Springer Collected Works in Mathematics

For further volumes:
http://www.springer.com/series/11104

Edmund Hlawka

Selecta

Editors

Peter M. Gruber · Wolfgang M. Schmidt

Reprint of the 1990 Edition

Springer

Author
Edmund Hlawka (1916–2009)
University of Technology
Vienna
Austria

Editors
Peter M. Gruber
University of Technology
Vienna
Austria

Wolfgang M. Schmidt
University of Colorado
Boulder, CO
USA

ISSN 2194-9875
ISBN 978-3-642-34622-4 (Softcover)
978-3-540-50623-2 (Hardcover)
DOI 10.1007/978-3-642-35384-0
Springer Heidelberg New York Dordrecht London

Library of Congress Control Number: 2012954063

Mathematics Subject Classification (1980): 10Hxx, 10Jxx, 10Kxx

Printed on acid-free paper

Springer is part of Springer Science+Business Media (www.springer.com)

Preface

The scholarly work up to now of Professor Hlawka has had strong influence on various areas of number theory, as well as other branches of mathematics. But many of his papers are not easily available. It therefore seemed appropriate to collect and to publish some of his most important works in the present Selecta. The idea for this endeavour came from Professor Chandrasekharan, who also established contact with the publisher. He deserves our sincere gratitude. The selection of the papers was done by Professor Hlawka himself after extensive consultation with other experts. The commentary is also his own.

Each of the papers was thoroughly checked. Since a number of small errors and misprints was found, but none which could not be remedied with a little effort, we decided to provide corrections of the main mistakes of only a few of the manuscripts. We are indebted to many colleagues and friends for the painstaking and laborious checking of the papers. Special thanks are due to Professors Harald Niederreiter and Karl Prachar, and to the Dozents and Doctors Michael Drmota, Johann Linhart, Gerhard Ramharter, Johannes Schoißengeier, Rudolf Taschner, Robert Tichy and Gerhard Turnwald. We wish to express our gratitude to Professor Walter Philipp for translating the commentary.

In our editorial work we received thorough help from Dr. Hartwig Sorger, and Dr. Christa Binder prepared the list of publications. Our sincere thanks are due to the staff of Springer-Verlag for their readiness to accommodate our wishes, as well as for the pleasant way in which they helped us along.

Vienna and Boulder
Autumn 1989

Peter M. Gruber
Wolfgang M. Schmidt

Preface

Table of contents

* The numbers in brackets refer to the List of publications at the end of this volume.

Notes on the papers on geometry of numbers and on Diophantine approximations

The first paper in the Selecta [3]* deals with Minkowski's theorem on the product of two inhomogeneous linear forms which he developed in his treatise "Diophantische Approximationen" [1907]. In contrast to Minkowski, the coefficients are complex numbers and the variables now are Gaussian integers. The paper is based on a suggestion of N. Hofreiter and was submitted as a Ph.D. thesis in 1938. I was able to obtain the exact constant. Later different proofs of this theorem were given by Mahler, by Chalk [1956] who also dealt with the case of other quadratic number fields, further by Perron and by Mordell [1956], see also L. Schmetterer [1952a]. The simplest proof appears to be the one given in Niven's [1963] book whose Theorem 5.3 is similar to Hauptsatz 1 of my paper. This area is known as "Complex Geometry of Numbers" since it involves integers in complex quadratic number fields. This subject was mainly studied by Perron (viz. [85]) in Munich and by Hofreiter's research group in Vienna. More recent work in this area is due to A. Schmidt (Copenhagen).

The second paper [4]* in the Selecta is an improvement of the results in my thesis. Special linear forms are considered here. The paper genuinely belongs to the area of Diophantine approximation. An analogue in the real case was established by Grace [1918]. However Grace used continued fractions, not available in the complex domain. Hence a new method had to be developed. Hilfssatz 1 was rediscovered by Davenport [1946] who observed that "this is of some interest in itself, so it is probably not new". Compare L. Schmetterer [1952b].

Paper [5]* on complex homogeneous linear forms has a longer history. The result on complex homogeneous linear forms established here can already be found in Minkowski's book "Diophantische Approximationen". The proof given there is extraordinarily long and as far as I know has not yet been checked in detail. It was desirable to find a simpler proof of this theorem. I first tried to find an analytic proof using Siegel's transcendental method. This involves the summation of certain infinite series containing Bessel functions. Because of my work on differential equations those were very familiar to me (see [1] and [2]). In a talk I gave in Göttingen in November 1938 I mentioned this possibility. Siegel who was in the audience observed that it would be very difficult to carry out this program – and he was absolutely right. In my paper the theorem is then proved by reducing it to the investigation of two Hermite forms. Mordell called this proof a proof with little algebra. (He meant to say elementary algebra.) A different proof was given by Cassels, Ledermann and Mahler

---

* Numbers in brackets refer to the List of publications at the end of this volume; an asterisk indicates a paper contained in these Selecta and citations like Mordell [1956] are given in the References of these Notes.

[1951]. Until today it is not known how an analogous theorem on three or more complex linear forms should look like. One cannot even conjecture what the constant should be. Applications of the papers on complex inhomogeneous linear forms can be found in [13] and [14].

I returned to the area of Diophantine approximations on several occasions, however mostly in the real case. I mention papers [38] and [84] (the latter one is dedicated to Hans Hornich and uses an idea of Landau) and particularly the papers [89], [94]*, [101] and [126]. Papers [89] and [101] address the problem of introducing analytic methods into the theory of Diophantine approximation of several irrational numbers. For the approximation of a single irrational number this is done in the papers by Hardy and Littlewood, Hecke and Behnke in the 1920's (cf. Koksma [1936], Chaps. 4 and 5).

In [94]* the approximation of irrational numbers is investigated under side conditions on the approximating rational numbers. For example it is required that they belong to a Pythagorean triple. In a qualitative way this appears to have been considered for the first time by Sierpinski [1954]. Paper [126] represents an application of Diophantine approximation to differential equations.

Paper [7]* entitled "Zur Geometrie der Zahlen" was my Habilitationsschrift. Submitted in 1941/42 it appeared in 1943. The theorem which got the most attention is Satz 4. It contains the proof of a conjecture of Minkowski. Minkowski never published a proof of this. But for the case of the sphere a proof can be constructed from his last paper. (This theorem was announced for the world fair in Chicago. This was kindly pointed out to me by Prof. Gruber. The conjecture was also stated in Hilbert's eulogy of Minkowski.) A more general statement is contained in Satz 2 which involves arbitrary Jordan measurable sets. This result was not even posed as a conjecture in Minkowski's work. The proof is analytic and is based on a generalization, namely on the so called deformation theorem. Another essential ingredient is Hilfssatz 1. It is suggested by the reduction theory of binary quadratic forms. The proof is unnecessarily indirect. It is possible to carry it out to a large extent in a constructive way. It then immediately yields a measure on the set of lattices of determinant 1. This theorem precipiated an extensive body of work, compare the book by Gruber and Lekkerkerker [1987] and the papers [19]* and [49]*. I only emphasize the paper by Siegel [1945] who in the proof uses the complete reduction theory of quadratic forms due to Minkowski. I also refer to the proof by Santaló [1950] using integral geometry. The simplest proof is due to Rogers [1947]. Later improvements of this theorem were obtained, in particular by Rogers and by W. Schmidt. (In [7]* a sharpening of this theorem for convex sets was in the planning stage. However this was abandoned due to the hardships of these times.) Here I also refer to the literature and to my lecture in London in 1986 on the occasion of the retirement of C. A. Rogers. The main part of the paper deals with Siegel's transcendental method applied to the inhomogeneous problem in the geometry of numbers. Central to this paper is the so-called Alternativsatz. This theorem has been further sharpened by means of Blichfeldt's method. Moreover, instead of convex bodies, more general regions have been investigated - called $M(k)$-regions in the paper - and the problem of the existence of k lattice points, different from the origin, is considered. Minkowski only treated the case $k=1$. This generalization is called k-theory. Although many problems remain unsolved this case hardly has been investigated during the last years.

The Alternativsatz also can be proved without transcendental methods. I refer to the papers [10]* and [22] and to the survey article [27] which, in particular, surveys the theory of Mahler obtained during the war and its continuation by Davenport, Rogers and Cassels. Although this is primarily a survey article, it still contains some new ideas, e.g. on p. 45 a proof of a theorem involving maximal lattices of arbitrary sets. A conjecture stated in this paper has been disproved in the aforementioned papers of Rogers and W. Schmidt in a brilliant fashion. The survey deals with the homogeneous problem, whereas the inhomogeneous problem is treated in [30]*. This was contained in a lecture at the International Mathematical Congress in Amsterdam in 1952.

In the subsequent papers [8]* and [12]* power sums of linear forms are considered by means of Blichfeldt's method. This was parallel to the investigations of Rankin. From 1947 on I kept a keen interest in Siegel's method. Siegel developed this method in a letter to Mordell, in order to prove in the conjecture on inhomogeneous linear forms at least the existence of an absolute constant. It was conjectured that this constante is $1/2^n$ for products of n linear forms. This approach of Siegel was no longer pursued since Tschebotareff [1940] proved the same result as Siegel but with a different method and with the relatively good bound $1/2^{n/2}$. The method of Siegel has been at first applied to complex inhomogeneous linear forms in [9]. This paper belongs to an area which was later called geometry of numbers in algebraic number fields. Siegel's method has been investigated further in [18]*. Here it is applied to the converse problem of Mordell. At the same time an assertion of Blichfeldt, stated without proof, was verified. Morever, Minkowski's theorem on successive minima is extended to the k-theory.

Paper [17] is a generalization of the problem of Mordell and was later significantly sharpened by Rogers. In [16]* the geometric aspect of the geometry of numbers is extensively treated. Here, inspired by a paper of Rankin, some assertions of Minkowski's are proven. Those investigations have been continued by several authors. In particular I mention the papers by Rogers, Bambah, Woods, Groemer, L. and G. Fejes Tóth, A. Florian, Kuperberg and Gruber. I also refer to paper [15] which appeared at the same time, as well as to [28] and [29] which were inspired by an idea of Davenport and Cassels. Papers [34]* and [51] represent to some extent the conclusion to these investigations. However, [51] led to further work by several authors. In [23] geometry of numbers is applied to the theory of additive set functions.

Siegel's transcendental method has many other applications. I only want to mention paper [99]. The difficulty of the method is based on the fact that the integrals occuring there are very hard to evaluate. A series of investigations was launched to remedy the situation. In the special case of the plane the integrals to be evaluated simply are the Fraunhofer integrals occuring in theoretical optics. The asymptotic expansion of these integrals extended over convex bodies is obtained by the method of the stationary phase as developed by van der Corput. Van der Corput's method had to be complemented by estimates of the error terms and the assumption that the convex bodies be very smooth had to be made. As I discovered later F. John [1957] obtained a similar formula in a different way, but without estimates of the error terms. These formulae have been applied to sharpen Minkowski's fundamental theorem in k-theory. However, they have the disadvantage that they only yield an imprrovement for larg values of k. The main reason for this is the fact that in the estimates of the error terms the constants cannot be given explicitly. (Recently I succeeded in obtaining upper bounds for these constants). This formula was used in a generalization of a method of Landau, in order to obtain the asymptotic number of lattice points in convex bodies.

In paper [20]* the asymptotic expansions in the case of the plane have been carried a step further. Here in the asymptotic formula a further term was obtained. Thanks to a course given by Strubecker this quantity had a familiar look and in fact it turned out to be the affine curvature. In [26]* I continued to follow this train of thought and I succeeded in obtaining the complete asymptotic expansion of the Fraunhofer integrals. This expansion involves the curvature and its derivatives. These expansions were already used in paper [20]* to prove several assertions by von Laue in theoretical optics (in the *green* Handbook of Physics). In the third part [24] the general expansions have been used to extend an analogue of Hankel's integral theorem in the theory of Bessel functions to general convex bodies. Moreover, a theorem in probability theory, established by Esseen [1944] for the classical case of balls, was extended to smooth convex sets. Also the case of lattice distributions is explicitly investigated. In paper [21] these asymptotic expansions have been applied to the analytic continuation of the zeta-function for convex bodies. This zeta-function was considered by Minkowski in the case of the plane and later by Herglotz.

Paper [25]* "Zur Theorie des Figurengitters" represents the passage from geometry of numbers to the theory of uniform distribution. There were several reasons why this paper, which is based on lectures in Munich, Göttingen, Hamburg and Erlangen, was written. First there was an invitation to write two articles for the Mathematische Enzyklopädie, one on Diophantine equations and the other on Diophantine inequalities (these surveys actually have been written, but they were considered as too long; today they are outdated to a large extent.) Second as I noticed at that time Siegel's method also can be applied to the theory of uniform distribution. C. L. Siegel himself in 1935 in a letter to Koksma (cf. Koksma [1936]) established a formula; under close scrutiny I noticed that this formula was a one dimensional special case of the general formula of Siegel which he had developed in the geometry of numbers. I don't know whether Siegel was aware of this connection. On p.185 of [25]* a property U is considered and it is conjectured that it is satisfied for all groups without elements of finite order. It would be interesting to prove or disprove this conjecture. Paper [25]* is very abstract, perhaps too abstract, so much so, that perhaps the basic ideas cannot be easily recognized. Many things are only sketched and only now (1987/88) the details have been worked out.

In paper [32]* following Eckmann [1943/44] a theory of uniform distribution on compact groups is developed and it is shown that the main theorems of the theory of uniform distribution carry over to the general case. However is should be noted, that it was always tacitly assumed that the boundary of the sets under consideration has measure zero. The introduction of this paper could have been simplified, but at that time there was some hope that, with the machinery developed here, it would be easily possible to develop a theory of a discrepancy i.e. a quantitative theory of uniform distribution on compact groups. However, that turned out not to be very simple. In [61], a joint paper with H. Niederreiter, such a concept was established for abelian groups. Later this has been pursued further by Niederreiter. We shall return to the concept of discrepancy later.

In [33]* the theory of uniform distribution is more generally developed for compact metrizable spaces. Under the imprint of the integration theory, as presented in Bourbaki (which just had appeared in print) the point of view of functional analysis is assumed. In this way measure theoretic difficulties are avoided which still had encumbered paper [32]*. Paper [37] is a continuation and presents a generalization of the theory of normal numbers. A further continuation containing the theory of absolutely normal numbers had been planned, but was never carried out.

Paper [35] is a preparatory note to [37] perhaps with a clearer presentation of focus. Concerning the foundations of the theory of uniform distribution I refer to [74]*, [76], [91] and [102]*, which also deal with the concept of the discrepancy. For the history of the theory of uniform distribution, in particular, as developed by its founder H. Weyl, I refer to [120] a joint paper with Ch. Binder.

Papers [46]* and [55] on rhythmic sequences on compact groups (the second paper, jointly with W. Henhapl, contains his Ph.D. thesis) represent a supplement to the general theory of uniform distribution. A theory of rhythmic sequences on compact groups is treated here. Van der Corput [1932] developed such a theory for the special case of the unit interval mod 1. Here some methods are used as presented in A. Weil's [1951] book "Integration on topological groups".

Rhythmic sequences $x(j)(j \in Z)$ on a compact group G have two nice properties:

1) If M is an open subset of G containing one element of the sequence then M contains infinitely many of them.
2) If $(x(j+1)x^{-1}(j))$ denotes the difference sequence of a sequence $x(j)$ then $x(j)$ is rhythmic if and only it its difference sequence is rhythmic.

This is not true for uniformly distributed sequences, but the following holds: Let $x(k)$ be a sequence (now $k \in N$) and define a matrix with entries $A_{ik} = x(k)x(k+1) \ldots x(i+k-1)$ then the following theorem holds: If the rows are uniformly distributed then the columns are also uniformly distributed.

As pointed out earlier the concept of discrepancy plays a fundamental role in the classical theory of uniform distribution. (The name was coined by van der Corput; when he used it for the first time cannot be determined, perhaps in a manuscript often quoted by Koksma, but never published, see [120].) Up to this point in time mostly the one dimensional case was treated in the literature. For instance Koksma, in connection with a paper of Behnke, had developed an important inequality. In [43]* the higher dimensional case is treated by making use of functions of bounded variation in the sense of Hardy and Krause. Hobson's [1927] book was taken as a basis for this paper. This paper got considerable attention and its proof has been simplified by several authors. One of the applications presented in the paper needs a certain supplement in its proof. It is the matter of the estimate of $N(C)$ where a formula of Davenport is used. In his "Lectures on Uniform Distribution" W. Schmidt gave such a supplement. In the last display (15) in [43]* a term is estimated which, for the case that B is convex, later was named isotropic discrepancy. This was done by Zaremba [1970] who showed that in the case of a tetrahedron the bound in (15) cannot be improved. In paper [69]* the connection between discrepancy and Riemann integration is made clear. In the classical discrepancy subintervals of the unit cube are taken as basis – in the paper [70] balls are taken instead. This definition has been investigated by Cassels and W. Schmidt but – as far as I know – never been published. Other definitions of discrepancy, such as the polynomial discrepancy – are studied in [78], [79], [82] and [118]. The theorems developed there permit a variety of applications. Of course these theorems served the purpose of studying the discrepancy of uniformly distributed sequences. My idea was now to use sequences with small discrepancy to evaluate integrals, i.e. to establish quadrature formulae. This, above all, is true for the first paper in this series [45]*.

The starting point is the following known fact. Given an integral equation

$$\Phi(x)-\lambda\int_0^1 K(x,y)\Phi(y)dy=f(x)$$

it is well-known that the solution (for sufficiently small λ) is given by the Neumann series

$$\Phi(x)=f(x)+\sum_{n=1}^{\infty}\lambda^n\int_0^1 K^{(n)}(x,y)f(y)dy$$

where

$$K^{(n)}(x,y)=\int_0^1\dots\int_0^1 K(x,x_1)K(x_1,x_2)\dots K(x_{n-1},y)dx_1\dots dx_{n-1}$$

are the iterated kernels of K. That means the solution requires the calculation of multiple integrals of dimension increasing to infinity. It is now almost obvious to apply approximation formulae for such integrals to the Neumann series.

In these papers the quadrature formulae developed by me are applied to the iterated kernels. Of course they also can be applied to Fredholm determinants. Independently similar formulae have been developed by Korobov [1963]. For this area I proposed the name number theoretic analysis. For the further development I only mention the book by Hua and Wang [1981], Lecture Notes 1114 and 1262 edited by me (Springer-Verlag) and the book "Zahlentheoretische Methoden in der numerischen Mathematik" (jointly with F. Firneis und P. Zinterhof). This method also can be applied to other types of integral equations as well as to linear systems of equations (I only refer to the papers [47] and [50] – the second paper is jointly with K. Kreiter), to interpolation [60]*), to numerical analytic continuation [81]; this paper has been continued by R. Taschner [1976]), to approximation formulae ([86]), to mathematical linguistics ([90]) to the kinetic theory of gases (building on a paper of Egerváry and Turán [1951]; [57], [64], [65] and [75]*). In a conversation P. Turán proposed for this theory the name deterministic kinetic theory of gases.

In order to extend the applicability to numerical mathematics in papers [71] and [73] (jointly with R. Mück) a method was developed with which sequences can be constructed which are uniformly distributed with respect to an arbitrary density. For this purpose sequences uniformly distributed in the classical sense were used. This uniform distribution method has also been used to treat problems in pure mathematics, such as the distribution of the zeros of the Riemann zeta function [83]*. Moreover this method helped in the investigation of the convergence properties of power series on the boundary of the circle of convergence [108]. Paper [96] also is perhaps of interest. Further applications are given in [109]* and [111].

In paper [102]* a quantitative theorem of uniform distribution on spheres was developed, inspired by the number theoretic papers by Malyšev [1957] and

Pommerenke [1959] and by the geodesical paper of Freeden [1979]. Paper [102] was applied immediately to establish approximate solutions of partial differential equations in mathematical physics [113]. Here the determination of solutions of linear partial differential equations with given initial and boundary conditions is at stake. This is a very important task and a preliminary stage in the treatment of nonlinear partial differential equations. (It is perhaps worth mentioning that in the very recent past also other mathematicians are interested in the construction of uniformly distributed sequences, in particular on the two-dimensional sphere.)

There are integral representations of solutions of partial differential equations. However, these integrals are volume integrals or surface integrals, particularly over spheres, – that is integrals not easily calculable. I only remind the reader of the volume integrals for retarded potentials or of the spherical mean values occurring in the wave equations whose importance was stressed by Courant and by John [1957]. This is the point of entry of the above mentioned paper. First the Poisson equation is considered; in the center of the investigation is the evaluation of the solutions of the equation of Euler – Poisson – Darboux then the wave equation in 2,3 and higher dimensions. The results are then for instance applied to the Maxwell equation and also to the telegraph equation.

I still want to point out paper [72] "Ein metrisches Gegenstück zu einem Satz von V. A. Veech" written in 1971.

A metric theorem in the theory of *C*-uniform distribution was given in [67] using the Wiener measure. In this context also an inequality for integrals of independent interest was proved.

Outside these investigations on geometry of numbers and on the theory of uniform distribution paper [6]* published by H. Hornich should be mentioned. It contains an inequality for lengths of vectors in finite dimensional Euclidean space. Hornich told me about this paper before it appeared in print and he stated that it could not be reduced to the triangle inequality. To both our satisfaction I was able to disprove this statement with the help of an identity now known as Hlawka's identity. Hornich included all these arguments in his paper. Hence this inequality is now named after me and Hornich. It has been included in many text books.

In 1942 I became interested in the calculation of integrals and investigated difference equations. A recent contribution to the latter is [107]. A letter to L. Koschmieder in 1935 who then was in Brno was reviewed in Koschmieder's paper [1946]. The papers [54], [58] deal with trigonometric interpolation in more than one variable.

References

J. W. S. Cassels, W. Ledermann, K. Mahler: Farey section in $k(i)$ and $k(g)$. Phil. Trans. Roy. Soc. London A **243** (1951), 585–628

J. H. H. Chalk: Rational Approximations in the complex plane II. J. London Math. Soc. **31** (1956), 216–221

J. G. van der Corput: Diophantische Ungleichungen II. Rhythmische Systeme A, B, Acta Math. **59** (1932), 209–328

H. Davenport: Non-homogeneous binary quadratic forms, Proc. Ned. Akad. Wet. **49** (1946), 815–823

B. Eckmann: Über monothetische Gruppen, Comment. Math. Helv. **16** (1943/44), 249–263

E. Egerváry, P. Turán: On a certain point of the kinetic theory of gases. Studia Math. **12** (1951), 170–180; Közleményei Magyar Tud. Akad. Mat. Fiz. Oszt. **1** (1951), 303–314

C.-G. Esseen: Fourier analysis of distribution functions. A mathematical study of the Laplace - Gaussian Law. Acta Math. **77** (1944), 1–125

W. Freeden: Über eine Klasse von Integralformen der mathematischen Geodäsie. Geodät. Inst., Aachen 1979

J. H. Grace: Note on a Diophantine approximation. Proc. London Math. Soc. (2) **17** (1918), 316–319

P. M. Gruber; C. G. Lekkerkerker: Geometry of Numbers. North-Holland, Amsterdam 1987

E. W. Hobson: The theory of functions of a real variable and the theory of Fourier Series. Vol. I. Cambridge Univ. Press, London 1927

L. K. Hua; Y. Wang: Applications of number theory to numerical analysis. Springer, Berlin Heidelberg New York, 1981

F. John: Bestimmung einer Funktion aus ihren Integralen über gewisse Mannigfaltigkeiten. Math. Ann. **109** (1957), 488–520

J. F. Koksma: Diophantische Approximationen. Springer, Berlin 1936

N. M. Korobov: Zahlentheoretische Methoden in der Näherungsanalysis (russ.). Fizmatgiz, Moscow 1963

L. Koschmieder: Beispiele des Gebrauchs gewisser Ableitungsformeln von Liouville, Spitzer und Schlömilch. Rev. Math. Hisp.-Amer. (4) **6** (1946)

A. V. Malyšev: The distribution of integer points in a fourdimensional sphere. Doklady Akad. Nauk SSSR **114** (1957), 25–28

H. Minkowski: Diophantische Approximationen. Teubner, Leipzig Berlin 1907

L. J. Mordell: Diophantine inequalities in complex quadratic fields. Publ. Math. Debrecen **4** (1956), 242–255

J. Niven: Diophantine Approximations. Interscience Publishers, Wiley, New York 1963

C. Pommerenke: Über die Gleichverteilung von Gitterpunkten auf m-dimensionalen Ellipsoiden. Acta Arithm. **5** (1959), 227–257

C. A. Rogers: Existence theorems in the geometry of number. Ann. Math. **48** (1947), 994–1002

L. A. Santaló: Integral geometry in projective and affine spaces. Ann. Math. **51** (1950), 739–755

L. Schmetterer: Notiz zu einem Satz über diophantische Approximationen. Monatsh. Math. **56** (1952), 253–255

L. Schmetterer: Über das Produkt zweier inhomogener Linearformen. Monatsh. Math. **56** (1952), 339–343

C. L. Siegel: A mean value theorem in geometry of numbers. Ann. Math. **46** (1945), 340–347

W. Sierpinski: Pythagoreische Tripel (poln.). Panstwowe wydow nictwo naukowe, Warszawa 1954

R. J. Taschner: Probleme der numerischen analytischen Fortsetzung. Sitzungsber. Österr. Akad. Wiss., Math.-naturwiss. Kl. II **185** (1976), 459–484

N. Tschebotareff: Beweis des Minkowskischen Satzes über lineare inhomogene Formen. Vierteljschr. Naturforsch. Ges. Zürich **85**, Beiblatt (1940), 27–30

A. Weil: L'intégration dans les groups topologiques et ses applications. Hermann, Paris 1939, 1951

S. K. Zaremba: La discrépance isotrope et l'integration numérique. Ann. Mat. Pura Appl. (IV) **87** (1970), 125–136

Über die Approximation von zwei komplexen inhomogenen Linearformen

Monatsh. Math. Phys. **46** (1938). 324–334

H. Minkowski[1]) hat den Satz bewiesen:

Ist $(\xi-\xi_0)(\eta-\eta_0)=(\alpha x+\beta y-\xi_0)(\gamma x+\delta y-\eta_0)$ eine zerlegbare, nichthomogene Form, in der $\alpha\delta-\beta\gamma=1$, $\alpha, \beta, \gamma, \delta, \xi_0, \eta_0$ sonst beliebig reell, so gibt es ganze rationale Zahlen x, y, so daß $|(\xi-\xi_0)(\eta-\eta_0)|\leqq\frac{1}{4}$.

Seither sind für diesen Satz mehrere Neubeweise mit verschiedenen Methoden geführt worden.

Ich beweise den Satz:

Ist $(\xi-\xi_0)(\eta-\eta_0)=(\alpha x+\beta y-\xi_0)(\gamma x+\delta y-\eta_0)$ eine zerlegbare nichthomogene Form, in der $\alpha\delta-\beta\gamma=1$, $\alpha, \beta, \gamma, \delta, \xi_0, \eta_0$ sonst beliebig komplex, so gibt es ganze Zahlen x, y aus $k(i)$, so daß

$$|(\xi-\xi_0)(\eta-\eta_0)|\leqq\frac{1}{2}.$$

Das Gleichheitszeichen gilt dann und nur dann, wenn die Form arithmetisch äquivalent ist, mit der Form

$$\left(x-\frac{1+i}{2}\right)\left(y-\frac{1+i}{2}\right).$$

Liegt $\frac{\alpha}{\beta}$ nicht in $k(i)$, so gibt es unendlichviele x, y, die die Ungleichung befriedigen und gleichzeitig $|\xi-\xi_0|$ beliebig klein machen.

Hilfssatz 1: Es seien p und q zwei komplexe Zahlen. Dann gibt es stets eine ganze Zahl z aus $k(i)$, so daß

$$|z-p|\,|z-q|\leqq\frac{\sqrt{5}}{4} \text{ oder } <\frac{|p-q|}{\sqrt{2}}$$

ist.

Beweis: *O. B. d. A.* sei $0\leqq q_1\leqq 1$, $0\leqq q_2\leqq 1$, wo $q=q_1+iq_2$ und $p=p_1+ip_2$.

[1]) H. Minkowski: Diophantische Approximationen.

Fall I: $0 \leqq p_1 \leqq 1,\ 0 \leqq p_2 \leqq 1.$

Es sei $z_1=0,\ z_2=1,\ z_3=i,\ z_4=1+i$. Wir behaupten: Es ist

$$\prod_{i=1}^{4} |z_i-p|\ |z_i-q| \leqq \frac{5^2}{2^8}.$$

Um dies zu beweisen, genügt es zu zeigen, daß

$$\prod_{i=1}^{4} |z_i-p| \leqq \frac{5}{2^4},\quad \prod_{i=1}^{4} |z_i-q| \leqq \frac{5}{2^4}.$$

Wir beschränken uns auf p. Wir betrachten

$$P=\prod_{i=1}^{4} |z_i-p|^2$$

und setzen $p=\varepsilon+\frac{1}{2}(1+\ )$, wo $\varepsilon=\varepsilon_1+i\varepsilon_2$.

Dabei ist $|\varepsilon_i| \leqq \frac{1}{2}$. Für P erhalten wir:

$$P=\left(b^2+\frac{1}{4}\right)^2-aR,$$

mit

$$R=3b-\frac{1}{2}a-a^3-4a^2b-6ab^2-4b^3 \geqq 5b\left(\frac{1}{2}-3b^2\right) \geqq 0,$$

wenn wir $a=\varepsilon_1^2,\ b=\varepsilon_2^2$ setzen und $a \leqq b$ annehmen, was wir *O.B.d.A.* voraussetzen können. Also ist

$$P \leqq \left(b^2+\frac{1}{4}\right)^2 \leqq \frac{5^2}{2^8}$$

womit alles gezeigt ist.

Fall II: $p_1 \leqq 0,\ p_2 \leqq 0,\ p_1+p_2 \neq 0.$

Damit behandeln wir auch die Gebiete

$$p_1 \leqq 0,\ p_2 \geqq 1;\ p_1 \geqq 1,\ p_2 \leqq 0;\ p_1 \geqq 1,\ p_2 \leqq 1,$$

denn diese Gebiete gehen aus $p_1 \leqq 0,\ p_2 \leqq 0$ durch Drehung um $\frac{\Pi}{2}$ hervor. Wir werden zeigen: Es gibt ein z_i, so daß

$$|z_i-p|\ |z_i-q| < \frac{1}{\sqrt{2}}\,|p-q|. \tag{1}$$

Ist $q=0$ oder $\frac{1+i}{2}$ so ist, wie man sofort sieht, die Behauptung mit $z=z_1$ richtig. Wir können daher $q \neq 0$ und $\neq \frac{1+i}{2}$ annehmen. Auch im Falle $|q| \leqq \frac{1}{\sqrt{2}}$ ist sie mit $z=z_1$ richtig, denn dann ist

$$|p-q|-\sqrt{2}\,|p|\,|q|\geqq|p-q|-|p|>0.$$

Wir können also noch $|q|>\frac{1}{\sqrt{2}}$ voraussetzen. Dann gibt es nur folgende Fälle:

$$a)\ \frac{1}{2}\leqq q_1\leqq 1,\ 0\leqq q_2\leqq\frac{1}{2};\quad b)\ 0\leqq q_1\leqq 1,\ \frac{1}{2}\leqq q_2\leqq 1;$$

$$c)\ \frac{1}{2}\leqq q_1\leqq 1,\ \frac{1}{2}\leqq q_2\leqq 1.$$

Wir setzen noch zur Abkürzung

$$e_i=|z_i-q|^2,\quad d_i=|z_i-p|^2,\quad 2e_1-1=e.$$

a) In diesem Fall gilt (1) mit z_1 oder z_2. Nehmen wir an, es sei $|p|\,|q|\geqq\frac{1}{\sqrt{2}}|p-q|$ oder anders geschrieben

$$ed_1\geqq e_1-2\,(p_1q_1+p_2q_2), \tag{2}$$

so betrachten wir

$$I_1=|p-q|^2-2e_2d_2.$$

Ersetzen wir in diesem Ausdruck, als Funktion von p_1, p_2, q_1, q_2 betrachtet, d_1 nach (2), so verkleinern wir I_1, da der Koeffizient von d_1 $1-2e_2$ größer als 0 ist. Es kommt

$$I_1\geqq 2q_1-1-2p_2q_2+\frac{2}{e}\,(R_1-p_1R_2),\ \text{wo}$$

$$R_1=e_1(2q_1-1)-ee_2=(2q_1-1)^2-2e_2\left(q_1-\frac{1}{2}\right)^2+2e_2\left(\frac{1}{4}-q_2^2\right)$$

$$\geqq 4\left(q_1-\frac{1}{2}\right)^2-\left(q_1-\frac{1}{2}\right)^2\geqq 0\ \text{ist}$$

und

$$R_2=4ee_2-8q_2^2-4q_1+4\geqq 2e-8q_2^2-4q_1+4>0$$

ist. Ist $q_2>0$, so ist $I_1>0$. Ist dagegen $q_2=0$, so ist $R_1\geqq 3\left(q_1-\frac{1}{2}\right)^2>0$, da dann $q_1>\frac{1}{\sqrt{2}}$, also wieder $I_1>0$.

b) erledigt sich analog *a)* mit z_1 und z_3.

c) In diesem Fall gilt (1) mit z_2 und z_4. Angenommen, es gälte wieder (2), so betrachten wir $I_2=|p-q|^2-2e_4d_4$ und gehen wie bei I_1 vor und erhalten

$$I_2\geqq-4e_4+\frac{4}{e}\,(q_1+q_2-1)\,e_1+I_3$$

mit

$$I_3=-\frac{4}{e}(p_1R_3+p_2R_4),$$

wo

$$R_{i+2}=2q_i(q_1+q_2-1)-ee_4 \qquad i=1,2$$

ist. R_{i+2} ist größer als 0, denn betrachten wir z. B. R_3 und setzen

$$q=\varepsilon+\frac{1}{2}(1+i),\ 0\leqq\varepsilon_i\leqq\frac{1}{2},\ \varepsilon_1+\varepsilon_2\neq 0,$$

so erhalten wir

$$R_3=3\varepsilon_1^2+\varepsilon_2^2+6\varepsilon_1\varepsilon_2-2(\varepsilon_1^2+\varepsilon_2^2)^2>0.$$

Also ist auch $I_3>0$. Aus (2) folgt, daß wenigstens für ein i

(3) $$-ep_i>2q_i$$

etwa p_1. Benützen wir dies, so folgt weiter

$$I_2>-4e_4+\frac{4}{e}(q_1+q_2-1)e_1-\frac{4}{e}p_1R_3$$

$$>-4e_4+\frac{4}{e}(q_1+q_2-1)e_1+\frac{8q_1}{e^2}R_3.$$

Ersetzt man wieder q durch ε wie oben, so sieht man unmittelbar

$$e^2I_2>0.$$

Fall III. $\qquad 0<p_1\leqq 1,\ p_2<0.$

Damit erledigen wir die Fälle

$$0<p_2\leqq 1,\ p_1<0;\quad 0<p_1\leqq 1,\ p_2>1;\quad 0<p_2\leqq 1,\ p_1>1,$$

denn sie gehen aus $0<p_1\leqq 1,\ p_2<0$ wieder durch Drehung um $\frac{\Pi}{2}$ hervor.

1) $$0\leqq q_1\leqq 1,\ 0\leqq q_2\leqq\frac{1}{2};\quad -\frac{1}{2}\leqq p_2<0.$$

Dann gilt für z_1 oder z_2

$$|z_i-p|\,|z_i-q|\leqq\frac{\sqrt{5}}{4}.$$

Zu diesem Zweck zeigen wir

$$|p|\,|p-1|\leqq\frac{\sqrt{5}}{4},\ |q|\,|q-1|\leqq\frac{\sqrt{5}}{4}.$$

Wir beweisen es für p. Wir setzen

$$p_1=\varepsilon_3+\frac{1}{2},\ p_2=-\varepsilon_4,\ |\varepsilon_3|\leqq\frac{1}{2},\ 0<\varepsilon_4\leqq\frac{1}{2}$$

und erhalten

$$I_4=d_1d_2=\left(\varepsilon_1^2-\frac{1}{2}\right)^2+\varepsilon_2^4+2\varepsilon_1^2\varepsilon_2^2+\frac{\varepsilon_2^2}{2}\leqq\varepsilon_1^4+\frac{1}{4}\leqq\frac{5}{16}.$$

In den weiteren Fällen können wir uns auf das Gebiet

$$0<p_1\leqq\frac{1}{2},\ p_2<0$$

beschränken, denn $\frac{1}{2}<p_1\leqq 1,\ p_2<0$ geht aus ihm durch Spiegelung an $|z|=|z-1|$ hervor.

2) $$0\leqq q_1\leqq\frac{1}{2},\quad 0\leqq q_2\leqq\frac{1}{2},\ p_2<-\frac{1}{2}.$$

In diesem Fall gilt (1) mit z_1, denn es ist

$$|p-q|^2-2e_1d_1=p_2^2e+(p_1-q_1)^2+q_2^2-2p_2q_2-2p_1^2e_1>\frac{e}{4}+$$
$$+\left(p_1-\frac{1}{2}\right)^2+q_2^2-\frac{e_1}{2}\geqq 0.$$

3) $$\frac{1}{2}<q_1\leqq 1,\quad 0\leqq q_2\leqq\frac{1}{2},\ p_2<-\frac{1}{2}.$$

Ist $p_2<-1$, so gilt (1) mit z_2, denn es ist

$$|p-q|^2-2e_2d_2\geqq p_2^2e-2p_2q_2+q_2^2-2e_2\geqq 1-4(1-q_1)^2-3q_2^2+$$
$$+2q_2>0.$$

Ist $-1\leqq p_2<-\frac{1}{2}$ so gilt (1) mit z_1 oder z_2. Um dies zu zeigen, bemerken wir, daß

$$q_2^2-2p_2q_2-2d_1q_2^2\geqq q_2^2-2\left(d_1-\frac{1}{4}\right)^2q_2^2\geqq q_2-2q_2^2\geqq 0$$

und $$q_2^2-2p_2q_2-2d_2q_2^2=$$
$$=q_2\left(-2p_2-2q_2\left(d_2-\frac{1}{2}\right)\right)\geqq q_2\left(-2p_2-d_2+\frac{1}{2}\right)\geqq 0.$$

Es genügt nun zu zeigen

(4) $$I_5=(p_1-q_1)^2+p_2^2-2q_1(1-q_1)\sqrt{d_1d_2}\geqq 0,$$

denn dann ist

$$(p_1-q_1)^2+p_2^2>2(q_1-z_i)\,d_i$$

für ein i, also

$$(p_1-q_1)^2+p_2^2+q_2^2-2p_2q_2-2d_iq_2^2>2(q_1-z_i)^2d_i,$$

d. h. die Behauptung. Um nun (4) zu zeigen, bemerken wir, daß

$$d_1d_2=(d_1-p_1)^2+p_2^2,$$

wo $$d_1-p_1\geqq p_1^2-p_1+\frac{1}{4}\geqq 0.$$

Es ist also

$$I_5 > \left(\frac{1}{2}-p_1\right)^2 + p_2^2 - 2q_1(1-q_1)(d_1-p_1-p_2) =$$
$$= \frac{1}{2}\left(p_1-\frac{1}{2}\right)^2 + \frac{1}{2}\left(p_2+\frac{1}{2}\right)^2 \geqq 0.$$

4) $$0 \leqq q_1 \leqq \frac{1}{2}, \quad \frac{1}{2} < q_2 \leqq 1; \quad +p_2 < 0.$$

Ist $|q| \leqq \frac{1}{\sqrt{2}}$, so gilt (1) mit z_1, denn

$$|p-q|^2 - 2e_1 d_1 \geqq e_1 - q_1 > q_1^2 - q_1 + \frac{1}{4} \geqq 0.$$

Ist nun $|q| > \frac{1}{\sqrt{2}}$ und $-1 \leqq p_2$ so gilt wieder (1) mit z_1, denn dann ist

$$|p-q|^2 - 2e_1 d_1 =$$
$$e_1 - 2p_1 q_1 - 2p_2 q_2 - e d_1 \geqq e_1 - q_1 - 2p_2 q_2 - \frac{1}{4}e - e p_2^2$$
$$= \left(q_1 - \frac{1}{2}\right)^2 + \frac{1}{2}(q_2^2 - q_1^2) - 2p_2\left(q_2 + \frac{e}{2}p_2\right) > 0,$$

denn $$q_2 + \frac{e}{2}p_2 \geqq q_2(1-q_2) + \frac{1}{2} - q_1^2 > 0.$$

Es sei also nur $p_2 < -1$ und $|q| > \frac{1}{\sqrt{2}}$. Ist nun $-p_2 e \leqq 2q_2$, so gilt wieder (1) mit z_1, denn dann ist

$$|p-q|^2 - 2e_1 d_1 =$$
$$e_1(1-2p_1^2) + p_1^2 - 2p_1 q_1 - 2p_2 q_2 - p_2^2 e > \frac{1}{2} - p_1^2 + p_1^2 - q_1 \geqq 0.$$

Ist nun aber $-p_2 e > 2q_2$, dann gilt (1) mit z_3 statt z_1, denn dann ist

$$I_6 = |p-q|^2 - 2e_3 d_3 = p_1^2(1-2e_3) + e_1 - 2p_1 q_1 + p_2^2(1-2e_3) +$$
$$+ 4p_2 e_3 - 2e_3 - 2p_2 q_2,$$

wenn man p_1^2 durch 0, ebenso $e_1 - 2p_1 q_1$ durch 0 und p_2^2 durch $-\frac{2p_2 q_2}{e}$ ersetzt.

$$I_6 \geqq -4p_2\left(\frac{q_2}{e}(2q_2-1) - e_3\right) - 2e_3.$$

Weiter ist $$\frac{q_2}{e}(2q_2-1) - e_3 \geqq \frac{1}{e}\left(q_2 - \frac{1}{2}\right)^2 > 0.$$

Ersetzt man also $-p_2$ durch $\frac{2q_2}{e}$, so folgt

$$e^2 I_6 > R_5,$$

mit $\qquad R_5=8q_2^2(2q_2-1)-2e_3e(4q_2+e).$

Verkleinert man nun R_5, indem man q_1 durch $\frac{1}{2}$ ersetzt und setzt $q_2=\varepsilon+\frac{1}{2}$ mit $0<\varepsilon\leqq\frac{1}{2}$, so sieht man sofort $R_5>0$, also $I_6>0$, womit alles gezeigt.

5) $$\frac{1}{2}<q_1\leqq 1,\ \frac{1}{2}<q_2\leqq 1.\quad p_2<0.$$

Ist $-\frac{1}{2}\leqq p_2<0$, dann gilt (1) mit z_1, denn

$$|p-q|^2-2e_1d_1\geqq e_1-q_1-2p_2q_2-\frac{1}{4}e-ep_2^2=$$
$$=\left(q_1-\frac{1}{2}\right)^2+\frac{1}{2}(q_2^2-q_1^2)-2p_2\left(q_2+\frac{e}{2}p_2\right)>0,$$

denn $q_2+\frac{e}{2}p_2\geqq q_2-\frac{e}{4}>\frac{1}{2}\left(1-\frac{q_2}{2}\right)+\frac{1}{4}-\frac{q_1^2}{2}\geqq 0.$

Es sei nun also $p_2<-\frac{1}{2}$.

Dann gilt (1) mit z_1, z_2 oder z_4. Angenommen, es sei

$$|p|\ |q|\geqq\frac{1}{\sqrt{2}}|p-q|$$

oder anders geschrieben

(5) $$ep_2^2\geqq-2p_2q_2+(p_1^2+e_1(1-2p_1^2)-2p_1q_1).$$

Ist $e_2\leqq\frac{1}{2}$, dann gilt (1) mit z_2, denn dann ist

$$|p-q|^2-2e_2d_2\geqq|p-q|^2-d_2>0.$$

Es sei jetzt $e_2>\frac{1}{2}$, dann ist der zweite Term in (5) größer als 0, denn er ist etwas umgeformt

$$\frac{e}{2}(1-2p_1^2)-2p_1q_1+\frac{1}{2}\geqq\frac{e_1}{2}+\frac{1}{4}-q_1>0,$$

wenn man benützt, daß $q_2^2>\frac{1}{2}-(1-q_1)^2$. Es gelte also jetzt (5) mit $e_2>\frac{1}{2}$. Dann gilt (1) mit $i=4$. Zum Beweise betrachten wir

$$I_2=|p-q|^2-2e_4d_4.$$

Geht man wie unter II*c* vor, ersetzt p_1^2 durch 0 und benützt (5), so kommt

$$I_2\geqq-\frac{4p_1}{e}R_3+4\frac{e_1}{e}(q_1+q_2-1)-4e_4-\frac{4p_2}{e}R_4.$$

Nun wissen wir, daß $R_3>0$, $R_4>0$. Wir verkleinern also I_2 weiter, wenn wir p_1 durch $\frac{1}{2}$ ersetzen. Da $e_2>\frac{1}{2}$ ist, so ist der zweite Term in (5) größer als 0, es gilt daher $-p_2 e>2q_2$. Berücksichtigt man dies, so kommt $e^2 I_2 \geqq F$, wo

$$F=4q_2^2(q_1+q_2-1)-\frac{e}{2}(4q_2+e)e_4+(q_1+q_2-1)(e_1-q_1)e.$$

Setzt man wieder $q=\varepsilon+\frac{1}{2}(1+i)$, wie oben, so sieht man nach kurzer Rechnung $F>0$.

Damit ist der Beweis des Hilfssatzes zu Ende geführt.

Hilfssatz 2: Sind α, β, γ, δ beliebige komplexe Zahlen, wobei $\alpha\delta-\beta\gamma=1$ und $\frac{\alpha}{\beta}$ nicht in $k(i)$ liegt, so gibt es unendlich viele ganze Zahlen p, $q\neq 0, 0$, wobei p zu q relativ prim ist, so daß

$$|(\alpha p+\beta q)(\gamma p+\delta q)|<\frac{1+\varepsilon}{\sqrt{3}}, \tag{6}$$

wobei $\varepsilon>0$ ist. Dabei ist es noch möglich, $|\gamma p+\delta q|$ beliebig groß zu machen.

Beweis: Es gibt unendlich viele Zahlen p, q, wobei $(p, q)=1$, so daß

$$-\frac{\beta}{\alpha}=\frac{p}{q}-\frac{\vartheta}{q^2}, \quad |\vartheta|\leqq\frac{1}{\sqrt{3}}\ ^{2)}.$$

Daraus folgt

$$|\gamma p+\delta q|=\left|\gamma\left(-\frac{\beta}{\alpha}q+\frac{\vartheta}{q}\right)+\delta q\right|=\left|\frac{q}{\alpha}+\frac{\gamma\vartheta}{q}\right|\geqq\left|\frac{|q|}{|\alpha|}-\frac{|\gamma\vartheta|}{|q|}\right|,$$

und

$$|(\alpha p+\beta q)(\gamma p+\delta q)|=\left|\frac{\alpha\vartheta}{q}\right|\left|\frac{q}{\alpha}+\frac{\gamma\vartheta}{q}\right|\leqq|\vartheta|+\frac{|\alpha\gamma|\,|\vartheta|^2}{|q|^2},$$

also kann man durch geeignete Wahl von q erreichen, daß $|\gamma p+\delta q|$ beliebig groß wird und (6) gilt.

Beweis des Satzes:

Wir bestimmen zwei Zahlen x_0, y_0 so, daß

$$\alpha x_0+\beta y_0=\xi_0$$
$$\gamma x_0+\delta y_0=\eta_0$$

was stets möglich ist, da $\alpha\delta-\beta\gamma=1$ ist. Dann wird

[2]) O. Perron, Math. Ann. **103** oder L. R. Ford, Trans. Amer. Math. Soc. **27**

$$(\xi-\xi_0)(\eta-\eta_0)=[\alpha(x-x_0)+\beta(y-y_0)]\,[\gamma(x-x_0)+\delta(y-y_0)]$$
$$=a(x-x_0)^2+b(x-x_0)(y-y_0)+c(y-y_0)^2,$$

wo $a=\alpha\gamma,\ b=\beta\gamma+\alpha\delta,\ c=\beta\delta$ und $b^2-4ac=(\alpha\delta-\beta\gamma)^2=1$.

Weiter setzen wir noch $f_0(x,y)=ax^2+bxy+cy^2$. Wir nehmen zuerst an, $\frac{\alpha}{\beta}$ liege nicht in $k(i)$. Dann gibt es nach Hilfssatz 2 unendlich viele ganze Zahlen $p, q, (p, q)=1$, so daß

$$|f_0(p,q)|<\frac{1+\varepsilon}{\sqrt{3}}.$$

Dabei können wir noch erreichen, daß $|\mu|=|\gamma p+\delta q|$ beliebig groß wird. Wir setzen noch $\alpha p+\beta q=\lambda$. Nun bestimmen wir eine ganze Zahl n so, daß

$$\mu\xi_0-\lambda\eta_0=n+k$$

und $|k|\leqq\frac{1}{\sqrt{2}}$ ist. Nun ist die Gleichung

(7) $$\mu\xi-\lambda\eta=qx-py=n$$

in ganzen Zahlen x, y lösbar. Die Lösung laute

$$\bar{x}_r=\bar{x}-rp,\quad \bar{y}_r=\bar{y}-rq. \qquad (r \text{ ganz}).$$

Wir bestimmen nun r, so daß

(8) $$\gamma\bar{x}_r+\delta\bar{y}_r-\eta_0=a\mu$$

wird, mit $a=a_1+ia_2,\ 0\leqq a_i\leqq 1$. Die allgemeine Lösung von (7) können wir dann so schreiben, wenn wir $\bar{x}_r=x'_0,\ \bar{y}_r=y'_0$ setzen, wo r gemäß (8) gewählt ist:

$$x'_s=x'_0-sp,\quad y'_s=y'_0-sq.$$

Wir setzen nun

$$\bar{\xi}_s=\alpha x'_s+\beta y'_s=\alpha x'_0+\beta y'_0-s\lambda=\bar{\xi}_0-s\lambda$$
$$\bar{\eta}_s=\gamma x'_s+\delta y'_s=\gamma x'_0+\delta y'_0-s\mu=\bar{\eta}_0-s\mu.$$

Dann ist

$$\bar{\eta}_s-\eta_0=(a-s)\mu \quad\text{und}\quad \mu(\bar{\xi}_s-\xi_0)=\lambda\mu(a-s)-k,$$

also

(9) $$|(\bar{\xi}_s-\xi_0)(\bar{\eta}_s-\eta_0)|=|a-s|\,\left|a-\frac{k}{\lambda\mu}-s\right|\,|\lambda\mu|$$

und

(10) $$|\bar{\xi}_s-\xi_0|=\frac{1}{|\mu|}\left|\lambda\mu(a-s)-k\right|.$$

Aus Hilfssatz 1 folgt, daß stets $(9) < \frac{1}{2}$ ist, und aus dem Beweis dieses Hilfssatzes ersieht man noch, daß man mit $s = 0, 1, i$ oder $1+i$ auskommt. Daraus folgt noch für (10):

$$|\xi_s - \xi_0| < \frac{1}{|\mu|}\left(\sqrt{\frac{2}{3}}(1+\varepsilon) + \frac{1}{\sqrt{2}}\right).$$

Man kann also durch geeignete Wahl von p, q stets erreichen, daß

$$|\xi - \xi_0|\ |\eta - \eta_0| < \frac{1}{2}$$

und gleichzeitig $|\xi - \xi_0|$ beliebig klein wird. Aus Hilfssatz 2, (9) und (10) folgt noch, daß es unendlich viele ganze Zahlen x, y gibt, für die der Satz gilt.

Es liege jetzt $\frac{\alpha}{\beta}$ in $k(i)$. Dann gibt es ganze Zahlen p, q so daß

$$f_0(p, q) = 0,$$

und $(p, q) = 1$. Dann bestimmen wir ganze Zahlen m und n so, daß

$$pn - mq = 1.$$

Wir transformieren nun die Form

$$a(x - x_0)^2 + b(x - y_0)(y - y_0) + c(y - y_0)^2,$$

mittels der Transformation

$$x = px' + my'. \quad y = qx' + ny' \text{ (entsprechend für } x_0,\ y_0).$$

Dann wird $a' = f_0(p, q) = 0$, also

$$f'(x', y') = (y' - y_0')\,[b'(x' - x_0') + c(y - y_0')], \quad D = b'^2.$$

Es wird nun y' ganzzahlig so bestimmt, daß

$$|y' - y_0'| \leqq \frac{1}{\sqrt{2}}$$

und sodann x' so, daß

$$\left|x' - x_0' + \frac{c'}{b'}(y' - y_0')\right| \leqq \frac{1}{\sqrt{2}}.$$

Daraus folgt

$$|f'(x' y')| \leqq \frac{|b'|}{2} = \frac{1}{2}$$

und daraus die Behauptung. Das Gleichheitszeichen gilt nur, wenn

$$|y' - y_0'| = \frac{1}{\sqrt{2}} \text{ und } \left|x' - x_0' + \frac{c'}{b'}(y' - y_0')\right| = \frac{1}{\sqrt{2}}.$$

Aus der ersten Bedingung folgt $y_0'=z_1+\frac{1+i}{2}$, aus der zweiten

$$x_0'=z_2-\frac{c'}{b'}\frac{1+i}{2}+\frac{1+i}{2},$$

wobei z_1 und z_2 ganze Zahlen sind, dabei muß $\frac{c'}{b'}$ ganz sein. Daraus folgt, daß $f'(x', y')$ der Form

$$\left(x-\frac{1+i}{2}\right)\left(y-\frac{1+i}{2}\right)$$

äquivalent sein muß. Diese Form stellt $\frac{1}{2}$ als kleinste Zahl dar. Die Schranke $\frac{1}{2}$ kann also nicht herabgedrückt werden. Damit ist der Satz vollständig bewiesen.

(Eingegangen: 22. XI. 1937.)

Über die Approximation von inhomogenen Linearformen

Monatsh. Math. Phys. **47** (1938). 181–185

Es seien Θ und β zwei beliebige komplexe Zahlen, wobei nur nicht Θ in $k(i)$ liegen und β keine ganze Zahl aus $k(i)$ sein soll. Dann betrachten wir die Approximation

$$|\Theta y - x - \beta| \leqq \frac{c}{|y|}, \tag{1}$$

wo x, y ganze Zahlen aus $k(i)$ sind und c eine gegebene Konstante ist. Ich habe gezeigt[1]): Ist $c = \frac{1}{2}$, dann besitzt (1) unendlich viele Lösungen in ganzen x und y. Man kann nun fragen, ob $\frac{1}{2}$ die kleinste Konstante dieser Art ist, d. h. ob (1) für ein $c < \frac{1}{2}$ nur mehr endlich viele Lösungen besitzt. Dies ist nun der Fall, wie ich zeigen will. Das entsprechende Problem für $k(1)$, wo Θ und β reell sind, Θ nicht in $k(1)$ und β nicht ganz rational ist, wurde bereits von Grace[2]) und Morimoto gelöst, indem sie zeigten, daß

$$|\Theta y - x - \beta| \leqq \frac{c}{|y|} \tag{1'}$$

(x, y ganz in $k(1)$ für $c < \frac{1}{4}$ nur endlich viele Lösungen besitzt. Sie benützten dazu die Theorie der Kettenbrüche. Ich vermeide diese Theorie und gebe zugleich einen neuen Beweis für das Problem in $k(1)$. Am Schluß der Arbeit beweise ich eine Verallgemeinerung eines Satzes von Mordell: Sind A, B, C, D, P, Q beliebige komplexe Zahlen, wobei $\Delta = AD - BC \neq 0$, $\overline{BC}\Delta$ reell und größer als 0 ist,

---

[1]) E. Hlawka, „Über die Approximation von zwei inhomogenen Linearformen" (Monatsh. Bd. **46**, 1938).

[2]) J. H. Grace, „Note on a Diophantine Approximation" (Proceedings of the London M. S. II, **17**, 1918).

so gibt es ganze x, y aus $k(i)$, so daß

$$|Ax+By+P| \leqq \frac{|A|}{\sqrt{2}}$$

$$|Cx+Dy+Q| \leqq \frac{|D|}{\sqrt{2}}$$

ist.

Hilfssatz 1:

Es sei $f(x, y)$ die quadratische Form $bx^2-2(b-1)xy-2y^2$, wo b eine ganz rationale Zahl ist. Dann ist für alle ganzen rationalen Zahlen x, y mit $x\equiv 1\,(2)$

$$|f(x,y)| \geqq b.$$

Für $x\equiv 0\,(2)$ ist dies für $b\geqq 3$ nicht richtig.

Beweis: $f(x, y)$ kann, außer im trivialen Falle, sicher nicht die Null darstellen, es müßte sonst ganze rationale Zahlen x, y geben, so daß $f(x, y)=0$ und $b\frac{x}{y}=b-1\pm\sqrt{b^2+1}$ wäre, was einen Widerspruch bedeutet. Im Falle $b=1$ oder 2 ist der Hilfssatz sicher richtig. Wir können also $b\geqq 3$ und außerdem noch $y\geqq 0$ annehmen. Wir betrachten zuerst den Fall $x>0$. Wenn $x>2y$ oder $1\leqq x\leqq y$ ist, so ist sicher

$$|f(x, y)|=|bx(x-2y)+2y(x-y)|\geqq b.$$

Ebenso, wenn $b(2y-x)\geqq y$ ist, denn dann ist wieder

$$|f(x, y)|=|b(2y-x)^2+2(x-y)(b(2y-x)-y)|\geqq b.$$

Es sei nun $b(2y-x)<y$, $y<x<2y$. Dann setzen wir

$$x_1=2y-x_1 \quad y_1=bx-(2b-1)y$$

und erhalten

$$|f(x, y)|=|f(x_1, y_1)|$$

mit $x_1\equiv 1\,(2)$, $1\leqq x_1<y$, $0\leqq y_1<y$. Es ist nun entweder $b(2y_1-$ $-x_1)\geqq y_1$, dann ist

$$|f(x, y)|=|f(x_1, y_1)|\geqq b$$

oder es ist wieder $b(2y_1-x_1)<y_1$. Dann setzen wir den Prozeß fort und kommen schließlich zu

$$|f(x,y)|=|f(x_1, y_1)|=\ldots=|f(x_{n-1}, y_{n-1})|=|f(1, 0)|=b.$$

Wir nehmen jetzt $x<0$ an und ersetzen x durch $-x$, betrachten also

$$f(-x, y)=bx^2+2(b-1)xy-2y^2$$

mit $x>0$. Wenn $y \geqq bx$ ist, so ist

$$|f(-x, y)| = |2y(y-bx) + x(2y-bx)| \geqq b,$$

analog wenn $y \leqq (b-1)x$, denn dann ist

$$|f(-x, y)| = |bx^2 + 2y((b-1)x-y)| \geqq b.$$

Es sei also $(b-1)x < y < bx$. Wenn nun $(2b-1)x > 2y$ ist, so ist

$$|f(-x, y)| = |bx((2b-1)x-2y) + 2(bx-y)(y-(b-1)x)| \geqq b$$

Es sei nun $(2b-1)x < 2y$. Setzen wir nun

$$x_1 = -(2b-1)x + 2y,\quad y_1 = bx - y,$$

so ist
$$|f(-x, y)| = |f(-x_1, y_1)|$$

mit $x_1 \equiv 1(2)$, $1 \leqq x_1 < x$, $0 \leqq y_1 < x$. Wir gehen nun so wie oben vor und erhalten schließlich

$$|f(-x, y)| \geqq b.$$

Damit ist alles gezeigt.

Hilfssatz 2: Es sei $F(x, y)$ die quadratische Form

$$F(x, y) = bx^2 - 2i(b-1)xy + 2y^2,$$

wo b eine ganze rationale Zahl ist. Wenn x, y ganze Zahlen aus $k(i)$ sind, wobei $x \equiv 1 + i\,(2)$ ist, so ist

$$|F(x, y)| \geqq 2b.$$

Beweis: Wir setzen $x = x_1 + ix_2$, $y = y_1 + iy_2$, wobei $x_1 \equiv x_2 \equiv$ $\equiv 1\ (2)$ ist. Dann ist

$$|F(x, y)|^2 = (f(x_1, -y_2) - f(x_2, y_1))^2 + 4\Phi(x_1, x_2, y_1, y_2),$$

wo $f(x, y)$ die Form aus Hilfssatz 1 ist und

$$\begin{aligned}\Phi &= [bx_1x_2 + 2y_1y_2 - (b-1)(x_1y_1 - x_2y_2)]^2\\ &= f(x_1, -y_2)f(x_2, y_1) + (b^2+1)(x_1y_1 + x_2y_2)^2.\end{aligned}$$

Es sei zuerst $\varphi = x_1y_1 + x_2y_2 \neq 0$, dann ist

$$|F(x, y)|^2 = f(x_1, -y_2)^2 + f(x_2, y_1)^2 + 2\Phi + 2(b^2+1)\varphi^2,$$

also nach Hilfssatz 1

$$|F(x, y)|^2 > 4b^2.$$

Ist $\varphi = 0$, dann ist

$$|F(x, y)|^2 = (f(x_1, -y_2) + f(x_2, y_1))^2.$$

Weiter ist nun $\operatorname{sgn} f(x_1, -y_2) = \operatorname{sgn} f(x_2, y_1)$

da
$$f(x_1, -y_2)\,f(x_2, y_1) = \Phi > 0$$

ist. Also ist wieder

$$|F(x, y)|^2 \geqq 4b^2.$$

Wir gehen nun zu dem Beweis des Satzes über. Wir wollen zeigen: Genügt Θ der Gleichung

$$2b\Theta^2 - 2c(b-1)\Theta - c^2 = 0 \qquad (1'')$$

wo $c = 1$, wenn wir $k(1)$ betrachten und $c = i$ im Falle $k(i)$, so hat die Ungleichung

$$|\Theta y' - x' - \beta| < \frac{k}{|y'|}$$

nur endlich viele Lösungen im ganzen x' y' aus $k(1)$ bzw. aus $k(i)$, wenn im ersten Falle $\beta = \frac{1}{2}$, $k < \frac{1}{4}$ ist, im zweiten Falle $\beta = \frac{1+i}{2}$, $k < \frac{1}{2}$. Um dies zu zeigen, schreiben wir die Ungleichung so:

$$\left|\Theta - \frac{2x' + \beta'}{2y'}\right| < \frac{hk'}{|2y'|^2} \qquad h < 1,$$

wo $\beta' = 1$, $k' = 1$ bzw. $\beta' = 1 + i$, $k' = 2$ ist. Setzen wir also $\xi = 2x' + \beta'$, $2y' = \eta'$, so haben wir zu zeigen: zu jedem $h < 1$ gibt es ein $\Theta(h)$, so daß

$$\left|\Theta - \frac{\xi}{\eta}\right| < \frac{hk'}{|\eta|^2}, \qquad (2)$$

wo $\xi \equiv 1\,(2)$ bzw. $1 + i\,(2)$ und $\eta \equiv 0(2)$ ist, nur endlich viele Lösungen hat.

Beweis: Ich bestimme eine ganze, rationale Zahl b so, daß $h \leqq 1 - \frac{1}{b}$ ist und ein Θ, daß $(1'')$ genügt. Angenommen (2) habe unendlich viele Lösungen. Ich setze

$$\Theta = \frac{\xi}{\eta} + \frac{h'k'}{\eta^2}, \qquad |h'| \leqq h \qquad (3)$$

und gehe damit in $(1'')$ ein, ordne und erhalte

$$[2b\xi^2 - 2c(b-1)\xi\eta - c^2\eta^2] + 2\left[\pm k'h'c\sqrt{b^2+1} - \frac{bh'^2k'^2}{\eta^2}\right] = 0.$$

Nun ist der 2. Term für genügend großes $|\eta|$ dem Betrage nach kleiner als

$$2k'h\sqrt{b^2+1} \leqq 2k'\sqrt{b^2+1}\left(1 - \frac{1}{b}\right) < 2k'b.$$

Die erste Klammer ist aber nach Hilfssatz 1 bzw. 2 dem Betrage nach größer oder gleich $2k'b$, wenn man $\xi = x$, $\eta = 2y$ setzt. Das ist aber ein Widerspruch und damit ist die Behauptung bewiesen.

Zum Schlusse beweisen wir noch den Satz: Es gibt stets ganze ξ, η aus $k(i)$, so daß

$$|A\xi + B\eta + P| \leqq \frac{|A|}{\sqrt{2}}$$

$$|C\xi + D\eta + Q| \leqq \frac{|D|}{\sqrt{2}}$$

wenn nur $\Delta = AD - BC \neq 0$, $\overline{B}\,\overline{C}\Delta$ reell und $\geqq 0$ ist.

Beweis: Da $\overline{B}\,\overline{C}\Delta$ reell ist, so ist es auch

$$d = \frac{D\overline{B}A}{C}.$$

Setzen wir noch $a = A\overline{A}$, $b = B\overline{A}$, $f = P\overline{A}$, $g = \frac{Q\overline{B}A}{C}$, so folgt

$$b\overline{b} - ad = -\frac{A\overline{A}}{C\overline{C}} BC\Delta < 0.$$

Dann besitzt aber

$$\Phi(x, y) = ax\overline{x} + \overline{b}x\overline{y} + b\overline{x}y + dy\overline{y} + f\overline{x} + \overline{f}x + g\overline{y} + \overline{g}y$$

ein Minimum ξ, η. Es sei $e = \pm 1$, $\pm i$, so ist

$$\Phi(\xi + e, \eta) \geqq \Phi(\xi, \eta),$$

oder

$$ea\overline{\xi} + \overline{e}a\xi + a + b\overline{e}\eta + \overline{b}e\overline{\eta} + f\overline{e} + \overline{f}e \geqq 0.$$

Nehmen wir zuerst $e = \pm 1$ und dann $e = \pm i$, so erhalten wir

$$|\varphi_1| = |a(\xi + \overline{\xi}) + b\eta + \overline{b}\,\overline{\eta} + f + \overline{f}| \leqq a$$

$$|\varphi_2| = |i[(\overline{\xi} - \xi)a + \overline{b}\,\overline{\eta} - b\eta - f + \overline{f}]| \leqq a,$$

also

$$|a\xi + b\eta + f| = \frac{1}{2}|\varphi_1 + i\varphi_2| \leqq \frac{|a|}{\sqrt{2}}.$$

Wenn wir $\Phi(\xi, \eta + e)$ betrachten, erhalten wir

$$|\overline{b}\xi + d\eta + g| \leqq \frac{|c|}{\sqrt{2}}.$$

Ersetzen wir a, b, d, f, g durch A, B, C, D, P, Q, so erhalten wir die Behauptung.

(Eingegangen: 18. X. 1938.)

Über komplexe homogene Linearformen

Monatsh. Math. Phys. 49 (1941), 321–326

Minkowski[1]) hat gezeigt: Es gibt stets ganze Zahlen $(x,y)\neq(0,0)$ aus $k(i)$, so daß

$$(1)\qquad \begin{aligned} |\alpha x+\beta y|^2 &\leqq \frac{1}{\sqrt{2}}\left(1+\frac{1}{\sqrt{3}}\right)\\ |\gamma x+\delta y|^2 &\leqq \frac{1}{\sqrt{2}}\left(1+\frac{1}{\sqrt{3}}\right)\end{aligned}$$

ist, wobei $\alpha, \beta, \gamma, \delta$ komplexe Zahlen mit $|\alpha\delta-\beta\gamma|=1$ sind.

Für diesen Satz gebe ich einen neuen Beweis.

Sind $L_1(x,y)$ und $L_2(x,y)$ die obigen Linearformen, so bilden wir

$$|L_1(x,y)|^2-(-1)^k\,|L_2(x,y)|^2=F_k(x,y).\qquad (k=1,2)$$

Die $F_k(x,y)$ sind hermitesche Formen mit der Diskriminante $(-1)^k$:

$$F_k(x,y)=a_k|x+b_ky|^2+\frac{(-1)^{k+1}}{a_k}|y|^2,\ b_k=p_{2k-1}+ip_{2k}.\qquad (k=1,2)$$

Die a_k sind reell und die Linearformen $L_1(x,y)$ und $L_2(x,y)$ seien so numeriert, daß $a_1\geqq a_2\geqq 0$ ist. Die Formen

$$F_1(x,y)+(-1)^{k+1}F_2(x,y)=2\,|L_k(x,y)|^2$$

müssen die Diskriminante Null haben oder

$$(2)\qquad |b_1-b_2|^2=\frac{1}{a_2^{\,2}}-\frac{1}{a_1^{\,2}}$$

O. B. d. A. sei $a_1\leqq\sqrt{2},\ 0\leqq p_1\leqq\frac{1}{2},\ 0\leqq p_2\leqq\frac{1}{2}$[2]).

Die Behauptung (1) schreibt sich jetzt so: Es gibt ganze Zahlen $(x,y)\neq(0,0)$ aus $k(i)$, so daß

$$(3)\qquad f(x,y)=F_1(x,y)+|F_2(x,y)|\leqq K=\sqrt{2}\left(1+\frac{1}{\sqrt{3}}\right)$$

ist. Im besonderen ist $f(1,0)=a_1+a_2$. Ist nun $a_1+a_2\leqq K$ oder

---

[1]) Minkowski, Diophantische Approximationen.

[2]) Perron, Mathematische Zeitschrift, Bd. 36, S. 148. — Blaha, Monatshefte Bd. 47, S. 195.

$a_2 \leqq \sqrt{\frac{2}{3}}$ oder $a_1 \leqq \frac{K}{2}$ so sind wir fertig. Ebenso wenn $|b_2 - b_1|^2 \geqq 1$, denn dann ist nach (2) $\frac{1}{a_2^2} > 1 + \frac{1}{a_1^2} > \frac{3}{2}$, also wieder $a_2 \leqq \sqrt{\frac{2}{3}}$.

Wir können also annehmen:

(4) $$|b_1 - b_2|^2 < 1,\ a_2 > \sqrt{\tfrac{2}{3}},\ a_1 + a_2 > K,\ a_1 > \frac{K}{2},$$

und außerdem noch

(5) $$f(0,1) = F_1(0,1) + |F_2(0,1)| > K.$$

Es sei $p_1 \geqq p_2$, dann wollen wir zeigen:

(6) $$f(-1,1) = F_1(-1,1) + |F_2(-1,1)| < K.$$

Der Fall $p_2 > p_1$ erledigt sich analog.

Es ist $F_1(-1,1) = a_1(1-2p_1) + F_1(0,1)$, also

(6') $$f(-1,1) = a_1(1-2p_1) + 2F_1(0,1) + |F_2(-1,1)| - F_1(0,1),$$

daher nach (5)

(7) $$f(-1,1) < a_1(1-2p_1) + 2F_1(0,1) + |F_2(0,1)| + |F_2(-1,1)| - K.$$

Wir unterscheiden nun mehrere Fälle:

I) $$F_2(0,1) < 0.$$

Dann muß $p_1 > \frac{1}{2\sqrt{1+a_1^2}}$ sein, denn sonst wäre $|b_1|^2 \leqq 2p_1^2 \leqq \frac{1}{2(1+a_1^2)}$ und daher

(8) $$F_1(0,1) - F_2(0,1) =$$
$$= a_1|b_1|^2 - a_2|b_2|^2 + \frac{1}{a_1} + \frac{1}{a_2} \leqq \frac{a_1}{2(1+a_1^2)} + \frac{1}{a_1} + \frac{1}{a_2}.$$

Nun ist $\frac{1}{a_1} + \frac{1}{a_2} < \frac{1}{a_1} + \frac{1}{K-a_1} = 4K\,(K^2-(2a_1-K)^2)^{-1} \leqq (1+\sqrt{3})\frac{\sqrt{2}}{2}$ nach (4) und $\frac{a_1}{1+a_1^2} \leqq \frac{1}{2}$, also ist (8) kleiner als K, in Widerspruch zu (5).

I *a)* $$F_2(-1,1) < 0.$$

Dann schreibt sich (7) so:

(7') $$f(-1,1) < a_1[1-2(p_1 - |b_1|^2)] -$$
$$- a_2[(1-p_3)^2 + p_3^2 + 2p_4^2] + 2\left(\frac{1}{a_1} + \frac{1}{a_2}\right) - K.$$

Ist nun $|b_2 - b_1|^2 \leqq \frac{1}{4}$, so ist $a_2 > 1$, denn nach (2) wäre für $a_2 \leqq 1$

$$1 \leqq \frac{1}{a_2^2} \leqq \frac{1}{4} + \frac{1}{a_1^2}$$

also $a_1 \leqq \frac{2}{\sqrt{3}}$ und daher $a_1 + a_2 \leqq 1 + \frac{2}{\sqrt{3}} < K$. Dann ist aber (7′) kleiner als

$$a_1 - \frac{a_2}{2} + 2\left(\frac{1}{a_1} + \frac{1}{a_2}\right) - K \leqq \frac{3}{2} a_1 + 2\left(\frac{1}{a_1} + \frac{1}{a_2}\right) - \frac{3}{2} K,$$

da $p_1 > |b_1|^2$ und $(1-p_3)^2 + p_3^2 \geqq \frac{1}{2}$. Der letzte Ausdruck ist aber kleiner als K, denn er ist in a_1 konvex und sowohl für $a_1 = 1$ wie $\sqrt{2}$ kleiner als K, da $a_2 > \sqrt{\frac{2}{3}}$. Es sei also $|b_1 - b_2|^2 > \frac{1}{4}$. Dann werden wir zeigen:

$$|b_2|^2 - p_3 \geqq \frac{1}{a_2^2} - \frac{1}{a_1^2} - \sqrt{\frac{1}{a_2^2} - \frac{1}{a_1^2}}. \tag{9}$$

Ist $p_3 \geqq \sqrt{\frac{1}{a_2^2} - \frac{1}{a_1^2}}$, so ist (9) richtig. Im entgegengesetzten Fall schließen wir so: Es ist

$$|p_4 - p_2| = \sqrt{\frac{1}{a_2^2} - \frac{1}{a_1^2} - (p_3 - p_1)^2},$$

also wenn wir in (9) für p_4 einsetzen:

$$|b_2|^2 - p_3 > -(1 - 2p_1)p_3 + p_2^2 - p_1^2 + \frac{1}{a_2^2} - \frac{1}{a_1^2} - 2p_2\sqrt{\frac{1}{a_2^2} - \frac{1}{a_1^2}}$$

Nun ist der letzte Ausdruck in p_2 monoton abnehmend und da $p_2 < p_1$, so ist

$$|b_2|^2 - p_3 > \frac{1}{a_2^2} - \frac{1}{a_1^2} - \sqrt{\frac{1}{a_2^2} - \frac{1}{a_1^2}} + (1 - 2p_1)\left(\sqrt{\frac{1}{a_2^2} - \frac{1}{a_1^2}} - p_3\right).$$

Damit ist (9) bewiesen. Benützen wir dies, so wird (7′) kleiner als

$$2\left(a_1 + \frac{1}{a_1} + \sqrt{1 - \left(\frac{a_2}{a_1}\right)^2} + \frac{a_2}{a_1}\frac{1}{a_1}\right) - 2K,$$

denn es ist $(1-p_3)^2 + p_3^2 + 2p_4^2 = 1 + 2(|b_2|^2 - p_3)$. Dieser Ausdruck ist in $\frac{a_2}{a_1}$ monoton abnehmend, denn ihre Ableitung nach $\frac{a_2}{a_1}$ ist $\frac{1}{a_1}\left(1 - \left(\frac{1}{a_2^2} - \frac{1}{a_1^2}\right)^{-\frac{1}{2}}\right) < 0$ und $\frac{a_2}{a_1}$ nach (4) größer als $\frac{1}{\sqrt{3}}$ also (7′) kleiner als $2\left(a_1 + \frac{K}{\sqrt{2}\,a_1} + \sqrt{\frac{2}{3}}\right) - 2K$. Dieser Term ist aber in a_1 konvex und sowohl für $a_1 = \frac{K}{2}$ wie $\sqrt{2}$ gleich K.

I *b*) $$F_2(-1, 1) > 0.$$

Dann wird (7):

$$(7'') \qquad a_1[1-2(p_1-|b_1|^2)]+a_2(1-2p_3)+\frac{2}{a_1}-K.$$

Ist $p_3 \geqq \frac{1}{2}$, so ist (7″) kleiner als $a_1+\frac{2}{a_1}-K$. Dieser Ausdruck ist aber in a_1 monoton abnehmend und für $a_1>\frac{K}{2}$ kleiner als $\sqrt{2}<K$. Wenn aber $p_3<\frac{1}{2}$ ist, dann ist $(1-2p_3)\left(\frac{1}{a_1^2}+(p_1-p_3)^2\right)^{-\frac{1}{2}}$ wegen $|p_3-p_1|<1$ und $a_1\leqq\sqrt{2}$ monoton abnehmend, denn ihre Ableitung beträgt bis auf einen quadratischen Faktor:

$$(1-2p_3)\left((1-2p_1)(p_1-p_3)-\frac{2}{a_1^2}\right)<0.$$

Nun ist nach (2) $\frac{1}{a_2^2}-\frac{1}{a_1^2}=(p_1-p_3)^2+(p_2-p_4)^2>(p_1-p_3)^2$ und $p_3>-1+p_1$, also

$$a_2(1-2p_3)<(1-2p_3)\left(\frac{1}{a_1^2}+(p_1-p_3)^2\right)^{-\frac{1}{2}}<(3-2p_1)\left(1+\frac{1}{a_1^2}\right)^{-\frac{1}{2}}$$

daher, da $|b_1|^2\leqq 2p_1^2$, ist (7″) kleiner als

$$a_1+\frac{2}{a_1}+\frac{2a_1}{\sqrt{1+a_1^2}}-(1-2p_1)a_1\left(2p_1-\frac{1}{\sqrt{1+a_1^2}}\right)-K<K,$$

da $p_1>\frac{1}{2\sqrt{1+a_1^2}}$ und $a_1+\frac{2}{a_1}+2a_1(1+a_1^2)^{-\frac{1}{2}}$ für $1<a_1\leqq\sqrt{2}$ konvex ist und für 1 und $\sqrt{2}$ kleiner K ist.

II) $$F_2(0,1)>0.$$

Ist $|b_2|^2\leqq\frac{1}{2}$, so ist

$$(5') \quad F_1(0,1)+F_2(0,1)= \\ =a_1|b_1|^2+a_2|b_2|^2+\frac{1}{a_1}-\frac{1}{a_2}\leqq\frac{a_1+a_2}{2}+\frac{1}{a_1}-\frac{1}{a_2}\leqq\sqrt{2},$$

da $a_2\leqq a_1$. Es sei also $|b_2|^2>\frac{1}{2}$. Nun ist nach (2)

$$|b_2|^2=\frac{1}{a_2^2}-\frac{1}{a_1^2}-|b_1|^2+2(p_1p_3+p_2p_4),$$

also wird (5′):

$$(5'') \quad (a_1-a_2)|b_1|^2+2a_2(p_1p_3+p_2p_4)+ \\ +\frac{1}{a_1}-\frac{a_2}{a_1^2}\leqq\frac{a_1}{2}+\frac{1}{a_1}-\frac{a_2}{a_1^2}-\frac{a_2}{2}+2a_2(p_1p_3+p_2p_4),$$

da $|b_1|^2\leqq\frac{1}{2}$ ist. Nun ist

$$(10)\quad \frac{a_1}{2}+\frac{1}{a_1}-\frac{a_2}{a_1{}^2}=$$

$$=\frac{1}{2}\left(a_1+\frac{1}{a_1}\right)-\frac{1}{a_1{}^2}\left(a_2-\frac{a_1}{2}\right)\leqq\frac{3}{2}\sqrt{2}-\frac{1}{2}\left(a_2-\frac{a_1}{2}\right)\leqq\sqrt{2}-\frac{a_2}{2},$$

da $a_2>\frac{a_1}{2}$ und $\frac{1}{a_1{}^2}\geqq\frac{1}{2}$ ist, also (5″) kleiner als

$$(10')\qquad \sqrt{2}+2\,a_2\left(p_1\,p_3+p_2\,p_4-\frac{1}{2}\right).$$

Es sei jetzt z. B. $p_4<0$, und bemerken wir, daß $p_3\leqq\frac{3}{2}$ ist, so ist (10′) kleiner als $\sqrt{2}+2\,a_2\left(p_1 p_3-\frac{1}{2}\right)\leqq\sqrt{2}+\frac{a_2}{2}<K$. Also können wir p_3 und p_4 größer als Null annehmen. Ist jetzt z. B. $p_4<\frac{1}{2}$, dann ist (10′) kleiner als $\sqrt{2}+a_2\left(p_3-\frac{1}{2}\right)$. Nun ist aber stets für $\frac{3}{2}>p_k>\frac{1}{2}$.

$$(11)\quad a_2\left(p_k-\frac{1}{2}\right)<\left(p_k-\frac{1}{2}\right)\left((p_k-\frac{1}{2})^2+\frac{1}{a_1{}^2}\right)^{-\frac{1}{2}}\leqq\left(1+\frac{1}{a_1{}^2}\right)^{-\frac{1}{2}}\leqq\sqrt{\frac{2}{3}},$$

da ja $\frac{1}{a_2{}^2}-\frac{1}{a_1{}^2}=(p_3-p_1)^2+(p_4-p_2)^2\geqq\left(p_k-\frac{1}{2}\right)^2$ ist, also

$$\sqrt{2}+a_2\left(p_3-\frac{1}{2}\right)<K.$$

Wir nehmen also p_3 und p_4 größer $\frac{1}{2}$ an.

II *a*) $\qquad F_2(-1,1)<0.$

Dann schreibt sich (7):

$$(12)\qquad a_1(1-2(p_1-|b_1|^2))+2\,a_2\left(p_3-\frac{1}{2}\right)+\frac{2}{a_1}-K$$
$$<a_1+\frac{2}{a_1}+\frac{2\,a_1}{\sqrt{1+a_1{}^2}}-K<K$$

wegen (11) (Vgl. I *b*)

II *b*) $\qquad F_2(-1,1)>0.$

Dann wird (7)

$$(12a):\quad a_1(1-2(p_1-|b_1|^2))+a_2((1-p_3)^2+p_3{}^2+2\,p_4{}^2)+\frac{2}{a_1}-\frac{2}{a_2}-K$$
$$<a_1+2\,a_2\left((p_3-\frac{1}{2})^2+(p_4-\frac{1}{2})^2\right)+\frac{2}{a_1}-\frac{2}{a_2}+2\,a_2\,p_4-K.$$

Nun ist nach (2):

$$\frac{1}{a_2{}^2}-\frac{1}{a_1{}^2}=(p_3-p_1)^2+(p_4-p_2)^2\geqq\left(p_3-\frac{1}{2}\right)^2+\left(p_4-\frac{1}{2}\right)^2,$$

also wird (12′) kleiner als

$$a_1 - \frac{2a_2}{a_1^2} + \frac{2}{a_1} - K + 2\,a_2 p_4 \leqq 2\sqrt{2} + 2a_2\left(p_4 - \frac{1}{2}\right) - K < K$$

wegen (10) und (11).

Damit ist der Beweis vollendet. Aus dem Beweis folgt noch, daß dann und nur dann das Gleichheitszeichen gilt, wenn

$$a_1 = \sqrt{2},\ b_1 = \frac{1+i}{2},\ b_2 = \frac{1-i}{2},\ a_2 = \sqrt{\tfrac{2}{3}}$$

und für die dazu äquivalenten Formen.

(Eingegangen: 7. 10. 1940.)

1

Eine Ungleichung für Vektorlängen

Enthalten in: H. Hornich: Eine Ungleichung für Vektorlängen
Math. Z. **48** (1942). 268–274

In dieser Arbeit soll eine weitgehende Verallgemeinerung der Dreiecksungleichung gezeigt werden, die trotz ihrer Einfachheit und Allgemeinheit nicht bekannt zu sein scheint. Ihr hier gegebener Beweis erfordert einige Umwege und Hilfssätze.

Sind $\mathfrak{a}_1 \ldots \mathfrak{a}_n$, $\mathfrak{b}_1 \ldots \mathfrak{b}_m$ *beliebige* $n+m$ *Vektoren mit* $m \geqq 1$ *und mit*

$$(1) \qquad \sum_{i=1}^{n} \mathfrak{a}_i = \sum_{j=1}^{m} \mathfrak{b}_j$$

und ist $\mathfrak{c}$ *ein Einheitsvektor eines* R_k, *so gilt die folgende Ungleichung*:

$$(2) \qquad \sum_{i=1}^{n} (|\mathfrak{a}_i + \mathfrak{c}| - |\mathfrak{a}_i|) \leqq \sum_{j=1}^{m} (|\mathfrak{b}_j + \mathfrak{c}| - |\mathfrak{b}_j|) + n + m - 2.$$

Als einfachster Spezialfall ergibt sich für $m = 1$ und $n = 2$ mit drei beliebigen Vektoren $\mathfrak{a}\,\mathfrak{b}\,\mathfrak{c}$ die Ungleichung:

$$|\mathfrak{a}| + |\mathfrak{b}| + |\mathfrak{c}| + |\mathfrak{a} + \mathfrak{b} + \mathfrak{c}| \geqq |\mathfrak{a} + \mathfrak{b}| + |\mathfrak{b} + \mathfrak{c}| + |\mathfrak{c} + \mathfrak{a}|,$$

welche wieder die Dreiecksungleichung als Spezialfall enthält.

Zum Beweis von (2) setzen wir

$$\sum_{i=1}^{n} |\mathfrak{a}_i| + \sum_{j=1}^{m} |\mathfrak{b}_j| = R$$

und zeigen den Hilfssatz:

Es ist gleichmäßig für alle $\mathfrak{a}_i$ und $\mathfrak{b}_j$, welche (1) erfüllen:

$$(3) \qquad \limsup_{R \to \infty} \Big(\sum_{i=1}^{n} (|\mathfrak{a}_i + \mathfrak{c}| - |\mathfrak{a}_i|) - \sum_{j=1}^{m} (|\mathfrak{b}_j + \mathfrak{c}| - |\mathfrak{b}_j|) \Big) \leqq n + m - 2.$$

Zunächst ist:

$$|\mathfrak{a}_i + \mathfrak{c}| - |\mathfrak{a}_i| = \frac{2\,\mathfrak{a}_i\,\mathfrak{c} + 1}{|\mathfrak{a}_i + \mathfrak{c}| + |\mathfrak{a}_i|}$$

und analog für die $\mathfrak{b}_j$, so daß wir (3) auch schreiben können:

$$(4) \qquad \sum_{i=1}^{n} \frac{2\,\mathfrak{a}_i\,\mathfrak{c} + 1}{|\mathfrak{a}_i + \mathfrak{c}| + |\mathfrak{a}_i|} - \sum_{j=1}^{m} \frac{2\,\mathfrak{b}_j\,\mathfrak{c} + 1}{|\mathfrak{b}_j + \mathfrak{c}| + |\mathfrak{b}_j|} \leqq m + n - 2 + o(1)$$

Wir betrachten zunächst den Fall, daß für ein festes $R > 0$ alle $|\mathfrak{a}_i|$ und $|\mathfrak{b}_j| > \frac{R}{2(m+n)}$ sind. Dann können wir $|\mathfrak{a}_i + \mathfrak{c}| = |\mathfrak{a}_i|(1 + \varepsilon_i)$ setzen, wo ε_i absolut unterhalb einer nur von R abhängigen und mit $R \to \infty$ gegen Null strebenden Schranke liegt; analoges gilt für die $|\mathfrak{b}_j + \mathfrak{c}|$. Es ist also:

$$(5)\quad \sum_{i=1}^{n} \frac{2\mathfrak{a}_i\mathfrak{c}+1}{|\mathfrak{a}_i+\mathfrak{c}|+|\mathfrak{a}_i|} - \sum_{j=1}^{m} \frac{2\mathfrak{b}_j\mathfrak{c}+1}{|\mathfrak{b}_j+\mathfrak{c}|+|\mathfrak{b}_j|} = \sum_{i=1}^{n} \frac{\mathfrak{a}_i\mathfrak{c}}{|\mathfrak{a}_i|} - \sum_{j=1}^{m} \frac{\mathfrak{b}_j\mathfrak{c}}{|\mathfrak{b}_j|} + o(1).$$

$$\leqq \left| \sum_{i=1}^{n} \frac{\mathfrak{a}_i}{|\mathfrak{a}_i|} - \sum_{j=1}^{m} \frac{\mathfrak{b}_j}{|\mathfrak{b}_j|} \right| + o(1).$$

Wir bestimmen nun das Maximum der Größe $\left|\sum_{i=1}^{m+n} \mathfrak{e}_i\right|$, wo die $\mathfrak{e}_i$ Einheitsvektoren sind und eine Relation

$$\sum_{i=1}^{m+n} \alpha_i \mathfrak{e}_i = 0$$

mit $\alpha_i \geqq 0$ bestehen soll. Sei etwa $\alpha_1 = \operatorname{Max}_i \alpha_i = 1$; dann ist:

$$\left|\sum_{i=1}^{n+m} \mathfrak{e}_i\right| = \left|\sum_{i=2}^{n+m} (1-\alpha_i)\,\mathfrak{e}_i\right| \leqq \sum_{i=2}^{n+m} (1-\alpha_i);$$

wegen $\mathfrak{e}_1 = -\sum_{i=2}^{n+m} \alpha_i \mathfrak{e}_i$ ist aber $\sum_{i=2}^{n+m} \alpha_i \geqq 1$ und $\left|\sum_{i=1}^{n+m} \mathfrak{e}_i\right| \leqq n+m-2$. Diese Zahl ist natürlich auch das Maximum.

Es ist also stets:

$$\left| \sum_{i=1}^{n} \frac{\mathfrak{a}_i}{|\mathfrak{a}_i|} - \sum_{j=1}^{m} \frac{\mathfrak{b}_j}{|\mathfrak{b}_j|} \right| \leqq n+m-2,$$

und da alle $|\mathfrak{a}_i|$, $|\mathfrak{b}_j| > \frac{R}{2(m+n)}$ sind, so ist

$$(6)\quad \sum_{i=1}^{n} \frac{2\mathfrak{a}_i\mathfrak{c}+1}{|\mathfrak{a}_i+\mathfrak{c}|+|\mathfrak{a}_i|} - \sum_{j=1}^{m} \frac{2\mathfrak{b}_j\mathfrak{c}+1}{|\mathfrak{b}_j+\mathfrak{c}|+|\mathfrak{b}_j|} \leqq n+m-2+o(1).$$

Seien nun einige der $|\mathfrak{a}_i|$ oder $|\mathfrak{b}_j| \leqq \frac{R}{2(m+n)}$; es muß stets eines der $|\mathfrak{a}_i|$ oder eines der $|\mathfrak{b}_j| > \frac{R}{2(m+n)}$ sein. Wir beachten weiter, daß für ein λ mit $0 \leqq \lambda \leqq 1$ und einem beliebigen Vektor $\mathfrak{a}$ gilt:

$$\frac{(\mathfrak{a}-\lambda\mathfrak{c})\,\mathfrak{c}}{|\mathfrak{a}-\lambda\mathfrak{c}|} \leqq \frac{(\mathfrak{a}+\mathfrak{c})\,\mathfrak{c}+\mathfrak{a}\,\mathfrak{c}}{|\mathfrak{a}+\mathfrak{c}|+|\mathfrak{a}|} \leqq \frac{(\mathfrak{a}+\mathfrak{c}+\lambda\mathfrak{c})\,\mathfrak{c}}{|\mathfrak{a}+\mathfrak{c}+\lambda\mathfrak{c}|}.$$

Für alle $|\mathfrak{a}_i| \leqq \frac{R}{2(m+n)}$ ersetzen wir $\frac{2\mathfrak{a}_i\mathfrak{c}+1}{|\mathfrak{a}_i+\mathfrak{c}|+|\mathfrak{a}_i|}$ auf der linken Seite von (4) durch die größeren Werte $\frac{(\mathfrak{a}_i+\mathfrak{c}+\lambda\mathfrak{c})\,\mathfrak{c}}{|\mathfrak{a}_i+\mathfrak{c}+\lambda\mathfrak{c}|}$; es seien deren etwa insgesamt n'. Ist nun ein Glied $|\mathfrak{a}_i|$, etwa $|\mathfrak{a}_n| > \frac{R}{2(m+n)}$, so ersetzen wir das entsprechende Glied auf der linken Seite von (4) durch $\frac{(\mathfrak{a}_n-n'\mathfrak{c}-n'\lambda\mathfrak{c})\,\mathfrak{c}}{|\mathfrak{a}_n-n'\mathfrak{c}-n'\lambda\mathfrak{c}|}$, was für dieses Glied nur eine Änderung $o(1)$ — und zwar gleichmäßig für alle $\mathfrak{a}_i$ — ergibt. Gibt es aber kein Glied $|\mathfrak{a}_i| > \frac{R}{2(m+n)}$, so muß es ein $\mathfrak{b}_j$, etwa $|\mathfrak{b}_m| > \frac{R}{2(m+n)}$ geben; dann ersetzen wir das entsprechende Glied auf der linken Seite von (4) durch $\frac{(\mathfrak{b}_m + n'\mathfrak{c} + n'\lambda\mathfrak{c})\mathfrak{c}}{|\mathfrak{b}_m + n'\mathfrak{c} + n'\lambda\mathfrak{c}|}$, was für dieses Glied wieder nur eine Änderung $o(1)$ bedeutet.

Analog ist mit den $\mathfrak{b}_j$ zu verfahren, für welche $|\mathfrak{b}_j| \leqq \frac{R}{2(m+n)}$ ist: wir ersetzen $\frac{2\mathfrak{b}_j\mathfrak{c}+1}{|\mathfrak{b}_j+\mathfrak{c}|+|\mathfrak{b}_j|}$ auf der linken Seite von (4) durch die kleineren Werte $\frac{(\mathfrak{b}_j - \lambda\mathfrak{c})\mathfrak{c}}{|\mathfrak{b}_j - \lambda\mathfrak{c}|}$; es seien derer etwa m'. Gibt es ein $\mathfrak{b}_j$ mit $|\mathfrak{b}_j| > \frac{R}{2(m+n)}$, so ersetzen wir das entsprechende Glied auf der linken Seite von (4) durch $\frac{(\mathfrak{b}_j + m'\lambda\mathfrak{c})\mathfrak{c}}{|\mathfrak{b}_j + m'\lambda\mathfrak{c}|}$, was für dieses Glied nur eine Änderung $o(1)$ ergibt. Gibt es aber kein $|\mathfrak{b}_j| > \frac{R}{2(m+n)}$, so gibt es ein $|\mathfrak{a}_i| > \frac{R}{2(m+n)}$ und wir ersetzen das entsprechende Glied auf der linken Seite von (4) durch $\frac{(\mathfrak{a}_i - m'\lambda\mathfrak{c})\mathfrak{c}}{|\mathfrak{a}_i - m'\lambda\mathfrak{c}|}$, was wieder nur eine Änderung $o(1)$ gibt.

Für die so abgeänderte und — abgesehen von Größen $o(1)$ — vergrößerte linke Seite von (4) ist dieselbe Überlegung wie oben anzuwenden: es ist dann wieder

$$\sum_{i=1}^{n} \frac{2\mathfrak{a}_i\mathfrak{c}+1}{|\mathfrak{a}_i+\mathfrak{c}|+|\mathfrak{a}_i|} - \sum_{j=1}^{m} \frac{2\mathfrak{b}_j\mathfrak{c}+1}{|\mathfrak{b}_j+\mathfrak{c}|+|\mathfrak{b}_j|} \leqq m+n-2+o(1).$$

Damit ist (3) vollständig nachgewiesen.

Dem weiteren Beweis von (2) schicken wir folgende Bemerkung voraus: Wir können für diesen Beweis alle Vektoren $\mathfrak{a}_i$, $\mathfrak{b}_j$, $\mathfrak{a}_i + \mathfrak{c}$, $\mathfrak{b}_j + \mathfrak{c} \neq 0$ voraussetzen. Ist ein $\mathfrak{a}_i + \mathfrak{c} = 0$ oder ein $\mathfrak{b}_j = 0$, so reduziert sich (2) auf eine Summe von Dreiecksungleichungen. Die beiden anderen Fälle erledigen sich durch Rekursion:

Ist ein $\mathfrak{a}_i = 0$, so kommt man auf den Fall von $m + n - 1$ Vektoren, indem man den betreffenden $\mathfrak{a}$-Vektor streicht.

Ist ein $\mathfrak{b}_j + \mathfrak{c} = 0$, etwa $\mathfrak{b}_m = -\mathfrak{c}$, so lassen wir das Glied mit $\mathfrak{b}_m$ fort, und da dadurch $\Sigma\,\mathfrak{b}_j$ um $\mathfrak{c}$ vermehrt wird, fügen wir zu den $\mathfrak{a}$-Vektoren noch $\mathfrak{a}_{n+1} = \mathfrak{c}$ hinzu; die Gesamtzahl der $\mathfrak{a}$- und $\mathfrak{b}$-Vektoren bleibt $m + n$. Es ist dann:

$$\sum_{i=1}^{n+1} (|\mathfrak{a}_i+\mathfrak{c}| - |\mathfrak{a}_i|) - \sum_{j=1}^{m-1} (|\mathfrak{b}_j+\mathfrak{c}| - |\mathfrak{b}_j|)$$
$$= \sum_{i=1}^{n} (|\mathfrak{a}_i+\mathfrak{c}| - |\mathfrak{a}_i|) - \sum_{j=1}^{m} (|\mathfrak{b}_j+\mathfrak{c}| - |\mathfrak{b}_j|).$$

Es bleibt dann noch der Fall, daß alle m $\mathfrak{b}$-Vektoren $= -\mathfrak{c}$ sind. Führt man den angegebenen Vorgang m-mal durch, so erhalten wir die zu beweisende Ungleichung:

$$\sum_{i=1}^{n+m}(|\mathfrak{a}_i+\mathfrak{c}|-|\mathfrak{a}_i|)\leqq n+m-2,$$

wo $\sum_{i=1}^{n+m}\mathfrak{a}_i=0$ und $\mathfrak{a}_{n+m}=\mathfrak{c}$ ist.

Es ist dann $\mathfrak{a}_{n+m-1}=-\mathfrak{c}-\mathfrak{a}_1-\mathfrak{a}_2-\cdots-\mathfrak{a}_{n+m-2}$ und die zu beweisende Ungleichung:

$$\sum_{i=1}^{n+m-2}(|\mathfrak{a}_i+\mathfrak{c}|-|\mathfrak{a}_i|)-(|\mathfrak{a}_1+\mathfrak{a}_2+\cdots+\mathfrak{a}_{n+m-2}+\mathfrak{c}|-$$
$$-|\mathfrak{a}_1+\mathfrak{a}_2+\cdots+\mathfrak{a}_{n+m-2}|)\leqq n+m-3,$$

also die Ungleichung (2) mit einem $\mathfrak{b}$- und $n+m-2$ $\mathfrak{a}$-Vektoren.

Für $n+m=2$ ist die Ungleichung (2) ohne weiteres zu verifizieren; sonst wird an keiner Stelle des Beweises Rekursion verwendet. Wir können daher fortan die Ungleichung (2) für die genannten Fälle als bewiesen annehmen.

Aus unserer Ungleichung (3) ergibt sich:

Soll die Funktion

$$f(\mathfrak{a}_i,\mathfrak{b}_j)=\sum_{i=1}^{n}(|\mathfrak{a}_i+\mathfrak{c}|-|\mathfrak{a}_i|)-\sum_{j=1}^{m}(|\mathfrak{b}_j+\mathfrak{c}|-|\mathfrak{b}_j|)$$

auch Werte $>m+n-2$ annehmen, so müßten auch Extremwerte $>m+n-2$ auftreten; wir gehen also daran, die Extreme von $f(\mathfrak{a}_i,\mathfrak{b}_j)$ aufzusuchen. Wir ersetzen dabei $\mathfrak{b}_m=\sum_{i=1}^{n}\mathfrak{a}_i-\sum_{j=1}^{m-1}\mathfrak{b}_j$; wir bemerken ferner, daß mit Ausnahme der Stellen, an denen einer der Vektoren $\mathfrak{a}_i$, $\mathfrak{a}_i+\mathfrak{c}$, $\mathfrak{b}_j$, $\mathfrak{b}_j+\mathfrak{c}$ verschwindet, f stets differenzierbar ist; in den ausgeschlossenen Fällen konnten wir (2) als bewiesen annehmen.

Für die Extreme von f gilt:

$$\frac{\partial f}{\partial\mathfrak{a}_i}=\frac{\mathfrak{a}_i+\mathfrak{c}}{|\mathfrak{a}_i+\mathfrak{c}|}-\frac{\mathfrak{a}_i}{|\mathfrak{a}_i|}-\frac{\mathfrak{b}_m+\mathfrak{c}}{|\mathfrak{b}_m+\mathfrak{c}|}+\frac{\mathfrak{b}_m}{|\mathfrak{b}_m|}=0\qquad(i=1,2,\ldots,n)$$

$$-\frac{\partial f}{\partial\mathfrak{b}_j}=\frac{\mathfrak{b}_j+\mathfrak{c}}{|\mathfrak{b}_j+\mathfrak{c}|}-\frac{\mathfrak{b}_j}{|\mathfrak{b}_j|}-\frac{\mathfrak{b}_m+\mathfrak{c}}{|\mathfrak{b}_m+\mathfrak{c}|}+\frac{\mathfrak{b}_m}{|\mathfrak{b}_m|}=0\qquad(j=1,\ldots,m-1)$$

oder es ist für $i=1\ldots n$, $j=1\ldots m$:

$$\frac{\mathfrak{a}_i+\mathfrak{c}}{|\mathfrak{a}_i+\mathfrak{c}|}-\frac{\mathfrak{a}_i}{|\mathfrak{a}_i|}=\frac{\mathfrak{b}_j+\mathfrak{c}}{|\mathfrak{b}_j+\mathfrak{c}|}-\frac{\mathfrak{b}_j}{|\mathfrak{b}_j|}=\mathfrak{A}.\tag{7}$$

Wir können annehmen, daß für höchstens einen der $\mathfrak{a}$- und $\mathfrak{b}$-Vektoren

$$|\mathfrak{a}_i+\mathfrak{c}|=|\mathfrak{a}_i|\quad\text{bzw.}\quad|\mathfrak{b}_j+\mathfrak{c}|=|\mathfrak{b}_j|\tag{8}$$

gilt: andernfalls folgte (2) schon aus Dreiecksungleichungen. Gibt es einen

$\mathfrak{a}$- oder $\mathfrak{b}$-Vektor, für den (8) gilt, so ist der Vektor $\mathfrak{A} = \mathfrak{c} \cdot \alpha$ ($\alpha \neq 0$) und alle $\mathfrak{a}_i$ und $\mathfrak{b}_j$, für welche (8) nicht gilt, sind linear von $\mathfrak{c}$ abhängig: unter den Vektoren $\mathfrak{a}_i$ und $\mathfrak{b}_j$ sind also höchstens zwei linear unabhängig. Gibt es endlich keinen Vektor, für den (8) gilt, so sind alle Vektoren $\mathfrak{a}_i$ und $\mathfrak{b}_j$ linear aus $\mathfrak{A}$ und $\mathfrak{c}$ zusammenzusetzen.

Die Vektoren $\mathfrak{a}_i$ und $\mathfrak{b}_j$ liegen also auf einer ebenen Vektormannigfaltigkeit $\mathfrak{E}$. Bezeichnet man mit $(\mathfrak{v}, \mathfrak{w})$ den orientierten Winkel zweier Vektoren $\mathfrak{v}$ und $\mathfrak{w}$ aus $\mathfrak{E}$, so folgt mit einem beliebigen Vektor $\mathfrak{e} \neq 0$ aus $\mathfrak{E}$ aus (7):

$$\begin{aligned}(9)\qquad &\sin\frac{(\mathfrak{a}_i+\mathfrak{c},\mathfrak{e})+(\mathfrak{a}_i,\mathfrak{e})}{2}\sin\frac{(\mathfrak{a}_i+\mathfrak{c},\mathfrak{e})-(\mathfrak{a}_i,\mathfrak{e})}{2}\\ &=\sin\frac{(\mathfrak{b}_j+\mathfrak{c},\mathfrak{e})+(\mathfrak{b}_j,\mathfrak{e})}{2}\sin\frac{(\mathfrak{b}_j+\mathfrak{c},\mathfrak{e})-(\mathfrak{b}_j,\mathfrak{e})}{2}.\end{aligned}$$

Denkt man sich den Vektor $\mathfrak{e}$ um $\frac{\pi}{2}$ gedreht, quadriert und addiert die so entstehenden Gleichungen (9), so wird für alle i und j:

$$(10)\qquad \sin^2\frac{(\mathfrak{a}_i+\mathfrak{c},\mathfrak{e})-(\mathfrak{a}_i,\mathfrak{e})}{2}=\sin^2\frac{(\mathfrak{b}_j+\mathfrak{c},\mathfrak{e})-(\mathfrak{b}_j,\mathfrak{e})}{2}=\varrho.$$

Ist $\varrho = 0$, so ist $(\mathfrak{a}_i + \mathfrak{c}, \mathfrak{e}) - (\mathfrak{a}_i \mathfrak{e}) = (\mathfrak{a}_i + \mathfrak{c}, \mathfrak{a}_i) \equiv 0\ (2\pi)$, also alle $\mathfrak{a}_i$ und ebenso alle $\mathfrak{b}_j$ sind linear von $\mathfrak{c}$ abhängig: dieser Fall wird am Ende des Beweises behandelt.

Ist $\varrho \neq 0$, so folgt aus (9) weiter:

$$(11)\qquad \sin[(\mathfrak{a}_i+\mathfrak{c},\mathfrak{e})+(\mathfrak{a}_i,\mathfrak{e})]=\sin[(\mathfrak{b}_j+\mathfrak{c},\mathfrak{e})+(\mathfrak{b}_j,\mathfrak{e})]$$

und wegen der Willkür von $\mathfrak{e}$:

$$(12)\qquad (\mathfrak{a}_i+\mathfrak{c},\mathfrak{e})+(\mathfrak{a}_i,\mathfrak{e})\equiv(\mathfrak{b}_j+\mathfrak{c},\mathfrak{e})+(\mathfrak{b}_j,\mathfrak{e})\ (2\pi).$$

Aus (10) folgt wieder entweder

$$(13\,a)\qquad (\mathfrak{a}_i+\mathfrak{c},\mathfrak{e})-(\mathfrak{a}_i,\mathfrak{e})\equiv(\mathfrak{b}_j+\mathfrak{c},\mathfrak{e})-(\mathfrak{b}_j,\mathfrak{e})\ (2\pi)$$

oder

$$(13\,b)\qquad (\mathfrak{a}_i+\mathfrak{c},\mathfrak{e})-(\mathfrak{a}_i,\mathfrak{e})\equiv(\mathfrak{b}_j,\mathfrak{e})-(\mathfrak{b}_j+\mathfrak{c},\mathfrak{e})\ (2\pi).$$

Aus (12) und (13a) folgt

$$\begin{aligned}(\mathfrak{a}_i,\mathfrak{e})&\equiv(\mathfrak{b}_j,\mathfrak{e})\ (\pi),\\ (\mathfrak{a}_i+\mathfrak{c},\mathfrak{e})&\equiv(\mathfrak{b}_j+\mathfrak{c},\mathfrak{e})\ (\pi).\end{aligned}$$

Es sind also $\mathfrak{a}_i$ und $\mathfrak{b}_j$ linear abhängig und ebenso $\mathfrak{a}_i + \mathfrak{c}$ und $\mathfrak{b}_j + \mathfrak{c}$. Dann ist entweder $\mathfrak{a}_i = \mathfrak{b}_j$ oder es sind sowohl $\mathfrak{a}_i$ als $\mathfrak{b}_j$ linear von $\mathfrak{c}$ abhängig. Den ersteren Fall können wir gleich ausscheiden, da dann (2) sofort aus Dreiecksungleichungen folgt.

Aus (12) und (13b) folgt

$$\begin{aligned}(\mathfrak{a}_i+\mathfrak{c},\mathfrak{e})&\equiv(\mathfrak{b}_j,\mathfrak{e})\ (\pi),\\ (\mathfrak{a}_i,\mathfrak{e})&\equiv(\mathfrak{b}_j+\mathfrak{c},\mathfrak{e})\ (\pi).\end{aligned}$$

Es sind also $\mathfrak{a}_i + \mathfrak{c}$ und $\mathfrak{b}_j$ linear abhängig und ebenso $\mathfrak{a}_i$ und $\mathfrak{b}_j + \mathfrak{c}$. Dann sind entweder sowohl $\mathfrak{a}_i$ als $\mathfrak{b}_j$ von $\mathfrak{c}$ linear abhängig, oder wenn weder $\mathfrak{a}_i$ noch $\mathfrak{b}_j$ linear von $\mathfrak{c}$ abhängen, muß $\mathfrak{a}_i + \mathfrak{b}_j + \mathfrak{c} = 0$ gelten. Wir zeigen, daß dieser letztere Fall nicht eintreten kann, sondern daß alle $\mathfrak{a}_i$ und $\mathfrak{b}_j$ linear von $\mathfrak{c}$ abhängig sein müssen. Wäre etwa $\mathfrak{a}_1$ nicht von $\mathfrak{c}$ linear abhängig, so müßte, und zwar mit jedem $\mathfrak{b}_j$ gelten: $\mathfrak{a}_1 + \mathfrak{b}_j + \mathfrak{c} = 0$, d. h. alle $\mathfrak{b}_j$ wären einander gleich und nicht von $\mathfrak{c}$ linear abhängig; dann müßten aber auch wieder alle $\mathfrak{a}_i$ der Gleichung genügen $\mathfrak{a}_i + \mathfrak{b}_j + \mathfrak{c} = 0$ und ebenfalls alle $\mathfrak{a}_i$ einander gleich und nicht von $\mathfrak{c}$ linear abhängig sein. Es wäre aber dann:

$$\sum_{i=1}^{n} \mathfrak{a}_i = \sum_{j=1}^{m} \mathfrak{b}_j = n\mathfrak{a}_1 = m\mathfrak{b}_1.$$

Weiter müßte gelten $\mathfrak{a}_1 + \mathfrak{b}_1 + \mathfrak{c} = 0$, also $\mathfrak{a}_1 = -\frac{m}{m+n}\mathfrak{c}$, also $\mathfrak{a}_1$ entgegen der Voraussetzung von $\mathfrak{c}$ linear abhängig.

Es ist also schließlich nur der Fall zu untersuchen, daß alle $\mathfrak{a}_i$ und $\mathfrak{b}_j$ linear von $\mathfrak{c}$ abhängig sind.

Sei also $\mathfrak{a}_i = \alpha_i \mathfrak{c}$ und $\mathfrak{b}_j = \beta_j \mathfrak{c}$; dann ist die Ungleichung zu zeigen:

$$\sum_{i=1}^{n} (|\alpha_i + 1| - |\alpha_i|) - \sum_{j=1}^{m} (|\beta_j + 1| - |\beta_j|) \leqq n + m - 2,$$

wo also die α_i und β_j beliebige reelle Zahlen mit $\sum_{i=1}^{n} \alpha_i = \sum_{j=1}^{m} \beta_j$ und $m \geqq 1$ sei.

Wir zerlegen die n Zahlen α_i in drei Klassen: mit α_i' bezeichnen wir die Zahlen α_i mit $\alpha_i' > 0$, mit α_i'' die mit $-1 \leqq \alpha_i'' \leqq 0$ und mit α_i''' die mit $\alpha_i''' < -1$; die Anzahlen der α_i in diesen drei Kla sen seien n', n'', n''', also $n' + n'' + n''' = n$. Analog zerlegen wir die m Zahlen β_j; deren Anzahlen seien m', m'', m''' und $m' + m'' + m''' = m$. Je nachdem α_i der ersten, zweiten oder dritten Klasse angehört, ist dann $|\alpha_i + 1| - |\alpha_i| = 1$, $1 - 2|\alpha_i''|$, -1; ebenso für die β_j.

Die zu beweisende Ungleichung hat dann die folgende Gestalt:

$$-\Sigma |\alpha_i''| + \Sigma |\beta_i''| \leqq n''' + m' + m'' - 1.$$

Wegen $\Sigma |\beta_i''| \leqq m''$ ist diese Ungleichung sicher erfüllt, wenn n''' oder $m' > 0$ sind. Wir nehmen also weiterhin $n''' = m' = 0$ an und hätten dann zu zeigen:

$$\Sigma |\beta_i''| \leqq \Sigma |\alpha_i''| + m'' - 1.$$

Nun ist

$$-\Sigma |\beta_i''| + \Sigma \beta_i''' = \Sigma \alpha' - \Sigma |\alpha_i''|,$$

also ist zu zeigen

$$-\Sigma \alpha_i' + \Sigma \beta_i''' \leqq m'' - 1;$$

wäre nun

$$-\Sigma \alpha_i' + \Sigma \beta_i''' > m'' - 1,$$

so müßte, weil links eine Zahl $\leqq 0$ steht, $m'' = 0$ sein und weiter

$$|-\Sigma \alpha_i' + \Sigma \beta_i'''| < 1$$

und

$$\Sigma |\beta_i'''| < 1$$

und wegen $|\beta_i'''| > 1$ ist auch $m''' = 0$. Also wäre $m = m' + m'' + m''' = 0$ entgegen unserer Voraussetzung.

Bemerkenswert ist die Voraussetzung $m \geqq 1$, während die Anzahl der $\mathfrak{a}$-Vektoren nicht beschränkt ist.

Für den Spezialfall $m = 1$, $n = 2$ hat Herr HLAWKA mir einen rein algebraischen Beweis angegeben:

Es ist, wenn man die Relation

$$|\mathfrak{a}|^2 + |\mathfrak{b}|^2 + |\mathfrak{c}|^2 + |\mathfrak{a} + \mathfrak{b} + \mathfrak{c}|^2 = |\mathfrak{a} + \mathfrak{b}|^2 + |\mathfrak{b} + \mathfrak{c}|^2 + |\mathfrak{c} + \mathfrak{a}|^2$$

berücksichtigt:

$$\begin{aligned}
&|\mathfrak{a}| + |\mathfrak{b}| + |\mathfrak{c}| - |\mathfrak{a} + \mathfrak{b}| - |\mathfrak{b} + \mathfrak{c}| - |\mathfrak{c} + \mathfrak{a}| + |\mathfrak{a} + \mathfrak{b} + \mathfrak{c}| \\
&\quad = (|\mathfrak{a}| + |\mathfrak{b}| - |\mathfrak{a} + \mathfrak{b}|)\left(1 - \frac{|\mathfrak{a}| + |\mathfrak{b}| + |\mathfrak{a} + \mathfrak{b}|}{|\mathfrak{a}| + |\mathfrak{b}| + |\mathfrak{c}| + |\mathfrak{a} + \mathfrak{b} + \mathfrak{c}|}\right) \\
&\quad + (|\mathfrak{b}| + |\mathfrak{c}| - |\mathfrak{b} + \mathfrak{c}|)\left(1 - \frac{|\mathfrak{b}| + |\mathfrak{c}| + |\mathfrak{b} + \mathfrak{c}|}{|\mathfrak{a}| + |\mathfrak{b}| + |\mathfrak{c}| + |\mathfrak{a} + \mathfrak{b} + \mathfrak{c}|}\right) \\
&\quad + (|\mathfrak{c}| + |\mathfrak{a}| - |\mathfrak{c} + \mathfrak{a}|)\left(1 - \frac{|\mathfrak{c}| + |\mathfrak{a}| + |\mathfrak{c} + \mathfrak{a}|}{|\mathfrak{a}| + |\mathfrak{b}| + |\mathfrak{c}| + |\mathfrak{a} + \mathfrak{b} + \mathfrak{c}|}\right) \geqq 0.
\end{aligned}$$

Wir geben schließlich noch einige Ungleichungen, die sich als Spezialfälle von (2) ergeben.

Für $m = 2$, $n = 1$ ist mit drei willkürlichen Vektoren $\mathfrak{a}$, $\mathfrak{b}$, $\mathfrak{c}$:

$$|\mathfrak{a}| + |\mathfrak{b}| - |\mathfrak{c}| + |\mathfrak{a} + \mathfrak{b} + \mathfrak{c}| \leqq |\mathfrak{a} + \mathfrak{b}| + |\mathfrak{b} + \mathfrak{c}| + |\mathfrak{c} + \mathfrak{a}|.$$

Für $m = 3$, $n = 0$ ist:

$$|\mathfrak{a}| + |\mathfrak{b}| + |\mathfrak{a} + \mathfrak{b}| \leqq |\mathfrak{a} + \mathfrak{c}| + |\mathfrak{b} + \mathfrak{c}| + |\mathfrak{a} + \mathfrak{b} - \mathfrak{c}| + |\mathfrak{c}|.$$

Für beliebige Vektoren $\mathfrak{b}_1 \ldots \mathfrak{b}_m$ und $\mathfrak{c}$ gilt ($n = 1$):

$$\sum_{j=1}^{m} |\mathfrak{b}_j| - \left|\sum_{j=1}^{m} \mathfrak{b}_j\right| \leqq \sum_{j=1}^{m} |\mathfrak{b}_j + \mathfrak{c}| - \left|\sum_{j=1}^{m} \mathfrak{b}_j + \mathfrak{c}\right| + (m - 1)|\mathfrak{c}|.$$

(Eingegangen am 9. Februar 1942.)

Corrections

268.1 Prof.Prachar wird einen Kommentar zu dieser Arbeit bezüglich des von H.Hornich stammenden Teils verfassen, der bei der Österreichischen Akademie der Wissenschaften veröffentlicht wird.

Zur Geometrie der Zahlen

Math. Z. 49 (1943) 285–312

§ 1.

Einleitung.

Die folgenden Ausführungen beziehen sich stets auf den n-dimensionalen euklidischen Raum R_n. Zugrunde gelegt wird ein orthogonales Koordinatensystem. Der Punkt mit den Koordinaten $x_1, \ldots, x_n$ wird mit $\mathfrak{x}$ bezeichnet. Sind die Koordinaten alle ganzzahlig, so bezeichnen wir sie mit $\mathfrak{g}$, $\mathfrak{l}$ oder $\mathfrak{r}$. Es sei $n \geqq 2$.

Wir betrachten nun im R_n einen konvexen Körper K mit dem Mittelpunkt im Koordinatenursprung $\mathfrak{o}$. Seine Distanzfunktion sei $\mathfrak{f}(\mathfrak{x})$. Diese ist bekanntlich homogen vom ersten Grad, symmetrisch in bezug auf $\mathfrak{o}$ und sie erfüllt die Ungleichung

$$\text{(a)} \qquad \mathfrak{f}(\mathfrak{x} + \mathfrak{y}) \leqq \mathfrak{f}(\mathfrak{x}) + \mathfrak{f}(\mathfrak{y}).$$

Dann lautet der Fundamentalsatz von Minkowski*) [1]: Besitzt K ein Volumen $V \geqq 2^n$, so liegt in K mindestens ein von $\mathfrak{o}$ verschiedenes Gitterpunktpaar $\pm\,\mathfrak{g}$ oder anders ausgesprochen: Es gibt zu jeder Zahl $M_h \geqq \left(\frac{V}{J}\right)^{\frac{1}{n}}$ ein $\mathfrak{g} \neq \mathfrak{o}$, so daß

$$\text{(b)} \qquad \mathfrak{f}(\mathfrak{g}) \leqq M_h.$$

Dabei ist J das Volumen des Eichkörpers $\mathfrak{f}(\mathfrak{x}) = 1$. Man sieht sofort, daß es konvexe Körper gibt mit einem Volumen $V < 2^n$, ja bei denen sogar $V \leqq \varepsilon$ ist ($\varepsilon > 0$, sonst aber beliebig) und trotzdem (b) gilt. Betrachten wir z. B. das Parallelepiped $P: |x_1| \leqq \frac{\varepsilon}{2 \cdot 4^{n-1}}$, $|x_i| \leqq 2$ ($i = 2, \ldots, n$), dessen Volumen $V = \varepsilon$ ist und welches den Gitterpunkt mit den Koordinaten $g_1 = 0$, $g_2 = \cdots = g_n = 1$ enthält. Die Sache wird aber anders, wenn wir mit Minkowski die Klasse $\Gamma(K)$ aller konvexen Körper betrachten, die aus K durch volumentreue affine Transformationen, bei denen $\mathfrak{o}$ festbleibt, hervorgehen. Von dieser Klasse $\Gamma(K)$ hat Minkowski folgendes behauptet: Ist

---

*) Die eckigen Klammern verweisen auf das Schriftenverzeichnis am Ende der Arbeit.

das Volumen $V < 2 \sum_{l=1}^{\infty} \frac{1}{k^n}$, so gibt es in $\Gamma(K)$ stets einen konvexen Körper K', der außer $\mathfrak{o}$ keine weiteren Gitterpunkte mehr enthält oder anders ausgesprochen: Es gibt stets eine unimodulare Matrix $\mathfrak{A}$, so daß für alle $\mathfrak{g} \neq \mathfrak{o}$

$$f(\mathfrak{A}\mathfrak{g}) > \left(\frac{V}{J}\right)^{\frac{1}{n}}.$$

Dabei ist es nicht notwendig vorauszusetzen, daß die Distanzfunktion von K die Eigenschaft (a) besitzt. Man nennt solche Körper Strahlkörper. MINKOWSKI hat für diesen Satz keinen Beweis veröffentlicht und es ist auch seither, soviel ich weiß, keiner veröffentlicht worden. Es ist mir nun gelungen, einen Beweis dafür zu erbringen, indem ich zunächst in § 2 folgenden Satz (Deformationssatz) beweise: Ist M eine beschränkte, im JORDANschen Sinne meßbare Menge und $\varphi(\mathfrak{x})$ eine auf M definierte beschränkte, im RIEMANNschen Sinne integrierbare Funktion, so gibt es zu jedem $\varepsilon > 0$ eine unimodulare Matrix $\mathfrak{A}$, so daß

$$\sum_{\mathfrak{g} \neq \mathfrak{o}} \varphi(\mathfrak{A}\mathfrak{g}) \leqq \int_M \varphi(\mathfrak{x})\, d\mathfrak{x} + \varepsilon$$

ist. Dabei erstreckt sich die Summe links über alle Gitterpunkte $\mathfrak{g} \neq \mathfrak{o}$. Aus diesem Satz wird dann in § 3 die Behauptung von MINKOWSKI hergeleitet.

Wir haben bis jetzt das homogene Problem (b) betrachtet. Das inhomogene Problem besteht darin, eine Zahl M_J so zu bestimmen, so daß es zu jedem $\mathfrak{x}_0$ des R_n wenigstens ein $\mathfrak{g}$ gibt, wobei jetzt auch $\mathfrak{o}$ zugelassen wird, so daß

$$f(\mathfrak{g} + \mathfrak{x}_0) \leqq M_J$$

wird. M_J hängt im Gegensatz zu M_n nicht vom Volumen V von K ab; z. B. gibt es keinen Gitterpunkt $\mathfrak{g}$, so daß das Parallelepiped

$$|x_1 + g_1| \leqq \frac{1}{4}, \quad |x_2 + g_2| \leqq V, \quad |x_i + g_i| \leqq \tfrac{1}{2} \quad (i = 3, \ldots, n)$$

mit dem Volumen V den Punkt $\mathfrak{x}_0$ mit den Koordinaten $x_1 = \frac{1}{2}$, $x_2 = \cdots = x_n = 0$ enthält, wie groß auch V sein mag. Das inhomogene Problem ist in der letzten Zeit mehrfach behandelt worden. MAHLER hat folgendes gezeigt: Ist $\overline{M}$ das erste Minimum des polaren Körpers $\overline{K}$ von K, so ist $M_J \cdot \overline{M} \geqq [(n+1)!]^2$. Mit dem Fall $n = 2$ hat sich besonders HEINHOLD [6] beschäftigt.

Ich beweise in § 6 folgenden Alternativsatz, der dem Alternativsatz bei linearen Gleichungssystemen analog ist: Besitzt der konvexe Körper K ein

Volumen $V \geqq 2^{n-1} \cdot k_0$ (k_0 natürliche Zahl) und gibt es höchstens $k_0 - r - 1$ Gitterpunktpaare $\mathfrak{g} \neq \mathfrak{o}$ $\left(0 \leqq r \leqq k_0 - \left[\frac{k_0}{2}\right] - 1\right)$, so daß

$$\text{(c)} \qquad f(\mathfrak{g}) < M_h \quad M_h \geqq \left(\frac{V}{J}\right)^{\frac{1}{n}},$$

so gibt es zu jedem $\mathfrak{x}_0$ mindestens $r + 1$ Gitterpunkte $\bar{\mathfrak{g}}$, so daß

$$\text{(d)} \qquad f(\bar{\mathfrak{g}} + \mathfrak{x}_0) \leqq M_h$$

ist. Der Satz gilt nicht mehr, wenn $V < 2^{n-1} k_0$. Betrachten wir das Parallelepiped $|x_1| \leqq \frac{1}{2}(1-\varepsilon)^{\frac{1}{n}}$, $|x_2| \leqq k_0(1-\varepsilon)^{\frac{1}{n}}$, $|x_i| \leqq (1-\varepsilon)^{\frac{1}{n}}$ $(i = 3, \ldots, n)$, dessen Volumen $V = 2^{n-1} k_0 (1-\varepsilon)$ ist. Es enthält die $k_0 - 1$ von $\mathfrak{o}$ verschiedenen Gitterpunktpaare $\pm \mathfrak{g}_j$ $(j = 1, \ldots, k_0 - 1)$, deren Koordinaten $g_{j1} = 0$, $g_{j2} = \pm j$, $g_{ji} = 0$ $(i = 3, \ldots, n)$ sind. Trotzdem gibt es keinen Gitterpunkt $\mathfrak{g}$ mit den Koordinaten $g_1, \ldots, g_n$, so daß

$$|g_1 + \tfrac{1}{2}| \leqq \tfrac{1}{2}(1-\varepsilon)^{\frac{1}{n}}, \quad |g_2| \leqq k_0(1-\varepsilon)^{\frac{1}{n}}, \quad |g_i| \leqq (1-\varepsilon)^{\frac{1}{n}} \quad (i = 3, \ldots, n)$$

ist. Ist das Volumen $V \geqq 2^{n-1} k_0$, so gibt es im allgemeinen auch nicht mehr als $r + 1$ Gitterpunkte, die (d) erfüllen, wie man am Parallelepiped $|x_1| \leqq \frac{6}{7}$, $|x_2| \leqq \frac{2}{3}$, $|x_i| \leqq \left(\frac{7}{8}\right)^{\frac{1}{n-2}}$ $(i = 3, \ldots, n)$ sieht. Sein Volumen ist 2^{n-1} und es enthält keinen von $\mathfrak{o}$ verschiedenen Gitterpunkt. Es gibt aber nur einen Gitterpunkt $\mathfrak{g}$, nämlich $\mathfrak{o}$, so daß

$$|g_1 + \tfrac{1}{8}| \leqq \tfrac{6}{7}, \quad |g_2 + \tfrac{1}{8}| \leqq \tfrac{2}{3}, \quad |g_i| \leqq \left(\tfrac{7}{8}\right)^{\frac{1}{n-2}} \qquad (i = 3, \ldots, n)$$

ist. Der Alternativsatz ist in einem allgemeineren Satz enthalten. Um ihn einfach aussprechen zu können, führe ich folgende Bezeichnungen ein: $T_{\mathfrak{g}}$ sei die Translation, die $\mathfrak{o}$ in $\mathfrak{g}$ überführt, $t_{\mathfrak{g}} M$ sei die Menge, die aus einer Menge M bei dieser Translation hervorgeht. Dann lautet der Satz: Sind K_1 und K_2 zwei konvexe Körper mit dem Mittelpunkt in $\mathfrak{o}$ und ist die Summe ihrer Volumina $V_1 + V_2 \geqq 2^n \cdot k_0$ und enthalten die Körper K_i $(i = 1, 2)$ höchstens $k_0 - r_i - 1$ Gitterpunktpaare $\neq \mathfrak{o}$, so erfüllen die konvexen Körper $t_{\mathfrak{g}} K_3$, wo $K_3 = \frac{1}{2}(K_1 + K_2)$ den R_n lückenlos, und zwar ist die Überdeckung mindestens $r + 1$-fach, wo $r = \left[\frac{r_1 V_1 + r_2 V_2}{2\sqrt{V_1 V_2}}\right]$ ist. Setzt man $K_1 = K_2 = K$, so ist $V_1 = V_2$, $r_1 = r_2 = r$ $K_3 = K$, und man erhält den Alternativsatz. Der obige Satz enthält auch noch den folgenden Satz von Mordell [7]: Es seien $L_i(\mathfrak{x})$ n Linearformen in $\mathfrak{x}$ mit der Determinante 1. Weiter seien

$\mu_1, \ldots, \mu_n, \nu_1, \ldots, \nu_n$ positive Zahlen mit $\mu_1 \ldots \mu_n + \nu_1 \ldots \nu_n \geqq 1$. Dann gibt es wenigstens zu einem der drei Systeme einen Gitterpunkt $\mathfrak{g}$, so daß

$$|L_i(\mathfrak{g})| \leqq \mu_i \ (\mathfrak{g} \neq \mathfrak{o}), \quad |L_i(\mathfrak{g})| \leqq \nu_i \ (\mathfrak{g} \neq \mathfrak{o}), \quad |L_i(\mathfrak{g} + \mathfrak{x}_0)| \leqq \tfrac{1}{2} (\mu_i + \nu_i).$$

Die soeben ausgesprochenen Sätze sind nicht auf konvexe Körper beschränkt. Ich betrachte gleich allgemeiner folgende Mengen: Es sei M eine beschränkte abgeschlossene, im JORDANschen Sinne meßbare Menge und $\mathfrak{K}$ irgendeine Diagonalmatrix

$$\begin{pmatrix} k_1 & & \\ & \ddots & \\ & & k_n \end{pmatrix},$$

wo die $k_i > 0$ sind. Dann lassen sich alle Sätze auf die Menge $M(\mathfrak{K})$ aller Punkte der Form $\mathfrak{K}^{-1}(\mathfrak{x} - \mathfrak{y})$ ($\mathfrak{x}$ und $\mathfrak{y}$ in M) übertragen. Die Sätze lassen sich für engere Klassen von Mengen verschärfen. In § 7 betrachte ich Mengen, ich nenne sie ω-konvex, die eine Distanzfunktion besitzen, bei denen statt (a) folgendes gilt: Sind k und k_0 irgendwelche natürliche Zahlen, aber $k > k_0$, so kann man aus k Punkten $\mathfrak{x}_1, \ldots, \mathfrak{x}_n$ stets $k_0 + 1$ Punkte $\mathfrak{x}'_0, \ldots, \mathfrak{x}'_{k_0}$ so herausgreifen, daß

$$f^\sigma(\mathfrak{x}'_0 - \mathfrak{x}'_i) \leqq \frac{c}{k - k_0} \sum_{j=1}^{k} f^\sigma(\mathfrak{x}_j) \qquad (i = 1, \ldots, k_0)$$

gilt. Dabei sind σ und c feste, positive Zahlen. Diese Eigenschaft ist stets dann vorhanden, wenn für jedes System von k Punkten $\mathfrak{x}_1, \ldots, \mathfrak{x}_k$

$$\sum_{i=1}^{k} \sum_{j=1}^{k} f^\sigma(\mathfrak{x}_i - \mathfrak{x}_j) \leqq c\, k \sum_{j=1}^{k} f^\sigma(\mathfrak{x}_j)$$

gilt. Dies ist der Fall für $f^\sigma(\mathfrak{x}) = \sum_{i=1}^{n} |L_i(\mathfrak{x})|^\sigma$ ($\sigma \geqq 2$), wo die $L_i(\mathfrak{x})$ Linearformen in $\mathfrak{x}$ sind mit einer von Null verschiedenen Determinante, und zwar ist $c = 2^{\sigma-1}$ (v. d. CORPUT-SCHAAKE [5]). Es wird in § 8 gezeigt, daß für ω-konvexe Mengen (c) und (d) mit $V \geqq \frac{k_0}{2} \frac{n+\sigma}{\sigma} c^{\frac{n}{\sigma}}$ gilt. Für die Ellipsoide ($\sigma = 2$, $c = 2$) ist wirklich $2^{-2+\frac{n}{2}} (n+2)\, k_0 < 2^{n-1} k_0$ ($n > 2$).

§ 2.

Deformationssatz.

Hilfssatz 1. *Es sei $F(x_1, x_2)$ eine reelle, im RIEMANNschen Sinne integrierbare Funktion, welche außerhalb des Quadrates Q:*

$$|x_1| \leqq K, \quad |x_2| \leqq K$$

verschwindet und in Q beschränkt ist: $|F(x_1, x_2)| \leqq C$. Dann gibt es zu jedem $\varepsilon > 0$ und zu jedem $\beta > 0$, welches nur kleiner als

$$c(\varepsilon) = \operatorname{Min}\left(K, \frac{\varepsilon}{42\,KC}, \frac{1}{4K}\left(\frac{\varepsilon}{21\,KC}\right)^2\right)$$

ist, ein α im offenen Intervall $\left(0, \frac{1}{\beta}\right)$, so daß

$$(1) \qquad \sum_g{}' \int\limits_{-\infty}^{+\infty} F\left(\beta g + \left(\alpha - \frac{1}{\beta}\right) x_2, \beta g + \alpha x_2\right) d x_2$$

$$\leqq \frac{1}{\beta} \iint\limits_{-\infty}^{+\infty} F\left(x_1 - \frac{x_2}{\beta}, x_1\right) d x_1\, d x_2 + \varepsilon \ldots$$

Dabei durchläuft g alle ganzen Zahlen mit Ausschluß der Null.

Wir verwenden folgenden Satz: Ist $f(x, y)$ über ein Rechteck im Riemannschen Sinne integrierbar, so sind die Funktionen

$$F_1(x) = \overline{\int} f(x, y)\, d y \quad \text{und}$$

$$F_2(y) = \overline{\int} f(x, y)\, d x$$

im Riemannschen Sinne integrierbar und es ist

$$\iint f(x, y)\, d x\, d y = \int F_1(x)\, d x = \int F_2(y)\, d y$$

(s. A. Schoenfliess, Mengenlehre, S. 196; Haupt-Aumann, Differential- und Integralrechnung III). Wir schreiben stets

$$F_1(x) = \int f(x, y)\, d y, \quad F_2(y) = \int f(x, y)\, d x.$$

In demselben Sinne sind auch die mehrfachen Integrale in Hilfssatz 2 und Satz 1 aufzufassen.

Bemerkung. Die Reihe ist endlich und die Integrale sind eigentlich, weil F außerhalb Q verschwindet. Für den Beweis von (1) ist es notwendig, Schranken für die x_1, x_2, y zu bestimmen. Wenn F nicht verschwinden soll, so müssen die Argumente von F stets im Quadrat Q liegen; auf der linken Seite von (1) ist also

$$(2) \qquad -K \leqq -\beta g + \left(\frac{1}{\beta} - \alpha\right) x_2 \leqq K, \quad -K \leqq \alpha x_2 + \beta g \leqq K.$$

Daraus folgt durch Addition $|x_2| \leqq 2 K \beta$. Multiplizieren wir nun die erste Ungleichung mit α, die zweite mit $\frac{1}{\beta} - \alpha$ und addieren dann die erste (zweite) Hälfte der ersten Ungleichung zur zweiten (ersten) Hälfte der zweiten, so erhalten wir $|\beta g| \leqq K$, also zusammen:

$$(3\,\text{a}) \qquad |x_2| \leqq 2 K \beta, \quad |g| \leqq K_1 = \frac{K}{\beta}.$$

Für die Integrationsgrenzen auf der rechten Seite von (1) erhalten wir analog

(3b) $$|x_2| \leqq 2K\beta, \quad |x_1| \leqq K.$$

Wir setzen zur Abkürzung

$$F\left(\beta g + \left(\alpha - \frac{1}{\beta}\right)x_2, \beta g + \alpha x_2\right) = F_1(\alpha, \beta, g, x_2); \; F\left(x_1 - \frac{x_2}{\beta}, x_1\right) = F(x_1, x_2, \beta)$$

und beachten noch, daß für alle $\beta > 0$, welche kleiner als $c(\varepsilon)$ sind,

$$3KC\beta \ln \frac{2K}{\beta} < \frac{\varepsilon}{7}$$

ist, da $e^{\frac{\varepsilon}{21KC\beta}} > \frac{1}{2}\left(\frac{\varepsilon}{21KC\beta}\right)^2 > \frac{4K\beta}{2\beta^2} = \frac{2K}{\beta}$ ist.

Wir führen den Beweis des Hilfssatzes indirekt. Wir nehmen an, das Gegenteil von (1) sei richtig. Dann gibt es ein $\varepsilon > 0$ und ein positives $\beta < c(\varepsilon)$, so daß für alle α in $\left(0, \frac{1}{\beta}\right)$

$$\sum_g{}' \int_{-\infty}^{+\infty} F_1(\alpha, \beta, g, x_2)\, dx_2 > \frac{1}{\beta} \int_{-\infty}^{+\infty} dx_1 \int_{-\infty}^{+\infty} F_2(x_1, x_2, \beta)\, dx_2 + \varepsilon$$

ist. Dann ist aber auch*)

$$\beta \int_0^{\frac{1}{\beta}} d\alpha \sum_g{}' \int_{-2K\beta}^{2K\beta} F_1(\alpha, \beta, g, x_2)\, dx_2 > \frac{1}{\beta} \iint_{-\infty} F_2(x_1, x_2, \beta)\, dx_1\, dx_2 + \varepsilon.$$

Nun ist

$$\left|\beta \int_0^{\frac{1}{\beta}} d\alpha \int_{-2K\beta}^{2K\beta} F_1(\alpha, \beta, 0, x_2)\, dx_2 \right| \leqq 4KC\beta \leqq < \frac{4KC\varepsilon}{42KC} < \frac{\varepsilon}{7},$$

also

$$\beta \int_0^{\frac{1}{\beta}} d\alpha \sum_g \int_{-\infty}^{+\infty} F_1(\alpha, \beta, g, x_2)\, dx_2 > \frac{1}{\beta} \int_{-\infty}^{+\infty} dx_1 \int_{-\infty}^{+\infty} F_2(x_1, x_2, \beta)\, dx_2 + \frac{6}{7}\varepsilon.$$

Dann ist aber entweder (1. Fall):

$$\beta \int_0^{\frac{1}{\beta}} d\alpha \sum_g \int_0^{\infty} F_1\, dx_2 > \frac{1}{\beta} \int_{-\infty}^{+\infty} dx_1 \int_0^{\infty} F_2\, dx_2 + \frac{3}{7}\varepsilon$$

---

*) Denn ist auf einem Intervall die Funktion $\varphi(x)$ stets > 0 so, so ist auch $\int \varphi\, dx > 0$.

oder (2. Fall):

$$\beta \int_0^{\frac{1}{\beta}} d\alpha \sum_g \int_{-\infty}^{0} F_1\, dx_2 > \frac{1}{\beta} \int_{-\infty}^{+\infty} dx_1 \int_{-\infty}^{0} F_2\, dx_2 + \frac{3}{7}\, \varepsilon.$$

Wir nehmen an, der erste Fall liege vor. Der zweite Fall wird analog behandelt. Es ist nun

$$\left| \beta \int_0^{\frac{1}{\beta}} d\alpha \sum_{-K \leqq g \leqq K_1} \int_0^{\beta^2} F_1\, dx_2 \right| \leqq (2\,K_1 + 1)\, C \beta^2$$
$$= (2\,K + \beta)\, C\beta \leqq 3\, K C \beta < \frac{\varepsilon}{14}$$

und ebenso

$$\left| \frac{1}{\beta} \int_{-K}^{K} dx_1 \int_0^{\beta^2} F_2\, dx_2 \right| \leqq 2\, K C \beta < \frac{\varepsilon}{14},$$

also

$$\beta \sum_g \int_{\beta^2}^{\infty} dx_2 \int_0^{\frac{1}{\beta}} F_1\, d\alpha > \frac{1}{\beta} \int_{-\infty}^{+\infty} dx_1 \int_{\beta^2}^{\infty} F_2\, dx_2 + \frac{2}{7}\, \varepsilon.$$

Es ist nun

$$\left| \beta \sum_{-K_1 \leqq g \leqq K_1} \int_{\beta^2}^{2K\beta} d\,x_2 \int_{\left[\frac{x_2}{\beta^2}\right] \frac{\beta}{x_2}}^{\frac{1}{\beta}} F_1\, d\alpha \right| \leqq \beta \left(\frac{2\,K}{\beta} + 1 \right) C \int_{\beta^2}^{2K\beta} d\,x_2 \left(\frac{1}{\beta} - \left[\frac{x_2}{\beta^2}\right] \frac{\beta}{x_2} \right)$$
$$= 3\, K C \beta \int_{\beta^2}^{2K\beta} \frac{d\,x_2}{x_2} \left(\frac{x_2}{\beta^2} - \left[\frac{x_2}{\beta^2}\right] \right) \leqq 3\, K C \beta \int_{\beta^2}^{2K\beta} \frac{d\,x_2}{x_2} = 3\, K C \beta \ln \frac{2\,K\beta}{\beta^2} < \frac{\varepsilon}{7},$$

also ist

$$\beta \sum_g \int_{\beta^2}^{\infty} d\,x_2 \int_0^{\left[\frac{x_2}{\beta^2}\right] \frac{\beta}{x_2}} F_1(\alpha, \beta, g, x_2)\, d\alpha > \frac{1}{\beta} \int_{-K}^{K} d\,x_1 \int_{\beta^2}^{\infty} F_2\,(x_1, x_2, \beta)\, d\,x_2 + \frac{\varepsilon}{7}.$$

Es ist nun

$$\frac{1}{\beta} \int_{-K}^{K} d\,x_1 \int_{\beta^2}^{\infty} F_2\,(x_1, x_2, \beta)\, d\,x_2 = \int_{-K}^{K} d\,x_1 \int_{\beta^2}^{\infty} d\,x_2 \int_0^{\frac{1}{\beta}} F_2\,(x_1, x_2, \beta)\, d\alpha$$

und

$$\left|\int_{-K}^{K} d x_1 \int_{\beta^2}^{2K\beta} d x_2 \int_{\left[\frac{x_2}{\beta^2}\right]\frac{\beta}{x_2}}^{\frac{1}{\beta}} F_2(x_1, x_2, \beta)\, d\alpha\right| \leqq 2KC \int_{\beta^2}^{2K\beta} d x_2 \left(\frac{1}{\beta} - \left[\frac{x_2}{\beta^2}\right]\frac{\beta}{x_2}\right) < \frac{\xi}{7},$$

also

$$\beta \int_{\beta^2}^{\infty} d x_2 \sum_g \int_0^{\left[\frac{x_2}{\beta^2}\right]\frac{\beta}{x_2}} F_1\, d\alpha > \int_{\beta^2}^{\infty} d x_2 \int_{-\infty}^{+\infty} d x_1 \int_0^{\left[\frac{x_2}{\beta^2}\right]\frac{\beta}{x_2}} F_2\, d\alpha.$$

Dann muß es aber ein ξ geben, so daß

$$\beta \sum_g \int_0^{\left[\frac{\xi}{\beta^2}\right]\frac{\beta}{\xi}} F_1\, d\alpha > \int_{-\infty}^{+\infty} d x_1 \int_0^{\left[\frac{\xi}{\beta^2}\right]\frac{\beta}{\xi}} F_2\, d\alpha$$

oder anders geschrieben

$$\sum_{l=0}^{\left[\frac{\xi}{\beta^2}\right]-1} \int_{\frac{\beta l}{\xi}}^{\frac{\beta(l+1)}{\xi}} d\alpha \sum_g F_1 > \frac{1}{\beta} \sum_{l=0}^{\left[\frac{\xi}{\beta^2}\right]-1} \int_{\beta l}^{\frac{\beta(\lambda+1)}{\xi}} d\alpha \int_{-\infty}^{+\infty} F_2\, d x_1.$$

Dann muß es aber ein l geben, so daß

$$\sum_g \int_{\frac{\beta l}{\xi}}^{\frac{\beta(l+1)}{\xi}} F\left(\alpha\xi + \beta g - \frac{1}{\beta}\, \xi, \beta g + \alpha\xi\right) d\alpha > \frac{1}{\beta} \int_{\frac{\beta l}{\xi}}^{\frac{\beta(l+1)}{\xi}} d\alpha \int_{-\infty}^{+\infty} F_2\, d x_1.$$

Wir machen nun im linksstehenden Integral die Substitution $\alpha\xi + \beta g = x_1$, dann kommt

$$\frac{1}{\xi} \sum_g \int_{(l+g)\beta}^{(l+g+1)\beta} F\left(x_1 - \frac{\xi}{\beta}, x_1\right) d x_1 = \frac{1}{\xi} \int_{-\infty}^{+\infty} F\left(x_1 - \frac{\xi}{\beta}, x_1\right) d x_1$$

$$> \frac{1}{\xi} \int_{-\infty}^{+\infty} F\left(x_1 - \frac{\xi}{\beta}, x_1\right) d x_1.$$

Widerspruch.

Ich führe nun einige Abkürzungen ein. Ist

$$\mathfrak{x} = \begin{pmatrix} x_1 \\ \vdots \\ x_n \end{pmatrix},$$

so setze ich

$$\mathfrak{x}_1 = \begin{pmatrix} x_1 \\ 0 \\ \vdots \\ 0 \end{pmatrix},\ \mathfrak{x}_1' = \begin{pmatrix} x_1 \\ 0 \end{pmatrix},\ \mathfrak{x}_2 = \begin{pmatrix} 0 \\ x_2 \\ 0 \\ \vdots \\ 0 \end{pmatrix},\ \mathfrak{x}_2' = \begin{pmatrix} 0 \\ x_2 \end{pmatrix},$$

$$\mathfrak{x}_{n-1} = \mathfrak{x} - \mathfrak{x}_1,\ \mathfrak{x}_{n-2} = \mathfrak{x} - \mathfrak{x}_1 - \mathfrak{x}_2,$$

$$\mathfrak{x}_{n-1}' = \begin{pmatrix} x_2 \\ \vdots \\ x_n \end{pmatrix},\ d\mathfrak{x} = d x_1 \ldots d x_n,\ d\mathfrak{x}_1 = d\mathfrak{x}_1' = d x_1,$$

$$d\mathfrak{x}_2 = d\mathfrak{x}_2' = d x_2,\ d\mathfrak{x}_{n-1} = d x_2 \ldots d x_n$$

und $d\mathfrak{x}_{n-2} = d x_3 \ldots d x_n$, wenn $n \geqq 3$.

Dann kann ich den Hilfssatz 1 so aussprechen:

Es gibt zu jedem $\varepsilon > 0$ und zu jedem $\beta > 0$, welches kleiner als $c(\varepsilon)$ ist, eine Matrix

$$\mathfrak{A}' = \begin{pmatrix} \beta\alpha & -\frac{1}{\beta} \\ \beta & \alpha \end{pmatrix}, \qquad 0 < \alpha < \frac{1}{\beta},$$

so daß

$$\text{(1')} \qquad \sum_{g_1}{}' \int_{-\infty}^{+\infty} F(\mathfrak{A}'(\mathfrak{g}_1' + \mathfrak{x}_2'))\, d\mathfrak{x}_2' \leqq \iint_{-\infty}^{+\infty} F(\mathfrak{x}_1' + \mathfrak{x}_2')\, d\mathfrak{x}_1'\, d\mathfrak{x}_2' + \varepsilon$$

ist.

Es ist ja $\frac{1}{\beta} \iint_{-\infty}^{+\infty} F\left(x_1 - \frac{x_2}{\beta}, x_1\right) d x_1\, d x_2 = \iint_{-\infty}^{+\infty} F(x_1, x_2)\, d x_1\, d x_2$.

Hilfssatz 2. *Es sei $\varphi(\mathfrak{x})$ eine reelle, beschränkte, im* RIEMANN*schen Sinne integrierbare Funktion, die außerhalb des Würfels $|x_i| \leqq K$ $(i = 1, \ldots, n)$ verschwindet. Dann gibt es einen Vektor $\mathfrak{b}_{n-1}$*

$$\mathfrak{b}_{n-1} = \begin{pmatrix} 0 \\ b_2 \\ \vdots \\ b_n \end{pmatrix} = \begin{pmatrix} 0 \\ \mathfrak{b}_{n-1}' \end{pmatrix},$$

so daß

$$\text{(4)} \qquad \sum_{g_1}{}' \sum_{\mathfrak{g}_{n-1}} \varphi(\mathfrak{g}_{n-1} + \mathfrak{b}_{n-1} g_1 + \mathfrak{g}_1) \leqq \sum_{g_1}{}' \int_{-\infty}^{+\infty} \varphi(\mathfrak{x}_{n-1} + \mathfrak{g}_1)\, d\mathfrak{x}_{n-1}.$$

Dabei durchläuft $\mathfrak{g}_{n-1}$ alle Gitterpunkte, deren erste Koordinate Null ist, und g_1 durchläuft alle ganzen Zahlen mit Ausschluß der Null. Den Beweis von (4) führen wir indirekt. Nehmen wir an, (4) sei für alle $\mathfrak{b}_{n-1}$ falsch, insbesondere für alle $\mathfrak{b}_{n-1}$, für die

$$W\colon \quad 0 \leqq b_i \leqq 1 \qquad (i = 2, \ldots n)$$

ist! Dann muß auch

$$\sum_{g_1}' \int_W d\mathfrak{b}_{n-1} \sum_{\mathfrak{g}_{n-1}} \varphi(\mathfrak{g}_{n-1} + \mathfrak{b}_{n-1} g_1 + \mathfrak{g}_1) > \sum_{g_1}' \int_W d\mathfrak{b}_{n-1} \int_{-\infty}^{+\infty} \varphi(\mathfrak{x}_{n-1} + \mathfrak{g}_1)\, d\mathfrak{x}_{n-1}$$

sein. Dann gibt es ein $g_1 \neq 0$, so daß

$$\int_W d\mathfrak{b}_{n-1} \sum_{\mathfrak{g}_{n-1}} \varphi(\mathfrak{g}_{n-1} + \mathfrak{b}_{n-1} g_1 + \mathfrak{g}_1) > \int_W d\mathfrak{b}_{n-1} \int_{-\infty}^{+\infty} \varphi(\mathfrak{x}_{n-1} + \mathfrak{g}_1)\, d\mathfrak{x}_{n-1}.$$

Wir teilen nun jede Seite des Würfels W in g_1 gleiche Teile und damit W in g_1^{n-1} Würfel W_j mit den Kantenlängen $\frac{1}{g_1}$. Dann ist also

$$\sum_{j=1}^{g_1^{n-1}} \int_{W_j} d\mathfrak{b}_{n-1} \sum_{\mathfrak{g}_{n-1}} \varphi(\mathfrak{g}_{n-1} + \mathfrak{b}_{n-1} g_1 + \mathfrak{g}_1)$$

$$> \sum_{j=1}^{g_1^{n-1}} \int_{W_j} d\mathfrak{b}_{n-1} \int_{-\infty}^{+\infty} \varphi(\mathfrak{x}_{n-1} + \mathfrak{g}_1)\, d\mathfrak{x}_{n-1}.$$

Dann muß es aber einen Würfel W_j geben, so daß

$$\int_{W_j} d\mathfrak{b}_{n-1} \sum_{\mathfrak{g}_{n-1}} \varphi(\mathfrak{g}_{n-1} + \mathfrak{b}_{n-1} g_1 + \mathfrak{g}_1) > \int_{W_j} d\mathfrak{b}_{n-1} \int_{-\infty}^{+\infty} \varphi(\mathfrak{x}_{n-1} + \mathfrak{g}_1)\, d\mathfrak{x}_{n-1}$$

ist. Wir machen nun links die Substitution $\mathfrak{x}_{n-1}^* = g_1 \mathfrak{b}_{n-1}$, dann geht W_j in einen Würfel W^* mit der Kantenlänge 1 über:

$$\int_{W^*} d\mathfrak{x}_{n-1}^* \sum_{\mathfrak{g}_{n-1}} \varphi(\mathfrak{g}_{n-1} + \mathfrak{x}_{n-1}^* + \mathfrak{g}_1) > \int_{-\infty}^{+\infty} \varphi(\mathfrak{x}_{n-1} + \mathfrak{g}_1)\, d\mathfrak{x}_{n-1}.$$

Wir setzen schließlich links $\mathfrak{x}_{n-1} = \mathfrak{g}_{n-1} + \mathfrak{x}_{n-1}^*$ und erhalten

$$\int_{-\infty}^{+\infty} d\mathfrak{x}_{n-1}\, \varphi(\mathfrak{x}_{n-1} + \mathfrak{g}_1) > \int_{-\infty}^{+\infty} \varphi(\mathfrak{x}_{n-1} + \mathfrak{g}_1)\, d\mathfrak{x}_{n-1}.$$

Widerspruch.

Hilfssatz 2 läßt sich noch so aussprechen: *Es gibt eine Matrix $\mathfrak{A}_3$:*

$$\mathfrak{A}_3 = \begin{pmatrix} 1 & \mathfrak{O}_3 \\ \mathfrak{b}'_{n-1} & \mathfrak{E}_{n-1} \end{pmatrix}, \tag{4'}$$

wo $\mathfrak{O}_3$ Nullmatrix und $\mathfrak{E}_{n-1}$ die $n-1$-zeilige Einheitsmatrix ist, so daß

$$\sum_{\mathfrak{g}}' \varphi(\mathfrak{A}_3 \mathfrak{g}) \leq \sum_{g_1=-\infty}^{+\infty}{}' \int_{-\infty}^{+\infty} \varphi(\mathfrak{x}_{n-1} + \mathfrak{g}_1)\, d\mathfrak{x}_{n-1} \tag{4''}$$

ist. Dabei sind links alle Gitterpunkte auszuschließen, deren erste Koordinate Null ist.

Es ist ja

$$\mathfrak{A}_3\mathfrak{g} = \begin{pmatrix} 1 & \mathfrak{O}_3 \\ \mathfrak{b}'_{n-1} & \mathfrak{E}_{n-1}\end{pmatrix}\begin{pmatrix} g_1 \\ \mathfrak{g}'_{n-1}\end{pmatrix} = \begin{pmatrix} g_1 \\ \mathfrak{g}'_{n-1} + \mathfrak{b}'_{n-1}\, g_1\end{pmatrix}$$

$$= \begin{pmatrix} g_1 \\ \mathfrak{o}'_{n-1}\end{pmatrix} + \begin{pmatrix} 0 \\ \mathfrak{g}'_{n-1} + \mathfrak{b}'_{n-1}\, g_1\end{pmatrix} = \mathfrak{g}_1 + \mathfrak{g}_{n-1} + \mathfrak{b}_{n-1} g_1 .$$

Wir bemerken noch, daß $\mathfrak{A}_3\mathfrak{g}_{n-1} = \mathfrak{g}_{n-1}$ ist.

Satz 1 (Deformationssatz).

Es sei $\varphi(\mathfrak{x})$ reell, beschränkt, im RIEMANN*schen Sinne integrierbar und außerhalb des Würfels W: $|x_i| \leqq \frac{K}{2}$ $(i = 1, \ldots n)$ identisch Null. Dann gibt es zu jedem $\varepsilon > 0$ eine unimodulare Matrix $\mathfrak{A}$, so daß*

$$\sum_{\mathfrak{g} \neq \mathfrak{o}} \varphi(\mathfrak{A}\,\mathfrak{g}) \leqq \int_{R_n} \varphi(\mathfrak{x})\, d\mathfrak{x} + \varepsilon. \tag{5}$$

Dabei erstreckt sich die Integration rechts über den ganzen Raum R_n.

Beweis*). Ist $n \geqq 3$, so betrachte ich die Matrizen

$$\mathfrak{A}_1 = \begin{pmatrix} 1 & 0 & \\ 0 & \frac{1}{K^{n-2}} & \mathfrak{O}_1 \\ & \mathfrak{O}_2 & K\mathfrak{E}_{n-2}\end{pmatrix}, \quad \mathfrak{A}_2 = \begin{pmatrix} \mathfrak{A}' & \mathfrak{O}_1 \\ \mathfrak{O}_2 & \mathfrak{E}_{n-2}\end{pmatrix}, \quad \mathfrak{A}_3.$$

Dabei sind $\mathfrak{O}_1$, $\mathfrak{O}_2$ Nullmatrizen und $\mathfrak{A}'$ und $\mathfrak{A}_3$ genau wie in (1″) bzw. (4″) definiert. Die $b_2, \ldots b_n$ in $\mathfrak{A}_3$ sind beliebig, und von den α, β in $\mathfrak{A}'$ setzen wir nur voraus: $0 < \alpha < \frac{1}{\beta}$, $0 < \beta < \frac{1}{K(1+K^{n-2})}$. Dann ist $\frac{1}{K^{n-2}}\left(\frac{1}{\beta} - \alpha\right) + \frac{\alpha}{K^{n-2}} = \frac{1}{\beta K^{n-2}} > \left(\frac{1}{K^{n-2}} + 1\right)K$. Also muß entweder $\frac{1}{\beta} - \alpha$ oder $\frac{\alpha}{K^{n-2}}$ größer als K sein. Dann liegt aber der Punkt mit den Koordinaten $\left(\alpha - \frac{1}{\beta}\right) g_2$, $\frac{\alpha g_2}{K^{n-2}}$, $K g_3, \ldots, K g_n$, bei denen die $g_2, \ldots, g_n$ ganzzahlig sind und nicht alle zugleich verschwinden, stets außerhalb des Würfels W, d. h. es ist für alle $\mathfrak{g}_{n-1} \neq \mathfrak{o}_{n-1}$

$$\varphi(\mathfrak{A}_1\mathfrak{A}_2\mathfrak{g}_{n-1}) = \varphi\left(\left(\alpha - \frac{1}{\beta}\right) g_2, \frac{\alpha g_2}{K^{n-2}}, K g_3, \ldots, K g_n\right) = 0.$$

Ist $n = 2$, so betrachte ich die Matrizen

$$\mathfrak{A}_1 = \begin{pmatrix} 1 & 0 \\ 0 & \frac{1}{K^{n-2}}\end{pmatrix}, \quad \mathfrak{A}_2 = \mathfrak{A}', \quad \mathfrak{A}_3,$$

wo wieder $0 < \alpha < \frac{1}{\beta}$, $0 < \beta < \frac{1}{K(1+K^{n-2})}$ und $\varphi(\mathfrak{A}_1\mathfrak{A}_2\,\mathfrak{g}_{n-1}) = 0$ ist für alle $\mathfrak{g}_{n-1} \neq \mathfrak{o}_{n-1}$, d. h. für alle $g_2 \neq 0$.

*) Es würde genügen Satz 1 für stetiges φ zu zeigen, da es zu integrablen φ stets Polynome P_1, P_2 gibt, so daß auf W $P_1 < \varphi < P_2$, $\int (P_2 - P_1)\, d\mathfrak{x} < \varepsilon$ ist.

Um nun im folgenden die Fälle $n \geqq 3$ und $n = 2$ nicht immer unterscheiden zu müssen, so setze ich noch fest, daß unter dem Integral $\int\limits_{-\infty}^{+\infty} d\mathfrak{x}_{n-2}$ im Falle $n = 2$ die Zahl 1 zu verstehen ist. Wir führen nun die Funktion $F(\mathfrak{x}_1' + \mathfrak{x}_2') = \int\limits_{-\infty}^{+\infty} \varphi(\mathfrak{A}_1 \mathfrak{x})\, d\mathfrak{x}_{n-2} = \int\limits_{-\infty}^{+\infty} \varphi(\mathfrak{A}_1(\mathfrak{x}_{n-2} + \mathfrak{x}_1 + \mathfrak{x}_2))\, d\mathfrak{x}_{n-2}$ ein. Es ist

$$F(\mathfrak{A}'(\mathfrak{x}_1' + \mathfrak{x}_2')) = \int\limits_{-\infty}^{+\infty} \varphi(\mathfrak{A}_1 \mathfrak{A}_2 \mathfrak{x})\, d\mathfrak{x}_{n-2}.$$

Wir wählen bei gegebenen ε nun $\beta < \mathrm{Min}\left(c(\varepsilon), \frac{1}{K(1+K^{n-2})}\right)$ und α entsprechend dem Hilfssatz 1. Es erfüllt ja $F(\mathfrak{x}_1' + \mathfrak{x}_2')$ die Voraussetzungen des Hilfssatzes 1 mit $C = K^{n-2} C_1$, wenn $|\varphi(\mathfrak{x})| \leqq C_1$ ist. Nachdem wir α und β gewählt haben, wählen wir $\mathfrak{b}_{n-1}$ entsprechend dem Hilfssatz 2 für die Funktion $\overline{\varphi}(\mathfrak{x}) = \varphi(\mathfrak{A}_1 \mathfrak{A}_2 \mathfrak{x})$. Setzen wir nun $\mathfrak{A} = \mathfrak{A}_1 \mathfrak{A}_2 \mathfrak{A}_3$, so haben wir nach (1″) und (4″)

$$\varepsilon + \int\limits_{R_n} \varphi(\mathfrak{x})\, d\mathfrak{x} = \varepsilon + \int\limits_{R_n} \varphi(\mathfrak{A}_1 \mathfrak{x})\, d\mathfrak{x} = \varepsilon + \iint\limits_{-\infty}^{+\infty} F(\mathfrak{x}_1' + \mathfrak{x}_2')\, d\mathfrak{x}_1'\, d\mathfrak{x}_2'$$

$$\geqq \sum_{\mathfrak{g}_1}{}' \int\limits_{-\infty}^{+\infty} F(\mathfrak{A}'(\mathfrak{g}_1' + \mathfrak{x}_2'))\, d\mathfrak{x}_2' = \sum_{\mathfrak{g}_1}{}' \int\limits_{-\infty}^{+\infty} d\mathfrak{x}_2' \int\limits_{-\infty}^{+\infty} d\mathfrak{x}_{n-2}\, \varphi(\mathfrak{A}_1 \mathfrak{A}_2 (\mathfrak{g}_1 + \mathfrak{x}_2 + \mathfrak{x}_{n-2}))$$

$$= \sum_{\mathfrak{g}_1}{}' \int\limits_{-\infty}^{+\infty} d\mathfrak{x}_{n-1}\, \varphi(\mathfrak{A}_1 \mathfrak{A}_2 (\mathfrak{g}_1 + \mathfrak{x}_{n-1})) \geqq \sum_{\mathfrak{g}}{}' \varphi(\mathfrak{A}_1 \mathfrak{A}_2 \mathfrak{A}_3 \mathfrak{g}).$$

Nun ist $\varphi(\mathfrak{A}_1 \mathfrak{A}_2 \mathfrak{A}_3\, \mathfrak{g}_{n-1}) = \varphi(\mathfrak{A}_1 \mathfrak{A}_2\, \mathfrak{g}_{n-1}) = 0$ für $\mathfrak{g}_{n-1} \neq \mathfrak{o}$, also

$$\varepsilon + \int\limits_{R_n} \varphi(\mathfrak{x})\, d\mathfrak{x} \geqq \sum_{\mathfrak{g} \neq \mathfrak{o}} \varphi(\mathfrak{A}\, \mathfrak{g}).$$

§ 3.

Die Minkowskische Behauptung

Wir nennen zwei Mengen M' und M *äquivalent*, wenn es eine unimodulare Matrix $\mathfrak{A}$ gibt, so daß folgendes gilt: Zu jedem Punkt $\mathfrak{x}'$ aus M' gibt es einen Punkt $\mathfrak{x}$ aus M mit $\mathfrak{x}' = \mathfrak{A}\mathfrak{x}$, und umgekehrt zu jedem $\mathfrak{x}$ aus M ein $\mathfrak{x}'$ aus M', so daß $\mathfrak{x} = \mathfrak{A}^{-1}\mathfrak{x}'$. Wir schreiben kurz $M' = \mathfrak{A} M$. Dann lautet Satz 2:

Satz 2. *Ist M im Jordanschen Sinne meßbar und ist sein Volumen $V < k$ (k natürliche Zahl), so gibt es stets eine zu M äquivalente Menge, die höchstens $k - 1$ Gitterpunkte $\neq \mathfrak{o}$ enthält.*

Beweis. Ich betrachte die charakteristische Funktion $\varphi(\mathfrak{x})$ von M — ($\varphi(\mathfrak{x}) = 1$, wenn $\mathfrak{x}$ in M, $\varphi(\mathfrak{x}) = 0$, wenn $\mathfrak{x}$ außerhalb M). Es gibt nun sicher ein $\varepsilon > 0$, so daß auch $V + \varepsilon < k$ ist. Nun wenden wir (5) an und erhalten

$$\sum_{\mathfrak{g} \neq \mathfrak{o}} \varphi(\mathfrak{A}\mathfrak{g}) \leqq \int_{R_n} \varphi(\mathfrak{x})\, d\mathfrak{x} + \varepsilon = V + \varepsilon < k.$$

Links steht die Anzahl der von $\mathfrak{o}$ verschiedenen Gitterpunkte, die in der Menge $\mathfrak{A}^{-1} M$ liegen, und damit ist der Satz schon bewiesen.

Der Satz läßt sich für Strahlkörper S verschärfen.

Es sei $f(\mathfrak{x})$ eine reelle Funktion mit den Eigenschaften

a) $f(\mathfrak{x}) > 0$ für $\mathfrak{x} \neq \mathfrak{o}$, $f(\mathfrak{o}) = 0$.

b) $f(t\mathfrak{x}) = t f(\mathfrak{x})$ für $t \geqq 0$.

c) $\int\limits_{f(\mathfrak{x}) \leqq 1} d\mathfrak{x} = J$ vorhanden und endlich.

Dann heißt der Bereich $f(\mathfrak{x}) \leqq t$ *Strahlkörper*.

Satz 3. *Ist das Volumen V eines Strahlkörpers S kleiner als* $\sum\limits_{k=1}^{\infty} \frac{1}{k^n}$, *so gibt es stets einen äquivalenten Strahlkörper, der außer dem Nullpunkt keine weiteren Gitterpunkte mehr enthält.*

Beweis. Es ist $\sum\limits_{k=1}^{\infty} \frac{1}{k^n} \sum\limits_{k=1}^{\infty} \frac{\mu(k)}{k^n} = 1$. Dabei ist $\mu(k)$ die Möbiussche Funktion: $\mu(1) = 1$, $\mu(k) = 0$, wenn k durch ein Quadrat (außer 1) teilbar ist und $\mu(k) = (-1)^r$, wenn dies nicht der Fall ist und r die Anzahl der verschiedenen Primfaktoren von k ist. Unsere Voraussetzung lautet also

$$V \sum_{k=1}^{\infty} \frac{\mu(k)}{k^n} < 1.$$

Nun kann ich sicher ein $\varepsilon > 0$ so wählen, daß auch

$$V \sum_{k=1}^{\infty} \frac{\mu(k)}{k^n} + 2\varepsilon < 1$$

ist. Ist p_j die j-te Primzahl ($p_1 = 2$, $p_2 = 3, \ldots$), so ist

$$\prod_{j=1}^{\infty} \left(1 - \frac{1}{p_j^n}\right) = \sum_{k=1}^{\infty} \frac{\mu(k)}{k^n}.$$

Wegen der Konvergenz des linksstehenden Produktes kann ich ein $m \geqq$. so wählen, daß

$$V \prod_{j=1}^{m} \left(1 - \frac{1}{p_j^n}\right) + \varepsilon < V \sum_{k=1}^{\infty} \frac{\mu(k)}{k^n} + 2\varepsilon < 1$$

wird. Setzen wir $N = p_1 \dots p_m$, so ist das linksstehende Produkt nichts anderes als $\sum_{k=1}^{N}{}' \frac{\mu(k)}{k^n}$, wo k im Intervall $[1, N]$ alle ganzen Zahlen der Form $p_1^{e_1} \dots p_m^{e_m}$ $(e_j \geqq 0)$ durchläuft. Es ist also $V \sum_{k=1}^{N}{}' \frac{\mu(k)}{k^n} + \varepsilon < 1$

Ich betrachte nun auf S die Funktion

$$\psi_N(\mathfrak{x}) = \sum_{k=1}^{N}{}' \mu(k)\, \varphi(k\mathfrak{x}),$$

wo $\varphi(\mathfrak{x})$ die charakteristische Funktion von S ist und wende (5) an. Es ist

$$\int\limits_S \psi_n(\mathfrak{x})\, d\mathfrak{x} = \sum_{k=1}^{N}{}' \mu(k) \int\limits_{f(\mathfrak{x}) \leqq t} \varphi(k\mathfrak{x})\, d\mathfrak{x} = \sum_{k=1}^{N}{}' \mu(k) \int\limits_{f(\mathfrak{x}) \leqq \frac{t}{k}} \varphi(\mathfrak{x})\, d\mathfrak{x}$$

$$= \sum_{k=1}^{N}{}' \frac{\mu(k)}{k^n} \cdot V$$

Also ist

$$\sum_{\mathfrak{g} \neq \mathfrak{o}} \psi_N(\mathfrak{A}\mathfrak{g}) = \sum_{k=1}^{N}{}' \mu(k) \sum_{\mathfrak{g} \neq \mathfrak{o}} \varphi(\mathfrak{A} k \mathfrak{g}) \leqq V \sum_{k=1}^{N}{}' \frac{\mu(k)}{k^n} + \varepsilon < 1.$$

Verstehen wir unter dem größten gemeinschaftlichen Teiler eines Gitterpunktes den größten gemeinschaftlichen Teiler seiner Koordinaten, so ist

$$\sum_{k=1}^{N}{}' \mu(k) \sum_{\mathfrak{g} \neq \mathfrak{o}} \varphi(\mathfrak{A} k \mathfrak{g})$$

nichts anderes als die Anzahl der von $\mathfrak{o}$ verschiedenen Gitterpunkte in $\mathfrak{A}^{-1} S$ deren größte gemeinschaftliche Teiler durch keine der Primzahlen $p_1, \dots, p_n$ teilbar sind. Diese Anzahl ist nach obigem kleiner als 1, also Null. Im besonderen können daher in $\mathfrak{A}^{-1} S$ auch keine Gitterpunkte $\neq \mathfrak{o}$ liegen, deren größter gemeinschaftlicher Teiler 1 ist. Dann können im Strahlkörper $\mathfrak{A}^{-1} S$ überhaupt keine Gitterpunkte liegen; denn nehmen wir an, in $\mathfrak{A}^{-1} S$ liege ein Gitterpunkt $\mathfrak{g} \neq \mathfrak{o}$, so muß sein größter gemeinschaftlicher Teiler $d > 1$ sein und $\frac{\mathfrak{g}}{d}$ ist ebenfalls ein Gitterpunkt. Da $f(\mathfrak{A}\mathfrak{g}) \leqq t$ ist, so ist $f\left(\mathfrak{A}\frac{\mathfrak{g}}{d}\right) \leqq \frac{t}{d} < t$ also liegt $\frac{\mathfrak{g}}{d}$ im Strahlkörper $\mathfrak{A}^{-1} S$ und sein größter gemeinschaftlicher Teiler ist 1. Dies ist aber nicht möglich. Damit ist der Satz vollständig bewiesen.

Satz 4. *Ist der Strahlkörper S in bezug auf $\mathfrak{o}$ symmetrisch, d. h. ist $f(\mathfrak{x}) = f(-\mathfrak{x})$ und ist das Volumen V von S kleiner als $2 \sum_{k=1}^{\infty} \frac{1}{k^n}$, so gibt es stets einen zu S äquivalenten Strahlkörper, welcher ebenfalls in bezug auf $\mathfrak{o}$ symmetrisch ist, und der außer $\mathfrak{o}$ keine weiteren Gitterpunkte enthält.*

Der Beweis verläuft genau so wie bei Satz 3; man hat nur zu beachten, daß die Anzahl der Gitterpunkte, die in einem in bezug auf $\mathfrak{o}$ symmetrischen Strahlkörper liegen, stets gerade ist.

Satz 3 und Satz 4 stellen die MINKOWSKIsche Behauptung dar.

§ 4.

Die $M(\mathfrak{K})$-Mengen.

Wir stellen nun eine Verallgemeinerung des MINKOWSKIschen Fundamentalsatzes auf.

Es sei M eine beschränkte, abgeschlossene, im JORDANschen Sinne meßbare Menge vom Volumen V. Sie liege im Würfel $|x_i| \leqq K$ $(i = 1, \ldots, n)$. Auf M sei eine beschränkte, im RIEMANNschen Sinne integrierbare Funktion $\varphi(\mathfrak{x})$ definiert; außerhalb M setzen wir $\varphi(\mathfrak{x}) = 0$. Weiter sei $\mathfrak{K}$ die Diagonalmatrix

$$\mathfrak{K} = \begin{pmatrix} k_1 & & \\ & \ddots & \\ & & k_n \end{pmatrix}, \quad k_i > 0, \tag{6}$$

$|\mathfrak{K}| = k_1 \ldots k_n$. Dann betrachten wir die Funktion

$$f(\mathfrak{x}) = \sum_{\mathfrak{g}} \varphi(\mathfrak{K}(\mathfrak{x} - \mathfrak{g})). \tag{7}$$

Dabei durchläuft $\mathfrak{g}$ alle Gitterpunkte des R_n. Die Anzahl der nicht verschwindenden Summanden in der rechtsstehenden Summe ist endlich. $f(\mathfrak{x})$ ist in den x_i periodisch mit der Periode 1 und ebenfalls integrierbar. Wir bestimmen nun die Fourierkoeffizienten $a_{\mathfrak{l}}$ von $f(\mathfrak{x})$ oder gleich allgemeiner von

$$F(\mathfrak{x}) = \sum_{\mathfrak{g}}' [\varphi(\mathfrak{K}(\mathfrak{x} - \mathfrak{g})) + \varphi_1(\mathfrak{K}(\mathfrak{x} - \mathfrak{g}) - \mathfrak{x}_0)].$$

Dabei ist $\varphi_1(\mathfrak{x})$ auf einer Menge M_1 definiert und $\mathfrak{x}_0$ ein Punkt des R_n. Es sei W der Würfel $0 \leqq x_i \leqq 1$ $(i = 1, \ldots, n)$ und $\mathfrak{l}\mathfrak{x} = \sum_{i=1}^{n} l_i x_i$. Dann berechnen sich die Fourierkoeffizienten $c_{\mathfrak{l}}$ von $F(\mathfrak{x})$ so:

$$c_{\mathfrak{l}} = \int_W F(\mathfrak{x}) e^{-2\pi i \mathfrak{l}\mathfrak{x}} d\mathfrak{x} = \sum_{\mathfrak{g}} \int_W [\varphi(\mathfrak{K}(\mathfrak{x} - \mathfrak{g})) + \varphi_1(\mathfrak{K}(\mathfrak{x} - \mathfrak{g}) - \mathfrak{x}_0)] e^{-2\pi i \mathfrak{l}\mathfrak{x}} d\mathfrak{x}$$

$$= \int_{R_n} [\varphi(\mathfrak{K}\mathfrak{x}) + \varphi_1(\mathfrak{K}\mathfrak{x} - \mathfrak{x}_0)] e^{-2\pi i \mathfrak{l}\mathfrak{x}} d\mathfrak{x}.$$

Führen wir noch die Substitution $\mathfrak{x}' = \mathfrak{K}\mathfrak{x}$ aus, so lautet $c_{\mathfrak{l}}$:

$$c_{\mathfrak{l}} = |\mathfrak{K}|^{-1} \int_{R_n} [\varphi(\mathfrak{x}) + \varphi_1(\mathfrak{x} - \mathfrak{x}_0)] e^{-2\pi i \mathfrak{K}^{-1}\mathfrak{l}\cdot\mathfrak{x}} d\mathfrak{x} \tag{8}$$

und insbesondere für $\varphi_1(\mathfrak{x}) \equiv 0$

(8a) $$a_{\mathfrak{l}} = |\mathfrak{K}|^{-1} \int_{R_n} \varphi(\mathfrak{x})\, e^{-2\pi i \mathfrak{K}^{-1}\mathfrak{l}\cdot\mathfrak{x}}\, d\mathfrak{x}.$$

Ist $\varphi(\mathfrak{x})$ die charakteristische Funktion von M, so ist

$$a_0 = |\mathfrak{K}|^{-1} V$$

und wenn $\varphi_1(\mathfrak{x})$ die charakteristische Funktion von M_1 ist:

$$c_0 = |\mathfrak{K}|^{-1}(V + V_1),$$

wo V_1 das Volumen von M_1 ist.

Ist $\psi(\mathfrak{x})$ eine Funktion mit denselben Eigenschaften wie $\varphi(\mathfrak{x})$, so bilden wir wieder

$$g(\mathfrak{x}) = \sum_{\mathfrak{g}} \psi(\mathfrak{K}(\mathfrak{x} - \mathfrak{g}))$$

und die Fourierkoeffizienten von $g(\mathfrak{x})$ seien $b_{\mathfrak{l}}$.

Dann ist $\sum_{\mathfrak{l}} a_{\mathfrak{l}} \bar{b}_{\mathfrak{l}}$ konvergent und es gilt die Parsevalsche Gleichung

$$\int_W f(\mathfrak{x})\, \bar{g}(\mathfrak{x})\, d\mathfrak{x} = \sum_{\mathfrak{l}} a_{\mathfrak{l}} \bar{b}_{\mathfrak{l}}.$$

Die linke Seite ist gleich

$$\int_W \sum_{\mathfrak{g}} \varphi(\mathfrak{K}(\mathfrak{x} - \mathfrak{g})) \sum_{\mathfrak{r}} \bar{\psi}(\mathfrak{K}(\mathfrak{x} - \mathfrak{r}))\, d\mathfrak{x} = \sum_{\mathfrak{g}} \int_W \varphi(\mathfrak{K}(\mathfrak{x} - \mathfrak{g})) \sum_{\mathfrak{r}} \bar{\psi}(\mathfrak{K}(\mathfrak{x} - \mathfrak{r}))\, d\mathfrak{x}$$
$$= \int_{R_n} \varphi(\mathfrak{K}\mathfrak{x}) \sum_{\mathfrak{r}} \bar{\psi}(\mathfrak{K}(\mathfrak{x} - \mathfrak{r}))\, d\mathfrak{x},$$

also

(9) $$|\mathfrak{K}|^{-1} \sum_{\mathfrak{r}} \int_{R_n} \varphi(\mathfrak{x})\, \bar{\psi}(\mathfrak{x} - \mathfrak{K}\mathfrak{r})\, d\mathfrak{x} = \sum_{\mathfrak{l}} a_{\mathfrak{l}} \bar{b}_{\mathfrak{l}}.$$

Ebenso ist für $F(\mathfrak{x})$:

$$\int_W |F(\mathfrak{x})|^2\, d\mathfrak{x} = \int_W d\mathfrak{x} \sum_{\mathfrak{g}} [\varphi(\mathfrak{K}(\mathfrak{x} - \mathfrak{g})) + \varphi_1(\mathfrak{K}(\mathfrak{x} - \mathfrak{g}) - \mathfrak{x}_0)] \sum_{\mathfrak{r}} [\bar{\varphi}(\mathfrak{K}(\mathfrak{x} - \mathfrak{r})) +$$
$$+ \bar{\varphi}_1(\mathfrak{K}(\mathfrak{x} - \mathfrak{r}) - \mathfrak{x}_0)$$
$$= \sum_{\mathfrak{r}} \int_{R_n} [\varphi(\mathfrak{K}\mathfrak{x}) + \varphi_1(\mathfrak{K}\mathfrak{x} - \mathfrak{x}_0)]\, [\bar{\varphi}(\mathfrak{K}(\mathfrak{x} - \mathfrak{r})) + \bar{\varphi}_1(\mathfrak{K}(\mathfrak{x} - \mathfrak{r}) - \mathfrak{x}_0)]\, d\mathfrak{x},$$

also

(9') $$|\mathfrak{K}|^{-1} \sum_{\mathfrak{r}} \int_{R_n} [\varphi(\mathfrak{x}) + \varphi_1(\mathfrak{x} - \mathfrak{x}_0)]\, [\bar{\varphi}(\mathfrak{x} - \mathfrak{K}\mathfrak{r}) + \bar{\varphi}_1(\mathfrak{x} - \mathfrak{x}_0 - \mathfrak{K}\mathfrak{r})]\, d\mathfrak{x} = \sum_{\mathfrak{l}} |c_{\mathfrak{l}}|^2.$$

Ich bilde nun die Menge $M(\mathfrak{K}) = \overline{M}$ aller Punkte der Form $\mathfrak{K}^{-1}(\mathfrak{x} - \mathfrak{y})$, wo $\mathfrak{x}$ und $\mathfrak{y}$ der Menge angehören. Daraus folgt, daß $\overline{M}$ $\mathfrak{o}$ enthält und in bezug auf $\mathfrak{o}$ symmetrisch ist. Ist $\overline{M} = M$ und $k_i = 2$ $(i = 1, \ldots, n)$, so ist M konvex. Ich beweise nun den

Hilfssatz 3. Sind $\mathfrak{x}$ und $\mathfrak{y}$ Punkte von M und ist $\mathfrak{x}$ innerer Punkt von M, so ist $\mathfrak{K}^{-1}(\mathfrak{x} - \mathfrak{y}) = \mathfrak{z}$ innerer Punkt von $M(\mathfrak{K}) = \overline{M}$.

Beweis. Wir bezeichnen mit $d(\mathfrak{x})$ den Abstand des Punktes $\mathfrak{x}$ von $\mathfrak{o}$. Wir führen den Beweis indirekt. Wir nehmen an, $\mathfrak{z}$ ist kein innerer Punkt von $\overline{M}$. Dann enthält jede Kugel $d(\mathfrak{z}-\mathfrak{z}') \leqq \varrho \operatorname{Min}\left(\frac{1}{k_1}, \ldots, \frac{1}{k_n}\right)$ ($\varrho > 0$, sonst beliebig) einen Punkt $\mathfrak{z}'$, der nicht zu $\overline{M}$ gehört. Nun ist $d(\mathfrak{R}^{-1}(\mathfrak{x}-\mathfrak{y})-\mathfrak{z}') = d(\mathfrak{R}^{-1}(\mathfrak{x}-\mathfrak{y}-\mathfrak{R}\mathfrak{z}')) \geqq \operatorname{Min}\left(\frac{1}{k_1}, \cdots, \frac{1}{k_n}\right) d(\mathfrak{x}-\mathfrak{y}-\mathfrak{R}\mathfrak{z}')$, also $d(\mathfrak{x}-\mathfrak{y}-\mathfrak{R}\mathfrak{z}') \leqq \varrho$. Der Punkt $\mathfrak{y}+\mathfrak{R}\mathfrak{z}'$ liegt nicht in M, denn sonst liegt $z' = \mathfrak{R}^{-1}(\mathfrak{R}\mathfrak{z}'+\mathfrak{y}-\mathfrak{y})$ in $\overline{M}$. Also enthält jede Kugel von $\mathfrak{x}$ einen Punkt, der nicht in M liegt. Widerspruch.

Wir wollen nun die Voraussetzung machen, daß in $\overline{M}$ *höchstens* $k_0 - 1$ verschiedene Gitterpunktpaare $\neq \mathfrak{o}$ *innere* Punkte von M sind. Dabei sei k_0 eine natürliche Zahl. Setzen wir in (9) $\psi(\mathfrak{x}) \equiv \varphi(\mathfrak{x})$ auf M, so hat die Summe links in (9) höchstens k_0 nichtverschwindende Summanden. Wenn wir nämlich annehmen, die Anzahl dieser Summanden ist größer als k_0, so gibt es mindestens k_0 voneinander verschiedene Punkte $\mathfrak{r}_m \neq \mathfrak{o}$ ($m = 1, \ldots, k_0$), für die

$$\int_M \varphi(\mathfrak{x})\,\overline{\varphi}(\mathfrak{x}-\mathfrak{R}\mathfrak{r}_m)\,d\mathfrak{x} \neq 0$$

ist. Dann gibt es ein $\mathfrak{x}$, so daß $\mathfrak{x}$ und $\mathfrak{x}-\mathfrak{R}\mathfrak{r}_m$ innere Punkte von M sind; dann liegt aber $\pm\,\mathfrak{r}_m$ in $\overline{M}$ und zwar ist $\mathfrak{r}_m$ nach Hilfssatz 3 innerer Punkt von $\overline{M}$. Das heißt aber, $\overline{M}$ enthält k_0 voneinander verschiedene, von $\mathfrak{o}$ verschiedene, Gitterpunktpaare, welche innere Punkte von $\overline{M}$ sind, was nicht sein soll. Es gilt also wegen (9) und (8)

$$(10)\qquad \sum_{m=0}^{G} \int_M \varphi(\mathfrak{x})\,\overline{\varphi}(\mathfrak{x}-\mathfrak{R}\mathfrak{r}_m)\,d\mathfrak{x} = |\mathfrak{R}|^{-1} \sum_{\mathfrak{l}} \left|\int_M \varphi(\mathfrak{x})\,e^{-2\pi i \mathfrak{R}^{-1}\mathfrak{l}\cdot\mathfrak{x}}\,d\mathfrak{x}\right|^2,$$

wo $0 \leqq G \leqq k_0 - 1$ und $\mathfrak{r}_0 = \mathfrak{o}$ ist. Nun ist

$$\left|\int_M \varphi(\mathfrak{x})\,\overline{\varphi}(\mathfrak{x}-\mathfrak{R}\mathfrak{r}_m)\,d\mathfrak{x}\right|^2 \leqq \int_M |\varphi(\mathfrak{x})|^2\,d\mathfrak{x} \int_M |\overline{\varphi}(\mathfrak{x}-\mathfrak{R}\mathfrak{r}_m)|^2\,d\mathfrak{x},$$

also ist die Summe links dem Betrage nach kleiner als

$$(G+1)\int_M |\varphi(\mathfrak{x})|^2\,d\mathfrak{x} \leqq k_0 \int_M |\varphi(\mathfrak{x})|^2\,d\mathfrak{x}.$$

Also folgt aus (10)

$$(11)\qquad k_0 \int_M |\varphi(\mathfrak{x})|^2\,d\mathfrak{x} \geqq |\mathfrak{R}|^{-1} \sum_{\mathfrak{l}} \left|\int_M \varphi(\mathfrak{x})\,e^{-2\pi i \mathfrak{R}^{-1}\mathfrak{l}\cdot\mathfrak{x}}\,d\mathfrak{x}\right|^2.$$

Es sei nun wieder $\varphi(\mathfrak{x})$ die charakteristische Funktion von M. Dann folgt aus (11)

$$(12)\qquad k_0\,|\mathfrak{R}|\,V \geqq V^2 + \sum_{\mathfrak{l}\neq\mathfrak{o}} \left|\int_M e^{-2\pi i \mathfrak{R}^{-1}\mathfrak{l}\cdot\mathfrak{x}}\,d\mathfrak{x}\right|^2.$$

Es gilt also

Satz 5. *Erfüllt das Volumen V von M die Ungleichung*

$$(13)\qquad V > k_0|\mathfrak{K}| - \frac{1}{V}\sum_{\mathfrak{l}\neq\mathfrak{o}}\left|\int_M e^{-2\pi i\mathfrak{K}^{-1}\mathfrak{l}\cdot\mathfrak{x}}\,d\mathfrak{x}\right|^2,$$

so besitzt $M(\mathfrak{K})$ *mindestens* k_0 *von* $\mathfrak{o}$ *verschiedene Gitterpunktpaare.*

Für konvexe Körper und $k_0 = 1$ stammt der Satz und sein Beweis von C. Siegel [3]. Insbesondere gilt Satz 5, wenn $V > k_0 k_1 \ldots k_n$. Dies wurde auf andere Weise zuerst von v. d. Corput [4] gezeigt. Durch einen einfachen Grenzübergang zeigt man in üblicher Weise (v. d. Corput [4]) den

Satz 5'. *Ist das Volumen V von M größer oder gleich* $k_0|\mathfrak{K}|$, *so besitzt* $M(\mathfrak{K})$ *mindestens* k_0 *von* $\mathfrak{o}$ *verschiedene Gitterpunktpaare.*

§ 5.

Ein Grenzfall.

Wir wollen nun in diesem Paragraphen die Voraussetzung machen, daß M das Volumen $V = k_0|\mathfrak{K}| = k_0 k_1 \ldots k_n$ besitzt. Nach Satz 5' besitzt also sicher $M(\mathfrak{K})$ mindestens k_0 von $\mathfrak{o}$ verschiedene Gitterpunktpaare.

Wir wollen nun voraussetzen, daß in $M = M(\mathfrak{K})$ *höchstens* $k_0 - 1$ *innere Punktepaare Gitterpunktpaare* $\neq \mathfrak{o}$ *sind.* Wir betrachten nun neben M die um $\mathfrak{K}\mathfrak{g}$ verschobenen Mengen $M_{\mathfrak{g}}$ $(M_{\mathfrak{o}} = M)$. Dann behaupten wir: *Ein Punkt des* R_n *kann von höchstens* k_0 *solchen Mengen* $M_{\mathfrak{g}}$ *innerer Punkt sein.* Nehmen wir an, es gibt einen Punkt $\mathfrak{x}_0$, der mindestens $k_0 + 1$ solchen $M_{\mathfrak{g}}$ als innerer Punkt angehört, d. h. es gibt $k_0 + 1$ Gitterpunkte $\mathfrak{g}_j$ $(j = 1, \ldots, k_0 + 1)$, so daß $\mathfrak{x}_0 - \mathfrak{K}\mathfrak{g}_j$ innere Punkte von M sind! Dann liegen die Gitterpunktpaare $\pm(\mathfrak{g}_j - \mathfrak{g}_1)$ in $\overline{M}$ und sind innere Punkte von $\overline{M}$, nach dem obigen Hilfssatz 3. Diese k_0 Gitterpunktpaare sind alle voneinander und von $\mathfrak{o}$ verschieden und dies ist ein Widerspruch zur Voraussetzung. Wir behaupten nun

Satz 6: *Ist M eine beschränkte, abgeschlossene Menge vom Volumen* $V = k_0|\mathfrak{K}|$ *und besitzt* $M(\mathfrak{K})$ *höchstens* $k_0 - 1$ *Gitterpunktpaare als innere Punktepaare, so liegt jeder Punkt des* R_n *in mindestens einer der Mengen* $M_{\mathfrak{g}}$.

Für konvexe Mengen und $k_0 = 1$ wurde dies von Minkowski bewiesen (Geometrie der Zahlen, § 32).

Beweis. Wir nehmen an, es gibt ein $\mathfrak{x}_0$, welches keinem der $M_{\mathfrak{g}}$ angehört. Dieser Punkt $\mathfrak{x}_0$ kann auch kein Häufungspunkt der Vereinigungsmenge $S = \sum_{\mathfrak{g}} M_{\mathfrak{g}}$ sein. Um dies zu zeigen, denke ich mir alle Gitterpunkte in irgendeiner Weise so als Folge geschrieben $\mathfrak{g}_1, \mathfrak{g}_2, \ldots$, so daß $\lim_{i\to\infty} d(\mathfrak{K}\mathfrak{g}_i) = \infty$.

st nun $\mathfrak{x}_0$ Häufungspunkt von S, so kann ich aus S eine unendliche Folge $(\mathfrak{x}_k)$ ɩerausgreifen mit $\lim\limits_{k\to\infty} \mathfrak{x}_k = \mathfrak{x}_0$ und daher $d(\mathfrak{x}_k - \mathfrak{x}_0) < C_1$ für alle k. Nun ;ibt es sicher einen Index j, so daß die Folge $(\mathfrak{x}_k)$ bereits in $\sum\limits_{i=1}^{j} M_{\mathfrak{g}_i}$ liegt. Denn onst gibt es eine unendliche Folge von Gitterpunkten $\mathfrak{g}'_k$ mit $d(\mathfrak{R}\mathfrak{g}'_k) \to \infty$ ür $k \to \infty$ und eine Teilfolge (φ'_k) von $(\mathfrak{x}_k)$, so daß $\mathfrak{x}'_k$ in $M_{\mathfrak{g}_{k'}}$ liegt. Nun iegt die Folge $(\varphi^*_k) = (\varphi'_k - \mathfrak{R}\mathfrak{g}'_k)$ in der beschränkten Menge $M = M\mathfrak{o}$, also $l(\mathfrak{x}'_k - \mathfrak{R}\mathfrak{g}'_k) < C_2$ für alle k. Dann ist aber $d(\mathfrak{R}\mathfrak{g}'_k) = d(\mathfrak{R}\mathfrak{g}'_k - \varphi'_k + \mathfrak{x}'_k - \mathfrak{x}_0)$ $\leqq d(\mathfrak{R}\mathfrak{g}'_k - \mathfrak{x}'_k) + d(\mathfrak{x}'_k - \mathfrak{x}_0) \leqq C_1 + C_2 = C_3$ für alle k, was unmöglich st. Es gibt also einen Index j, so daß die Folge $(\mathfrak{x}_k)$ in $\sum\limits_{i=1}^{j} M_{\mathfrak{g}_i}$ liegt. Dann ;ibt es aber eine unendliche Teilfolge $(\mathfrak{x}'_k)$ von $(\mathfrak{x}_k)$, die in einem deı $M_{\mathfrak{g}_i}$ $(1 \leqq i \leqq j)$ liegt, z. B. $M_{\mathfrak{g}_1}$. Nun ist $M_{\mathfrak{g}_1}$ abgeschlossen und $\mathfrak{x}_0$ Häu- ungspunkt der Folge $(\mathfrak{x}'_k)$, die ganz auf $M_{\mathfrak{g}_1}$ liegt, also muß $\mathfrak{x}_0$ auch in $M_{\mathfrak{g}_1}$ iegen, in Widerspruch zur Voraussetzung. Es gibt daher eine Kugel K um len Punkt $\mathfrak{x}_0$ mit einem Radius $\varrho > 0$, welche zu allen $M_{\mathfrak{g}}$ punktfremd ist.

Dann sind auch die Kugeln $K_{\mathfrak{g}}$ mit den Mittelpunkten $\mathfrak{x}_0 + \mathfrak{R}\mathfrak{r}$ ($\mathfrak{r}$ Gitter- ɔunkt) zu allen $M_{\mathfrak{g}}$ punktfremd. Ist $\varphi_1(\mathfrak{x})$ die charakteristische Funktion ler Kugel $\overline{K}$ mit dem Radius ϱ und dem Mittelpunkt im Ursprung, so ist also

$$\varphi_1(\mathfrak{x} - \mathfrak{x}_0 - \mathfrak{R}\mathfrak{r})\ \varphi(\mathfrak{x} - \mathfrak{R}\mathfrak{g}) = 0,$$

vo $\varphi(\mathfrak{x})$ die charakteristische Funktion von M ist. Wir wählen noch ϱ sc klein, daß die Kugeln $K_{\mathfrak{r}}$ zueinander punktfremd sind, d. h.

$$\varphi_1(\mathfrak{x} - \mathfrak{x}_0)\ \varphi_1(\mathfrak{x} - \mathfrak{x}_0 - \mathfrak{R}\mathfrak{r}) = 0$$

ür alle $\mathfrak{r}$. Dann ist nach (9')

$$(14)\quad |\mathfrak{R}|^{-1} \sum_{\mathfrak{r}} \int_{R_n} [\varphi(\mathfrak{x}) + \varphi_1(\mathfrak{x} - \mathfrak{x}_0)]\,[\varphi(\mathfrak{x} - \mathfrak{R}\mathfrak{r}) + \varphi_1(\mathfrak{x} - \mathfrak{x}_0 - \mathfrak{R}\mathfrak{r})]\,d\mathfrak{x} = \sum_{\mathfrak{l}} |c_{\mathfrak{l}}|^2.$$

Der linksstehende Ausdruck lautet ausmultipliziert

$$\sum_{\mathfrak{r}} \int_{R_n} \varphi(\mathfrak{x})\ \varphi(\mathfrak{x} - \mathfrak{R}\mathfrak{r})\,d\mathfrak{x} + \sum_{\mathfrak{r}} \int_{R_n} [\varphi_1(\mathfrak{x} - \mathfrak{x}_0)\ \varphi(\mathfrak{x} - \mathfrak{R}\mathfrak{r}) +$$
$$+ \varphi(\mathfrak{x})\ \varphi_1(\mathfrak{x} - \mathfrak{x}_0 - \mathfrak{R}\mathfrak{r})]\,d\mathfrak{x} + \sum_{\mathfrak{r}} \int_{R_n} \varphi_1(\mathfrak{x} - \mathfrak{x}_0)\ \varphi_1(\mathfrak{x} - \mathfrak{x}_0 - \mathfrak{R}\mathfrak{r})\,d\mathfrak{x}.$$

Nach Voraussetzung verschwindet die zweite Summe. Auf die erste Summe können wir die Überlegungen von (10) bis (12) anwenden; bei der dritten Summe verschwinden alle Glieder außer dem Glied mit $\mathfrak{r} = \mathfrak{o}$. Bezeichnen wir das Volumen von $\overline{K}$ mit V_1, so erhalten wir

$$|\mathfrak{R}|^{-1}(k_0 V + V_1) \geqq |c_{\mathfrak{o}}|^2 + \sum_{\mathfrak{l} \neq \mathfrak{o}} |c_{\mathfrak{l}}|^2.$$

Da nun $c_{\mathfrak{o}} = |\mathfrak{R}|^{-1}(V + V_1)$, so ist

$$(15)\qquad |\mathfrak{R}|\,(k_0 V + V_1) \geqq (V + V_1)^2.$$

Nun ist nach Voraussetzung $V = k_0 k_1 \ldots k_n = k_0 = k_0 |\mathfrak{R}|$, also

$$|\mathfrak{R}|\, V_1 = \frac{V V_1}{k_0} \geqq 2\, V V_1 + V_1^2,$$

dann muß aber $V_1 = 0$ sein, also auch $\varrho = 0$. Widerspruch.

§ 6.

Die Alternativsätze.

Satz 7. *Besitzt M ein Volumen $V \geqq \frac{k_0 |\mathfrak{R}|}{2}$ und sind höchstens $k_0 - r - 1$ Gitterpunktpaare, welche von $\mathfrak{o}$ verschieden sind, innere Punkte von $M(\mathfrak{R})$ $\left(0 \leqq r \leqq k_0 - \left[\frac{k_0}{2}\right] - 1\right)$, so erfüllen die Mengen $\mathfrak{t}_{\mathfrak{g}} M(\mathfrak{R})$ den R_n lückenlos, und die Überdeckung ist mindestens $r + 1$-fach, d. h. es gibt zu jedem $\mathfrak{x}_0$ mindestens $r + 1$ Gitterpunkte $\mathfrak{g}$, so daß $\mathfrak{x}_0 + \mathfrak{g}$ in $M(\mathfrak{R})$ liegt.*

Satz 7 ist ein Spezialfall des folgenden allgemeineren Satzes:

Satz 7a. *Es seien M_1 und M_2 zwei beschränkte, abgeschlossene, im* Jordanschen *Sinne meßbare Mengen und die Summe ihrer Volumina $V_1 + V_2 \geqq k_0 |\mathfrak{R}| = k_0 k_1 \ldots k_n$ $(V_1 > 0,\ V_2 > 0)$. Weiter seien höchstens $k_0 - r_i - 1$ $\left(i = 1, 2,\ 0 \leqq r_i \leqq k_0 - \left[\frac{V_i}{\mathfrak{R}}\right] - 1\right)$ von $\mathfrak{o}$ verschiedene Gitterpunktpaare innere Punkte von M_i. Ist dann $M_3(\mathfrak{R})$ die Menge aller Punkte der Form $\pm \mathfrak{R}^{-1}(\mathfrak{x} - \mathfrak{y})$, wo $\mathfrak{x}$ in M_1 und $\mathfrak{y}$ in M_2 liegt, so erfüllen die $\mathfrak{t}_{\mathfrak{g}} M_3(\mathfrak{R})$ den R_n lückenlos, und zwar ist die Überdeckung mindestens $r + 1$-fach mit*

$$r = \left[\frac{r_1 V_1 + r_2 V_2}{2\sqrt{V_1 V_2}}\right].$$

Es muß $r_i \leqq k_0 - \left[\frac{V_i}{\mathfrak{R}}\right] - 1$ vorausgesetzt werden, da nach Satz 5' $M_i(\mathfrak{R})$ mindestens $\left[\frac{V_i}{|\mathfrak{R}|}\right]$ von $\mathfrak{o}$ verschiedene Gitterpunktpaare enthält.

Wir führen folgende Bezeichnungen ein: Die charakteristische Funktion von M_i sei $\varphi_i(\mathfrak{x})$. Ist $\mathfrak{x}_0$ ein Punkt des R_n, so setzen wir $\psi_i(\mathfrak{x}) = \varphi_i(\mathfrak{x} - \mathfrak{R}\mathfrak{x}_0)$. Mit M_i^0 bezeichnen wir die Menge aller inneren Punkte von M_i.

Ist $\mathfrak{x}_0$ Gitterpunkt und $M_1 = M_2 = M$, so läßt sich Satz 7a sofort beweisen. Es ist dann $V_1 = V_2 = V$, $r_1 = r_2 = r$, $M_1(\mathfrak{R}) = M_2(\mathfrak{R}) = M_3(\mathfrak{R})$. In $M^0(\mathfrak{R})$ liegen nach Voraussetzung höchstens $k_0 - r - 1$ von $\mathfrak{o}$ verschiedene Gitterpunktpaare. Dabei ist $r \leqq k_0 - \left[\frac{k_0}{2}\right] - 1$, d. h. $M(\mathfrak{R})$ enthält nach Satz 5' mindestens $s = \left[\frac{k_0}{2}\right]$ Gitterpunktpaare $\pm \mathfrak{g}_1, \ldots, \pm \mathfrak{g}_s \neq \mathfrak{o}$. Dann liegt der Gitterpunkt $\mathfrak{x}_0$ in den $2s + 1$ Mengen $\mathfrak{t}_{\mathfrak{x}_0} M(\mathfrak{R})$, $\mathfrak{t}_{\mathfrak{x}_0 \pm \mathfrak{g}_j} M(\mathfrak{R})$. Nun ist $2s + 1 = 2\left[\frac{k_0}{2}\right] + 1 \geqq k_0 - \left[\frac{k_0}{2}\right] \geqq r + 1$, was zu zeigen war.

Wir können also im folgenden den Fall ausschließen, daß $M_1 = M_2$ und $\mathfrak{x}_0$ Gitterpunkt ist.

Hilfssatz 4: Für jedes $\mathfrak{y}$ ist stets

$$\int_{R_n} \varphi_1(\mathfrak{x})\, \varphi_2(\mathfrak{x}-\mathfrak{y})\, d\mathfrak{x} + \int_{R_n} \varphi_2(\mathfrak{x})\, \varphi_1(\mathfrak{x}-\mathfrak{y})\, d\mathfrak{x} \leqq 2\sqrt{V_1 V_2},$$

und das Gleichheitszeichen tritt nur dann ein, wenn $M_1 = M_2$, $\mathfrak{y} = 0$ ist.

Beweis. Es ist $\varphi_1(\mathfrak{x})\, \varphi_2(\mathfrak{x}-\mathfrak{y})$ dann und nur $\neq 0$, wenn $\mathfrak{x}$ in M_1 und $\mathfrak{x}-\mathfrak{y}$ in M_2 liegt. Es ist also das erste Integral das Volumen des Durchschnittes D_1 von M_1 und $M_2 + \mathfrak{y}$ und daher $\leqq \operatorname{Min}(V_1, V_2) \leqq \sqrt{V_1 V_2}$, ebenso das zweite Integral das Volumen des Durchschnittes D_2 von M_2 und $M_1 + \mathfrak{y}$. Soll nun bei beiden Integralen das Gleichheitszeichen gelten, so muß $V_1 = V_2 =$ Volumen von $D_1 =$ Volumen von D_2 sein. Also muß, da D_1 Durchschnitt von M_1 und $M_2 + \mathfrak{y}$ ist*), $D_1^0 = M_1^0 = M_2^0 + \mathfrak{y}$ sein und ebenso $D_2^0 = M_2^0 = M_1^0 + \mathfrak{y}$, also $M_1^0 = M_1^0 + 2\,\mathfrak{y}$. Dies ist, da M_1^0 beschränkt ist, nur so möglich, daß $\mathfrak{y} = 0$, $M_1^0 = M_2^0$ und daher $M_1 = M_2$ ist, w. z. b. w.

Beweis von Satz 7a. Die Fourierkoeffizienten von $\varphi_i(\mathfrak{x})$ seien $a_{i\mathfrak{l}}$; dann ist

$$(16)\quad |\mathfrak{K}|^{-1} \sum_{\mathfrak{r}} \int_{R_n} [\varphi_1(\mathfrak{x})\, \varphi_2(\mathfrak{x} - \mathfrak{K}(\mathfrak{x}_0 + \mathfrak{r})) + \varphi_2(\mathfrak{x})\, \varphi_1(\mathfrak{x} - \mathfrak{K}(\mathfrak{x}_0 + \mathfrak{r}))]\, d\mathfrak{x}$$
$$= \sum_{\mathfrak{l}} (a_{1\mathfrak{l}} \overline{a_{2\mathfrak{l}}} + \overline{a_{1\mathfrak{l}}}\, a_{2\mathfrak{l}})\, e^{-2\pi i \mathfrak{l}\mathfrak{x}_0} = 2\,|\mathfrak{K}|^{-2} V_1 V_2 + 2 \sum_{\mathfrak{l} \neq 0} R(a_{1\mathfrak{l}} \overline{a_{2\mathfrak{l}}})\, e^{-2\pi i \mathfrak{l}\mathfrak{x}_0}.$$

Nun folgt aus (12) wegen der gemachten Voraussetzungen

$$(k_0 - r_i)\, V_i\, |\mathfrak{K}|^{-1} \geqq |\mathfrak{K}|^{-2} V_i^2 + \sum_{\mathfrak{l} \neq 0} |a_{i\mathfrak{l}}|^2 \qquad (i = 1, 2).$$

Dann ist

$$\Big|2 \sum_{\mathfrak{l} \neq 0} R(a_{1\mathfrak{l}} \overline{a_{2\mathfrak{l}}})\, e^{-2\pi i \mathfrak{l}\mathfrak{x}_0}\Big| \leqq \sum_{\mathfrak{l} \neq 0} (|a_{1\mathfrak{l}}|^2 + |a_{2\mathfrak{l}}|^2)$$
$$\leqq |\mathfrak{K}|^{-1} \sum_{i=1}^{2} (k_0 V_i - |\mathfrak{K}|^{-1} V_i^2 - r_i V_i),$$

also ist

$$(16')\quad \Big|2\,|\mathfrak{K}|^{-2} V_1 V_2 + 2 \sum_{\mathfrak{l} \neq 0} R(a_{1\mathfrak{l}} \overline{a_{2\mathfrak{l}}})\, e^{-2\pi i \mathfrak{l}\mathfrak{x}_0}\Big|$$
$$\geqq 2\,|\mathfrak{K}|^{-2} V_1 V_2 - 2\,\Big|\sum_{\mathfrak{l} \neq 0} R(a_{1\mathfrak{l}} \overline{a_{2\mathfrak{l}}})\, e^{-2\pi i \mathfrak{l}\mathfrak{x}_0}\Big|$$
$$\geqq \Big(\sum_{i=1}^{2} V_i\Big)^2 |\mathfrak{K}|^{-2} - k_0 \sum_{i=1}^{2} V_i\, |\mathfrak{K}|^{-1} + \sum_{i=1}^{2} r_i V_i\, |\mathfrak{K}|^{-1} \geqq \sum_{i=1}^{2} r_i V_i\, |\mathfrak{K}|^{-2}.$$

*) Ist nämlich z. B. $D_1^0 \neq M_1^0$, so gibt es eine ganz in M_1 liegende Kugel K, welche mit D_1 keinen inneren Punkt gemeinsam hat und es ist dann $V_1 \geqq \text{Vol.}\, D_1 + \text{Vol.}\, K > \text{Vol.}\, D_1$.

Daraus folgt, daß die Summe auf der linken Seite von (16) mindestens $r+1$ nicht verschwindende Summanden haben muß. Sind nämlich nur höchstens r solche Summanden vorhanden, so ist der Betrag dieser Summe *kleiner* als $\sum_{i=1}^{2} r_i V_i |\mathfrak{K}|^{-2}$ in Widerspruch zu (16'). Es ist nämlich nach Hilfssatz 4

$$\int_{R_n} [\varphi_1(\mathfrak{x})\, \varphi_2(\mathfrak{x} - \mathfrak{K}(\mathfrak{x}_0 + \mathfrak{r})) + \varphi_2(\mathfrak{x})\, \varphi_1(\mathfrak{x} - \mathfrak{K}(\mathfrak{x}_0 + \mathfrak{r}))]\, d\mathfrak{x} < 2\sqrt{V_1 V_2},$$

da wir ja den Fall $M_1 = M_2$, $\mathfrak{x}_0 + \mathfrak{r} = 0$ als bereits erledigt, ausgeschlossen hatten. Also ist, wenn die fragliche Summe höchstens r nichtverschwindende Summanden hat, die Summe $< 2r\sqrt{V_1 V_2}\, |\mathfrak{K}|^{-1}$ $= 2|\mathfrak{K}|^{-1}\sqrt{V_1 V_2}\left[\frac{r_1 V_1 + r_2 V_2}{2\sqrt{V_1 V_2}}\right] \leqq |\mathfrak{K}|^{-1} \sum_{i=1}^{2} r_i V_i$. Es gibt also $r+1$ Punkte $\mathfrak{r}_m$, so daß

$$\int_{R_n} [\varphi_1(\mathfrak{x})\, \varphi_2(\mathfrak{x} - \mathfrak{K}(\mathfrak{x}_0 + \mathfrak{r}_m)) + \varphi_2(\mathfrak{x})\, \varphi_1(\mathfrak{x} - \mathfrak{K}(\mathfrak{x}_0 + \mathfrak{r}_m))]\, d\mathfrak{x} > 0$$

ist. Dann gibt es aber $r+1$ Punkte $\mathfrak{x}_m$, so daß

$$\varphi_1(\mathfrak{x}_m)\, \varphi_2(\mathfrak{x}_m - \mathfrak{K}(\mathfrak{x}_0 + \mathfrak{r}_m)) + \varphi_2(\mathfrak{x}_m)\, \varphi_1(\mathfrak{x}_m - \mathfrak{K}(\mathfrak{x}_0 + \mathfrak{r}_m)) > 0$$

ist. Dann muß aber entweder $\mathfrak{x}_m$ in M_1 und $\mathfrak{x}_m - \mathfrak{K}(\mathfrak{x}_0 + \mathfrak{r}_m)$ in M_2 liegen oder umgekehrt. Also liegen auf jeden Fall die Punkte $\mathfrak{x}_0 + \mathfrak{r}_m$ ($m = 1, \ldots, r+1$) in $M_3(\mathfrak{K})$, w. z. b. w.

§ 7.

Die ω-konvexen Mengen.

Es sei $f(\mathfrak{x})$ eine reelle Funktion mit folgenden Eigenschaften:

a) $f(\mathfrak{x}) > 0$ für alle $\mathfrak{x} \neq \mathfrak{o}$, $f(\mathfrak{o}) = 0$,

b) $f(t\mathfrak{x}) = t f(\mathfrak{x})$ für alle $t \geqq 0$.

c) Es gibt zwei positive Zahlen c und σ, so daß folgendes gilt: Sind k und k_0 beliebige natürliche Zahlen ($k_0 < k$), so kann man stets aus k Punkten $\mathfrak{x}_1, \ldots, \mathfrak{x}_k$ $k_0 + 1$ Punkte $\overline{\mathfrak{x}}_0, \overline{\mathfrak{x}}_1, \ldots, \overline{\mathfrak{x}}_{k_0}$ herausgreifen, so daß

$$f^\sigma(\overline{\mathfrak{x}}_i - \overline{\mathfrak{x}}_0) \leqq \frac{c}{k - k_0} \sum_{j=1}^{k} f^\sigma(\mathfrak{x}_j) \quad (i = 1, \ldots, k_0)$$

ist.

d) $f(\mathfrak{x})$ besitzt auf der Kugel $\mathfrak{x}^2 = 1$ eine positive untere Schranke.

Dann nenne ich die durch $f(\mathfrak{x}) \leqq c^{\frac{1}{\sigma}}\, t$ definierte Menge K ω-konvex. Im weiteren setze ich noch, nur der Einfachheit halber, voraus

e) $f(-\mathfrak{x}) = f(\mathfrak{x})$.

Wir zeigen zunächst

Satz 8. *Erfüllt* $f(\mathfrak{x})$ *außer den Eigenschaften* a) b), d) *noch* c_1): *Sind* $\mathfrak{x}_1, \ldots, \mathfrak{x}_k$ *irgendwelche Punkte des* R_n, *und ist*

$$\sum_{i=1}^{k} \sum_{j=1}^{k} f^\sigma(\mathfrak{x}_i - \mathfrak{x}_j) \leqq c\,k \sum_{j=1}^{k} f^\sigma(\mathfrak{x}_j), \tag{17}$$

so hat $f(\mathfrak{x})$ *auch die Eigenschaft* c).

Wir müssen also zeigen: Es gibt unter den $\mathfrak{x}_1, \ldots, \mathfrak{x}_k$ stets $k_0 + 1$ Punkte $\bar{\mathfrak{x}}_0, \bar{\mathfrak{x}}_1, \ldots, \bar{\mathfrak{x}}_{k_0}$ (k_0 beliebig $< k$), so daß

$$f^\sigma(\bar{\mathfrak{x}}_i - \bar{\mathfrak{x}}_0) \leqq \frac{K}{k - k_0} = \frac{c}{k - k_0} \sum_{j=1}^{k} f^\sigma(\mathfrak{x}_j).$$

Wir schreiben für die Summe links in (17) kurz $\sum_{i \neq j} (i\,j)$*). Sie enthält $k(k-1)$ nichtverschwindende Summanden; zu jeder Zahl i ($1 \leqq i \leqq k$) gibt es $k-1$ Paare $(i\,j)$. Aus diesen können wir auf $\binom{k-1}{k_0}$ Arten Systeme von k_0 Paaren $(i, i_1), \ldots (i, i_{k_0})$ herausgreifen. Wir nennen sie die Systeme der i-ten Zeile. Führen wir dies für $i = 1, \ldots, k$ durch, so erhalten wir $k\binom{k-1}{k_0}$ solche Systeme. Jedes System enthält k_0 Paare, also die Gesamtheit aller Systeme $k\,k_0\binom{k-1}{k_0}$ Paare; nun haben wir aber nur $k(k-1)$ Paare, also kommt jedes Paar

$$\frac{k\,k_0\binom{k-1}{k_0}}{k(k-1)} = \binom{k-2}{k_0-1}\text{-mal}$$

vor. Also ist die Summe über alle Paare aller Systeme gleich

$$\binom{k-2}{k_0-1} \sum_{i \neq j} (i\,j)$$

und daher

$$\leqq K\,k\binom{k-2}{k_0-1}.$$

Nehmen wir nun an der Satz sei nicht richtig, dann gibt es in jedem System mindestens ein Paar $(i\,j)$, welches $> \frac{K}{k-k_0}$ ist; ich nenne sie kurz b-Paare. Wir zeigen nun: In jeder Zeile gibt es mindestens $k - k_0$ voneinander verschiedene b-Paare. Ich zeige es für die erste Zeile. Ich greife ein System aus der ersten Zeile heraus und aus ihm ein b-Paar $(1, i^1)$. Dieses b-Paar kommt gleichzeitig in $\binom{k-2}{k_0-1}$ Systemen vor. Diese Systeme denke ich mir gestrichen. Ich nehme eines der übriggebliebenen Systeme und greife wieder ein b-Paar

*) O. B. d. A. $(i\,j) \neq (i'\,j')$, wenn $(i - i')^2 + (j - j')^2 \neq 0$.

heraus $(1, i^2)$, welches natürlich $\neq (1, i^1)$ ist. Dieses kommt in $\binom{k-3}{k_0-1}$ Systemen vor. Auch diese denke ich mir gestrichen, und so gehe ich weiter. Nach $k - k_0$ Schritten ist dieser Prozeß zu Ende, alle Systeme sind gestrichen, und ich habe $k - k_0$ voneinander verschiedene b-Paare in der 1-ten Zeile gefunden. Jedes Paar kommt $\binom{k-2}{k_0-1}$-mal vor, also ist die Summe der b-Paare der 1-Zeile

$$> \binom{k-2}{k_0-1} (k - k_0) \frac{K}{k - k_0} = \binom{k-2}{k_0-1} K.$$

Dies geht natürlich für jede Zeile, und da wir k Zeilen haben, so ist die Summe der b-Paare sicher größer als $Kk\binom{k-2}{k_0-1}$; dies ist aber ein Widerspruch.

Ich setze nun in (9) $\mathfrak{K} = \mathfrak{x}$, $\varphi(\mathfrak{x}) = \psi(\mathfrak{x}) = t^\sigma - f^\sigma(\mathfrak{x})$, wenn $t^\sigma > f^\sigma(\mathfrak{x})$ und $\varphi(\mathfrak{x}) = \psi(\mathfrak{x}) = 0$ sonst. Es gilt noch nach Corput-Schaake [5]

$$\int\limits_{R_n} \varphi(\mathfrak{x})\, d\mathfrak{x} = \frac{n}{n+\sigma} t^{n+\sigma} J,$$

wo $J = \int\limits_{f(\mathfrak{x}) \leqq 1} d\mathfrak{x}$ ist. Dann folgt aus (9)

$$\int\limits_K \sum_{\mathfrak{r}} \overline{\varphi}(\mathfrak{x}) (t^\sigma - f^\sigma(\mathfrak{x}+\mathfrak{r}))\, d\mathfrak{x} = \sum_{\mathfrak{l}} |a_{\mathfrak{l}}|^2.$$

Bezeichnen wir die Anzahl der Summanden der linken Summe mit l, so ist

$$\int\limits_K \sum_{\mathfrak{r}} \overline{\varphi}(\mathfrak{x}) f^\sigma(\mathfrak{x}+\mathfrak{r})\, d\mathfrak{x} = l\, t^\sigma \int\limits_K \overline{\varphi}(\mathfrak{x})\, d\mathfrak{x} - \sum_{\mathfrak{l}} |a_{\mathfrak{l}}|^2.$$

Wir beweisen nun

Satz 9. *Ist $f(\mathfrak{x})$ ω-konvex und*

$$\sum_{\mathfrak{l}} |a_{\mathfrak{l}}|^2 > k_0\, t^\sigma \int\limits_K \overline{\varphi}(\mathfrak{x})\, d\mathfrak{x} \tag{18}$$

oder schärfer

$$|a_0|^2 > k_0\, t^\sigma \int \overline{\varphi}(\mathfrak{x})\, d\mathfrak{x}, \tag{18a}$$

so gibt es mindestens k_0 von $\mathfrak{o}$ verschiedene Gitterpunktpaare $\pm \mathfrak{g}_1, \ldots, \pm \mathfrak{g}_{k_0}$, so daß

$$f^\sigma(\mathfrak{g}_i) \leqq c\, t^\sigma$$

ist.

Bemerkung. Es ist $a_0 = \int\limits_K \varphi(\mathfrak{x})\, d\mathfrak{x}$, also lautet (18a)

$$\int\limits_K \varphi(\mathfrak{x})\, d\mathfrak{x} = \frac{n}{n+\sigma} t^{n+\sigma} J > k_0\, t^\sigma$$

oder

$$t > t_0 = \left(k_0 \frac{u+\sigma}{\sigma J}\right)^{\frac{1}{n}}.$$

Beweis. Benutzen wir (18), so folgt

$$\int_K \sum_{\mathfrak{r}} \overline{\varphi}(\mathfrak{x}) f^{\sigma}(\mathfrak{x}+\mathfrak{r})\, d\mathfrak{x} < (l-k_0)\, t^{\sigma} \int_K \overline{\varphi}(\mathfrak{x})\, d\mathfrak{x}.$$

Es gibt daher ein $\mathfrak{x}'$, so daß

$$\sum_{\mathfrak{r}} f^{\sigma}(\mathfrak{x}'+\mathfrak{r}) < (l-k_0)\, t^{\sigma}$$

ist, also muß $l > k_0$, d. h. $l \geqq k_0 + 1$ sein. Dann kann ich aus den l Punkten $\mathfrak{x}' + \mathfrak{r}_j$ $k_0 + 1$ Punkte $\mathfrak{x}' + \mathfrak{r}_j'$ $(j = 0, 1, \ldots k_0)$ so herausgreifen, daß

$$f^{\sigma}(\mathfrak{r}_j' - \mathfrak{r}_0') \leqq \frac{c}{l-k_0} \sum_{s=1}^{l} f^{\sigma}(\mathfrak{x}' + \mathfrak{r}_s) < c\, t^{\sigma} \qquad (j = 1, \ldots, k_0)$$

ist. Die Punkte $\mathfrak{r}_j' - \mathfrak{r}_0'$ sind alle voneinander verschieden, denn aus $\mathfrak{r}_j' - \mathfrak{r}_0' = \mathfrak{r}_i' - \mathfrak{r}_0'$ $(i \neq j)$ folgt $\mathfrak{r}_j' = \mathfrak{r}'$.

Satz 9a. *Satz 9 bleibt richtig, wenn in* (18a) *das Gleichheitszeichen gilt.* Den Beweis führt man in bekannter Weise so:

Ist q die positive untere Schranke von $f(\mathfrak{x})$ auf der Kugel $\mathfrak{x}^2 = 1$, so ist stets

$$f^{\sigma}(\mathfrak{x}) \geqq q\, (\mathfrak{x}^2)^{\frac{\sigma}{2}}.$$

Liegt ein Gitterpunkt $\mathfrak{g}$ in $f^{\sigma}(\mathfrak{x}) < c\, t^{\sigma}$, dann liegt er auch in der Kugel $\mathfrak{x}^2 < \left(\frac{c\,t}{q}\right)^{\frac{2}{\sigma}}$. Nun gilt Satz 9 für alle $t_{\varepsilon} = (t_0^{\sigma} + \varepsilon)^{\frac{1}{\sigma}}$ $(\varepsilon > 0)$, d. h. zu diesen t_{ε} gibt es k_0 Gitterpunktpaare, die in $f^{\sigma}(\mathfrak{x}) < c\, t^{\sigma}$ und zugleich in der Kugel $\mathfrak{x}^2 < \left(\frac{c\,t}{q}\right)^{\frac{2}{\sigma}}$ liegen. Ich betrachte nun eine Nullfolge (ε_m). Zu jedem t_{ε_m} gehören nun k_0 Gitterpunktpaare $\neq \mathfrak{o}$. Nun liegen in einer Kugel nur endlich viele Gitterpunkte, also gibt es sicher eine Teilfolge (ε_m'), zu der dieselben k_0 Gitterpunktpaare $\pm\, \mathfrak{r}_i$ gehören. Es ist also

$$f^{\sigma}(\mathfrak{r}_i) < c\, t_0^{\sigma} + \varepsilon_m' c \qquad (i = 1, \ldots, k_0).$$

Nun ist die linke Seite von den ε_m' unabhängig, also ist

$$f^{\sigma}(\mathfrak{r}_i) \leqq c\, t_0^{\sigma} \qquad (i = 1, \ldots, k_0).$$

Satz 9 kann also so ausgesprochen werden: Ist $f(\mathfrak{x})$ ω-konvex, so gibt es wenigstens k_0 Gitterpunktpaare $\pm\, \mathfrak{g}_j$, so daß

$$f^{\sigma}(\mathfrak{g}_j) \leqq c\, k_0^{\frac{\sigma}{n}} \left(\frac{n+\sigma}{\sigma J}\right)^{\frac{\sigma}{n}} \qquad (j = 1, \ldots, k_0).$$

Für $k_0 = 1$ wurde der Satz von v. d. Corput-Schaake [5] bewiesen.

§ 8.

Alternativsatz für ω-konvexe Mengen.

Wir beweisen nun eine Verschärfung von Satz 7.

Satz 10. *Ist $f(\mathfrak{x})$ ω-konvex und gibt es höchstens $k_0 - r - 1$ von $\mathfrak{o}$ verschiedene Gitterpunktpaare* $\pm\, \mathfrak{r}_j$ $\left(0 \leqq r \leqq k_0 - \left[\frac{k_0}{2}\right] - 1\right)$, *so daß*

$$f^{\sigma}(\mathfrak{r}_j) < c\left(\frac{k_0}{2}\right)^{\frac{\sigma}{n}} \left(\frac{n+\sigma}{\sigma J}\right)^{\frac{\sigma}{n}} \quad j = 1, \ldots, k_0 - r - 1,$$

so gibt es zu jedem $\mathfrak{x}_0$ mindestens $r+1$ Gitterpunkte $\mathfrak{r}_{j'}$, so daß

$$f^{\sigma}(\mathfrak{r}_{j'} - \mathfrak{x}_0) \leqq c\left(\frac{k_0}{2}\right)^{\frac{\sigma}{n}} \left(\frac{n+\sigma}{\sigma J}\right)^{\frac{\sigma}{n}} \qquad j' = 1, \ldots, r+1.$$

Es ist also die Bedingung $|a_0|^2 \geqq t^{\sigma} k_0 \int\limits_K \varphi(\mathfrak{x})\, d\mathfrak{x}$ durch $|a_0|^2 \geqq \frac{k_0 t^{\sigma}}{2} \int \varphi(\mathfrak{x})\, d\mathfrak{x}$ ersetzt.

Beweis. Wir wenden wieder (9) an, und zwar einmal auf

$$\varphi(\mathfrak{x}) = \psi(\mathfrak{x}) = \begin{cases} t^{\sigma} - f^{\sigma}(\mathfrak{x}) & \text{wenn } t^{\sigma} > f^{\sigma}(\mathfrak{x}) \\ 0 & \text{sonst} \end{cases}$$

und einmal auf

$$\psi(\mathfrak{x}) = \begin{cases} t^{\sigma} - f^{\sigma}(\mathfrak{x} + \mathfrak{x}_0) & \text{wenn } f^{\sigma}(\mathfrak{x} + \mathfrak{x}_0) < t^{\sigma} \\ 0 & \text{sonst} \end{cases}$$

und das obige $\varphi(\mathfrak{x})$.

Dann erhalten wir

$$\text{(19 a)} \qquad \sum_{\mathfrak{r}} \int\limits_K \varphi(\mathfrak{x})\,(t^{\sigma} - f^{\sigma}(\mathfrak{x} + \mathfrak{r}))\, d\mathfrak{x} = \sum_{\mathfrak{l}} |a_{\mathfrak{l}}|^2$$

und

$$\text{(19 b)} \qquad \sum_{\mathfrak{r}'} \int\limits_K \varphi(\mathfrak{x})\,(t^{\sigma} - f^{\sigma}(\mathfrak{x} + \mathfrak{x}_0 + \mathfrak{r}'))\, d\mathfrak{x} = \sum_{\mathfrak{l}} |a_{\mathfrak{l}}|^2\, e^{2\pi i \mathfrak{l}\mathfrak{x}_0}.$$

ie Anzahl der Summanden in (19a) sei l, die in (19b) l', also ist

$$20\,\mathrm{a})\qquad \sum_{\mathfrak{r}} \int_K \varphi(\mathfrak{x})\, f^{\sigma}(\mathfrak{x}+\mathfrak{r})\, d\mathfrak{x} = l\, t^{\sigma} \int_K \varphi(\mathfrak{x})\, d\mathfrak{x} - \sum_{\mathfrak{l}} |a_{\mathfrak{l}}|^2,$$

$$20\,\mathrm{b})\qquad \sum_{\mathfrak{r}'} \int_K \varphi(\mathfrak{x})\, f^{\sigma}(\mathfrak{x}+\mathfrak{x}_0+\mathfrak{r}')\, d\mathfrak{x} = l'\, t^{\sigma} \int_K \varphi(\mathfrak{x})\, d\mathfrak{x} - \sum_{\mathfrak{l}} |a_{\mathfrak{l}}|^2 e^{2\pi i \mathfrak{l}\mathfrak{x}_0}.$$

s ist daher

$$\int_K \varphi(\mathfrak{x})\, [\sum_{\mathfrak{r}} f^{\sigma}(\mathfrak{x}+\mathfrak{r}) + \sum_{\mathfrak{r}'} f^{\sigma}(\mathfrak{x}+\mathfrak{r}+\mathfrak{x}_0)]\, d\mathfrak{x} = (l+l')\, t^{\sigma} \int_K \varphi(\mathfrak{x})\, d\mathfrak{x} -$$
$$- \sum_{\mathfrak{l}} |a_{\mathfrak{l}}|^2 (1+e^{2\pi i \mathfrak{l}\mathfrak{x}_0}).$$

un ist

$$\left|\sum_{\mathfrak{l}} |a_{\mathfrak{l}}|^2 (1+e^{2\eta i \mathfrak{l}\mathfrak{x}_0})\right| = \left|\, 2\,|a_0|^2 + \sum_{\mathfrak{l}\neq 0} |a_{\mathfrak{l}}|^2 + \sum_{\mathfrak{l}\neq 0} |a_{\mathfrak{l}}|^2 e^{2\mu i \mathfrak{l}\mathfrak{x}_0}\right|$$
$$\geqq 2\,|a_0|^2 + \sum_{\mathfrak{l}\neq 0} |a_{\mathfrak{l}}|^2 - \sum_{\mathfrak{l}\neq 0} |a_{\mathfrak{l}}|^2 = 2\,|a_0|^2 \geqq k_0\, t^{\sigma} \int_K \varphi(\mathfrak{x})\, d\mathfrak{x},$$

lso

$$\int_K \varphi(\mathfrak{x})\, (\sum_{\mathfrak{r}} f^{\sigma}(\mathfrak{x}+\mathfrak{r}) + \sum_{\mathfrak{r}'} f^{\sigma}(\mathfrak{x}+\mathfrak{x}_0+\mathfrak{r}')) \leqq (l+l'-k_0)\, t^{\sigma} \int_K \varphi(\mathfrak{x})\, d\mathfrak{x}.$$

un gibt es wieder ein $\mathfrak{x}'$, so daß

$$\sum_{\mathfrak{r}} f^{\sigma}(\mathfrak{x}'+\mathfrak{r}) + \sum_{\mathfrak{r}'} f^{\sigma}(\mathfrak{x}'+\mathfrak{x}_0+\mathfrak{r}') \leqq (l+l'-k_0)\, t^{\sigma}.$$

st $\mathfrak{x}_0$ kein Gitterpunkt, so ist die linksstehende Summe größer als Null. st sie nämlich Null, so ist $\mathfrak{x}'+\mathfrak{r} = \mathfrak{x}'+\mathfrak{x}_0+\mathfrak{r}' = 0$, also $\mathfrak{x}_0$ Gitterpunkt. ür einen Gitterpunkt $\mathfrak{g}$ können wir aber sofort den Satz 10 beweisen. Es ibt ja nach Satz 9a $\left[\frac{k_0}{2}\right]$ Gitterpunktpaare $\pm\, \mathfrak{g}_j \left(j = 1, \ldots \left[\frac{k_0}{2}\right]\right)$, wenn $_0 > 1$, so daß

$$f^{\sigma}(\mathfrak{g}_j) \leqq c \left(\frac{k_0}{2}\right)^{\frac{\sigma}{n}} \left(\frac{n+\sigma}{\sigma J}\right)^{\frac{\sigma}{n}}.$$

s gibt daher $2\left[\frac{k_0}{2}\right]+1$ Gitterpunkte $\mathfrak{g}'$, nämlich $\mathfrak{g}$, $\mathfrak{g}\pm\mathfrak{g}_j$, so daß

$$f^{\sigma}(\mathfrak{g}-\mathfrak{g}') \leqq c \left(\frac{k_0}{2}\right)^{\frac{\sigma}{n}} \left(\frac{n+\sigma}{\sigma J}\right)^{\frac{\sigma}{n}}$$

st. Nun ist $2\left[\frac{k_0}{2}\right]+1 \geqq k_0 - \left[\frac{k_0}{2}\right] \geqq r+1$ und damit Satz 10 für alle itterpunkte bewiesen. Ist $\mathfrak{x}_0$ kein Gitterpunkt, so ist also $l+l'-k_0 > 0$, lso $l+l' \geqq k_0+1$. Aus den $l+\bar{l}$ Punkten $\mathfrak{x}'+\mathfrak{r}_j$ $(j = 1, \ldots, l)$ und $' + \mathfrak{x}_0 + \mathfrak{r}'_{j'}$ $(j' = 1, \ldots, l')$, die wir kurz mit $\mathfrak{y}_i$ $(i = 1, \ldots, l+l')$ beeichnen, lassen sich k_0+1 Punkte $\mathfrak{y}'_0, \mathfrak{y}'_1, \ldots, \mathfrak{y}'_{k_0}$ herausgreifen, so daß

$$f^{\sigma}(\mathfrak{y}_i - \mathfrak{y}'_0) \leqq \frac{c}{l+l'-k_0}\left(\sum_{\mathfrak{r}} f^{\sigma}(\mathfrak{x}'+\mathfrak{r}) + \sum_{\mathfrak{r}'} f^{\sigma}(\mathfrak{x}'+\mathfrak{x}_0+\mathfrak{r}')\right) < c\, t^{\sigma}$$

ist. Nehmen wir an, $\mathfrak{y}_0'$ sei von der Form $\mathfrak{x}' + \mathfrak{r}$. Dann können höchsten[s] $k_0 - r - 1$ Punkte $\mathfrak{y}_j'$ $(j \geqq 1)$ von der Form $\mathfrak{x}' + \mathfrak{r}_j$ sein. Denn sonst liege[n] mindestens $k_0 - r$ Gitterpunktpaare $\pm(\mathfrak{r}_j - \mathfrak{r}_0)$ in $f^\sigma(\mathfrak{x}) < c\,t^\sigma$. Also müsse[n] $r + 1$ Punkte $\mathfrak{y}_j'$ $(j \geqq 1)$ von der Form $\mathfrak{x}' + \mathfrak{r}' + \mathfrak{x}_0$ sein. Dann gibt e[s] also $r + 1$ Gitterpunkte $\mathfrak{g}_j = \mathfrak{r}_j' - \mathfrak{r}$, so daß $f^\sigma(\mathfrak{x}_0 + \mathfrak{g}_j) < c\,t^\sigma$ ist. Analo[g] schließt man, wenn $\mathfrak{y}_0'$ von der Form $\mathfrak{x}' + \mathfrak{x}_0 + \mathfrak{r}'$ ist.

Literatur.

[1] H. Minkowski, Über Geometrie der Zahlen. Ges. Abhandlungen 1, S. 264

[2] K. Mahler, Ein Übertragungsprinzip für konvexe Körper. Časopis **68**.

[3] C. Siegel, Über Gitterpunkte in konvexen Körpern. Acta Mathem. **65**.

[4] J. G. v. d. Corput, Verallgemeinerung einer Mordellschen Beweismethode in der Geometrie der Zahlen I, II. Acta Arithm. 1, S. 2.

[5] J. G. v. d. Corput und G. Schaake, Anwendung einer Blichfeldtschen Beweismethode in der Geometrie der Zahlen. Acta Arithm. **2**.

[6] J. Heinhold, Zur Geometrie der Zahlen. Math. Zeitschr. **47** (1941), S. 198.

[7] L. J. Mordell, An arithmetical theorem of linear forms. Acta Arithm. **2**.

(Eingegangen am 2. Dezember 1942.)

Ausfüllung und Überdeckung konvexer Körper durch konvexe Körper

Monatsh. Math. **53** (1949) 81–131

(Eingelangt am 3. August 1948.)

Die Arbeit beschäftigt sich mit der Ausfüllung und Überdeckung eines konvexen Körpers B im R_n mit endlichem Volumen $V(B) > 0$ durch Körper, die aus einem konvexen Körper $K : \mathfrak{f}(x) \leqq r$ mit Mittelpunkt durch Parallelverschiebung hervorgehen. Bei der Ausfüllung verlangen wir, daß diese Körper nicht übereinandergreifen und das Gesamtvolumen S ein Maximum ist. Bei der Überdeckung soll das Gesamtvolumen Σ der überdeckenden Körper ein Minimum sein. Man kann zunächst verlangen, daß die maximale Ausfüllung, bzw. minimale Überdeckung regulär ist, d. h. daß die Mittelpunkte der Körper stets zu einem Gitter gehören. Die entsprechenden Gesamtvolumina seien S^*, bzw. Σ^*. Es ist natürlich $S \geqq S^*$, $\Sigma \leqq \Sigma^*$. Es ist, wie aus der Definition folgt, stets SV^{-1} und $S^* V^{-1} \leqq 1$, ΣV^{-1} und $\Sigma^* V^{-1} \geqq 1$. Es wird gezeigt, daß die Limiten ($r \to 0$) von $S V^{-1}$, $S^* V^{-1}$, ΣV^{-1}, $\Sigma^* V^{1-}$ stets existieren und von B unabhängig sind. Diese nennen wir die Dichten $\Delta, \Delta^*, \vartheta, \vartheta^*$. Für Ausfüllungen wurde dies bei Kugeln von Rankin[1] gezeigt. Für Δ^*, bzw. ϑ^* (§ 6, § 13) wird eine Abschätzung nach unten und oben gegeben, die von B abhängt und die für $r \to 0$, $\Delta^* = J (\frac{1}{2} M)^n$, $\vartheta^* = J \varepsilon^n$ liefert. Dabei ist J das Volumen von $\mathfrak{f}(x) \leqq 1$, M das absolute Minimum, eine Größe, die von *Minkowski*[2] in der Geometrie der Zahlen eingeführt wurde. ε ist ebenfalls eine zahlengeometrische Größe, die ich absolute Exzentrizität nenne, da eine analoge Größe in der Theorie der Wabenzellen auftritt. (Remak, Hofreiter[3]). Der Zusammenhang, der in der ersten Formel aufscheint, findet sich anschaulich schon bei *Minkowski*. Für die Berechnung von M gibt es

---

[1] *Rankin*, Annals of Math. 48.
[2] *Minkowski*, Geometrie der Zahlen.
[3] *Hofreiter*, Monatshefte 40.

(wenigstens für $n \leqq 3$) Methoden. Insbesonders ist $\Delta^* < 1$, wenn K kein Polyeder, genauer kein $\frac{1}{2}$ M-Körper ist. ε ist nun (abgesehen vom Würfel) für Kugeln bis $n = 4$ bestimmt, und zwar von *Remak* und *Hofreiter*. Es wird in der Arbeit gezeigt, daß $\vartheta^* > 1$, wenn K kein $\frac{1}{2}$ M-Körper, insbesondere kein Polyeder ist. Die Bestimmung von Δ und ϑ liegt bedeutend tiefer und ist allgemein ungelöst. Für $n = 2$ und für den Kreis ist $\Delta = \Delta^*$, $\vartheta = \vartheta^*$ wie *Fejes*, *Hadwiger* und *Kerschner* gezeigt haben, doch lassen sich die Beweise auf den R_n nicht übertragen. Ihnen verdanken wir auf diesem Gebiet mit dem sich schon *Bravais*, *Kelvin* und andere beschäftigt haben, die ersten exakt bewiesenen Sätze. Für $n \geqq 4$ ist es sehr wahrscheinlich, daß es Körper gibt, für die $\Delta > \Delta^*$, $\vartheta < \vartheta^*$ ist. Es ist mir aber zu zeigen gelungen, daß trotzdem noch immer $\Delta < 1$, $\vartheta > 1$ ist, wenn K kein Polyeder von höchstens $3^n - 1$ Seiten ist, und überhaupt eine Abschätzung für diese Dichten, $S\,V^{-1}$, $\Sigma\,V^{-1}$ zu liefern. Es wird weiter gezeigt, daß $\Delta = 1$ und $\vartheta = 1$ auseinanderfolgen. Die allgemeine Bestimmung der genauen Werte erscheint heute als unangreifbares Problem, nachdem selbst im gitterförmigen Fall erst sehr wenige Fälle erledigt sind.

In der Arbeit werden noch verschiedene Verallgemeinerungen behandelt (vgl. § 11, § 12 und § 16). Besitzt z. B. K keinen Mittelpunkt, so ist immer $\Delta < 1$, $\vartheta > 1$. Für Δ^* findet sich das bereits bei *Minkowski*[8]. Der letzte Paragraph beschäftigt sich mit einem Problem, der mit dem Vitalischen Überdeckungssatz aus der Theorie des *Lebesgue*schen Maßes zusammenhängt. Es handelt sich um die Frage, wieviele nicht übereinandergreifende Körper aus einer Überdeckung eines Bereiches B herausgegriffen werden können. Wir haben uns in der Arbeit hauptsächlich auf konvexe Bereiche B beschränkt. Dadurch wurde es ermöglicht, für die Abschätzungen von $S\,V^{-1}$, $\Sigma\,V^{-1}$ usw. die *Brun-Minkowski*sche Theorie herauszuziehen. Es wäre wünschenswert, dies für allgemeine Bereiche durchzuführen. Ist der Rand von B genügend oft differenzierbar, so läßt sich dies, wenn K eine Kugel ist, mit Hilfe der Entwicklungen von *Weyl*, Journal of Math. 61, 1939 leicht durchführen.

---

[4] Für die Literatur, die mir nur zum Teil zugänglich war, vgl. *Fejer L.* Acta Sceged 11, 1946, Bulletin Amerc. Soc. 54 (1948). — *Hadwiger*, Mathem. Zeitschrift 49.

Anmerkung bei der Korrektur: Hiezu kommt noch die wichtige Arbeit von *Segre* und *Mahler*, Amer. Monthly **51** (1944) S. 261. Dort wird auch auf eine Arbeit von *A. Thue*, Norske Skrifter 1910 hingewiesen.

Inhalt.

§ 1. Figurengitter.

Ist z ein Punkt des R_n und sind $a_1, \ldots, a_n$ linear unabhängige Vektoren des R_n, so heißt die Gesamtheit der Punkte $p = z + \sum_{i=1}^{n} \lambda_i a_i$, $(\lambda_i = 0, \pm 1, \pm 2, \ldots)$ Punktgitter $\Gamma_z [a_1, \ldots, a_n]$. Die $a_1, \ldots, a_n$ heißen Basisvektoren des Gitters. Die Verbindungsvektoren zweier Gitterpunkte heißen Gittervektoren. Punktgitter, die aus $\Gamma_z [a_1, \ldots, a_n]$ durch Parallelverschiebung hervorgehen, rechnen wir in dieselbe Klasse und nennen diese Klasse Gitter $\Gamma [a_1, \ldots, a_n]$ erzeugt von den Gittervektoren $\sum_{i=1}^{n} \lambda_i a_i$ $(\lambda_i = 0, \pm 1, \pm 2, \ldots)$. Wird ein spezielles Punktgitter betrachtet $\Gamma_z [a_1, \ldots, a_n]$, so sagen wir das Gitter Γ befindet sich in der Lage z.

Wir bezeichnen Det $(a_1, \ldots, a_n)$ mit $D(\Gamma)$, es ist das Volumen des Elementarparallelepipeds $P [a_1, \ldots, a_n] = P_a$, das von den Basisvektoren $a_1, \ldots, a_n$ aufgespannt wird. Gehen n Gittervektoren $b_1, \ldots, b_n$ aus $a_1, \ldots, a_n$ durch eine ganzzahlige, unimodulare Transformation hervor, so erzeugen sie dasselbe Gitter. Besitzt ein Gitter Γ' die Basisvektoren $\lambda a_1, \ldots, \lambda a_n$, $(\lambda \neq 0)$, so schreiben wir $\Gamma' = \lambda \Gamma$ und es ist $D(\Gamma') = \lambda_n D(\Gamma)$. Es sei nun K ein konvexer Körper mit Mittelpunkt. Wir sprechen von einem Figurengitter[5] $[\Gamma, K]$, wenn in jeder Lage von Γ, jeder Gitterpunkt, Mittelpunkt des konvexen Körpers ist, der aus K durch Parallelverschiebung hervorgeht. Ist Γ in der Lage z, so sagen wir, das Figurengitter befindet sich in der Lage z und schreiben $[\Gamma, K]_z$. Ist K ein Punkt, so ist $[\Gamma, K] = \Gamma$; ist K das Parallelepiped

[5] Der Name stammt von *Hadwiger*.

P_a, so überdeckt $[\Gamma, P_a]$ in jeder Lage den R_n einfach und lückenlos, bei entsprechender Wahl der Randpunkte von P_a.

§ 2. Minima eines konvexen Körpers.

Es sei $f(x)$ die Distanzfunktion von K. Wir nennen $f(x)$ die f-Länge von x. Wir betrachten nun die i-ten Minima $M_i(\Gamma, f)$ $(i = 1, 2, \ldots n)$.

Sie sind so definiert:

$$M_i = f(m_i) = \operatorname{Min} f(g) \tag{1}$$

unter allen Gittervektoren g, die linear unabhängig von $m_0 = 0, m_1, \ldots, m_{i-1}$ sind $(i = 1, \ldots, n)$.

Aus der Definition folgt sofort:

$$M_1 \leqq M_2 \leqq \ldots \leqq M_n \tag{2}$$

Weiters gilt:

Hilfssatz 1: [*Minowski*[2], § 50]

$$\frac{2^n}{n!} D(\Gamma) \leqq J\, M_1\, M_2 \ldots M_n \leqq 2^n D(\Gamma) \tag{3}$$

wo J das Volumen des Eichkörpers $f(x)$ W 1 ist.

Durch Adaption des Gitters Γ können wir aus $m_1, \ldots, m_n$ neue Gittervektoren $l_1, \ldots, l_n$ bilden, so daß die l_i eine Linearkombination von $m_1 \ldots, m_i$ sind und[6]

$$N_i = f(l_i) \leqq \sum_{k=1}^{i} f(m_k) \leqq i\, M_i \tag{4}$$

ist. Diese l_i sind Basisvektoren des Gitters.

Wir wollen kurz die l_i ausgezeichnete Basisvektoren des Gitters Γ nennen. Es ist $\sup M_1(\Gamma, f) = M\, D^{\frac{1}{n}}$ über alle Gitter Γ mit festem $D(\Gamma) = D$, wo M von D unabhängig ist. M heißt das absolute Minimum von f.

Hilfssatz 2: Es gibt stets ein Gitter $\overline{\Gamma}$, wir nennen es maximales Gitter, mit $D(\overline{\Gamma}) = D$, so daß

$$M_1(\overline{\Gamma}, f) = M\, D^{\frac{1}{n}} \text{ ist.}$$

Beweis: Nach *K. Mahler*[7], gibt es stets ein kritisches Gitter Γ^* in Bezug auf den Körper $f(x) \leqq 1$. Dann ist $\overline{\Gamma} = M\, D^{\frac{1}{n}}\, \Gamma^*$ ein Gitter der gewünschten Eigenschaft.

---

[6] *K. Mahler*, Proceedings Amsterdam, 41.

[7] *K. Mahler*, Revista Tucuman (A) 5, 1946.

Hilfssatz 3: Ist $\bar{\Gamma}$ ein maximales Gitter, so ist $M_1(\bar{\Gamma}, f) = M_i(\bar{\Gamma}, f)$ $i = 1, \ldots n$ (*Minowski*[8] § 3).

Hilfsastz 4: (*E. Hlawka*[9])

$$J\, M^n(f) \geqq 2\,\zeta(n) \text{ wo } \zeta(n) = \sum_{k=1}^{\infty} \frac{1}{k^n} \text{ ist.}$$

Wir wollen jetzt den Begriff des 1. Minimums $M_1(\Gamma, f)$ verallgemeinern. Unter $\overset{k}{M}_1(\Gamma, f)$ verstehen wir folgendes:

Es gibt höchstens k—1 Gittervektorenpaare $\pm\, g \neq 0$ aus Γ, so daß $f(g) < \overset{k}{M}_1(\Gamma, f)$ ist, aber $f(g) \leqq \overset{k}{M}_1$ mindestens k-Gitterpunktpaare enthält.

Für $k = 1$ erhalten wir das 1. Minimum.

Hilfssatz 5: Es ist $\overset{k}{M}_1 \leqq k\, M_1$ und

$$J\, \overset{k}{M}{}_1^n \leqq 2^n\, k\, D(\Gamma) \tag{5}$$

Die 1. Behauptung ist trivial, die zweite entnimmt man z. B. aus (*E. Hlawka*[9], Satz 5).

Das sup $\overset{k}{M}_1(\Gamma, f)$ über alle Γ mit $D(\Gamma) = D$ hat wieder die Gestalt $\overset{k}{M}\, D^{\frac{1}{n}}$ $\overset{k}{M}$ heißt das absolute Minimum.

Hilfssatz 6: (*E. Hlawka*[9], Satz 2)

$$J\, \overset{k}{M}{}^n \geqq 2\, k \tag{6}$$

§ 3. Exzentrizität.

Definition 1: Es werde die Menge aller Zahlen d betrachtet, für die das Figurengitter $[\Gamma, f(x) \leqq d]$ den R_n überdeckt. Dann heißt das Minimum aller dieser Zahlen d die Exzentrizität $\varepsilon(\Gamma, f)$.

Aus der Definition folgt, daß es zu jedem x des R_n einen Gitterpunkt g aus Γ_z gibt, z beliebig, so daß

$$f(x - g) \leqq \varepsilon(\Gamma, f) \tag{1}$$

ist.

Hilfssatz 7: Es ist stets

---

[8] Abhandlungen II, S. 3ff.

[9] *E. Hlawka*, Math. Zeitschrift 49.

Nach *K. Mahler* Duke Journal 13 gilt sogar $J M^n \geqq 2\,\zeta(n) + {}^1/_6$.

Anmerkung bei der Korrektur: *Davenport* und *Rogers* haben diese Abschätzung zu $J\, M^n \geqq 4{,}921$ verschärft (Duke Journal 14).

$$\tfrac{1}{2} M_n (\Gamma, f) \leqq \varepsilon (\Gamma, f) \leqq \tfrac{1}{2} \sum_{i=1}^{n} M_i (\Gamma, f) \tag{2}$$

Beweis: Die linke Seite der Ungleichung folgt aus *Minowski*[2], Seite 226. Die rechte Seite zeigen wir so:

Die Vektoren $m_1, \ldots, m_n$ für die $f(m_i) = M_i$ ist, sind linear unabhängig. Jeder Punkt x läßt sich daher in der Gestalt $x = z + \sum_{i=1}^{n} \lambda_i m_i$ darstellen. Es seien nun γ_i ganze Zahlen, so daß $|\lambda_i - \gamma_i| \leqq \tfrac{1}{2}$.

Der Punkt $g = z + \Sigma \gamma_i m_i$ gehört sicher Γ_z an. Dann ist

$$f(x-g) \leqq \Sigma \,|\lambda_i - \gamma_i|\, f(mi) \leqq \tfrac{1}{2} \Sigma M i$$

w. z. b. w.

Aus Hilfssatz 7 folgt

$$\varepsilon (\Gamma, f) \geqq \tfrac{1}{2} M_1 (\Gamma, f) \tag{3}$$

Wir wollen jetzt schärfer folgendes zeigen:

Hilfssatz 8: Es ist

$$J \varepsilon^n (\Gamma, f) \geqq D\,[1 - \tfrac{1}{2} (1 - \tfrac{1}{2} z)^n]^{-1}, \; z^n = J M_1^n D^{-1} \tag{4}$$

Beweis: Wir betrachten das Figurengitter $[\Gamma, f \leqq \varepsilon]$ und greifen einen Körper $\overline{K}$ heraus; sein Mittelpunkt p_1 ist Anfangspunkt eines Gittervektors g mit $f(g) = M_1 (\Gamma, f)$. Der Körper des Figurengitters, dessen Mittelpunkt p_2, Endpunkt g ist, hat mit $\overline{K}$ einen Körper $\overline{K}'$ gemeinsam, nämlich $f(x - \tfrac{1}{2}(p_1 + p_2)) \leqq \varepsilon - \tfrac{1}{2} M_1$. Wir definieren nun in K eine Funktion $\varphi(x, \overline{K})$. Es ist

$$\varphi(x, \overline{K}) = \tfrac{1}{2}, \text{ wenn } x \text{ in } \overline{K}' \text{ und}$$
$$= 1 \text{ sonst in } K. \text{ Dann sei}$$

$\varphi(x) = \Sigma\, \varphi(x, \overline{K})$ ($\overline{K}$ durchlaufe alle Körper des Figurengitters).

Nun ist $\varphi(x) \geqq 1$ wegen der lückenlosen Überdeckung.

Wir betrachten jetzt das Parallelepiped $t\, P_a$ $(t > 1)$, wo P_a ein Elementarparallelepiped von Γ ist. Es enthält t^n Gitterpunkte von Γ. Die Körper von $[\Gamma, \overline{K}]$, deren Mittelpunkte in $P = t\, P_a$ liegen, überdecken dann ein Parallelotop $(t - c)\, P_a = \overline{P}$, wo c von K abhängt.

1 Sein Volumen ist $(t-c)^n D$. Dann ist $I = \int \varphi(x)\, d\,x \geqq (t-c)^n D$.

Nun ist

2 $$I = \sum_{\overline{K} \subset \overline{P}} \int_{\overline{K}} \varphi(x, \overline{K})\, d\,x = \sum_{\overline{K} \subset \overline{P}} (J \varepsilon^n - \tfrac{1}{2} (\varepsilon - \tfrac{1}{2} M_1)^n) =$$

3 $$= t^n (J \varepsilon^n - \tfrac{1}{2} (\varepsilon - \tfrac{1}{2} M_1)^n).$$

Daraus folgt für $t \to \infty$

$$J \varepsilon^n - \tfrac{1}{2} J (\varepsilon - \tfrac{1}{2} M_1)^n \geqq D$$

Also ist zunächst $J\,\varepsilon^n \geqq D$. Daraus folgt:

$$J\,\varepsilon^n\,[1 - \tfrac{1}{2}(1 - \tfrac{1}{2}z)^n] \geqq J\,\varepsilon^n\,[1 - \tfrac{1}{2}\,(1 - \frac{J^{\frac{1}{n}} M_1}{J^{\frac{1}{n}}\,\varepsilon})^n] \geqq D \text{ w. z. b. w.}$$

Aus Hilfssatz 8 folgt, daß $J\,\varepsilon^n \geqq D$ ist. Bezeichnen wir mit *Minkowski*[2], § 32, einen Körper $f \leqq \tfrac{1}{2}\,M_1$ als $\frac{M}{2}$-Körper, wenn $J\,M_1{}^n = 2^n\,D$ ist, so folgt aus Hilfssatz 8, daß $J\,\varepsilon^n > D$, wenn f zu keinem $\tfrac{1}{2}\,M$-Körper gehört.

Definition 2:

$\varepsilon\,(f) = D^{-\frac{1}{n}}$ inf $\varepsilon\,(\Gamma,\,f)$ über alle Γ, mit $D\,(\Gamma) = D$ heißt absolute Exzentrizität.

Satz 1: Es gibt stets ein Gitter $\bar{\bar{\Gamma}}$ mit $D\,(\bar{\bar{\Gamma}}) = D$, so daß $\varepsilon\,(\bar{\bar{\Gamma}},\,f) = \varepsilon\,(f)\,D^{\frac{1}{n}}$ ist. Wir nennen $\bar{\bar{\Gamma}}$ minimales Gitter. Es genügt, den Fall $D = 1$ zu beweisen.

Beweis: Aus der Definition von $\varepsilon\,(f)$ folgt, daß es sicher eine Folge $\Gamma_1, \Gamma_2, \ldots$ mit $D\,(\Gamma_i) = 1$ gibt, so daß $\lim \varepsilon\,(\Gamma_i,\,f) = \varepsilon\,(f)$. Wir können annehmen, daß $\varepsilon\,(\Gamma_i,\,f) < 2\,\varepsilon$. Nun ist stets $\varepsilon\,(\Gamma_i,\,f) \geqq \tfrac{1}{2}\,M_n\,(\Gamma_i,\,f)$, also $\varepsilon > \tfrac{1}{4}\,M_i$. $(i = 2, 3, \ldots n)$. Aus Hilfssatz 1 folgt dann, daß

$$4^{n-1}\,J\,\varepsilon^{n-1}\,M_1\,(\Gamma_i,\,f) \geqq \frac{2^n}{n!}$$

also $M_1\,(\Gamma_i,\,f) \geqq c$, $c =$ Konstante, die nur von ε und f abhängt. Daraus folgt, daß für jedes g aus Γ_i $(i = 1, 2, \ldots)$, $f\,(g) \geqq c$ ist. Also auch $|g| \geqq c_1$. Aus dem Fundamentallemma von *K. Mahler*[6], S. 116, folgt, daß es eine unendliche Teilfolge $\{\Gamma'_i\}$ gibt, die gegen ein Gitter $\bar{\bar{\Gamma}}$ konvergiert, d. h. die Gittervektoren von Γ'_i streben gegen Gittervektoren von $\bar{\bar{\Gamma}}'_i$. Man sieht sofort, daß $\varepsilon\,(\bar{\bar{\Gamma}},\,f) = \varepsilon\,(f)$.

Satz 2: Für die absolute Exzentrizität $\varepsilon\,(f)$ gilt:

4 $$[1 - \tfrac{1}{2}\,(1 - \tfrac{1}{2}\,V_0^{\frac{1}{n}})^n]^{-1} \leqq J\,\varepsilon^n \leqq (\frac{2}{n})^n\,V_0 \leqq n^n \tag{5}$$

Dabei ist $V_0 = J\,M^n\,(f)$, wo $M\,(f)$ das absolute Minimum von f ist.

Beweis: Die linke Seite der Ungleichung folgt aus Hilfssatz 8. Die rechte Seite folgt aus Hilfssatz 7 und Hilfssatz 3, wenn für Γ das maximale Gitter genommen wird. Aus Hilfssatz 1 folgt dann der Rest der Behauptung.

Aus diesem Satz folgt, daß
$J\,\varepsilon^n\,(f) > 1$ ist, wenn f zu keinem ½ M-Körper gehört.

Nach *Minkowski*, § 32—§ 34 ist für einen ½ M-Körper $\varepsilon = 1$, weiters ist der ½ M-Körper ein Polyeder von höchstens $2\,(2^n—1)$ Seiten.

§ 4. Ein Satz über Figurengitter.

Wir betrachten nun, außer dem Figurengitter $[\Gamma, K]$ in einer Lage z, eine abgeschlossene, im *Jordan*schen Sinne meßbare Menge B. Weiters setzen wir voraus, daß $B + K$ ebenfalls einen *Jordan*schen Inhalt besitzt. Dabei verstehen wir unter $A_1 + A_2$, wo A_1, A_2 Mengen des R_n sind, die Menge der Punkte $x_1 + x_2$, wo x_1 in A_1, x_2 in A_2 liegt.

Satz 3: Ist $N\,([\Gamma, K]z, B)$ die Anzahl der Körper aus dem Figurengitter $[\Gamma, K]$ in der Lage z, die mit B einen inneren Punkt gemeinsam haben, so ist

$$\int\limits_P N\,([\Gamma, K]_z, B)\,dz = V\,(B + K)$$

dabei ist P irgendein Elementarparallelepiped des Gitters Γ.

Beweis: Es sei $\alpha\,(x)$ folgende Funktion des $R_n : \alpha\,(x) = 1$, wenn B mit dem Körper K_x, der aus K durch Parallelverschiebung hervorgeht und x als Mittelpunkt hat, einen inneren Punkt gemeinsam hat, sonst sei $\alpha\,(x) = 0$. Jeder Gitterpunkt von Γ_z hat die Gestalt $z + g$, wo g Gittervektor von Γ ist. Dann ist, da $[\Gamma, P]$ den R_n einfach und lückenlos überdeckt

$$\int\limits_P \sum_\Gamma \alpha\,(z + g)\,dz = \int_{Rn} \alpha\,(x)\,dx$$

Nun ist, nach Definition von $\alpha\,(x)$, $\alpha\,(x)$ dann und nur dann ungleich Null, wenn B mit K_x einen inneren Punkt gemeinsam hat. Das heißt, da K symmetrisch ist, wenn x innerer Punkt von $B + K$ ist. Es ist also

$$\int_{Rn} \alpha\,(x)\,dx = V\,(B + K)$$

Weiters ist $\Sigma\,\alpha\,(z + g)$ die Anzahl $N\,([\Gamma, K]z, B)$ der Körper aus dem Figurengitter $[\Gamma, K]$ in der Lage z, die mit B einen inneren Punkt gemeinsam haben. Damit ist die Formel gezeigt.

Folgerung: Ist $N^0 = N^0\,([\Gamma, K], B)$ das Maximum aller Zahlen N, ebenso $N_u = N_u\,([\Gamma, K], B)$ das Maximum, dann ist

$$D\,(\Gamma)\,N^0 \geqq V\,(B + K) \tag{1}$$

$$D\,(\Gamma)\,Nu \leqq V\,(B + K) \tag{2}$$

Ist K ein Punkt, so stellt (1) im wesentlichen den *Blichfeldt*schen Satz dar (*Koksma*)[10].

Hilfssatz 9 (*Bonnesen-Fenchel*[11] 29, 32):

Ist B konvex, so ist $V(B+K) = \sum_{i=0}^{n} \binom{n}{i} V_i(B, K)$, wo Vi die gemischten Volumina von B und K sind. Insbesondere ist V_0 das Volumnen $V(B)$ von B. Ist K die Einheitskugel E, so sind die $V_i(B, E)$ die Quermaßintegrale $W_i(B)$ von B.

§ 5. Extremale Punktverteilungen in konvexen Körpern.

Es sei B ein konvexer Körper, $f(x)$ die Distanzfunktion eines konvexen Körpers mit Mittelpunkt und d eine positive Zahl.

Definition 3: Eine Punktmenge P von B[12] heißt (f, d) Konfiguration von B, wenn

$$f(p_1 - p_2) \geqq d \tag{1}$$

ist, für alle p_1, p_2 von P.

Definition 4: Ist $s(P)$ die Anzahl der Punkte einer (f, d)-Konfiguration von B, so heißt P_0 extremale (f, d)-Konfiguration von B, wenn $s(P_0) = Max\ s(P)$ über alle (f, d)-Konfigurationen P von B ist. Es werde $s(P_0) = s(B, f, d)$ gesetzt.

Satz 4: Es ist stets $s(B, f, d) \,.\, Jd^n \leqq 2^n\, V(B + K_1)$, wo J das Volumen des Eichkörpers ($f(x) \leqq 1$ und K_1 der Körper $f \leqq 1/2\, d$ ist.

Beweis: Es sei P irgendeine (f, d)-Konfiguration von B. Dann greifen die Körper $f(x - p) \leqq 1/2\, d$ nicht übereinander, wenn $p \,\varepsilon\, P$ ist und sie liegen alle in $B + K_1$. Damit ist Satz 4 gezeigt.

Folgerung: Ist B ein Körper $f(x) \leqq t\, d$, so ist

$$s(B, f, d) \leqq (2\,t + 1)^n \tag{2}$$

Definition 5: Es sei Γ ein Punktgitter in der Lage z. Dann heißt
5 P eine (f, d)-Konfiguration von B eine Γ-Konfiguration, wenn alle Punkte von P, Γ_z angehören.

Eine (f, d)-Konfiguration P heißt extremale Γ-Konfiguration, wenn $s(\overline{P}) = max\ s(P')$ ist über alle Γ-Konfigurationen der Form (f, d) von B. Man setzt $s(\overline{P}) = s(B, \Gamma, f, d)$.

Es soll nun $s(B, \Gamma, f, d)$ nach oben und unten abgeschätzt werden.

Satz 5: Ist $M_1(\Gamma, f)$ das 1. Minimum von Γ, $D(\Gamma)$ die Determinante des Gitters Γ, so gilt für $s = s(B, \Gamma, f, d)$

[10] *Koksma*, Diophantische Approximationen, Springer 1935.

$$J\, d^n\, D^{-1}\, V(B) \leqq s\, J\, d^n \leqq D^{-1}\, J\, M_1{}^n\, V(B) + 2^n n!\, (V(B + n\, d\, F) - V(B)) \qquad (3)$$

dabei ist F der Eichkörper $f \leqq 1$ und J sein Volumen.

Beweis: Ist P eine Γ-(f, d)-Konfiguration von B, so gilt nach *Definition 1:* $f(p_1 - p_2) \geqq d$, $(p_1 \neq p_2)$. Da p_1, p_2 zu einem Γ_z gehören, so muß $M_1(\Gamma, f) \geqq d$ sein.

Nach § 4 (1) gibt es ein z, so daß mindestens $V(B)\, D^{-1}(\Gamma)$ Punkte von Γ_z in B liegen (*Blichfeldt*scher Satz). Diese Punkte bilden sicher eine (f, d)-Konfiguration von B. Damit ist die untere Abschätzung von s bewiesen.

Wir wollen nun s nach oben abschätzen. Für die im folgenden benützten Sätze über gemischte Volumina verweisen wir auf *Bonnesen-Fenchel*, 29. Es sei $l_1, \ldots, l_n$ eine ausgezeichnete Basis, wie in § 2, von Γ und P_l das dazugehörige Parallelepiped. $P_l{}^*$ sei das Parallelepiped, welches aus P_l durch Verkürzung um $\lambda = d\, M_1{}^{-1} \leqq 1$ hervorgeht. Dann greifen die Parallelepipede aus dem Figurengitter $[\Gamma, P_l{}^*]$ sicher nicht übereinander, da dies auch beim Figurengitter $[\Gamma, P_l]$ nicht der Fall ist. Also ist der Gesamtinhalt dieser Parallelepipede, deren Mittelpunkte, Punkte der extremalen Γ-Konfiguration P sind, gleich $s\, D \lambda^n$. Sie liegen alle im konvexen Körper $B + P_l{}^*$, also

$$s\, D\, \lambda^n \leqq V(B + P_l{}^*) = \sum_{i=0}^{n} \binom{n}{i}\, V_1(B, P_l{}^*) \qquad (4)$$

Es soll nun $V_i(B, P_l{}^*)$ für $1 \leqq i \leqq n$ abgeschätzt werden.

Wenn jetzt die Vektoren $l_1, \ldots, l_n$ als konvexe Körper, die von 0 ausgehen, aufgefaßt werden, so können wir schreiben

$$P_l{}^* = \lambda \sum_{i=1}^{n} l_i$$

Dann ist

$$V_i(B, P_l{}^*) = \lambda^i \sum_{\sigma_1 \cdots \sigma_i} V(\underbrace{B, \ldots, B}_{n-i}, l\,\sigma_1, \ldots\, l\,\sigma_i) \qquad (5)$$

wo $\sigma_1, \ldots, \sigma_l$ alle Zahlen von 1 bis n durchlaufen. Nun verschwinden in dieser Summe alle Glieder, für die $l\,\sigma_i = l\,\sigma_j$ ist, also erstreckt sich die Summe nur über alle Variationen aus $\{1, \ldots, n\}$ zur i-ten Klasse. Wir setzen nun $l_i = t_i\, N_i$. Dann ist

$$V_i(B, P_l{}^*) = \lambda^i\, \Sigma' \, \Sigma\, N\sigma_1 \ldots N\sigma_i\, V(B, \ldots, B, t\sigma_1, \ldots, t\sigma_i) \qquad (6)$$

wo die Summation sich über alle verschiedenen σ_i erstreckt.

Nun ist nach § 2 (4)

$$N\sigma_1 \ldots N\sigma_i \leqq \sigma_1 \ldots \sigma_i \, M\sigma_1 \ldots M\sigma_i \leqq n!\, M\sigma_1 \ldots M\sigma_i \, (\sigma_1 \neq \sigma_j) \tag{7}$$

Nun ist nach Hilfssatz 1, § 2

$$M_1^{\,n-i} M\sigma_1 \ldots M\sigma_i \leqq 2^n D J^{-1}$$

also: $V_i(B, P_l^*) \leqq \lambda^i \, 2^n D J^{-1} M_1^{\,i-n} n! \sum\limits_{\sigma_1 \ldots \sigma_i} V(B, \ldots, B, t\sigma_1 \ldots t\sigma_i)$.

Setzen wir also $Q = \sum\limits_{i=1}^{n} t_i$, wo Q das Parallelelotop ist, erzeugt von $t_1, \ldots, t_n$, so folgt

$$V_i(B, P_l^*) \leqq \lambda^i \, 2^n n! \, D J^{-1} M_1^{\,i-n} V_i(B, Q)$$

also

$$s D \lambda^n \leqq V(B) + 2^n n! \, J^{-1} D M_1^{-n} \sum_{i=1}^{n} \binom{n}{i} (\lambda M_1)^i V_i(B, Q) =$$

6 $$= V(B) + \frac{2^n n!}{J M_1^{\,n}} D \left(V(B t \lambda M_1 Q) - V(B)\right) \tag{8}$$

Nun liegt Q im Körper $f(x) \leqq n$, also wenn wir mit F den Eichkörper $f \leqq 1$ bezeichnen, so ist

7 $$s D \lambda^n \leqq V(B) + (2 M_1)^n n! \, D J^{-1} \left(V(B + \lambda n M_1 F) - V(B)\right) \tag{9}$$

Beachten wir nun, daß $\lambda = d M_1^{-1}$ ist, so folgt daraus die Behauptung.

Definition 6 (*Bonnesen-Fenchel* 35):

Es sei T ein konvexer Körper, dann heißt die größte Kugel, die ganz in T liegt, Inkugel, ihr Radius sei $\tau(T)$. Die kleinste Kugel, die T enthält, heißt Umkugel von T, ihr Radius sei $\varrho(T)$.

Definition 7: Ist B konvex, sind W_i die Quermaßintegrale von B, so heiße $\sum\limits_{i=1}^{n} \binom{n}{i} W_i(B) t^i$, $t > 0$ Quermaßsumme $O(B, t)$.

Es ist $O(B, 0) = 0$.

Gelten dieselben Voraussetzungen wie bei Satz 5 und ist $\varrho(F)$ der Umkugelradius des Eichkörpers F, so können wir s noch weiter abschätzen.

Es ist nach Definition des absoluten Minimums von f (§ 2):

$$M_1^n \leqq M^n(f) \, D(\Gamma).$$

Weiters ist nach Definition 6, F in seiner Umkugel enthalten, also wenn E die Einheitskugel ist, liegt F in $\varrho(F) E$. Daraus folgt aber, daß

$$V(B + n d F) \leqq V(B + n d \varrho E) = \sum_{i=0}^{n} \binom{n}{i} W_i(B) (n d \varrho)^i =$$

$$= V(B) + O(B, n d \varrho) \tag{10}$$

Berücksichtigen wir, daß bei der rechten Ungleichung von Satz 5, so erhalten wir:

Satz 5': Es ist

$$J\, d^n D^{-1}(\Gamma)\, V(B) \leqq s\, J\, d^n \leqq J\, M^n\, V(B) + 2^n n!\, O(B, n\, d\, \varrho). \quad (11)$$

Definition 8: $s^*(B, f, d) = \max(B, \Gamma, f, d)$ über alle Gitter Γ.

Satz 6: Ist B ein konvexer Bereich, $f(x)$ die Distanzfunktion eines konvexen Körpers K, ϱ der Umkugelradius, J das Volumen des Eichkörpers, $M(f)$ das absolute Minimum, so ist:

$$J\, M^n\, V(B) \leqq s^*(B, f, d)\, J\, d^n \leqq J\, M^n\, V(B) + 2^n n!\, O(B, n\, d\, \varrho) \quad (12)$$

dabei ist $O(B, n\, d\, \varrho)$ nach Definition 5 die Quermaßsumme.

Beweis: Die rechte Seite der Umgleichung folgt aus Satz 5'. Wir wählen nun für Γ das maximale Gitter $\overline{\Gamma}$ mit der Determinante $(d M^{-1})^n$ für welches also $M_1(\overline{\Gamma}, f) = M\, D^{\frac{1}{n}} = d$ ist (§ 2, Hilfssatz 2).

Dann ist nach (11): $s^*(B, \overline{\Gamma}, f, d)\, J\, d^n \geqq J\, M^n\, V(B)$.

Damit ist Satz 6 gezeigt.

Folgerung: $\lim\limits_{d \to 0} \dfrac{J\, d^n\, s^*(B, f, d)}{V(B)}$ existiert und ist gleich $J\, M^n(f)$.

Wir wollen das eben betrachtete Problem noch verallgemeinern.

Definition 9: Eine Punktmenge P vom Bereich B heißt (k, f, d)-Konfiguration von B, wenn es zu einem Punkt $p_1 \varepsilon P$ höchstens $2(k-1)$ Punkte p_2 aus P gibt; so daß $f(p_1 - p_2) < d$ ist.

Gehören alle Punkte von P zugleich einem Gitter Γ_z an, so heißt P eine (Γ, k, f, d)-Konfiguration. Die Anzahl von P werde mit $\overset{k}{s}(P)$ bezeichnet. Die (k, f, d), bzw. (Γ, k, f, d)-Konfiguration P heißt extremal, wenn $\overset{k}{s}(P) = \max \overset{k}{s}(P')$ über alle (k, f, d), bzw. (Γ, k, f, d)-Konfigurationen ist. Wir setzen $\overset{k}{s}(P) = \overset{k}{s}(B, f, d)$, bzw. $\overset{k}{s}(B, \Gamma, f, d)$. Es soll nun $\overset{k}{s}(B, \Gamma, f, d)$ abgeschätzt werden. Zunächst muß $M_1(\Gamma, f) \geqq d$ sein. Es gilt dann:

Satz 7:

$$J\, d^n D^{-1}(\Gamma)\, V(B) \leqq \overset{k}{s}\, J\, d^n \leqq J\, \overset{k}{M}{}_1^n\, V(B) + 2^n n!\, k^n O(B, n\, d\, \varrho) \quad (13)$$

Beweis: Er erfolgt analog wie bei Satz 5, bzw. 5'. Es wird das Figurengitter $[\Gamma, P_l'']$, wo jetzt P_l'' aus P_l durch Verkürzung um $\bar{\lambda} = d\, \overset{k}{M}{}_1^{-1} \leqq 1$

---

[11] *Bonnesen-Fenchel*, Theorie der konvexen Körper, Springer 1934. Die Zahlen bedeuten die Abschnitte des Buches.

hervorgeht. Die obigen Überlegungen bis (9) bleiben ungeändert, wenn λ durch $\bar{\lambda}$ ersetzt wird. Multiplizieren wir die so entstehende Gleichung mit $\overset{k}{M}_1$, so kommt

8 $$J\, D\, d^n\, \overset{k}{s} \leqq J\, \overset{k}{M}{}_1^n\, V(B) + 2^n n!\, D\,.\,(V(B + n\, d\, \overset{k}{M}_1\,.\, M_1^{-1} F) - V(B)) \tag{14}$$

Nun ist nach Hilfssatz 5

$\overset{k}{M}_1 \leqq k\, M_1$, also der zweite Summand in (14)

$$\leqq 2^n n!\, D\,.\,(V(B + n\, d\, k\, F) - V(B))$$

Die weitere Abschätzung erfolgt wie bei Satz 5'.

Definition 10:

$$\overset{k}{s}(B, f, d) = \max \overset{k}{s}(B, \Gamma, f, d) \text{ über alle } \Gamma.$$

Satz 8: Es ist

$$J\, \overset{k}{M}{}^n\, V(B) \leqq \overset{k}{s}(B, f, d)\, J\, d^n \leqq J\, \overset{k}{M}{}^n\, V(B) + 2^n n!\, k^n\, O(B, n\, d\, \varrho) \tag{15}$$

Dabei ist $\overset{k}{M}$ das absolute k-Minimum von f.

Beweis: Die rechte Seite der Ungleichung folgt aus Satz 7. Die linke Seite erhalten wir, wenn man in Satz 7, ein Gitter Γ mit $D^{\frac{1}{n}} = \mathrm{d}\, \overset{k}{M}{}^{-1}$ nimmt.

§ 6. Gitterförmige Ausfüllung konvexer Körper durch konvexe mit Mittelpunkt.

Es sei ein konvexer Bereich B gegeben, der jetzt innere Punkte besitzen soll. Der Inkugelradius sei $\beta(B)$. Der Ursprung o soll Mittelpunkt einer Inkugel sein. Es sei r eine positive Zahl und $K = K(r)$ der Körper $f(x) \leqq r$. Wir setzen stets für r voraus, daß der Umkugelradius $\varrho(K) < \beta$ ist. Das Volumen $V(K) = J\, r^n$.

Definition 11: Eine Punktmenge A von B[12] heißt eine $K(r)$-Ausfüllungskonfiguration, wenn:

1. A eine $(f, 2\,r)$-Konfiguration ist, f ist Distanzfunktion von K;
2. die Körper $f(x - p) \leqq r$, wo p in A, ganz in B liegen.

Ausfüllungskonfigurationen sollen immer mit A - $K(r)$ bezeichnet werden. Diese Definition besagt, daß die Körper $f(x - p) \leqq r$, wo p in A, nicht übereinandergreifen und ganz in B liegen. Die Anzahl der Punkte von A werde mit $a(A)$ bezeichnet.

---

[12] B braucht nicht konvex zu sein.

Definition 12: Eine Ausfüllungskonfiguration $A_0 - K(r)$ heißt maximal, wenn $a(A_0) = \max a(A)$ ist, über alle $A - K(r)$. $a(A_0)$ bezeichnen wir mit $a(B, K(r))$, die „Maximalanzahl der Körper K, die in B eingelagert werden können." $a(B, K(r))\, J\, r^n = S(B, K(r))$ ist das Gesamtvolumen der Körper K, die in B eingelagert werden können.

Aus der Definition folgt sofort:

$$S(B, K(r)) \leqq V(B) \tag{1}$$

Definition 13: Ist Γ ein Punktgitter in der Lage z, dann heißt eine Ausfüllungskonfiguration $A\ K(r)$ von B eine Γ-Ausfüllungskonfiguration, wenn alle Punkte von A, Γ_z angehören. Eine Konfiguration $A^* - K(r)$ heißt maximale, gitterförmige Ausfüllungskonfiguration, wenn $a(A') = \max a(A)$ über alle Γ-Konfigurationen der Form $A - K(r)$

Wir setzen: $a(A'') = a''(B, \Gamma, K(r))$.

Definition 14: $a^*(B, K(r)) = \max a^*(B, \Gamma, K(r))$ über alle Gitter Γ.

a^* nennen wir die Maximalanzahl der Körper, die in B gitterförmig eingelagert sind, und nicht übereinandergreifen. $S^*(B, K(r)) = a^*\, J\, r^n$ ist das Gesamtvolumen der gitterförmig eingelagerten Körper.

Aus der Definition 14 und § 5, 8 folgt sofort

$$a^*(B, K(r)) \leqq s^*(B, f, 2r) \tag{2}$$

Es soll nun a^* nach unten abgeschätzt werden.

Hilfssatz 10: Ist $B_1 = (1 - \frac{\varrho}{\beta})\, B$, so ist

$$a^*(B, K(r)) \geqq s^*(B_1, f, 2r) \tag{3}$$

Beweis: Es sei P eine $(f, 2r)$-Konfiguration von B_1. Dann liegen die Körper $f(x - p) \leqq r$ ganz in B. Zunächst liegen sie ganz in $B_1 + K$. $B_1 + K$ liegt aber in B. Es ist nämlich seine Stützfunktion $H_1(n)$ in irgendeiner Richtung (u = Einheitsvektor)

$$\leqq (1 - \frac{\varrho}{\beta})\, H(u) + \overline{H}(u)$$

Dabei sei $H(u)$ die Stützfunktion von B und $\overline{H}(u)$ die von K.

Daraus folgt

$$H_1(u) \leqq H(u) - \frac{\varrho}{\beta}\beta + \varrho = H(u)$$

Dies gilt für jedes u. Dabei wurde benützt, daß

$$H(u) \geqq \beta \text{ ist.}$$

Für die benützten Eigenschaften der Stützfunktion sei auf *Bonnesen-Fenchel*, 15, verwiesen.

Satz 9: Besitzt der Körper K das Volumen $J\,r^n$ und ist sein Umkugelradius ϱ kleiner als der Inkugelradius β des konvexen Bereiches B, so gilt für das Gesamtvolumen $S^*(B, K(r))$ der Körper K, die in B, maximal gitterförmig eingelagert werden können:

$$(1-\frac{\varrho}{\beta})^n J\,(\frac{M}{2})^n V(B) \leqq S^*(B, K(r)) \leqq \mathrm{Min}\,[V(B), J(\frac{M}{2})^n V(B) + \\ + n!\,O(B, 2n\varrho)], \qquad (4)$$

dabei ist M das absolute Minimum der Distanzfunktion von K, J das Volumen des Eichkörpers und

$$O(B, 2n\varrho) = \sum_{i=1}^{n} \binom{n}{i} W_i(B)\,(2n\varrho)^i \qquad (5)$$

Die $W_i(B)$ sind die Quermaßintegrale von B.

Beweis: Der Satz folgt aus Satz 6, § 5 (1) und Hilfssatz 10.

$$\text{Folgerung: } \lim_{r\to 0} \frac{S^*(B, K(r))}{V(B)} = J\,(\tfrac{1}{2} M(f))^n \qquad (6)$$

Eine analoge Formel gilt für $S^*(B, \Gamma, K(r)) = a^*(B, \Gamma, K(r))\,J\,r^n$, nämlich

9 $$(1-\frac{\varrho}{\beta})^n J\,V.D^{-1} r^n \leqq S^*(B, \Gamma, K) \leqq J\,(\tfrac{1}{2}M)^n V.D^{-1} + n!\,O(B, 2n\varrho) \qquad (7)$$

wie sofort aus Satz 5′ folgt.

Nehmen wir jetzt für Γ das maximale Gitter $\overline{\Gamma}$ mit $D = (\frac{2r}{M})^n$ und $M_1(\overline{\Gamma}, f) = 2r$, so gelten für $S^*(B, \overline{\Gamma}, K)$ dieselben Schranken, wie für $S^*(B, K)$.

Bezeichnen wir die Schranken mit C_1 und C_2, so gilt

$$C_1 \leqq S^*(B, \overline{\Gamma}, K) \leqq S^*(B, K) \leqq C_2$$

Daraus folgt:

Satz 10: $S^*(B, \overline{\Gamma}, K)$ ist asymptotisch gleich $S^*(B, K)$ für $r \to 0$

Definition 15:

$\Delta^*(f) = \lim\limits_{r\to 0} \dfrac{S^*(B,K)}{V(B)}$ heißt Dichte der gitterförmigen, maximalen Lagerung durch Körper mit der Distanzfunktion f.

Es ist daher aus der Folgerung von Satz 9

$$\Delta^* (f) = J (\tfrac{1}{2} M)^n < 1$$

wenn f zu keinem $\tfrac{1}{2}$ M-Körper gehört (vgl. § 3).

Nach Hilfssatz 4, § 2 ist $\Delta^* (f)$ immer $\geqq 2^{1-n} \zeta (n)$.

Wir können nun nach dem inf $\Delta^* (f)$ über alle konvexen Körper fragen. Da $\Delta^* (f)$ nur von der Eichfunktion abhängt, können wir auch sagen:

Welche Eichfunktion liefert die kleinste Dichte oder anders ausgesprochen, „die dünnste, dichteste gitterförmige Lagerung?" Für $n = 2$ wurde dies von *Reinhardt*[13] und *Mahler*[14] behandelt.

Definition 16: Eine Punktmenge A_k von B heißt eine $k - K(r)$-Ausfüllungskonfiguration, wenn sie

1. eine $(k, f, 2r)$-Konfiguration von B ist und
2. die Körper $f(x - p) \leqq r$, wo p in A_k, ganz in B liegen.

Gehören alle Punkte von A_k zugleich einem Gitter Γ_z an, so heißt A_k eine Γ-Ausfüllungskonfiguration $A_k - \Gamma - K(r)$.

Die Anzahl der Punkte von A_k werde mit $a(A_k)$ bezeichnet. Eine Ausfüllungskonfiguration A_k o, bzw. A^*_{k0} heißt eine maximale, bzw. eine maximale gitterförmige Ausfüllungskonfiguration, wenn $a(A_{k0}) =$
10 $\max a(A_k)$, bzw. $a(A^*_{k0}) = \max (A_k)$ über alle gitterförmigen $A_k - K(r)$.

Wir setzen $a^*_k = a(A^*_{k0}$ und $S^*_k (B, K(r)) = a^*_k J r^n$ (8)

Nach Definition 16 dürfen die in B eingelagerten Körper jetzt übereinandergreifen, aber nur mit höchstens $2(k-1)$.

Dann gilt für die Maximalanzahl $a^*_k (B, K(r))$, bzw. für das Gesamtvolumen $S^*(B, K(r))$ der gitterförmig eingelagerten Körper:

Satz 11:

$$(1 - \frac{\varrho}{\beta})^n J (\tfrac{1}{2} \overset{k}{M})^n V(B) \leqq S^*_k (B, K(r)) \leqq J (\tfrac{1}{2} \overset{k}{M})^n V(B) +$$

11 $$+ n!\, k^n O(B, 2nr\varrho) \qquad (9)$$

Dabei ist M das absolute k-Minimum von f (§ 2).

Der Satz wird genau so bewiesen wie Satz 9, unter Heranziehung des Satzes 8, § 5.

Folgerung:

$$\Delta^*_k (f) = \lim_{r \to 0} \frac{S^*_k (B, K(r))}{V(B)} = J (\tfrac{1}{2} \overset{k}{M})^n \qquad (10)$$

---

[13] Hamburger Abhandlungen 10.

[14] Proceedings Amsterdam 50 (1947).

§ 7. Ausfüllung konvexer Körper durch konvexe mit Mittelpunkt.

Es soll jetzt $a\ (B, K\ (r))$, bzw. $S\ (B, K\ (r)) = a\ (B, K)\ J\ r^n$, die in Definition 12 definiert wurden, abgeschätzt werden. Es wird also jetzt nicht mehr vorausgesetzt, daß die Ausfüllung von B durch Körper $K\ (r)$ gitterförmig ist. Aus Definition 14 folgt, daß

$$a\ (B, K) \geqq a^*\ (B, K),\ S\ (B, K) \geqq S^*\ (B, K) \tag{1}$$

ist. Wir wollen jetzt a ,bzw. S nach oben abschätzen.

Definition 17: Ist R ein Polyeder von kleinstem Volumen, welches

1. dem Körper $K : f \leqq r$ umschrieben ist,
2. von höchstens z^n-Seiten ist,

dann sei $\varphi_z\ (f) = \dfrac{V\ (R)}{V\ (K)}$.

φ hängt, wie man sofort sieht, nur von der Distanzfunktion f ab.

Satz 12: Es ist stets $\varphi_z \geqq 1$. Ist K kein Polyeder von höchstens $2\ [\frac{z^n}{2}]$ Seiten, so ist $\varphi_z > 1$. Dies ist stets dann der Fall, wenn K kein Polyeder ist.

Beweis: Die 1. Behauptung ist trivial. Die 2. Behauptung begründet man folgendermaßen: Ist $\varphi_z = 1$, so muß K ein Polyeder von höchstens
2 z^n-Seitenflächen sein. Da aber $2\ [\frac{z^n}{2}]$ symmetrisch ist, kann es höchstens $2\ [\frac{z^n}{2}]$ sein.

Definition 18: Ist $\varkappa\ (K)$ der Inkugelradius von $K : f \leqq r$, $\varrho\ (K)$ der Umkugelradius von K, δ eine beliebige positive Zahl, χ_m das Volumen der m-dimensionalen Einheitskugel, so sei

$$\omega\ (f, \delta) = 1 + \frac{\chi_{m-1}}{\chi_m}\ \frac{\delta^n}{(l + \delta)^{n-1}}\ \left(\frac{\varkappa}{\varrho}\right)^n \tag{2}$$

Hilfssatz 11: Es sei $K : f \leqq r$, $\overline{K} : f \leqq (1 + \delta)\ r$, $\delta > 0$.

Weiters sei T ein konvexer Körper, der K enthält und in $\overline{K}$ liegt. Haben T und $\overline{K}$ Randpunkte gemeinsam, so ist

$$V\ (T) \geqq \omega\ (f, \delta)\ V\ (K) \tag{3}$$

Beweis: Die Stützfunktion $H\ (u)$ von K mit $u^2 = 1$ ist nach Definition von K stets $\geqq \varkappa$, die von $\overline{K}$ stets $\geqq (1 + \delta)\ \varkappa$. Als Inkugel I

nehmen wir die, welche O als Mittelpunkt besitzt. Nun soll T mit dem Rand von $\overline{K}$ mindestens einen Punkt gemeinsam haben. Es sei $\bar{p}$ ein solcher. Dann betrachten wir den Durchstoßpunkt p des Halbstrahles $\overrightarrow{o\,\bar{p}}$ mit K und die zugehörige Stützebene ε (oder eine von denen). Der Abstand $\bar{p}$ von ε ist gleich δ. $\varepsilon\, o$, wo $\varepsilon\, o$ der Abstand der Ebene ε von o ist. Wir bringen nun den Kappenkörper von K (*Bonnesen-Fenchel*, 12) mit $\bar{p}$ als Eckpunkt zum Schnitt mit ε. Der so entstandene Kegel hat sicher eine Höhe $\delta\,.\,\bar{\varepsilon}\, o \geqq \delta\, \varkappa$. Nun schneidet jene Ebene, welche parallel zu ε ist und durch o geht, aus dem Kappenkörper einen Bereich aus, der sicher eine $(n-1)$-dimensionale Kugel mit dem Radius $\varkappa$ enthält. Also schneidet die Ebene ε aus dem Kegel einen Bereich aus, der eine $(n-1)$-dimensionale Kugel vom Radius $\frac{\delta}{1+\delta}\,\varkappa$ enthält. Dabei ist das Volumen des Kegels mindestens $\frac{1}{n}\left(\frac{\delta}{1+\delta}\varkappa\right)^{n-1} \chi_{n-1}\,\delta\,\varkappa$. Dieser Kegel hat mit K keinen Punkt gemeinsam, da die beiden Körper durch ε getrennt werden.

Es ist also

$$V(T) - V(K) \geqq \frac{1}{n}\ \chi_{n-1}\ \frac{\delta^n}{(1+\delta)^{n-1}}\ \varkappa^n$$

Nun ist $V(K) \leqq \chi_n\, \varrho^n$, also ist

$$V(T) \geqq \omega(f, \delta)\, V(K) \text{ w.z.b.w.}$$

Satz 13: Ist $0 < \delta < (2\,.\,3^n n^2)^{-1}$,

$$\tau(f) = \operatorname{Min}(\omega(f, \delta), \varphi_3(f)) \tag{4}$$

$$\text{so ist } S(B, K) \leqq \frac{1}{\tau}\, V(B + \delta K) \leqq \frac{1}{\tau}\,(V(B) + O(B, \delta\varrho)) \tag{5}$$

Dabei ist die Quermaßsumme O in Definition 7, § 5, ω, φ_3 in Definition 17 und 18 festgelegt.

Beweis: Die Mittelpunkte der Körper $K(r)$, die in B liegen, bezeichnen wir mit p_i und die zugehörigen Körper $f(x - p_i) \leqq r$ mit K_i $f(x - p_i) \leqq (1 + \delta)\, r$ mit $\overline{K}_i$. Diese liegen alle in $B + \delta K$.

Die Punkte $q_{ij} = \frac{1}{2}(p_i + p_j)$ nennen wir Halbierungspunkte. Sie liegen wieder in B. Wir wollen zunächst die Anzahl $s + 1$ der Halbierungspunkte q_{ij}, die in $\overline{K}_i$ liegen, abschätzen. Nun ist, da die K_i nicht übereinandergreifen, $f(p_i - p_j) \geqq 2\, r$, also $f(q_{i\,j} - q_{ik}) \geqq r$.

Die q_{ij} bilden also eine (f, r)-Konfiguration von $\overline{K}_i$. Also ist nach. der Folgerung von Satz 4 mit $t = 1 + \delta$, $s + 1 \leqq (3 + 2\,\delta)^n \leqq 3^n + 1$[15] Daher ist die Anzahl s der q_{ij} mit $j \neq i$, höchstens 3^n. Der Körper $K : f \leqq r$ besitzt in jedem seiner Randpunkte eine Stützebene. Ist der Randpunkt singulär, so sei eine seiner Stützebenen willkürlich ausgewählt, aber so, daß in Bezug auf o symmetrisch gelegene Randpunkte von K parallele Stützebenen besitzen. Dies ist stets möglich. Ist nun q ein Halbierungspunkt q_{ij} $(j \neq i)$ in $\overline{K}_i$, so betrachten wir jene Ebene durch q, die zu der Stützebene parallel ist, welche der Durchstoßpunkt des Halbstrahles durch o mit dem Richtungsvektor $\overrightarrow{p_{ij}\, q}$ mit K besitzt. Dies wird für alle q_{ij} $(j \neq i)$ durchgeführt. Wir erhalten so höchstens 3^n Ebenen. Der Durchschnitt der Halbräume dieser Ebenen, die K_i enthalten, mit $\overline{K}$ bildet einen konvexen Körper L_i, der nicht notwendigerweise einen Mittelpunkt besitzen muß. Es liegt L_i in $\overline{K}_i$ und er enthält K_i. Diese L_i greifen nicht übereinander. Hätten nämlich L_i und L_j $(i \neq j)$ einen Punkt gemeinsam, dann auch $\overline{K}_i$ und $\overline{K}_j$. Diese hätten dann auch q_{ij} gemeinsam. Dann können wir oben eine Ebene durch q_{ij} legen, welche sowohl Stützebene von L_i wie von L_j ist, also L_i und L_j trennt.

Es ist also

$$\Sigma\, V(L_i) \leqq V(B + \delta K) \tag{6}$$

Wir müssen nun zwei Fälle unterscheiden:

1. L_i hat mit dem Rand von $\overline{K}_i$ einen Punkt gemeinsam. Dann ist nach Hilfssatz 11: $V(L_i) \geqq \omega\, V(K_i)$ oder
2. dies ist nicht der Fall, dann ist L_i ein Polyeder von höchstens 3^n Seiten. Dann ist nach Definition 17: $V(L_i) \geqq \varphi_3\, V(K_i)$.

Daher haben wir in beiden Fällen

$$V(L_i) \geqq \tau\, V(K) = \tau\, J\, r^n$$

Also ist

$$S(B, K) = a(B, K)\, J\, r^n \leqq \frac{1}{\tau}\, V(B + \delta K) \tag{7}$$

Nun liegt K im Körper $\varrho\, E$ (E Einheitskugel).

Also ist $V(B + \delta K) \leqq V(B) + O(B, \delta\, \varrho)$.

[15] Man benütze die Ungleichungen

$$\left(1 + \frac{a}{n^2}\right)^n \leqq 1 + a \ (a < 1),\ (1 - a)^{-1} < 1 + 2a \ (a < 1/2).$$

Damit ist der Satz bewiesen.

Mit Hilfe des eben gewonnenen Satzes läßt sich Satz 4 verschärfen.

Satz 14: Mit denselben Bezeichnungen wie in Satz 4 gilt:

13 $$s\,(B, f, d)\,J\,d^n \leqq \frac{2^n}{\tau\,(f)}\,V\,(B + (1 + \delta)\,K_1) \qquad (8)$$

Beweis: Er erfolgt wie bei Satz 4, nur wird jetzt statt der trivialen Abschätzung $s\,J\,(\tfrac{1}{2}\,d)^n \leqq V\,(B + K_1)$ Satz 13 benützt.

Wir stellen jetzt eine Verallgemeinerung von Satz 13 auf.

Satz 15: Es sei $a > 0$, $a \leqq 1$,

$$l = \frac{1}{\alpha},\ L = (2\,l^2 + 1)^n,\ h^n = [L],\ 0 < \delta_l < \frac{1}{(l+1)\,n^2\,L} \qquad (9)$$

$$\tau_l\,(f) = \mathrm{Min}\,(\omega\,(f, \delta_l), \varphi\,n\,(f)) \qquad (10)$$

dann gilt für $S_\alpha(B, K\,(r))$, = Supremum der Gesamtinhalte aller Körper von der Form $f\,(x - p) \leqq \bar{r}$ wo $\alpha\,r \leqq \bar{r} \leqq r$, die ganz in B liegen

$$S_\alpha\,(B, K\,(r)) \leqq \frac{1}{\tau_l}\,V\,(B + \delta_l\,r\,F)$$

wo F der Eichkörper $f \leqq 1$ ist.

Ist $\alpha = 1$, so erhalten wir Satz 13.

Beweis: Wir gehen wie in Satz 13 vor. Die Körper $f\,(x - p_i) \leqq r_i$, die in B liegen, bezeichnen wir mit K_i und $f\,(x - p_i) \leqq (1 + \delta_l)\,r_i$ mit $\bar{K}_i$. Diese liegen alle in $B + \delta_l\,.\,r\,F$. Jetzt betrachten wir die Punkte $q_{ij} = \dfrac{r_i\,p_j + r_j\,p_i}{r_i + r_j}$. Es soll nun die Anzahl $s + 1$ der $q_{i\,j}$

14 die in $\bar{K}_i$ liegen, abgeschätzt werden. Zunächst folgt aus $f\,(q_{i\,j} - p_i) \leqq r_i$, wo $c = 1 + \delta_l$ ist: $f\,(p_j - p_i) \leqq c\,(r_i + r_j$. Nun greifen die K_i nicht übereinander, also gilt $f\,(p_j - p_k) \geqq r_j + r_k\ (j \neq k)$.

Daraus folgt durch einfache Abschätzung[16]

$$f\,(q_{i\,j} - q_{i\,k}) \geqq \frac{\alpha^2}{2}\,(1 + \alpha - c\,(1 - \alpha))\,r = \gamma\,r$$

Die Folgerung von Satz 4 mit $t = (1 + \delta_l)\,\dfrac{r_i}{\gamma\,r}$, angewendet auf $\bar{K}_i$ liefert[14] $s + 1 \leqq L + 1$, also ist die Anzahl s der $q_{i\,j}$ mit $j \neq i$ höchstens h^n. Nun geht der Beweis wie bei Satz 13 zu Ende.

---

[16] Es ist ja $2\,(q_{ij} - q_{ik})\,r_{ij}\,r_{ik} = (r_{ij} + r_{ik})\,(p_i - p_k) - (r_i - r_k)\,(p_k + p_j - 2p_i)$ wo $r_{il} = r_i + r_l$ und $a\,r \leqq r_i \leqq r$ ist.

§ 8. λ-Körper.

Es soll φ_z (*f*) gegeben durch Definition 17 für eine spezielle Klasse von konvexen Körpern nach unten abgeschätzt werden.

Definition 19: Ein konvexer Körper K heißt λ-Körper, wenn es ein $\lambda > 0$ gibt, so daß folgendes gilt[17]: Es sei F eine Ebene, die K schneidet, F die Stützebene von K, welche zu F parallel und ihm am nächsten ist. Es sei $d(F)$ der Abstand der beiden Ebenen und $T(F)$ der n—1 dimensionale Inhalt des Durchschnittes von K und F. Dann sei stets für jede solche Ebene F . $T(F) \leqq [\lambda\, d(F)\, \varrho(K)]^{\frac{n-1}{2}}$. Dabei sei $\varrho(K)$ der Umkugelradius von K.

Satz 15: Ist K ein λ Körper und $\varkappa(K)$ sein Inkugelradius, dann ist $\varphi_z \geqq 1 + c\left(\frac{\varkappa}{\varrho}\right)$, wo c nur von λ und z abhängt.

Beweis: Es sei L ein Polyeder von höchstens z^n Seiten. Die Seiten bezeichnen wir mit L und ihre n—1 dimensionale Inhalte mit $J(\mathsf{L})$. Aus der isoperimetrischen Ungleichung folgt

$$(\Sigma J(\mathsf{L}))^n\, V(L)^{1-n} \geqq n^n \chi_n = c_2$$

also $\Sigma J(\mathsf{L}) \geqq c_3\, V(L)^{1-\frac{1}{n}}$.

Nun ist die Anzahl der L höchstens z^n. Es gibt also mindestens eine Seitenfläche L, so daß

$$J(\mathsf{L}) \geqq c_4\, V^{1-\frac{1}{n}}(L) \tag{2}$$

Wir betrachten nun die zu L am nächsten gelegene Stützebene E an K, die zu L parallel ist. Ihr Abstand von o sei l ebenso der Abstand von L an o sei s. Dann ist $s \geqq l \geqq \varkappa$. Wir betrachten weiter die Pyramide P mit L als Basis und o als Spitze, die sicher in L liegt.

Wir unterscheiden nun zwei Fälle:

1. $(\lambda\, l\, \varrho)^{\frac{n-1}{2}} \leqq \frac{1}{2n} J(\mathsf{L})$.

---

[17] Eine eng verwandte Definition bei *van der Corput* Davenport Proc., Amsterdam 49, S. 701. Die Verfasser betrachten den gitterförmigen Fall. Hier haben die Polyeder höchstens 2 $(2^n$—1) Seitenflächen und Mittelpunkt.

Wichtig wäre die Lösung des folgenden Problems: Gibt es zu einem konvexen Körper K mit Mittelpunkt 0 stets ein Polyeder mit $2s$ Seitenflächen (s vorgegeben) mit minimalen Volumen, welches K umgeschrieben ist und 0 als Mittelpunkt hat. Für den R_2 wurde das Problem von *Dowker* (Bull. Amer. Soc. 50 (1944) S. 120) gelöst.

Es ist sicher $V(L) - V(K) \geqq V(P) - V(P \cap K)$ (3)

Zunächst ist $V(P) = \frac{1}{n} J(\mathsf{L})\, s \geqq \frac{1}{n} J(\mathsf{L})\, l.$

Wir müssen nun $V(P \cap K)$ nach oben abschätzen. Wir betrachten die Ebene durch o parallel zu L, die K in zwei volumsgleiche Teile K' und K'' teilt.

Dann ist $V(P \cap K) \leqq V(K')$. Nun ist der Abstand $d(F)$ jeder Ebene F, welche parallel zu L ist und K schneidet $\leqq l$.

Also ist

$$V(K') \leqq \frac{1}{2n} J(\mathsf{L})\, l \tag{4}$$

Es ist also $V(L) - V(K) \geqq \frac{1}{2n} J(\mathsf{L})\, l \geqq \frac{C_4}{2n}\, l\, V^{1-\frac{1}{n}}(L) \geqq \frac{C_4}{2n} \varkappa V^{1-\frac{1}{n}}(K).$

Nun ist wieder $V(K) \leqq \chi_n \varrho^n$, also

15 $$V(L) \geqq V(K)\left(1 + c_5 \left(\frac{\varkappa}{\varrho}\right)^n\right) \tag{5}$$

2. Es sei nun $(\lambda\, l\, \varrho)^{\frac{n-1}{2}} > \frac{1}{2n} J(\mathsf{L})$.

Wir setzen $T = \left(\frac{1}{2n} J(\mathsf{L})\right)^{\frac{2}{n-1}} \frac{1}{\lambda \varrho}$. Dann ist zunächst $T < l$.

Wir betrachten noch die Ebene $\bar{E}$, welche zu E und L parallel ist,

16 und von o den Abstand T hat. Die Ebene $\bar{E}$ schneidet von der Pyramide P einen Pyramidenstumpf P_1 ab, mit der Basis L und der Deckfläche $\bar{\mathsf{L}}$.

Dann ist wieder

17 $$V(L) - V(K) \geqq V(P_1) - V(P \cap K).$$

Es sei P_2 die Pyramide mit $\bar{\mathsf{L}}$ als Basis und der Höhe $l - T$. Dann ist

$$V(P_2) = \left(\frac{l-T}{s}\right)^n V(P)$$

also $V(P_1) = V(P)\left(1 - \left(\frac{l-T}{s}\right)^n\right) \geqq V(P) \frac{T}{l} = \frac{J(\mathsf{L})}{n} T.$

Weiter ist ganz analog wie unter 1. $V(P \cap K) \leqq \frac{1}{2n} J(\mathsf{L})\, T$, da

18 wieder $T(F) \leqq (\lambda\, \varrho\, d(F))^{\frac{n-1}{2}} \leqq (\lambda\, \varrho\, T)^{\frac{n-1}{2}} = \frac{1}{2n} J(\mathsf{L})\, T$ ist, da ja nur jene F in Frage kommen, für die $d(F) \leqq T$ ist.

Also erhalten wir

$$V(L) - V(K) \geqq \frac{J(\mathsf{L})\, T}{2n} \geqq \frac{J(\mathsf{L})}{2n} \left(\frac{1}{2n} J(\mathsf{L})\right)^{\frac{2}{n-1}} \frac{1}{\lambda \varrho}$$

also $V(L) - V(K) \geqq c_c\, V^{1+\frac{1}{n}}(L) \dfrac{1}{\lambda \varrho}$.

Nun ist $V(L)^{1+\frac{1}{n}} \geqq V(K)\, \chi_n^{\frac{1}{n}}\, \varkappa$, da $V(K) \leqq \chi_n\, \varkappa^n$,

daher
$$\frac{V(L)}{V(K)} \geqq 1 + c_7 \left(\frac{\varkappa}{\varrho}\right)^n.$$

Zusammenfassend sehen wir, daß für alle L stets

$$\frac{V(L)}{V(K)} \geqq 1 + c \left(\frac{\varkappa}{\varrho}\right)^n \tag{6}$$

also gilt auch
$$\varphi_z \geqq 1 + c \left(\frac{\varkappa}{\varrho}\right)^n \tag{7}$$

Dies ist die Abschätzung, die wir erhalten wollten.

§ 9. Ausfüllung durch Zylinder.

Unter einen Zylinder verstehen wir einen konvexen Körper K mit der Distanzfunktion $f(x) = \operatorname{Max}(f_1(x), f_2(x))$, wo $f_1 = (x_1^2 + \dots + x_m^2)^{\frac{1}{2}}$, $f_2 = \operatorname{Max}(|x_{m-1}|, \dots |x_n|)$, dabei sind die x_i die Koordinaten von x.

Für $m = n$ erhalten wir die Kugel.

Hier lassen sich bedeutend schärfere Abschätzungen[18] erzielen.

Satz 17: Ist K ein Zylinder, so gilt:

$$S(B, K(r)) \leqq 2^{-\frac{m}{2}} \frac{m+2}{2} \left(V(B) + O(B, \sqrt{n-m+2}\, r)\right) \tag{1}$$

Es sei K ein Zylinder $f \leqq r$. Die Mittelpunkte der Zylinder, die in B eingelagert werden, seien wieder mit p_i bezeichnet. Sie haben die Form $f(x - p_i) \leqq r$. Wir betrachten jetzt weiter die Zylinder $\overline{K}: f_1 \leqq \sqrt{2}\, r$, $f_2 \leqq r$. Dieser Zylinder liegt sicher in einer Kugel vom Radius $\sqrt{n-m+2}\, r = \bar{r}$. Diese Zylinder liegen also alle im Bereich

[18] Der Gedankengang stammt von *Blichfeldt*, Math. Annalen 101.

$B^* = B + r\,E$ (E = Einheitskugel). Wir definieren nun auf B^* folgende Funktion:

1. Liegt x in l-Zylindern $\overline{K}$, so sei $\psi(x) = \sum_{i=1}^{l} (2\,r^2 - f_1^2(x - p_i))$,

2. liegt x in keinen solchen Körper, dann sei $\psi(x) = 0$. Es ist nun bekanntlich, vergleiche z. B. *Blichfeldt*[18],

$$\sum_{i=1}^{l} f_1^2(p_i - p_j) \leqq 2\,l \sum_{i=1}^{l} f_1^2(x - p_i).$$

Die Körper K greifen nach Voraussetzung nicht übereinander, also gilt $f(p_i - p_j) \geqq 2\,r$. Greifen also zwei Körper $\overline{K}$ übereinander, so muß also $f_1(p_i - p_j) \geqq 2\,r$ sein, da nach Definition von $\overline{K}$ dann $f_2(p_i - p_j) \leqq 2\,r$ ist. Daraus folgt also, daß $\sum_{i=1} f_1^2(x - p_i) \geqq 2\,r^2(l-1)$ ist, wenn der Punkt x in l-Zylindern $\overline{K}$ liegt. Daraus folgt sofort:

$$\psi(x) \leqq 2\,r^2(l - 1)) = 2\,r^2. \tag{2}$$

Es ist also stets $\psi(x) \leqq 2\,r^2$.

Daraus folgt (für die Berechnung der auftretenden Integrale sei auf *Blichfeldt*[18] verwiesen):

$$2\,r^2\,V(B^* \geqq \int_{B^*} \psi(x)\,dx = \sum_{i=1}^{a} \int_{\overline{K}_i} (2\,r^2 - f_1^2(x - p_i))\,dx =$$
$$= a \int_{\overline{K}} (2\,r^2 - f_1^2(x))\,dx = 2^{\frac{m}{2}+1} \frac{2}{m+2}\,a\,J\,r^n$$

also $$S = a\,J\,r^n \leqq \frac{m+2}{2}\,2^{-\frac{m}{2}}\,V(B^*) \quad \text{w.z.b.w.}$$

Diese Abschätzung läßt sich noch verschärfen. Schärfere Abschätzungen für die Kugel stammen von *Blichfeldt*[19] und *Rankin*[1]. Letzterer hat auch den Fall betrachtet, wo die eingelagerten Kugel, nicht alle den gleichen Radius haben müssen. Diese Betrachtungen lassen sich auch übertragen auf Körper von der Distanzfunktion $(\Sigma\,|\,x_i\,|^{\alpha})^{\frac{1}{\alpha}}$ für $\alpha \geqq 1$ (*E. Hlawka*)[19].

---

[19] Sitzungsberichte 1946, 1947. Die Sätze sind dort nur für den gitterförmigen Fall ausgesprochen, gelten aber auch im allgemeinen.

Anmerkung bei der Korrektur: *Mahlev* (Quarterly Journal **17**) hat gezeigt, daß das Problem des Zylinders stets auf das analoge Problem der Basis zurückgeführt werden kann. Vgl. dazu *E. Hlawka*, Anzeiger der math. naturw. Kls. Akademie 1948 S. 116.

Weitere Untersuchungen über Potenzsummen bei *Rankin*. Proc. Amsterdam **51**. Dort auch weitere Literaturangaben.

§ 10. Dichte einer Ausfüllung.

Es soll jetzt gezeigt werden, daß der Limes $\frac{S(B, K(r))}{V(B)}$ existiert und unabhängig vom Bereich B ist. Wir nennen ihn die Dichte. Für Kugeln und B-Würfel wurde dies von *Rankin* gezeigt[1]. Der Gedanken-
19 gang der Hilfssätze 3 bis 5 ist von der Arbeit von *Rankin* beeinflußt.

Hilfssatz 12: Es sei W ein Würfel von der Kantenlänge w und T ein konvexer Körper mit Mittelpunkt. Es sei $\varrho\,(T)$ der Umkugelradius und es sei $4\,\varrho < w$. Dann gilt:

1. Hat T mit W einen Punkt gemeinsam, so liegt er in den zu W konzentrischen Würfel W_1 mit der Kantenlänge $w + 4\,\varrho$.
2. Liegt T nicht ganz in W, so hat er mit dem Würfel W_2, der konzentrischen zu W ist und die Kantenlänge $w - 4\,\varrho$ hat, keinen Punkt gemeinsam.

Beweis: T liegt ganz in seiner Umkugel mit dem Radius ϱ, also ganz in einem Würfel W_3 mit demselben Mittelpunkt wie T und der Kantenlänge $2\,\varrho$. Daraus folgt sofort die Behauptung.

Hilfssatz 13: Es sei A eine $K\,(r)$-Konfiguration von B, $a\,(A)$ die Anzahl der Punkte von A, $S\,(B, A, K\,(r)) = a\,(A)\,J\,r^n$ wie in Definition 12. Weiter sei $\lambda > 0$ eine positive Zahl. Ist $A_1 = \lambda\,A$, $r_1 = \lambda\,r$ dann gilt: A_1, ist eine $K\,(r_1)$-Konfiguration von $B_1 = \lambda\,B$

$$a\,(A_1) = a\,(A),\; S\,(B_1, A_1, K\,(r_1)) = \lambda^n\,S\,(B, A, K\,(r)).$$

Hilfssatz 14: Es sei W ein achsenparalleler Würfel von der Kantenlänge w mit Mittelpunkt 0. Weiter seien λ, r_1, r_2 positive Zahlen mit $r_2 \geqq r_1$, $\lambda \geqq 1$. Dann gibt es Zahlen $C_1, C_2 > 0$, so daß, wenn $\varrho_2 \leqq \frac{1}{4}\,w$ wo ϱ_2 der Umkugelradius von $K_2 : f \leqq r_2$ ist.

$$|S\,(\lambda\,W, K\,(r_1)) - S\,(W, K\,(r_2))| \leqq C\,(\lambda), C\,(\lambda) = \left(C_1\left(\lambda - \frac{r_1}{r_2}\right) + C_2\,r_1\right)\lambda^{n-1} \tag{1}$$

Dabei ist S in Definition 12 definiert.

Beweis: Es sei $\mu = \lambda\frac{r_2}{r_1}$ und A_2 eine $K\,(r_2)$-Ausfüllungskonfiguration von $W_2 = \mu\,W$ und $\Sigma_2 = a\,(A_2)\,J\,r_2{}^n$. Es sei nun A'_2 die Menge jener Punkte von A_2, die eine $K\,(r_2)$-Konfiguration von W bilden, $\Sigma' = a\,(A_2')\,J\,r_2{}^n$. $S_2 = S\,(W, K\,(r_2))$, $A_2'' = A_2 - A_2'$, $\Sigma'' = a\,(A_2'')\,J\,r_2{}^n$. Dann ist $\Sigma_2 = \Sigma' + \Sigma''$. Nun ist nach Definition 12 $\Sigma' \leqq S_2$. Es soll nun Σ'' abgeschätzt werden.

Wir betrachten die Körper $\overline{K}p: f(x-p) \leqq r_2$, für die p zu A_2'' gehört. Das Gesamtvolumen dieser $\overline{K}\,p$ ist gerade Σ''. Sie liegen alle $W_2 = \mu\, W$. Da sie nicht ganz in W liegen, so haben sie nach Hilfssatz 12 mit dem Würfel $W' = W\left(1 - \frac{4\,\varrho_2}{w}\right)$ keinen Punkt gemeinsam.

Es ist also

$$\Sigma'' \leqq V(W_2) - V(W') \leqq (\mu\, w)^n - (w - 4\,\varrho_2)^n \leqq n\,(w\,\mu)^{n-1}\,(w\,(n-1) + 4\,\varrho_2) \qquad (2)$$

Nun ist $\mu = \lambda\,\frac{r_2}{r_1}$, $\varrho_2 = r_2\,\varrho^*$, wo ϱ^* der Umkugelradius von $f \leqq 1$ ist.

Setzen wir $C_1 = n\,w^n$, $C_2 = 4\,\varrho\,.\,n\,w^{n-1}$

$$C(\lambda) = \left(C_1\left(\lambda - \frac{r_1}{r_2}\right) + C_2\,r_1\right)\lambda^{n-1}$$

so ist also

$$\Sigma'' \leqq \left(\frac{r_2}{r_1}\right)^n C(\lambda).\ \Sigma \leqq S_2 + \left(\frac{r_2}{r_1}\right)^n C(\lambda) \qquad (3)$$

Dies gilt für jedes A_2, also auch für $S_{2\,2} = S(W_2, K(r_2))$. Setzen wir noch $S_{1\,1} = S(\lambda\,W, K(r_1))$, so ist nach Definition 12 und Hilfssatz 13, da $\mu = \lambda\,\frac{r_2}{r_1}$, $S_{2\,2} = \left(\frac{r_2}{r_1}\right)^n S_{1\,1}$. Wir erhalten also wegen (3) und $r_1 \leqq r_2$

$$D = S_{1\,1} - S_2 = \left(\frac{r_1}{r_3}\right)^n S_{2\,2} - S_2 \leqq C(\lambda)$$

Weiter ist $S_{2\,2} \geqq S_2$, $S_2 \leqq V(W)$,

also $\quad D \geqq -\left(1 - \left(\frac{r_1}{r_2}\right)^n\right) S_2 \geqq -C(\lambda)$

Damit erhalten wir zusammenfassend: $|D| \leqq C(\lambda)$. w. z. b. w.

Hilfssatz 15: Es sei W der Würfel von Hilfssatz 14. Dann gilt für jedes $r > 0$ mit $\varrho(K) < \frac{1}{4}\,W$ und für jede natürliche Zahl g

$$|S(W, K(r\,g^{-1})) - S(W, K(r))| \leqq C_3\,r \qquad (4)$$

wo C_3 nur von w und f abhängt.

Beweis: Nach Hilfssatz 13 ist $a(W, r\,g^{-1}) = a(Wg, r)$.

Weiter ist

$$a(Wg, r) \geqq g^n\,a(W, r),\ \text{also}\ S(W, K(r\,g^{-1})) \geqq S(W, K(r)).$$

Es sei nun A eine $K(r)$-Konfiguration von $\overline{W} = Wg$. Wir teilen nun $\overline{W}$ in g^n achsenparallele Würfel W_j mit der Kantenlänge W, die also aus W durch Parallelverschiebung hervorgehen. Mit A_j bezeichnen wir die Punkte von A, die in W_j liegen. Dann ist also $A = \Sigma A_j$. Nach Hilfssatz 12 ist A_j eine $K(r)$-Konfiguration des Würfels $\overline{W}_j$, konzentrisch zu W_j mit der Kantenlänge $w + 4\varrho$. Setzen wir also $W_4 = W(1 + \frac{4\varrho}{w})$, dann ist

$$a(Aj) \leqq a(W_4, K(r)) = \bar{a},$$

also $a(A) \leqq g^n \bar{a}$. Dies gilt für jedes A, daraus folgt $a(\overline{W}, K(r)) \leqq g^n \bar{a}$, also ist nach Hilfssatz 13

$$S(W, K(r g^{-1})) \leqq S_4$$

Weiter ist nach Hilfssatz 14 mit $\lambda = 1 + \frac{4\varrho}{w}$, $r_1 = r_2 = r$,

$$S_4 \leqq S + C'$$

wo
$$C' = (1 + \frac{4\varrho}{w})^{n-1} [\frac{4\varrho}{w} C_1 + C_2 r]$$

Nun ist $4\varrho \leqq w$, $\varrho \leqq \varrho^* r$, wo ϱ^* der Umkugelradius vom Eichkörper ist. Setzen wir also

$$C_3 = 2^{n-1} (\frac{4 C_1}{w} + C_2)$$

so folgt die Behauptung.

Hilfssatz 16: Ist W der Würfel von Hilfssatz 14, $S(r) = S(W, K(r))$, dann existiert

$$\Delta = \lim_{r \to 0} \frac{S(r)}{w^n} \tag{5}$$

und ist unabhängig von w.

Beweis: Es seien r_1, r_2 zwei positive Zahlen, die $< \frac{1}{2}$ und $\frac{w}{8\varrho}$. Es ist $\frac{1}{r_3}$, die nächst kleinere ganze Zahl an $\frac{1}{r_1}$, ebenso $\frac{1}{r_4}$ von $\frac{1}{r_2}$. Weiter sei $r_5 = r_3 r_4$, $S_i = S(r_i)$. Dann wollen wir zunächst $S_1 - S_2$ abschätzen. Es ist

$$|S_1 - S_2| \leqq |S_1 - S_3| + |S_3 - S_4| + |S_4 - S_2| \tag{6}$$

Es ist

$$|S_3 - S_4| \leqq |S_5 - S_3| + |S_5 - S_4|$$

Nach Hilfssatz 15 mit $g = r_4^{-1}$ ist $|S_5 - S_i| \leqq C_3 r_j$ $(i = 3, 4)$.

Nun ist $\frac{1}{r_3} \geqq \frac{1}{r_1} - 1$, also $r_3 < 2\, r_1$, ebenso $r_4 < 2\, r_2$, also $|S_3 - S_4| < 2\, C_3\,(r_1 + r_2)$. Weiter ist nach Hilfssatz 14 mit $\lambda = 1$ angewendet auf r_1 und r_3

$$|S_1 - S_3| \leqq C_1\left(1 - \frac{r_1}{r_3}\right) + C_2\, r_1 \leqq (C_1 + C_2)\, r_1 = C_5\, r_1$$

ebenso $|S_2 - S_4| \leqq C_5\, r_2$, also

$$|S_1 - S_2| \leqq (2\, C_3 + C_5)\,(r_1 + r_2) = C_6\,(r_1 + r_2).$$

Es sei nun $\varepsilon > 0$ eine positive Zahl. Dann ist für alle $r_1, r_2 < \frac{\varepsilon}{2\, C_6}$ $|S_1 - S_2| < \varepsilon$, d. h. $\lim\limits_{r \to 0} \frac{S(r)}{w^n}$ existiert. Es muß noch gezeigt werden, daß er von w nicht abhängt. Es sei W_1 ein achsenparalleler Würfel von der Kantenlänge w_1, der o. B. d. A. zu W konzentrisch angenommen werden kann. Setzen wir $\lambda = \frac{w_1}{w}$, $r_1 = \frac{r}{\lambda}$, so ist nach Hilfssatz 13

$$S\,(W_1, K\,(r)) = \lambda^n\, S\,(W, K\,(r_1)).$$

Aus dem eben bewiesenen erfolgt sofort die Behauptung. Es kann zugelassen werden, daß w_1 von r abhängt, aber muß in r beschränks sein.

Satz 18: Es sei B ein im *Jordan*schen Sinne meßbarer Bereich, mit $0 < V\,(B) < \infty$. $S\,(r) = S\,(B, K\,(r))$ das Maximum der Gesamtinhalte der konvexen Körper mit Mittelpunkt $f\,(x - p) \leqq r$, die, ohne übereinanderzugreifen, in B liegen. Dann existiert

$$\lim_{r \to 0} \frac{S\,(r)}{V\,(B)} = \Delta$$

und ist von B unabhängig.

Beweis: Da B einen Inhalt besitzt, so gibt es bei vorgegebenen $\varepsilon > 0$ eine Menge von achsenparallelen Würfeln W_i mit der Kanten-
21 länge $w = w\,(\varepsilon)$, die nicht übereinandergreifen, so daß

1. der Gesamtinhalt V_1 der W_i, die nicht zu B fremd sind, $\leqq V\,(B) + \varepsilon$ ist,
2. der Gesamtinhalt V_2 der W_i, welche ganz zu B gehören, wir bezeichnen sie mit W_i', $\geqq V\,(B) - \varepsilon$ ist.

Wir setzen $S\,(W_i, K\,(r))$, bzw. $S\,(W_i', K\,(r))$, S_i, bzw. S_i'. Nun ist nach Definition von S: $S \geqq \Sigma\, S_i$.

22 Nun existiert nach Hilfssatz 16 $\lim\limits_{r\to 0} S_i' = \Delta\, w''$, also

$$\underline{\lim} \frac{S}{V(B)} \geqq \Delta' \frac{\Sigma' w^n}{V(B)} \geqq \Delta \left(1 - \frac{\varepsilon}{V(B)}\right)$$

Weiter wird B von den W_i mit der Eigenschaft 1 überdeckt. Ist nun A eine $K(r)$-Konfiguration von B, so sei A_i die Menge der Punkte, die in W_i liegen. Dann ist $A = \Sigma A_i$. Dann bilden nach Hilfssatz 12 die A_i eine $K(r)$-Konfiguration von $\overline{W}_i$, wo die $\overline{W}_i$ zu W_i konzentrisch sind und die Kantenlänge $\overline{w} = w + 4\,\varrho$ besitzen. Also

$$a(A_i) \leqq a(\overline{W}_i, K(r)),\ S \leqq \Sigma\, S(\overline{W}_i)$$

Daraus folgt $\overline{\lim} \dfrac{S}{V} \leqq \Delta \dfrac{\Sigma W''}{V} \leqq \Delta \left(1 + \dfrac{\varepsilon}{V}\right)$

also $\Delta \left(1 - \dfrac{\varepsilon}{V}\right) \leqq \underline{\lim} \dfrac{S}{V} \leqq \overline{\lim} \dfrac{S}{V} \leqq \Delta \left(1 + \dfrac{\varepsilon}{V}\right)$.

Dies gilt für jedes $\varepsilon > 0$, also existiert $\lim \dfrac{S}{V}$ und ist gleich Δ w. z. b. w.

Δ heißt Dichte der Ausfüllung durch konvexe Körper mit der Distanzfunktion $f(x)$.

Satz 19: Es ist

$$\frac{1}{2^{n-1}} \sum_{k=1}^{\infty} \frac{1}{k^n} \leqq J \left(\frac{M}{2}\right)^n \leqq \Delta \leqq \frac{1}{\tau} \tag{9}$$

Die Dichte ist stets kleiner als < 1, wenn K kein Polyeder von höchstens 3^{n-1}-Seiten ist. Dabei wurden τ, bzw. M, § 7, Satz 13, bzw. § 2 definiert. J ist Volumen des Eichkörpers $f \leqq 1$.

Beweis: Die Behauptungen folgen sofort aus § 7, Satz 13, § 7 (1), Satz 9, § 6 und Hilfssatz 4, § 2.

Ist K ein Zylinder nach § 9, so gilt für die Dichte nach Satz 17, § 9

$$\Delta \leqq 2^{-\frac{m}{2}} \frac{m+2}{2}$$

Daraus folgt $\lim\limits_{n\to\infty} \Delta = 0$, wenn m mit $n \to \infty$ strebt. Dies gilt insbesondere für die Kugel $(m = n)$. Man kann zeigen, daß diese Beziehung auch gilt für alle Körper mit einer Distanzfunktion von der Form $f = (\Sigma\, |x_i|^\alpha)^{\frac{1}{\alpha}}$ vgl. § 9 Schluß.

Satz 20: Es sei $\alpha \leqq 1$ eine positive Zahl, $l = \dfrac{1}{\alpha}$. Es sei $S_a(B, K(r))$

und τ_l wie in Satz 15 definiert. Dann existiert $\Delta_\alpha = \lim\limits_{r \to 0} \frac{S_\alpha}{V(B)}$ ist unabhängig von B und es ist

$$\Delta_\alpha \leqq \frac{1}{\tau_l} \tag{10}$$

Insbesondere ist $\Delta_\alpha < 1$, wenn K kein Polyeder ist. Weiter ist $\lim\limits_{\alpha \to 0} \Delta_\alpha = 1$.

Beweis: Die Existenz von Δ_α zeigt man in analoger Weise wie bei Δ. Für die Kugel vgl. *Rankin*[1]. Die zweite Behauptung entnimmt man § 7, Satz 15. Die letzte Behauptung folgt aus § 12, Satz 25.

Man kann *Rankin*[1], Satz 3 entnehmen, daß für die Kugel $\lim\limits_{n \to \infty} \Delta_\alpha = 0$ ist.

Es soll noch eine weitere Bemerkung angeführt werden.

Ist $S_k(B, K(r))$ das Maximum aller Gesamtvolumina von k - $K(r)$-Ausfüllungen, wie dies in Definition 16 definiert wurde, so kann man zeigen, daß $\Delta_k = \lim\limits_{r \to 0} \frac{S_k}{V(B)}$ existiert. Weiters kann gezeigt werden, analog wie bei Satz 17, daß für eine Kugel $\Delta_k \leqq k \cdot 2^{-\frac{n}{2}} \frac{n+2}{2}$, allgemein für einen Zylinder $\leqq k \cdot 2^{-\frac{m}{2}} \frac{m+2}{2}$ (vgl. *E. Hlawka*[19]). Daraus folgt $\lim\limits_{n \to \infty} \Delta_k = 0$, wenn $k = O(2^{\frac{n}{2}(1-\varepsilon)})$ z. B. $k = n$ ist, d. h. die Dichte Δ_k geht auch dann gegen 0, wenn zugelassen wird, daß die eingelagerten Kugeln übereinandergreifen, allerdings dürfen, grob gesprochen, nicht zuviele einen inneren Punkt gemeinsam haben.

§ 11. Ausfüllung durch konvexe Körper ohne Mittelpunkt.

Bis jetzt wurde vorausgesetzt, daß die in einem Bereich B eingelagerten Körper einen Mittelpunkt besitzen. Diese Voraussetzung wird jetzt fallen gelassen. Es sei K ein konvexer Körper $f(x) \leqq r$, der Koordinatenursprung O ein innerer Punkt von K, $f(x)$ seine Distanzfunktion, $H(u)$ $(u^2 = 1)$ seine Stützfunktion.

Definition 20: Es sei $a(B, K(r))$ die größte Anzahl der Körper, die aus K durch Parallelverschiebung hervorgehen, nicht übereinandergreifen und ganz in B liegen, $S(B, K(r))\, J\, r^n$ sein Gesamtvolumen.
23

Hilfssatz 17: Zwei Körper $K_a : f\,(x - a) \leqq r$, $K_b : f\,(x - b) \leqq r$ greifen dann und nur dann übereinander, wenn

$$(b - a)\,u < H\,(u) + H\,(-u) \tag{1}$$

ist, für jedes u mit $u^2 = 1$.

Beweis: 1. Angenommen K_a und K_b greifen übereinander. Dann gibt es einen Punkt p, so daß $(p - a)\,u < H\,(u)$, $(p - b)\,u < H\,(u)$.

Daraus folgt sofort die Behauptung.

2. Angenommen, es gelte $(b - a)\,u < H\,(u) + H\,(-u)$ für jedes u, dann müssen wir zeigen, daß K_a und K_b übereinandergreifen. Nehmen wir an, sie greifen nicht übereinander. Dann gibt es eine Ebene ε mit der Richtung u_0, die sie trennt. Dann ist der Abstand $\overline{a\,\varepsilon} = -a\,u_0 > H\,(u_0$, ebenso $\overline{b\,\varepsilon} = b\,u_0 > H\,(u_0)$, also wäre $(b - a)\,u > H\,(u_0) + H\,(-u_0)$ in Widerspruch zur Voraussetzung. Zum Beweis vgl. man *Minkowski*[8].

Wir können Hilfssatz 17 noch etwas anders aussprechen.

Definition 21: (*Bonnesen-Fenchel* 42). Es sei $\hat{K}$ der Konvexe Körper, der aus K durch Zentralsymmetrisierung in bezug auf 0 aus K entsteht. Seine Stützfunktion ist $\frac{1}{2}\,(H\,(u) + H\,(-u))$ und er besitzt O als Mittelpunkt.

Dann folgt sofort:

Hilfssatz 18: Zwei Körper K_a und K_b greifen dann und nur dann übereinander, wenn dies $\hat{K}_a$ und $\hat{K}_b$ tiun.

Satz 21: Es sei ϱ der Umkugelradius von K, β der Inkugelradius von B, $\varrho < \beta$, dann gilt

$$S\,(B_1, \hat{K}) \leqq \frac{V\,(\hat{K})}{V\,(K)}\, S\,(B, K\,(r)) \leqq S\,(B + \hat{K}, \hat{K}) \tag{2}$$

Dabei ist $B_1 = (1 - \frac{\varrho}{\beta})\;B$, wenn der Ursprung, Mittelpunkt einer Inkugel ist.

Beweis: Betrachten wir eine Ausfüllung von B_1 durch Körper, die aus $\hat{K}$ durch Parallelverschiebung hervorgehen. Diese greifen nach Definition nicht übereinander. Jetzt ersetzen wir die eingelagerten Körper $\hat{K}$ durch die zugehörigen Körper K. Nach Hilfssatz 18 greifen diese nicht übereinander. Nach einer Überlegung wie in Hilfssatz 10, § 6 liegen diese alle in B, also $a\,(B_1, \hat{K}) \leqq a\,(B, K)$ nach Definition 12 und 20. Daraus folgt die linke Hälfte der Ungleichung. Jede Ausfüllung von B durch Körper K ist, wenn die eingelagerten Körper durch $\hat{K}$ ersetzt werden, eine Ausfüllung von $B + \hat{K}$ durch Körper $\hat{K}$, also

$$a\,(B, K) \leqq a\,(B + \hat{K}, \hat{K})$$

Damit ist alles bewiesen.

Daraus folgt

Satz 22: $\Delta = \lim \dfrac{S\,(B, K)}{V\,(B)}$ existiert, ist unabhängig von B und $\Delta = \dfrac{V(K)}{V(\hat{K})}\,\hat{\Delta}$, wo $\hat{\Delta}$ Dichte von $\hat{K}$

Hilfssatz 19: (*Godbersen*[20], *Bonnesen-Fenchel* 53).

Besitzt K keinen Mittelpunkt, so ist

$$V\,(K) < V\,(\hat{K}) \leqq \mu\, V\,(K)$$

wo $$\mu = \frac{1}{2^n} \sum_{i=0}^{n} \binom{n}{i}\ {}_n\mathrm{Min}\,(i, n-i)$$

Ist K ein Simplex, so ist $V\,(\hat{K})\; V^{-1}\,(K) = \dfrac{1}{2^n}\binom{2n}{n}$

Daraus folgt

Satz 23: Es ist $\Delta < \hat{\Delta} \leqq 1$. Ist K ein Simplex, so ist $\lim\limits_{n\to\infty} \Delta = 0$.

§ 12. Ausfüllung mittels einer Menge $\{K\}$.

Es sei ein konvexer Bereich B mit Inkugelradius β vorgegeben. Weiters sei $\{K\}$ eine unendliche Menge von konvexen Körpern mit Mittelpunkt gegeben. Dabei brauchen die Körper nicht alle voneinander
24 verschieden sein, die Durchmesser sollen aber alle $\tfrac{1}{2}\,\beta$ sein und das inf $\varkappa\,(K)$ über alle Inkugelradien soll > 0 sein. Wir betrachten nun konvexe Körper, die aus der Menge $\{K\}$ durch Translationen hervorgehen, nicht übereinandergreifen und ganz in B liegen.

Wir wollen sagen, es werden Körper aus $\{K\}$ in B eingelagert. Das Gesamtvolumen dieser eingelagerten Körper sei S.

Definition 22: $S\,(B, \{K\})$ ist das Supremum von S für alle Einlagerungen von $\{K\}$ in B.

[20] Dissertation, Göttingen 1938, vgl. Zentralblatt 20, S. 77.

Satz 24: Es ist

$$S = S(B, \{K\}) \geqq C\, V(B) \quad (1)$$

wo $$C = 2^{1-n}\left(1 - \sqrt{\tfrac{n}{2n+2}}\right)^n \zeta(n) \quad (2)$$

Es können also soviele Körper aus $\{K\}$ in B eingelagert werden, so daß der C-te Teil von B ausgefüllt wird.

Beweis: Die Menge $\{K\}$ ist beschränkt. Nach dem *Blaschke*schen Auswahlsatz (*Bonnesen-Fenchel* 25) gibt es eine Teilfolge $\{K_i\}$, die gegen einen konvexen Körper $\overline{K}$ konvergiert. Ist $1 > \varepsilon > 0$ beliebig, so liegen fast alle Körper K_i im Körper $T = \overline{K} + \frac{1}{8}\,\beta\,\varepsilon\,E$ (E Einheitskugel), und $\overline{K}$ in $K_i + \frac{1}{8}\,\varepsilon\,\beta\,E$. Nun gilt nach § 7 (1), Satz 9, § 6

$$S_1 = S(B, T) \geqq \left(1 - \frac{\varrho_1}{\beta}\right)^n \overline{J}\,(\tfrac{1}{2}\,\overline{M})^n\, V(B) \quad (3)$$

wo ϱ_1 Umkugelradius von T, $\overline{M}$ das absolute Minimum und $\overline{J}$ das Volumen des Eichkörpers von T ist. Nach Hilfssatz 4 ist $\overline{J}\,\overline{M}^n \geqq 2\,\zeta(n)$. Die Anzahl der eingelagerten Körper T bei dieser maximalen Ausfüllung sei a. Weiters ist (*Bonnesen-Fenchel* 44, Seite 78) $\varrho_1 \leqq D(T)\sqrt{\frac{n}{2n+2}}$, wo $D(T)$, $D(K_i)$ die Durchmesser der zugehörigen Körper sind. Weiter ist

$$D(T) < D(K_i + \tfrac{1}{4}\,\beta\,\varepsilon\,E) < \frac{\beta}{2}(1 + \varepsilon)$$

nach Voraussetzung über die Durchmesser von K_i.

Es ist also $$S_1 \geqq C\,(1 + \varepsilon)^n\, V(B) \quad (4)$$

Es sei nun jedem eingelagerten Körper T eindeutig ein K_i zugeordnet und wir ersetzen den betreffenden, eingelagerten Körper T durch den Körper, der aus K_i durch jene Translation hervorgeht, welcher den Mittelpunkt von K_i mit dem des zugeordneten, eingelagerten Körper T zur Deckung bringt. Dann sind diese K_i, da sie in T liegen, in B eingelagert. Das Gesamtvolumen deser K_i sei S'. Dann ist

$$S' = S_1 + S' - S_1$$

Nun ist $$|S_1 - S'| = \left|\sum_{i=1}^{a} (V(T) - V(K_i))\right|$$

Nun liegt T in $K_i + \frac{1}{4}\,\beta\,\varepsilon\,E = \overline{K}_i$,

also $$|S_1 - S'| \leqq \sum_{i=1}^{a} (V(\overline{K}_i) - V(K_i)) \leqq C_1 \cdot \varepsilon,$$

25 wo C_1 noch von a und ε abhängt. Da $\inf \varkappa(K) > 0$, so ist $D(K) \geqq S > 0$ für alle K aus $\{K\}$. Es liegt also a unter einer Schranke, die von den Ki und von ε nicht abhängt. Also gilt $|S_1 - S'| \leqq C_2 \varepsilon$, wo C_2 unabhängig von a und ε ist.

Daraus folgt

$$S' \geqq S_1 - C_2 \varepsilon$$

also $$S \geqq S' \geqq (1 + \varepsilon)^n C V(B) - C_2 \varepsilon$$

Dies gilt für jedes ε und daraus folgt die Behauptung.

Erweiterungen: 1. In Satz 24 kann die Voraussetzung, daß alle K aus $\{K\}$ einen Mittelpunkt besitzen, fallen gelassen werden, wenn man C durch $\overline{C} = C \mu^{-1}$ ersetzt, wo μ in Hilfssatz 19 definiert ist. Es ist nur nötig, die linke Seite von Satz 21 heranzuziehen.

2. Wir können Satz 24 noch dadurch verallgemeinern, daß wir nur verlangen, daß ein Punkt von B höchstens $2(k-1)+1$ eingelagerten Körpern angehört (k natürliche Zahl). Dann ist C durch $\left(1 - \sqrt{\frac{n}{2n+2}}\right)^n k\, 2^{1-n}$ zu ersetzen. Es ist nur nötig, § 6, Satz 11 und § 2, Hilfssatz 6 heranzuziehen.

Satz 25: Es sei ein Bereich B vom *Jordan*schen Inhalt $0 < V(B) < \infty$ gegeben. Weiters sei noch eine Menge $\{K\}$ von konvexen Körpern mit Mittelpunkt gegeben, so beschaffen, daß es zu jedem $\varepsilon > 0$ eine unendliche Menge $M(\varepsilon)$ aus $\{K\}$ mit Durchmesser ε gibt. Die Körper aus $\{M\}$ brauchen nicht alle voneinander verschieden zu sein. Es sei weiter η eine beliebige, positive Zahl. Dann gibt es $b = \left[\frac{2V}{C\eta}\right]$ verschiedene Mengen $M(\varepsilon)$, die so in B eingelagert werden können, ohne daß sie übereinandergreifen, so daß für ihr Gesamtvolumen Σ gilt

$$V(B) - \eta \leqq \Sigma \leqq V(B) \tag{5}$$

Dabei ist C in Satz 24 definiert.

Ist W ein Würfel von der Kantenlänge w, so gibt es nach Voraussetzung eine Menge $M(\frac{1}{4} w)$ aus $\{K\}$ mit dem Durchmesser $\varepsilon = \frac{1}{4} w$. Nach Satz 24 können dann, da $\beta(W) = \frac{1}{2} w$ mindestens soviele Körper aus $M(\frac{1}{4} w)$ eingelagert werden, so daß ihr Gesamtinhalt $\geqq C V(W)$ ist.

Ist B^* eine im *Jordan*schen Sinne meßbare Menge vom Inhalt $V(B^*)$, dann gibt es zu jedem $\delta > 0$ eine Menge achsenparalleler Würfel W_i mit der Kantenlänge $w(B^*, \delta)$, die nicht übereinandergreifen, ganz in B^* liegen und

$$\Sigma V(W_i) \leqq V(B^*) \leqq \Sigma V(W_i) + \delta$$

ist. In jedem W_i werden jetzt Körper aus M ($\frac{1}{2}w$) eingelagert. Dann gilt für das Gesamtvolumen S (B^*, δ) der eingelagerten Körper aus M ($\frac{1}{4}w$) nach obigem:

$$V(B^*) \geq S(B^*, \delta) \geqq C \Sigma V(W_i) \geqq C(V(B^*) - \delta)$$

Die Menge der eingelagerten Körper bezeichnen wir mit L (B^*. δ), Wir definieren nun $B_0 = B$, $L_0 = L(B_0, \delta)$

allgemein $B_{i+1} = B_i - L_i$, $Li = L(B_i, 2^{-i}\delta)$ $V(L_i) = S(B, 2^{-i}\delta)$.
Nach obigem ist

$$V(L_i) \geqq C(V(B_i) - 2^{-i}\delta) \tag{6}$$

Weiters ist

$$V(B_i) \leqq V(B)(1 - C)^i + C\delta \sum_{j=0}^{i-1} 2^{-j} \tag{7}$$

wie man durch vollständige Induktion zeigt.

Denn für $i = 1$ ist es richtig, allgemein

$$V(B_{i+1}) = V(B_i) - V(L_i) \leqq V(B_i)(1 - C) + 2^{-i}\delta C$$

und daraus folgt sofort (7).

Aus (7) folgt weiter

$$V(B_i) \leqq V(B)(1 + iC)^{-1} + 2C\delta$$

Setzen wir $i = b$, $\delta = \eta(4C)^{-1}$, so folgt $V(B_b) \leqq \eta$. Setzen wir $\sum_{j=i}^{b} L_j = L$, so gilt für $\Sigma = V(L)$ tatsächlich $V(B) - \eta \leqq \Sigma \leqq V(B)$.

Satz 25 kann verallgemeinert werden, wenn man die Erweiterungen von Satz 24 heranzieht.

§ 13. Gitterförmige Überdeckung konvexer Körper durch konvexe mit Mittelpunkt.

Es sei $K: f(x) \leqq r$ wieder ein konvexer Körper mit Mittelpunkt, B ein Bereich vom Inhalt $V(B)$.

Definition 23: Eine Punktmenge U heißt K (r)-Überdeckungskonfiguration von B, wenn jeder Punkt von B in einem Körper $K\,p$: $f(x - p) \leqq r$ liegt, wo $p \varepsilon U$ und jeder Körper K_p mit $p \varepsilon U$, mit B einen inneren Punkt gemeinsam hat. Die Anzahl werde mit $u(U)$ bezeichnet. Eine K (r)-Überdeckungskonfiguration U_0 heißt minimal, wenn $u(U_0) = \min u(U)$ über alle U. Wir setzen $u(U_0) = u(B, K(r))$, $\Sigma(B, K(r)) = u\,J\,r^n$.

Definition 24: Ist Γ ein Punktgitter in der Lage z, so heißt eine Punktmenge eine Γ-K (r)-Überdeckungskonfiguration, wenn alle Punkte von U zu Γ_z gehören und U eine K (r)-Überdeckungskonfiguration ist. Die Γ-Konfiguration U_0^* mit u $(U_0^*) = \min u$ (U) über alle Γ_z-Konfigurationen heißt minimale Γ-Überdeckungskonfiguration. Ihre Anzahl bezeichnen wir mit u^* $(B, \Gamma, K(r))$.

Definition 25: u^* $(B, K(r)) = \min u^*$ $(B, \Gamma, K(r))$ über alle Gitter

$$\Gamma, \Sigma^* (B, K(r)) = u^* J r^n$$

u^* nennen wir die Minimalzahl der Körper $K(r)$ die B in gitterförmiger Lagerung überdecken. Σ^* ist das Gesamtvolumen der überdeckenden Körper.

Satz 26: Es sei B konvex, $\beta(B)$, $\varrho(K)$ wie früher definiert, $\varrho < \frac{1}{2n^2}\beta$

Wieder sei O der Mittelpunkt einer Inkugel, $B_1 = (1 - 2n^2 \frac{\varrho}{\beta})$ B.

Dann ist

$$\operatorname{Max}(J \varepsilon^n V(B_1), V(B)) \leqq \Sigma^* (B, K(r)) \leqq J \varepsilon^n (V(B) + O(B, \varrho)) \quad (1)$$

Dabei ist ε (f) die absolute Exzentrizität nach Definition 2, § 3 O (B, ϱ) die Quermaßsummen nach Definition 7, § 5.

Beweis: Es sei U eine Γ-K (r)-Überdeckungskonfiguration. Dann muß nach Definition 1, § 3 ε $(\Gamma, f) \leqq r$ sein. Wir betrachten nun das Figurengitter $[\Gamma, P^*]$, wo $P^* = P_l \frac{r}{\varepsilon}$ ist. Dabei sei Γ auf seine ausgezeichneten Basisvektoren bezogen und P_l das zugehörige Parallelepiped. $[\Gamma, P^*]$ überdeckt wegen $r \geqq \varepsilon$, den R_n lückenlos, da dies
26 $[\Gamma_l, P]$ tut. P_l liegt ganz im Körper $f(x) \leqq c$, wo $c_1 \leqq \frac{1}{2} \sum_{i=1}^{n} N_i \leqq \frac{1}{2} n^2 M_n$ (man vgl. § 2) also P^* in $\overline{K}: f(x) \leqq c_2 = c_1 \frac{r}{\varepsilon}$. Nun ist nach Hilfssatz 7, § 3 $\varepsilon \leqq \frac{1}{2} M u$, also $c_2 \leqq n^2 r$.

Wir betrachten nun B_1. Es liegt $B_2 = B_1 + 2P^*$ in $B_3 = B_1 + 2\overline{K}$. Für die Stützfunktion H_3 von B gilt

$$H_3 \leqq H(1 - 2n^2 \frac{\varrho}{\beta}) + 2\overline{H}$$

27 wo H, $\overline{H}$ die Stützfunktionen von $\frac{3}{4}$, bzw. $\overline{K}$ sind.

Nun ist $\overline{H} \leqq \frac{1}{2} n^2 \varrho$, $H \geqq \beta$ also $H_3 \leqq H$.

Daraus folgt, daß B_2 in B liegt.

Die Parallelotope aus $[\Gamma, P^*]$, deren Mittelpunkte in B liegen, müssen also B_1 überdecken. Daraus folgt

$$u(U)\,D\,(\frac{r}{\varepsilon})^n \geqq V(B_1), \tag{2}$$

wo D das Volumen von P_l ist, also

$$u(U)\,J\,r^n \geqq D^{-1}\,J\,\varepsilon^n(\Gamma, f)\,V(B_1) \geqq J\,\varepsilon^n(f)\,V(B_1) \tag{3}$$

Dies gilt für jedes Γ, also

$$\Sigma^*(B, K(r)) \geqq J\,\varepsilon^n\,V(B_1). \tag{4}$$

Es soll nun Σ^* nach oben abgeschätzt werden.

Wir wählen ein Gitter Γ mit $\varepsilon(\Gamma, f) \leqq r$.

Dann überdeckt $[\Gamma, K]$ den R_n lückenlos. Nach § 4 (2) kann Γ in eine solche Lage gebracht werden, so daß die Anzahl der Körper, die B treffen $\leqq V(B + K)\,D^{-1}(\Gamma)$ ist. Diese müssen aber B überdecken, also ist

$$u(B, K(r)) \leqq V(B + K)\,D^{-1}(\Gamma)$$

Wir wählen nun für Γ ein minimales Gitter, nach § 3, Satz 1 existiert ein solches, so daß $D(\Gamma) = (\frac{r}{\varepsilon})^n$. Daraus folgt

$$\Sigma^* = u^*(B, K(r))\,J\,r^n \leqq V(B + K)\,J\,\varepsilon^n$$

Damit ist alles gezeigt.

Folgerung: $\vartheta^*(f) = \lim\limits_{r \to 0} \frac{\Sigma^*}{V(B)}$ existiert und ist $J\,\varepsilon^n(f)$.

Definition 25: ϑ^* heißt Dichte der Überdeckung bei gitterförmiger Lagerung.

Nach § 3, Satz 2 ist $\vartheta^* > 1$, wenn K kein $\frac{1}{2}\,M$-Körper ist, also sicher dann, wenn K kein Polyeder ist. Weiters ist nach denselben Satz $\vartheta^* \leqq n^n$. Wir können nach dem sup ϑ^* über alle konvexen Körper fragen, d. h. welche Körper liefern die dichtstee, minimale Überdeckung. Für $n = 2$ wurde von *Sas*[21] gezeigt, daß für die Kugel ϑ^* den größten Wert annimmt.

14. Dichte einer minimalen Überdeckung.

Es soll gezeigt werden, daß $\vartheta = \lim\limits_{r \to 0} \frac{\Sigma(B, K(r))}{V(B)}$ existiert und unabhängig von B ist. Wir nennen sie die Dichte der Überdeckung.

Wie sofort aus Definition 23 folgt, ist analog zu Hilfssatz 13, § 10

$$u(\lambda B, K(\lambda r)) = u(B, K(r)), \Sigma(\lambda B, K(\lambda r)) = \lambda^n\,\Sigma(B, K(r)) \tag{1}$$

[21] Compositio 6.

für $\lambda > 0$.

Hilfssatz 20: Es sei W der Würfel von Hilfssatz 14. Weiter seien λ, r_1, r_2 positive Zahlen mit $\lambda \leqq 1$, $r_1 \geqq r_2$ und $\varrho^* r_1 \leqq \frac{1}{4} \lambda w$, wo ϱ^* Umkugelradius von $f \leqq 1$ ist.

Dann gilt $$|\Sigma(\lambda W, K(r_1)) - \Sigma(W, K(r_2))| \leqq \overline{C}(\lambda)$$

wo $$\overline{C}(\lambda) = n(4\varrho^* r_1 + w)^{n-1}\left(\left(1 - \lambda \frac{r_2}{r^1}\right) w + 8\varrho^* r_2\right)\left(\frac{r_1}{r_2}\right)^n$$

Beweis: Wir setzen $\mu = \lambda \dfrac{r_2}{r_1}$, dann ist $\mu \leqq 1$.

Es sei U eine $K(r_2)$-Überdeckungskonfiguration von W. Es sei U' die Menge der Punkte von U, für welche die zugehörigen K_p ganz in $W_2 = \mu W$ liegen, ihr Gesamtvolumen sei Σ'. Es ist $\Sigma' \leqq \Sigma(W_2, K(r_2)) = \Sigma_2$. Die Körper K_p für die p zu $U - U' = U''$ gehört, überdecken $W - W_2$. Das Gesamtvolumen sei Σ''. Die K_p mit $p \varepsilon U''$ liegen nach § 10, Hilfssatz 12, ganz in $W_3 = W\left(1 + \dfrac{4\varrho_2}{w}\right)$ und haben mit $W_4 = W_2$

28 $\left(1 - \dfrac{4\varrho}{\mu n}\right)$ keinen Punkt gemeinsam. Es ist also

$$\Sigma'' \leqq V(W_3) - V(W_4) \leqq n(w + 4\varrho_2)^{n-1}((1-\mu) w + 8\varrho_2) \leqq \overline{C}(\lambda)$$

Daraus folgt, da $\Sigma' + \Sigma'' = \Sigma'''$ das Gesamtvolumen der K_p mit $p \varepsilon U$ ist $\Sigma''' \leqq \Sigma_2 + \overline{G}(\lambda)$. Dies gilt für jedes U, also auch

$$\Sigma_1 = \Sigma(W, K(r_2)) \leqq \Sigma_2 + \overline{C}(\lambda).$$

Setzen wir $\Sigma_3 = \Sigma(\lambda W, K(r_1))$, so folgt

$$\Sigma_3 - \Sigma_1 = \left(\frac{r_1}{r_2}\right)^n \Sigma_2 - \Sigma_1 \geqq \Sigma_2 - \Sigma_1 \geqq -\overline{C}$$

Weiter ist $\Sigma_2 \leqq \Sigma_1 \leqq V(W_3)$, also

$$\Sigma_3 - \Sigma_1 \leqq \left(\left(\frac{r_1}{r_2}\right)^n - 1\right)\Sigma_1 \leqq \overline{C} \text{ w.z.b.w.}$$

Hilfssatz 21: Für jedes $r > 0, < 1$ mit $\varrho(K(r)) < \frac{1}{4} w$ und für jede natürliche Zahl g ist

$$|\Sigma(W, K(r g^{-1})) - \Sigma(W, K(r))| \leqq C r \qquad (2)$$

wo C nur von W und f abhängt.

Beweis: Es ist $u(W, r g^{-1}) = u(Wg, r)$. Weiter ist $u(Wg, r) \leqq g^n u(W_j r)$, also $\Sigma(W, K(r g^{-1})) \leqq \Sigma(W, K(r))$.

Es sei U eine $K(r)$-Überdeckung von $\overline{W} = Wg$. Wir teilen $\overline{W}$ in g^n achsenparallele Würfel W_j mit der Kantenlänge w. Mit U_j bezeichnen wir die Punkte von U, die in W_j liegen. Dann ist also U_j eine Überdeckung des Würfels W_j, konzentrisch zu W_j mit der Kantenlänge $w - 4\,\varrho$. Setzen wir also $W_4 = W\left(1 - \frac{4\,\varrho}{w}\right)$, dann ist

$$u\,(U_j) \geqq u\,(W_4, K\,(r)) = \bar{u},$$

also $u\,(U) \geqq g^n\,\bar{u}$. Dies gilt für jedes U, also $u\,(\overline{W}, K\,(r)) \geqq g^n\,\bar{u}$. Setzen wir $\Sigma_4 = \Sigma\,(W_4, K\,(r))$, so ist also

$$\Sigma\,(W, K\,(r\,g^{-1})) \geqq \Sigma_4$$

Weiter ist nach Hilfssatz 20, mit $\lambda = 1 - \frac{4\,\varrho}{w}$, $r_1 = r_2 = r$, da $r < 1$

$$\Sigma_4 \geqq \Sigma - C\,r$$

9 wo $C = 12\,n\,\varrho^*\,(w + 4\,\varrho^*)^{n-1}$ w. z. b. w.

Hilfssatz 22: Ist $\Sigma\,(r) = \Sigma\,(W, K\,(r))$, so existiert

$$\vartheta = \lim_{r \to 0} \frac{\Sigma\,(r)}{w^n} \tag{3}$$

und ist unabhängig von w.

Der Beweis ist analog zum Beweis des Hilfssatzes 16, § 10, wenn man Hilfssatz 20 und 21 heranzieht.

Satz 27: Es sei B ein im *Jordan*schen Sinne meßbarer Bereich mit $0 < V\,(B) < \infty$, $\Sigma = \Sigma\,(B, K\,(r))$ das Minimum der Gesamtinhalte der konvexen Körper mit Mittelpunkt $f\,(x - p) \leqq r$, die B überdecken, dann existiert

$$\vartheta = \lim_{r \to 0} \frac{\Sigma}{V}$$

und ist von B unabhängig. ϑ heißt Dichte der Überdeckung durch konvexe Körper mit der Distanzfunktion f.

Der Beweis erfolgt analog eu Satz 18. Es ist nur der Unterschied, daß man statt den A_i, die U_i zu nehmen hat, wo die U_i die Menge der Punkte einer Überdeckung U sind, welche in den W_i liegen. Diese bilden dann nach Hilfssatz 12 eine Überdeckung der W_i' mit der Kantenlänge $w - 4\,\varrho$, welche konzentrisch zu W_i liegen.

Man zeigt analog

Satz 28: Es sei B ein Bereich wie in Satz 27. Es sei $\Sigma_\alpha = \Sigma_\alpha\,(B, K\,(r))$ das Infimum der Gesamtinhalte der konvexen Körper mit Mittelpunkt $f\,(x - p) \leqq \bar{r}$, $\alpha\,r \leqq \bar{r} \leqq r$, $(\alpha > 0)$ die B überdecken, dann existiert

$$\lim_{r \to 0} \frac{\Sigma\, \alpha}{V(B)} = \vartheta_\alpha \tag{4}$$

und ist von B unabhängig. ϑ_α heißt wieder Dichte der Überdeckung.

Es sei noch bemerkt, daß man in Satz 27, die Voraussetzung, daß die Körper einen Mittelpunkt besitzen, fallen lassen kann.

§ 15. Überdeckung konvexer Körper durch konvexe mit Mittelpunkt.

Es soll jetzt $\Sigma(B, K(r))$ von Satz 27, bzw. gleich allgemein $\Sigma_\alpha = \Sigma_\alpha (B, K(r))$ von Satz 28 für konvexe Körper B nach unten abgeschätzt werden.

Die Abschätzung nach oben haben wir nach Definition 23 und 25

$$\Sigma \leqq \Sigma^*. \tag{1}$$

Satz 29: Es sei B ein konvexer Körper mit $V(B) > 0$. Es sei der Ursprung 0 Mittelpunkt einer Inkugel von B mit Radius β, ϱ der Umkugelradius von K, $\varrho < \beta$, $B_1 = (1 - \frac{\varrho}{\beta})\, B$, $V_1 = V(B_1)$ $0 < \lambda < 1$ beliebig, dann ist

$$\Sigma_\alpha (B, K(r)) \geqq \operatorname{Max}(V, V_1(1 + C)) \tag{2}$$

30 $$C = \frac{\alpha^n}{4} \operatorname{Min}\left(\frac{2^{1-e}}{(2n)^n}\left(1 - \frac{S_1}{V_1}\right)^{n+e}, \left(\frac{1}{\lambda} - 1\right)^n \left(1 - \frac{S_1}{V_1}\right)\right) \quad e = 1,$$

wenn n gerade, 0 sonst,

$$S_1 = S_{\lambda\,\alpha}(B, K(r)).$$

Dabei ist $S_{\lambda\alpha}$ in Satz 15 definiert (§ 7).

Weiter ist

$$\vartheta_\alpha \geqq 1 + \tfrac{1}{4}\left(\frac{\alpha}{2\,n}\right)^n C_1 \tag{3}$$

wo

$$C_1 = \left(\frac{1}{\lambda} - 1\right)^n (1 - \Delta_{\alpha\,\lambda})^{n+1}, \; (\tfrac{1}{2} \leqq \lambda < 1).$$

Beweis: Die Körper K_p der Überdeckung von B, die B_1 überdecken, liegen ganz in B. Dies zeigt man genau so, wie in Hilfssatz 10. Es werden jetzt in folgendem nur diese Körper $K_i : f(x - p_i) \leqq r_i$ mit $\alpha\, r \leqq r_i \leqq r$ in Betracht gezogen, die B_1 überdecken. Es sei v die Anzahl dieser Körper $\Sigma_1 = \sum^{v} J\, r_i^n$. Dann ist $\Sigma_\alpha \geqq \Sigma_1$.

Wir bilden nun

$$\operatorname*{Min}_{j \neq i} (r_i + r_j)^{-1} f(p_i - p_j) = \lambda_i, \; \hat{r}_i = \lambda_i r_i \tag{4}$$

also
$$f(p_i - p_j) \geqq \lambda_i (r_i + r_j) \; (i \neq j)$$

Es wird in Hilfssatz 23 gezeigt werden, daß $\lambda_j \leqq 1$ also $\hat{r}_i \leqq r_i$.

Es sei $\overline{K}_i$ jener Körper, für welchen

$$f(p_i - \overline{p}_i) = \lambda_i (r_i + \overline{r}_i), \; q_i = (r_i + \overline{r}_i)^{-1} (p_i \overline{r}_i + \overline{p}_i r_i) \tag{5}$$

Dann definieren wir auf K_i eine Funktion $\Phi_i(x)$ so:

$$\Phi_i(x) = 1/2, \text{ wenn } f(x - q_i) \leqq \alpha (r_i - \hat{r}_i) \tag{6}$$
$$1, \text{ sonst in } K_i$$

Wir definieren weiter

$$\Phi(x) = \sum_{i=1}^{\nu} \Phi_i(x) \tag{7}$$

31 Dann ist in $B: \Phi(x) \geqq 1$. Denn jedes $x \varepsilon B_1$ liegt in einem K_i. Liegt es sogar in $f(x - q_i) \leqq \alpha (r_i - \hat{r}_i)$, dann liegt es auch in $\overline{K}_i'$, denn

$$f(x - \overline{p}_i) \leqq f(x - q_i) + f(q_i - \overline{p}_i) \leqq \alpha r_i (1 - \lambda_i) + \lambda_i \overline{r}_i \leqq \overline{r}_i \tag{8}$$

Es ist also

$$J \sum_{i=1}^{\nu} [r_i^n - \frac{\alpha^n}{2} (r_i - \hat{r}_i)^n] = \int_{B_1} \Phi(x) \, d\, x \geqq V(B_1) = V_1 \tag{9}$$

Es seien die K_i $(i \leqq m)$ jene Körper, für die $\lambda_i \leqq \lambda$, also $\hat{r}_i \leqq \lambda r_i$. Für die K_i mit $m > i$ ist dann $\lambda_i > \lambda$.

Aus (9) folgt

$$\Sigma_1 \geqq V_1 + 1/2\, \alpha^n J \sum_{i=1}^{\nu} (r_i - \hat{r}_i)^n$$

Wir unterscheiden nun zwei Fälle:

1.
$$\Sigma_2 = J \sum_{i=1}^{m} \hat{r}_i^{\,n} \geqq k V_1, \; k = 1/2 \left(1 - \frac{S_1}{V_1}\right) \tag{10}$$

Dann ist

$$\Sigma_1 \geqq V_1 + \sum_{i=1}^{m} (r_i - \hat{r}_i)^n \geqq V_1 [1 + 1/2\, \alpha^n (\frac{1}{\lambda} - 1)^n k] \geqq V_1 (1 + C) \tag{11}$$

2.
$$\Sigma_2 < k V_1. \tag{12}$$

Es ist $(\sum_i^{\nu} (r_i^n - \hat{r}_i^{\,n}))^n \leqq n^n (\Sigma r_i^n)^{n-1} (\Sigma (r_i - \hat{r}_i)^n)$,

also folgt aus (9)

32 $$\Sigma_1 \geqq V_1 + \frac{\alpha^n}{2n^n \alpha^n} \Sigma_1^{1-n} (\Sigma_1 - \Sigma_3)^n \geqq V_1 [1 + \tfrac{1}{2} (\frac{\alpha}{n})^n (1 - \Sigma_3 V_1^{-1})^{n+e}] \tag{13}$$

Dabei ist $\Sigma_3 = \overset{\nu}{\Sigma} J \hat{r}_i^n = \Sigma_2 + \Sigma_4$, wo $\Sigma_4 = \underset{m+1}{\overset{\nu}{\Sigma}} J \hat{r}_i^n$. Für die K_i, die zu Σ_4 beitragen, gilt nun folgendes: Wir betrachten die Körper $\hat{K}_i : f(x - p_i) \leqq \hat{r}_i$. Diese greifen nicht übereinander, da $f(p_i - p_j) \geqq \lambda_i (r_i + r_j) \geqq \hat{r}_i + \hat{r}_j$ und liegen ganz in B, da dies die K_i tun.

Weiter ist
$$\alpha \lambda r \leqq \hat{r}_i = \lambda_i r_i \leqq r. \tag{14}$$

Also ist nach Definition $\Sigma_4 \leqq S_{\lambda\alpha}(B, K) = S_1$, daher $\Sigma_3 \leqq k V_1 + S_1 = \tfrac{1}{2}(V_1 + S_1)$ und daraus folgt wieder $\Sigma_1 \geqq V_1 (1 + C)$ w.z.b.w.

Hilfssatz 23: Es ist $\lambda_i \leqq 1$.

Beweis: Der Körper $K_i' : f(x - p_i) \leqq r_i (1 + \varepsilon)$, wo $\varepsilon > 0$ genügend klein, hat mit B_1 sicher einen Punkt x gemeinsam, da dies K_1 tut. Es gibt nun in B_1 einen Punkt y, der im Innern liegt und nicht zu K_i' gehört, denn sonst wäre

$$\chi_n (\varrho (1+\varepsilon))^n \geqq J (r (1+\varepsilon))^n \geqq (1 - \frac{\varrho}{\beta})^n V(B) \geqq \chi_n (1 - \frac{\varrho}{\beta})^n \beta^n,$$

was für genügend kleines ε auf einen Widerspruch führt. Die Verbindungsstrecke x, y wird den Rand von K_i' in einen Punkt $\bar{x}$ treffen, der ebenfalls in B_1 liegt. Es ist also $f(\bar{x} - p_i) = r_i (1 + \varepsilon)$. Es liegt aber nicht in K_i. Da aber B_1 von den K_j überdeckt wird, so muß es ein j geben, so daß $f(\bar{x} - p_j) \leqq r_j$. Daraus folgt $f(p_i - p_j) \leqq r_i (1 + \varepsilon) + r_j$, daher $\lambda_i \leqq (r_i + r_j)^{-1} f(p_i - p_j) \leqq 1 + \varepsilon$. Dies gilt für jedes $\varepsilon > 0$, also $\lambda_i \leqq 1$. w.z.b.w.

Es sei $l = \alpha^{-1}$. $L = (2 l^2 + 1)^n$, $\nu = L - [L]$ und τ_l sei wie in Satz 15 definiert. Dann gilt

Satz 30: Es ist
$$\vartheta_\alpha \geqq 1 + \tfrac{1}{2} (\frac{\alpha}{2n} (\frac{1}{\lambda} - 1)(1 - \frac{1}{\tau_l}))^n \tag{15}$$

für jedes $\lambda \geqq 1 - \frac{1}{4 n^2 L}(1 - \nu)$. Es ist also $\vartheta_\alpha > 1$, wenn K kein Polyeder ist. Weiter ist $\lim\limits_{\alpha \to 0} \vartheta_\alpha = 1$.

Beweis: Wir haben nach Satz 29 nur $\Delta_{\lambda\alpha}$ abzuschätzen. Es ist
33 nach § 10, Satz 20, $\Delta_{\lambda\alpha} \leqq \tau_r^{-1}$, wo $r = l\,\alpha^{-1}$.

Nun ist nach Satz 15, § 7

$$\tau_r = \mathrm{Min}\,(\omega\,(f,\,\delta_r),\ \varphi_s\,(f))$$

34 Dabei ist $s^n = [(2\,r + 1)^n]$. Man rechnet leicht nach[15], daß $[L] \leqq s^n < [L] + 1$, also $s = h$, wenn $h^n = [L]$, daher ist $\tau_r = \tau_l$, da $\delta_s < \delta_l$ ein passendes δ_l ist. Die letzte Behauptung wird sich aus § 16, Satz 32, ergeben.

Benützt man *Rankin*[1], Satz 3, so kann man für Kugeln ein etwas besseres Resultat erzielen. Man kann die Abschätzungen von ϑ verbessern, wenn man Polyeder heranzieht, die K eingeschrieben sind. Doch soll darauf nicht weiter eingegangen werden. Wir haben bis jetzt vorausgesetzt, daß K einen Mittelpunkt besitzt. Wir wollen diese Voraussetzung jetzt fallen lassen. Es sei $\hat{K}$, der nach Definition 21 K zugeordnete Körper. Dabei sei o der Schwerpunkt von K. Dann gilt, wenn V_1 wie im Satz 29 definiert ist:

Satz 29':

$$\Sigma\,(B,\,K\,(r)) \geqq \mathrm{Max}\,(V,\ V_1\,(1 + C)) \tag{16}$$

wo

$$C = \tfrac{1}{4}\,\beta^n\,q\ \mathrm{Min}\left(\frac{2^{1-e}}{(2\,n)^n}\Bigl(1 - \frac{S_1}{2\,V_1}\Bigr)^{n+e},\ \Bigl(\frac{1}{\lambda} - 1\Bigr)^n\Bigl(1 - \frac{S_1}{2\,V_1}\Bigr)\right)$$

$$q = \frac{V\,(\hat{K})}{V\,(K)},\ \beta = \frac{2}{n+1}\,S_1 = S_\lambda\,(B,\,K\,(r)),\ \ 0 < \lambda < 1.$$

Weiter ist

$$\vartheta \geqq 1 + \tfrac{1}{4}\,\Bigl(\frac{\beta}{2\,n}\Bigr)^n\,C_1 \tag{17}$$

wo

35 $$C_1 = \Bigl(\frac{1}{\lambda} - 1\Bigr)^n\,\Bigl(1 - \frac{\Delta\,\lambda}{g}\Bigr)^{n+1} \quad (\lambda \geqq \tfrac{1}{2})$$

Daraus folgt, da nach Hilfssatz 19, $q > 1$ ist, wenn K keinen Mittelpunkt besitzt: Es ist stets $\vartheta > 1$, wenn K keinen Mittelpunkt besitzt. Der Beweis verläuft weitgehend analog zu Satz 29. Die Körper K_p $(x - p)\,u \leqq H\,(u)$ der Überdeckung von B, die B_1 überdecken, liegen in B. Es sollen wieder nur solche in Betracht gezogen werden, $K_i: (x - p_i)\,u \leqq H\,(u)\ (i = 1, \ldots v)$, $\Sigma_1 = \sum^{v} V\,(K)$. Dann ist $\Sigma \geqq \Sigma_1$.

Es habe $\hat{K}: xu \leqq \frac{1}{2}(H(u) + H(-u))$ die Distanzfunktion $f(x)$, also $\hat{K}: f(x) \leqq r$.

Dann bildet man

$$\operatorname*{Min}_{j \neq i} \frac{1}{2r} f(p_i - p_j) = \lambda_i, \hat{r}_i = \lambda_{ir} \tag{18}$$

Dann ist wieder analog zu Hilfssatz 23 $\lambda_i \leqq 1$, denn $K_i': (x - p_i) u \leqq H(u)(1 + \varepsilon)$ hat mit B_1 sicher einen Punkt gemeinsam. Dann gibt es wieder ein $\bar{x}$ und ein p_i, so daß

$$(\bar{x} - p_i) u \leqq H(u)(1 + \varepsilon), (\bar{x} - p_j) u \leqq H(u),$$

also $(p_j - p_i) u \leqq H(u)(1 + \varepsilon) + H(-u)$, $f(p_i - p_j) \leqq 2r + \varepsilon_1$, also $\lambda_i \leqq 1$. Es sei nun $\overline{K}_i$ jener Körper, für welchen $f(p_i - \bar{p}_i) = 2r\lambda_i$, daher $(p_i - \bar{p}_i) u \leqq \lambda_i (H(u) + H(-u))$. Dann gibt es einen Punkt q_i, so daß

$$(q_i - p_i) u \leqq \lambda_i H(u), (q_i - \bar{p}_i) u \leqq \lambda_i H(u)$$

wie man sich analog zu Hilfssatz 17 überlegt.

Der Körper $f(x - q_i) < \beta(r - \hat{r}_i)$, d. h. $(x - q_i) u < \frac{1}{n+1} (H(u) + H(u))(1 - \lambda_i)$ liegt sowohl in K_i wie in $\overline{K}_i$, denn es ist zum Beispiel, für alle diese x $(x - p_i) u = (x - q_i) u + (q_i - p_i) u \leqq H(u)$, denn nach *Bonnesen-Frenchel* 34, S. 53 ist $H(-u) \leqq n H(u)$.

Nun definieren wir wieder auf K_i, $\Phi_i(x)$ mit

$$\Phi_i(x) = \begin{matrix} \frac{1}{2} & f(x - q_i) \leqq \beta(r - \hat{r}_i) \\ 1 & \text{sonst in } K_i \end{matrix}$$

und $\Phi(x) = \Sigma \Phi_i(x)$. Dann ist $\Phi(x) \geqq 1$ in B_1, und

$$\sum_i [V(K) - \tfrac{1}{2}\beta^n J(r - \hat{r}_i)^n] \geqq V_1$$

wo J das Volumen von $f \leqq 1$. Es ist daher

$$\Sigma_1 \geqq V_1 + \tfrac{1}{2}\beta^n \Sigma J (r - \hat{r}_i)^n.$$

Jetzt geht der Beweis wie bei Satz 29 zu Ende. Es ist nur zu beachten, daß jetzt $\Sigma J r^n = q\Sigma$ und für k $\frac{1}{2}(q - \frac{S_1}{V_1})$ zu nehmen ist.

§ 16. Überdeckungen durch eine Menge $\{K\}$.

Es sei wie in § 12 $\{K\}$ eine Menge von konvexen Körpern und B ein konvexer Bereich. Es mögen dieselben Bezeichnungen und Voraussetzungen wie dort gelten. Dann sei Σ der Gesamtinhalt der Körper, die aus Körpern von $\{K\}$ durch Translation hervorgehen und B überdecken. Weiter sei $\Sigma\,(B, \{K\}) =$ Inf. aller dieser Σ. Dann gilt[22]

Satz 31:
$$\Sigma\,(B, \{K\}) \leqq n^n\,(V\,(B) + O\,(B, \beta \sqrt{\tfrac{n}{2n+2}})) \tag{1}$$

Dabei ist $O\,(B, t)$ die Quermaßsumme nach Definition 7. Der Beweis ist analog wie bei Satz 24. Es ist nur zu beachten, daß nach § 15 (1), Satz 26, für einen Körper

$$\Sigma\,(B, K) \leqq J\,\varepsilon^n\,(V\,(B) + O\,(B, \varrho)) \tag{2}$$

und nach Satz 2, § 3, $J\,\varepsilon^n \leqq n^n$ ist

Erweiterungen: 1. Setzt man nicht voraus, daß alle Körper aus $\{K\}$ einen Mittelpunkt besitzen, so ist in Satz 31, der Ausdruck rechts mit $\frac{1}{2}\,(n+1)$ zu multiplizieren.

Dies folgt aus folgenden

Hilfssatz (*Bonnesen-Fenchel* 34, Seite 53):

Es sei K ein konvexer Körper und sein Schwerpunkt liege in O. Dann enthält er den Körper $\frac{2}{n+1}\,\hat{K}$, wo $\hat{K}$ der Körper aus Definition 21 ist.

2. Verlangt man, daß B mindestens l^n-fach bedeckt wird ($l \geqq 1$), so hat man noch mit l^n zu multiplizieren. Man geht bei der Abschätzung wie in Hilfssatz 7 und 8 bei ε vor (§ 3).

Satz 32: Es sei B im *Jordan*schen Sinne meßbar mit $V\,(B) > O$. Weiters sei $\{K\}$ eine Menge von konvexen Körpern, welche die Voraussetzungen von Satz 25 erfüllen. Dann gibt es eine Konstante C_n, die nur von n abhängt mit folgender Eigenschaft:

Ist $\eta_1 > 0$ eine beliebige Zahl, $\eta = \frac{1}{2}\,\eta_1\,C_n^{-1}$, so gibt es $b + 1$ Mengen M_ε, welche B so überdecken, daß ihr Gesamtinhalt

$$\Sigma_1 \leqq V\,(B) + \eta_1 \tag{3}$$

Dabei hat b dieselbe Bedeutung wie in Satz 25 (§ 12).

Beweis: Nach Satz 25 können b-Mengen M_ε so in B eingelagert werden, daß sie nicht übereinandergreifen und für das Volumen Σ dieser Menge L gilt

36
$$V\,(B) - \eta \leqq \mathrm{T} \leqq V\,(B)$$

also $V\,(B - L) \leqq \eta$

[22] Für $n = 2$ wurde für kleine r ein ähnlicher Satz von *Fejes* aufgestellt, vgl. die erste Arbeit in

Nun ist $B - L$ meßbar, kann also bei einem $\delta > 0$ so durch eine Menge von achsenparallelen Würfeln von einer Kantenlänge w überdeckt werden, daß

$$\Sigma\, V(W_i) \leqq V(B - L) + \delta$$

Nach Satz 31 kann jedes W_i durch Körper aus einer Menge M_ε so überdeckt werden, so daß

$$\Sigma(W_i, M_\varepsilon) \leqq n^n \left(V(W_i) + O\left(W_i, \beta\sqrt{\tfrac{n}{2n+2}}\right)\right) = C_n . w^n \quad (4)$$

Es wird daher $B - L$, so durch Körper aus M_ε überdeckt, daß

$$\Sigma_2 = \Sigma(B - L, M_\varepsilon) \leqq C_n \Sigma\, V(W_i) \leqq C_n(\eta + \delta) = \eta_1,$$

wenn $\delta = \eta$ genommen wird. Die Menge $L + M_\varepsilon = L_1$ überdeckt B und es gilt für $\Sigma_1 = V(L_1)$

$$\Sigma_1 = \Sigma + \Sigma_2 \leqq V(B) + \eta_1 \text{ w.z.b.w.}$$

Es soll noch folgende Ergänzung zu Satz 29 bewiesen werden: Es ist stets

$$\vartheta_\alpha \leqq \Delta_\alpha + C_n(1 - \Delta_\alpha) \quad (5)$$

Beweis: Es liege eine Ausfüllung von B durch Körper $f(x - p) \leqq r'$ $\alpha r \leqq r' \leqq r$ $(\alpha \leqq 1)$ vor. Es sei L die Menge von B, die dabei überdeckt wird. $B - L$ kann nun durch Körper $f(x - p) \leqq r$ so überdeckt werden, daß das Gesamtvolumen $\Sigma(B - L)$ dieser Körper $\leqq C_n V(B\text{-}L)$ ist. Dies zeigt man wie in Satz 32. Damit ist B überdeckt und da $V(B - L) = V(B) - V(L)$, so ist

$$\Sigma_\alpha(B, K) \leqq V(L) + C_n(V(B) - V(L))$$

also auch

$$\vartheta_\alpha \leqq \Delta_\alpha + C_n(1 - \Delta_\alpha) \text{ w.z.b.w.}$$

Daraus folgt jetzt:

Ist ein konvexer Körper K mit Mittelpunkt so beschaffen, daß $\Delta_\alpha = 1$ ist, dann ist auch $\vartheta_\alpha = 1$ und umgekehrt.

Beweis: Ist $\Delta_\alpha = 1$, so folgt sofort aus (5), daß $\vartheta_\alpha \leqq 1$ also $\vartheta_\alpha = 1$. Es sei nun $\vartheta_\alpha = 1$. Dann folgt aus Satz 29, daß $\Delta_{\alpha\lambda} = 1$ für $\frac{1}{2} \leqq \lambda < 1$. Aber es ist $\Delta_{\alpha\lambda} \leqq \lambda^{-n} \Delta_\alpha$.

Dies sieht man so ein: Es liege eine Ausfüllung von B durch Körper $K_i : f(x - p_i) \leqq r_i$ und $\lambda \alpha r \leqq r_i \leqq r$ vor. Jeder der Körper K_i für die $\lambda r \leqq r_i \leqq r$ ist, ersetze man durch $K_i' : f(x - p_i) \leqq \lambda r$, die anderen bleiben ungeändert. Dies stellt wieder eine Ausfüllung von B dar, mit $\lambda \alpha r \leqq \bar{r}_i \leqq \lambda r$.

Das Gesamtvolumen der eingelagerten Körper der ursprünglichen Ausfüllung Σ können wir so schreiben: $\Sigma = \Sigma_1 + \Sigma_2$, wo $\Sigma_1 = \sum_{i=1}^{t} J\, r_i^n$ mit $\lambda r \leqq r_i \leqq r$. Es ist nun $\Sigma_1 \leqq \lambda^{-n} \sum_{i=1}^{t} J\, (\lambda r)^n$, daher

$$S_{\lambda\alpha}(B, K(r)) \leqq \lambda^{-n} S_\alpha (B, K(\lambda r)), \text{ also } \Delta_{\alpha\lambda} \leqq \lambda^{-n} \Delta_\alpha.$$

Da nun $\Delta_{\alpha\lambda} = 1$, so folgt $\Delta_\alpha \geqq \lambda^n$ für jedes $\lambda \geqq 1/2$, daher $\Delta_\alpha \geqq 1$, d. h. $\Delta_\alpha = 1$.

§ 17. Zwei Sätze über Konfigurationen.

Es sollen einfache Eigenschaften bei Ausfüllungen und Überdeckungen bewiesen werden.

Es liege eine maximale Ausfüllung eines konvexen Körpers B durch Körper K vor, mit $\varrho < \frac{1}{3}\beta$. Dabei seien ϱ und β wie vorher definiert. Es sei $K: f(x - p) \leqq r$ irgendein Körper dieser Ausfüllung von B,
37 der mit $B_3\ (1 - \frac{3\varrho}{\beta})\ B$ Punkte gemeinsam hat. Dann liegt K_3: $f(x - p) \leqq 3r$ ganz in B. Dann gilt

Satz 33:

Die Menge der Mittelpunkte p_i aller Körper K_i der Ausfüllung, die mit K_3 innere Punkte gemeinsam haben, liegt in keiner Ebene (R^{n-1}).

Beweis: Angenommen der Satz wäre falsch. Dann liegen die p_i in einer Ebene durch p. Wir betrachten den Streifen ε der beiden Stützebenen an K, welche parallel zur Ebene sind. Dann liegen alle diese K_i in ε. Es gibt nun sicher Punkte von $f(x - p) = 2r$, welche nicht in ε liegen. Es sei $\bar{p}$ ein solcher. Dann hat $K': f(x - \bar{p}) \leqq r$ mit ε keinen inneren Punkt gemeinsam, da sonst $f(p - \bar{p}) < 2r$ wäre. Weiters liegt K' in K_3, da für jedes x in $K': f(x - p) \leqq f(x - \bar{p}) + f(p - \bar{p}) \leqq 3r$ K' hat aber mit keinem Körper $\bar{K}$ der Ausfüllung innere Punkte gemeinsam, denn sonst wäre $\bar{K}$ ein K_i, was nicht geht. Dies ist ein Widerspruch zur Annahme, daß die Ausfüllung maximal ist.

Es liege nun eine Überdeckung von B durch Körper K vor. Es sei $B_2 = (1 - \frac{2\varrho}{\beta})\ B$. Es sei $K: f(x - p) \leqq r$ ein Körper der Überdeckung, der mit B_2 Punkte gemeinsam hat ($\varrho < 1/2\,\beta$).

Dann gilt

Satz 34: Die Menge der Mittelpunkte p_i der K_i der Überdeckung, die mit K Punkte gemeinsam haben, liegen in keiner Ebene.

Beweis: Es genügt zu zeigen, daß dies richtig ist für die p_i, deren
38 K_i mit $\overline{K}: f(x-p) \leqq r(1+\varepsilon)$, $(\varrho > 0)$ Punkte gemeinsam haben, nicht in einer Ebene liegen.

Angenommen, dies wäre falsch. Die p_i würden also in einer Ebene durch p liegen. Dann betrachten wir den Streifen ε, gebildet wie im Beweis von Satz 33. Die K_i würden wieder alle im Streifen liegen. Nun gibt es sicher innere Punkte von $\overline{K}$, die nicht in ε liegen. Nun liegt $\overline{K}$ in B, daher gäbe es Punkte von B, die nicht überdeckt werden. Das ist ein Widerspruch.

§ 18. Weitere Sätze über Überdeckungen.

Es sei B ein im *Jordan*schen Sinne meßbarer Bereich mit $V(B) = V$. Er werde überdeckt von konvexen Körpern, die aus einem konvexen Körper $K: F(x) \leqq 1$ mit der Stützfunktion $H(u)$ durch Parallelverschiebung hervorgehen. Es sei $\hat{K}$ der Körper von Definition 21. Dabei ist vorausgesetzt, daß o in K liegt. Es sei $\hat{K}: f(x) \leqq r$ und J das Volumen des Eichkörpers $f(x) \leqq 1$.

Satz 35: Ist $M(f)$ das absolute Minimum, dann gibt es in der Überdeckung von B soviele Körper, die nicht übereinandergreifen, so daß ihr Gesamtinhalt

$$S \geqq \operatorname{Max}\left(\frac{1}{4^n} j\, M^n \frac{V(K)}{V(\hat{K})}, \frac{V(K)}{V(K+2\hat{K})} V(B)\right) \qquad (1)$$

ist.

Für K-Würfel wurde dies von *Sokolin*[23] gezeigt.

Beweis: Es sei $\overline{\Gamma}$ das maximale Gitter mit $D(\overline{\Gamma}) = (4\, r\, M^{-1})^n$ (vgl. zur Definition § 2). Es ist also $M_1(\overline{\Gamma}, f) = 4\, r$. Dann kann man nach § 4 (1) $\overline{\Gamma}$ in eine solche Lage z bringen, so daß B mindestens $V \cdot D^{-1}$ Gitterpunkte enthält. Jeder dieser Gitterpunkte p muß in einen K_α: $(p-a)\, u \leqq H(u)$ liegen. Kein K kann zwei dieser Gitterpunkte enthalten, sonst wäre $(p'-p'')\, u \leqq H(u) + H(-u)$, d. h. $f(p'-p'') \leqq 2\, r$, was nicht geht. Also liegt jeder Gitterpunkt nur in einem solchen K_a. Diese K_α können nicht übereinandergreifen, denn dann wäre

[23] C. R. Acad. ScURSS BNr. 20 (1940).

$$(b - a)\, u < H(u) + H(-u) \text{ also } (p_a - p_b)\, u < 2\,(H(u) + H(-u))$$

also $f(p_a - p_b) < 4\,r$, was nicht geht.

Es ist daher

$$S \geqq V . D^{-1} V(K) = \tfrac{1}{4} J M^n \frac{V(K)}{V(\hat{K})} V(B) \qquad (2)$$

Der weitere Teil der Behauptung ist enthalten in den allgemeineren

Satz 36: Es sei B überdeckt von den Körpern $K_j : F(x - p_j) \leqq t_j$. Dann gilt für das Supremum $\overline{S}$ der Gesamtvolumina der K_i, die nicht übereinandergreifen

$$\overline{S} \geqq C\, V(B),\, C = \frac{V(K)}{V(K + 2\,\hat{K})} \qquad (3)$$

Besitzt K einen Mittelpunkt, so ist $K = \hat{K}$, daher $\overline{S} \geqq 3^{-n}\, V(B)$, sonst $\geqq \gamma_n^{-1}\, V(B)$, wo $\gamma_n = \sum_{i=0}^{n} \binom{n}{i}\, 2^{n-1}\, {}_n\mathrm{Min}\,(i, n - i)$.

Der Beweis ist *Zygmund-Marcinkiewicz*[24] nachgebildet.

Beweis: Es sei $\tau_0 = \sup t_j$. Ist $\tau_0 = \infty$, so ist nichts zu beweisen. Es sei also $\tau_0 < \infty$. Dann gibt es ein t_0 mit $t_0 > \tau_0\,(1 - \varepsilon)$, $(\varepsilon > 0)$. Es sei nun L_0 die Menge aller K_j, $L_1' \subset L_0$ die Menge der K_j, welche mit K_0 übereinandergreifen, L_1 die übrigen. Es sei allgemein $\tau_i = \sup t_j$ aller K_j die zu L_i gehören, $t_i > \tau_i\,(1 - \varepsilon)$, L_{i+1}, die Menge der K_j, die mit K_i-Punkte gemeinsam haben, L_{i+1} die übrigen also $L_i = L_{i+1} + L_{i+1}$. Die τ_i bilden eine im schwächeren Sinne abnehmende Folge. Wir unterscheiden zwei Fälle:

1. Es ist stets $\tau_i \neq 0$.
2. Es gibt ein m, so daß $\tau_m = 0$.

Ad 1. Ist $\lim\limits_{i \to \infty} \tau_i = \alpha > 0$, so sind wir fertig, denn dann ist, da die K_i nach Konstruktion nicht übereinandergreifen, für genügend großes j

$$\sum_{i}^{j} V(K_i) > J\,(1 - \varepsilon)^n \sum^{j} \tau_i{}^n > C\, V(B) \qquad (4)$$

Es kann daher $\lim \tau_i = 0$ angenommen werden.

Wir bilden die Vereinigungsmenge $S = \Sigma\, K_i$, ebenso $S_j = \Sigma\, K_i$ mit $K_i \subset L_j'$. Dann ist $S = \Sigma\, S_j$. Denn ist $\overline{K}$ ein Körper im L_0 mit einem $\overline{t}$, dann kann $\overline{K}$ nicht in allen L_j liegen, da es sicher ein j mit $t_j < \overline{t}$ gibt. Dann muß $\overline{K}$ in S_j liegen. Die umgekehrte ist klar. Da $B \subset S$ so gilt also

[24] Fundamenta Mathematica 32.

$$\sum_{j=1}^{\infty} V(S_j) \geqq V(B)$$

Es sei nun K aus einem L_j', dann muß $K \subset L_{j-1}$ sein, greift also mit K_{j-1} übereinander. Es ist also $t \leqq \tau_{j-1}$, $t_{j-1} \geqq \tau_{j-1}(1-\varepsilon)$, daher $t \leqq \dfrac{1}{1-\varepsilon} t_{j-1}$, d. h. $(p - p_{j-1}) u \leqq H(u) t + H(-u) t_{j-1}$ (vgl. dazu § 12) Daraus folgt für jedes x von K

$$(x - p_{j-1}) u = (x - p) u + (p - p_{j-1}) u$$

$$\leqq \left(H(u) + H(-u) + \frac{1+\varepsilon}{1-\varepsilon} H(u)\right) t_{j-1} = H_1 t_{j-1} \qquad (6)$$

Es ist daher H_1 die Stützfunktion von $\overline{K} = 2\hat{K} + \dfrac{1+\varepsilon}{1-\varepsilon} K$. Es
39 liegt also K in $\overline{K}_{j-1}: (x - p_{j-1}) u \leqq H_1 t_{j-1}$, daher auch $L_j \subset \overline{K}_{j-1}$. Daraus folgt $V(S_j) \leqq V(\overline{K}_{j-1})$, also nach (5)

$$\frac{V(\overline{K})}{V(K)} \sum_{j=1}^{\infty} V(K_{j-1}) \geqq V(B)$$

daher

$$S \geqq \frac{V(K)}{V(\overline{K})} V(B)$$

Dies gilt für jedes ε und daraus folgt die Behauptung.

Es ist $V(\overline{K}) \leqq V(K) \sum_{i=0}^{n} \binom{n}{i} 2^{n-i} {}_n\mathrm{Min}\,(i, n-i)$, wie man analog zu *Bonnesen-Fenchel* 53 nachweist.

Der Fall 2 wird genau so behandelt.

Wir wollen in einem Spezialfall den Sachverhalt noch genauer untersuchen. Wir setzen jetzt voraus, daß die K_i einen Mittelpunkt besitzen und $F(p_i - p_0) \geqq \mathrm{Max}\,(t_i, t_j)$ ist.

Dann gilt zunächst folgender

Hilfssatz: Es sei eine endliche Menge von Körpern K_i mit obiger Voraussetzung gegeben. Es sei $t_0 \leqq t \leqq \ldots \leqq t_j$. Dann können höchstens $5^n - 1$ Körper K_j mit K_0 innere Punkte gemeinsam haben.

Beweis: Die Anzahl der K_j, die mit K_0 übereinandergreifen, sei c. K_i kann nur dann mit K_0 übereinandergreifen, wenn $F(p_i - p_0) < t_0 + t_i$ ist. Ordnen wir p_i den Punkt

$$\overline{p}_i = (t_0 + t_i)^{-1} (p_0 (t_i - t_0) + 2 p_i t_0) \qquad (7)$$

zu, so greifen K_0 und K_i nur dann übereinander, wenn $F(\overline{p}_i - p_0) < 2 t_0$ ist. Weiter ist stets für zwei solche Punkte $\overline{p}_i$, $\overline{p}_j$ $F(\overline{p}_i - p_j) > t_0$, denn es ist

$$\bar{p}_i - \bar{p}_j = \frac{2\,t_0}{t_0 + t_j}\,[p_j - p_i + \frac{t_i - t_j}{t_0 + t_i}\,(p_i - p_0)]$$

also

$$F\,(\bar{p}_i - \bar{p}_j) \geqq \frac{2\,t_0}{t_0 + t_j}\,[\mathrm{Max}\,(t_i, t_j) - |\,t_i - t_j\,|] \geqq t_0.$$

Die $\bar{p}_i$ bilden also eine F - t_0-Konfiguration von $F\,(x - p_0) \leqq 2\,t_0$. Daher ist nach Satz 4, Folgerung

$$c + 1 \leqq (2 + \tfrac{1}{2})^n$$

daher $\quad c \leqq 5^n - 1$ w.z.b.w.

Daraus folgt

Satz 37: Liegt eine beschränkte Menge von Körpern K_j vor, mit $F\,(p_i - p_j) \geqq \mathrm{Max}\,(t_i, t_j)$, so kann sie in höchstens 5^n-Teilmengen so aufgespalten werden, daß die K_i, die in einer Teilmenge liegen, nicht übereinandergreifen.

Dabei verstehen wir unter einer beschränkten Menge, eine Menge, die im endlichen liegt.

Beweis: Zunächst ist die Anzahl der K_j höchstens abzählbar. Wir ordnen sie so an, daß $t_1 \geqq t_2 \geqq t_3 \geqq \ldots$

Wir bezeichnen die gesuchten Teilmengen mit T_i $(i \leqq 5^n)$. In T_i nehmen wir zunächst die K_i $(i \leqq 5^n)$ auf. Jetzt wenden wir vollständige Induktion an. Es sei schon gezeigt, daß K_1, K_j in einen dre T_i liegen. Dann muß dies auch bei K_{j+1} gelten. Denn sonst gäbe es in jedem T_i einen Körper, der mit K_{j+1} übereinandergreifen würde, was nach dem obigen Hilfssatz nicht geht. w.z.b.w.

Ist die Menge der K_i nicht beschränkt, so muß 5^n durch $2 \cdot 5^n$ ersetzt werden.

Der Satz gilt auch, wenn nur

$$F\,(p_i - p_j) \geqq \mathrm{Max}\,(t_i, t_j) - \varepsilon\,\mathrm{Min}\,(t_i, t_j)$$

für $O < \varepsilon < 5^{-n}$ vorausgesetzt wird.

Bei Kugeln kann 5^n unter Benutzung von § 9 durch $[\frac{n+2}{2}(2\sqrt{2}+1)^n]$ ersetzt werden.

Aus Satz 35 folgt wieder Satz 34, allerdings mit 5^{-n} statt 3^{-n}. Aus Satz 34, bzw. 35 können der Vitalische Überdeckungssatz und seine Verallgemeinerungen gefolgert werden (*Besicovitsch*[25]).

---

[25] Für $n = 2$ und K = Kreis Besicovitch, Proc. Cambridge 41 (1945).

E. Hlawka:

Corrections

86.1 $I = \int\limits_{\bar{\bar{P}}} \varphi(x)\,dx$

86.2 $J(\varepsilon^n$

86.3 wie 86.2

87.4 $\leq (\frac{n}{2})^n V_0$

89.5 heißt eine (f,d) – Konfiguration P von B

91.6 $D(V(B+\lambda M_1 Q) - V(B))$

91.7 $(2M_1^{-1})^n$

93.8 $nd\overset{k}{M_1^{-1}} \cdot M_1 F$

95.9 $\leq J(\frac{1}{2}M)^n V + n!O(B, 2n\rho)$

96.10 gitterförmigen $A_k - \Gamma - K(r)$.

96.11 $O(B, 2n\rho)$

97.12 Da aber K symmetrisch ist, kann es höchstens $2[\frac{z^n}{2}]$ Seitenflächen haben.

100.13 $V(B + \delta K_1)$

100.14 $f(q_{ij} - p_i) \leq cr_i$,

102.15 $(1 + c_5(\frac{\kappa}{\rho}))$

102.16 von E den Abstand

102.17 $\geq V(P_1) - V(P_1 \cap K)$.

103.18 $= \frac{1}{2n} J(L)$ ist,

105.19 Hilfssätze 14 bis 16 ist

106.20 $(w(\mu - 1)+$

108.21 so daß $B \subseteq \bigcup W_i$ und

109.22 $\Delta\, w^n$

110.23 $S(B, K(r))$ das Gesamtvolumen

112.24 alle $< \frac{1}{2}\beta$

114.25 $D(K) \geq C_3 > 0$

116.26 $[\Gamma, P_l]$

116.27 von B, bzw.

118.28 $W_4 = W_2(1 - \frac{4\rho_2}{\mu w})$

119.29 $C = (\frac{4}{w} + 8)n\rho^*(w + 4\rho^*)^{n-1}$

120.30 $), e = 1,$

121.31 in B_1 :

122.32 $V_1 + \frac{\alpha^n}{2n^n} \sum_1^{1-n}$

123.33 $r = l\lambda^{-1}$.

123.34 $[(2r^2+1)^n]$

123.35 $C_1 = q(\frac{1}{\lambda} - 1)^n (1 - \frac{\Delta_\lambda}{2})^{n+1}$

125.36 $\leq \Sigma \leq$

127.37 $B_3 = (1 - \frac{3\rho}{\beta})B$

128.38 $\leq r(1+\eta), (\eta > 0)$

130.39 auch $\bigcup L_j' \subseteq \overline{K}_{j-1}$.

Über Gitterpunkte in Parallelepipeden

J. Reine Angew. Math. **187** (1950). 246–252

§ 1.

Einleitung. Es soll folgender Satz bewiesen werden:

Es seien im R_n n linear unabhängige Vektoren $a_1, \ldots, a_n$ und eine positive Zahl V vorgegeben. Dann gibt es stets ein Parallelepiped P mit dem Volumen V, mit dem Mittelpunkt im Koordinatenursprung 0, *dessen Seitenflächen die Normalenrichtungen $a_1, \ldots, a_n$ haben, so daß die Anzahl der verschiedenen Gitterpunktpaare $\pm g \neq 0$, welche in P liegen, höchstens $A_n V$ ist, wo $A_n = \frac{1}{n}(n!)^2 2^{\frac{n(n-1)}{2}}$ ist.*

Es ist also A_n von V und den $a_1, \ldots, a_n$ unabhängig. Es ist klar, daß nicht jedes Parallelepiped mit vorgegebenen Volumen und Normalenrichtungen diese Eigenschaft hat. Dies zeigt schon das Rechteck mit den Seitenflächen $NF, \frac{1}{N}$ (N, F natürliche Zahlen), welches mindestens N Gitterpunkte enthält und den Inhalt F hat, wo also die Anzahl der Gitterpunkte mit N wächst. Ist $V < A_n^{-1}$, so enthält das obige Parallelepiped P keine Gitterpunkte $\neq 0$. Dies ist ein Satz, der zuerst von Mordell[1]) formuliert wurde. Es wäre wichtig, das kleinste A_n zu kennen, für welches der Satz noch gilt.

Der Beweis verwendet die Methode, die C. L. Siegel[2]) für den inhomogenen Linearformensatz angewendet hat.

Die Hilfssätze, welche weiter unten entwickelt werden, sind in weiterer Fassung ausgesprochen, als es für den Beweis des Satzes notwendig wäre, da sie auch eine selbständige Bedeutung haben.

Am Schlusse der Arbeit wird noch eine Verallgemeinerung des Satzes für beliebige konvexe Körper mit Mittelpunkt besprochen.

§ 2.

Hilfssatz 1. *Es sei K ein konvexer Körper vom Volumen V, der den Ursprung* 0 *im Innern enthält. Weiter enthalte K eine Menge $\mathfrak{M}$ von k Punkten $\neq 0$ mit folgenden Eigenschaften:*

1. *$\mathfrak{M}$ liege nicht in einem R_{n-1} durch* 0,
2. *Sind x' und x Punkte von $\mathfrak{M}$ und ist $x' = \lambda x$ $(\lambda > 1)$, so gehört auch $x' - x$ der Menge an.*

---

[1]) *L. J. Mordell*, Note on an arithmetical problem on linear forms, Journal London Math. Soc. **12** (1937).

[2]) *H. Davenport*, Note on a result of Siegel, Acta Arithmetica **2** (1937).

Es sei $D = \text{Min}\,|\text{Det.}(x_1, \ldots, x_n)|$ *erstreckt über alle n-tupel linear unabhängiger Punkte von* $\mathfrak{M}$. *Dann ist*

$$V \geq \frac{k-n+1}{n!} D. \tag{1}$$

Bemerkung. Sind die x Gitterpunkte, so ist (2) von selbst erfüllt und $V > \frac{k-n+1}{n!}$. Für diesen Spezialfall wurde der Satz von Blichfeldt[3]) ohne Beweis aufgestellt.

Ist $k = n$ und K symmetrisch, so gilt schärfer, wie allgemein bekannt,

$$V \geq \frac{2^n}{n!}. \tag{2}$$

Eine entsprechende Verschärfung des Satzes wäre wünschenswert.

Beweis. Die k Punkte seien auf s Halbstrahlen $g_1, \ldots, g_s$ durch 0 verteilt. Es muß $s \geq n$ sein. Auf g_i mögen k_i Punkte liegen. Dann ist $\sum_{i=1}^{s} k_i = k$. Ist $\bar{x}_i$ der Punkt auf g_i, der 0 am nächsten liegt, so gehören, wegen Eigenschaft 2, genau die k_i Punkte $j\bar{x}_i$ $(1 \leq j \leq k_i)$ von g_i zu $\mathfrak{M}$. Wir schlagen nun um 0 eine so kleine Kugel $\bar{K}$, daß sie ganz in K liegt und betrachten die Schnittpunkte $\bar{p}_i$ der g_i mit $\bar{K}$. Es werde nun die konvexe Hülle $\bar{P}$ der $\bar{p}_i$ und 0 gebildet. $\bar{P}$ ist ein Polyeder und alle $\bar{p}_i$ sind Eckpunkte, denn sie sind extreme Punkte von $\bar{P}$. Außer den $\bar{p}_i$ kann höchstens noch 0 Eckpunkt sein. Die Seitenflächen von $\bar{P}$, die 0 nicht enthalten, bezeichnen wir allgemein mit $\bar{S}$. Jedes $\bar{S}$ kann nun so in Simplizes S' zerlegt werden, daß die Eckpunkte der Simplizes Eckpunkte von $\bar{S}$ sind und alle Eckpunkte von $\bar{S}$ dabei auftreten, wie man durch Induktion nach den Dimensionen zeigt. Wir bilden nun die Simplizes $0S' = S''$, welche eine Zerlegung Z von $\bar{P}$ bilden. Aus diesen Simplizes S'' können wir nun $l = s - n + 1$ Simplizes $S_1'', \ldots, S_l'$ so herausgreifen, daß $S_1'', \ldots, S_j''$ für jedes $j \leq l$ genau $j + n - 1$ verschiedene Eckpunkte $\bar{p}$ besitzen. Dies zeigt man durch vollständige Induktion. Für $j = 1$ erfüllt dies jedes Simplex S''. Sind schon $S_1'', \ldots, S_j''$ $(j < l)$ gefunden, so überlegen wir so: Es muß in $\bar{P}$ sicher mindestens ein weiteres Simplex auftreten, sogar eines, welches einen Eckpunkt besitzt, der nicht schon Eckpunkt der S_i'' $(i \leq j)$ war, denn die Anzahl der verschiedenen Eckpunkte der S_i'' ist $j + n - 1 < s$. Es muß daher Simplizes S'' geben, die eine Seitenfläche mit einem der S_i'' $(i \leq j)$ gemeinsam haben, da $\bar{P}$ konvex ist. Hat eines davon einen Eckpunkt, der noch nicht vorkommt, so nehmen wir dieses S'' als S_{j+1}''. Wenn nicht, so betrachten wir die Simplizes (welche sicher vorhanden sind), die mit diesen S'' eine Seitenfläche gemeinsam haben. Nach endlich vielen Schritten muß dann ein Simplex auftreten, so beschaffen, daß $n-1$ seiner Eckpunkte $\bar{P}$ unter denen von S_i'' $(i \leq j)$ vorkommen und ein weiterer von diesen verschieden ist. Dieses nehmen wir als S_{j+1}''.

Wir betrachten nun die Simplizes S_j $(1 \leq j \leq l)$, die aus den S_j' dadurch hervorgehen, daß wir die Eckpunkte $\bar{p}_i$ von S_j'' durch $p_i = k_i \bar{x}_i$ ersetzen, welche alle in K liegen.

Nun gilt für das Volumen $V(S)$ eines solchen Simplex mit den Eckpunkten p_i $(1 \leq i \leq n)$:

$$V(S) \geq \frac{1}{n!} k_1 \cdots k_n D,$$

und es ist

$$k_1 \cdots k_n \geq k_1 + \cdots + k_n - n + 1 \geq \text{Max}\,(k_1, \ldots, k_n),$$

wie man leicht durch Induktion nach n zeigt, da $k_i \geq 1$ ist. Folglich ist

$$n!\,V(S_1 + \cdots + S_l) = n!\,(V(S_1) + \sum_{i=2}^{l} V(S_i)) \geq k_1 + \cdots + (k_n - n + 1 + \sum_{i=2}^{l} k_{n+i-1}) D,$$

also $V(K) \geq \frac{k-n+1}{n!} D$, w. z. b. w.

---

[3]) Vgl. *Koksma*, Diophantische Approximationen, Ergebnisse der Math. 4 (1936), 22.

§ 3.

Es sei nun K ein konvexer Körper mit Mittelpunkt in 0. Es sei $f(x)$ seine Distanzfunktion und J das Volumen des Eichkörpers $f \leq 1$. Weiter sei $l \geq 1$ eine beliebige natürliche Zahl. Wir definieren nun n Zahlen M_{li} so: Ist q eine Zahl mit $0 \leq q < M_{li}$, so besitze der durch $f \leq q$ bestimmte Körper K_q mindestens eine der beiden Eigenschaften:

a) die Anzahl der Gitterpunktpaare $\neq 0$ in K_q ist $< l$,

b) die Anzahl der linear unabhängigen Gittervektoren in K_q ist $< i$.

Ist q eine Zahl mit $q \geq M_{li}$, so enthalte K_q mindestens l Gitterpunktpaare $\neq 0$ und die Anzahl der linear unabhängigen ist $\geq i$. Dadurch sind die M_{li} eindeutig bestimmt und es muß $M_{li} \leq M_{l,i+1}$ sein. Für M_{l1} ist Eigenschaft b) in a) enthalten, also enthält $f \leq M_{l1}$ im Innern höchstens $l-1$ Gitterpunktpaare $\neq 0$, aber mit Einschluß des Randes mindestens l Paare. Die Anzahl der linear unabhängigen, die in $f \leq M_{l1}$ liegen, sei s. Es seien dies die Punkte $m_1, \ldots, m_s$. Dann ist $M_{l1} = \cdots = M_{ls}$. Für alle Körper $f \leq q < M_{lj}$ $(j > s)$ gilt dann stets die Eigenschaft b), wenn $q \geq M_{l1}$. Wir definieren nun weitere Gitterpunkte $m_{s+1}, \ldots, m_n$ so, daß alle Gitterpunkte im Innern von $f \leq M_{lj}$ $(j > s)$ von $m_1, \ldots, m_{j-1}$ abhängig sind, während m_j ein Gitterpunkt am Rande ist, welcher linear unabhängig von $m_1, \ldots, m_{j-1}$ ist. Ein solcher muß nach Definition existieren. Diese n Vektoren $m_1, \ldots, m_n$ sind linear unabhängig. (Genauer müßte m_{li} $(i = 1, 2, \ldots, n)$ geschrieben werden.) Diese m_i sind aber nicht eindeutig bestimmt. Wir beweisen nun den folgenden

Hilfssatz 2. *Es ist*

$$\frac{l}{n!} < J M_{l1} \cdots M_{ln} \leq 2^n l. \tag{3}$$

Bemerkung. Ist $l = 1$, so ist dies der Minkowskische Satz über die sukzessiven Minima. Hier kann $\frac{1}{n!}$ durch $\frac{2^n}{n!}$ ersetzt werden. Auch für $l > 1$ könnte die untere Abschätzung leicht verbessert werden, doch verzichten wir darauf. Die obere Abschätzung wird der Vollständigkeit halber hergeleitet.

Beweis. Zuerst soll die untere Abschätzung bewiesen werden. Wir betrachten den Eichkörper $\bar{K}$ mit $f \leq 1$. Sind $\pm g$ die Gitterpunktpaare, die in $f \leq M_{l1}$ liegen, so liegen die Punkte $\pm \frac{1}{M_{l1}} g$ in $\bar{K}$. Ebenso liegen $\pm \frac{1}{M_{li}} m_i$ $(i > s)$ in $\bar{K}$. Die Gesamtzahl dieser beiden Punktgattungen sei $k = 2r$. Es ist $r \geq l + n - s$. Weiter enthält die Menge dieser Punkte n linear unabhängige Punkte, nämlich $\frac{1}{M_{li}} m_i$ $(i = 1, \ldots, n)$. Die Punkte der ersten Art sind nur linear abhängig von $\frac{1}{M_{l1}} m_i$ $(i = 1, \ldots, s)$, die Punkte zweiter Art voneinander linear unabhängig. Ist $x' = \lambda x$ $(\lambda > 1)$, so müssen also beide von erster Art sein und es ist

$$f(x' - x) = (\lambda - 1) f(x) < f(x') \leq 1,$$

also, da diese x, x' von der Gestalt $\frac{1}{M_{l1}} g$ sind, so ist $x' - x$ ebenfalls von dieser Gestalt, also $\pm g$ in $f(x) \leq M_{l1}$. Die Menge erfüllt also beide Voraussetzungen des Hilfssatzes 1. Daraus folgt

$$J \geq \frac{2r - n + 1}{n!} D.$$

Nun ist $D^{-1} \leq M_{l1} \cdots M_{ln}$. Weiters ist $r \geq l$ und $\geq n$, also

$$J M_{l1} \cdots M_{ln} > \frac{l}{n!}, \qquad \text{w. z. b. w.}$$

Jetzt soll die rechte Seite von (3) bewiesen werden. Wir gehen so vor, wie dies Davenport[4]) für den Fall $l=1$ getan hat. Wir führen neue Gittervektoren $l_1, \ldots, l_n$ ein, durch Adaption aus den m_i gebildet. Es ist Det. $(l_1, \ldots, l_n)=1$. Durch eine unimodulare Transformation gehen die l_i in die i-ten Einheitsvektoren e_i über. Nach Definition enthält K_i, d. h. $f \leq M_{li}$, für $i>s$ mindestens einen Gitterpunkt am Rande, der von $m_1, \ldots, m_{i-1}$ linear unabhängig ist, aber keinen solchen im Innern. Also gilt für die Koordinaten $(x^1, \ldots, x^n)$ aller Gitterpunkte mit $f<M_{li}$ $(i>s)$:

$$x^i = \ldots = x^n = 0.$$

Dies kann auch so ausgesprochen werden: Sind x und y zwei beliebige Punkte im Innern von K_i und ist $x \equiv y\,(2)$, d. h. $x^j \equiv y^j\,(2)$ $(j=1, \ldots, n)$, so ist $x^j = y^j$ $(j \geq i)$, denn dann ist $\frac{1}{2}(x-y)$ Gitterpunkt. Wir definieren nun weitere (nicht notwendig konvexe) Körper L_i rekursiv $(i=s, \ldots, n)$ so: Es sei $L_s = K_s$. Ist L_{i-1} $(i>s)$ bereits definiert, so bestimmen wir uns $i-1$ stetige Funktionen $f_j(x^i, \ldots, x^n)$, $1 \leq j \leq i-1$ so, daß $(f_1, \ldots, f_{i-1}, x^i, \ldots, x^n)$ in K_{i-1} liegt. Dann sei L_i die Menge aller Punkte $\bar{x}$ von der Gestalt

$$\text{(4)} \qquad \bar{x}^j = \begin{cases} x^j + \left(\dfrac{M_{li}}{M_{l,i-1}} - 1\right) f_j(x^i, \ldots, x^n), & \text{wenn } 1 \leq j \leq i-1 \\ \dfrac{M_{li}}{M_{l,i-1}}\, x^j, & \text{wenn } i \leq j \leq n, \end{cases}$$

wo die $(x^1, \ldots, x^n)$ alle Punkte von L_{i-1} durchlaufen. Von diesen Körpern L_i stellt man sofort folgendes durch vollständige Induktion fest: $L_i \subset K_i$ $(i \geq s)$. Denn mit $(x^1, \ldots, x^n)$ und $(f_1, \ldots, f_{i-1}, x^i, \ldots, x^n)$, welche in L_{i-1}, also im konvexen Körper K_{i-1} liegen, liegt $(\bar{x}^1, \ldots, \bar{x}^n)$ in K_i. Außerdem ist das Volumen

$$V(L_i) = \left(\frac{M_{li}}{M_{l,i-1}}\right)^{n-i+1} V(L_{i-1}) \quad (i>s)$$

also $V(L_n) = J M_{l1} \cdots M_{ln}$, da $V(L_s) = V(K_s) = J M_{l1}^n$.

Weiter gilt folgende Eigenschaft E: Sind $\bar{x}$, $\bar{y}$ zwei Punkte von L_i und x, y die Punkte aus L_{i-1}, die nach (4) zu $\bar{x}$, $\bar{y}$ gehören, und ist $\bar{x}^j = \bar{y}^j$ $(j=i, i+1, \ldots, n)$, so ist

$$\bar{x} - \bar{y} = x - y.$$

Jetzt kann die obere Abschätzung von (3) leicht gezeigt werden. Wir führen den Beweis indirekt. Es werde also angenommen, es sei

$$\text{(5)} \qquad J M_{l1} \cdots M_{ln} > 2^n l.$$

Nun ist die linke Seite von (5) gerade das Volumen von K_n. Dann enthält nach einem bekannten Satz von Blichfeldt[5]) L_n im Innern $l+1$ Punkte x_j, y $(j=1, \ldots, l)$, so daß

$$\text{(6)} \qquad x_j \equiv y\,(2) \quad (j=1, \ldots, l).$$

Daraus folgt $x_j^n = y^n$ nach Definition von K_n. Dann gibt es, nach der Eigenschaft E, $l+1$ Punkte im Innern von L_{n-1}, für die ebenfalls (6) gilt. Für diese muß dann aber $x_j^{n-1} = y^{n-1}$, $x_j^n = y^n$ sein. Setzt man dies bis zu $L_s = K_s$ fort, so folgt, daß K_s im Innern $l+1$ solche Punkte mit (6) enthalten sollte, also l verschiedene Gitterpunktpaare $\neq 0$ im Widerspruch zur Definition von K_s, w. z. b. w.

---

[4]) *H. Davenport*, Minkowski's inequality for the minima associated with a convex body. Quarterly Journal **10** (1939), 119—121. Für beliebige Körper läßt sich ein ähnlicher Satz aufstellen, wie dies für $l=1$ *Jarník*, *Rogers* und *Chabauty* getan haben; vgl. z. B. C. R. Acad. Sci. Paris **227** (1948).

[5]) Vgl. *Koksma*, a. a. O., S. 20ff.

§ 4.

Nach diesen Vorbereitungen wenden wir uns zum Beweis des Satzes (§ 1). Sind $a_1, \ldots, a_n$ die vorgegebenen Vektoren (o. B. d. A. Det. $(a_1, \ldots, a_n) = 1$), so bilden wir die Linearformen $L_i(x) = a_i^* x$ (a^* transponierter Vektor zu a) und die Distanzfunktion $f = \operatorname{Max}_i |L_i|$. Weiter sei $l = [A_n V] + 1$. Die zugehörigen M_{li}, m_i, s $(i = 1, \ldots, n)$ seien wie in § 3 definiert. Es seien nun $\pm g_j$ $(j = 1, \ldots, l)$ beliebige l Gitterpunktpaare $\neq 0$. Dann muß es unter ihnen mindestens ein g mit $f(g) \geq M_s$ geben. Es sei g jenes g_i unter diesen l Gitterpunkten, für welches $f(g)$ maximal ist. Dann gibt es ein t so, daß $M_{l,t+l} > f(g) \geq M_{l,t}$ (ist $t = n$, so fällt die linke Ungleichung fort).

Dann bestehen zwischen den $L_i(g)$, wenn $t < n$ ist, $n-t$ Relationen

$$\sum_{k=1}^{n} A_{ik} L_k = 0 \quad (i = 1, \ldots, n-t).$$

Man kann durch elementare Umformungen und Umnumerierung erreichen, daß diese Relationen die Gestalt

$$\sum_{k=1}^{n-i+1} A_{ik} L_k = 0 \quad (i = 1, \ldots, n-t)$$

annehmen, wo $|A_{i,n-i+1}| = \operatorname{Max} |A_{ik}|$. Daraus folgt sofort

$$|L_j| \leq \sum_{k=1}^{j-1} |L_k| \quad (j = n-i+1;\ i = 1, \ldots, n-t)$$

und hieraus $|L_j| \leq 2^{j-t-1} \sum_{k=1}^{t} |L_k|$ durch vollständige Induktion. Denn für $j = t+1$ ist die Behauptung richtig. Gilt sie auch bis $j-1$, so folgt aus

$$|L_j| \leq \sum_{k=1}^{j-1} \leq \left(1 + \sum_{k=t+1}^{j-1} 2^{k-t-1}\right) \sum_{k=1}^{t} \leq 2^{j-t-1} \sum_{k=1}^{t}$$

die Behauptung.

Dabei ist $j \geq t+1$. Es ist also auch, da $j \leq n$ ist,

$$|L_j| \leq 2^{n-t+1}\, t \operatorname{Max}(|L_1|, \ldots, |L_t|) \; (j = 1, \ldots, n),$$

also auch

$$M_{l,t} \leq f(g) \leq 2^{n-t+1}\, t \operatorname{Max}(|L_1|, \ldots, |L_t|).$$

Daraus folgt, wenn wir $r_i = \frac{2^{i+1-n}}{i} M_{li}$ $(i \leq n-1)$, $r_n = M_{ln}$ setzen, folgendes:

Betrachten wir die Distanzfunktion

$$\bar{f} = \operatorname{Max}_i \left(\frac{|L_i|}{r_i}\right)$$

und sind l beliebige Gitterpunktpaare $\pm g_i \neq 0$ $(i = 1, \ldots, l)$ vorgelegt, so muß für mindestens eines davon $\bar{f}(g) \geq 1$ sein. Denn ist t jene Zahl, für die $f(g) \geq M_{l,t}$, so ist

$$\bar{f}(g) \geq \frac{1}{r_t} \operatorname{Max}(|L_1|, \ldots, |L_t|) \geq 1,$$

da die r_i monoton zunehmen. Folglich gilt, daß im Innern des Parallelepipeds

$$|L_i| \leq r_i \quad (i = 1, \ldots, n)$$

höchstens $l-1$ Gitterpunktpaare $\neq 0$ liegen. Wir setzen nun $(r_1 \cdots r_n)^{\frac{1}{n}} = r$ und $\frac{r}{r_i} = \varepsilon_i$. Dann ist $\varepsilon_1 \cdots \varepsilon_n = 1$. Weiter ist nach Hilfssatz 2, da $J = 2^n$ ist,

$$r_1 \cdots r_n = \frac{1}{(n-1)!}\, 2^{-\frac{(n-1)(n-2)}{2}} M_{l1} \cdots M_{ln} > \frac{n\, l}{(n!)^2}\, 2^{-\frac{n^2-n+2}{2}} = C\, l.$$

Daraus folgt, daß die Anzahl der Gitterpunktpaare $\neq 0$ im Parallelepiped P_1:

$$\varepsilon_i |L_i| \leq \sqrt[n]{Cl} \quad (i=1,\ldots,n)$$

höchstens gleich $l-1=[A_n V] \leq A_n V$ ist.

Das Volumen von P_1 ist aber $2^n Cl > \frac{1}{A_n}([A_n V]+1) > V$. Also enthält auch das Parallelepiped P:

$$\varepsilon_i |L_i| \leq \frac{1}{2} V^{\frac{1}{n}} \quad (i=1,\ldots,n)$$

vom Volumen V höchstens $A_n V$ Gitterpunktpaare $\neq 0$, w. z. b. w.

§ 5.

Wir können den eben bewiesenen Satz auch so aussprechen:

Liegt ein Würfel W vom Volumen V vor, mit 0 als Mittelpunkt, und ist A eine beliebige Matrix mit Det. $A=1$, so gibt es eine Diagonalmatrix D mit Det. $D=1$, so daß die Anzahl der Punktepaare $\pm DAg$ (g Gitterpunkt), die in W liegen, $\leq A_n V$ ist.

Dieser Satz läßt sich auf beliebige konvexe Körper K mit Mittelpunkt in 0 und Volumen V verallgemeinern.

Satz. *Es sei K ein konvexer Körper mit Volumen V und Mittelpunkt in* 0. *Dann gibt es stets einen konvexen Körper $\overline{K}$ mit denselben Eigenschaften, der aus K durch eine volumentreue Affinität hervorgeht, so daß folgendes gilt: Ist A eine beliebige Matrix mit* Det. $A=1$, *so gibt es stets eine Diagonalmatrix D mit* Det. $D=1$, *so daß die Anzahl der Punktepaare $\pm DAg$ (g Gitterpunkt), die in $\overline{K}$ liegen, kleiner als $\overline{A}_n V$ ist, wo*

$$\overline{A}_n = A_n (n!)^{4n+1} = \frac{1}{n}(n!)^{4n+3} 2^{\frac{n(n-1)}{2}}.$$

Zieht man einen Satz von Davenport und Rogers heran[6]), dann kann die Schranke um $\frac{4{,}921}{n!}$ herabgedrückt werden.

Beweis. Es sei $f(x)$ die Distanzfunktion von K. Es seien die $M_i=f(m_i)$ gewöhnliche Minima (also $l=1$). Durch Adaption des Zahlengitters erhält man neue Gittervektoren[7]) $u_1,\ldots,u_n$ aus $m_1,\ldots,m_n$, so daß Det. $U=1$ für die Matrix $U=(u_1,\ldots,u_n)$ gilt und

$$\frac{2^n}{n!} \leq f(u_1)\cdots f(u_n) J \leq 2^n n!$$

ist (es ist wieder J das Volumen des Eichkörpers). Dann sei $\overline{K}$ der Körper $f(Ux)\leq t$, wenn K der Körper $f\leq t$ ist. Sind die e_i wieder die Einheitsvektoren des R_n, so ist also $F(e_i)=f(u_i)$, wenn $F(x)=f(Ux)$ gesetzt wird, und es ist daher

$$\frac{2^n}{n!} \leq F(e_1)\cdots F(e_n) J \leq 2^n n! \tag{7}$$

Man betrachte nun die polare Distanzfunktion $G(y)$ zu $F(x)$, für die also

$$|x^* y| \leq F(x)\, G(y) \tag{8}$$

ist. Eine leichte Rechnung zeigt[8]), daß

$$F(e_i)\, G(e_i) \leq (n!)^4,$$

[6]) *Davenport* und *Rogers*, Hlawka's theorem in the geometry of numbers, Duke Journal **14** (1947).

[7]) *Mahler*, A theorem on inhomogenous diophantine inequalities, Proc. Acad. Amsterdam **41** (1938).

[8]) *Mahler*, Ein Übertragungsprinzip für konvexe Körper, Casopis **68** (1939), 93ff.; *Hlawka*, Über eine Verallgemeinerung des Satzes von Mordell, Annals of Math. **50** (1949).

also, wenn man (7) beachtet,

$$JG(e_1)\cdots G(e_n)\leq 2^{-n}(n!)^{4n+1}=B$$

ist. Es wurde nun in § 4 gezeigt, daß das Parallelepiped

$$\mathrm{Max}\,|\varepsilon_i L_i|\leq (Cl)^{\frac{1}{n}}$$

mit $C=\frac{n}{(n!)^2}2^{-\frac{n^2-n+2}{2}}$ höchstens $l-1$ Gitterpunktpaare $\neq 0$ enthält. (Dabei ist l eine beliebige natürliche Zahl.) Die spezielle Wahl von l in § 4 wurde an dieser Stelle noch nicht verwendet. Setzen wir nun

$$G(e_i)=G_i,\quad G_i(G_1\cdots G_n)^{\frac{1}{n}}=b_i,\quad \lambda_i=b_i\varepsilon_i\quad (i=1,\ldots,n),$$

so ist

$$F(\lambda_1|L_1|,\ldots,\lambda_n|L_n|)\geq \mathrm{Max}\,(\varepsilon_1|L_1|,\ldots,\varepsilon_n|L_n|)\,(G_1\cdots G_n)^{-\frac{1}{n}},$$

denn es ist doch $G_iF(\lambda_1|L_1|,\ldots,\lambda_n|L_n|)\geq\lambda_i|L_i|$.

Nun ist $JG_1\cdots G_n\leq B$, also enthält

$$F(\lambda_1|L_1|,\ldots,\lambda_n|L_n|)\leq\left(\frac{Cl}{BJ}\right)^{\frac{1}{n}}$$

höchstens $l-1$ Gitterpunktpaare $\neq 0$, und es ist $\lambda_1\cdots\lambda_n=1$, also das Volumen dieses Körpers $C\,B^{-1}l$. Wählen wir daher $l=[\bar{A}_n V]+1$, so erhalten wir die gewünschte Behauptung.

Eingegangen 20. Juni 1949.

Über Integrale auf konvexen Körpern I

Monatsh. Math. **54** (1950). 1–36

(Eingelangt am 8. Juni 1949.)

Einleitung.

Es sei B ein konvexer Körper im R_m ($m \geqq 2$), der den Ursprung o im Innern enthält. Es sei $f(x)$ seine Distanzfunktion und B sei ihr Eichkörper. (Wir bezeichnen die Punkte $(x_1, \ldots, x_m)$ mit x und fassen sie als Spaltenvektoren auf). Dann führen Probleme aus der Geometrie der Zahlen auf die Integrale $G_1(l) = \int\limits_B (m + i\,l\,x)\, e^{ilx}\, d\,x$, $G(l) = \int\limits_B e^{ilx}\, d\,x$, allgemeiner auf $\int\limits_B \Phi(f(x))\, e^{ilx}\, d\,x$ ($d\,x$ Volumselement, $l\,x = \Sigma\, l_j\, x_j$). Ist B die Kugel vom Radius 1, so lassen sich G_1 und G durch Besselfunktionen ausdrücken:

$$G_1 = (2\pi)^{\frac{m}{2}} J_{\frac{m}{2}-1}(|l|)/|l|^{\frac{m}{2}-1}, \quad G = (2\pi)^{\frac{m}{2}} J_{\frac{m}{2}}(|l|)/|l|^{\frac{m}{2}}.$$

Es interessiert vor allem das Verhalten dieser Integrale für großes $|l|$. Hier ist, soviel ich weiß, außer für $m = 2$ nichts bekannt, wenn B nicht gerade ein Parallelepiped oder ein Ellipsoid ist.* Unter Voraussetzung weiterer Eigenschaften von B gelingt es nun asymptotische Ausdrücke für G_1 und G anzugeben (§ 3, § 5). Der Rand $S(B)$ von B muß sich durch gleichsinnig-parallele Stützebenen umkehrbar eindeutig und $6\,m$-mal stetig auf die Oberfläche der Einheitskugel abbilden lassen. Weiter muß, und dies ist die wesentliche Einschränkung, $\inf K > 0$ auf $S(B)$ sein, wo K das Produkt der Hauptkrümmungsradien ist.

Zur Herleitung der asymptotischen Entwicklungen benützen wir die Methode der stationären Phase, wie sie van der Corput[1] in strenger und umfassender Weise begründet hat. Das von uns Benötigte wird in § 2, Hilfssatz 1 hergeleitet.

---

* Vgl. aber *F. John*, Math. Anm. 109, S. 488.

[1] Compositio Mathematica I, III. Proceedings Amsterdam 51, S. 650.

Die Integrale $\int\limits_B \Phi(f)\, e^{ilx}\, d\, x$ lassen sich auf die einfachen Integrale $\int\limits_0^1 \Phi(u)\, u^{m-1}\, G_1(u\, l)$ zurückführen (§ 1). Dies ist für die Kugel bekannt (vgl. Bochner[2]), aber im allgemeinen Fall scheint sie neu zu sein. Diese Formel ermöglicht es, auch für diese Integrale asymptotische Ausdrücke anzugeben. Dies wird insbesondere für $\int\limits_B (1 - f^2)^\delta\, e^{ilx}\, d\, x$ ($\delta \geqq 0$) durchgeführt (§ 4, § 5). Für die Kugel ist der Wert bekanntlich $(2\pi)^{\frac{m}{2}}\, 2^\delta\, \Gamma(\delta + 1)\, J_{\frac{m}{2}+\delta}(|l|) \,/\, |l|^{\frac{m}{2}+\delta}$.

Die weiteren Paragraphen bringen Anwendungen auf die Theorie der Gitterpunkte. Es wird z. B. in § 8 gezeigt, daß der „Gitterrest" $\Phi(0, u) - \mathfrak{J} u^{\frac{m}{2}}$, wo Φ die Anzahl der Gitterpunkte in $f \leqq \sqrt{u}$ und $\mathfrak{J}$, das Volumen von B ist, $O\left(u^{\frac{m(m-1)}{2(m+1)}}\right)$ und $\Omega\left(u^{\frac{m-1}{4}}\right)$ ist, Abschätzungen, wie sie bei der Kugel bekannt sind und die sich noch verschärfen ließen. In § 9 und § 10 verallgemeinern und verschärfen wir die interessanten Untersuchungen von Kendall[3]. § 11 beschäftigt sich mit einem Problem aus der Geometrie der Zahlen: Es ist durch Blichfeldt, van der Corput[4] u. a. bekannt, daß für die Anzahl $2\,S$ der Gitterpunkte $\neq 0$ in einen konvexen Körper $B_t : f(x) \leqq t$, $\mathfrak{J}\, t^m \leqq 2^m (S + 1)$ gilt. Es gelingt uns, wenn $m \not\equiv 1\ (4)$, diese Abschätzung für großes t zu verschärfen, wenn B die obigen Voraussetzungen erfüllt. § 12 bringt die Abschätzung der Zahl $2\,S$ in Zusammenhang mit dem analogen Problem für den polaren Körper von B. In einen Spezialfall (B Parallelepiped, $S = 0$) wurde die Methode dieses Paragraphens bereits von *Gelfond*[5] angewandt. Weitere Anwendungen und Verschärfungen der asymptotischen Entwicklungen werden in der Fortsetzung dieser Arbeit gegeben werden.

§ 1.

Es sei $f(x)$ stetig, dann gilt folgender

Satz 1: Es sei $\Phi(u)$ stetig auf dem Intervall $[0, T]$, $T > 0$, dann ist

$$\int\limits_{f \leqq T} e^{ilx}\, \Phi(f)\, dx = \int\limits_0^T \Phi(u)\, u^{m-1}\, du \int\limits_{f(x) \leqq 1} e^{iulx}\, (m + i\, u\, l\, x)\, dx \tag{1}$$

---

[2] Vorlesungen über Fouriersche Integrale, S. 186, Satz 56.

[3] Quarterly Journal 19 (1948).

[4] Acta Arithmetica 2.

[5] C. R. URSS 1937, XVII, S. 447.

Beweis: $O . B . d . A .$ kann vorausgesetzt werden, daß $\Phi(u)$ ein Polynom ist. Dann ist

$$\int\limits_{f\leqq T} e^{ilx}\Phi(f)dx=\Phi(T)\int\limits_{f\leqq T} e^{ilx}dx-\int\limits_{f\leqq T} e^{ilx}(\Phi(T)-\Phi(f))\,dx=I_1-I_2.$$

Es ist $I_1=\Phi(T)\,T^m\int\limits_{f\leqq 1} e^{ilx}\,dx$ und

$$I_2=\int\limits_{f\leqq T} e^{ilx}\,dx\int\limits_{f(x)}^{T}\Phi'(u)\,du=\int\limits_{0}^{T}\Phi'(u)\,du\int\limits_{f\leqq u} e^{ilx}\,dx=$$

$$=\int\limits_{0}^{T}\Phi'(u)\,u^m\,du\int\limits_{f\leqq 1} e^{iulx}\,dx=\Phi(T)\,T^m\int\limits_{f\leqq 1} e^{ilx}\,dx-$$

$$-\int\limits_{0}^{T}\Phi(u)\,du\int\limits_{f\leqq 1}\frac{d}{du}\left(e^{iulx}\,u^m\right)dx \tag{2}$$

und daraus folgt die Behauptung.

Folgerung. Ist $\Phi(T)=0$, $\Phi'(u)$ vorhanden und existiert $\int\limits_{0}^{T}\Phi'(u)\,u^m\,du\int\limits_{f\leqq 1} e^{iulx}\,dx$, so ist

$$\int\limits_{f\leqq T} e^{ilx}\Phi(f)\,dx=-\int\limits_{0}^{T}\Phi'(u)\,u^m\,du\int\limits_{f\leqq 1} e^{iulx}\,dx \tag{3}$$

Dies gilt auch dann noch, wenn das Integral rechts in (3) ein uneigentliches ist.

§ 2.

Für die weiteren Entwicklungen benötigen wir einen Hilfssatz über die Methode der stionären Phase. Wir benützen Methoden, die *van der Corput*[1] entwickelt hat.

Hilfssatz 1: Auf einem Intervall $[a, b]$ seien zwei Funktionen $f(x)$ und $g(x)$ definiert. Sie seien $3(k+1)$-mal stetig differenzierbar (k natürliche Zahl beliebig). Es sei weiter

$$f'(a)=f'(b)=0,\ f''(a)<0,\ f''(b)>0,\ f'(x)\neq 0 \text{ in } (a, b) \tag{1}$$

Dann ist für $\omega\geqq 1$

$$\int\limits_{a}^{b} g(x)\,e^{i\omega f(x)}\,dx=\Sigma_k(a)+\Sigma_k(b)+0\left(\frac{1}{\omega^{\frac{k}{2}+1}}\right) \tag{2}$$

wo

$$\left\{\begin{aligned} \Sigma_k(a) &= \tfrac{1}{2} e^{i\omega f(a)} \sum_{j=0}^{k} A_j \, e^{-\frac{i(j+1)\pi}{4}} \omega^{-\frac{j+1}{2}} \\ \Sigma_k(b) &= \tfrac{1}{2} e^{i\omega f(b)} \sum_{j=0}^{k} (-1) \, B_j \, e^{i\frac{(j+1)\pi}{4}} \omega^{-\frac{j+1}{2}} \end{aligned}\right. \tag{3}$$

dabei ist

$$\left\{\begin{aligned} A_j &= \sum_{t=0}^{j} \frac{1}{j!} \, \Gamma\left(\frac{j+1}{2} + t\right) (\tfrac{1}{2} |f''(a)|)^{-\frac{j+1}{2} - t} \, l_{t, j+2t}(a) \\ l_{t\,s}(a) &= \sum_{n=3t}^{s} \binom{s}{n} g^{(s-n)}(a) \, K_{t\,n}(a) \\ K_{00} &= 1, \; K_{0\,n} = 0 \; (u \geqq 1), \; K_{t+1,\,n+1} = \sum_{r=3t}^{n-2} \binom{n}{r} f^{(n-r+1)}(a) \, K_{t\,r} \end{aligned}\right. \tag{A}$$

Um die B_j zu erhalten ist in (A) a durch b zu ersetzen.

Beweis: Wegen (1) gibt es ein $c_1 > 0$, so daß in $[a, a + c_1]$, $|f'(x)|$ monoton wachsend und in $[b - c_1, b]$ monoton abnehmend ist. In $[a + c_1, b - c_1]$ ist $\operatorname{Min} |f'(x)| = C_1 > 0$. Es sei weiter $\operatorname{Max} |f'''(x)| = C_2$

$$C = \frac{1}{2(C_2 + 1)} \operatorname{Min}(|f''(a)|, |f''(b)|, C_1(C_2 + 1)) \text{ und } \varepsilon = C\,\omega^{-\frac{1}{3}}.$$

Unter $w(x)$ verstehen wir die Funktion

$$\left\{\begin{array}{ll} \int_x^1 e^{-\frac{1}{t} - \frac{1}{1-t}} dt \Big/ \int_0^1 e^{-\frac{1}{t} - \frac{1}{1-t}} dt & \text{wenn } 0 \leqq x \leqq 1 \\ 0 & \text{für } x > 1. \end{array}\right.$$

Es besitzt $w(x)$ alle Ableitungen, es ist $w(0) = 1$, $w(1) = 0$ und $w^{(j)}(0) = w^{(j)}(1) = 0$ für $j \geqq 1$.

Es sei $N_a(x) = w\left(\frac{x - a}{\varepsilon}\right)$, $N_b(x) = w\left(\frac{b - x}{\varepsilon}\right)$. Dann ist also

$$N_a(a) = 1, N_a^{(j)}(a) = N_a^{(j)}(a + \varepsilon) = 0, (j \geqq 1), N_a^{(j)}(x) = 0\left(\frac{1}{\varepsilon^j}\right) (j \geqq 0) \tag{4}$$

und $N_a(x) = 0$ für $x > a - \varepsilon$. Das analoge gilt für $N_b(x)$.

Es ist nun weiter

$$|f'(x)| \geqq \tfrac{1}{2} |f''(a)| (x - a) \text{ für } a \leqq x \leqq a + \varepsilon, \tag{5}$$

denn es ist ja $f'(x) = (x - a)[f''(a) + (x - a) f'''(a + \vartheta(x - a))]$, also $|f'(x)| \geqq (x - a)[|f''(a)| - \varepsilon C_2]$ w. z. b. w. Ebenso gilt

$$|f'(x)| \geqq \tfrac{1}{2} |f''(x)| (b-x) \text{ für } b-\varepsilon \leqq x \leqq b. \qquad (5')$$

Weiters ist in $[a+\varepsilon, b-\varepsilon]$, $|f'(x)| \geqq \mathrm{Min}\,(\tfrac{1}{2}|f''(a)|\,\varepsilon, \tfrac{1}{2}|f''(b)|\,\varepsilon, C_1)$ also $\geqq C_3\,\varepsilon$. Es ist nun

$$\int_a^b g\, e^{i\omega f}\, dx = \int_a^b N_a g\, e^{i\omega f}\, dx + \int_a^b N_b g\, e^{i\omega f}\, dx + \int_a^b g H\, e^{i\omega f}\, dx, \qquad (6)$$

wo $H = g\,(1 - N_a - N_b)$.

Es ist für $0 \leqq j \leqq 3\,(k+1)$, $H^{(j)}(a) = H^{(j)}(b) = 0$, $H^{(j)}(x) = 0\,(\varepsilon^{-j})$.

Wir definieren $\dfrac{H^{(j)}(a)}{(f')^{2i-j}(a)} = 0$, wenn $i \geqq j$, analog an der Stelle b.

Es sei weiters

$H_0 = H$, $H_{i+1}(x) = \dfrac{d}{dx}\left(\dfrac{H_i}{f'}\right)$. Es ist, wie man mittels vollständiger Induktion zeigt

$$H_i(x) = \sum_{j=0}^{i} H^{(j)}(x)\,(f')^{j-2i}\, C_j(f', \ldots, f^{(i+1)}),$$

wo C_j ein Polynom in seinen Argumenten ist. Es ist $H_i(a) = H_i(b) = 0$, $H_i(x)$ ist stetig, differenzierbar und $H_i(x) = 0\,(\varepsilon^{-2i})$. Es ist nämlich $H^{(j)}(x) = \dfrac{(x-a)^{2i-j}}{(2i-j)!} H^{(2i)}(a + \vartheta\,(x-a))$. Nun ist für x in $[a, a+\varepsilon]$, $|f'(x)| \geqq C_4\,(x-a)$, $H^{(2i)}(x) = 0\,(\varepsilon^{-2i})$ und die C_j sind beschränkt. Daraus folgt diese Behauptung für $[a, a+\varepsilon]$. Ebenso wird dies für $[b-\varepsilon, b]$ gezeigt. Für $[a+\varepsilon, b-\varepsilon]$ ist die Behauptung klar.

Es ist für $j \leqq 3\,k+2$

$$\int_a^b H\, e^{i\omega f}\, dx = 0\,((\omega\,\varepsilon^2)^{-j}) \quad (j \geqq 0) \qquad (7)$$

Denn es ist, wie man durch mehrfache partielle Integration feststellt $\int_a^b H\, e^{i\omega f}\, dx = \dfrac{1}{(i\,\omega)^j} \int_a^b H_j\, e^{i\omega f}\, dx$. Aus $H_j = 0\,(\varepsilon^{-2j})$ folgt sofort die Behauptung.

Auch (7) folgt wegen $\varepsilon = C\,\omega^{-\frac{1}{3}}$ mit $j = [\frac{3}{2}\,(k+2)] + 1$

$$\int_a^b H\, e^{i\omega f}\, dx = 0\,(\omega^{-\frac{k}{2}-1}) \qquad (7')$$

In (6) sind also nur mehr die ersten beiden Integrale zu betrachten. Es sei

$I_1 = \int\limits_a^{a+\varepsilon} B\, e^{i\omega f}\, dx$, wo $B = N_a\, g$. Es ist $B^{(j)}(x) = 0\,(\varepsilon^{-j})(j \leqq 3\,(k+1))$.

Wir setzen $p\,(x) = f\,(a) + \dfrac{(x-a)^2}{2} f''\,(a)$. Nun ist

$$f\,(x) = p\,(x) + \frac{1}{3!}\,(x-a)^3\, f'''\,(a + \vartheta\,(x-a))$$

also für $0 \leqq x - a \leqq \varepsilon$

$$f^{(s)}\,(x) - p^{(s)}\,(x) = 0\,(\varepsilon^{3-s}),\ (0 \leqq s \leqq 3),\ f^{(s)}\,(x) - p^{(s)}\,(x) = 0\,(1)\,(s > 3). \tag{8}$$

Setzen wir $L = B\, e^{i\omega(f-p)}$, so wird $I_1 = \int\limits_a^{a+\varepsilon} L\, e^{i\omega p}\, dx$. Nun ist

$\dfrac{d}{dx}\,(e^{i\omega(f-p)}) = i\,\omega\, e^{i\omega(f-p)}\,(f' - p') = 0\,(\varepsilon^2\,\omega) = 0\,(\omega^{\frac{1}{3}})$, allgemein

$$\frac{d^j}{dx^j}\,(e^{i\omega(f-p)}) = 0\,(\omega^{\frac{j}{3}}), \tag{9}$$

wie man mittels vollständiger Induktion zeigt, denn es ist ja

$$\frac{d^{j+1}}{dx^{j+1}}\,(e^{i\omega(f-p)}) = i\,\omega\, \Sigma \binom{j}{k} \frac{d^k\, e^{i\omega(f-p)}}{dx^k}\,(f^{(j-k+1)} - p^{(j-k+1)}).$$

Weiter ist

$$L^{(j)} = 0\,(\omega^{\frac{j}{3}}) \qquad (j \leqq 3\,(k+1)), \tag{10}$$

denn nach (9) und wegen $B^{(j)} = 0\,(\varepsilon^{-j})$ ist

$$L^{(j)} = \Sigma \binom{j}{k} B^{(k)} \frac{d^{j-k}}{dx^{j-k}}\, e^{i\omega(f-p)} = \Sigma\, 0\,(\varepsilon^{-k}\,\omega^{\frac{j-k}{3}}) = (\omega^{\frac{j}{3}}).$$

Wir setzen nun für $x \geqq 0$

$$\begin{cases} \psi_0\,(x) = e^{i\omega p} \\ \psi_{\mu+1}\,(x) = \dfrac{1}{\mu!} \displaystyle\int\limits_{C(x)} (\zeta - x)^\mu\, e^{i\omega p(\zeta)}\, d\,\zeta \quad (\mu \geqq 0) \end{cases} \tag{11}$$

wo $C\,(x)$ der Weg $\zeta\,(x, t) = \dfrac{e^{-\frac{i\pi}{4}}}{\sqrt{2}}\,(t + \dfrac{x^2}{t}\, i)\ (x \leqq t \leqq \infty)$. Für

$x = 0$ sei stets $\dfrac{x^2}{t} = 0$. Es ist $\zeta\,(x, x) = x$. Weiter ist, da

$t^2 - \dfrac{x^4}{t^2} \geqq (t-x)^2$

$$|e^{i\omega p(\zeta)}| = e^{-\frac{\omega}{4}|f''(a)|\left(t^2-\frac{x^4}{t^2}\right)} \leqq e^{-\frac{\omega}{4}|f'(a)|(t-x)^2}, \qquad (11')$$

also ist das in (11) auftretende Integral konvergent.

Weiters ist

$$\psi_{\mu+1}(x) = -\psi_\mu(x) \qquad (12)$$

denn $\frac{1}{h}(\psi_{\mu+1}(x+h) - \psi_{\mu+1}(x)) = \frac{1}{h}\Big[\int\limits_{c(x+h)} (\zeta-x-h)^\mu e^{i\omega p}\, d\zeta -$

$$-\int\limits_{c(x)} (\zeta-x-h)^\mu e^{i\omega p}\, d\zeta\Big] + \int\limits_{c(x)} \frac{(\zeta-x-h)^\mu - (\zeta-x)^\mu}{h}\, d\zeta.$$

Der erste Teil ist $\frac{1}{h}\int\limits_x^{x+h} (\zeta-x-h)^\mu e^{i\omega p}\, d\zeta$, geht also mit $h \to 0$ gegen 0. Der zweite Teil geht aber mit $h \to 0$ gegen $-\psi_\mu(x)$ w. z. b. w.

Es ist, wenn $\tau = \frac{t}{\sqrt{2}}$ gesetzt wird

$$\psi_{\mu+1}(0) = \frac{e^{i\omega f(a)\frac{i(\mu+1)\pi}{4!}}}{\mu!}\int\limits_0^\infty \tau^\mu e^{-\frac{\omega}{2}|f''(a)|\tau^2}\, d\tau \qquad (13)$$

also

$$\psi_{\mu+1}(0) = \frac{e^{i\omega f(a)}}{2\left(\frac{\omega}{2}|f''(a)|\right)^{\frac{\mu+1}{2}}}\,\frac{\Gamma\left(\frac{\mu+1}{2}\right)}{\mu!}\, e^{-\frac{i(\mu+1)\pi}{4}} \qquad (13')$$

Aus (11′) und (11) folgt noch

$$\psi_{\mu+1}(x) = 0\left(\omega^{-\frac{\mu+1}{2}}\right). \qquad (14)$$

Wir gehen jetzt auf I_1 zurück und setzen $M(x) = L(x-a)$, $q(x) = p(x-a)$. Dann ist

$$f(a) = p(a) = q(0),\ M^{(j)}(0) = L^{(j)}(a) = g^{(j)}(a),\ M^{(j)}(\varepsilon) = L^{(j)}(a+\varepsilon) = 0 \qquad (14')$$

und nach (10) $L^{(j)}(x) = 0\left(\omega^{\frac{i}{3}}\right)$.

Dann erhalten wir

$f(a) = p(a) = q(0)$, $M^{(j)}(0) = L^{(j)}(a)$, $M^{(j)}(\varepsilon) = L^{(j)}(a+\varepsilon)$ und nach (10) $L^{(j)}(x) = 0\left(\omega^{\frac{j}{3}}\right)$.

Dann folgt durch partielle Integration

$$I_1 = \int_0^\varepsilon M e^{i\omega q}\, dx = \int_0^\varepsilon M \psi_0\, dx = M(0)\,\psi_1(0) + \int_0^\varepsilon M' \psi_1\, dx =$$

$$= \sum_{s=0}^{K} L^{(s)}(a)\,\psi_{s+1}(0) + \int_0^\varepsilon M^{(K+1)}\,\psi_{K+1}\, dx \tag{15}$$

wo $K = 3(k+1)$. Es ist

$$\int_0^\varepsilon M^{(K+1)}\,\psi_{K+1}\, dx = 0\,(\omega^{\frac{K+1}{3}})\; 0\,(\omega^{-\frac{K+1}{2}})\,\varepsilon = 0\,(\omega^{-\frac{k}{2}-1})$$

Nun muß $L^{(s)}(a)$ betrachtet werden. Setzen wir für die n-te Ableitung von $e^{i\omega(f-p)}$ an der Stelle a, P_n, dann ist $P_0 = 1$, $P_1 = P_2 = 0$ und es ist (vgl. den Beweis von (9))

$$P_{n+1} = i\,\omega \sum_{r=0}^{n-2} \binom{n}{r} P_r\, f^{(n-r+1)}(a) \tag{16}$$

Die P_n sind also Polynome in $i\,\omega$, und zwar nach (9) vom Grad $\left[\frac{n}{3}\right]$. Wir setzen daher $P_n = \sum_{t=0}^{n} K_{tn}\,(i\,\omega)^t$ mit $K_{00} = 1$, $K_{0n} = 0$ für $n \geqq 1$ und $K_{tn} = 0$, wenn $t > \left[\frac{n}{3}\right]$. Dann folgt aus (16)

$$K_{t+1,\,n+1} = \sum_{r=3t}^{n-2} \binom{n}{r} f^{(n-r+1)}(a)\, K_{t\,r} \tag{17}$$

also z. B. $K_{1n} = f^{(n)}(a)$. Weiter ist

$$L^{(s)}(a) = \sum_{n=0}^{s} \binom{s}{n} g^{(s-n)}(a)\, P_n = \sum_{t=0}^{s} (i\,\omega)^t\, l_{ts} \quad \text{wo}$$

$$l_{ts} = \sum_{n=3t}^{s} \binom{s}{n} g^{(s-n)}(a)\, K_{tn} \tag{18}$$

also $l_{ts} = 0$ für $t > \frac{s}{3}$ und $l_{0s} = g^{(s)}(a)$.

Jetzt betrachten wir die Summe rechts in (15), wobei wir $\psi_{j+1}(0) = \Psi_{j+1}\,\omega^{-\frac{j+1}{2}}$ setzen, welche sich jetzt wegen (18) schreibt:

$$\sum_{s=0}^{K} \sum_{t=0}^{s} i^t\, \omega^{t-\frac{s+1}{2}}\, l_{ts}\, \Psi_{s+1} \tag{19}$$

Wir ordnen nach Potenzen von $\omega^{-\frac{1}{2}}$. Um den Koeffizienten A_j der $j+1$-ten Potenz zu erhalten ($j \geqq 0$) müssen wir alle Glieder in (19) nehmen, wo $s - 2t = j$ ist und erhalten daher $A_j = \sum_{t=0}^{j} i^t\, l_{t,\,j+2t}$

Ψ_{j+2t+1}, da für $t > j$, $l_{t,j+2t} = 0$. Dieser Ausdruck ist aber gerade der in (17) angegebene. Behalten wir in (19) nur die Potenzen mit $j \leqq k$ bei, so bekommen wir gerade $\Sigma_k(a)$, es ist also

$$I_1 = \Sigma_k(a) + 0\,(\omega^{-\frac{k}{2}-1}).$$

Bei dem Integral $\int\limits_{b-\varepsilon}^{b} N_b\, g e^{i\omega f}\, dx$ geht man analog vor. Es ist nur $\psi_{\mu+1}$ zu ersetzen durch

$$\frac{1}{\mu!} \int\limits_{x}^{\infty} (\zeta - x)^{\mu}\, e^{i\omega(f(b) - {}^1/_2 f''(b)\zeta^2)} \frac{\partial \zeta}{\partial t}\, dt$$

wo $\zeta = \frac{1}{\sqrt{2}}\, e^{\frac{i\pi}{4}} (t + \frac{x^2}{t}\, i)$. Beachtet man noch (7'), so ist alles gezeigt.

§ 3.

Es sei im R_m ($m \geqq 2$) B ein konvexer Körper, der den Koordinatenursprung o im Innern enthält. Der Rand $S(B)$ von B sei durch gleichsinnig parallele Stützebenen umkehrbar eindeutig und $6\,m$-mal stetig differenzierbar auf die Einheitskugeloberfläche abgebildet. Die Distanzfunktion $f(x)$ und Stützfunktion $H(u)$ ($|u| = 1$) sind also für $x \neq 0$, $u \neq 0$ $6\,m$-mal stetig differenzierbar. B selbst habe die Darstellung $f(x) \leqq 1$. Die Ableitungen von H nach den u_i, wo $u = (u_1, \ldots, u_m)$ sind, wenn ihre Ordnung $\leqq 6\,m$, auf der Kugel $|u| = 1$ gleichmäßig beschränkt, da sie stetig sind. Weiters ist $\inf H(u) > 0$ auf $|u| = 1$, da o innerer Punkt von B ist. Ist nun $K = r_1 \ldots r_{m-1}$ das Produkt der Hauptkrümmungsradien r_i von $S(B)$, so machen wir die folgende Voraussetzung

$$\inf K = \varrho > 0 \text{ auf } S(B) \tag{1}$$

Man vergleiche zum Folgenden stets *Bonnesen-Fenchel*, Konvexe Körper (kurz *B. F.*).

Es soll nun

$$G = \int\limits_B e^{i\omega x_1}\, dx, \text{ bzw. } G_1 = \int\limits_B (m + i\omega x_1)\, e^{i\omega x_1}\, dx \tag{2}$$

für $\omega \geqq 1$ asymptotisch entwickelt werden.

Da sich der Rand von B auf $|u| = 1$ abbilden läßt, so ist auch K eine Funktion von u. Dann gilt für $k \leqq m + 3$

Satz 2: Es ist

$$G = \frac{(2\pi)^{\frac{m-1}{2}}}{\omega^{\frac{m+1}{2}}} \left[e^{i\omega H_1} \sum_{j=0}^{k} \frac{a_j}{\omega^{\frac{j}{2}}} + e^{-i\omega H_{-1}} \sum_{j=0}^{k} \frac{b_j}{\omega^{\frac{j}{2}}} \right] + 0\,(\omega^{-\frac{m+k}{2}-1}) \tag{3}$$

wo

$$H_{\pm 1} = H(\pm 1, 0, \ldots 0),\; K_{\pm 1} = K(\pm 1, 0, \ldots 0),$$

$$a_0 = \sqrt{K_1}\, e^{-i\frac{(m+1)\pi}{4}}\; b_0 = \sqrt{K_{-1}}\, e^{i\frac{(m+1)\pi}{4}} \tag{4}$$

$$G_1 = \left(\frac{2\pi}{\omega}\right)^{\frac{m-1}{2}} \left[e^{i\omega H_1} \sum_{j=0}^{k} \frac{a'_j}{\omega^{\frac{j}{2}}} + e^{-i\omega H_{-1}} \sum_{j=0}^{k} \frac{b'_j}{\omega^{\frac{j}{2}}} \right] + 0\,(\omega^{-\frac{m+k}{2}}) \tag{5}$$

wo

$$a_0' = H_1 \sqrt{K_1}\, e^{-\frac{(m-1)\pi}{4}},\; b_0' = H_{-1} \sqrt{K_{-1}}\, e^{i\frac{(m-1)\pi}{4}} \tag{6}$$

Zuerst schicken wir einen einfachen Hilfssatz voraus.

Hilfssatz 2: Es sei $Q = \sum_{i,k=1}^{s} A_{ik}\, l_i\, l_k$ eine positiv-definite Form. Es sei $D = \operatorname{Det} A_{ik}$ und $C = \operatorname{Max}_{i,k} A_{ik}$. Dann ist für alle $(l_1, \ldots, l_s)$ mit $\Sigma\, l_i^2 = 1$

$$Q \geqq \frac{D}{(s\,C)^s} \tag{7}$$

Beweis: Die Eigenwerte von Q seien $\lambda_1 \leqq \lambda_2 \leqq \ldots \leqq \lambda_s$. Dann ist für alle $(l_1, \ldots l_s)$ mit $\Sigma\, l_i^2 = 1$

$$Q \geqq \lambda_1 = \frac{\lambda_1 \ldots \lambda_s}{\lambda_2 \ldots \lambda_s} \geqq \frac{D}{(\lambda_1 + \ldots + \lambda_s)^s} \geqq \frac{D}{(s\,C)^s}$$

da $\lambda_1 + \ldots + \lambda_s = A_{11} + \ldots + A_{ss} \leqq s\,C$ w. z. b. w.

Beweis von Satz 2: Nach dem *Gauß*schen Integralsatz ist

$$G = \frac{1}{\omega i} \int_{S(B)} e^{i\omega x_1}\, u_1\, do. \tag{8}$$

Dabei ist do das Oberflächenelement von $S(B)$ und u_1 die erste Komponente des normierten, nach außen gerichteten Normalenvektors u im Punkt $x = (x_1 \ldots x_m)$. Es sei gleich $m > 2$. Es ist nun x auf $S(B)$ eine Funktion von u, da $S(B)$ auf $|u| = 1$ abgebildet ist. Es ist weiter $do = K\, d\sigma$, wo $d\sigma$ das Oberflächenelement von $E_m : |u| = 1$ (vgl.

B. F. S. 63, Z. 6 v. u.) ist. Weiters sei $d\bar{\sigma}$ das Oberflächenelement der $m-1$-dimensionalen Einheitskugel E_{m-1}. Wir setzen nun

$$u_1 = \cos\vartheta,\ u_j = \sin\vartheta\, a_j\ (j \geqq 2),\ 0 \leqq \vartheta \leqq \pi,\ \sum_{j=2}^{m} a_j^2 = 1 \tag{9}$$

Dann folgt aus (8)

$$i\,\omega\, G = \int_{E_{m-1}} d\bar{\sigma} \int_0^{\pi} e^{i\,\omega\, x_1}\, K \cos\vartheta \sin^{m-2}\vartheta\, d\vartheta \tag{10}$$

Wir halten zunächst $a_2, \ldots a_m$ fest und betrachten nur das innere Integral in (10).

Es ist nun nach *B. F.* S. 58 (1) $x_1 = \dfrac{\partial H}{\partial u_1}$, also

$$\frac{\partial x_1}{\partial \vartheta} = \sum_{i=1}^{m} \frac{\partial^2 H}{\partial u_1\, \partial u_i}\, \frac{\partial u_i}{\partial \vartheta} = -\frac{\partial^2 H}{\partial u_1^2} \sin\vartheta + \cos\vartheta \sum_{i=2}^{m} \frac{\partial^2 H}{\partial u_1\, \partial u_i}\, a_i \tag{11}$$

Nun gilt nach *B. F.* S. 59, Z. 1 v. u. wegen der Homogenität von H

$$\sum_{i=1}^{m} u_i \frac{\partial H}{\partial u_i} = H,\ \sum_{i=1}^{m} u_i \frac{\partial^2 H}{\partial u_i\, \partial u_j} = 0\ (j = 1, \ldots, m) \tag{12}$$

Daraus folgt zunächst

$$\frac{\partial H}{\partial u_1}(\pm 1, 0, 0 \ldots 0) = \pm H(\pm 1, 0, \ldots 0),$$

$$\frac{\partial^2 H}{\partial u_1\, \partial u_j}(\pm 1, 0 \ldots 0) = 0 \tag{13}$$

und weiter

$$\frac{\partial^2 H}{\partial u_1\, \partial u_i} \cos\vartheta = \frac{\partial^2 H}{\partial u_1\, \partial u_i}\, u_1 = -\sum_{k=2}^{m} \frac{\partial^2 H}{\partial u_2\, \partial u_k}\, u_k,$$

also folgt aus (11)

$$\frac{\partial x_1}{\partial \vartheta} = -\sin\vartheta\, F(\vartheta, a_i) \tag{14}$$

wo

$$F(\vartheta, a_i) = \frac{\partial^2 H}{\partial u_1^2} + \sum_{i,k=2}^{m} \frac{\partial^2 H}{\partial u_i\, \partial u_k}\, a_i\, a_k\ \Big(\sum_{i=2}^{m} a_i^2 = 1\Big) \tag{15}$$

Es ist nun nach *B. F.* S. 62 (5) K die Summe aller $m-1$-reihigen Hauptminoren h der Matrix $(H_{ik}) = \left(\dfrac{\partial^2 H}{\partial u_i\, \partial u_k}\right)$. Nun ist nach (1)

$\inf K = \varrho > 0$, also $K \geqq \varrho$ auf $|u| = 1$. Es muß also einer der Hauptminoren $h \geqq \frac{1}{m} \varrho > 0$ sein. Dann kann die quadratische Form $Q(z)$ in den Variablen z, deren Koeffizienten von den Elementen dieses Hauptminors gebildet werden, sicher nicht ausgeartet sein. Nun ist H konvex, also nach *B. F.* S. 18, Z. 1 v. u. $\sum_{i,k=1}^{m} H_{ik} z_i z_k \geqq 0$, also muß das obige Q sogar positiv-definit sein. Nun sind nach Voraussetzung die H_{ik} auf $|u| = 1$ beschränkt, also $\leqq C$.

Dann gilt nach Hilfssatz 2

$$Q(z) \geqq \frac{\varrho}{(m-1)(mC)^{m-1}} = \varrho_1 \tag{16}$$

auf der $m-1$-dimensionalen Einheitskugel, da $s = m-1$. Wir müssen nun zwei Fälle unterscheiden

1. Es sei Q gerade die Form $\sum_{i,k=2}^{m} H_{ik} z_i z_k$, dann ist, da $H_{11} \geqq 0$

$$F \geqq \varrho_1 \tag{17}$$

für alle $a_2, \ldots, a_m$ mit $\Sigma a_i^2 = 1$.

2. Ist das obige Q nicht von dieser Gestalt, so muß H_{11} als ein Koeffizient in Q auftreten und es ist dann wegen (16) $H_{11} \geqq \varrho_1$, also gilt wieder (17).

Es besitzt also nach (14) $\frac{\partial x_1}{\partial \vartheta}$ nur die Nullstellen $\vartheta = 0$ und $\vartheta = \pi$, d. h. für $u = (1, 0 \ldots 0)$ und $u = (-1, 0, \ldots, 0)$.

Weiters ist

$$\frac{\partial^2 x_1}{\partial \vartheta^2} \Big|_0 = -F(0, a_i) \leqq -\varrho_1 < 0, \quad \frac{\partial^2 x_1}{\partial \vartheta^2} \Big|_\pi = F(\pi, a_i) \geqq \varrho_1 > 0$$

wo ϱ_1 von den a_i nicht abhängt.

Wir können also Hilfssatz 1 aus § 2 anwenden mit $a = 0$, $b = \pi$, $f(\vartheta) = x_1$, $g = K \sin^{m-2} \vartheta \cos \vartheta$. In $\vartheta = 0$, bzw. $\vartheta = \pi$ besitzt g die Entwicklungen

$$g(\vartheta) = \begin{cases} \vartheta^{m-2} \left[K_1 + \left(\frac{\partial K}{\partial \vartheta}\right)_0 \vartheta + \ldots\right] \\ -(\pi - \vartheta)^{m-2} \left[K_{-1} + \left(\frac{\partial K}{\partial \vartheta}\right)_\pi (\pi - \vartheta) + \ldots\right], \end{cases}$$

also

$$g^{(j)}(\vartheta) = 0 \text{ für } \vartheta = 0 \text{ und } \pi \, (0 \leqq j \leqq m-2), \; g^{(m-2)}(0) = (m-2)!\, K_1,$$
$$g^{(m-2)}(\pi) = (-1)^{m-1}(m-2)!\, K_{-1}. \tag{19}$$

Setzen wir noch $F(0, a_i) = F_1$, $F(\pi, a_i) = F_{-1}$, so ist also

$$\int_0^\pi e^{i\omega x_1} K \cos\vartheta \sin^{m-2}\vartheta \, d\vartheta$$
$$= \tfrac{1}{2} e^{i\omega x_1(1,0,\ldots,0)} \sum_{j=m-2}^{k+m-2} A_j e^{-\frac{i(j+1)}{4}} \omega^{-\frac{j+1}{2}}$$
$$+ \tfrac{1}{2} e^{i\omega x_1(-1,0,\ldots,0)} \sum_{j=m-2}^{k+m-2} (-1)^j B_j e^{\frac{i(j+1)\pi}{4}} \omega^{-\frac{j+1}{2}} + 0\left(\frac{1}{\omega^{\frac{m+k}{2}}}\right) \tag{20}$$

Dabei gilt $0\,(\omega^{-\frac{m+k}{2}})$ gleichmäßig in den a_i.

Daraus folgt, wenn wir beachten, daß $x_1(\pm 1, 0 \ldots 0) = \frac{\partial H}{\partial u_1}(\pm 1, 0 \ldots 0) = \pm H_{\pm 1}$ nach (13),

$$i\omega G = \left(\frac{2\pi}{\omega}\right)^{\frac{m-1}{2}} \sum_{j=0}^{k} \frac{a_j^* e^{i\omega H_1} + b_j^* e^{-i\omega H_{-1}}}{\omega^{\frac{j}{2}}} + 0\,(\omega^{-\frac{m+k}{2}}). \tag{21}$$

Dabei ist

$$\left\{\begin{aligned} a_0^* &= \left(\frac{1}{2\pi}\right)^{\frac{m-1}{2}} \frac{\Gamma\left(\frac{m-1}{2}\right)}{2\,(m-2)!}\, 2^{\frac{m-1}{2}} e^{-\frac{i(m-1)\pi}{4}} \int_{E_{m-1}} \frac{g^{(m-2)}(0)}{F_0^{\frac{m-1}{2}}}\, d\bar\sigma \\ b_0^* &= \left(\frac{1}{2\pi}\right)^{\frac{m-1}{2}} \frac{\Gamma\left(\frac{m-1}{2}\right)}{2\,(m-2)!} (-1)^{m-2}\, 2^{\frac{m-1}{2}} e^{\frac{i(m-1)\pi}{4}} \int_{E_{m-1}} \frac{g^{(m-2)}(\pi)}{F_0^{\frac{m-1}{2}}}\, d\bar\sigma \end{aligned}\right. \tag{22}$$

Es soll nun a_0^* und b_0^* bestimmt werden. Nach (13) ist $F(0, a_2, \ldots a_m) = \sum_{i,j=2}^{m} H_{ik}(1, 0 \ldots 0)\, a_i a_k$. Wegen (9) haben wir nur $\int_{E_{m-1}} \frac{d\bar\sigma}{F_1^{\frac{m-1}{2}}}$ zu berechnen. Durch eine orthogonale Transformation können wir F_1 überführen in $\sum_{i=2}^{m} \lambda_i a_i^2$ $(\lambda_i > 0)$. Dabei ist $\lambda_2 \ldots \lambda_m = K_1 = K(1, 0, \ldots, 0)$, da die Hauptminoren von (H_{ik}) für $(1, 0 \ldots 0)$ verschwinden bis auf Det. H_{ik} $(2 \leqq i, k \leqq m)$.

Wir gehen nun von dem absolutkonvergenten Integral

$$I = \int_{R_{m-1}} e^{-\sum_{i=2}^{m} \lambda_j x_j^2} d x_2 \ldots d x_m \tag{23}$$

aus. Es ist einerseits

$$I = \prod_{i=2}^{m} \int_{-\infty}^{+\infty} e^{-\lambda_i x_i^2} d x_i = \frac{\pi^{\frac{m-1}{2}}}{\sqrt{\lambda_2 \ldots \lambda_m}} = \frac{\pi^{\frac{m-1}{2}}}{\sqrt{K_1}}$$

Führen wir andererseits in I Polarkoordinaten ein: $x_j = r a_j$ $(j = 2, \ldots m)$, $\Sigma a_j^2 = 1$, so kommt

$$I = \int_0^{\infty} r^{m-2} d r \int_{E_{m-1}} e^{-r^2 \sum_{i=2}^{m} \lambda_i x_i^2} d\bar{\sigma} = \tfrac{1}{2} \Gamma\left(\frac{m-1}{2}\right) \int_{E_{m-1}} \frac{d\bar{\sigma}}{(\Sigma \lambda_i x_i^2)^{\frac{m-1}{2}}},$$

also erhalten wir wegen (19) aus (22)

$$a_0^* = \sqrt{K_1}\, e^{-i\frac{(m-1)\pi}{4}}, \quad b_0^* = -\sqrt{K_{-1}}\, e^{i\frac{(m-1)\pi}{4}}.$$

Damit erhalten wir für G (3) und (4), wenn wir $a_i = -i a_i^*$, $b_i = i b_i^*$ setzen. Damit ist der erste Teil des Satzes bewiesen.

Bei G_1 gehen wir analog vor. Es ist nach dem *Gauß*schen Integralsatz

$$G_1 = \frac{1}{i \omega} \int_S e^{i \omega x_1} (m - 1 + i \omega x_1)\, u_1\, d o \tag{24}$$

Man braucht also nur $G_3 = \int_S u_1 x_1 e^{i \omega x_1} d o$ zu betrachten und erhält dann (5) und (6).

Man stellt sofort fest, daß der Satz und sein Beweis auch für $m = 2$ gilt, wenn man in (10) unter $\int_{E_{m-1}} d\sigma$ die Zahl 2 versteht.

Bemerkung: Betrachtet man das Integral $G_4^{\pm} = \int_B e^{\pm \omega x_1} d x$, so zeigt eine ganz analoge Rechnung, wo statt Hilfssatz 1 aus § 2 die *Laplace*sche Methode zu verwenden ist, daß

$$G_4^{\pm} \sim \frac{(2\pi)^{\frac{m-1}{2}}}{\omega^{\frac{m+1}{2}}} \sqrt{K(\pm 1, 0 \ldots 0)}\, e^{\omega H(\mp 1, 0, \ldots 0)}, \quad (\omega \to \infty) \tag{25}$$

§ 4.

Es soll nun mittels Satz 1 und Satz 2

$$I_\delta = \int_B (1 - f^2)^\delta e^{i \omega x_1} d x \tag{1}$$

für $\delta \geqq 0$ asymptotisch dargestellt werden. Über B mögen wieder dieselben Voraussetzungen wie in § 3 gelten. Insbesondere ist $B\colon f(x) \leqq 1$. Dann gilt:

Satz 3:

$$I_\delta = \frac{(2\pi)^{\frac{m-1}{2}}\, 2^\delta\, \Gamma(\delta+1)}{\omega^{\frac{m+1}{2}+\delta}} \left[\frac{\sqrt{K_1}}{H_1^\delta}\, e^{i\omega H_1 - \frac{i\pi}{2}\left(\delta + \frac{m+1}{2}\right)} + \right.$$

$$\left. + \frac{\sqrt{K_{-1}}}{H_{-1}^\delta}\, e^{-i\omega H_{-1} + \frac{i\pi}{2}\left(\delta + \frac{m+1}{2}\right)} \right] + 0\left(\omega^{-\frac{m}{2}-\delta-1}\right) + 0\left(\omega^{-m-2}\right) \tag{2}$$

Dabei ist $H_{\pm 1}, = H(\pm 1, 0, \ldots 0)$, ebenso ist $K_{\pm 1}$ erklärt (wie in § 3). Die Abschätzungen gelten in δ gleichmäßig in jedem Intervall $[0, \delta_0]$.

Unter stärkeren Voraussetzungen kann in (2) das zweite Fehlerglied weggelassen werden. Dies wird in der Fortsetzung dieser Arbeit gezeigt werden, wo die Integrale (1) nach einer anderen Methode behandelt werden.

Beweis: Für $\delta = 0$ wurde (2) bereits in § 3 Satz 2 bewiesen, da $I_0 = G$. Es kann also $\delta > 0$ vorausgesetzt werden. Nach § 1 (3) ist für $\delta > 0$

$$I_\delta = 2\delta \int_0^1 (1-u^2)^{\delta-1}\, u^{m+1}\, G(u)\, du, \tag{3}$$

wo $G(u) = \int_B e^{iu\omega x_1}\, dx$. Da $|G| \leqq V(B)$ (Volumen von B), so ist $\int_0^{\frac{1}{\omega}} (1-u^2)^{\delta-1}\, u^{m+1}\, G(u)\, du = 0\left(\omega^{-(m+2)}\right)$ Wir erhalten weiter, wenn wir § 3 (3) mit $k = m+3$ verwenden.

$$I_\delta = 2\delta\, 0\left(\frac{1}{\omega^{m+2}}\right) + \frac{2\delta\,(2\pi)^{\frac{m-1}{2}}}{\omega^{\frac{m+1}{2}}} \int_{\frac{1}{\omega}}^{1} (1-u^2)^{\delta-1}\, u^{\frac{m+1}{2}}\, du$$

$$\left(\sum_{j=0}^{m+3} \frac{a_j\, e^{i\omega u H_1} + b_j\, e^{-i\omega u H_{-1}}}{(u\omega)^{\frac{j}{2}}}\right) + 0\left(\frac{2\delta}{\omega^{m+\frac{5}{2}}} \int_{\frac{1}{\omega}}^{1} (1-u^2)^{\delta-1}\, u^{-\frac{3}{2}}\, du\right) \tag{4}$$

Der letzte 0-Term in (4) ist $0\left(\frac{\delta+1}{\omega^{m+2}}\right)$, denn es ist

$$\int_{\omega^{-1}}^{1} (1-u^2)^{\delta-1}\, u^{-\frac{3}{2}}\, du = \int_{\omega^{-1}}^{1/2} + \int_{1/2}^{1} \leqq 4/3\, \sqrt{\omega} + \frac{2\sqrt{2}}{\delta}.$$

Wir haben nun weiter die Integrale

$$I_{\pm} = 2\delta \int_0^1 (1-u^2)^{\delta-1} u^{\frac{l}{2}} e^{\pm i\omega u H \pm 1} du$$

zu betrachten, wo $l = m+1-j$, $0 \leqq j \leqq m+3$. Wir beschränken uns auf I_+ und setzen $u = 1 - v$, dann kommt

$$I_t = 2^\delta \delta e^{i\omega H_1} \int_0^{1-\frac{1}{\omega}} v^{\delta-1} (1 - \frac{v}{2})^{\delta-1} (1-v)^{\frac{l}{2}} e^{-i\omega H_1 v} dv. \quad (5)$$

Wir betrachten nun in der komplexen v-Ebene den Bereich begrenzt durch die Strecke $v = t$ $(0 \leqq t \leqq 1 - \frac{1}{\omega})$ auf der positiven reellen Achse, durch $v = t e^{\frac{-i\pi}{2}}$ $(0 \leqq t \leqq 1)$ auf der negativ-imaginären Achse und den Viertelkreis $V: v = (1 - \frac{1}{\omega}) e^{-it}$ $(0 \leqq t \leqq \frac{\pi}{2})$. Außerdem ist der Punkt $v = 0$ durch einen Kreisbogen V_ϱ vom Radius ϱ $(\varrho > 0)$ um diesen Punkt auszuschließen. In den so begrenzten Bereich ist der Integrand von (5) eindeutig und regulär, also das Integral über den Rand dieses Bereichs 0. Man sieht sofort, daß $\lim_{\varrho \to 0} \int_{V_\varrho} = 0$.

Wir erhalten also

$$\int_0^{1-\frac{1}{\omega}} = e^{-\frac{i\pi}{2}\delta} \int_0^{1-\omega^{-1}} t^{\delta-1} (1 + \frac{t}{2} i)^{\delta-1} (1+it)^{\frac{l}{2}} e^{-\omega H_1 t} dt + \int_V. \quad (6)$$

Nun ist

$$\left|\int_V\right| \leqq (1 - \frac{1}{\omega})^\delta \, 5 \, (\frac{8}{5})^{\delta-1} \int_0^{\frac{\pi}{2}} \left| 1 - (1 - \frac{1}{\omega}) e^{-it} \right|^{\frac{l}{2}} e^{-\omega H_1 \sin t} dt.$$ Da

$$\frac{1}{\omega^2} \leqq \left| 1 - (1 - \frac{1}{\omega}) e^{-it} \right|^2 = 1 + (1 - \frac{1}{\omega})^2 - 2 (1 - \frac{1}{\omega}) \cos t \leqq$$

$$\leqq \frac{1}{\omega^2} + t^2 \quad (7)$$

wegen $\cos t \geqq 1 - \frac{t^2}{2}$, so ist

$$\left|\int_V\right| \leqq 5 \, (\frac{8}{5})^{\delta-1} \int_0^{\frac{\pi}{2}} (\frac{1}{\omega^2} + C \,.\, t^2)^{\frac{l}{4}} e^{-\omega H_1 \sin t} dt \leqq 5 \, (\frac{8}{5})^{\delta-1}$$

$$\int_0^\infty (\frac{1}{\omega^2} + C\, t^2)^{\frac{l}{4}} e^{-\frac{2}{\pi}\omega H_1 t} dt$$

wo C zwischen 0 und 1 liegt. Wir setzen nun $t = \frac{\tau}{\omega}$ und erhalten

$$|\int_v| \leqq \frac{25_c{}^\delta}{8\omega^{\frac{l}{2}+1}} \int_0^\infty (1 + C\tau^2)^{\frac{l}{4}} e^{-\frac{2}{\pi} H_1 t} dt = 0\left(\frac{1}{\omega^{\frac{l}{2}+1}}\right).$$ Dabei

wurde $c = \frac{8}{5}$ gesetzt und benutzt, daß H auf $|u| = 1$ nach unten beschränkt ist.

Nun betrachten wir das Integral rechts in (6). Entwickeln wir $(1 + \frac{t}{2} i)^{\delta-1} (1 + t i)^{\frac{l}{2}}$ bis zur ersten Potenz in t, so erhalten wir

$$e^{\frac{i\pi}{2}\delta} \int_0^{1-\frac{1}{\omega}} = \int_0^{1-\frac{1}{\omega}} t^{\delta-1} e^{-\omega H_1 t} dt + 0\left(\int_0^{1-\frac{1}{\omega}} t^\delta e^{-\omega H_1 t} dt\right) + $$

$$+ 0\left(\frac{1}{\omega^{\frac{l}{2}+1}}\right) \tag{8}$$

Nun ist für $\gamma \geqq 0$, $\int_0^{1-\frac{1}{\omega}} t^{\gamma-1} e^{-\omega H_1 t} dt = \int_0^\infty - \int_{1-\frac{1}{\omega}}^\infty$. Nun ist

$$\int_{1-\frac{1}{\omega}}^\infty t^{\gamma-1} e^{-\omega H_1 t} dt \leqq \frac{1}{(\omega H_1)^\gamma} \int_{\omega H_1}^\infty t^{\gamma-1} e^{-t} dt \leqq$$

$$\frac{1}{(\omega H_1)^{\gamma+1}} \int_0^\infty t^\gamma e^{-t} dt.$$

Es ist also

$$\int_0^{1-\frac{1}{\omega}} t^{\gamma-1} e^{-\omega H_1 t} dt = \frac{\Gamma(\gamma)}{(\omega H_1)^\gamma} + 0\left(\frac{1}{(\omega H_1)^{\gamma+1}}\right).$$

Wenden wir dies für $\gamma = \delta$ und $\delta + 1$ an, so erhalten wir

$$e^{\frac{i\pi}{2}\delta} \int_0^{1-\frac{1}{\omega}} = \frac{\Gamma(\delta)}{(\omega H_1)^\delta} + 0\left(\frac{1}{(\omega H_1)^{\delta+1}}\right) + 0\left(\frac{1}{\omega^{\frac{l}{2}+1}}\right) \tag{9}$$

also

$$I_\pm = \frac{2^\delta \Gamma(\delta+1) e^{\pm i\omega H \pm 1}}{(\omega H_{\pm 1})^\delta} \left(1 + 0\left(\frac{1}{\omega}\right)\right) + 0\left(\frac{1}{\omega^{\frac{l}{2}+1}}\right) \tag{10}$$

Bei I_- ist einfach der Bereich in der komplexen v-Ebene zu nehmen, der aus dem bei I_+ verwendeten durch Spiegelung an der reellen Achse entsteht. Setzen wir dies alles in (4) ein, so kommt

$$I_\delta = \frac{2^\delta \Gamma(\delta+1)(2\pi)^{\frac{m-1}{2}}}{\omega^{\frac{m+1}{2}}} \sum_{j=0}^{m+3} \frac{1}{\omega^{\delta+\frac{j}{2}}} \left(\frac{a_j e^{i(\omega H_1 - \frac{\pi}{2}\delta)}}{H_1^\delta} + \right.$$

$$\left. + \frac{b_j e^{-i(\omega H_{-1} - \frac{\pi}{2}\delta)}}{H_{-1}^\delta} \right) + \frac{1}{\omega^{\frac{m+1}{2}}} \left| \sum_{j=0}^{m+3} 0 \left(\frac{1}{\omega^{\delta+1+\frac{j}{2}}} \right) + \right.$$

$$+ \sum_{j=0}^{m+3} 0 \left(\frac{1}{\omega^{\frac{j}{2} + \frac{m+1-j}{2} + 1}} \right) + 0 \left(\frac{1}{\omega^{m+2}} \right) \tag{11}$$

Die Glieder in der Summe für $j \geqq 1$ und der erste 0-Term sind $0\left(\frac{1}{\omega^{\delta+\frac{1}{2}}}\right)$, der zweite 0-Term ist $0\left(\frac{1}{\omega^{m+2}}\right)$. Setzt man für a_0 und b_0 ihre Werte nach § 3 (4) ein, so erhält man (2) w. z. b. w.

Man kann mittels der Sätze 1 und 2 auch asymptotische Entwicklungen für $\int_D \Phi(f) e^{i\omega x_1} dx$ aufstellen, wenn $\Phi(u)$ genügend oft differenzierbar ist. Wir beschränken uns, um ein Beispiel zu geben, auf den Fall, daß $\Phi(1) \neq 0$ ist. Dann gilt

Satz 4

$$I = \int_B \Phi(f) e^{i\omega x_1} dx = \Phi(1) \int_B e^{i\omega x_1} dx + 0(\omega^{-\frac{m}{2}-1}) \tag{12}$$

wenn $\Phi(u)$ mindestens zweimal differenzierbar ist.

Beweis: Aus § 1 (1) folgt durch partielle Integration

$$I = \Phi(1) G(1) - \int_0^1 \Phi'(u) u^m G(u) du. \tag{13}$$

Dann folgt aus § 3 (3) mit $k = 0$, wenn wir beachten, daß

$$\int^{\omega^{-\frac{1}{2}}} \Phi'(u) u^m G(u) du = 0(\omega^{-\frac{m}{2}-1}),$$

$$\int_0^1 \Phi'(u) u^m G(u) du = \frac{(2\pi)^{\frac{m-1}{2}}}{\omega^{\frac{m+1}{2}}} [a_0 \int_{\omega^{-\frac{1}{2}}}^1 \Phi'(u) u^{\frac{m-1}{2}} e^{i\omega u H_1} du +$$

$$+ b_0 \int\limits_{\omega^{-\frac{1}{2}}}^{1} \Phi'(u)\, u^{\frac{m-1}{2}} e^{-i\omega u H_1}\, du] + 0\left(\frac{1}{\omega^{\frac{m}{2}+1}}\right) \int\limits_{\omega^{-\frac{1}{2}}}^{1} \Phi'(u)\, u^{\frac{m}{2}-1}\, du)$$

$$+ 0\left(\frac{1}{\omega^{\frac{m}{2}+1}}\right) = 0\left(\frac{1}{\omega^{\frac{m}{2}+1}}\right),$$

da $m \geqq 2$. Man sieht die Abschätzung ein, wenn man die auftretenden Integrale partiell integriert. Damit ist bereits alles gezeigt.

§ 5.

Bis jetzt wurden nur Integrale von der Gestalt $\int\limits_B \Phi(f)\, e^{i\omega x_1}\, dx$ betrachtet. Für die Anwendungen benötigen wir aber Integrale von der allgemeinen Gestalt $\int\limits_{B_t} \Phi(f)\, e^{ilx}\, dx$, wo $B_t : f(x) \leqq t$ für großes $|l|\, t$, insbesondere für $\Phi(f) = (t^2 - f^2)^\delta$ $(\delta \geqq 0)$. Wir wollen uns gleich auf diese, also auf

$$I_\delta = \int\limits_{f \leqq t} (t^2 - f^2)^\delta\, e^{ilx}\, dx \quad (\delta \geqq 0) \tag{1}$$

beschränken. Dabei ist $B_1 = B : f \leqq 1$. Für B mögen wieder die Voraussetzungen von § 3 gelten. Wir können schreiben

$$I_\delta = t^{2\delta+m} \int\limits_B (1 - f^2)^\delta\, e^{it|l|\, e_l x}\, dx, \text{ wo } e_l = \frac{l}{|l|}.$$

Durch eine orthogonale Drehung A können wir erreichen, daß e_l übergeht in $(1, 0, \ldots 0)$. Dabei geht B über in $B' = A\,B$ und es wird

$$I_\delta = t^{2\delta+m} \int\limits_B (1 - f'^2(x))^\delta\, e^{it|l|x_1}\, dx, \tag{2}$$

wo f' Distanzfunktion von B'. Jetzt können wir mit $\omega = t\,|l|$ auf (2) Satz 3, § 4 anwenden. Dies ist möglich, da B' dieselben Voraussetzungen wie B erfüllt. Denn zunächst ist inf K auf $|u| = 1$ eine Drehinvariante, also auch inf K'. Weiter sind alle $6\,m$ Ableitungen von H', wo H' Stützfunktion von B', beschränkt, denn diese sind Linearkomposita in den Ableitungen von H mit Koeffizienten, welche Produkte der Elemente von A sind. Da A orthogonal, so sind diese Produkte und damit auch die Ableitungen von H' beschränkt. Machen wir nun die Drehung wieder rückgängig, so ist in (2) überall der Vektor $(1, 0, \ldots 0)$ durch e_l zu ersetzen und wir erhalten also

Satz 5: Es ist für $\delta \geqq 0$

$$\int_{(x)\leqq t} (t^2 - f^2(x))^\delta e^{i l x}\, d x = \frac{(2\pi t)^{\frac{m-1}{2}} (2t)^\delta \Gamma(\delta+1)}{|l|^{\frac{m+1}{2}+\delta}} [C_1 e^{i|l| t H(e_l)-\alpha} +$$

$$+ C_2 e^{-i|l| t H(e_{-l})+\alpha}] + 0\left(\frac{t^{\frac{m}{2}+\delta-1}}{|l|^{\frac{m}{2}+\delta+1}}\right) + 0\left(\frac{t^{2\delta-2}}{|l|^{m+2}}\right) \qquad (3)$$

Dabei ist

$$e_l = \frac{l}{|l|}, \alpha = \frac{i\pi}{2}(\delta + \frac{m+1}{2}), C_1 = \frac{\sqrt{K(e_l)}}{(H(e_l))^\delta}, C_2 = \frac{\sqrt{K(-e_l)}}{(H(-e_l))^\delta} \qquad (3')$$

Die Abschätzung ist gleichmäßig in δ in jedem Intervall $[0, \delta_0]$. Ist insbesondere B symmetrisch mit o als Mittelpunkt, so ist das Hauptglied in (3), wenn $t = 1$

$$2\frac{(2\pi)^{\frac{m-1}{2}} 2^\delta \Gamma(\delta+1)}{|l|^{\frac{m+1}{2}+\delta}} \frac{\sqrt{K(e_l)}}{(H(e_l))^\delta} \cos(|l| H(e_l) - \frac{\pi}{2}(\delta + \frac{m+1}{2})) \qquad (3'')$$

Dies ist für die Kugel wohlbekannt, da $I_\delta = (2\pi)^{\frac{m-1}{2}} \dfrac{J_{\frac{m}{2}+\delta}(|l|)}{|l|^{\frac{m}{2}+\delta}}$

ist, wo J rechts die *Bessel*sche Funktion 1. Art ist. Hier kann das Restglied bedeutend schärfer angegeben werden. Für das Integral $G_4(\pm l) = \int_B e^{\pm l x}\, d x$ erhält man mittels § 3 (25)

$$G_4(\pm l) = \int_B e^{\pm l x}\, d x \sim \frac{(2\pi)^{\frac{m-1}{2}}}{|l|^{\frac{m+1}{2}}} \sqrt{K(\pm e_l)}\, e^{|l| H(\pm e_l)} \qquad (4)$$

Daraus folgt

$$\overline{\lim_{|l|\to\infty}} \frac{\ln G_4(\pm l)}{|l|} = H(\pm l) \qquad (5)$$

Dies steht mit den allgemeinen Untersuchungen von *Plancherel-Polya*[6] in Einklang. Dort wird (5) für beliebige konvexe Körper bewiesen.

Es sollen noch Abschätzungen für $G = \int_B e^{i\omega x_1}\, d x$ und $G_1 = \int_B e^{i\omega x_1}(m + i\omega x_1)\, d x$ gegeben werden, die auch dann gelten, wenn nur auf B der *Gauß*sche Integralsatz anwendbar ist.

[6] Commentarii Helvetici 9.

Satz 6: Es ist, wenn $\omega \neq 0$

$$|G| \leqq \frac{S(B)}{|\omega|},\ |G_1| \leqq C(B) \tag{6}$$

Dabei ist $S(B)$ die Oberfläche von B und $C(B)$ eine Konstante, die nur von B abhängt.

Beweis: Die Abschätzung von G folgt sofort aus § 3 (8). Bei G_1 sieht man dies so: Für $|\omega| \leqq 1$ ist

$$|G_1| \leqq \int_B (m + |x_1|)\, d\, x = C_1(B).$$

Ist $|\omega| \geqq 1$, so folgt aus § 3 (24)

$$|G_1| \leqq (m-1)\, S(B) + \int_{S(B)} |x_1|\, d\, o = C_2(B) \text{ w.z.b.w.}$$

§ 6.

Es sei wieder B_t der Bereich $f(x) \leqq t$, $B_1 = B$ ist dann der Eichkörper. Sein Volumen bezeichnen wir mit $\mathfrak{J}$. Weiter sei A eine Matrix mit Det. $A = 1$. $\varphi(x)$ sei eine integrierbare Funktion auf B_t, die außerhalb B_t verschwindet und c eine Zahl > 0.

Mit g, bzw. l sollen im folgenden stets Gitterpunkte des R_m bezeichnet werden.

Wir bilden uns die in g periodische Funktion

$$\Phi(x) = \sum_g \varphi(c\, A\, (g - x)) \tag{1}$$

und entwickeln sie formal in eine Fouriersche Reihe

$$\Phi(x) \sim \sum_l a_l\, e^{2\pi i l x} \tag{2}$$

Ist E der Würfel $0 \leqq x_i \leqq 1$ $(i = 1, \ldots m)$, so ist

$$a_l = \int_E \Phi(x)\, e^{-2\pi i l x}\, d\, x = \frac{1}{c^m} \int_{R_m} \varphi(-A\, x)\, e^{-\frac{2\pi i}{c} l x}\, d\, x$$

also

$$a_l = \frac{1}{c^m} \int_{B_t} \varphi(x)\, e^{\frac{2\pi i k x}{c}}\, d\, x \quad (k = A^{*-1}) \tag{3}$$

Nach dem Vollständigkeitssatz ist $\Sigma\, |a_l|^2$ konvergent und es ist

$$\int_E \Phi^2\, d\, x = \sum_l |a_l|^2 \tag{4}$$

In (4) spezialisieren wir jetzt, indem wir $\varphi(x) = 1$ setzen und $c = 2$ nehmen. Dann ist $a_0 = \frac{1}{2^m} \int\limits_{B_t} dx = \frac{V}{2^m}$, wo $V = \Im t^m$ das Volumen von B_t. Weiter ist

$$\int\limits_E \Phi^2 dx = \int\limits_{Rm} \varphi(-2Ax) \sum_g \varphi(2A(g-x))dx = \frac{1}{2^m} \sum_g \int\limits_{B_t} \varphi(-x+2Ag)\, dx,$$

also wird aus (4)

$$2^m [V + \sum_g{}' \int\limits_{B_t} \varphi(-x + 2Ag)\, dx] = V^2 + \sum_l{}' \left| \int\limits_{B_t} e^{\pi i k x} dx \right|^2 \qquad (5)$$

Dabei bedeutet Σ', daß bei der Summation $g = 0$, bzw. $l = 0$ auszulassen ist. Diese Formel stammt von *C. Siegel*[7].

§ 7.

Wir nehmen für $\varphi(x)$ wieder $\varphi(x) = 1$ auf B_t, 0 sonst, setzen aber $c = 1$. Dann stellt

$$\Phi(y) = \sum_g \varphi(A(g-y)) \sim \sum_l a_l e^{2\pi i l y} \qquad (1)$$

die Anzahl der Gitterpunkte in $f(A(x-y)) \leqq t$ dar.

Es genügt, wenn y auf den Würfel

$$E: 0 \leqq y_i \leqq 1 \quad (i = 1, \ldots, m) \qquad (2)$$

beschränkt wird.

Wir betrachten gleich allgemeiner, wenn wir $t^2 = u$ setzen, die Funktion $\varphi_\delta(x) = \frac{1}{\Gamma(\delta+1)} (u - f^2(x))^\delta$ auf B_t. Damit ist, wenn $f(Ax) = F(x)$ gesetzt wird, nach § 6 (2) und (3)

$$\Phi_\delta(y, u) = \frac{1}{\Gamma(\delta+1)} \sum_{F^2(g-y) \leqq u} (u - F^2(g-y))^\delta \sim \sum_l L_\delta(l, u) e^{2\pi i l y} \qquad (3)$$

wo

$$L_\delta(l, u) = \frac{1}{\Gamma(\delta+1)} \int\limits_{f^2(x) \leqq u} (u - f^2(x))^\delta e^{2\pi i k x} dx. \qquad (4)$$

Es ist

$$L_{\delta+1}(l, u) = \int\limits_0^u L_\delta(l, v)\, dv, \qquad (5)$$

[7]) Acta Mathematica 64.

denn es ist

$$\int_0^u L_\delta(l, v)\, dv = \frac{1}{\Gamma(\delta+1)} \int_0^u dv \int_{f \leqq \sqrt{v}} (v - f^2)^\delta e^{2\pi i k x}\, dx =$$

$$= \frac{1}{\Gamma(\delta+1)} \int_{f \leqq \sqrt{u}} e^{2\pi i k x}\, dx \int_{f^2}^u (v - f^2)^\delta\, dv =$$

$$= \frac{1}{\Gamma(\delta+2)} \int_{f \leqq \sqrt{u}} e^{2\pi i k x}\, (u - f^2)^{\delta+1}\, du = L_{\delta+1}.$$

Weiter ist

$$L_\delta(0, u) = \mathfrak{J}\, u^{\frac{m}{2}+\delta} \frac{\Gamma(\frac{m}{2}+1)}{\Gamma(\delta + \frac{m}{2}+1)} \tag{6}$$

denn nach § 1 (1) ist

$$\int_{f \leqq \sqrt{u}} (u - f^2(x))^\delta\, dx = m I \int_0^{\sqrt{u}} (u - v^2)^\delta v^{m-1}\, dv =$$

$$= \frac{m}{2} \mathfrak{J}\, u^{\frac{m}{2}+\delta} \frac{\Gamma(\delta+1)\,\Gamma(\frac{m}{2})}{\Gamma(\delta + \frac{m}{2} + 1)}$$

Wie (5) zeigt man

$$\Phi_\delta(l, u) = \frac{1}{\Gamma(\delta)} \int_0^u \Phi(v, y)\,(u - v)^{\delta-1}\, dv \tag{7}$$

wo $\Phi_0 = \Phi$ aus (1) ist

Daraus folgt

$$\Phi_{\delta+1}(y, u) = \int_0^u \Phi_\delta(y, v)\, dv \tag{7'}$$

Es wird jetzt gezeigt werden, daß in (3) für $\delta > \frac{m-1}{2}$ das $=$ Zeichen steht. Wir zeigen gleich folgenden allgemeineren Satz:

Satz 7: Ist h ein beliebiger Punkt, aber kein Gitterpunkt $\neq 0$, so gilt

$$\sum_{F^2(g-y) \leqq u} (u - F^2(g-y))^\delta e^{-2\pi i h g} =$$

$$= \sum_l e^{2\pi i (l-h) y} \int_{f^2 \leqq u} e^{-2\pi i (k-\bar{h}) x} (u - f^2)^\delta\, dx, \tag{8}$$

wo $\bar{h} = A^{*-1} h$, wenn

1. $\delta > \frac{m-1}{2}$, und zwar für alle y in E. Die Reihe rechts in (8) konvergiert absolut und gleichmäßig in jeden Intervall $[u_1, u_2]$ wo $u_1 > 0$ und wenn

2. $\delta \geqq 0$, $h = 0$ für alle in E bis auf eine Menge vom Lebesgueschen Maße 0.

Dabei ist vorausgesetzt, daß $B_1 = B$ alle Voraussetzungen von § 3 erfüllt.

Beweis: Es werde gesetzt

$$\Phi_\delta^* (y, u) = \frac{1}{\Gamma(\delta+1)} \sum_{F^2(g-y) \leqq u} (u - F^2 (g-y))^\delta \, e^{-2\pi i h g}, \quad L_\delta^* (l, u) =$$

$$= \int_{f^2 \leqq u} e^{2\pi i (k-\bar{h}) x} (u - f^2)^\delta \, d\, x.$$

Dann gilt auch für diese Größen (5) und (7').

Es soll zunächst der erste Teil der Behauptung bewiesen werden. Es sei also $\delta > \frac{m-1}{2}$. Dann ist die Reihe rechts in (8) absolut und in jedem Intervall $[u_1, u_2]$ in u gleichmäßig konvergent, denn es ist nach § 5 (3)

$$L_\delta^* (l, u) = u^{\frac{m-1}{4} + \frac{\delta}{2}} \, 0 \left(\frac{1}{|k-\bar{h}|^{\frac{m+1}{2}+\delta}} \right) + u^{\delta-1} \, 0 \left(\frac{1}{|k-\bar{h}|^{m+2}} \right). \tag{9}$$

Da $k - \bar{h} = A^{*-1} (l - h) \neq 0$, da h kein Gitterpunkt und $\sum_l \frac{1}{|k-\bar{h}|^\sigma}$ für $\sigma > m$ konvergent ist, so ist damit diese Behauptung gezeigt.

Wir betrachten die Funktion $\varphi_\lambda (x) = (u - F^2 (x-y))^\delta \, e^{-2\pi i h x}$ auf $F^2 (x-y) \leqq u$. Außerhalb dieses Bereiches sei $\varphi_\lambda = 0$. Man sieht weiter, daß alle Voraussetzungen des Satzes von *Bochner*[8] über die *Poisson*sche Summenformel erfüllt sind und wir erhalten nach diesem Satz (8).

Für $u = 0$ gilt sie selbstverständlich auch.

Es soll nun der zweite Teil der Behauptung gezeigt werden. Nach (3) ist die rechte Seite von (8) für $h = 0$ die Fouriersche Reihe der linken

---

[8] Mathematische Annalen 106.

Seite. Nun ist aber nicht nur $\sum_l |L_\delta|^2$ konvergent, wie es nach dem Vollständigkeitssatz sein muß (vgl. § 6 (4)), sondern sogar $\sum_l |L_\delta|^2 |l|^\eta$ für jedes η mit $0 < \eta < 1$. Denn es ist zunächst nach Hilfssatz 2 $|l| \leqq C(A)|k|$, wo C von A abhängt. Weiters ist nach (9) $\sum_l |L_\delta|^2 |k|^\eta$ konvergent. Daraus folgt aber nach einem Satz von *Kaczmarz*[9], daß die Reihe rechts in (8) für fast alle y in E konvergiert und gleich der linken Seite ist. Damit ist der Satz vollständig bewiesen.

Satz 8: Ist $f(x)$ die Distanzfunktion, $H(u)$ die Stützfunktion eines konvexen Körpers B, welcher den Voraussetzungen des § 3 genügt, und sind außerdem noch diese Funktionen für $x \neq 0$, $u \neq 0$ sogar analytisch, so gibt es eine periodische Funktion $X(x)$ (Periode 1 in $x_1, \ldots x_m$), so beschaffen, daß für seine Fourierkoeffizienten a_l gilt:

$$\overline{\lim_{R \to \infty}} \left| \sum_{f(g) \leqq R} \left(1 - \frac{f^2(g)}{R^2}\right)^{\frac{m-1}{2}} a_g \right| = +\infty \tag{10}$$

Für die Kugel wurde dies von *Bochner*[10] gezeigt. Für den Beweis ist folgendes zu beachten: $H(u)$ ist sicher nicht analytisch in $u = 0$ und dies ist auch die einzige Stelle nach Voraussetzung. Weiters haben wir in § 5 (3) eine asymptotische Formel für L_δ^*, welche in δ im Intervall $[\frac{m-1}{2}, \delta_0]$ ($\delta_0 > \frac{m-1}{2}$, beliebig) gilt. Dann kann der Beweis wie bei *Bochner* geführt werden.

§ 8.

Es soll jetzt für die Anzahl der Gitterpunkte $\Phi(y, u)$ von $f(A(x-y)) \leqq \sqrt{u}$ eine asymptotische Abschätzung gegeben werden:

Satz 9:

$$\Phi(y, u) = \mathfrak{J} u^{\frac{m}{2}} + O\left(u^{\frac{m(m-1)}{2(m+1)}}\right) \tag{1}$$

Für das Ellipsoid wurde dies von *Landau*[11] gezeigt. Wir werden weitgehend seiner Methode folgen. Satz 7, § 7 gilt sicher für $\delta = [\frac{m+1}{2}] > \frac{m-1}{2}$ und $h = 0$. Es ist nach (5)

$\frac{d^\delta L^\delta}{d u^\delta} = L_0 = \int_{f^2(x) \leqq u} e^{2\pi i k x}\, dx$ und nach § 5 (3)

---

[9] Studia Mathematica 2.

[10] Transactions 40, S. 193.

$$L_\delta = u^{\frac{m-1}{4} + \frac{\delta}{2}}\, 0 \left(\frac{1}{|k|^{\frac{m+1}{2}+\delta}} \right), \tag{2}$$

da das zweite 0-Glied wegen $\frac{m+1}{2} + \delta \leqq m + 1$ wegfällt. Wir setzen kurz $L_\delta = L$ und betrachten mit *Landau*

$$\Delta L = \sum_{\gamma=0}^{\delta} (-1)\, l^{\delta-\gamma} \binom{\delta}{\gamma} L(u + \nu\tau, y)$$

wo $\tau = u^{\frac{1}{m+1}}$, also $0 < \tau \leqq u$ ist. Dann ist zunächst nach (2)

$$\Delta L = u^{\frac{m-1}{4} + \frac{\delta}{2}}\, 0 \left(\frac{1}{|k|^{\frac{m+1}{2}+\delta}} \right)$$

Nun gibt es sicher ein τ' mit $u \leqq \tau' \leqq u + \delta\tau$, so daß $\Delta L = \tau^\delta L^{(\delta)}(\tau')$ Nun ist $L^{(\delta)} = L_0$, also folgt aus (2) mit $\delta = 0$

$$\Delta L = \tau^\delta u^{\frac{m-1}{4}}\, 0 \left(\frac{1}{|k|^{\frac{m+1}{2}}} \right).$$

Zusammenfassend haben wir

$$\Delta L = \frac{u^{\frac{m-1}{4}}}{|k|^{\frac{m+1}{2}}} \operatorname{Min}\left(\left(\frac{\sqrt{u}}{|k|} \right)^\delta, \tau^\delta \right). \tag{3}$$

Nach § 7 (8) und § 7 (6) gilt

$$\Phi_\delta(y, u) = \frac{\Gamma(\frac{m}{2}+1)}{\Gamma(\delta+\frac{m}{2}+1)}\, u^{\frac{m}{2}+\delta} + \sum_l{}' e^{2\pi i l y} L_\delta(l, u).$$

Wenden wir auf Φ_δ die obige Δ-Operation an, welche wir rechts gliedweise anwenden können, so kommt

$$\Delta \Phi_\delta = \frac{\Gamma(\frac{m}{2}+1)}{\Gamma(\delta+\frac{m}{2}+1)}\, \Delta\left(u^{\frac{m}{2}+\delta}\right) + \sum_l{}' e^{2\pi i l y} \Delta L_\delta(l, u) \tag{4}$$

Dann ist nach (3)

$$\left| \sum{}' e^{2\pi i l y} \Delta L \right| \leqq C\, u^{\frac{m-1}{4}} \sum{}' \frac{1}{|k|^{\frac{m+1}{2}}} \left(\operatorname{Min}\left(\frac{\sqrt{u}}{|k|}, \tau \right) \right)^\delta =$$

$$= C \Big[u^{\frac{m-1}{4}} \tau^\delta \sum_{|k| \leqq \frac{\sqrt{u}}{\tau}} |k|^{-\frac{m+1}{2}} + u^{\frac{m-1}{4}+\frac{\delta}{2}} \sum_{|k| > \frac{\sqrt{u}}{\tau}} |k|^{-\frac{m+1}{2}-\delta}$$

Nun ist die erste Summe rechts $0\left(\left(\frac{\sqrt{u}}{\tau}\right)^{\frac{m-1}{2}}\right)$, die zweite

$$0\left(\left(\frac{\sqrt{u}}{\tau}\right)^{\frac{m-1}{2}-\delta}\right)$$

also folgt

$$\sum_l{}' e^{2\pi i l y}\, \Delta L = 0\,(\tau^\delta\, u^{\frac{m(m-1)}{2(m+1)}}) \tag{5}$$

Weiter ist

$$\Delta\,(u^{\frac{m}{2}+\delta}) = \tau^\delta\, \frac{d^\delta}{d\,u^\delta}\,(u^{\frac{m}{2}+\delta})\,\Big|_{u=\tau'} = \tau^\delta\, \frac{\Gamma(\delta+\frac{m}{2}+1)}{\Gamma(\frac{m}{2}+1)}\, \tau'^{\frac{m}{2}} =$$

$$= \tau^\delta\, \frac{\Gamma(\delta+\frac{m}{2}+1)}{\Gamma(\frac{m}{2}+1)}\,(u^{\frac{m}{2}} + 0\,(u^{\frac{m}{2}-1}\,\tau)),$$

wo τ' eine Zwischenstelle in $[u, u+\delta\,\tau]$. Es ist also

$$\frac{\Gamma(\frac{m}{2}+1)}{\Gamma(\delta+\frac{m}{2}+1)}\,\Delta\,(u^{\frac{m}{2}+\delta}) = \tau^\delta\,(u^{\frac{m}{2}} + 0\,(u^{\frac{m(m-1)}{2(m+1)}})) \tag{6}$$

Daraus folgt, wenn wir (5) und (6) zusammenfassen

$$\Delta\,\Phi_\delta = \tau^\delta\,(\mathfrak{J}\,u^{\frac{m}{2}} + 0\,(u^{\frac{m(m-1)}{2(m+1)}})) \tag{7}$$

Nun ist, wenn wir das Argument y unterdrücken,

$$\Delta\,\Phi_\delta = \int_u^{u+\tau} d\,u_1 \int_{u_1}^{u_1+\tau} du_2 \ldots \int_{u_{\delta-1}}^{u_{\delta-1}+\tau} \Phi\,(u_\delta)\,d\,u_\delta \tag{8}$$

Nun ist

$$\Phi\,(u) \leqq \Phi\,(u_\delta) \leqq \Phi\,(u+\delta\,\tau), \text{ also nach (8)}$$

$$\tau^\delta\,\Phi\,(u) \leqq \Delta\,\Phi \leqq \tau^\delta\,\Phi\,(u+\delta\,\tau). \tag{9}$$

Dann ist nach (7) einerseits $\Phi\,(u) \leqq \mathfrak{J}\,u^{\frac{m}{2}} + 0\,(u^{\frac{m(m-1)}{2(m+1)}})$, andererseits $\Phi\,(u+\delta\,u^{\frac{1}{m+1}}) \geqq \mathfrak{J} u^{\frac{m}{2}} + 0\,(u^{\frac{m(m-1)}{2(m+1)}})$.

Setzen wir $u+\delta\,u^{\frac{1}{m+1}} = v$, dann ist $u = v - \delta\,u^{\frac{1}{m+1}} = v + 0\,(v^{\frac{1}{m+1}})$ also $\Phi\,(v) \geqq \mathfrak{J}\,(v + 0\,(v^{\frac{1}{m+1}}))^{\frac{m}{2}} + 0\,(v^{\frac{m(m-1)}{2(m+1)}}) = \mathfrak{J}\,v^{\frac{m}{2}} + 0\,(v^{\frac{m(m-1)}{2(m+1)}})$ also

$$\Phi(u, y) = \mathfrak{J} u^{\frac{m}{2}} + 0\,(u^{\frac{m(m-1)}{2(m+1)}}) \text{ w.z.b.w.}$$

Wir wollen jetzt, und zwar für $y = 0$ weiter zeigen (vgl. *Landau*[11])

Satz 10: Es ist

$$\Phi(0, u) = \sum_{F^2(g) \leqq u} 1 = \mathfrak{J} u^{\frac{m}{2}} + \Omega\,(u^{\frac{m-1}{4}}) \tag{10}$$

Beweis: Wir setzen $R_\delta(0, u) = \Phi_\delta(0, u) - \mathfrak{J} u^{\frac{m}{2}+\delta} \frac{\Gamma(\frac{m}{2}+1)}{\Gamma(\delta+\frac{m}{2}+1)}$,

dann gilt wegen § 7 (7)

$$R_\delta = \frac{1}{\Gamma(\delta)} \int_0^u R_0(v, 0)\,(u - v)^{\delta-1}\,d\,v. \tag{11}$$

Weiter ist nach § 7 (8) und § 7 (6), da R_δ reell ist,

$$R_\delta = \sum_l{}' \text{Real } L_\delta(l\,u), \tag{12}$$

wenn $\delta > \frac{m-1}{2}$. Wir verwenden (12) zunächst für $\delta_1 = \delta + 1$, wo $\delta = [\frac{m+1}{2}]$. Dann ist nach (2), da $\delta_1 \leqq m + 2$, auf (12) angewendet

$$R_{\delta_1} = 0\,(u^{\frac{m-1}{4}+\frac{\delta_1}{2}}). \tag{13}$$

Weiter ist nach § 5 (3)

$$\text{Real } L_\delta = \frac{u^{\frac{m-1}{4}+\frac{\delta}{2}}}{|\,k\,|^{\frac{m+1}{2}+\delta}} [A_l \cos\alpha_l + B_l \cos\beta_l] + 0\left(\frac{u^{\frac{m-1}{4}+\frac{\delta-1}{2}}}{|\,k\,|^{\frac{m}{2}+1+\delta}}\right) \tag{14}$$

wo $A_l = \frac{\Gamma(\delta+1)}{2\pi^{\delta+1}} \frac{\sqrt{K(e_k)}}{(H(e_k))^\delta}$, $B_l = \frac{\Gamma(\delta+1)}{2\pi^{\delta+1}} \frac{\sqrt{K(-e_k)}}{(H(-e_k))^\delta}$. Sie sind also positiv. Weiter ist

$$\alpha_l = 2\pi\sqrt{u}\,H(k) - \frac{\pi}{2}(\delta + \frac{m+1}{2}), \beta_l = 2\pi\sqrt{u}\,H(-k) - \frac{\pi}{2}(\delta + \frac{m+1}{2})$$

Weiter ist, da $\delta = [\frac{m+1}{2}]$, $\cos\alpha_l = (-1)^j \cos(2\pi\sqrt{u}\,H(k) - \frac{\pi}{4}\varepsilon)$, wo $j = [\frac{m+1}{2}]$ und $\varepsilon = 1$ wenn m gerade und sonst 0. Dasselbe gilt

[11] Preußische Sitzungsberichte 1915, Göttinger Nachrichten, S. 137ff.

für $\cos \beta_l$. Setzen wir noch $2\pi\sqrt{u}\,H(k)-\frac{\pi}{4}\varepsilon=\alpha_l'$, $2\pi\sqrt{u}\,H(-k)-\frac{\pi}{4}\varepsilon=$ $=\beta_l'$ so wird (12)

$$R_\delta = (-1)^j\, u^{\frac{m-1}{4}+\frac{\delta}{2}}\, \Sigma' \frac{A_l \cos\alpha_l' + B_l \cos\beta_l'}{|k|^{\frac{m+1}{2}+\delta}} + 0\,(u^{\frac{m-1}{4}+\frac{\delta-1}{2}}) \tag{15}$$

Es sei nun K eine natürliche Zahl. Dann kann man zu jedem $\eta > 0$ und zu jedem u_0 ein $u\,(\eta, K, u_0)$ so bestimmen, so daß

$$\cos\alpha_l' = \cos\,(2\,\pi\sqrt{u}\,H\,(\pm\,k) - \frac{\pi}{4}\,\varepsilon) > \cos\frac{\pi}{4}\,\varepsilon - \eta \geqq \frac{1}{\sqrt{2}} - \eta$$

für alle k mit $|\,k\,| \leqq K$ (Dirichletscher Approximationssatz. Dann ist aber, wenn Σ' die Summe in (15) rechts bedeutet

$$\Sigma' \geqq \sum_{|k|\leqq K}{}' - \sum_{|k|>K} \frac{A_l + B_l}{|k|^{\frac{m+1}{2}+\delta}}, \text{ also}$$

$$\overline{\lim_{u\to\infty}}\ \Sigma' \geqq (\frac{1}{\sqrt{2}} - \eta) \sum_{|k|\leqq K}{}' \frac{A_l + B_l}{|k|^{\frac{m+1}{2}+\delta}} - \sum_{|k|>K} \frac{A_l + B_l}{|k|^{\frac{m+1}{2}+\delta}},$$

also für $\eta \to 0$, $K \to \infty$

$$\overline{\lim_{u\to\infty}}\ \Sigma' \geqq \frac{1}{\sqrt{2}} \sum_{k}{}' \frac{A_l + B_l}{|k|^{\frac{m+1}{2}+\delta}} > 0 \tag{16}$$

Daraus folgt, daß

$$R_\delta = \Omega\,(u^{\frac{m-1}{4}+\frac{\delta}{2}}) \tag{17}$$

Wäre nun (10) falsch, also

$$R_0 = o\,(u^{\frac{m-1}{4}}) \tag{18}$$

so wäre z. B. nach *M. Riesz*[12], wegen (13) und $0 < \delta < \delta_1$

$$R_\delta = o\,(u^{\frac{m-1}{4}(1-\frac{\delta}{\delta_1})}\, u^{(\frac{m-1}{4}+\frac{\delta_1}{2})\frac{\delta}{\delta_1}}) = o\,(u^{\frac{m-1}{4}+\frac{\delta}{2}}) \tag{19}$$

in Gegensatz zu (17). Damit ist alles gezeigt.

[12] Acta Szeged 1.

§ 9.

Es soll nun Satz 9 in gewisser Hinsicht für fast alle η aus E (§ 7 (2)) verschärft werden.

Satz 11: Es sei $\lambda(u)$ eine monoton wachsende Funktion mit $\lambda(u) \to \infty$, δ eine beliebige Zahl mit $0 \leqq \delta < \frac{2}{m-1}$ und $\{u_\nu\}$ eine solche Folge, so daß

$$\Sigma \frac{1}{\lambda_\nu^{2+\delta}},$$

wo $\lambda_\nu = \lambda(u_\nu)$, konvergiert. Dann ist für fast alle y aus E

$$\Phi(y, u) = \mathfrak{J} u^{\frac{m}{2}} + 0\,(u^{\frac{m-1}{4}} \lambda(u)), \tag{1}$$

wo u die Folge $\{u_\nu\}$ durchläuft.

Für $\delta = 0$ wurde dieser Satz von *Kendall*[3] gezeigt, und zwar für $m = 2$.

Beweis: Nach § 7 (1) ist $\sum\limits_l L_0\, e^{2\pi i l y}$ die *Fourier*sche Reihe von $\Phi(y, u)$, also

$$R_0 = \Phi - \mathfrak{J} u^{\frac{m}{2}} \sim \sum_l{}' L_0(l, u)\, e^{2\pi i l y} \tag{2}$$

Es ist nach § 6 (4) $\sum\limits_l{}' |L_0|^2$ konvergent. Da aber nach § 8 (2) $L_0 = 0 \left(\frac{u^{\frac{m-1}{4}}}{|k|^{\frac{m+1}{2}}}\right)$, so ist sogar $\Sigma' |L_0|^\alpha$ mit $\alpha = 1 + \frac{1}{\delta+1}$ konvergent, da $\Sigma |k|^{-\frac{m+1}{2}\alpha}$ wegen $\frac{m+1}{2}\alpha > m$ konvergent. Dann folgt nach dem Satz von *Hausdorff-Yonny*[13], daß wegen $1 < \alpha < 2$

$$\sigma_\beta = [\int_E |R_0(y, u)|^\beta\, d\,y]^{\frac{1}{\beta}} \leqq [\Sigma' |L_0|^\alpha]^{\frac{1}{\alpha}}, \tag{3}$$

wo $\frac{1}{\alpha} + \frac{1}{\beta} = 1$, also $\beta = 2 + \delta$. Wir erhalten also

$$\sigma_\beta^\beta = 0\,(u^{\frac{m-1}{4}\beta}). \tag{4}$$

---

[13] F. Riesz, Math. Zeitschrift 18.

Nun gehen wir wie *Kendall* vor. Ist, u fest, M die Teilmenge von E, wo $|R_0| \geqq \lambda\,\sigma_\beta$, dann ist

$$\sigma_\beta^\beta \geqq \int_M |R_0|^\beta\, d\,y \geqq \lambda^\beta\,\sigma_\beta^\beta\, V(M),$$

wo $V(M)$ das Maß von M. Es ist also

$$V(M) \leqq \lambda^{-\beta}. \tag{5}$$

Um (1), also $R_0(y,u) \leqq C(y)\, u^{\frac{m-1}{4}}\,\lambda(u)$ für $u = u_\nu \to \infty$ und fast alle y zu zeigen, gehen wir so vor: Es ist nur zu zeigen, daß die Menge $\overline{M}$, für die $|R_0(y,u_\nu)| < \lambda_\nu\,\sigma_\beta^{(\nu)}$ $(\sigma_\beta^{(\nu)} = \sigma_\beta\,(u_\nu))$ für alle ν, mit $\sigma_\beta^{(\nu)} = 0\,(u_\nu^{\frac{m-1}{4}})$ das Maß 1 hat. Mit M^* soll stets das Komplement von M in E bezeichnet werden.

Dann ist $\overline{M} = \Sigma\, S_\nu$, wo $S_\nu = \sum_{j=\nu}^{\infty} M_j^*$. Dabei ist

$$M_j = \text{Menge}\,\{\,|R_0(y,u)| \geqq \lambda_\nu\,\sigma_\beta^{(\nu)}\}$$

Es ist also $V(S_\nu^*) \leqq \sum_{j=\nu}^{\infty} V(M_j) \leqq \sum_{j=\nu}^{\infty} \lambda_j^{-\beta} < \eta$, wenn $\nu \geqq \nu_0(\eta)$ für jedes $\eta > 0$, also $V(\overline{M}) > 1 - \eta$ für alle η w.z.b.w.

§ 10.

Für $\beta = 2$ gilt nach § 6 (4) in § 9 (3) das Gleichheitszeichen, also wenn $\sigma = \sigma_2$ gesetzt wird

$$\sigma^2 = \Sigma'\,|L_0|^2 \tag{1}$$

Wir nennen mit *Kendall* σ die Streuung. Es ist

$$\frac{\sigma^2}{u^{\frac{m-1}{2}}} = \frac{1}{4\pi^2}\,\Sigma'\,\frac{1}{|k|^{m+1}}\,\Big(K(e_k) + K(-e_k) + 2\sqrt{K(e_k)\,K(-e_k)}\cos\Big(|k|\sqrt{u}\,(H(e_k)+H(-e_k)) - \frac{\pi}{2}(m+1)\Big)\Big) + 0\,(u^{-\frac{1}{4}}) \tag{2}$$

Daraus folgt

$$\frac{1}{U}\int_1^U \frac{\sigma^2}{u^{\frac{m-1}{4}}}\,d\,u = \frac{1}{4\pi^2}\,\Sigma'\,\frac{K(e_k)+K(-e_k)}{|k|^{m+1}} + 0\,(U^{-\frac{1}{4}}) \tag{3}$$

also

$$\lim_{U \succ \infty} \frac{1}{U} \int_1^U \frac{\sigma^2}{u^{\frac{m-1}{2}}} \, d\,u = \frac{1}{4\pi^2} \Sigma' \frac{K(e_k) + K(-e_k)}{|k|^{m+1}} \tag{4}$$

Aus (3) folgt sofort

Satz 12: Es ist die Streuung $\sigma = \Omega(u^{\frac{m-1}{4}})$.

Wäre nämlich $\sigma = o(u^{\frac{m-1}{4}})$, so wäre die linke Seite von (3), $o(1)$ in Gegensatz zur rechten Seite.

Aus (3) läßt sich noch eine weitere Folgerung ziehen (vgl. dazu für $m = 2$ *Kendall*, S. 22 (31)): Es war ja σ, wie B_0 usw., von der Matrix A mit Det. $A = 1$ abhängig. Es sei nun A orthogonal. Dann bilden wir uns den Mittelwert $\bar{\sigma}^2$ aller σ^2, also

$$\bar{\sigma}^2 = \frac{1}{\{A\}} \int \sigma^2 \, d\,\{A\}. \tag{5}$$

Dabei bedeutet d $\{A\}$ das Volumselement in der Gruppe der Drehungen und $\{A\}$ das Gesamtvolumen.

Nun ist $\int K\,d\,\sigma = S(B)$ (vgl. B. F. *S.* 63, Z. 6 v. u.), wo $d\,\sigma$ wie in § 3 das Oberflächenelement der m-dim. Einheitskugel $E\,m$ bedeutet. Es sei ω die Gesamtoberfläche. Dann folgt aus (3)[13a]

$$\lim_{U \succ \infty} \int_1^U \frac{\bar{\sigma}^2}{u^{\frac{m-1}{2}}} \, d\,u = \frac{S(B)}{2\pi^2\omega} \sum_l{}' \frac{1}{|L|^{m+1}} \tag{6}$$

Dabei ist, wie immer $B: f(x) \leqq 1$ und $\omega = \dfrac{2\pi^{\frac{m}{2}}}{\Gamma(\frac{m}{2})}$.

§ 11.

Wir wollen nun unsere asymptotischen Formeln auf die Geometrie der Zahlen anwenden.

Es soll jetzt vorausgesetzt werden, daß B symmetrisch in bezug auf 0 ist. Weiter soll $m \not\equiv 1$ (4) sein, wo m die Dimension unseres Raumes ist. Die Anzahl der Gitterpunkte $\neq 0$ in $f(A\,x) \leqq t$, welche gerade ist, sei $2\,S$. Es ist $V = \mathfrak{J}\,t^n$ das Volumen von $B_t: f \leqq t$ und es werde $B_1 = B$ gesetzt. Dann gilt

Satz 13: Es gibt ein t_0, welches nur von B abhängt, so daß für alle $t > t_0$ und für alle Matrizen A mit Det. $A = 1$

$$V \leqq 2^{m-1}(S+1) + \sqrt{4^{m-1}(S+1)^2 - \varrho_t C}. \tag{1}$$

[13a] B besitze Mittelpunkt.

Dabei ist $\varrho_t = \inf_{B_t} K$ und $C > 0$ eine absolute Konstante, die nur von m abhängt.

Dieser Satz stellt für große t eine Verschärfung von

$$V \leqq 2^m (S+1) \tag{2}$$

dar. (Verallgemeinerter Minkowkischer Satz.)

Beweis: Nach Satz 5, § 5 (3) und (3') ist für $\delta = 0$

$$I = t^{\frac{1-m}{2}} \int_{B_t} e^{i\pi k x}\, dx = \frac{1}{\pi}\left(\frac{2}{|k|}\right)^{\frac{m+1}{2}} \sqrt{K(e_k)} \left(\cos (\pi k t H(e_k) - \right.$$
$$\left. - \frac{(m+1)\pi}{4}\right) + 0\left(\frac{1}{\sqrt{t\,|k|}}\right) \tag{3}$$

wo $k = A^{\nu-1} l = \bar{A} l$. Es ist Det. $\bar{A} = 1$. Nun wählen wir ein $l_0 \neq 0$ so, daß

$$1 \leqq |k_0| \leqq \sqrt{m} \tag{4}$$

Dies geht nach dem *Minkowski*schen Linearformensatz. Dann ist für $k = k_0 s$ (s natürliche Zahl $\varrho = \inf_B K$)

$$|I| \geqq \frac{1}{\pi}\, \varrho^{\frac{1}{2}} \left(\frac{2}{\sqrt{m\, s}}\right)^{\frac{m+1}{2}} \left|\, \left|\cos\left(\pi\, t\, s\, H(k_0) - \frac{(m+1)\pi}{4}\right)\right| - \right.$$
$$\left. - \frac{|C_1(B)|}{\sqrt{t\, s\, k_0}}\, \right|$$

Wir wählen nun t_0 so groß, daß $\dfrac{|C_1(B)|}{\sqrt{t_0}} < 2$ wird. Weiter wählen wir s, bei gegebenen $t > t_0$, so daß $\left|\frac{1}{2} s\, t\, H(k_0) - g\right| < \frac{1}{12}$ und $(32)^2 \leqq s \leqq 1\,2\,(32)^2$ (g ganz). Dies ist nach den *Dirichlet*schen Approximationssatz stets möglich. Dann ist, wenn $\alpha = \pi\,(s\, t\, H(k_0) - 2g)$

$$|\cos \alpha| \geqq \cos\frac{\pi}{6},\ \sin\frac{\pi}{6} \geqq |\sin \alpha| \text{ und } \left|\cos\left(\alpha - \frac{(m+1)\pi}{4}\right)\right| \geqq$$

$$\geqq |\cos \alpha|\, \left|\cos\frac{(m+1)\pi}{4}\right| - |\sin \alpha| \sin\frac{(m+1)\pi}{4} \geqq \frac{1}{2\sqrt{2}}\,(\sqrt{3}-1) > \frac{1}{8},$$

da $\left|\cos \pi\, \dfrac{m+1}{4}\right| \geqq \dfrac{1}{\sqrt{2}}$, $\left|\sin \pi\, \dfrac{m+1}{4}\right| \leqq \dfrac{1}{2\sqrt{2}}$ für $m \not\equiv 1\ (4)$.

Weiter ist $\frac{C_1(B)}{\sqrt{ts\,|k_0|}} \leqq \frac{2}{\sqrt{s}} = \frac{2}{32}$. Wir haben also, wenn $t > t_0$, uns ein k so konstruiert, daß

$$t^{\frac{1-m}{2}} \,|\int_{B_t} e^{\pi i k x}\, d\,x\,| \geqq \sqrt{C}\,\varrho^{\frac{1}{2}}, \tag{5}$$

wo $C = \frac{1}{(16\,\pi)^2}\,(4800\sqrt{m})^{-m-1}$. Daraus folgt nach § 6 (5)

$$2^m\,[V + \Sigma'_g \int_{B_t} \varphi\,(x - 2\,A\,g)\,d\,x] \geqq V^2 + C\,\varrho\,t^{m-1} \tag{6}$$

Nun ist $\int_{B_t} \varphi\,(x - 2\,A\,g)\,d\,x$ genau dann $\neq 0$, wenn $\pm\,A\,g$ in B_t, da B_t konvex und symmetrisch in 0 ist. Weiters sind natürlich diese Integrale $\leqq V$, also folgt aus (6), da $f\,(A\,x) \leqq t$, S Gitterpunktpaare $\neq 0$ enthält,

$$2^m\,V\,(S + 1) \geqq V^2 + C\,\varrho_t \tag{7}$$

da $\varrho_t = \varrho\,t^{m-1}$. Daraus folgt zunächst (2) und weiter $C\,\varrho_t \leqq 2^m\,V\,(S+1)$. Ist nun $V \leqq 2^{m-1}\,(S + 1)$, so ist (1) richtig. Ist aber $V > 2^{m-1}\,(S + 1)$, so folgt aus (7)

$$[V - 2^{m-1}\,(S+1)]^2 \leqq 4^{m-1}\,(S+1)^2 - C\,\varrho_t,$$

also wieder (1) w. z. b. w.

§ 12.

Die Verschärfung von § 11 (2) wurde unter stark einschränkenden Voraussetzungen über B_t erhalten ($m \not\equiv 1$ (4), Differenzierbarkeit von $S\,(B)$ usw.). Es soll nun eine andere Verschärfung angegeben werden, die nicht diese Voraussetzungen macht. (Für Parallelepipede wurde diese Verschärfung schon von *Gelfond*[5] angegeben.)

Satz 14: Ist N die Anzahl der Gitterpunktpaare $\neq 0$ in

$$H\,(A^{-1}\,l) \leqq \frac{\lambda}{2\,t}, \tag{1}$$

wo $0 < \lambda < 1$ beliebig, H Stützfunktion von $B: f\,(x) \leqq 1$, so ist

$$V \leqq \frac{2^m\,(S + 1)}{1 + N\cos^2\frac{\pi}{2}\lambda} \tag{2}$$

Dabei ist natürlich wieder B symmetrisch und V, S haben dieselbe Bedeutung wie in § 11.

Beweis: Es ist, da $H \leqq 1$ der polare Körper zu $B: f \leqq 1$ (*B.F.* S. 28, *Mahler*[14]), $|x\,y| \leqq f(x)\,H(y)$, also $|k\,x| \leqq t\,H(A^{-1}\,l) \leqq \frac{\lambda}{2}$, wo $k = A^{*-1}\,l$, nach Voraussetzung (1). Für diese l ist dann

$$\left|\int e^{\pi i k x}\,d\,x\right| \geqq V \cos\frac{\pi}{2}\lambda$$

Also folgt aus § 6 (5) analog zu § 11 (7)

$$2^m\,(S+1)\,V \geqq V^2\left(1 + N\cos^2\frac{\pi}{2}\lambda\right) \quad \text{w.z.b.w.}$$

Daraus folgt

Satz 15: Besitzt für

$$\tau < \frac{\lambda}{8}\left(\frac{\mathfrak{J}}{S+1}\right)^{\frac{1}{m}} \quad (0 < \lambda < 1) \tag{3}$$

$$H\,(A^{-1}\,l) \leqq \tau \tag{4}$$

eine Lösung und ist

$$\mathfrak{J}\,t^m + \frac{\lambda}{2\,\tau}\,t^{m-1}\cos^2\frac{\pi}{2}\lambda > 2^m\,(S+1) \tag{5}$$

so besitzt $f(A\,x) \leqq t$ mehr als S Gitterpunktpaare $\neq 0$.

Beweis: Ist $\mathfrak{J}\,t^m > 2^m\,(S+1)$, so ist dies nach § 11 (2), welches immer gilt, richtig. Ist $\mathfrak{J}\,t^m \leqq 2^m\,(S+1)$, dann ist

$$(t\,\tau)^m \leqq \frac{2^m\,(S+1)}{\mathfrak{J}}\left(\frac{\lambda}{4}\right)^m \frac{I}{2^m\,(S+1)} = \left(\frac{\lambda}{4}\right)^m,$$

also $\frac{\lambda}{2\,t\,\tau} \geqq 2$. Dann hat $H\,(A^{-1}\,l) \leqq \frac{\lambda}{2\,t} = \tau\,\frac{\lambda}{2\,t\,\tau}$ mindestens $2\left[\frac{\lambda}{2\,t\,\tau}\right]$ Gitterpunkte $\neq 0$, also $N \geqq 2\left[\frac{\lambda}{2\,t\,\tau}\right] \geqq \frac{\lambda}{2\,t\,\tau}$, da $\frac{\lambda}{2\,t\,\tau} \geqq 2$. Dann würde nach (2), wenn $f(A\,x) \leqq t$ höchstens S Gitterpunktpaare $\neq 0$ enthielte,

$$\mathfrak{J}\,t^m \leqq \frac{2^m\,(S+1)}{1 + \frac{\lambda}{2\,t\,\tau}\cos^2\frac{\pi}{2}\lambda}$$

---

[14] Casopis 68, S. 93ff.

in Gegensatz zur Voraussetzung (5).

Bermerkung: (3) ist sicher erfüllt, wenn $\mathfrak{J}_1 \tau^m < (\frac{\lambda}{2})^m \frac{1}{(S+1)(m!)^2}$ wo $\mathfrak{J}_1$ das Volumen von $H(y) \leqq 1$ ist, denn nach *Mahler*[14] ist $\mathfrak{J}_1 \geqq \geqq \frac{4^m}{(m!)^2}$.

Über eine Klasse von mehrfachen Integralen

Abh. Math. Semin. Univ. Hamb. **18** (1952), 53–69

Zu der Klasse der Integrale, die wir hier betrachten wollen, kommt man durch folgendes Problem: Es liege im euklidischen m-dimensionalen Raum R_m ein Bereich B mit JORDANschem Inhalt vor. Wir fragen nun nach der Anzahl der Punkte g mit ganzzahligen Koordinaten, welche in B liegen. Es sei $\varphi(x)$ die charakteristische Funktion von B, dann ist $\sum_g \varphi(g)$ diese Anzahl. Wir bilden $\psi(x) = \sum_g \varphi(x+g)$. Diese Funktion ist periodisch in g. Sie kann daher formal in eine FOURIERsche Reihe $\sum a_\iota e^{-2\pi i \iota x}$ entwickelt werden. (ιx bedeute das Skalarprodukt $\sum_{k=1}^{m} \iota_k x_k$ wenn $\iota_1, \ldots, \iota_m$ bzw. $x_1, \ldots, x_m$ die Koordinaten von ι bzw. x sind; unter dx wollen wir das m-dimensionale Volumselement verstehen.) Die FOURIER-koeffizienten a_ι sind (E sei der m-dimensionale Einheitswürfel $0 \leqq x_i \leqq 1$ $(i = 1, \ldots, m)$).

$$a_\iota = \int_E \psi(x)\, e^{2\pi i \iota x}\, dx = \sum_g \int_E \varphi(x+g)\, e^{2\pi i \iota x}\, dx = \sum_g \int_{E-g} \varphi(x)\, e^{2\pi i \iota x}\, dx = \int_{R_m} \varphi(x)\, e^{2\pi i \iota x}\, dx$$

daher $a_\iota = \int_B e^{2\pi i \iota x}\, dx$.

Wir wollen also die Integrale

$$G(B; \iota) = \int_B e^{i x \iota}\, dx$$

betrachten. Auch physikalische Probleme führen auf diese Integrale. Betrachten wir die FRAUENHOFERschen Beugungserscheinungen an einer beugenden Öffnung bzw. Scheibe B. Dann ist bekanntlich die Lichterregung u_P in einem „weit entfernten" Punkte P, wenn auch die Lichtquelle Q „weit von B entfernt" ist

$$u_P = c \iint_B e^{ik[(\alpha-\alpha_0)x_1 + (\beta-\beta_0)x_2]}\, dx_1\, dx_2$$

wo $k = \frac{2\pi}{\lambda}$ (λ Wellenlänge) „groß", α_0, β_0 bzw. α, β die Richtungskosinusse von OQ bzw. OP mit der x_1, x_2-Achse sind. Dabei liegt der Koordinatenursprung O in B und die Koordinatenachsen x_1, x_2 in der Ebene von B. Es liegt hier wieder und zwar für $m = 2$ ein solches Integral vor. Ebenso kommt in der Theorie der FRAUENHOFERschen Beugungserscheinungen der Elektronenoptik, G für $m = 3$ vor. Hier interessiert man sich insbesondere für die Punkte ι für die $G(B;\iota)$ verschwindet. Es wäre leicht, noch weitere Probleme anzugeben, die auf diese Integrale führen.

Liegt ein Würfel vor, so läßt sich $G(B;\iota)$ leicht berechnen. Ist B eine Kugel, so kommt man auf BESSELfunktionen. Hat die Kugel den Koordinatenursprung 0 als Mittelpunkt, und den Radius 1, so ist $G = (2\pi)^{\frac{m}{2}} \frac{I_{\frac{m}{2}}(|\iota|)}{|\iota|^{\frac{m}{2}}}$ ($|\iota|$ Entfernung des Punktes ι von 0). Es ist hier nötig Polarkoordinaten einzuführen; dann schreibt sich σ

$$G = O_{m-1}\int_0^1 r^{m-1}\,dr\int_0^{\pi} e^{i|\iota|\cos\vartheta}\sin^{m-1}\vartheta\,d\vartheta$$

(O_{m-1} Oberfläche der m—1-dimensionalen Einheitskugel). Das Integral nach ϑ ist aber, von Faktoren abgesehen, $\frac{I_{\frac{m}{2}-1}(|\iota|r)}{(|\iota|r)^{\frac{m}{2}-1}}$ und geläufige Formeln über BESSEL-Funktionen führen zum angegebenen Resultat. Sonst liegt in der Literatur über $G(B;\iota)$ wenig vor. Für den Fall, daß B ein Polygon ($m = 2$) bzw. ein Polyeder ($m = 3$) ist, findet man in der physikalischen Literatur Berechnungen. In neuerer Zeit hat sich besonders LAUE damit beschäftigt. Dieser Fall ist grundsätzlich einfach zu behandeln, aber allgemeine Resultate dürften, auch wenn die Polyeder konvex sind, nicht leicht zu erhalten sein.

Nun sind einfache Resultate zu erwarten, wenn $|\iota|$ groß ist und wenn B ein konvexer Körper ohne Ecken und Kanten ist. Man stellt sich also die Frage nach asymptotischen Entwicklungen unter obigen Voraussetzungen. Genauer formuliert, wollen wir hier folgende Voraussetzungen über B machen: B sei ein konvexer Körper mit Mittelpunkt und sein Rand lasse sich durch gleichsinnig-parallele Stützebenen umkehrbar eindeutig und analytisch auf die Oberfläche der m-dimensionalen Einheitskugel E_m abbilden.

Diese Voraussetzungen ließen sich abschwächen, man könnte besonders die Annahme fallen lassen, daß B einen Mittelpunkt besitzt. Mit den asymptotischen Entwicklungen von $G(B;\iota)$ und verwandter Integrale habe ich mich in letzter Zeit beschäftigt[1]). Es sollen hier einige Ergebnisse dargelegt werden. Zunächst soll $m = 2$ betrachtet werden.

I.

Es liege in der Ebene ein konvexer Bereich B mit Mittelpunkt vor, dessen Rand C analytisch ist und der nur Berührungen erster Ordnung

[1]) Vgl. Über Integrale auf konvexen Körpern I, II, III. Monatshefte f. Mathematik (1950), 54; S. 1—36, S. 81—99; 55, S. 105—137.

besitzt. Wir betrachten das Integral

$$G(B;\iota) = \int\limits_B e^{\iota x} dx\,. \tag{1}$$

Es weist zwei Eigenschaften auf:

1. Wird auf B eine flächentreue, affine Transformation T (Det $T = 1$) ausgeübt, geht also B in $\bar{B} = TB$ über, so ist

$$G(\bar{B},\iota) = G(B, T^*\iota)\,, \tag{2}$$

denn wenn $\bar{x} = Tx$ gesetzt wird, so ist $\iota\bar{x} = (T^*\iota)x$.

2. Wird auf B die Translation $\bar{x} = x + t$ ausgeübt, wo t auf ι senkrecht steht, so ist

$$G(B;\iota) = G(B + t;\iota)\,. \tag{3}$$

Wir nehmen nun an, daß der Mittelpunkt von B auf der Geraden durch den Koordinatenursprung senkrecht zu ι liegt. Wir beziehen die Randkurve C von B auf die affine Bogenlänge σ. Wir bezeichnen die Ableitungen nach σ durch Punkte und verstehen unter (x, y), $x_1 y_2 - x_2 y_1$, wenn x_1, x_2 bzw. y_1, y_2 die Komponenten von x bzw. y sind. Dann ist also $(\dot{x}, \ddot{x}) = 1$. (Für alle verwendeten Sätze der affinen Differentialgeometrie verweisen wir auf W. BLASCHKE, Affine Differentialgeometrie.) Nun verwandeln wir (1) mit Hilfe des GAUSSschen Integralsatzes in ein Integral über C: Es ist $\Delta(e^{\iota x}) = -|\iota|^2 e^{\iota x}$ (Δ der LAPLACEsche Operator), also

$$G = -\frac{1}{|\iota|^2}\int\limits_B \Delta(e^{\iota x})\,dx = -\frac{i}{|\iota|^2}\int\limits_C e^{\iota x}(\iota_1 dx_2 - \iota_2 dx_1)$$

daher

$$G = -\frac{i}{\omega}\int\limits_C (n, \dot{x})\, e^{in\omega x} d\sigma \tag{4}$$

wo $\iota = \omega n$, $\omega = |\iota|$ gesetzt ist. Dabei sei die Kurvennormale ins Außengebiet von B gerichtet. Nun können wir nach unseren Voraussetzungen über C ein σ_0 und somit einen Punkt $X = x(\sigma_0)$ auf C so bestimmen, daß

$$n = \frac{1}{|\dot{X}|}(\dot{X}_2, \dot{X}_1) \tag{5}$$

und σ_0 durch n eindeutig bestimmt ist.

Es ist also

$$nx = \frac{1}{|\dot{X}|}(x, \dot{X})\,,$$

insbesondere ist die Stützfunktion

$$h(n) = nX = \frac{1}{|\dot{X}|}(X, \dot{X})$$

und der zugehörige Krümmungsradius

$$K(n) = |\dot{X}|^3, \text{ da } (\dot{X}, \ddot{X}) = 1\,.$$

Wir wollen noch

$$f(\sigma) = (x, \dot{X}), \quad \varrho = \frac{|\iota|}{|\dot{X}|} \tag{6}$$

setzen.

Wir können sicher n als Linearkombination von X und $\dot{X}$ ansetzen: $n = c_1 X + c_2 \dot{X}$, wo $c_1 = (nX)^{-1}$ ist. Dann ist σ nach (4) eine Summe von zwei Integralen.

$$G = -\frac{ic_1}{\omega}\int (X, \dot{x})\, e^{i\varrho f(\sigma)} d\sigma - \frac{ic_2}{\omega}\int (\dot{X}, \dot{x})\, e^{i\omega f(\sigma)} d\sigma$$

Das zweite Integral ist aber Null, denn der Integrand ist, abgesehen von konstanten Faktoren, gleich der Ableitung von $e^{i\omega f(\sigma)}$ nach σ. Wir erhalten daher für σ die Darstellung, die wir erreichen wollten:

$$G = -\frac{i}{\varrho}\int g(\sigma)\, e^{i\varrho f(\sigma)} d\sigma \tag{7}$$

wo, wenn wir alles zusammenstellen,

$$\varrho = \frac{|\iota|}{|\dot{X}|}, \quad g(\sigma) = \frac{(X, \dot{x})}{(X, X)}, \quad f(\sigma) = (x, \dot{X}) \text{ ist.}$$

Wir können noch erreichen, daß

$$\dot{g}(\sigma_0) = (X, \ddot{X}) = 0 \tag{8}$$

ist.

Denn ist das nicht der Fall, so üben wir die Translation $x = \bar{x} - c\dot{X}$ $(c = -(X, \ddot{X}))$ aus. Dabei ändert sich nach (3) σ nicht, da $\dot{X}$ auf ι senkrecht steht. Dann ist

$$(\bar{X}, \ddot{\bar{X}}) = (X, \ddot{X}) + c(\dot{X}, \ddot{X}) = 0\,.$$

Nun wollen wir mit Hilfe der Methode der „stationären Phase“, für das Integral einen asymptotischen Ausdruck für großes $|\iota|$, d. h. für großes ϱ aufstellen. Dazu müssen wir die Punkte auf C bestimmen, für die $\dot{f}(\sigma) = (x, X) = 0$ ist. Dann muß $\dot{x}$ parallel zu $\dot{X}$ sein, d. h. für X und für den in Bezug auf den Mittelpunkt von B spiegelbildlichen Punkt, dessen Parameter $\bar{\sigma}_0$ sei. Es ist natürlich $\dot{X}(\bar{\sigma}_0) = -\dot{X}(\sigma_0)$.

Wir benötigen noch die höheren Ableitungen von $f(\sigma)$ und $g(\sigma)$ für σ_0 bzw. $\bar{\sigma}_0$. Zunächst ist

$$\ddot{f}(\sigma_0) = (\ddot{X}, \dot{X}) = -1\,, \quad \ddot{f}(\bar{\sigma}_0) = (\dot{X}, \ddot{X}) = 1; \tag{9}$$

es ist also σ_0 die Stelle des Maximums, $\bar{\sigma}_0$ die Stelle des Minimums von $f(\sigma)$.

Weiters ist

$$f^{(k)}(\sigma_0) = (X^{(k)}, \dot{X})\,.$$

Nun ist $\dddot{X} + \varkappa\dot{X} = 0$, wo $\varkappa$ die affine Krümmung von $C(B)$ ist. Es ist also $f^{(3)}(\sigma_0) = 0$ und aus

$$X^{(k+3)} = -\sum_{j=0}^{k}\binom{k}{j}\varkappa^{(k-j)} X^{(j+1)} \tag{10}$$

folgt die Rekursionsformel für $f^{(k)}(\sigma_0)$

$$f^{(k+3)}(\sigma_0) = -\sum_{j=1}^{k} \binom{k}{j} \varkappa^{(k-j)} f^{(j+1)}(\sigma_0) \tag{11}$$

Wie man sieht, drücken sich die $f^{(k)}(\sigma_0)$ durch die affine Krümmung und ihre Ableitungen aus.

Weiter ist $g(\sigma_0) = 1$, $\dot{g}(\sigma_0) = 0$ nach (8) und für

$$g^{(k)}(\sigma_0) = (X, X^{(k+1)}) / (X, \dot{X})$$

folgt aus (10) die Rekursionsformel

$$g^{(k+2)}(\sigma_0) = -\sum_{j=0}^{k} \binom{k}{j} \varkappa^{(k-j)} g^{(k)}(\sigma_0) \tag{11'}$$

insbesondere ist $g^{(2)}(\sigma_0) = -\varkappa$. Es drücken sich also auch die $g^{(k)}(\sigma_0)$ $(k \geqq 0)$ durch $\varkappa$ und ihre Ableitungen aus.

Für die Stelle $\bar{\sigma}_0$ gilt dasselbe, da

$$f^{(k)}(\sigma_0) = -f^{(k)}(\bar{\sigma}_0), \quad g^{(k)}(\sigma_0) = -g^{(k)}(\bar{\sigma}_0) \quad \text{ist.}$$

Die Methode der stationären Phase sagt nun folgendes aus[1]): Liegt ein Integral $\int_a^b g(\sigma) e^{i\varrho f(\sigma)} d\sigma$ vor, wo $f'(a) = f'(b) = 0$, $f''(a) < 0$, $f''(b) > 0$ und $f'(\sigma) \neq 0$ in $(a < \sigma < b)$, so ist

$$\int_a^b g(\sigma) e^{i\varrho f(\sigma)} d\sigma = \sum_k (a) + \sum_k{}^* (b) + O\left(\varrho^{-\frac{k}{2}-1}\right) \tag{12}$$

wo

$$\sum_k (a) = \tfrac{1}{2} e^{i\varrho f(a)} \sum_{j=0}^{k} A_j(a) e^{-\frac{i(j+1)\pi}{4}} \varrho^{-\frac{j+1}{2}} \tag{13}$$

$$\sum_k{}^* (b) = \tfrac{1}{2} e^{i\varrho f(b)} \sum_{j=0}^{k} (-1)^j A_j^*(b) e^{\frac{i(j+1)\pi}{4}} \varrho^{-\frac{j+1}{2}} \quad \text{ist.}$$

Dabei ist

$$A_j(a) = \sum_{t=0}^{j} \frac{1}{(j+2t)!} \Gamma\left(\frac{j+1}{2} + t\right) \left(\frac{1}{2} / f''(a) /\right)^{-\frac{j+1}{2}-t} l_{j,j+2t}(a)$$

$$l_{t,s}(a) = \sum_{n=3t}^{s} \binom{s}{n} g^{(s-n)}(a) K_{tn}(a) \tag{14}$$

$$K_{00} = 1,\ K_{0n} = 0,\ (n \geqq 1),\ K_{t+1,\,n+1} = \sum_{r=3t}^{n-2} \binom{n}{r} f^{(n-r+1)}(a) K_{tr}. \tag{15}$$

Analog ist $A_j^*(b)$ gebildet. Es ist in (14) nur a durch b zu ersetzen und jedes Glied der Summe mit $(-1)^t$ zu versehen[2]).

Bemerkung: Gilt umgekehrt $f''(a) > 0$, $f''(b) < 0$, so lautet (12)

$$\int_a^b g(\sigma) e^{i\varrho f(\sigma)} d\sigma = S_k(a) + S_k^*(b) + O\left(\varrho^{-\frac{k}{2}-1}\right) \tag{12'}$$

---

[1]) Vgl. I S. 3—9.

[2]) Der Faktor $(-1)^t$ ist in I S. 4 irrtümlich vergessen.

wo
$$S_k(a) = \tfrac{1}{2} e^{i\varrho f(a)} \sum_{j=0}^{k} A_j^*(a) e^{\frac{i(j+1)\pi}{4}} \varrho^{-\frac{j+1}{2}}$$
$$S_k^*(b) = \tfrac{1}{2} e^{i\varrho f(a)} \sum_{j=0}^{k} (-1)^j A_j(b) e^{\frac{-i(j+1)\pi}{4}} \varrho^{-\frac{j+1}{2}}.$$

Um nun dies auf unser Integral (7) anzuwenden, haben wir zu berücksichtigen, daß
$$G = -\frac{i}{\varrho} \int_C g(\sigma) e^{i\varrho f(\sigma)} d\sigma = -\frac{i}{\varrho} \Big(\int_{\sigma_0}^{\bar\sigma_0} + \int_{\bar\sigma_0}^{\sigma_0+S} \Big) = -\frac{i}{\varrho}(I + II)$$
wo $\sigma = \sigma_0 + S$ denselben Punkt X auf C liefern soll. (Es ist ja C eine geschlossene Kurve.) Da B ein konvexer Bereich mit Mittelpunkt ist, so ist $II = -\bar I$.

Es ist also
$$G = +\frac{2}{\varrho} Jm \Big(\int_{\sigma_0}^{\bar\sigma_0} \Big).$$

Wir wenden (12) mit $a = \sigma_0$, $b = \bar\sigma_0$ an. Die $K_{rt}(\sigma_0)$ können hier, wo $\dddot f(\sigma_0) = 0$ ist, leicht explizit berechnet werden, und zwar ist
$$K_{tn}(\sigma_0) = \frac{n!}{t!} \sum_{k=0}^{n-4t} \binom{t}{k} L_0^{t-k} S_k(n-4t) \tag{16}$$

Dabei ist $L_i = \frac{f^{(i+4)}(\sigma_0)}{(i+4)!}$ für $i \geqq 0$ und $S_0(O) = 1$, $S_0(j) = 0$ für $j > 0$ und sonst
$$\sum_{\substack{\alpha_1 + \cdots + \alpha_k = j \\ \alpha_i \geqq 1}} L\alpha_1 \cdots L\alpha_k \tag{16'}$$

Den Beweis von (16) führen wir durch vollständige Induktion nach t. Für $t = 0$ ist (16) nach (15) richtig. Nun ist
$$\sum_{s=k}^{m} L_{m-s} S_k(s) = S_{k+1}(m) + L_0 S_k(m), \quad \sum_{s=k}^{m} (m-s) L_{m-s} S_k(s) = \frac{m}{k+1} S_{k+1}(m) \tag{16''}$$

Denn es ist z. B. für die zweite Summe, wenn wir $m - s = \alpha_{k+1}$ setzen
$$\sum_{s=k}^{m} = \sum_{\alpha_1 + \cdots + \alpha_{k+1} = m} \alpha_{k+1} L_{\alpha_1} \cdots L_{\alpha_{k+1}} = \frac{1}{k+1} \sum (\alpha_1 + \cdots + \alpha_{k+1}) L_{\alpha_1} \cdots L_{\alpha_{k+1}}$$
$$= \frac{m}{k+1} S_{k+1}(m)$$

Jetzt zum Induktionsbeweis: Es ist nach (15)
$$K_{t+1,n} = \sum_{r=3t}^{n-3} \binom{n-1}{r} f^{(n-r)} K_{tr}$$

Nun ist $f^{(3)}(\sigma_0) = 0$ und nach Induktionsannahme $K_{tr} = 0$ für $r < 4t$. Setzen wir noch $s = r - 4t$, $m = n - 4(t+1)$, so kommt nach (16) da $f^{(n-r)} = (m + 4 - s)!\, L_{m-s}$

$$K_{t+1,n} = \frac{(n-1)!}{t!} \sum_{s=0}^{m} (m-s+4) L_{m-s} \sum_{k=0}^{s} \binom{t}{k} L_0^{t-k} S_k(s)$$

$$= \frac{(n-1)!}{t!} \sum_{k=0}^{m} \binom{t}{k} L_0^{t-k} \sum_{s=k}^{m} (m-s+4) L_{m-s} S_k(s)$$

also nach (16″)

$$K_{t+1,n} = \frac{(n-t)!}{t!} \left(4 L_0^{t+1} S(m) + \sum_{k=1}^{m} L_0^{t+1-k} \left(4\binom{t}{k} + \frac{m+4k}{k}\binom{t}{k-1}\right) S_k(m)\right)$$

$$= \frac{n!}{(t+1)!} \sum_{k=0}^{m} \binom{t+1}{k} L_0^{t+1-k} S_k(m) \qquad \text{w. z. b. w.}$$

Dasselbe gilt auch für $K_{tn}(\overline{\sigma}_0)$ und zwar entnimmt man aus (16) sofort, da $f^{(k)}(\overline{\sigma}_0) = -f^{(k)}(\sigma_0)$, daß $K_{tn}(\overline{\sigma}_0) = (-1)^t K_{tn}(\sigma_0)$. Wenn man noch beachtet, das $g^{(k)}(\sigma_0) = -g^{(k)}(\sigma_0)$ ist, so folgt daraus, daß $A_j^*(\sigma_0) = -A_j(\sigma_0)$ ist.

Also erhalten wir nach (12)

$$\int_{\sigma_0}^{\overline{\sigma}_0} g(\sigma) e^{i\varrho f(\sigma)} d\sigma = \tfrac{1}{2} \sum_{j=0}^{k} A_j(\sigma_0) \left(e^{i\alpha_j} - (-1)^j e^{-i\alpha_j}\right) \varrho^{\frac{-j+1}{2}} + O\left(\varrho^{-\left(\frac{k}{2}+1\right)}\right)$$

wo $\alpha_j = \varrho f(\sigma_0) - \frac{(j+1)\pi}{4}$ gesetzt ist. Da uns der Imaginärteil interessiert, können wir alle Glieder mit ungeradem j weglassen.

Die $A_j(\sigma_0)$ hängen dabei nur von der affinen Krümmung und ihren Ableitungen ab.

Weiters ist $f(\sigma_0) = (X, \dot{X})$, also $\varrho f(\sigma_0) = |\iota|\, h(n) = h(\iota)$.

Wir erhalten daher

$$G = \frac{1}{\varrho^{\frac{3}{2}}} \left(\sum_{r=0}^{R} \frac{C_r(\varkappa)}{\varrho^r} \cos\left(h(\iota) - (2r+3)\frac{\pi}{4}\right) + O\left(\frac{1}{\varrho^{R+1}}\right)\right). \tag{17}$$

Dabei haben wir $j = 2r$, $k = 2R+1$, $A_j(\sigma_0) = C_r(\varkappa)$ gesetzt.

Es ist nun $\varrho = \frac{|\iota|}{|\dot{X}|} = \frac{|\iota|}{K^{\frac{1}{3}}} = \frac{\iota\,|\,h(n)}{(X, \dot{X})}$ und wenn wir nach $C_r = 2\sqrt{2\pi} D_r(\varkappa)$ setzen, so erhalten wir endgültig

$$G = 2\frac{\sqrt{2\pi K}}{|\iota|^{\frac{3}{2}}} \sum_{r=0}^{R} \left(\frac{(X, \dot{X})}{h(\iota)}\right)^r D_r(\varkappa) \cos\left(h(\iota) - (2r+3)\frac{\pi}{4}\right) + O\left(|\iota|^{-R-\frac{5}{2}}\right). \tag{18}$$

Die $D_r(\varkappa)$ hängen nur von der affinen Krümmung und ihren Ableitungen im Punkt $\sigma = \sigma_0$ ab. Die asymptotische Entwicklung baut sich also aus folgenden einfachen Bestandteilen auf: Aus dem Flächenelement $(X, \dot{X})$, der Stützfunktion $h(\iota) = \operatorname{Max} x\,\iota$ über alle x von $C(B)$ und der affinen Krümmung $\varkappa$ und ihren Ableitungen. Es ist sehr instruktiv, dies mit der Definition (1) von C und den Eigenschaften 1. und 2. zu vergleichen.

Wir wollen noch die ersten drei D angehen. Es ist

$$D_0 = 1,\ D_1 = \frac{3}{8}\varkappa, \quad D_2 = -\frac{3\cdot 5}{2\cdot 8^2}\left(\varkappa^2 + \frac{16}{15}\ddot{\varkappa}\right).$$

Ist B der Einheitskreis, so ist $(X, \dot{X}) = K = \varkappa = 1$, $h(\iota) = |\iota|$ und man erhält

$$G = \frac{2\sqrt{2\pi}}{|\iota|^{\frac{3}{2}}}\left(\cos\left(|\iota| - \frac{3\pi}{4}\right) - \frac{3}{8|\iota|}\sin\left(|\iota| - \frac{3\pi}{4}\right) + \frac{3\cdot 5}{2\cdot 8^2|\iota|^2}\cos\left(|\iota| - \frac{3\pi}{4}\right) + \cdots\right)$$

also die ersten Glieder der asymptotischen Entwicklungen von $2\pi \frac{J_1(|\iota|)}{|\iota|}$.

Wir wollen noch auf einige Folgerungen aus (18) hinweisen:

Es ist von Interesse, die Punkte ι zu betrachten, für die $G(B; \iota) = 0$ ist. Nun verschwindet das erste Glied in (18) für alle Kurven der Gestalt.

$$h(\iota) = \frac{3\pi}{4} + (2k+1)\frac{\pi}{2} \tag{19}$$

$(k = 0, 1, 2, \ldots)$. Die Eichkurve dieser Kurven ist polar zur Randkurve $C(B)$ von B. Man wird erwarten, daß für großes $|\iota|$, die Punkte für die $G = 0$ ist, auf Kurven (Nullkurven) liegen werden, die eine zu (19) analoge Gestalt haben. Man kann nun tatsächlich zeigen, daß es unendlich viele Kurven C_k gibt, auf welchen G verschwindet und welche die Gestalt

$$r = \left(\frac{3\pi}{4} + (2k+1)\frac{\pi}{2}\right) h^{-1}(u) + O\left(\frac{1}{k}\right)$$

$(|u. = 1)$ haben und daß für genügend großes ι alle Punkte, für welche G verschwindet, auf diesen Kurven liegen. Diese Kurven sind für genügend großes k konvex, mit 0 als Mittelpunkt, wenn der Mittelpunkt von B in 0 liegt[1]). Man kann nun fragen, ob die Nullkurven genau die Gestalt (19) haben können. Dies ist bekanntlich bei der Ellipse der Fall. Man kann unter Benützung der zweiten Näherung in (18) zeigen, daß für unendlich viele k, die Nullkurven von G nur dann die Gestalt (19) haben können, wenn B eine Ellipse ist. Dies läuft darauf hinaus, alle konvexen, geschlossenen Kurven zu betrachten. für die $(x, \dot{x})\varkappa$ konstant $= C$ für alle σ ist. Es muß $C \neq 0$ sein, sonst würde eine Parabel vorliegen. Dann folgt daraus, daß $(\varkappa^{-1})^{\cdot\cdot} = \frac{1-C}{C}$ sein muß, da $(x, \dddot{x}) = -\varkappa(x, \dot{x}) = -C$ ist, also $(x, \dot{x})^{\cdot\cdot} = (\dot{x}, \ddot{x}) + (x, \dddot{x}) = 1 - C$. Durch Integration folgt daraus $\varkappa^{-1} = \frac{1-C}{C}\sigma^2 + c_1\sigma + c_2$. Da eine geschlossene Kurve vorliegt, so muß $\varkappa = c_2$ sein und da die Hyperbel nicht in Frage kommt, muß die Kurve eine Ellipse sein.

Man kann die Voraussetzung, daß B einen Mittelpunkt besitzt, fallen lassen. Dann lautet die erste asymptotische Näherung von σ

$$G = \frac{1}{|\iota|^{\frac{3}{2}}}\left[\sqrt{K(e_\iota)}\, e^{i\left(h(\iota) - \frac{3\pi}{4}\right)} + \sqrt{K(-e_\iota)}\, e^{-i\left(h(-\iota) - \frac{3\pi}{4}\right)} + O\left(\frac{1}{|\iota|}\right)\right] (\iota = |\iota| e_\iota).$$

Man kann wieder fragen, ob es Richtungen ι gibt, auf welchen unendlich viele Punkte liegen, für die $G(B; \iota) = 0$ ist. Man überlegt sich leicht,

[1]) Vgl. dazu II § 3—6.

daß dies genau dann der Fall ist, wenn $K(e_\iota) = K(-e_\iota)$ ist. Nach W. Blaschke gibt es mindestens drei solche Richtungen. Fragt man nach der Existenz von unendlich vielen Nullkurven von G, so folgt aus dem eben Gesagten, daß für jede Richtung $e\, K(e) = K(-e)$ sein muß. Dann muß aber nach Minkowski B einen Mittelpunkt besitzen. Es können also dann und nur dann bei einem Bereich B unendlich viele Nullkurven auftreten, auf welchen G verschwindet, wenn B einen Mittelpunkt besitzt.

II.

Es sei nun B ein konvexer Körper im R_m (m beliebig $\geqq 2$). Es werde wieder vorausgesetzt, daß er einen Mittelpunkt besitzt, der Rand $S(B): x = x(u)$ ($|u| = 1$) analytisch ist und Berührungen erster Ordnung besitzt. Dann ist es auch hier möglich, das erste nichtverschwindende Glied der asymptotischen Entwicklung zu bestimmen[1]): Es ist

$$G(B;\iota) = 2\frac{(2\pi)^{m-1}}{|\iota|^{\frac{m+1}{2}}}\left[\sqrt{K(e_\iota)}\cos\left(H(\iota) - \frac{(m+1)\pi}{4}\right) + O\left(\frac{1}{|\iota|}\right)\right]. \qquad (20)$$

Dabei ist $\iota = |\iota| e_\iota$, $H(\iota)$ die Stützfunktion und $K(e_\iota)$ das zugehörige Produkt der Hauptkrümmungsradien. Eine zu (18) analoge Entwicklung ist noch nicht aufgestellt, aber im Prinzip möglich. Ich will auf den Beweis von (20) nicht eingehen, sondern will den Gedankengang an einem anderen Integral, das in der Theorie der Verteilungen auf konvexen Körpern auftritt, entwickeln.

Wir betrachten die Verteilungsfunktion (vollständig additive Mengenfunktion) $\varphi(M)$ definiert für alle Borelsche Mengen im R_m:

$$\varphi(M) = \frac{1}{\sigma_m}\int_u d\sigma .$$

Dabei ist $d\sigma$ das Oberflächenelement der m-dimensionalen Einheitskugel E_m, σ_m die Gesamtoberfläche von E_m und das Integral wird erstreckt über alle u für die die zugehörigen Punkte $x(u)$ des Randes $S(B)$ zu M gehören. Es ist $\varphi(R_m) = 1$. Die Fouriertransformierte ist dann

$$\Lambda(\iota;\varphi) = \int_{R_m} e^{i\iota x}\Phi(dR_x) = \frac{1}{\sigma_m}\int_{E_m} e^{i\iota x(u)} d\sigma . \qquad (21)$$

Wir wollen dieses Integral betrachten und zwar für $\iota = (\omega, 0, \ldots, 0)$.

Wir setzen $u_1 = \cos\vartheta$, $u_j = \sin\vartheta\, a_j$ $(j \geqq 2)$, $0 \leqq \vartheta \leqq \pi$, $\sum_{j=2}^{m} a_j^2 = 1$ (22)

Dann ist[2]), wenn $d\bar{\sigma}$ das Oberflächenelement von E_{m-1} bedeutet

$$J = \Lambda\sigma_m = \int_{E_{m-1}} d\bar{\sigma}\int_0^\pi e^{i\omega x_1}\sin^{m-2}\vartheta\, d\vartheta . \qquad (23)$$

---

[1]) I S. 10 und S. 20 (3) für $\delta = 0$.

[2]) Vg. dazu I § 3.

Wir halten zunächst die a_j fest und betrachten nur das innere Integral in (23). Nun ist, wenn $H(u)$ die Stützfunktion von B ist (die Ableitungen nach den Komponenten von u deuten wir durch Indices an), $x_1 = H_1$, also

$$\frac{\partial x_1}{\partial \vartheta} = - H_{11} \sin\vartheta + \cos\vartheta \sum_{i=2}^{m} H_{1i} a_i . \tag{23}$$

Weiter ist

$$\sum_{i=1}^{m} u_i H_i = H, \quad \sum_{i=1}^{m} u_i H_{ij} = 0 \quad (j = 1, \ldots, m) \tag{24}$$

Daraus folgt, wenn e_1 der Vektor $(1, 0, \ldots, 0)$ ist $H_1(\pm e_1) = \pm H(\pm e_1)$, $H_{1j}(\pm e_1 = O$ und da $u_1 = \cos\vartheta$, $H_{1i}\cos\vartheta = -\sum_{k=2}^{m} H_{ik} u_k$, daher wird (23)

$$\frac{\partial x_1}{\partial \vartheta} = - \sin\vartheta F(\vartheta, a_i) \tag{24}$$

wo

$$F(\vartheta, a_i) = H_{11} + \sum_{i,k=2}^{m} H_{ik} a_i a_k \qquad \left(\sum_{i=2}^{m} a_i^2 = 1\right). \tag{25}$$

Nun ist $K(u)$ die Summe aller m—1-zeiligen Hauptminoren der Matrix (H_{ik}). Da wegen unserer Voraussetzungen über $S(B)$ sicher für alle u, $K(u) \geqq \varrho > 0$ auf $|u| = 1$ ist, so folgt leicht, das $F(\vartheta, a_i) \geqq \varrho_1 > 0$ sein muß für alle $a_2, \ldots, a_m$ mit $\sum a_i^2 = 1$ und alle ϑ. Also hat $\frac{\partial x_1}{\partial \vartheta}$ nur die Nullstellen $\vartheta = 0$ und $\vartheta = \pi$, d. h. für alle $u = \pm e_1$ und es ist $\left.\frac{\partial^2 x_1}{\partial \vartheta^2}\right|_0 \leqq -\varrho_1 < 0$, $\left.\frac{\partial^2 x_1}{\partial \vartheta^2}\right|_\pi \geqq \varrho > 0$. Dann können wir das Prinzip der stationären Phase mit $f(\vartheta) = x_1$, $g = \sin^{(m-2)}\vartheta$ anwenden. Es ist $g^{(j)}(\vartheta) = 0$ für $\vartheta = 0$ und $\pi (0 \leqq j \leqq m-2)$, $g^{(m-2)}(0) = (m-2)!$, $g^{(m-2)}(\pi) = (-1)^{m-1}(m-2)!$ Also beginnt die asymptotische Entwicklung mit $\omega^{-\frac{m-1}{2}}$ und wir erhalten:

$$\Lambda\sigma_m = \left(\frac{2\pi}{\omega}\right)^{\frac{m-1}{2}} \left(a_0^* e^{i\omega H(e)} + b_0^* e^{-i\omega H(e)} + O\left(\omega^{-\frac{1}{2}}\right)\right)$$

wo

$$a_0^* = \left(\frac{1}{2\pi}\right)^{\frac{m-1}{2}} \frac{\Gamma\left(\frac{m-1}{2}\right)}{2(m-2)!} 2^{\frac{m-1}{2}} e^{\frac{-i(m-1)\pi}{4}} \int_{E_{m-1}} \frac{d\bar{\sigma}}{F_0^{\frac{m-1}{2}}}$$

und b_0^* analog gebaut ist. Nun ist

$$\int_{E_{m-1}} \frac{d\bar{\sigma}}{F_0^{\frac{m-1}{2}}} = \int_{E_{m-1}} \frac{d\bar{\sigma}}{(\Sigma H_{ik}(e) a_i a_k)^{\frac{m-1}{2}}} = 2\, \frac{\pi^{\frac{m-1}{2}}}{\Gamma\left(\frac{m-1}{2}\right)} \frac{1}{\sqrt{K(e)}} .$$

Wird alles eingesetzt, so erhält man, wenn man noch berücksichtigt, daß man durch eine Drehung ι in ωe überführen kann

$$\Lambda = \frac{2}{\sigma_m} \left(\frac{2\pi}{|\iota|}\right)^{\frac{m-1}{2}} \left[\frac{1}{\sqrt{K(e_\iota)}} \cos\left(H(\iota) - \frac{(m-1)\pi}{4}\right) + O\left(\frac{1}{|\iota|^{\frac{1}{2}}}\right)\right] \tag{26}$$

Für $m=2$ kann man die Entwicklung bis zum zweiten Glied führen. Setzt man $e_\iota=(\cos\varphi, \sin\varphi)$, und bezeichnet die Ableitung nach φ durch Striche, so lautet das nächste nichtverschwindende Glied in der Klammer von (24)

$$\frac{1}{|\iota| K^{\frac{3}{2}}}\left(\frac{5}{6}\left(\frac{K'}{K}\right)^2+\frac{1}{8}\frac{(K-3K'')}{K}\right)\sin\left(H(\iota)-\frac{\pi}{4}\right). \tag{27}$$

Für $m=2$ wurde (24) von WINTNER und JESSEN[1]) und von WINTNER und HAVILAND[2]) gezeigt.

Aus (24) kann man wie bei WINTNER und JESSEN folgendes schließen: Sind $\varphi_i (i=1,\ldots,n)$ die Verteilungen auf $B_i: f(x)=r_i$, so ist die Konvolution $\psi_n=\varphi_1*\varphi_2*\cdots*\varphi_n$ absolut stetig mit einer stetigen Dichte $D_n(x)$, wenn $n>\frac{2m}{m-1}$ und sie besitzt Ableitungen von einer Ordnung $\leq p$, wenn $n>\frac{2(m+p)}{m-1}$ ist. Neben dem Integral $\sigma(B;\iota)=\int\limits_B e^{\iota x}dx$ benötigt man in den Anwendungen noch Integrale von der Gestalt

$$J_\delta(B;\iota)=\int\limits_B (1-f^2)^\delta e^{\iota x}dx \tag{28}$$

$(\delta>0)$, wenn $B: f(x)\leq 1$ ist, allgemeiner $\int\limits_{f(x)\leq T}\Phi(f(x))e^{\iota x}dx$. Diese lassen sich unter gewissen Voraussetzungen auf $G(B,\iota)$ bzw. auf ein verwandtes Integral $G_1(B,\iota)$ (s. u.) zurückführen. Gehen wir nämlich allgemein von einem Integral $J=\int\limits_{f\leq T}\psi(x)\,dx$ aus, wo ψ stetig sei (über B mögen unsere üblichen Annahmen gelten, $B_T: f\leq T$), dann setzen wir $x=ry$, wo $y=y(u)$, $(|u|=1)$ Randpunkt von B ist. Ist die Oberfläche der Einheitskugel E_m in zwei Halbkugeln E', E'' zerlegt, so bedeutet dies eine Zerlegung von $C(B)$ und damit von $B_T: B_1+B_2$. Es genügt das Integral über B_1 zu betrachten. Dann lassen sich auf E' Parameter $\alpha_1,\ldots,\alpha_{m-1}$ einführen. Wir transformieren unser Integral auf $r,\alpha_1,\ldots,\alpha_{m-1}$. Die Funktionaldeterminante lautet

$$\Delta=\left|\frac{\partial x}{\partial r},\frac{\partial x}{\partial\alpha_1},\frac{\partial x}{\partial\alpha_2},\ldots\frac{\partial x}{\partial\alpha_{m-1}}\right|=r^{m-1}\left|y,\frac{\partial y}{\partial\alpha_1},\frac{\partial y}{\partial\alpha_{m-1}}\right|.$$

Da $y(u)-uH(u)$ eine Linearkombination von $\frac{\partial y}{\partial\alpha_1},\cdots\frac{\partial y}{\partial\alpha_{m-1}}$ ist, so ist

$$\Delta\, d\alpha_1\cdots d\alpha_{m-1}=r^{m-1}H(u)\left|u,\frac{\partial y}{\partial\alpha_1},\ldots,\frac{\partial y}{\partial\alpha_{m-1}}\right|d\alpha_1\cdots d\alpha_{m-1}=r^{m-1}H(u)\,do$$

(do das Oberflächenelement von B). Dasselbe gilt in bezug auf B_2 und es ist daher

$$J=\int\limits_{f\leq T}(x)\,dx=\int\limits_0^T r^{m-1}dr\int\limits_{C(B)}\psi(ry)H\,do. \tag{29}$$

Setzen wir $\psi(x)=\varphi(f(x))e^{\iota x}$, so erhalten wir

1) Transactions 38 (1935).

2) Duke Journal 2.

$$\int\limits_{f\leqq T} \varphi(f(x))e^{i\iota x}dx = \int\limits_0^T \varphi(r)r^{m-1}G_1(r\iota, B)dr \tag{30}$$

wo $G_1(\iota, B) = \int\limits_{C(B)} He^{i\iota y}do$ ist. G_1 läßt sich als Volumselement schreiben. Wir setzen in (29) $\psi(x) = (m + i\iota x)e^{i\iota x}$ und $T = 1$; dann lautet die rechte Seite

$$\int\limits_{C(B)} H\,do \int\limits_0^1 r^{m-1}(m + i\iota r y)e^{ir\iota y}dr\,.$$

Das innere Integral läßt sich aber ausführen, da der Integrand in r die Ableitung von $r^m e^{ir\iota y}$ nach r ist und man erhält $G_1(\iota, B)$. Es ist also

$$G_1(\iota, B) = \int\limits_B (m + i\iota x)e^{i\iota x}dx = \int\limits_{C(B)} He^{i\iota y}do\,. \tag{31}$$

Benützt man die erste Darstellung von $G_1(\iota, B)$ in (31), so kann man (30) leicht direkt unter sehr allgemeinen Voraussetzungen zeigen[1]). (Es genügt für B, daß er ein Sternkörper ist und für $\varphi(r)$ genügt, daß sie im RIEMANNschen Sinne absolut integrierbar sind.) Daraus folgt dann schon, daß $\varphi(f(x))$ im RIEMANNschen Sinne integrierbar ist und es gilt (30). Ist $\varphi(T) = 0$ und $\varphi'(r)$ vorhanden, so folgt aus (30) sofort, daß

$$\int\limits_{f\leqq T} \varphi(f)e^{i\iota x}dx = -\int\limits_0^T \varphi'(r)r^m G(r\iota, B)dr \tag{32}$$

Will man Abschätzungen des obigen Integrals nach $|\iota|$ bzw. asymptotische Entwicklungen, so kann man dies mittels der Formeln (30) und (32) aus jenen von G und G_1 herleiten. Für G_1 [2]) erhält man nach derselben Methode wie bei G und Λ

$$G_1(\iota, B) = 2\left(\frac{2\pi}{|\iota|}\right)^{\frac{m-1}{2}}\left[H(e_\iota)\sqrt{K(e_\iota)}\cos\left(H(\iota) - \frac{(m-1)\pi}{4}\right) + O\left(\frac{1}{|\iota|^{\frac{1}{2}}}\right)\right] \tag{33}$$

($\iota = He_\iota$). Für die Kugel ist übrigens $G_1 = (2\pi)^{\frac{m}{2}}\dfrac{J_{\frac{m}{2}-1}(|\iota|)}{|\iota|^{\frac{m}{2}-1}}$. Ebenso kann man zeigen, daß, wenn $|\iota| = R$ gesetzt wird,

$$G_2(\iota, B) = \frac{\partial\sigma_1}{\partial R} = \int\limits_C e_\iota y e^{i\iota y} H do = O\left(\frac{1}{|\iota|^{\frac{m-1}{2}}}\right) \quad \text{ist.} \tag{34}$$

(Wir begnügen uns hier mit der Angabe der Größenordnung.) Das Integral (28) läßt sich direkt aber einfacher behandeln. Wir wollen uns aber auf $\delta = \frac{k}{2}$ (k natürliche Zahl) beschränken und uns mit der Herleitung der Größenordnung zufrieden geben. Dazu betrachten wir in R_{m+k} den konvexen Körper $\bar{B}$: $f^2(x) + z_1^2 + \cdots + z_k^2 \leqq 1$. Nun ist

---

1) I S. 2, III S. 107 (1) mit $U(x) = e^{i\iota x}$.

2) I S. 10 (6).

$$\int\limits_{B} e^{\iota x} dx\, dz = \frac{\pi^{\frac{k}{2}}}{\Gamma\left(\frac{k}{2}+1\right)} \int\limits_{f\leqq 1} (1-f^2)^{\frac{k}{2}} e^{\iota x} dx,$$

also folgt sofort aus (20)

$$\int\limits_{f\leqq 1} (1-f^2)^{\delta} e^{\iota x} dx = O\left(\frac{1}{|\iota|^{\frac{m+1}{2}+\delta}}\right), \tag{35}$$

wenn $\delta = \frac{k}{2}$. (35) gilt, wie man zeigen kann, für jedes $\delta \geqq 0$.

Es ist auch leicht, die asymptotische Formel[1])

$$J_\delta = \frac{(2\pi)^{\frac{m-1}{2}} 2^\delta \Gamma(\delta+1)}{|\iota|^{\frac{m+1}{2}+\delta}} \left[\frac{\sqrt{K(e_\iota)}}{(H(e_\iota))^\delta} \cos\left(H(\iota) - \frac{\pi}{2}\left(\delta + \frac{m+1}{2}\right)\right) + O\left(\frac{1}{|\iota|}\right)\right] \tag{36}$$

herzuleiten. Für die Kugel ist $J_\delta = (2\pi)^{\frac{m-1}{2}} \frac{J_{\frac{m}{2}+\delta}(|\iota|)}{|\iota|^{\frac{m}{2}+\delta}}$. Wir wollen, um ein Beispiel für (30) zu geben, das wichtige Integral

$$G_3 = \lim_{T\to\infty} \int\limits_{f\leqq T} e^{-f^2+\iota x} dx \tag{37}$$

(wir schreiben kurz $\int_f$) abschätzen. Man schließt aus (30) leicht, analog wie der Beweis von (32)

$$G_3 = 2\int\limits_0^\infty e^{-r^2} r^m \sigma(B, \iota r)\, dr = \int\limits_0^\infty e^{-u} L_0(\iota, u)\, du, \tag{38}$$

wo allgemein $L_s(\iota, u) = \frac{1}{\Gamma(s+1)} \int\limits_{f^2\leqq u} (u-f^2)^s e^{\iota x} dx$ ist.

Unter Berücksichtigung, daß $L_{s+1}(\iota, u) = \int\limits_0^u L_s(\iota, r)\, dr$ ist, folgt aus (38) durch s-malige partielle Integration: $G_3 = \int\limits_0^\infty e^{-u} L_s(\iota, u)\, du$. Aus (35) folgt aber sofort $L_s = u^{\frac{m-1}{4}+\frac{s}{2}}\left(O|\iota|^{-\left(\frac{m+1}{2}+s\right)}\right)$. Es ist also für jede natürliche Zahl s

$$G_3 = \left(O|\iota|^{-\left(\frac{m+1}{2}+s\right)}\right). \tag{39}$$

Dies ist nicht die wahre Größenordnung, denn für die Kugel ist (37) bekanntlich, von Faktoren abgesehen $|\iota|^m e^{\frac{-|\iota|^2}{4}}$.

III.

Aus (30) kann man allgemein mittels eines Satzes von BOCHNER[2]) unter Benützung von (34) leicht folgenden Satz herleiten:

---

[1]) I S. 20 (3) und II S. 84 (1).

[2]) I S. 22 (5).

Ist $0 < q < \frac{m-1}{2}$ und

1. $\frac{1}{T^q}\int\limits_0^T |\Phi(r)| r^{\frac{m-1}{2}} dr$ beschränkt für alle r (40)

und

2. $\lim\limits_{T\to 0} \frac{1}{T^q}\int\limits_0^T \Phi(r) r^{m-1} dr$ vorhanden $= C$, (41)

dann ist

$$\lim_{|\iota|\to\infty} |\iota|^q \int_f \Phi(f) e^{i\iota x} dx = C q \int_f f^{q-m} e^{i e_\iota x} dx \tag{42}$$

$\left(\int\limits_f = \lim\limits_{T\to\infty} \int\limits_{f\leqq T}\right)$. Die Integrale sind absolut konvergent, wie man leicht aus (30) schließt. Dieser Satz macht also Aussagen über das asymptotische Verhalten von $\int \Phi(f) e^{i\iota x} dx$. Wir wollen auf den Beweis nicht näher eingehen, sondern uns mit der umgekehrten Fragestellung beschäftigen: Ist die Voraussetzung 2. notwendig, damit (42) für jedes ι, d. h. für jede Richtung e_ι gilt? Dies ist eine TAUBERsche Fragestellung und für die Kugel wurde dies, unter der Voraussetzung $\Phi \geqq 0$ von CHENG[1]) bejahend beantwortet. Es wurden dabei spezielle Eigenschaften der BESSELschen Funktion benützt. Wir wollen (unter den hier immer benutzten Annahmen über B) zeigen, daß, unter der Voraussetzung (40) und (42) für alle Richtungen e_ι und

$$\Phi \geqq 0, \tag{43}$$

(41) gilt. Dabei wollen wir dem Gedankengang von CHENG folgen.

Es genügt aber, statt (40) nur

$$\int_0^T |\Phi| r^{m-1} dr < A T^q \tag{40'}$$

vorauszusetzen.

Zunächst ist für jedes $\delta > 0$, $|\iota|^q \int\limits_{\delta\leqq f} \Phi(f) e^{i\iota x} dx = o(1)$ für $|\iota| \to \infty$. Denn es ist nach (30) und (33)

$$|\iota|^q \int_{\delta\leqq f} = O\left(|\iota|^q \int_\delta^\infty \Phi(r) r^{m-1} (|\iota| r)^{-\frac{m-1}{2}} dr\right) \tag{44}$$

und nach partieller Integration folgt, wenn wir $F(r) = \int\limits_0^r \Phi(r) r^{m-1} dr$ setzen, für (44), wegen (41)

$$O\left(|\iota|^{q-\frac{m-1}{2}} \left(F(r) r^{-\frac{m+1}{2}}\right)\Big|_\delta^\infty\right) + O\left(|\iota|^{q-\frac{m-1}{2}} \int_\delta^\infty F(r) r^{-\frac{m+1}{2}} dr\right) =$$

$$= o(1) + O\left(|\iota|^{q-\frac{m-1}{2}} \int_\delta^\infty r^{q-\frac{m+1}{2}} dr\right) = o(1) \quad \text{für} \quad |\iota| \to \infty.$$

[1]) Annals of Math. 50.

Es kann also o.B.d.A. angenommen werden, daß $\Phi(r) = 0$ für $r > \delta > 0$, also

$$\int_0^\infty |\Phi(r)| r^{m-1} dr < \infty \quad \text{ist.} \tag{45}$$

Nun gehen wir von der Funktion $e^{-f^2(x)}$ aus. Dann ist die zugehörige FOURIERtransformierte $\Psi(\iota) = \frac{1}{(2\pi)^m} G_3$ nach (39) für jedes natürliche s ein $O\left(|\iota|^{-\frac{m+1}{2}-s}\right)$. Daraus folgt, daß die FOURIERtransformierte von $\Psi(\iota)$ absolut konvergent ist und da alles stetig ist, können wir den FOURIERschen Integralsatz anwenden und erhalten

$$e^{-f^2(x)} = \int_{R_m} \Psi(\iota) e^{i\iota x} d\iota \,. \tag{46}$$

Dann ist, wenn $K(\iota) = q \int_f f^{q-m} e^{ie_\iota x} dx$ ist,

$$\int_{R_m} K(\iota) \frac{\Psi(\iota)}{|\iota|^q} d\iota = V(B) m \Gamma\left(\frac{q}{2} + 1\right), \tag{47}$$

denn es ist

$$q \int_{R_m} \frac{\Psi(\iota)}{|\iota|^q} d\iota \int_f f^{q-m} e^{ie_\iota x} dx = q \int_f f^{q-m} dx \int_{R_m} \Psi(\iota) e^{i\iota x} d\iota =$$

$$= q \int_f f^{q-m} e^{-f^2} dx = V(B) m q \int_0^\infty e^{-r^2} r^{q-1} dr = V(B) m \Gamma\left(\frac{q}{2} + 1\right).$$

Dabei haben wir (30) mit $\iota = 0$ benützt. Es ist nun nach (46)

$$U_R^* = \frac{R^q}{V(B) m} \int_f \Phi(f) e^{-R^2 f^2} dx = \frac{R_q}{V m} \int_{R_m} \Psi(\iota) d\iota \int_f \Phi(f) e^{iR\iota x} dx \,. \tag{48}$$

Wir setzen nun $U_R = \frac{1}{\Gamma\left(\frac{q}{2}+1\right)} U_R^*$ und beachten (47). Dann erhält man

$$U_R - C = \frac{1}{V m \Gamma\left(\frac{q}{2}+1\right)} \int_{R_m} \frac{\Psi(\iota)}{|\iota|^q} \left((|\iota| R)^q \int_f \Phi(f) e^{iR\iota x} dx - C K(\iota) \right) d\iota \,. \tag{49}$$

Nun ist nach Voraussetzung $\lim\limits_{R|\iota| \to \infty} (R|\iota|)^q \int_f \Phi(f) e^{i\iota R x} dx = C K(\iota)$.

Weiter ist, unter Benützung von (30) und (45)

$$\left| R^q \int_f \Phi(f) e^{iR\iota x} dx \right| = \left| (|\iota| R)^q \int_0^\infty \Phi(r) r^{m-1} G_1(R\iota r, B) dr \right| \leqq$$

$$\leqq (|\iota| R)^q \int_0^\infty |\Phi(r)| r^{m-1} dr = O\left((|\iota| R)^q\right).$$

Es werden nun M so groß gewählt, daß bei gegebenem $\varepsilon > o$, der Klammerausdruck in (49), für $R|\iota| \geqq M$, dem Betrage nach $< \varepsilon$ ist. Es ist

$$U_R - C \leqq \int_{R|\iota| \leqq M} + \int_{R|\iota| > M}$$

Dann ist also das zweite Integral $< \varepsilon \int\limits_{R|\iota|>M} \frac{|\Psi(\iota)|}{|\iota|^q} d\iota = O(\varepsilon)$, während das erste Integral

$$\int\limits_{|\iota|\leqq\frac{M}{R}} \frac{\Psi(\iota)}{|\iota|^q} d\iota (|\iota| R)^q \int\limits_f \Phi e^{iR\iota x} dx - CK(\iota) = O\Big(\int\limits_{R|\iota|\leqq M} \frac{|\Psi(\iota)|}{|\iota|^q} d\iota\Big) + O\Big(R^q \int\limits_{|\iota|\leqq\frac{M}{R}} d\iota\Big) = o(1)$$

ist. Damit ist gezeigt, daß $\lim\limits_{R\to\infty} U_R = C$ ist, wo $U(R)$ nach (48) und (30) $= \frac{R^q}{\Gamma\left(\frac{q}{2}+1\right)} \int\limits_0^\infty \Phi(t) t^{m-1} e^{-R^2 t^2} dt$ ist. Nach einem Satz von Hardy und Littlewood[1]) folgt aber daraus, daß dann $\lim\limits_{T\to 0} \frac{1}{T^q} \int\limits_0^T \Phi(r) r^{m-1} dr = C$ sein muß, w.z.b.w.

(28) gestattet eine weitere Anwendung. Man kann daraus folgern: Ist x ein beliebiger Punkt, aber kein Gitterpunkt $\neq 0$, so ist[2])

$$\sum_{f(g)\leqq R} (R^2 - f^2(g))^\delta e^{2\pi i x g} = \sum_\iota \int\limits_{f\leqq R} e^{-2\pi i(\iota - x)z} (R^2 - f^2)^\delta dz$$

für $\delta > \frac{m-1}{2}$. Für die Integrale rechts können wir nach (28)

$$R^{m+2\delta} J_\delta(B, R(x-\iota))$$

schreiben. Es sei nun $\Phi(x)$ eine periodische integrierbare Funktion mit der Periode 1 in jeder Komponente x und $\sum c_g e^{2\pi i g x}$ ihre Fouriersche Reihe. Dann ist bei festem x_0, $\Psi(x) = \Phi(x + x_0)$ ebenfalls periodisch in x mit den Fourierkoeffizienten $a_g = c_g e^{2\pi i g x_0}$. Wir multiplizieren nun (50) rechts und links mit $\Psi(x)$ und integrieren über den Einheitswürfel $W: 0 \leqq x_i \leqq 1$ $(i = 1, \ldots, m)$. Dann erhalten wir $\sum (R^2 - f^2(g))^\delta a_g$ $= R^{m+2\delta} \sum\limits_\iota \int\limits_W \Psi(x) J_\delta(B, R(x-\iota))\, dx$. Nun ist

$$\sum_\iota \int\limits_W = \sum_\iota \int\limits_{W+\iota} \Psi(x+\iota) J_\delta(B, Rx)\, dx = \int\limits_{R_m} \Psi(x) J_\delta(B, Rx)\, dx$$

da Ψ periodisch ist. Wenn wir x_0 durch x ersetzen, erhalten wir

$$\sum_{f(g)\leqq R} \left(1 - \frac{f^2(g)}{R^2}\right)^\delta c_g e^{2\pi i g x} = R^m \int\limits_{R_m} \Phi(x+z) J_\delta(B, Rz)\, dz \qquad (51)$$

Bei der Kugel ist diese Formel in der Bochnerschen Summierungstheorie der Fourierschen Reihen durch sphärische Mittel[3]) wichtig. Es erscheint möglich, durch (51) eine analoge Theorie aufzubauen[4]).

---

[1]) Proceedings London (2) 30 (1929) 23—37. Journal für reine u. angew. Math. 164 (1931) 27—39.

[2]) III S. 127 (29).

[3]) I S. 23 (8).

[4]) Transactions 40 (1936) S. 207—475.

Zum Schluß möchte ich noch auf folgendes Integral hinweisen[1]): Es existiert $\lim\limits_{T\to\infty}\int\limits_{H(x)\leqq T} e^{iax} G(B; x)\,dx$ für $a \neq 0$ ($H(x)$ Stützfunktion von B) und es ist

$$\frac{1}{(2\pi)^m}\lim_{T\to\infty}\int\limits_{H(x)\leqq T} e^{iax} G(B; x)\,dx = \begin{cases} 1 & \text{wenn } a \text{ im Innern von } B \text{ liegt} \\ \frac{1}{2} & \text{wenn } a \text{ am Rand von } B \text{ liegt} \\ 0 & \text{sonst} \end{cases}$$

Es ist dies eine Verallgemeinerung des bekannten diskontinuierlichen Integrals $\int\limits_0^\infty J_\mu(t) J_{\mu-1}(bt)\,dt$ aus der Theorie der Besselfunktionen. Man kann daraus folgende Verallgemeinerung eines Satzes von Berry[2]) folgern:

Liegt eine im Lebesqueschen Sinne quadratisch integrierbare Funktion $\Phi(x)$ im R_m vor, ist $\Psi(y) = \operatorname{l\cdot i\cdot m}\limits_{D\to\infty} \frac{1}{(2\pi)^{\frac{m}{2}}}\int\limits_D \Phi(x) e^{iyx}\,dx$ eine Fourier-transformierte, so läßt sich diese in der Gestalt

$$\lim_{r\to 0}\frac{1}{V(B)\,r^{\frac{m}{2}}}\int\limits_{R_m} \Phi(x) G(B, rx) e^{ixy}\,dx$$

darstellen.

---

[1]) Zum Beweis von (51) vgl. S. Minakshisundaram, Amer. Journal 71 (1949), S. 60—66.

[2]) Journal Massachusetts, Band 8.

Das inhomogene Problem in der Geometrie der Zahlen

Proceedings of the International Congress of Mathematicians, Amsterdam, vol. III, 1954, pp. 20–27

Der Fragenkreis, über welchen ich sprechen will, wurde in den vergangenen Jahren, besonders seit dem Bericht von Prof. H. Davenport [1]) auf dem letzten Kongreß, in vielen Arbeiten behandelt und es wurde dabei eine Reihe von wichtigen Ergebnissen erzielt. Es soll nun ein Überblick über die allgemeinen Begriffsbildungen und Sätze gegeben werden, ohne dabei den Anspruch auf Vollständigkeit zu erheben.

Zunächst einige Vorbemerkungen! Es sei $\mathfrak{G}$ ein n-dimensionales Gitter ($n \geqq 1$) im Vektorraum $\mathfrak{V}_n$ eines euklidischen Raumes R_n von gleicher Dimension über dem Körper der reellen Zahlen, besitzt also stets Basen, gebildet von freien Vektoren $\mathfrak{a}_1, \ldots, \mathfrak{a}_n$, so daß alle Gittervektoren $\mathfrak{g} \in \mathfrak{G}$ ganzzahlige Linearkomposita dieser Vektoren sind. Arithmetisch gesprochen, ist $\mathfrak{G}$ die Wertmenge von n homogenen reellen Linearformen $\xi_i = L_i(x) = \sum_{k=1}^{n} a_{ik}x_k$ mit Det $(a_{ik}) \neq 0$, wenn die $x_1, \ldots, x_n$ alle ganze Zahlen durchlaufen. Zwei Punkte p, p' im R_n heißen zueinander äquivalent modulo $\mathfrak{G}$($p \equiv p'$ (mod $\mathfrak{G}$)), wenn sie durch eine Gittertranslation von $\mathfrak{G}$ auseinander hervorgehen. Die Borelmeßbaren Fundamentalbereiche $F(\mathfrak{G})$ von $\mathfrak{G}$, für die es zu jedem $p \in R_n$ genau einen äquivalenten Punkt in F gibt, haben alle das gleiche Maß, $|$ Det. $(\mathfrak{a}_1, \ldots, \mathfrak{a}_n)|$, ($\mathfrak{a}_1, \ldots, \mathfrak{a}_n$ Basis von $\mathfrak{G}$), das Maß $m(\mathfrak{G})$ von $\mathfrak{G}$. Für reelles $\lambda \neq 0$ ist $m(\lambda\mathfrak{a}) = |\lambda|^n m(\mathfrak{a})$, ($\lambda\mathfrak{G}$ das Gitter der $\lambda\mathfrak{a}$ ($\mathfrak{a} \in \mathfrak{G}$). Ist T eine beliebige Menge im R_n, $T\mathfrak{a}$ die Menge, welche aus T durch eine Gittertranslation $\mathfrak{a} \in \mathfrak{G}$ hervorgeht, so heißt die Menge aller Figuren $T\mathfrak{a}$ mit H. Hadwiger [2]), das Figurengitter (T, $\mathfrak{G}$). Ist T ein Punkt p, so ist (p, $\mathfrak{G}$) die Menge aller zu p äquivalenten Punkte, das Punktgitter zum Vektorgitter $\mathfrak{G}$, welches p enthält.

Das inhomogene Problem lautet nun so: Wann überdeckt die Vereinigungsmenge der Figuren von (T, $\mathfrak{G}$), (das Innere T^0 von T nicht leer), den R_n, d.h. wann gibt es zu jedem p ein $p' \in T$ mit $p' \equiv p$ (mod $\mathfrak{G}$), symbolisch $p \in T$ (mod $\mathfrak{G}$). Es heißt dann $\mathfrak{G}$ Überdeckungsgitter zur Menge T, (Kurz *ÜG* zu T). Es ist $\mathfrak{G}$ stets *ÜG* zu jedem Fundamentalbereich $F(\mathfrak{G})$. Die Überdeckung ist

[1]) Proceedings of the International Congress of Math. (1950) I, 166–174.

[2]) Comment Math. Helvetici 11 (1938) 221–233.

hier einfach, denn die Figuren haben keinen Punkt gemeinsam. Ist $\mathfrak{G}$ kein $\ddot{U}G$ zu T, dann „verkleinern" wir das Gitter, d.h. wählen $\lambda \neq 0$ so klein, bis $\lambda\mathfrak{G}$ ein $\ddot{U}G$ zu T wird. Es heißt $E(T, \mathfrak{G}) = 1/\inf |\lambda|$ ($\lambda \in l$, die Menge[3]) aller λ), ($-\infty < \lambda < \infty$), für die $\lambda\mathfrak{G}$ kein $\ddot{U}G$ zu T ist), die Exzentrität zu $(T, \mathfrak{G})$. Es ist $E(T, \sigma\mathfrak{G}) = \sigma E(T, \mathfrak{G})$ $(\sigma > 0)$. Lassen wir jetzt $\mathfrak{G}$ fest und unterwerfen T von einem inneren Punkt z aus einer genügend großen Ähnlichkeitstransformation $T \to T_z(\lambda)$ (λ Vergrößerungsverhältnis), solange bis $\mathfrak{G}$ Überdeckungsgitter von $T_z(\lambda)$ wird! Es sei T aus der Klasse Σ_τ der Sternkörper S mit Mittelpunkt. Die S sind abgeschlossene Mengen mit einem Zentrum $z \in S^0$, so daß z Mittelpunkt ist und mit jedem Punkt p von S alle Punkte der Verbindungsstrecke zp (höchstens p ausgenommen), innere Punkte von S sind. Also $S \subset S_z(\lambda)$ für alle $\lambda \geqq 1$. Zu S gibt es eine stetige Indikatrix $f(p)$, so daß die Punkte von S^0 durch $f(p) < 1$ $(f(z) = 0)$ und jene von $S_z(\lambda)$ durch $0 \leqq f(p) \leqq \lambda$ charakterisiert sind. Beispiele für Sternkörper S liefert jede homogene Form $F(\xi) = F(\xi_1, \ldots, \xi_n)$ von einem Grad $d(F(t\xi) = t^d F(\xi))$ mit der Indikatrix $f(\xi) = |F|^{1/d}$. Ist $\mathfrak{G}$ definiert durch n homogene Linear-formen $\xi_i = L_i(x_1, \ldots, x_n)$ dann ist $\mathfrak{G}$ genau dann $\ddot{U}G$ zu $S_z(\lambda)$, wenn es zu jedem n Tupel $c = (c_1, \ldots, c_n)$ reeller Zahlen, ein n Tupel $g = (g_1, \ldots, g_n)$ ganzer Zahlen gibt, so daß

$$f(L_1(g-c), \ldots, L_n(g-c)) \leqq \lambda \tag{1}$$

ist. Da die Linearformen $L_i(x_1 - c_1, \ldots, x_n - c_n)$ inhomogen in $x_1, \ldots, x_n$ sind, heißt der ganze Fragenkreis inhomogenes Problem, im Gegensatz zum homogenen Problem, wo nach ganzzahligen Lösungen $g \neq 0$ von (1) für $c = 0$ gefragt wird.

Ist $p \in R_n$, dann sei $E(p) = E(p, S, \mathfrak{G}) = \inf \lambda$ ($\lambda \in L^*$, die Menge aller $\lambda > 0$, für die $p \in S_z(\lambda)$ (mod $\mathfrak{G}$) ist), also gleich $\inf f(p')$ ($p' \equiv p$ (mod $\mathfrak{G}$)) ist. Gehört $E(p)$ selbst zu L^*, so ist es ein angenommenes Minimum für p. Es ist $E(p) = E(p')$ für $p \equiv p'$ und ist in p oberhalb stetig. Die Exzentrizität $E = E(S, \mathfrak{G})$ ist dann nichts anderes, als $\sup E(p)$ über alle p, also ist $\mathfrak{G}$, für jedes $\lambda > E$, $\ddot{U}G$ zu $S_z(\lambda)$. Ist dies auch für $\lambda = E$ der Fall, dann heißt E ein angenommenes Minimum zu $(S, \mathfrak{G})$. Es gibt dann[4]) ein p, so daß $E(p)$ ein angenommenes Minimum und gleich E ist. Ist P die Menge aller p für die $E(p) = E$ und ist $E_2 = \sup E$ $(p \notin P)$ kleiner als E, dann heißt E isoliert, $E = E_1$ das erste Minimum, E_2 das zweite Minimum. Die Definition der Kette dieser Minima kann fortgesetzt werden, bis ein nicht isoliertes Minimum auftritt.

---

[3]) Ist T offen, dann ist das Komplement zu l in $-\infty < \lambda < \infty$ offen und enthält eine Umgebung von $\lambda = 0$. Vgl. H. P. F. Swinnerton-Dyer, Proc. Camb. Phil. Soc. 50 (1954) 20–25.

[4]) J. Heinhold, Math. Z. 44 (1939) 659–88.

Ist M die Menge aller $\mathfrak{G}$, welche $\ddot{U}G$ für S sind (nicht notwendig aus Σ_τ) und ist M^* ihr Komplement, dann sei $\vartheta(S) = \inf m(\mathfrak{G})(\mathfrak{G} \in M^*)$, $\Theta(S) = \sup m(\mathfrak{G})(\mathfrak{G} \in M)$ ($\vartheta(S) = \infty$, wenn M^* leer ist). Es heißt $\mathfrak{G}$ minimales bzw. maximales Gitter, wenn $m(\mathfrak{G}) = \vartheta(S)$ bzw. $\Theta(S)$ ist. Ist S abgeschlossen und meßbar mit einem Maß $m(S) < \infty$, dann gibt es stets [5]) maximale Gitter, welche, wenn S beschränkt und abgeschlossen ist [6]), (S^0 nicht leer), $\ddot{U}G$ von S sind; im allgemeinen Fall bedeckt aber doch $(S, \mathfrak{G})$ den R_n bis auf eine Menge vom Lebesqueschen Maße 0. Es ist

$$\vartheta(S) = E^{-n}(S),\ \Theta(S) = \varepsilon^{-n}(S),\ \varepsilon(S) = \inf E(S, \mathfrak{G})/m^{1/n},$$
$$E(S) = \sup E(S,\mathfrak{G})/m^{1/n}$$

($\mathfrak{G}$ durchläuft alle Gitter mit festem Maße m), (m beliebig). Wir nennen die Gitter $\mathfrak{G}$ mit Maß m, für welche $E(S, \mathfrak{G})$ angenommen und gleich $\varepsilon(S)$ ist, dünnste $\ddot{U}G$ des R_n durch Mengen S. Ist S abgeschlossen beschränkt, S^0 nicht leer, dann gibt es nach dem Vorhergehenden solche Gitter. $D(S) = m(S)\varepsilon^n(S) = m(S)/\Theta(S)$ heißt Dichte der dünnsten regulären Überdeckung des R_n durch Mengen S. Ist nämlich B eine beliebige abgeschlossene beschränkte konvexe Menge mit Mittelpunkt und positivem Volumen, (die Klasse dieser Mengen bezeichnen wir mit K), so heißt sie regulär überdeckt durch Mengen S, wenn es ein Gitter $\mathfrak{G}$ gibt, so daß B von den Figuren aus $(S, \mathfrak{G})$ lückenlos überdeckt wird und es ist $d = Nm(S)/m(B)$ die Dichte dieser Überdeckung, wenn N die Anzahl der Figuren ist, welche an der Überdeckung teilhaben. $D(B, S) = \min d$ ist die Dichte der dünnsten regulären Überdeckung von B durch Mengen S. Ich habe gezeigt [6]), daß $\lim\limits_{\lambda \to \infty} D(B_z(\lambda), S)$ existiert (z Mittelpunkt von B) und gleich $D(S)$ ist. Daraus folgt, daß $\Theta(S) \leqq m(S)$ ist.

Ist S aus der Klasse K(z Mittelpunkt von S), dann ist bekanntlich

$$\mu_n \leqq 2E(S, \mathfrak{G}) \leqq n\mu_n \leqq n2^n m(\mathfrak{G})/m(S)\mu_1{}^{n-1} \tag{2}$$

wo $\mu_i = \mu_i(S, \mathfrak{G}, z)$, $(i = 1, \ldots n)$ das i-te Minimum ist. Für beliebige Mengen S und Punkte p ist $\mu_i(S, \mathfrak{G}, p)$ $(1 \leqq i \leqq n) = \inf \lambda$ ($\lambda > 0$; dim $((p, \mathfrak{G}) \cap S_p(\lambda)) \geqq i$). Aus (2) folgert K. Mahler [7]) mit Hilfe seines Übertragungssatzes die wichtige Abschätzung [8]): $E(S, \mathfrak{G}) \ll \tau_1{}^{-1}$, mit $\tau_1 = \mu_1(T, \mathfrak{G}^*, z(S))$ wo T in Bezug auf z polar zu S und $\mathfrak{G}^*$ reziprok zu $\mathfrak{G}$ ist. Der Kroneckersche Approxi-

[5]) R. P. Bambah, Journ. Ind. (1953) 447–59.

[6]) Für konvexe Mengen bei E. Hlawka, Monatsh. Math. 53 (1949) 81–131.

[7]) Casopis Mat. a. Fys. 68 (1939) 93–102.

[8]) $A \ll B$ soll heißen: Es gibt eine positive Konstante c_n, die nur von n abhängt, daß $A \leqq c_n B$. Analog ist $A \gg B$ erklärt. Für die c_n sind in allen zu besprechenden Formeln, Abschätzungen bekannt. Wir begnügen uns hier mit der qualitativen Darstellung.

mationssatz ist eine unmittelbare Folgerung. Folgende Ungleichung wurde von mir [9]) gefunden:

$$2\,E(S,\mathfrak{G}) \leqq \{2^{n-1}(\nu+1)\,m\,(\mathfrak{g})/m(S)\} \tag{3}$$

wo $\nu(S,\mathfrak{G},z)$ die (ungerade) Anzahl der Gitterpunkte des Gitters $(z,\mathfrak{G})$ im Innern von S ist. ($\{x\}$ nächst größere ganze Zahl von x). Für den klassischen Fall $m(S)=2^n\,m(\mathfrak{g})$, $\nu=1$ liefert sie den genauen Wert $E=\frac{1}{2}$. S ist dann bekanntlich ein Polyeder, begrenzt von 2^n-1 Paaren von parallelen Seitenflächen. Ein anderes Beispiel ist folgendes. Es sei S ein Würfel W mit der Kantenlänge L und $\mathfrak{G}$ ein Gitter mit den Basisvektoren $\mathfrak{a}_1,\ldots,\mathfrak{a}_n$ parallel zu den Kanten und den Längen $2^{n-2}(\nu+1)/L^{n-1}$, $L/(\nu+1)$, $\frac{1}{2}L,\ldots,\frac{1}{2}L$, ($\nu$ ungerade, $2^{n-2}(\nu+1)/L^n>1$). Man sieht sofort:

$$\nu(W,\mathfrak{G},z)=\nu,\ m(W)=L^n,\ m(\mathfrak{G})=1, 2E(W,\mathfrak{G})=2^{n-1}(\nu+1)L^n$$

Da ν und L beliebig ist, so folgt: *Für jede beschränkte Menge S ist stets $E(S)=\infty$, also ist $\vartheta(S)=0$.*

Nach A. M. Macbeath [10]), ist dies auch dann noch der Fall, wenn S nicht beschränkt, aber vom endlichen Maß ist. Es genügt sogar anzunehmen, daß es ein Paar paralleler Hyperebenen gibt, so daß nur der Teil von S außerhalb dieses Streifens endliches Maß hat. Es ist [6]) für $S\in K$, $\Theta(S)\geqq(2/n)^{-n}\Delta_z(S)$ $\geqq n^{-n}\,m(S)$. Nach C. A. Rogers [11]) ist, sogar [12]) $\Theta\geqq 2^n/3^{n-1}\Delta\geqq 3^{-(n-1)}m$ ($\Delta_z(S)$ die Mahlersche Grenzdeterminante inf $m(\mathfrak{G})$ über alle $\mathfrak{g}$, für die $\nu(S,\mathfrak{G},z)$ $=1$ ist). Wenn man von dem ausführlich studierten Fall [13]) $n=2$ absieht, so ist im wesentlichen für $n\geqq 3$ nur die Kugel behandelt. Der genaue Wert von Θ ist für $n=3:32/5\sqrt{5}$, für $n\geqq 4$ ist bis jetzt nur bekannt [15]) $4/3-\varepsilon_n$ $<D(S)<(1{,}15)^n$ ($\varepsilon_n\to 0$ für $n\to\infty$). Diese Resultate werden mit Hilfe der konvexen Polyeder (Wabenzellen) Z erhalten, definiert als Menge aller Punkte

---

[9]) Für $\nu=1$ bei E. Hlawka, Math. Ann. 125 (1952) 183—207. Es kann (3) noch verschärft werden. Vgl. auch E. Hlawka, Monatsh. Math. 58 (1954) 287—91. M. Kneser, Math. Z. 61 (1954) 429—34.

[10]) Proc. Camb. Phil. Soc. 47 (1951) 627—28.

[11]) Journ. Lond. Math. Soc. 25 (1950) 328—31.

[12]) Besitzt S, n aufeinander senkrecht stehende Symmetrieebenen durch $z(S)$, dann ist nach R. P. Bambah und K. F. Roth, Journ. Ind. Math. Soc. 16 (1952) 7—12, $D(S)$ $\leqq \pi n^n/3\sqrt{3}\,n!\sim e^n(\pi/54n)^{\frac{1}{2}}$.

[13]) Vgl. das schöne Buch von L. Fejes Toth, Lagerungen in der Ebene, auf der Kugel und im Raum, Springer-Verlag, Berlin-Heidelberg-Göttingen 1953 X, 197 S. Hinzugefügt werde, J. Heinhold, Math. Z. (1942) 199—214.

[14]) Neuer Beweis bei R. P. Bambah, Proc. Nat. Inst. Sciences of India 20 (1954) 25—52.

[15]) R. P. Bambah and H. Davenport, Journ. Lond. Math. 27 (1952) 224—29, H. Davenport Rend. Palermo (2) 1 (1952) 92—107.

p mit $|p, z| \leqq |p, g|$, für alle Gitterpunkte g von $(z, \mathfrak{G})$, (z Mittelpunkt der Kugel). Es ist $\mathfrak{G}$, $\ddot{U}G$ zu Z mit $D(Z) = 1$, also besitzt Z höchstens 2^n-1 parallele Seitenflächenpaare und es ist $E(S, \mathfrak{G}) = \max |e, z|$ (erstreckt über alle Eckpunkte e von Z). Ist S konvex und offen (braucht nicht beschränkt zu sein), p_0 ein Punkt $\epsilon\, S(\bmod\, \mathfrak{G})$, so wird im allgemeinen p_0 in mehreren Figuren liegen, aber es gibt Teilmengen T von S, so daß $p \epsilon\, T(\bmod\, \mathfrak{G})$, aber nur mehr einer Figur angehört. Nach C. A. Rogers [16]) ist dies für $T = S \cap H(\neq 0)$ der Fall, wo H ein gewisser abgeschlossener Halbraum ist. Es liegt also in T *genau* ein zu p_0 äquivalenter Punkt p (C. A. Rogers nennt ihn äußeren Punkt zu p_0 in S mod $\mathfrak{G}$), also auch im Durchschnitt von S und dem Spiegelbild $S^+(p)$ an p außer ihm kein weiterer Gitterpunkt von $(p, \mathfrak{G})$. Nach dem Minkowskischen Fundamentalsatz gilt also

$$2^n m(\mathfrak{g}) \geqq 2^n \Delta_p(S \cap S^+(p)) \geqq m(S \cap S^+(p)) = F(p) \tag{4}$$

(Satz von A. M. Macbeath [17])). Es ist $\vartheta(S)$ genau dann ∞, wenn für jedes ϱ, S eine Kugel vom Radius ϱ enthält bzw. in keinem Streifen, begrenzt von parallelen Hyperebenen, liegt.

Ist S nicht konvex, so liegen in der Ebene, Überdeckungssätze von L. J. Mordell und K. Rogers für Mengen S von allgemeineren Charakter vor [18]). Für $n \geqq 2$ liefert der Satz von Macbeath (4): Ist $\mathfrak{G}$, $\ddot{U}G$ der konvexen Menge S, dann ist es auch $\ddot{U}G$ von der Menge $F(p) \leqq 2^n m(\mathfrak{G})$.

Ist S definiert durch $\xi_1 \geqq 0, \ldots, \xi_n \geqq 0$, $(F(p) = 2^n |p_1| \ldots |p_n|, p = (p_1, \ldots p_n))$ so folgt der Satz von J. H. W. Chalk [19]): Jedes $\mathfrak{G}$ ist $\ddot{U}G$ von $\xi_1 \ldots \xi_n \leqq m(\mathfrak{G})$ $\xi_1 \geqq 0, \ldots, \xi_n \geqq 0$. Die Menge S: $\xi_1 - (\xi_2^2 + \ldots + \xi_n^2) \geqq 0$ führt zu dem Satz, daß das gleiche gilt, für

$$0 \leqq \xi_1 - (\xi_2^2 + \ldots + \xi_n^2) \leqq (2^{n-2}(n+1) m(\mathfrak{G})/\omega_{n-1})^{2/n+1} \tag{5}$$

(ω_{n-1} Volumen der $(n-1)$–dim. Kugel).

Alle Mengen S, welche man genauer untersucht hat, sind automorph, d.h. 1) sie besitzen eine Gruppe $A = A_z$ von Automorphismen α in Bezug auf einen Punkt z, (der nicht zu S gehören braucht), d.h. von affinen Abbildungen α mit z als Fixpunkt, welche S in sich überführen.

---

16) Journ. London 29 (1954) 133–143.

17) Annals of Math. (2) 56 (1952) 269–93. Für $n = 2$ auch Quart. Journ. (2) 3 (1952) 268–81, Durch (4) ist der Inhalt dieser wichtigen Untersuchungen nicht erschöpft.

18) Duke Math. Journ. 19 (1952) 519–27, K. Rogers, Journ. London 28 (1953) 394–402. Dazu gehören die Untersuchungen von E. S. Barnes, Quart. J. (2) 1 (1950) 199–210, R. P. Bambah, Proc. Camb. Phil. Soc. 47 (1952) 457–60, J. H. W. Chalk ebenda 48 (1952).

19) Quart. J. Math. 18 (1947) 215–27. Ein weiterer Beweis bei A. M. Macbeath, Jour. London 23 (1948) 141–47; A. J. Cole, Quart. J. (2) 3 (1952) 56–62 zeigt dies für $\xi_1 > 0, \ldots, \xi_n > 0$, $|\xi_1 \ldots \xi_n| \leqq 1/2\, m(\mathfrak{G})$.

2) Es gibt eine beschränkte Menge $U(A)$, so daß es zu jedem p aus der Hülle von S ein $\alpha \in A$ gibt, so daß $\alpha p \in U$. Wichtige Beispiele sind: $S_1 : |\xi_1^2 + \ldots + \xi_r^2 - (\xi_{r+1}^2 + \ldots + \xi_n^2)| \leqq 1$ $(2 \leqq r < n)$, $S_2 : |\xi_1 \ldots \xi_n| \leqq 1$ $S(K_1, \ldots, K_t)$ (Verallgemeinerung von S_2): $f_1^{n_1} \ldots f_t^{n_t} \leqq 1$ $(n_1 + \ldots + n_t = n)$ wo f_i $(1 \leqq i \leqq t)$ die Indikatrix des konvexen Körpers K_i in Räumen L_i von der Dimension n_i ist, welche den Schnittpunkt z der t aufeinander senkrechten L_i zum Mittelpunkt hat. Ein Spezialfall mit $n_1 = l, n_2 = m$: (Max $(|x_1|, \ldots |x_l|))^l$ (Max $(|y_1|, \ldots |y_m|))^m$ (Diophantische Approximation!) Da die Einheitskugel E_z um z in S_1, S_2 und in $S(K_1, \ldots, K_t)$ liegt, wenn die K Kugeln sind, so erhält man nach (2) eine Abschätzung

$$E(S, \mathfrak{G}) \ll m(\mathfrak{G}) | \mu_1^{n-1}(E_z) \ll m(\mathfrak{G}) | \mu_1^{n-1}(S) \tag{6}$$

Da mit E_z auch αE_z in S liegt $(\alpha \in A)$, so kann in (6) $\mu_1(E_z)$ durch $\mu = \sup \mu(\alpha E_z)$ ersetzt werden. Je umfangreicher die Automorphismengruppe von S ist, desto größer wird μ sein. So lieferten die Methoden von C. L. Siegel und Tschebotarew (vgl. dazu [20]) für [21]) $S(K_1, \ldots, K_t)$: $E(S, \mathfrak{G}) \ll m(\mathfrak{G}) | \mu_1^{(n-t)}(S, \mathfrak{G})$: ($K_i$ Kugeln). In S_2 ist $t = n$, daher ist $\vartheta(S_2) > 0$. Das gleiche zeigte für S_1, H. Blaney [22]). Es gilt aber sogar:

$$E(S, \mathfrak{G}) \ll \mu_1^{(n-2)/2(n-1)}(S_4, \mathfrak{G})\, m(\mathfrak{G}), \text{wo } S_4 : 0 < \xi_1^2 + \ldots + \xi_r^2 - (\xi_{r+1}^2 + \ldots + \xi_n^2) < 1$$

Nun gibt es aber Gitter mit $m(\mathfrak{G}) = 1$ und beliebig kleinen μ_1; Z.B. für das Gitter: $\xi_i = x_i N^{-1/n}$ $(1 \leqq i < n)$, $\xi_n = N^{1-1/n} x_n$ (N natürliche Zahl), ist $\mu_1 = N^{-1/n}$. Daher ist für $n > 2$, $\Theta(S_1) = \infty$ also $\varepsilon(S_1) = 0$; d.h. es gibt für $n \geqq 3$ stets Überdeckungsgitter zu S_1 mit beliebig großem Maß [23]). Für $n = 2$ versagt der Schluß und es ist auch tatsächlich $\Theta(S) < \infty$. Allgemein läßt sich folgendes zeigen: Für $S(K_1, K_2)$ (K_1, K_2 nicht notwendigerweise Kugeln) ist $\Theta(S) \ll V \ll \Delta(S)$, wo V das Volumen von Max $(f_1, f_2) \leqq 1$ ist. Die Fälle, wo K_1, K_2 Intervalle oder Kreise sind, wurden zuerst von H. Davenport [1]) behandelt und von J. W. S. Cassels [24]) bedeutend vereinfacht. H. P. F. Swinnerton-Dyer [3]) ist es zuerst gelungen, ein allgemeines Kriterium dafür zu finden,

---

[20]) Der Fall $n = 2$ wird genau untersucht in Proc. Camb. Phil. Soc. 47 (1952) 266—73.

[21]) Der Exponent kann in vielen Fällen noch verschärft werden. Vgl. L. E. Clarke, Quart. J. of Math. (Oxford) (2) 2 (1951) 308—15; H. Davenport ebenda (2) 3 (1952) 32—41; E. S. Barnes, Proc. Camb. Phil. Soc. 49 (1953) 360—62; H. Davenport, ebenda 110—93.

[22]) Jour. London 23 (1948) 153—160; vgl. dazu C. A. Rogers, ebenda 27 (1952) 314—19, wo der oben herausgestellte Gesichtspunkt deutlich hervorgehoben ist.

[23]) Für $n = 3$ hat dies H. Davenport, Journ. London 23 (1948) 199—202 hervorgehoben und gibt eine schärfere Abschätzung für $E(S, \mathfrak{G})$.

[24]) Proc. Camb. Phil. Soc. 48 (1952) 72—86 und die dort angegebene Literatur. Der allgemeine Fall bei E. Hlawka, Monatsh. 58 (1954) 292—305.

daß $\vartheta(S)$, also $E(S) < \infty$ ist: Ist S offen, automorph in Bezug auf z und enthält jede Gerade, welche einen Punkt von $Fr_z(S)$ enthält, auch einen Punkt von S, dann ist $\vartheta(S) > 0$ und es gibt minimale Gitter. Dabei ist $Fr_z(S)$ die Hülle der äußeren Randpunkte von S, d.h. der Randpunkte r, welche Häufungspunkte von $zr \cap S$ sind.

Wenn S Sternkörper ist, $\vartheta(S) > 0$ ist und minimale Gitter $\mathfrak{G}'$ existieren, dann kann $\vartheta(S)$ isoliert sein, d.h. es kann $\vartheta_2 = \inf m(\mathfrak{G})$ erstreckt über alle Gitter, welche nicht Überdeckungsgitter sind, aber auch nicht von der Gestalt $\lambda\mathfrak{G}'$ ($\mathfrak{G}'$ minimal), $> \vartheta(S)$ sein. Diese Kette kann mann bis zu einem nicht isoliertem Minimum fortsetzen. Für $n = 2$ hat H. Davenport [25]) und für $n \geqq 2$ J. W. S. Cassels [26]) gezeigt, daß S_2 nicht isoliert ist, wenn die Minkowskische Vermutung richtig ist. (Bewiesen für $n \leqq 4$). Höhere Minima untersuchte für $0 \leqq \xi_1 - \xi_2 2 \leqq 1$, A. M. Macbeath. [27]) H. P. F. Swinnerton-Dyer und E. S. Barnes [28]) haben für $| \xi_1 \xi_2 | \leqq 1$ unter Benützung von Ideen J. W. S. Cassels gezeigt, wie $E(S, \mathfrak{G})$ und die weiteren Minima $E_k(S, \mathfrak{G})$ $(k = 2, \ldots)$ bestimmt werden können, wenn S automorph ist in Bezug auf eine Gruppe $A_z(\mathfrak{G})$, welche aber auch $\mathfrak{G}$ in sich selbst überführt. Zu jedem p gibt es dann ein p_1, so daß $E(p) = E(p_1)$ und $E(p_1)$ angenommen wird, Die Wertmenge von $E(p)$ ist abgeschlossen.

Gibt es ein α aus $A_z(\mathfrak{G})$ mit folgenden Eigenschaften: 1) Es besitzt keine Eigenwerte vom Betrage 1. 2) Sind R, R^* Mengen (R beschränkt), $\mathfrak{g} \in \mathfrak{G}$ so daß für jedes $p \in R$, entweder $\alpha p \in R^*$ (mod $\mathfrak{g}$) oder $\alpha p(-\mathfrak{g}) \in R$, $\alpha^{-1} p \in R \cup R^*$ dann ist $p \in R$ Fixpunkt f von α, $\alpha f = f\mathfrak{g}$, wenn 3) $\alpha^n p \in R^*$(mod $\mathfrak{g}$) für alle ganzen $n \neq 0$ ist. (Lemma von Cassels). Es sei nur eine wichtige Folgerung aus diesem Lemma hervorgehoben (wenn es $\alpha \in A(\mathfrak{G})$ mit der Eigenschaft 1 gibt): Gibt es nur endlich viele inäquivalente p, für die $E(p) = E$ ist, dann ist E isoliert. Dieser Satz gilt auch für die höheren Minima. Bisher wurde nur der Fall betrachtet, daß die Überdeckung mindestens einfach ist. Es kann auch mehrfache Überdeckung [30]) verlangt werden. In der Theorie der diophanti-

---

[25]) Proc. Kon. Ned. Wet. Amsterdam 49 (1946) 815–21.

[26]) Journ. London 27 (1952) 485–92. Für $\Delta(S)$ gilt das Gegenteil, vgl. dazu G.S. Barnes, Proc. Camb. Phil. Soc. 49 (1953) 59–62.

[27]) Proc. Camb. Phil. Soc. 47 (1951) 266–73.

[28]) Acta Math. 87 (1952) 259–323), 88 (1952) 279–316. Acta Math. 92 (1954) 199—234, E. S. Barnes ebenda 235—64 (Weiter Arbeiten: H. J. Godwin, Journ. London 36 (1955) 114—119, Quart. Journ. of Math. V(II)(1954) 28—46; K. Jukeri, Ann. Akad. Feunicae, (1952) Ser. A. 136, 1—16). Ternäre quadr. Farmen E. S. Barnes Acta Math. 92 (1954) 13—33.

[29]) Für $| \xi_1\xi_2\xi_3 | < 1$ ist die Methode angewendet bei P. A. Samet, Proc. Cam. 50 (1954) 372–390.

[30]) vgl. D. B. Sawyer Proc. Cam. Phil. S. 49 (1953) 156–57.

schen Approximationen interessiert man sich für die unendlich oftmalige Überdeckung. Hier hat eine Arbeit von L. J. Mordell [31]). Anlaß zu wichtigen Untersuchungen gegeben. Die Ergebnisse von J. M. H. Chalk [32]) und C. A. Rogers [33]) haben diesen Fragenkreis zu einem gewissen Abschluß gebracht.

---

[31]) J. London Math. Soc. 26 (1951) 93; vgl. auch Varnavides Proc. London Math. (3) 2 (1952), 234–244.

[32]) J. Lond. Math. Soc. 25 (1950) 46, Quart. J. of Math. (Oxford) 19 (1948) 67.

[33]) Journ. London (3) 4 (1954) 50–83.

Über einen Satz von van der Corput

Archiv Math. **6** (1955), 115–120

Von van der Corput [1] wurde folgender Satz bewiesen: *Ist α eine irrationale Zahl, und ist $(x_1, x_2, \ldots)$ eine Folge mit $\lim_{i\to\infty}(x_{i+1}-x_i)=\alpha$, dann ist diese Folge* mod 1 *gleichverteilt.*

Es soll nun eine Verallgemeinerung dieses Satzes gegeben werden, und zwar soll diese gleich für kompakte Gruppen ausgesprochen werden. Dazu schicken wir einige Begriffe aus der Theorie der Gleichverteilung auf einer kompakten Gruppe G voraus [2], [3]. Wir nehmen dabei das Haarsche Maß von G als zu 1 normiert an. Die irreduziblen unitären Darstellungen von G bezeichnen wir mit $D_j(x)$, $(x \in G, j=0, 1, 2, \ldots)$. Dabei sei D_0 die identische Darstellung.

Eine Folge $f=\{x_i\}$ heißt nun gleichverteilt auf G, wenn für jede auf G stetige Funktion $\Phi(x)$

$$\lim_{N\to\infty} M_N(\Phi, f) = M(\Phi) \tag{1}$$

gilt, wobei $M_N(\Phi, f) = \frac{1}{N}\sum_{i=1}^{N}\Phi(x_i)$ und $M(\Phi) = \int_G \Phi\, dx$ ist.

Es gilt das Weylsche Kriterium: f ist genau dann gleichverteilt, wenn für jede Darstellung $D_j(x)$ $(j>0)$

$$\lim_{N\to\infty} M_N(D_j, f) = 0$$

ist.

Mit f_k bezeichnen wir die Folge $\{x_{i+k}\} = (x_{k+1}, x_{k+2}, \ldots)$, welche für jedes feste k gleichverteilt ist, wenn dies für f zutrifft.

Wir führen nun den Begriff der gleichmäßig gleichverteilten Folge im schwachen Sinne, kurz der *sgg*-Folge, ein. Eine Folge f heiße *sgg*-Folge, wenn für jede stetige Funktion Φ

$$\lim_{N\to\infty}\ \overline{\lim_{H\to\infty}}\ \frac{1}{H}\sum_{r=0}^{H} \left| M_N(\Phi, f_{rN}) - M(\Phi) \right| = 0 \tag{2}$$

ist.

Jede *sgg*-Folge ist eine gleichverteilte Folge. Um nämlich für alle stetigen Φ die Formel (1) zu zeigen, genügt es, sich auf Φ mit $M(\Phi) = 0$ zu beschränken. Nun gilt für jede natürliche Zahl K und N die Identität

$$M_K(\Phi, f) = \frac{N}{K} \sum_{r=0}^{H} M_N(\Phi, f_{rN}) + \frac{1}{K} \sum_{i=N(H+1)+1}^{K} \Phi(x_i)$$

mit

$$H = \left[\frac{K}{N}\right] - 1.$$

Es sei nun ein festes $N(\varepsilon) > 0$ so gewählt, daß

$$\overline{\lim_{H\to\infty}} \frac{1}{H} \sum_{r=0}^{H} | M_N(\Phi, f_{rN}) | < \varepsilon$$

ist ($\varepsilon > 0$ beliebig vorgegeben). Dann gilt für alle $H \geqq H(N)$

$$\sum_{r=0}^{H} | M_N(\Phi, f_{rN}) | < 2\,\varepsilon H.$$

Für alle $K \geqq N(H(N) + 2)$ ist

$$H = \left[\frac{K}{N}\right] - 1 \geqq \frac{K}{N} - 2 \geqq H(N).$$

Es ist daher, wenn C eine obere Schranke von $\Phi(x)$ auf G ist:

$$| M_K(\Phi, f) | \leqq 2\varepsilon \frac{N}{K} H + \frac{C}{K}(K - N(H+1)) < 2\varepsilon + \frac{N}{K}.$$

Daher ist für jedes $\varepsilon > 0$

$$\overline{\lim_{K\to\infty}} | M_K(\Phi, f) | \leqq 2\varepsilon$$

und daher gilt (1).

Eine Folge f ist sicher eine *sgg*-Folge, wenn sie eine gleichmäßig gleichverteilte Folge, kurz *gg*-Folge ist [3]. Man versteht darunter eine Folge f, für welche es zu jedem ε und zu jeder stetigen Funktion Φ ein $N_0(\Phi, f, \varepsilon)$ gibt, so daß für alle ganzen Zahlen $k \geqq 0$ und alle $N > N_0$

$$| M_N(\Phi, f_k) | < \varepsilon \tag{3}$$

ist. Jede Folge $\{a^i\}$ ist eine *gg*-Folge, wenn sie gleichverteilt ist. Man zeigt in üblicher Weise das Analogon zum Weylschen Kriterium: Eine Folge f ist genau dann eine *sgg*-Folge, wenn für jede Darstellung $D_j (j > 0)$

$$\lim_{N\to\infty} \overline{\lim_{H\to\infty}} \frac{1}{H} \sum_{r=0}^{H} | M_N(D_j, f_{rN}) | = 0 \tag{4}$$

ist.

Dabei sei, wenn A eine k-zeilige quadratische Matrix ist, $(A \mid B) = \mathrm{Sp}(B^* A)$, $|A| = \sqrt{(A \mid A)}$. Bekanntlich gilt für jede unitäre Matrix $|AU| = |UA| = |A|$, also insbesondere $|U| = \sqrt{k}$. Nach diesen Vorbereitungen können wir die Verallgemeinerung des Satzes von VAN DER CORPUT, welche wir zeigen wollen, leicht aussprechen.

Satz 1: *Es sei auf G eine Folge $f = \{y_i\}$ eine sgg-Folge. Weiter sei $f' = \{x_i\}$ eine Folge so beschaffen, daß $\lim\limits_{i\to\infty} y_{i+1}^{-1} x_{i+1} x_i^{-1} y_i$ existiert und gleich dem Einheitselement e ist. Dann ist f wieder eine sgg-Folge, also insbesondere gleichverteilt*[1]).

Beweis: Wir setzen $y_{i+1}^{-1} x_{i+1} x_i^{-1} y_i = u_i$, dann ist für $s \geqq i$

$$x_s = y_s u_{s-1} u_{s-2} \ldots u_i y_i^{-1} x_i$$

und $\lim\limits_{i\to\infty} u_i = e$. Nach Voraussetzung gibt es zu jedem $\varepsilon > 0$ und zu $D = D_j (j > 0)$ ein $N_0(\varepsilon)$ so, daß für alle $N > N_0$, $\overline{\lim\limits_{H\to\infty}} \frac{1}{H} \sum\limits_{r=0}^{H} |M_N(D, f_{rN})| < \varepsilon$ gilt. Weiter gibt es, wegen der Stetigkeit von $D(x)$, ein k_0, so daß für alle $k \geqq k_0(N)$ gilt:

$$|D(u_k) - E| < \frac{1}{(N+1)^2} \tag{5}$$

(E ist d-zeilige Einheitsmatrix, wenn D vom Grade d ist). Dann ist für alle $i > k \geqq k_0$

$$|D(x_i) - D(y_i) D(y_k^{-1} x_k)| < \frac{i-k}{(N+1)^2}, \tag{6}$$

denn der Ausdruck links in (6) ist ja gleich $|D(u_{i-1}) \ldots D(u_k) - E|$, und mittels vollständiger Induktion folgert man aus (5) die Abschätzung (6). Daraus folgt weiter für jedes $k \geqq k_0$

$$\left|\sum_{i=1}^{N} D(x_{i+k})\right| \leqq \left|\sum_{i=1}^{N} D(y_{i+k})\right| + 1. \tag{7}$$

Dies gilt für $k = rN$, (r natürliche Zahl), wenn $r \geqq k_0 N^{-1}$ ist. Für $r < k_0 N^{-1}$ benützen wir die triviale Abschätzung

$$\left|\sum_{i=1}^{N} D(x_{i+k})\right| \leqq \left|\sum_{i=1}^{N} D(y_{i+k})\right| + N\sqrt{d}. \tag{8}$$

Es ist also für jedes H

$$\frac{1}{H} \sum_{r=0}^{H} |M_N(D, f'_{rN})| \leqq \frac{1}{H} \sum_{r=0}^{H} |M_N(D, f_{rN})| + \frac{\sqrt{d}\, k_0}{H} + N^{-1}. \tag{9}$$

Gehen wir mit $H \to \infty$, so ist für alle $N > N_0(\varepsilon)$

$$\overline{\lim} \frac{1}{H} \sum_{r=0}^{H} |M_N(D, f'_{rN})| < \varepsilon + N^{-1}$$

und daraus folgt alles.

Damit der Satz anwendbar ist, müssen wir zeigen, daß es auf G überhaupt *sgg*-Folgen gibt.

[1]) Ein schwächerer Satz findet sich in [3], Satz 4.

Wir betrachten dazu den Raum F aller Folgen f aus G. In diesem Raum läßt sich ein Maß μ einführen mit $\mu(F)=1$. Wir zeigen dann

Satz 2: ***Fast alle Folgen f sind sgg-Folgen.***

Dieser Satz könnte aus der Ergodentheorie gefolgert werden, wir wollen ihn aber direkt zeigen und erhalten dabei eine vertiefte Einsicht. Wir bezeichnen im Raum F das Differential mit df. Dann ist ja für jede Darstellung $D=D_j(x)$ $(j>0)$ (Grad d)

$$\int\limits_F D(x)df=\int\limits_G D(x)dx=0\,.$$

Daraus folgern wir nun für jede natürliche Zahl H und N

$$\int\limits_F \Big(\sum_{r=0}^{H}\Big|\sum_{i=1}^{N} D(x_{i+rN})\Big|^2-HNd\Big)^2 df\leqq HN^2d^2\,. \tag{10}$$

Beweis: Der Ausdruck unter dem Integral im Quadrat ist ja

$$\sum_{r=0}^{H}\,\Big|\,\sum_{i\neq j}(D(x_{i+rN})\,|\,D(x_{j+rN}))\,,$$

da für $i=j$ ein Term $(\ )=|D|^2=d$ ist. Also wird das Integral gleich

$$\sum_{r=0}^{H}\sum_{s=0}^{H}\ \sum_{i\neq j}\ \sum_{k\neq l}\int\limits_F (D(u_i)\,|\,D(u_j))\,(D(u_k)\,|\,D(u_l))\,df\,, \tag{11}$$

wo $u_a=x_{a+rN}$ für $a=i$ oder j und x_{a+sN} für $a=k$ oder l ist. Für $r\neq s$ sind die vier Variablen u_a im Integral alle voneinander verschieden, daher ist das Integral in (11) gleich Null. Für $r=s$ ist das Integral ebenfalls stets Null, außer wenn das Zahlenpaar (i,j), abgesehen von der Reihenfolge, mit (k,l) übereinstimmt. Im letzten Fall ist es aber $\leqq d^2$, da allgemeiner für Matrizen $|(A\,|\,B)|\leqq|A|\,|B|$ gilt.

Es ist also tatsächlich das Integral in (10) höchstens $2HN(N-1)d^2\leqq 2HN^2d^2$.

Aus (10) folgt nun für jedes N,H und $j(j>0)$

$$\int\limits_F\Big(\frac{1}{Hd}\sum_{r=0}^{H}|M_N(f_{rN})|^2-\frac{d}{N}\Big)^2 df\leqq\frac{1}{HN^2}\,. \tag{12}$$

Es sei der Ausdruck links unter dem Integral mit $R(f,j,N,H)$ bezeichnet, und wir wählen für H die Teilfolge h^2 $(h=1,2,\ldots)$ und gehen in üblicher Weise vor. Ist L_{jNh} die Menge in F, für die $R>(N^2h)^{-1}\log^{1+2\varepsilon}h$ gilt ($\varepsilon>0$ beliebig), dann ist das Maß nach (12) kleiner als $h^{-1}\log^{-(1+2\varepsilon)}h$. Also ist das Maß von $L_{jN}=\bigcap\limits_{h=1}^{\infty}\bigcup\limits_{k=h}^{\infty}L_{jNk}$ gleich Null. Daher ist auch das Maß der Vereinigungsmenge L der L_{jN} für alle j,N gleich Null. Also gilt für jede Folge f, welche nicht in L liegt, für alle j,N und für alle $h>h(j,N)$ $R(f,j,N,h^2)<\frac{1}{N^2h}\log^{1+2\varepsilon}h$, also für alle $H=h^2>H(j,N)$

$$\left|\frac{1}{H}\sum_{r=0}^{H}|M_N(D_j,f_{rN})|^2-\frac{d}{N}\right|\leqq K\,(NH^{\frac14})^{-1}\log^{\frac12+\varepsilon}H$$

($\varepsilon>0$, K Konstante, unabhängig von H und N).

Ist $H > H(j,N)$, so gibt es ein h mit $h^2 \leqq H < (h+1)^2$, so daß

$$\frac{1}{H}\sum_{r=0}^{H} \leqq \frac{1}{h^2}\left(\sum_{r=0}^{h^2} + \sum_{r=h^2+1}^{(h+1)^2}\right);$$

andererseits ist

$$\frac{1}{H}\sum_{r=0}^{H} \geqq \frac{1}{(h+1)^2}\left(\sum_{r=0}^{(h+1)^2} - \sum_{r=h^2+1}^{(h+1)^2}\right),$$

also gilt für alle $H > H(j,N)$

$$\frac{1}{H}\sum_{r=0}^{H} |M_N(D_j, f_{rN})|^2 = \frac{d}{N} + O(H^{-\frac{1}{2}}\log^{\frac{1}{2}+\varepsilon} H), \tag{13}$$

wo O unabhängig von N ist.

Daraus folgt, daß für fast alle Folgen f, für alle N und j

$$U_{Nj} = \lim_{H\to\infty} \frac{1}{H}\sum_{r=0}^{H} |M_N(D_j, f_{rN})|^2$$

existiert und gleich $\frac{d}{N}$ ist, und weiter, daß $\lim\limits_{N\to\infty} U_{Nj} = 0$ ist für fast alle Folgen und alle $j = 1, 2, \ldots$. Daraus folgt bereits unser Satz.

Aus (13) entnehmen wir noch genauer, daß für jedes $\varepsilon > 0$,

$$\frac{1}{H}\sum_{r=0}^{H} |M_N(D_j, f_{rN})| \leqq \sqrt{\frac{d}{N}} + O\left(\frac{\log^{\frac{1}{4}+\varepsilon} H}{H^{1/8}}\right) \tag{14}$$

ist. Diese Abschätzung ließe sich noch verbessern, doch gehen wir nicht darauf ein.

Wenn G das erste Abzählbarkeitsaxiom erfüllt, zeigt man leicht

Satz 3: *Jede Folge f, welche in G überall dicht liegt, läßt sich stets zu einer sgg-Folge umordnen.*

Der nächste Satz bringt eine Anwendung, die eine Verallgemeinerung des Satzes von FATOU [2]) ist.

Satz 4: *Es sei $\Phi(x)$ eine stetige Funktion aus G, und es sei $M(|\Phi|) > 0$. Weiter sei die Reihe $\sum\limits_{i=1}^{\infty} \alpha_i \Phi(x_i)$ für eine sgg-Folge $\{x_i\}$ absolut konvergent, und es gelte für alle i: $0 < \alpha_{i+1} \leqq \alpha_i$.*

Dann ist $\qquad \Sigma\,\alpha_i < \infty.$

Beweis: Wenn $\sum\limits_{r=0}^{\infty} \alpha_{(r+1)N}$ für ein N konvergiert, dann ist alles klar, da für jedes K

$$\sum_{j=1}^{K} \alpha_j \leqq \sum_{n=1}^{N} \alpha_n + \sum_{r=0}^{K} \sum_{i=1}^{N} \alpha_{(r+1)N+i} \leqq N(\alpha_1 + \sum_{r=0}^{\infty} \alpha_{(r+1)N})$$

gilt.

---

[2]) Vgl. auch die Arbeit [3].

Es kann also angenommen werden, daß $\sum_{r=0}^{\infty} \alpha_{(r+1)N}$ für alle N divergiert. Da die Folge $\{x_i\}$ eine *sgg*-Folge ist, so gibt es ein N, so daß für alle $H \geqq H(N) = H_0$

$$\frac{1}{H} \sum_{r=0}^{H} S_r < \frac{1}{2} C$$

ist, wo $M(|\Phi|) = 2C$ und $S_r = \left| \frac{1}{N} \sum_{i=1}^{N} \Phi(x_{rN+i}) - 2C \right|$ gesetzt wurde. Wir setzen $\beta_r = \alpha_{(r+1)N}$. Es ist natürlich β_r monoton abnehmend. Setzen wir $T_r = \frac{1}{r+1} \sum_{t=0}^{r} S_t$, dann ist $\sum_{r=0}^{H} \beta_r S_r = \beta_H (H+1) T_H + \sum_{r=0}^{H-1} (\beta_r - \beta_{r+1})(r+1) T_r$, und da für $r \geqq H_0$, $T_r < \frac{1}{2} C$ gilt, so ist für alle $H \geqq H_0$

$$\sum_{r=0}^{H} \beta_r S_r \leqq \left(L - \frac{1}{2} C\right) \sum_{r=0}^{H_0} \beta_r + \frac{C}{2} \sum_{r=0}^{H} \beta_r,$$

wo L eine obere Schranke von S_r ist, also, da $\Sigma \beta_r$ divergiert, für alle $H \geqq H_1$, $\sum_{r=0}^{H} \beta_r S_r \leqq C \sum_{r=0}^{H} \beta_r$ ist.

Es ist also auch $\sum_{r=0}^{H} \beta_r \sum_{i=1}^{N} |\Phi(x_{rN+i})| > C \sum_{r=0}^{H} \beta_r$ für alle $H \geqq H_1$.

Nun ist

$$\sum_{i=1}^{(H+1)N} \alpha_i |\Phi(x_i)| = \sum_{r=0}^{H} \sum_{i=1}^{N} \alpha_{rN+i} |\Phi(x_{rN+i})| \geqq \sum_{r=0}^{H} \beta_r \sum_{i=1}^{N} |\Phi(x_{rN+i})| > C \sum_{r=0}^{H} \beta_r,$$

also ist $\sum_{r=0}^{\infty} \beta_r$ doch konvergent, und damit ist alles gezeigt.

Literaturverzeichnis

[1] Van der Corput, Diophantische Ungleichungen. I. Zur Gleichverteilung modulo Eins. Acta Math. **56**, 373—456.

[2] B. Eckmann, Über monothetische Gruppen. Commentarii Math. Helvet. **16**, 249—263 (1943).

[3] E. Hlawka, Zur formalen Theorie der Gleichverteilung in kompakten Gruppen. Rend. Circ. Mat. Palermo. Im Druck.

Eingegangen am 17. 3. 1954

Zur formalen Theorie der Gleichverteilung in kompakten Gruppen

Rend. Circ. Mat. Palermo, Ser. II, 4 (1955) 1–15

B. Eckmann [1] hat den Begriff der Gleichverteilung auf kompakte Gruppen übertragen. Es soll nun die Theorie weiter ausgebaut werden und es soll hier die qualitative Seite betrachtet werden. In einer weiteren Arbeit soll auch die quantitative Theorie entwickelt werden.

§ 1. Es sei G eine kompakte Gruppe. Das Haarsche Mass von G sei zu 1 normiert. Die Klassen der irreduziblen Darstellungen von G seien $D_0, D_1, D_2, \ldots$ (D_0 sei die triviale irreduzible Darstellung). Die zugehörigen unitären Matrizen bezeichnen wir mit $D_i(x)$, ihre Grade mit r_i.

Weiter bedeute, wenn k-zeilige quadratische Matrizen betrachtet werden (k beliebige natürliche Zahl)

$$(A\,|\,B) = \mathrm{Sp}\,(B^* A), \quad (\mathrm{Sp} = \mathrm{Spur},\ B^* = \bar{B}^{\mathsf{T}}), \quad |A|^2 = (A\,|\,A).$$

Es ist bekanntlich

$$(B\,|\,A) = \overline{(A\,|\,B)}, \quad |(A\,|\,B)|^2 \leqq |A|^2 |B|^2$$

$|A + B| \leqq |A| + |B|$, $|AB| \leqq |A|\,|B|$ und es gilt für jede unitäre Matrix $(AU\,|\,BU) = (UA\,|\,UB) = (A\,|\,B)$, also insbesondere

$$|U| = \sqrt{k}, \quad |AU| = |A|.$$

§ 2. Es sei F die Menge aller Folgen $f = \{x_1, x_2, \ldots\}$ aus G. Es sei weiter eine unendliche Matrix $A = (a_{ik})$ von reellen Elementen $a_{ik} \geqq 0$ gegeben, so dass $\lim\limits_{i \to \infty} a_{ik} = 0$ fürjedes feste k gilt und es sei $A_i = \sum\limits_{k=1}^{\infty} a_{ik}$ konvergent und strebe

$\to 1$ für $i \to \infty$. Ist $\Phi(x)$ eine beschränkte Function auf G, dann sei

$$M_N(\Phi,\ f,\ A) = \sum_{k=1}^{\infty} a_{Nk}\Phi(x_k)$$

und $M(\Phi,\ f,\ A) = \lim\limits_{N\to\infty} M_N(\Phi,\ f,\ A)$, wenn vorhanden.

Eine Folge f auf G heisst dann A-gleichverteilt, wenn für jede stetige Funktion $\Phi(x)$, $M(\Phi,\ f,\ A)$ existiert und wenn

$$M(\Phi,\ f,\ A) = \int_G \Phi\,dx = M(\Phi) \tag{1}$$

ist. Ist S eine Teilmenge aus F, so heisst S, A-gleichmässig gleichverteilt auf G (kurz eine A g.g. Menge), wenn für alle $f \in S$ für jede stetige Funktion $\Phi(x)$, $M_N(\Phi,\ f,\ A)$ gleichmässig gegen $M(\Phi)$ strebt; d.h. wenn es zu jedem $\varepsilon > 0$, ein $N_0(\varepsilon,\ \Phi)$ gibt, so dass für alle $N > N_0$ und alle Folgen $f \in S$

$$|M_N(\Phi,\ f,\ A) - M(\Phi)| < \varepsilon \tag{2}$$

ist. Besteht S nur aus einer Folge, so geht (2) in (1) über. Nimmt man für Φ die Matrixfunktion $D_j(x)$, $(j > 0)$ und beachtet, dass dann $M(D_j) = 0$ ist, so folgt sofort als notwendige Bedingung, dass S eine A g.g. Menge ist, dass gleichmässig auf S

$$M_N(D_j,\ f,\ A) \to 0 \tag{3}$$

strebt. Es gilt auch hier wieder das Weylsche Kriterium, dass (3) auch hinreichend ist. Wir gehen dazu von der Parsevalschen Gleichung aus

$$\int \Lambda(s)\,\Phi(s)\,ds = \sum_{j=0}^{\infty}(L_j/C_j) \tag{4}$$

gültig für alle Funktionen von $L^2(G)$. Dabei sind die $C_j = \sqrt{r_j}\,M(\Phi \bar{D}_j)$ die Fourierkoeffizienten von $\Phi(s)$ (Analog sind die L_j definiert).

Es sei jetzt $M(\Phi) = 0$ vorausgesetzt, dann ist $C_0 = 0$. Wir betrachten nun statt $\Phi(s)$, die Funktion $M_N(\Phi,\ sf,\ A)$, wo $sf = s(x_1,\ x_2,\ \ldots)$ die Folge $(sx_1,\ sx_2,\ \ldots)$ bedeutet. Ihre Fourierkoeffizienten F_j lauten: $F_0 = 0$, $F_j = C_j S_j(N)$ wo $S_j = M_N(D_j,\ f)$. Also erhalten wir

$$\int \Lambda(s)\,M_N(\Phi,\ sf,\ A)\,ds = \sum_{j=1}^{\infty}(L_j/C_j S_j) \tag{5}$$

Es ist $C_j S_j = M_N(C_j D_j, f, A) = \sum_{i=1}^{\infty} a_{Ni} C_j D_j(x_i)$, also

$$|C_j S_j| \leqq \sum_{i=1}^{\infty} a_{Ni} |C_j D_j(x_i)| = |C_j| A_N. \tag{6}$$

Weiter ist $|(L_j|C_j S_j)| \leqq |L_j||C_j S_j| \leqq |L_j||C_j| A_N$ also gilt für jede natürliche Zahl K

$$\left|\int \Lambda(s) M_N(\Phi, sf, A) ds\right| \leqq \sum_{j=1}^{K} |L_j||C_j||S_j| + A_N \sum_{j=K+1}^{\infty} |L_j||C_j|. \tag{7}$$

Es sei nun $\Phi(s)$ stetig auf G. Dann gibt es zu jedem $\varepsilon > 0$ eine Funktion $\psi(s)$ aus $L(G)$ mit $M(\psi) = 1$, so dass auf G

$$|\Phi - \psi * \Phi| < \frac{\varepsilon}{4}$$

wo

$$\psi * \Phi = \int \psi(s) \Phi(s^{-1} x) ds = \int \psi(s^{-1}) \Phi(sx) ds.$$

Mit $\Lambda(s) = \psi(s^{-1})$ folgt aus (7)

$$|M_N(\Phi, f, A)| < \sum_{j=1}^{K} |L_j||C_j||S_j| + A_N\left(\frac{\varepsilon}{4} + \sum_{j=K+1}^{\infty} |L_j||C_j|\right). \tag{8}$$

Wird nun K so gross gewählt, dass $\sum_{j=K+1}^{\infty} |L_j||C_j| \leqq \frac{1}{2} \sum_{j=K+1}^{\infty} (|L_i|^2 + |C_j|^2 < \frac{\varepsilon}{8}$ und N so gross gewählt, dass für alle $f \in M$ und alle $j \leqq K$ $|S_j| < \frac{\varepsilon}{4}\left(1 + \sum_{j=1}^{K} |L_j||C_j|\right)^{-1}$ und $A_N < \frac{4}{3}$ gilt, so folgt $|M_N(\Phi, f, A)| < \frac{3\varepsilon}{4}$ gleichmässig auf S.

Ist nun $M(\Phi) \neq 0$, so wenden wir die Abschätzung auf $\Phi - M(\Phi)$ an und erhalten $|M_N(\Phi, f, A) - M(\Phi)| \leqq |1 - A_N||M(\Phi)| + \frac{3}{4}\varepsilon$.

Wählen wir noch N so gross, dass $|1 - A_N||M(\Phi)| < \frac{\varepsilon}{4}$, so folgt (2), für jede stetige Funktion Φ, w.z.b.w.

Ist Φ beliebige aus $L^2(G)$ und setzen wir in (5) $\Lambda(s) = M_N(\Phi, sf, A)$ so folgt,

$$\int M_N^2(\Phi, sf, A) ds = \sum_{j=1}^{\infty} |C_j S_j|^2$$

und aus (7) folgt wie vorher: Für jede A g.g. Menge S und fürjedes Φ aus $L^2(G)$, gilt gleichmässig auf S

$$\lim_{N\to\infty} \int (M_N(\Phi, sf, A) - M(\Phi))^2 ds = 0.$$

Diese Bemerkung verdanken wir J. F. Koksma und R. Salem [2].

Es ist leicht einzusehen, dass (2) nicht für jede Funktion $\Phi(s)$ zu gelten braucht. Der Gedankengang, welcher von (3) zu (2) geführt hat, zeigt, dass (2) für alle fastperiodischen Funktionen auf G gilt. Wir fügen gleich die Bemerkung an, dass sich der bisher entwickelte Gedankengang für jede Gruppe G, nicht nur für kompakte Gruppen durchführen lässt, wenn wir (2) für alle fastperiodischen Funktionen auf G verlangen und die Menge der D_j durch die Menge der beschränkten irreduziblen Darstellungen von G ersetzen. Dabei sind dann die $M(\Phi)$ die Mittelwerte der fastperiodischen Funktionen Φ. Von grösserer Wichtigkeit ist aber, dass (2) für alle beschränkten Funktionen $\Phi(x) \in L(G)$ richtig bleibt, zu welchen es zwei Folgen $\{\Phi_n(x)\}$, $\{\psi_n(x)\}$ von stetigen Funktionen gibt, so dass für alle $x \in G$ und alle n

$$\Phi_n(x) \leqq \Phi(x) \leqq \psi_n(x) \tag{9}$$

und

$$\lim_{n\to\infty} M(\Phi_n) = \lim_{n\to\infty} M(\psi_n) = M(\Phi) \text{ gilt.} \tag{10}$$

Es sei S eine A g.g. Menge. Dann gibt es zu jedem $\varepsilon > 0$ ein n, so dass

$$M(\Phi) - \frac{\varepsilon}{2} < M(\Phi_n) \leqq M(\psi_n) < M(\Phi) + \frac{\varepsilon}{2}$$

und zu diesem n ein N_0, so dass für alle $N > N_0$ und alle $f \in S$

$$M(\Phi_n) - \frac{\varepsilon}{2} < M_N(\Phi_n, f) \leqq M_N(\Phi, f) \leqq M_N(\psi_n, f) < M(\psi_n) + \frac{\varepsilon}{2}$$

gilt und daraus folgt alles.

Die Voraussetzungen (9), (10) sind erfüllt, wenn $\Phi(x)$ die charakteristische Funktion $\varphi_\tau(x)$ einer abgeschlossenen Menge T von G ist. Also gilt für jedes A g.g. Menge S

$$\lim_{N\to\infty} \sum_{i=1}^{\infty} a_{Ni} \varphi_\tau(x_i) = V(T) \tag{11}$$

gleichmässig auf S ($V(T)$ Mass von T).

Gilt (11) für alle abgeschlossenen Mengen in G, dann folgt daraus wieder (2). Zunächst gilt (11) auch für alle offenen Mengen in G.

Es sei nun $\Phi(x)$ eine stetige beschränkte nicht negative Funktion: $0 \leqq \Phi \leqq K$ (K natürliche Zahl). Ist $\Phi_m(x)$ die Treppenfunktion $= \frac{i}{m}$ für $\frac{i}{m} \leqq \Phi(x) < \frac{i+1}{m}$ ($i = 0, \ldots Km$), (m natürliche Zahl), so ist, wenn wir mit $\varphi_i = \varphi_{i,m}$ die charakteristische Funktion der offenen Menge $S_{i,m} : \Phi < \frac{i}{m}$ bezeichnen

$$M_N(\Phi_m, f, A) = \sum_{i=0}^{Km} \sum_{\frac{i}{m} \leq \Phi(x_j) < \frac{i+1}{m}} a_{Nj} \Phi_m(x_j) = \sum_{i=0}^{Km} \frac{i}{m} (M_N(\varphi_{i+1}, f, A) - M_N(\varphi_i, f, A)).$$

Nun ist nach Voraussetzung (11): $M_N(\varphi_{i,m}, f, A) = M(\varphi_{i,m}) + \Delta_{i,m}(N)$ wo $\lim\limits_{N\to\infty} \Delta_{j,m}(N) = 0$, also ist weiter

$$M_N(\Phi_m, f, A) - M(\Phi_m) = \frac{1}{m} \sum_{i=0}^{Km} i(\Delta_{i+1} - \Delta_i) = -\frac{1}{m} \sum_{i=1}^{Km} \Delta_i + K\Delta_{Km+1} \text{ und daher,}$$

da stets $0 \leqq \Phi(x) - \Phi_m(x) < \frac{1}{m}$

$$|M_N(\Phi, f, A) - M(\Phi)| \leqq \frac{1}{m} A_N + |M(\Phi) - M(\Phi_m)| + K \max_{1 \leqq i \leqq Km+1} |\Delta_{i,m}(N)| \quad (12)$$

(o.B.d.A $A_N < 2$).

Wird m so gross gewählt, dass die beiden ersten Terme zusammen $< \frac{\varepsilon}{2}$ sind und dann N so gross, dass der dritte Term ebenfalls $< \frac{\varepsilon}{2}$ wird, so ist alles gezeigt. Ist $\Phi(x)$ eine beliebige stetige Funktion, so wende man (12) auf $K - \Phi$ an, wo K obere Schranke von Φ ist.

Die bisher entwickelten Sätze sind in dem Spezialfall des arithmetischen Mittels $M_N(\Phi, f) = \frac{1}{N} \sum_{i=1}^{N} \Phi(x_i)$ bekannt ([1]) und werden nur deshalb hier entwickelt, da die Formeln (8) und (12) auch für die Aufstellung der quantitativen Theorie der Gleichverteilung, welche hier aber nicht entwickelt werden soll, von Nutzen sind. Die Nützlichkeit, allgemeinere Summierbarkeitsverfahren zugrunde zu legen, wurde zuerst von M. Tsuji [3] gezeigt. Nach einem Satz von A. Brudno [4] (vgl. [5]) genügt es aber, A auf normale Matrizen, d.h. auf Matrizen der Gestalt $(a_{\varkappa i})$ mit $a_{\varkappa i} = 0$ für $i > N$ und $a_{\varkappa\varkappa} > 0$, zu beschränken. Wird in Folgendem der Fall des arithmetischen Mittels betrachtet, welcher der wichtigste Fall ist, so soll die Bezeichnung A wegbleiben.

§ 3. Es sollen jetzt Sätze entwickelt werden, die es gestatten, aus Folgen, welche A gleichverteilt sind (kurz A g Folgen) weitere Folgen mit gleichen Eigenschaften herzuleiten. Klar ist folgendes: Ist $\{x_i\}$ eine A-g Folge, dann ist dies auch der Fall für die Folge $\{c_i x_i\}$ bzw. $\{x_i c_i\}$, wenn $\lim\limits_{i\to\infty} c_i = c$ existiert. Denn aus der Stetigkeit der $D_j(x)$ folgt, dass auch $\lim\limits_{i\to\infty} D_j(c_i)$ existiert und $D_j(c)$ ist und aus $D_j(c_i x_i) = D_j(c_i) D_j(x_i)$ folgt,

$$\Sigma a_{\varkappa i} D_j(c_i x_i) = D(c) \Sigma a_{\varkappa i} D_j(x_i) + \Sigma a_{\varkappa i} D_j(x_i)(D_j(c_i) - D_j(c)).$$

Es folgt daraus die Behauptung, da ja wegen der Beschränktheit der $D_j(x_i)$ die Folge im zweiten Term eine Nullfolge ist.

Mit $\{x_i\}$ ist also auch stets $\{c x_i\}$ eine A-Folge. Es gilt aber sogar, etwas schärfer

Satz 1: Die Menge M der Folgen $\{c x_i\}$, $\{x_i c\}$ ist, wenn c die Gruppe G durchläuft, A-g.g. verteilt auf G.

Zum Beweis ist nur zu beachten, dass für jedes c, stets

$$|\Sigma a_{\varkappa i} D_j(c x_i)| = |D_j(c)| |M_N(D_j, f)| \leqq \sqrt{r_j} |M_N(D_j, f, A)| \quad \text{gilt.}$$

Wir können unter etwas allgemeineren Voraussetzungen über die Folge $\{c_i\}$ (vgl. dazu [6]) zeigen, dass $\{c_i x_i\}$ bzw. $\{x_i c_i\}$ eine A-g Folge ist, wenn dies für $\{x_i\}$ zutrifft. Es seien $\{\alpha_\varkappa\}$, $\{\beta_\varkappa\}$ Folgen natürlicher Zahlen mit $\alpha_\varkappa \leqq \beta_\varkappa$, so dass $\lim\limits_{N\to\infty} \sum\limits_{h=1}^{\alpha_N} a_{\varkappa h} = 0$, $\lim\limits_{N\to\infty} \sum\limits_{h=\beta_N}^{\infty} a_{\varkappa h} = 0$ ist. Solche Folgen existieren stets vgl. [6]). Setzen wir nun voraus, dass für alle natürlichen Zahlen P_N im Intervall $[\alpha_\varkappa, \beta_\varkappa]$, $\lim\limits_{\varkappa\to\infty} c p_\varkappa c_{\alpha_\varkappa}^{-1} = e$ gilt, dann ist $\{c_i\}$ eine solche Folge. Zunächst ist natürlich für jedes $j > 0$

$$\left|\sum_{h=\alpha_\varkappa}^{\beta_\varkappa} a_{\varkappa i} D_j(x_i)\right| \leqq \left|\sum_{i=1}^{\infty} a_{\varkappa i} D_j(x_i)\right| + \sqrt{r_j}\left(\sum_{i=1}^{\alpha_\varkappa} a_{\varkappa i} + \sum_{i=\beta_\varkappa}^{\infty} a_{\varkappa i}\right)$$

strebt also gegen 0 für $N \to \infty$.

Da

$$D_j(c_i x_i) = (D_j(c_i) - D_j(c_{\alpha_\varkappa}))\, D(x_i) + D_j(c_{\alpha_\varkappa})\, D(x_i)$$

für $\alpha_\varkappa \leqslant i \leqslant \beta_\varkappa$, so ist $\Sigma a_{\varkappa i} D_j(c_i x_i)$ dem Betrage nach also höchstens gleich

$$\sqrt{r_j}\left(\sum_{i=1}^{\alpha_\varkappa} a_{\varkappa i} + \sum_{i=\beta_\varkappa}^{\infty} a_{\varkappa i} + \operatorname*{Max}_{\alpha_\varkappa \leqq i \leqq \beta_\varkappa} |D_j(c_i c_{\alpha_\varkappa}^{-1}) - E| + \left|\sum_{i=\alpha_\varkappa}^{\beta_\varkappa} a_{\varkappa i} D_j(x_i)\right|\right)$$

Wegen der Stetigkeit von D_j folgt wieder die Behauptung.

Im klassischen Fall, dass G das direkte Produkt von endlich vielen Kreisdrehungsgruppen ist, gilt bekanntlich: Ist $\{x_i\}$ gleichverteilt, dann ist für jedes natürliche r, auch $\{x_i^r\}$ gleichverteilt. Dies können wir allgemein nur behaupten, wenn G eine zusammenhängende kompakte abelsche Gruppe, also monothetisch ist. Denn dann ist mit D auch D^r eine Darstellung und diese ist ebenfalls die nichttriviale Darstellung, wenn dies für D der Fall ist.

Wir wollen uns im weiteren auf den Fall beschränken, dass A regulär ist, denn dann ist mit jeder Folge $\{x_i\}$ für jedes feste $h \geqq 1$, auch die Folge $\{x_{i+h}\} = (x_{h+1}, x_{h+2} \dots)$, A-gleichverteilt.

Wir nennen nun eine Folge $f = \{x_i\}$ gleichmässig gleichverteilt in Bezug auf A, kurz eine A g.g. Folge, wenn die Menge $S(f)$ der Folgen $f_h = \{x_{i+h}\}$ $(h = 0, 1, 2, \dots)$ A g.g. in G ist.

Ein Beispiel einer g.g. Folge wird durch jede Folge $f = \{a^i\}$ gegeben $(a \in G)$, wenn sie gleichverteilt ist. Denn nach Satz 1 ist die Menge der Folgen $cf = (ca^i)$ (c durchläuft G) g.g. auf G, also auch die Menge der Folgen $a^h f = f_h$ $(h = 0, 1, 2, \dots)$. Ist für ein $a \in G$, $D_j(x) \neq E$ für alle $j > 0$, so ist $\{a^i\}$ gleichverteilt (G ist dann monothetisch).

Ist eine Folge f ein A g.g. Folge dann folgt sofort aus (11):

Satz 2: Ist f eine A g.g. Folge, (A normal) dann gibt es zu jeder abgeschlossenen Menge S mit $V(S) > 0$ eine natürliche Zahl N, so dass für alle ganzen $h \geqq 0$, wenigstens eines der Elemente $x_{h+1}, x_{h+2}, \dots x_{h+n}$ in S liegt.

Es gibt ja dann ein N_0, so dass für alle $N > N_0$ und alle k

$$\sum_{i=1}^{N} a_{Ni} \varphi_s(x_{i+h}) - V(S) > -\frac{V(S)}{2}$$

ist.

Um die Bedeutung der A - g.g. Folgen zu beleuchten, sei folgende Verallgemeinerung eins Satzes von Fatou (vg. [7]) angegeben.

Satz 3: Es sei $\Phi(x)$ eine Funktion auf G, stetig bis auf eine Menge vom Masse 0 und es sei $M(|\Phi|) \neq 0$. Weiter sei $f = \{x_i\}$ eine A-g.g. Folge, (wo A die Bedingungen des Satzes 2 erfüllt), σ eine reelle Zahl > 0 (α_i) eine positive Zahlenfolge mit $0 < \alpha_{i+1} \leqq \sigma \alpha_i$ $(i = 1, 2, \dots)$ und es sei die Reihe $\sum_{i=1}^{\infty} \alpha_i \Phi(x_i)$ absolut konvergent. Dann ist $\sum_{i=1}^{\infty} |\alpha_i| < \infty$ also ist dann insbesondere, wenn Φ beschränkt ist, $\Sigma \alpha_i \Phi(z_i)$ für jede Folge $\{z_i\}$ absolut konvergent.

Beweis: Für $\sigma < 1$ ist der Satz trivial. Es sei also $\sigma \geqq 1$. Da Φ fast überall stetig ist, so gibt es eine abgeschlossenen Menge S in G mit $V(S) > 0$ und ein $C > 0$, so dass auf S, $|\Phi(x)| > C$ gilt.

Dann gibt es nach Satz 2 ein N, so dass für alle $h \geqq 0$ eines der Elemente $x_{h+1}, x_{h+2}, \ldots x_{h+N}$ in S liegt. Weiters ist für jede natürliche Zahl S und für alle $1 \leqslant n \leqslant N$,

$$\alpha_{s+n} \geqq \alpha_{s+N}\, \sigma^{n-N} \geqq \alpha_{s+N}\, \sigma^{-N}.$$

Dann ist

$$\sum_{i=1}^{\infty} \alpha_i |\Phi(x_i)| = \sum_{q=0}^{\infty} \sum_{n=1}^{N} \alpha_{qN+n} |\Phi(x_{qN+n})| \geqq C\sigma^{-N} \sum_{q=0}^{\infty} \alpha_{(q+1)N}$$

also für jedes $K \geqq N$, wenn $C_1 = \frac{\sigma^{N+1} - 1}{\sigma - 1}$ für $\sigma > 1$ und $C_1 = N$ für $\sigma = 1$ gesetzt wird

$$\sum_{i=1}^{K} \alpha_i \leqq \sum_{n=1}^{N} \alpha_n + \sum_{q=0}^{K} \sum_{n=1}^{N} \alpha_{(q+1)N+n} \leqq \alpha_1 C_1 + \sum_{q=0}^{\infty} \sum_{n=1}^{N} \alpha_{(q+1)N}\, \sigma^n$$

$$\leqq C_1 \Big(1 + \frac{\sigma^N}{C} \sum_{i=1}^{\infty} \alpha_i |\Phi(x_i)|\Big)$$

also

$$\sum_{i=1}^{\infty} \alpha_i < \infty.$$

§ 4. Wir zeigen jetzt, (wir beschränken uns auf g.g. Folgen. Für normale-Matrizen gilt der Satz auch).

Satz 4: Ist die Folge $f = \{y_i\}$, g.g. verteilt und besitzt weiter die Folge $f' = \{x_i\}$ die Eigenschaft, dass $\lim_{i\to\infty} y_{i+1}^{-1} x_{i+1} x_i^{-1} y_i$ existiert und gleich dem Einheitselement e ist, dann ist, auch die Folge f' eine g.g. Folge, also insbesondere gleichverteilt.

Ist die Folge $\{a^i\}$ gleichverteilt (G ist dann monothetisch) und ist $\lim_{i\to\infty} x_{i+1} x_i^{-1}$ vorhanden und gleich a, dann ist auch $\{x_i\}$ gleichverteilt, denn es strebt dann $a^{-(i+1)} x_{i+1} x_i^{-1} a^i = a^{-1} x_{i+1} x_i^{-1} \to e$ (Vgl. dazu I.G. van der Corput [8]).

Beweis: Wir setzen $y_{i+1}^{-1} x_{i+1} x_i^{-1} y_i = u_i$ dann ist für $s \geqq i$

$$x_s = y_s u_{s-1} u_{s-2} \ldots u_i y_i^{-1} x_i \tag{13}$$

Nach Vorraussetzung ist die Folge $\{y_i\}$, g.g. verteilt, es gibt also zu jedem $\varepsilon > 0$, ein $K_0(\varepsilon)$, so dass für alle $K > K_0(\varepsilon)$ und alle h $(j > 0)$

$$\Big|\frac{1}{K} \sum_{l=1}^{K} D_j(y_{l+h})\Big| < \frac{\varepsilon}{4}$$

Wir nehmen $K = \mathrm{Max}\left(\left[\frac{4}{\varepsilon}\right] + 1,\ K_0 + 1\right)$.

Weiters gibt es wegen der Stetigkeit von $D_j(x)$ und wegen $\lim u_i = e$, ein $k(\varepsilon)$, so dass für alle $k \geqq k_0 > 1$

$$|D_j(u_k) - E| < \frac{1}{(K+1)^2}$$

Dann ist für alle $i > k \geqslant k_0$

$$|D(x_i) - D(y_i) D(y_k^{-1} x_k)| < \frac{i-k}{(K+1)^2} \tag{14}$$

denn nach (13) ist ja die linke Seite von (14) $|D(u_{i-1}) \ldots D(u_k) - E|$ und für $i = k+1$ ist dies richtig. Durch vollständige Induktion zeigt man (14) denn es ist ja

$$|D(u_i) D(u_{i-1}) \ldots D(u_k) - E| = |D(u_i)(D(u_{i-1}) \ldots D(u_k) - E) + D(u_i) - E|$$
$$< \frac{i-k}{(K+1)^2} + \frac{1}{(K+1)^2}$$

Aus (12) folgt nun für jedes $k \geqq k_0$

$$\left|\sum_{i=k+1}^{k+K} D(x_i)\right| \leqq \left|D(y_k^{-1} x_k) \sum_{i=k+1}^{k+K} D(y_i)\right| + \frac{1}{(K+1)^2} \sum_{i=k+1}^{k+K} (i-k) < \frac{\varepsilon}{4} K + 1$$

Also gilt für $k \geqq k_0$ und für alle $N > N_0(\varepsilon) > \sqrt{r_j}(8\varepsilon^{-1} + 1)(K + k_0)$ mit $H = \left[\frac{N}{K}\right]$

$$\left|\frac{1}{N} \sum_{i=k+1}^{k+N} D(x_i)\right| \leqq \left|\frac{1}{N} \sum_{r=0}^{H-1} \sum_{i=k+rK+1}^{k+(r+1)K} D(x_i)\right| + \left|\frac{1}{N} \sum_{i=k+HK+1}^{k+N} D(x_i)\right|$$
$$< \frac{H}{N}\left(K \frac{\varepsilon}{4} + 1\right) + \left(1 - \frac{HK}{N}\right)\sqrt{r_j}$$

also

$$\left|\frac{1}{N} \sum_{i=k+1}^{k+N} D(x_i)\right| < \frac{\varepsilon}{4} + \frac{1}{K} + \frac{K}{N}\sqrt{r_j} < \frac{3\varepsilon}{4}$$

Ist $k < k_0$, dann ist nach (13)

$$\left|\frac{1}{N} \sum_{i=k+1}^{k+N} D(x_i)\right| = \left|\frac{1}{N} \sum_{i=k+1}^{k_0+1} + \sum_{i=k_0+1}^{N+k_0} - \sum_{i=k+N+1}^{N+k_0}\right|$$

$$\leqq N^{-1}(k_0 - k + 1)\sqrt{r_j} + \frac{3\varepsilon}{4} + (k_0 - k) r_j N^{-1} < \frac{2k_0\sqrt{r_j}}{N} + \frac{3\varepsilon}{4} = \varepsilon$$

also gibt es zu jedem $\varepsilon > 0$, ein $N_0(\varepsilon)$ so dass für alle $N > N_0(\varepsilon)$ und für alle $k \geqq 0$

$$|M_n(D_j, f_k)| < \varepsilon \text{ gilt, w. z. b. w.}$$

§ 5. Wir zeigen nun dass Analogon zur Fundamental-Ungleichung von van der Corput, welches auch aus dieser Ungleichung hergeleitet werden könnte (vg. [8], [3]).

Satz 5: Es seien $U_1, \ldots U_N$ quadratische Matrizen von gleicher Zeilenzahl. Es sei weiter q eine natürliche Zahl mit $1 \leqq q \leqq N$. Dann ist

$$q^2\left|\sum_{i=1}^{N} U_i\right|^2 \leqq (N+q-1)\left(q\sum_{i=1}^{N}\left|U_i\right|^2 + 2R\sum_{h=1}^{q-1}(q-h)\sum_{j=1}^{N-h}(U_j\,|\,U_{j+h})\right)$$

(R Realteil).

Beweis: Es werde definiert $U_i = 0$, für $i > N$ und $i < 1$, dann ist U_i für alle ganzen i definiert

$$q^2\left|\sum_{i=1}^{N} U_i\right|^2 = \left|\sum_{j=1}^{N+q-1}\sum_{h=0}^{q-1} U_{j-h}\right|^2 \leqq \left|\sum_{j=1}^{N+q-1}\left(\sum_{h=0}^{q-1} U_{j-h}\right)\right|^2$$

$$\leqq (N+q-1)\sum_{j=1}^{N+q-1}\left|\sum_{h=0}^{q-1} U_{j-h}\right|^2 = (N+q-1)\left[\sum_{j=1}^{N+q-1}\sum_{h=0}^{q-1}|U_{j-h}|^2\right.$$

$$\left.+2R\sum_{j=1}^{N+q-1}\sum_{l<h}(U_{j-h}\,|\,U_{j-l})\right] = (N+q-1)(\Sigma_1 + 2R\Sigma_2).$$

Wenn wir $j - k = m$ setzen, dann ist

$$\Sigma_1 = \sum_{j=1}^{N+q-1}\sum_{m=j-q+1}^{j}|U_m|^2 = \sum_{m=1}^{N}|U_m|^2\sum_{j=m}^{N+q-1}1 = q\sum_{m=1}^{N}|U_m|^2$$

und wenn $j - k = m$, $h = k - l$ geseztzt wird, also $j - l = m + h$, ist

$$\Sigma_2 = \sum_{h=1}^{q-1}\sum_{m=1}^{N+q-1}\sum_{j=m-q+1}^{m-h}(U_m\,|\,U_{m+h}) = \sum_{h=1}^{q-1}\sum_{m=1}^{N-h}(U_m\,|\,U_{m+h})\sum_{j=m-h}^{m+q-1}$$

$$= \sum_{h=1}^{q-1}\sum_{m=1}^{N-h}(q-h)(U_m\,|\,U_{m+h}).$$

Wir zeigen jetzt (vgl. [8], [3]) die Bemerkung von N. M. Korobow und A. G. Postnikow [9]).

Satz 6. Es sei f die Folge $(x_1, x_2 \ldots)$. Ist für jede natürliche Zahl h die Folge $f^{(h)} = (x_i^{-1}x_{i+h})$ gleichverteilt, so ist es auch die Folge f und für jede natürliche Zahl k und l auch die Folge $\{x_{kl+i}\}$. Ist jede der Folgen $f^{(h)}$ eine g.g.

Folge, so gilt dies auch für jede der Folgen $\{x_{ki+l}\}$ Beweis: Es sei $D(x)$ eine nichttriviale Darstellung vom Grade r. Dann ist (k, l fest)

$$\sum_{i=1}^{N} D(x_{ki+l}) = \frac{1}{k}\sum_{j=1}^{k}\sum_{s=1}^{kN} D(x_{s+l})\, e\left(\frac{js}{k}\right)$$

denn es ist $\frac{1}{k}\sum_{j=1}^{k} e\left(\frac{js}{k}\right) = \begin{matrix}1\\0\end{matrix}$ für $\begin{matrix} s \equiv 0 \pmod k \\ s \not\equiv 0 \pmod k\end{matrix}$ $(e(\alpha) = e^{2\pi\sqrt{-1}\alpha})$.

Daher ist

$$\left|\sum_{i=1}^{N} D(x_{ki+l})\right| \leqq \frac{1}{k}\sum_{j=1}^{k}\left|\sum_{s=1}^{kN} D(x_{s+l})\, e\left(\frac{js}{k}\right)\right| \leqq \operatorname*{Max}_{1\leqq j\leqq k}\left|\sum_{s=1}^{kN} D(x_{s+l})\, e\left(\frac{js}{k}\right)\right|.$$

Es sei nun q eine natürliche Zahl $\leqq kN$. Dann ist nach Satz 5

$$q^2\left|\sum_{s=1}^{kN} D(x_{s+l})\, e\left(\frac{js}{k}\right)\right|^2 \leqq (kN + q - 1)(qN + 2\sum_{k=1}^{q-1}(q-k)\,|\Sigma| \tag{15}$$

wo

$$\Sigma = \sum_{i=1}^{Nk-h}\left(D(x_{i+l})\, e\left(\frac{ji}{k}\right) \Big| D(x_{i+l+h})\, e\left(\frac{i+h}{k}j\right)\right)$$

$$= e\left(\frac{-hj}{k}\right)\sum_{i=1}^{Nk-h}(E\,|\,D(x_{i+l}^{-1}x_{i+l+h})) = e\left(\frac{-hj}{k}\right)\left(E\,|\sum_{i=1}^{Nk-h} D(x_{i+l}^{-1}x_{i+l+h})\right)$$

also gilt nach (15)

$$\left|\frac{1}{N}\sum_{s=1}^{kN} D(x_{s+l})e\left(\frac{js}{k}\right)\right|^2 \leqq \frac{kN+q-1}{N}\left(\frac{r}{q} + \frac{2\sqrt{r}}{q^2}\sum_{h=1}^{q-1}(q-h)\frac{Nk-h}{N}\left|\frac{1}{Nk-h}\sum_{i=1}^{Nk-h} D(x_{i+l}^{-1}x_{i+k+l})\right|\right). \tag{16}$$

Daraus folgt unmittelbar der erste Teil der Behauptung. Denn es ist nach Vorausetzung für jedes feste h, $x_{i+l+h}x_{i+l}^{-1}$ gleichverteilt, daher geht für $N \to \infty$, die rechte Seite von (16) gegen $\frac{kr}{q}$. Geht man nun mit $q \to \infty$, so folgt, dass der Limes der linken Seite existiert und 0 ist.

Es sei nun vorausgesetzt, dass die Folgen $(x_i^{-1} x_{i+h})$ g.g. verteilt sind. Dann gibt es also zu jedem $\varepsilon > 0$ ein $M(h, \varepsilon)$, so dass für alle l und alle $M > M(h)$

$$\left|\frac{1}{M}\sum_{i=1}^{M} D(x_{i+l}x_{i+l+h}^{-1})\right| < \frac{\varepsilon^2}{4\sqrt{r}\,k^2} \tag{17}$$

gilt.

Wir nehmen nun $q = q(\varepsilon) = \left[\frac{4(k+1)r}{\varepsilon^2}\right] + 1$ und $M''(\varepsilon) = \underset{1 \leqq h \leqq q-1}{\mathrm{Max}}\, M(h, \varepsilon)$.

Dann gilt für alle h mit $1 \leqq h \leqq q-1$ und alle l die Abschätzung (17). Ist $N(\varepsilon) = \left[\frac{1}{k}(M'(\varepsilon) + q(\varepsilon))\right] + 1$, so ist für alle $N > N(\varepsilon)$ und für jedes h $(1 \leqq h \leqq q-1)$, $Nk > q$ und $M_h = Nk - h > M'(\varepsilon) + q - h > M'(\varepsilon)$. Es folgt daher aus (16) für den Term links, die Abschätzung

$$\leqq 2k\left(\frac{r}{k} + \frac{2\sqrt{r}}{q^2}\sum_{h=1}^{q-1}(q-h)k\frac{\varepsilon^2}{4\sqrt{r}\,k(k+1)}\right) < \frac{\varepsilon^2}{2} + \frac{\varepsilon^2}{2} = \varepsilon^2$$

also auch für alle $N > N(\varepsilon)$ und für alle l,

$$\left|\frac{1}{N}\sum_{i=1}^{N} D(x_{ki+l})\right| < \varepsilon.$$

Daraus folgt alles, denn die Folgen $(x_{k(l+s)})$ sind ja Folgen der Gestalt x_{ki+l}.

Folgerung: Es sei $\mathfrak{A}$ die unendliche Matrix $(A_{ik})_{i,\,k=1,\ldots\infty}$, $A_{ik} = x_k \ldots x_{i+k-1}$ Sind die Zeilen gleichverteilt, dann sind es auch die Spalten.

Beweis: Die k-te Spalte ist ja $A_{ik} = x_k \ldots x_{i+k-1}$ $(i = 1, 2, \ldots)$. Dann ist für jedes h, $A_{ik}^{-1} A_{i+h,\,k} = x_{i+k} \ldots x_{i+k+h-1} = A_{h,\,i+k}$.

Da die Folge $\{A_{h,\,i+k}\}$ nach Voraussetzung gleichverteilt ist, ist damit alles gezeigt.

Sind mit (x_k) auch alle (x_k^r) gleichverteilt (wie dies bei zusammenhängenden, abelschen Gruppen der Fall ist) und existiert $\lim x_k^{-1} x_{k+1}$, so sind die Zeilen von $\mathfrak{A}$ bereits gleichverteilt, da $A_{ik} = x_k^i (x_k^{-1} x_{k+1})^{i-1} \ldots x_{i+k-2}^{-1} x_{i+k-1}$.

§ 6. Wir wollen kurz noch auf die Frage eingehen, ob es auf jeder Gruppe überhaupt gleichverteilte Folgen gibt.

Betrachten wir mit G das unendliche cartesische Produkt F der Punkte, die Folgen aus G sind. In diesem Raum lässt sich ein Mass μ einführen (vgl. dazu E. Hopf, Ergodentheorie) mit $\mu(F) = 1$. Dann folgt aus dem Ergodensatz, dass für jede stetige Funktion $\Phi(x)$ auf G für fast alle $f = \{x_i\}$,

$$M(\Phi, f) = \lim_{N \to \infty} \frac{1}{N}\sum_{i=1}^{N} \Phi(x_i) = M(\Phi)$$

gilt.

Wendet man dies auf alle $D_j (j > 0)$ an, so folgt, dass fast alle Folgen f gleichverteilt sind. Es gibt sogar eine Menge in F von einem Mass $> 1 - \delta$, welche g.g. verteilt ist. Denn nach dem Satz von Egoroff, gibt es zu jedem j, eine Menge $S_j \in F$ vom Mass $> 1 - \frac{\delta}{2^j}$ $(j = 1, 2, \ldots)$, in welcher $M_n(D_j, f)$ für

$N \to \infty$ gleichmässig konvergiert. Daher besitzt der Durchschnitt der S_i ein Mass $> 1 - \delta$. Wir können nun leicht den folgenden Satz beweisen:

Satz 7: Ist f eine beliebige Folge aus G, welche in G dicht liegt, so können die Glieder von f so umgeordnet werden, dass die neue Folge f^* in G gleichverteilt ist.

Beweis: Es sei $f_1 = (v_1, v_2, \ldots)$ irgendeine g. Folge. Da die Glieder x_i von f in G dicht liegen, so ist jedes v_i Häufungspunkt von f.

Es sei $U_n (n = 1, 2, \ldots)$ eine sich auf e (Einheitselement) zusammenziehende Umgebungsfolge. Dann gibt es zu jedem n, also sicher ein $y_n \in f$, so dass $y_n \in U_n v_n$, also $y_n v_n^{-1} \in U_n$ ist. Dann ist also $\lim\limits_{n \to \infty} y_n v_n^{-1} = e$ daher ist $y_n = C_n v_n$ wo $C_n \to e$ geht. Dann ist aber (vgl. § 3) die Teilfolge $\{y_n\}$ aus f gleichverteilt. Es seien z_j die restlichen Glieder der Folge f in irgendeiner Weise angeordnet. Dann betrachten wir die Folge $f^* = \{t_n\}$ definiert durch

$$\begin{aligned} t_k &= y_{k-i} \quad \text{für } i^2 + 1 \leqq k \leqq (i+1)^2 - 1 \\ &= z_i \qquad\qquad k = i^2 \end{aligned} \qquad (i = 1, 2, 3, \ldots)$$

f^* ist eine Umordnung von f. Die Anzahl der z in $t_1, \ldots t_N$ ist $\leqq \sqrt{N}$, also ist die Anzahl T der $y_1, y_2 \ldots$ in diesem Abschnitt $\geqq N - \sqrt{N}$, also

$$|M_N(D_j, f^*)| \leqq \frac{N}{T} \left| \frac{\sum\limits_{i=1}^{T} D_j(y_i)}{T} \right| + \frac{\sqrt{N}}{N} \sqrt{r_i}$$

Für $N \to \infty$ geht $T \to \infty$, $N/T \to 1$, also $M_N(D_j, f^*) \to 0$ w. z. b. w.

Wir haben bei dem Nachweis, dass fast alle Folgen gleichverteilt sind, den Ergodensatz benützt. Wir können aber elementar vorgehen und werden dabei gleich zeigen:

Satz 8: Für fast alle Folgen f ist für alle $j > 0$

$$M_N(D_j, f) = 0(N^{-\frac{1}{4}} \log^{\frac{1}{2}+\varepsilon} N)$$

($\varepsilon > 0$ bel.)

Beweis: Wir arbeiten wieder im Raum F mit dem zugehörigen Mass μ. Das Differential bezeichnen wir mit df. Dann ist ($j > 0$)

$$\int_F |D_j(x_1) + \ldots + D_j(x_N)|^2 df = N r_j$$

also

$$\int_F |M_N(D_j, f)|^2 df = r_j N^{-1}$$

für jedes N. Wir wählen nun für N die Teilfolge n^2 und gehen in üblicher Weise vor: Es sei $L_{j,n}$ die Menge in F, für die $|M_{n^2}(D_j, f)|^2 > r_j n^{-1} \log^{1+2\varepsilon} n$ ist, dann ist $\mu(L_{jn}) < n^{-1} \log^{-(1+2\varepsilon)} n$. Ist $V_{jn} = \bigcup_{k=n}^{\infty} L_{jk}$, dann ist bei festem j, die Mengenfolge V_{jn} monoton abnehmend, das äussere Mass von V_{jn}, wenn nur $n > n_0(\eta)$ kleiner als $\sum_{k=h}^{\infty} \frac{1}{k \log^{1+2\varepsilon} k}$, also $< \eta$ für jedes $\eta > 0$, so dass der Durchschnitt V_j der V_{jn} das Mass 0 hat.

Also hat die Vereinigungsmenge V der V_j ebenfalls das Mass 0. Es sei nun f eine Folge, welche nicht in V liegt. Dann liegt sie auch nicht in irgend einen V_j. Daher für ein $n_0 \geqq 1$ nicht in V_{j,n_0}, also für alle $n \geqq n_0$, in keinem $L_{j,n}$, Daher gilt für alle $n \geqq n_0$

$$|M_{n^2}(D_j, f)| \leqq r_j n^{-\frac{1}{2}} \log^{\frac{1}{2}+\varepsilon} n.$$

Ist jetzt $N \geqq n_0^2$, dann gibt es ein n mit $n^2 \leqq N < (n+1)^2$. Dann ist

$$|M_N(D_j, f)| \leqq \left|\frac{1}{n^2} \sum_{i=1}^{n^2} D_j(x_i)\right| + \left|\frac{1}{n^2} \sum_{i=n^2+1}^{N} D_j(x_i)\right| = O\left(\frac{\log^{\frac{1}{2}+\varepsilon} n}{\sqrt{n}}\right)$$

$$+ O\left(\frac{(n+1)^2 - 1}{n^2}\right) = O(N^{-\frac{1}{4}} \log^{\frac{1}{2}+\varepsilon} N) \qquad \text{w.z.b.w.}$$

Man zeigt ebenso leicht, dass für fast alle Folgen $\{x_n\}$, die Folgen $\{x_k \dots x_{i+k-1}\}$ für alle i und $\{x_1 \dots x_k\}$ gleichverteilt sind.

Aus dem obigen Satz folgt unmittelbar, dass auch fast alle Folgen g. Folgen in Bezug auf dass r-te Cesaromittel sind ($r > 0$).

Analog zeigt man, dass fast alle Folgen in Bezug auf das Eulersche Summierungsverfahren, also auch in Bezug auf das Borelsche Summierungsverfahren gleichverteilt sind.

Bemerkung: Wie oben gezeigt, gibt es in F stets eine Menge vom Mass $> 1 - \delta$ ($\delta > 0$), in welchem $M_N(D_j, f)$ gleichmässig gegen 0, für jedes $j > 0$, strebt.

Nun ist bekanntlich die Abbildung T, welche jeder Folge $\{x_i\}$ die Folge $\{x_{i+1}\}$ zuordnet, eine Massinvariante Abbildung in F und jede Folge f ist Häufungspunkt der Folgen $T^k f (k = 0, 1, 2, \dots)$. Daher folgt, dass es eine offene Menge U in F mit einem Mass beliebig nahe von 1 gibt, so dass es zu jedem $f \in U$ eine Folge natürlicher Zahlen $j_1, j_2, \dots$ gibt, so dass die Menge der Folgen $f_{jk} = \{x_{i+j_k}\}$ ($k = 0, 1, 2, \dots; j_0 = 0$) g.g. in G ist.

LITERATUR

[1] B. Eckmann, Commentarii math. Helv. 16, (1943), 249-263.
[2] J. F. Koksma und R. Salem, Acta Scient. Math. 12 (B), 1950, 87-96.
[3] M. Tsuji, Journ. Math. Soc, Japan, 4 (1952), 313-322.
[4] A. Brudno, Mat. Sbornik NS. 16 (1945), 191-247.
[5] A. M. Tropper, Proc. Amer. Math. Soc. 4, (1953), 671-677.
[6] R. P. Agnew, Annals Math. Princeton 46 (1949), 93-101.
[7] G. I. Sunouchi und S. Yano, Proc. Amer. Math. Soc. 2 (1951), 380-389.
[8] I. G. van der Corput, Acta Math. 56 (1931), 373-456.
[9] N. M. Korobow und A. G. Postnikow, Dokladi Acad. Nauk SSSR 84 (1952) 217-220.

Folgen auf kompakten Räumen

Wilhelm Blaschke zum 70. Geburtstag gewidmet

Abh. Math. Semin. Univ. Hamb. **20** (1956) 223–241

Eine Folge reeller Zahlen $x_1, x_2, \ldots$ heißt bekanntlich mod 1 gleichverteilt[1]), wenn für jede stetige periodische Funktion von der Periode 1

$$\lim_n \lambda_n(f) = \lim_n \frac{f(x_1) + \cdots + f(x_n)}{n} = \int_0^1 f(x)\,dx = \mu(f) \tag{0}$$

ist. Diese Definition überträgt sich sofort auf die Folge von Punkten im euklidischen Raum E_k von k Dimensionen, wenn man nur ihre Koordinaten mod 1 reduziert. Das Wesentliche an der Gleichverteilung ist das Vorhandensein der Verteilungswahrscheinlichkeit oder des Verteilungsmaßes $\mu(f) = \int_0^1 f\,dx$. Nun ist ja dieses Maß nicht das einzig mögliche und man hat für den eindimensionalen Fall in der Definition allgemeinere Verteilungswahrscheinlichkeiten oder Verteilungsmaße zugrunde gelegt und noch allgemeiner die Menge der Verteilungsmaße einer Folge studiert, d.h. man verlangt nicht mehr, daß $\lim \lambda_n$ existiert. Die weitgehendsten und tiefsten Untersuchungen stammen wohl von VAN DER CORPUT[2]). (Die Fragestellung ist bei VAN DER CORPUT noch wesentlich allgemeiner, als es hier angedeutet ist.) Der mehrdimensionale Fall wurde dagegen, so viel ich weiß, noch wenig untersucht. In dieser Arbeit soll nun mit einer systematischen Untersuchung der Anfang gemacht werden. Wir werden aber dabei noch zwei Verallgemeinerungen vornehmen. Wir werden nämlich nicht nur die Folgen von Gewichtssätzen $a_{nk} = 1/n$ für $k \leqq n$ und 0 für $k > n$ wie in (0) zugrunde legen, sondern gleich allgemeine Folgen a_{nk} zulassen. Wir werden $\lim_n \sum a_{nk} = 1$ verlangen. Daß dies auch für den eindimensionalen Fall sehr zweckmäßig ist, hat eine Arbeit von M. TSUJI[3]) gezeigt, welche im eindimensionalen Fall die Gewichte $a_{nk} = \gamma_k/(\gamma_1 + \cdots + \gamma_n)$ für $k \leqq n$ und sonst 0 zugrunde legt. Andererseits werden wir statt dem Torusraum $0 \leqq x_i < 1$

---

[1]) Literatur bei J. F. KOKSMA, Diophantische Approximationen.

[2]) Proc. Amsterdam 38 (1935) 813—21, 1058—1066, 39 (1936) 10—26, 149—153, 339—344, 489—494, 579—590.

[3]) Journ. Math. Soc. Japan 4 (1952) 313—322.

(mod 1), $j = 1, \ldots k$, gleich eine kompakte Gruppe[4]), allgemeiner einen kompakten Raum X mit abzählbarer Basis zugrunde legen. Diese Verallgemeinerung gestattet es, zu dem Gebiet der Summierbarkeit von Zahlenfolgen, welche nur aus 0 oder 1 bestehen und welche in der Summierungstheorie der Folgen eine fundamentale Stellung einnimmt, eine Brücke zu schlagen. Es seien besonders die bedeutenden Untersuchungen von J. D. Hill[5]) über die sogenannte Borelsche Eigenschaft solcher Folgen hervorgehoben, welchen die vorliegende Arbeit wesentliche Anregungen verdankt.

§ 1. Es sei X ein kompakter Raum mit abzählbarer Basis, $C(X)$ der Vektorraum der reellen stetigen Funktionen $f(x)$ auf X mit der Norm $\|f\| = \sup_{x \in X} |f(x)|$, welcher vom abzählbaren Typus ist. Unter einem Hauptsystem H auf X, verstehen wir eine abzählbare Menge von Funktionen aus $C(X)$, so daß der Vektorraum, erzeugt von H, dicht in $C(X)$ ist. Solche Hauptsysteme gibt es stets. Es sei $M(X)$ der Raum aller linearen stetigen Funktionale $\mu(f)$ auf $C(X)$, also der Raum aller Radonschen Maße μ mit der üblichen schwachen Topologie, definiert durch $\lim_n \mu_n = \mu$ für eine Folge von Maßen μ_n, wenn für jedes f aus $C(X)$ stets $\lim_n \mu_n(f) = \mu(f)$ gilt. Es ist wieder $M(X)$ vom abzählbaren Typus. Bedeutet wie üblich $\|\mu\| = \sup_{\|f\| \leq 1} |\mu(f)|$, $(f \varepsilon C(X))$ die Norm von μ, dann ist jede Menge K in M, für welche $\|\mu\|$ unter einer festen Zahl $a > 0$ liegt, relativ kompakt, d. h. es gibt eine Folge $\{\mu_n\}$ aus K, so daß $\lim \mu_n$ existiert. Wir schreiben auch $\mu(f) = \int_X f(x)\, d\mu$ und nennen $\mu(1)$ die Gesamtmasse von $\mu(f)$. Die Menge aller Maße mit $\mu(1) = 1$ bezeichnen wir mit W, die Menge aller positiven Maße mit M_+ und den Durchschnitt von W und M_+ mit V. Jedes μ aus M_+ definiert eine additive Mengenfunktion auf X, so daß alle Borelschen Mengen und alle Baireschen Funktionen in bezug auf μ meßbar, bzw. integrierbar sind.

Es sei $A = (a_{nk})$ eine Matrix von reellen[6]) Zahlen a_{nk}, wo n und k alle natürlichen Zahlen durchlaufen, so daß $\|A\| = \sup_n \sum_{k=1}^{\infty} |a_{nk}| < \infty$ und $\lim_n \sum_{k=1}^{\infty} a_{nk} = 1$ ist. Ist nun $\omega = (x_k)$ eine unendliche Folge von Elementen aus X, dann betrachten wir die Menge der Funktionale

[4]) Vgl. z.B. E. Hlawka, Rendiconti di Pal. 4 (2) (1955) 1—15 u. die dort angegebene Literatur.

[5]) Ann. of Math. 46 (1945) 556—562, Pacific. J. Math. 1 (1951) 390—409, 4 (1954) 227—242. Vgl. auch R. G. Cooke, Infinite Matrices and Sequence Spaces, Macmillan 1950.

[6]) Viele der nachfolgenden Sätze gelten unter geringfügigen Modifikationen, wenn man komplexe a_{nk} und komplexe Radonsche Maße in Betracht zieht.

$\lambda_n(\omega, A, f) = \sum_{k=1}^{\infty} a_{nk} f(x_k)$ aus $M(X)$. Wir wollen die Menge der Häufungsmaße der Folge $\lambda_n(\omega, A)$ in $M(X)$ untersuchen. Ist μ ein solches Häufungsmaß, so gibt es also eine Folge $\{n(t)\}$ natürlicher Zahlen, so daß $\lim_t \lambda n(t)\,(\omega, A) = \mu$ ist, d.h. für jede stetige Funktion $f(x)$, ist

$$\lim_{t\to\infty} \sum_{k=1}^{\infty} a_{n(t),k} f(x_k) = \mu(f) = \int_X f\, d\mu \tag{1}$$

Wir nennen jedes solche μ ein A-Maß der Folge ω. Da $\lim \sum a_{nk} = 1$ ist, so ist die Gesamtmasse $\mu(1)$ jedes A-Maßes stets 1. Sie liegen also alle in W. Liegt ein A-Maß in V, so nennen wir es ein A-Verteilungsmaß von ω. Die Menge aller Verteilungsmaße nennen wir $V(\omega, A)$. Sind alle $a_{nk} \geqq 0$, so besitzt ω nur A-Verteilungsmaße. Die Norm aller A-Maße ist stets $\leqq \|A\|$, denn es ist ja doch $\|\lambda_n(\omega, A)\| \leqq \|A\|$ für alle n. Da die Menge der λ_n relativ kompakt ist, so können wir sofort folgenden Satz aussprechen:

Satz 1: Jede Folge ω besitzt für jedes A stets mindestens ein A-Maß. Gibt es eine Folge (n), so daß für diese Folge $a_{nk} \geqq 0$, so besitzt es sogar ein A-Verteilungsmaß.

Wir zeigen sofort:

Satz 2: Gibt es ein f^* aus $C(X)$ und eine unendliche Folge natürlicher Zahlen $n(t)$ mit $n(t)\to\infty$, so daß $\lim_t \lambda_{n(t)}(\omega, A, f^*) = m^*$ existiert, so gibt es ein A-Maß μ^* von ω, so daß $\mu^*(f^*) = m^*$ gilt.

Beweis: Die Menge der $\lambda_{n(t)}(\omega, A, f)$ ist wieder relativ kompakt, es gibt also eine Teilfolge (n') der Folge $(n(t))$, so daß $\lim_{n'} \lambda_{n'}(\omega, A) = \mu^*$ existiert. Also ist μ^* ein A-Maß und da für jedes $f \in C(X)$, $\lim_{n'} \lambda_{n'}(\omega, A, f) = \mu^*(f)$ gilt, so ist tatsächlich $\mu^*(f^*) = m^*$.

Definition: Eine Folge ω heißt A-summierbar zum Maß μ, wenn sie nur ein A-Maß besitzt und dieses Maß μ ist. Sie heißt A-gleichverteilt, wenn μ aus V ist. Ist A die Matrix des arithmetischen Mittels, X die Torusgruppe, μ das Haarsche Maß auf X, so ist dies die klassische Definition von H. Weyl.

Satz 3: Ist eine Folge ω zum Maße μ, A-summierbar, so ist $\lim_n \lambda_n(\omega, A) = \mu$, d.h. für jede Funktion $f \in C(X)$, ist

$$\lim_n \lambda_n(\omega, A, f) = \mu(f) \tag{2}$$

Beweis: Würde (2) nicht für jedes f gelten, so gäbe es ein $f^* \in C(X)$ und eine Folge natürlicher Zahlen $n(t) \to \infty$, so daß $m^* = \lim_t \lambda_{n(t)}(\omega, A, f^*)$ existiert und $\neq \mu(f^*)$ ist. Dann gäbe es nach Satz 2 ein A-Maß μ^* von ω, so daß $\mu^*(f^*) = m^* \neq \mu(f^*)$ ist. Also würde ω zwei A-Maße besitzen.

Wir fragen jetzt nach einem notwendigen und hinreichenden Kriterium, damit eine Folge ω, A-summierbar ist.

Dazu zeigen wir:

Satz 4: Existiert für alle f aus einem Hauptsystem H, $\lim_n \lambda_n(\omega, A, f) = \mu_f$, dann existiert dieser Limes für alle f aus $C(X)$ und definiert damit ein Funktional $\mu(f)$; d.h. die Folge ω ist A-summierbar.

Das Kriterium des Satzes 4 ist nicht nur hinreichend, sondern klarerweise auch notwendig.

Beweis: Es gibt zu jedem f aus $C(X)$ und zu jedem $\varepsilon > 0$ ein Linearkompositum h aus endlich vielen Funktionen aus H, so daß $\|g\| = \|f - h\| < \varepsilon/(3\,\|A\| + 1)$ ist. Dies folgt aus der Definition von H. Für h gilt selbstverständlich unsere Behauptung. Daher gibt es ein n_0, so daß für alle $n > n_0$ und alle natürlichen Zahlen r, $|\lambda_{n+r}(h) - \lambda_n(h)| < \varepsilon/3$ ist, wo wir die Buchstaben ω, A weggelassen haben. Dann ist

$$|\lambda_{n+r}(f) - \lambda_n(f)| \leqq |\lambda_{n+r}(h) - \lambda_n(h)| + |\lambda_{n+r}(g) - \lambda_n(g)| < \varepsilon$$

w. z. b. w.

Aus Satz 4 folgt sofort

Satz 4': Eine Folge ω ist genau dann zum Maß μ, A-summierbar, wenn für alle f aus einem Hauptsystem H $\lim \lambda_n(\omega, A, f)$ existiert und gleich $\mu(f)$ ist.

Wenn μ ein A-Maß von ω ist, es braucht jetzt ω nicht A-summierbar sein, so gibt so nach Definition eine Folge $(n(t))$, so daß (1) gilt, d.h. nach Satz 3, daß die Folge ω in bezug auf die Matrix $A' = (a_{n(t),k})$ zum Maß μ, A'-summierbar ist. Aus Satz 4' folgt sofort

Satz 4'': Ein Maß μ ist genau dann ein A-Maß von ω, wenn es eine Folge natürlicher Zahlen $n(t)$ gibt, so daß für alle f aus einem Hauptsystem H die Relation (1) besteht.

Wir wollen den Sachverhalt, der durch die drei vorhergehenden Sätze ausgedrückt wird, das Weylsche Kriterium nennen.

Den Sachverhalt des Satzes 1 können wir jetzt auch so aussprechen, daß es zu jeder Folge ω und zu jedem A eine Teilmatrix A' gibt, so daß ω in bezug auf A' summierbar ist. Das können wir verallgemeinern zu

Satz 5: Es sei eine abzählbare Menge von Folgen $\omega_s = (x_{k,s})$, $s = 1, 2, \ldots$ und eine Matrix A gegeben. Dann gibt es eine Teilmatrix A^* von A, so daß alle diese Folgen in bezug auf A^* summierbar sind. Darüber hinaus sind noch alle Folgen $\omega = (x_k)$ in bezug auf A^* summierbar, für die es eine Folge $(s(t))$ natürlicher Zahlen $\to \infty$ gibt, so daß für jedes $f \in C(X)$ und alle k, $\lim_t f(x_{k,s(t)}) = f(x_k)$ gilt.

Beweis: Nach Satz 1 gibt es eine Teilmatrix $A_1 = (a_{n,k}^{(1)})$ von A, so daß ω_1 in bezug auf A_1 summierbar ist. Weiter gibt es eine Teilmatrix

$A_2 = (a^{(2)}_{n,k})$ von A_1, so daß ω_2 in bezug auf A_2 summierbar ist. Setzt man diesen Prozeß fort, so leistet $A^* = (a^{(n)}_{n,k})$ das Gewünschte. Bevor wir weiter gehen, wollen wir noch folgende Frage besprechen: Es sei ω, A-gleichverteilt zum Verteilungsmaß μ. Dann gilt (2) für alle f aus $C(X)$. Für welche Funktionen f gilt (2) noch? Wir begnügen uns mit folgendem Satz:

Satz 6: Es sei ω, A-gleichverteilt zum Verteilungsmaß μ und es sei für alle n, $a_{nk} \geqq 0$. Dann gilt (2) für alle Borel-meßbaren Funktionen, deren obere und untere Grenzfunktionen fast überall in bezug auf μ einander gleich sind. Insbesondere gilt also (2) für die charakteristische Funktion jeder Borel-meßbaren Menge, deren Rand vom μ-Maß 0 ist.

Beweis: Zu jeder solchen Funktion f und zu jedem $\varepsilon > 0$ gibt es Funktionen f_1, f_2 aus $C(X)$ mit $f_1 \leqq f \leqq f_2$ und $\mu(f_2) \leqq \mu(f_1) + \varepsilon$. Dann ist $\lambda_n(\omega, A, f_1) \leqq \lambda_n(\omega, A, f_2)$ also $\mu(f_1) \leqq \underline{\lim}\, \lambda_n(f) \leqq \overline{\lim}\, \lambda_n(f) \leqq \mu(f_2)$ und daher ist $\overline{\lim} - \underline{\lim} < \varepsilon$ und damit ist alles gezeigt.

Im wichtigen Fall einer kompakten Gruppe X, also insbesondere der Torusgruppe, und dem Haarschen Maß μ, ist für jede offene und abgeschlossene Menge, der obige Satz anwendbar.

§ 2. Wir zeigen jetzt

Satz 7: Die Menge $W(\omega, A)$ der A-Maße von ω bzw. die Menge $V(\omega, A)$ der Verteilungsmaße ist in $M(X)$ stets abgeschlossen.

Zu diesem Zwecke müssen wir zeigen: Ist (μ_t) eine Folge von A-Maßen von ω mit $\lim \mu_t = \mu$, dann ist auch μ ein A-Maß von ω. Es sei H ein Hauptsystem, die Funktionen aus H seien $f_1, f_2, \ldots$. Da μ_t ein A-Maß von ω ist, so gibt es zu jedem t ein $n(t)$, so daß

$$|\lambda_{n(t)}(\omega, A, f_j) - \mu_t(f_j)| < 1/t \qquad \text{für } j = 1, 2, \ldots t$$

Dabei kann noch angenommen werden, daß die Folge $(n(t)) \to \infty$ strebt. Dann ist für jedes f_r, $\lim\limits_t \lambda_{n(t)}(\omega, A, f_r)$, denn es ist bei festem r, für alle $t > r$

$$|\lambda_{n(t)}(\omega, A, f_r) - \mu(f_r)| \leqq |\mu_t(f_r) - \mu(f_r)| + 1/t$$

Aus Satz 4'' folgt dann die Behauptung.

Satz 8: Es gelte[7]) für A, $\lim\limits_n \sum\limits_k |a_{n+1,k} - a_{n,k}| = 0$. Es seien $g_1, \ldots, g_k$ Funktionen aus $C(X)$, dann ist die Menge S der Punkte $p = (p_1, \ldots, p_k)$ im euklidischen Raum E_k von k Dimension, für die es ein $\mu \in W(\omega, A)$ gibt, so daß $\mu(g_j) = p_j$ $(j = 1, \ldots k)$ ist, zusammenhängend, wenn sie aus mehr als einem Punkt besteht.

[7]) Die Voraussetzung ist für $a_{nk} = \gamma_k/(\gamma_1 + \cdots + \gamma_n)$ $(k \leqq n)$ und 0, sonst $(\gamma_k \geqq 0)$ erfüllt, wenn $\gamma_n = 0(S_n)$, $S_n = \gamma_1 + \cdots + \gamma_n$ also insbesondere für das arithmetische Mittel. Zu der Beweismethode von Satz 8 vgl. [2]).

Beweis: Um uns kurz ausdrücken zu können, sagen wir, daß μ dem Punkt $p \in S$ zugeordnet ist, wenn $\mu(g_j) = p_j$ $(j = 1, \dots k)$ ist. Nehmen wir nun an, S läßt sich in zwei nichtleere abgeschlossene Mengen T, U ohne gemeinsame Punkte zerlegen, so gibt es eine natürliche Zahl l, $a \in T$, $b \in U$, so daß der Abstand $d = d(a, b)$ der beiden Mengen T, $U > 2k/l$ ist. Es seien μ, ν den Punkten a, b zugeordnet. Dann gibt es Folgen $(n) = (n(s))$, $(m) = (m(s))$ natürlicher Zahlen $\to \infty$, so daß

$$\lim_{(m)} \lambda_m(\omega, A, f) = \mu(f), \ \lim_{(n)} \lambda_n(\omega, A, f) = \nu(f)$$

Dabei kann noch angenommen werden, daß für jedes s stets $n(s) > m(s)$ ist. Es sei nun $\varrho(n, f) = [l\lambda_n(f)]/l$ ($[\alpha]$-nächst kleinere ganze Zahl an α). Es ist $|l\varrho(n, f)| \leqq l\,||A||\,||f|| + 1$. Ist also K eine gemeinsame Schranke für die k Funktionen g_j, so nimmt das k-Tupel $(\varrho(n, g_1), \dots, \varrho(n, g_k))$ höchstens $K_1 = (l\,||A||\,K + 1)^k$ verschiedene Werte an. Es gibt also nur endlich viele verschiedene Punkte $q(n)$ mit den Koordinaten $\varrho(n, g_j)$ $(j = 1, \dots k)$. Wir definieren nun für jede natürliche Zahl s folgende natürliche Zahlen $s(1), \dots, s(t)$ mit einer Anzahl $t = t(s)$ so: Es sei $m(s(1)) = m(s)$ und allgemein sei $m(s(j))$ die größte natürliche Zahl m für die $m(s(j-1)) < m \leqq n(s)$, $q(m) = q(m(s(j-l)) + 1)$ gilt. Der Prozeß geht zu Ende, wenn $m(s(t)) = n(s)$ ist. Daraus folgt, daß die Punkte $q(m(s(j)))$ $(j = 2, \dots t(s))$ alle voneinander verschieden sind. Denn wäre z.B. $q(m(s(j)) = q(m(s(j-1))$, so wäre $q(m(s(j))) = q(m(s(j-2)) + 1)$ und es wäre $m(s(j-2)) < m(s(j-1)) < m(s(j)) < n(s)$ im Gegensatz zur Definition von $m(s(j-1))$. Da es aber nur endlich viele verschiedene Punkte q gibt, so liegt $t(s)$ unter einer festen Schranke. Es gibt also eine unendliche Folge (s)natürlicher Zahlen s mit gleichem $t(s)$. Da ja stets $||\lambda_n|| \leqq ||A||$ ist, so können wir aus (s) eine Teilfolge (s') so herausgreifen, so daß für $j = 1, 2, \dots t$, die Limiten $\lim_{(n'(j))} \lambda_n = \hat{\mu}_j$ ist, wo $n'(j)$ die Teilfolge $m(s'(j))$ durchläuft. Da $m(s(1)) = m(s)$, $m(s(t)) = n(s)$ ist, so gilt $\hat{\mu}_1 = \mu$, $\hat{\mu}_t = \nu$, d.h. die Punkte $(\hat{\mu}_j(g_1), \dots, \hat{\mu}_j(g_k)) = p(j)$ liegen für $j = 1$ bzw. t in T bzw. U. Es gibt daher ein r, so daß $p(r)$ in T und $p(r + 1)$ in U ist. Nun ist natürlich stets $0 \leqq \lambda_n(f) - \varrho(n, f) \leqq 1/l$, also ist für $f = g_j$ $(j = 1, \dots k)$, da $\varrho(m(s(r+1)) = \varrho(m(s(r)) + 1)$ für jedes s ist, (Wir lassen jetzt den Index f weg)

$$|\lambda_{m(s(r+1))} - \lambda_{m(s(r))}| \leqq |\lambda_{m(s(r+1))} - \varrho(m(s(r+1))| + |\varrho(m(s(r)) + 1) - \lambda m(s(r)) + 1| + |\lambda m(s(r)) + 1 - \lambda_{m(s(r))}|.$$

Lassen wir nun s die Folge (s') durchlaufen, so erhalten wir, da

$$|\lambda_{n+1} - \lambda_n| \leqq ||f|| \sum |a_{n+1,k} - a_{nk}|$$

$|\hat{\mu}_{r+1}(f) - \hat{\mu}_r(f)| \leqq 2/l$; also wäre doch $d(T, U) \leqq 2k/l$.

Die vorhergehenden Sätze haben nur Bedeutung, wenn es Folgen ω gibt, für die $W(\omega, A)$ und $V(\omega, A)$ aus mehr als einen Punkt bestehen. Dazu zeigen wir:

Satz 9: Ist A eine *reguläre* Matrix, d.h. ist $\lim_n a_{nk} = 0$ für jedes k, dann gibt es stets Folgen ω, welche mindestens zwei Verteilungsfunktionen besitzen, wenn X mehr als einen Punkt besitzt.

Beweis: Es seien x, y zwei verschiedene Punkte aus X, dann gibt es stets ein f_1 aus $C(X)$, so daß $f_1(x) \neq f_1(y)$. Wir definieren nun zwei unendliche monoton wachsende Folgen $(m(l))$, $(n(l))$ so, daß $m_0 = 0$ und bestimmen bei bereits gewählten $m(l-1)$, $n(l)$ so, daß $\sum_{k=1}^{m(l-1)} |a_{n(l),k}| < 1/l$ (das geht, da $\lim a_{nk} = 0$). Dann bestimmen wir $m(l)$ so, daß $\sum_{k=m(l)+1}^{\infty} |a_{n(l),k}| < 1/l$. Wir definieren dann die Folge $\omega = (x_k)$, so: Ist $m(l-1) < k \leqq m(l)$ und l ungerade, dann sei $x_k = x$ und wenn l gerade ist, dann sei $x_k = y$. Dann ist für jedes f aus $C(X)$, $\lim_n \lambda_n(\omega, A, f) = f(x)$, wenn n die Folge $n(l)$ mit ungeradem l durchläuft und gleich $f(y)$, wenn n die Folge $n(l)$ mit geradem l durchläuft. Ist nämlich $n = n(l)$ (l ungerade), so ist

$$\left|\sum\nolimits_n\right| = \left|\sum a_{nk}(f(x_n) - f(x))\right| \leqq \Big(\sum_{k=1}^{m(l-1)} + \sum_{k=m(l)+1}^{\infty}\Big) |a_{n,k}|\,|f(x) - f(y)|$$

daher ist $\lim \sum_n = 0$, also ist das gerade die Behauptung, da $\lim \sum a_{nk} = 1$ ist. Analog geht der Beweis für gerades l. Es besitzt also ω die Verteilungsmaße $f(x)$ und $f(y)$ und diese sind verschieden.

Analog zeigt man

Satz 9′: Unter den Voraussetzungen von Satz 9 gibt es stets Folgen, ω, welche mindestens N verschiedene Verteilungsmaße besitzen, welche Einzelmaße sind, wenn X mindestens N verschiedene Punkte enthält ($N = \infty$ zugelassen).

Wir stellen nun folgende Frage:

Kann jede vorgegebene beschränkte abgeschlossene Menge in W als ein $W(\omega, A)$ auftreten? Dazu zeigen wir

Satz 10: Es sei U eine beschränkte abgeschlossene Menge in W und ω eine beliebige Folge, welche in X überall dicht ist. Dann gibt es stets eine reguläre Matrix A, so daß $W(\omega, A) = U$ ist. Dabei kann noch A so gewählt werden, daß in jeder Zeile von A nur endlich viele Glieder von Null verschieden sind. Wir zeigen gleich etwas allgemeiner: Satz 10 bleibt richtig, wenn nur vorausgesetzt wird, daß ω dicht in allen Trägern der Maße aus U ist. Dabei verstehen wir unter dem Träger E_μ eines Maßes μ das Komplement der größten offenen Menge G in X, für welche $\mu = 0$ ist.

Wir nennen dabei eine Folge dicht in einer Menge E, wenn jeder Punkt von E Häufungspunkt der Folge ist.

Beweis: Es gibt in U eine abzählbar dichte Menge $\nu_1, \nu_2, \ldots$, also besitzt die Folge (μ_j), definiert durch $\nu_1, \nu_1, \nu_2, \nu_1, \nu_2, \nu_3, \ldots$ genau die Menge U als Menge ihrer Häufungsmaße. Da U beschränkt ist, gibt es ein $a > 0$, so daß für alle $\mu \in U$, also auch für die μ_j, $\|\mu\| \leqq a$, d.h. $|\mu(f)| \leqq a \|f\|$ für jedes $f \in C(X)$ ist.

Ist nun $E = E_\mu$ der Träger von $\mu = \mu_n$ $(n = 1, 2, \ldots)$, so ist $\mu(f) = \int\limits_X f d\mu = \int\limits_E f d\mu$, also gilt sogar $|\mu(f)| \leqq a \|f\|^*$, wo $\|f\|^* = \sup\limits_{x \in E} f(x)$ ist. Da die Folge $\omega = (x_k)$ dicht in E ist, so gibt es sicher ein x_l mit $l > n$, so daß $\|f\|^* \leqq 2\,|f(x_l)| \leqq 2 \sup\limits_{n \leqq k} |f(x_k)|$, also haben wir für jedes $f \in C(X)$

$$|\mu_n(f)| \leqq 2 \sup_{n \leqq k} |f(x_k)| \tag{3}$$

Es sei nun H ein Hauptsystem, ihre Funktionen seien $f_1, f_2, f_3 \ldots$ und es sei o. B. d. A. $\|f_j\| \leqq 1$ für alle j.

Dann behaupten wir zunächst folgenden[8])

Hilfssatz 1: Es gibt zu jedem n eine natürliche Zahl $k_n > n$, so daß für alle reellen Zahlen $h_1, \ldots, h_n$

$$\sup_{n \leqq k} |f_h^{(n)}(x_k)| \leqq 2 \sup_{n \leqq k \leqq k_n} |f_h^{(n)}(x_k)| \tag{4}$$

wo $f_h^{(n)} = \sum\limits_{j=1}^{n} h_j f_j(x_k)$ gesetzt wurde.

Wir zeigen dies so, indem wir gleich zeigen: Es gibt zu jeder natürlichen Zahl s eine natürliche Zahl $k_s \geqq n$, so daß $\sigma_s(h) = \sigma(h_1, \ldots, h_s) \leqq 2\sigma(k_s; h_1, \ldots, h_s) \leqq \sigma_s(k_s, h)$. Dabei wurde $\sigma_s(h) = \sup\limits_{n \leqq k} |f_h^{(s)}|$, $\sigma_s(l, h) = \sup\limits_{n \leqq k \leqq l} |f_h^{(s)}|$ gesetzt.

Wir zeigen dies durch vollständige Induktion nach s. Für $s = 1$ ist die Behauptung richtig, sie sei also schon für $s - 1$ gezeigt. Gibt es Zahlen $l_1, \ldots, l_s$, nicht alle Null, so daß $\sigma(l) = 0$, dann können wir $k_s = k_{s-1}$ nehmen, denn ist z.B. $l_s \neq 0$, dann ist für alle $k \geqq n$, $f_s(x_k)$ von der Gestalt $a_1 f_1(x_k) + \cdots + a_{s-1} f_{s-2}(x_k)$, wo die a nicht von k abhängen. Dann ist aber mit

$$h_r^* = h_r + a_r h_s \qquad (r = 1, \ldots s - 1)$$

$$\sigma_s(h) = \sigma_{s-1}(h^*) \leqq 2\sigma_{s-1}(k_{s-1}, h^*) = 2\sigma_s(k_{s-1}, h)\,.$$

Es kann also angenommen werden, daß für alle $h \neq (0, 0, \ldots 0)$, $\sigma_s(h) > 0$ ist. Wäre nun die Behauptung falsch, so gäbe es zu jedem

[8]) S. BANACH, Theorie des Operations lineaires (1932) S. 69. Ein Spezialfall von Satz 10, loc. cit. S. 217.

$k \geqq n$ Zahlen $h_1^{(k)}, \ldots, h_s^{(k)}$, so daß $\sigma_s(h^{(k)}) \geqq 2\sigma_s(k, h^{(k)})$ ist, wo o.B.d.A. $|h_1^{(k)}| + \cdots + |h_s^{(k)}| = 1$ angenommen werden kann. Dann gibt es eine Teilfolge (k_j) so daß die $h_r^{(k)}$ Grenzwerten $l_s (r = 1, \ldots s)$ zustreben, wo nicht alle l_s verschwinden. Dann folgt durch Grenzübergang $\sigma_s(l) \geqq 2\sigma_s(l)$, also doch $\sigma_s(l) = 0$. Nun kann der Beweis von Satz 10 leicht zu Ende geführt werden. Wir betrachten für jedes n das Gleichungssystem

$$\sum_{k=n}^{k_n} a_{nk} f_i(x_k) = \mu_n(f_i) \qquad (i = 1, \ldots, n) \qquad (5)$$

mit den Unbekannten $a_{nk} (n \leqq k \leqq k_n)$. Dieses System ist lösbar, da nach (3), (4) für alle reellen Zahlen $h_1, \ldots, h_n$

$$|\sum_{i=1}^{n} h_i \mu_n(f_i)| \leqq 2a \sup_{n \leqq k \leqq k_n} |\sum_{i=1}^{n} h_i f_i(x_k)|$$

ist. Dabei können die a_{nk} noch so gewählt werden, daß $\sum |a_{nk}| \leqq 2a$ ist. Dann leistet die Matrix $A = (a_{nk})$ mit $a_{nk} = 0$ für $k < n$ und $k \geqq k_n$ das Gewünschte. Sie ist regulär, $\|A\| \leqq 2a$.

Ist nun $\mu \in U$, dann gibt es eine Folge (n), so daß $\lim_{(n)} \mu_n = \mu$ ist. Für jedes feste i, wenn nur $n > i$, ist nach (5) $\lambda_n(\omega, A, f_i) = \mu_n(f_i)$, also ist $\lim_{(n)} \lambda_n(\omega, A, f_i) = \mu(f_i)$ für jedes f_i aus H. Es ist also μ in $W(\omega, A)$. Umgekehrt liegt jedes $\mu \in W(\omega, A)$ in U. Damit ist alles gezeigt.

§ 3. Wir fragen nun: Unter welchen Voraussetzungen gibt es Folgen ω, welche in bezug auf A gleichverteilt sind, mit dem Verteilungsmaß μ?

Wir wollen von nun an stets voraussetzen, daß alle Maße aus V sind, d.h. positiv und von der Gesamtmasse 1. Ist (μ_i) eine unendliche Folge von Maßen, so sei μ_P das Produktmaß $\otimes \mu_i$ auf dem Raum P aller Folgen $\omega = (x_k)$ aus X. Für jedes f aus $C(P)$, das nur von x_i abhängt, ist $\mu(f) = \mu_i(f)$. Dann behaupten wir zunächst[9])

Hilfssatz 2: Ist (f_i) eine Folge von Funktionen aus $C(X)$ mit $\|f_i\| \leqq K$ für alle i und ist $\mu_i(f_i) = 0$, dann ist für alle reellen Zahlen $h_1, \ldots, h_n$ und natürliche Zahlen m auf P

$$\mu_P((h_i f_i + \cdots + h_n f_n)^{2m}) \leqq (3mK^2 \sum_{i=1}^{n} h_i^2)^m. \qquad (5)$$

Bemerkung: Der Satz bleibt ungeändert richtig, wenn (X_i) eine Folge von kompakten Räumen, $P = \Pi X_i$, ist, die μ_i positive Maße auf den X_i mit der Gesamtmasse 1 sind und die f_i auf $C(X_i)$ definiert sind mit $\|f_i\|_i \leqq K$, wo sich die Norm auf X_i bezieht.

---

[9]) Vgl. die bekannte Untersuchung von A. Chincin, Math. Z. 18 (1922) 109—116.

Beweis: Unmittelbar klar ist die Abschätzung

$$\sum_{\beta_1+\cdots+\beta_n=s} ((2s)!/(2\beta_1)!\ldots(2\beta_n)!)h_1^{2\beta_1}\ldots h_n^{2\beta_n} \leqq ((2s)!/(2^s s!))\left(\sum_{i=1}^{n} h_i^2\right)^s \quad (6)$$

denn es ist doch $2^\beta \beta! \leqq (2\beta)!$

Wir können o.B.d.A. $\sum_{i=1}^{n} h_i^2 = 1$ und alle $h \geqq 0$ annehmen. Es ist die linke Seite von (5)

$$\leqq \sum_{\alpha_1+\cdots+\alpha_n=2m} ((2m)!/\alpha_1!\ldots\alpha_n!)\, h_1^{\alpha_1}\ldots h_n^{\alpha_n} J\alpha_1\ldots\alpha_n$$

wo die J die $\mu_\mathfrak{v}(f_1^{\alpha_1}\ldots f_n^{\alpha_n})$ sind, welche 0 sind, wenn ein $\alpha = 1$ und auf jedem Fall $\leqq K^{2m}$ sind. Wir haben also jetzt die Summe $\sum = \sum((2m)!/\alpha_1!\ldots\alpha_n!)\, h_1^{\alpha_1}\ldots h_n^{\alpha_n}$ abzuschätzen, wo $\sum\alpha_i = 2m$ ist und alle $\alpha_i \neq 1$ sind.

Wir fassen in dieser Summe alle Glieder zusammen, für welche die Anzahl der ungeraden α gleich ist, sagen wir $2q$. Die Summe dieser Glieder sei $\sum q$. Setzen wir für die ungeraden $\alpha = 2\beta + 3$ (es muß ja $\alpha \geqq 3$ sein) und für die geraden $\alpha = 2\beta$, dann ist $\sum\beta_i = m - 3q$, also $q \leqq m/3$. Weiter ist $\alpha_1!\ldots\alpha_n! \geqq 6^q(2\beta_1)!\ldots(2\beta_n)!$, daher wird

$$\sum q \leqq \sum \frac{(2m)!}{6^q(2\beta_1)!\ldots(2\beta_n)!}\, h_1^{2\beta_1}\ldots h_n^{2\beta_n}\, p_{2q}(h^3)$$

wo $p_l(k)$ die l-te elementarsymmetrische Funktion von $k_1,\ldots k_n$ ist. Nun ist z.B. $h_1\ldots h_{2q} \leqq (p_1(h^2)/2q)^q \leqq (2q)^{-q}$. Wenden wir dies z.B. auf $h_1^3\ldots h_{2q}^3$ an, so folgt $p_{2q}(h^3) \leqq (2q)^{-q} p_{2q}(h^2)$. Nun ist $p_{2q}(h^2) \leqq \binom{n}{2q}\left(\frac{p_1(h^2)}{n}\right)^{2q} \leqq 1/(2q)!$

Es wird also nach (5)

$$\sum \leqq \sum_q \sum q \leqq \frac{(2m)!}{2^m m!}\sum_{q=0}^{[m/3]}\binom{m}{3q} 2^q \leqq (3m)^m \qquad \text{w.z.b.w.}$$

Hilfssatz 3: Unter den Voraussetzungen von Hilfssatz 2 gilt für die reelle Zahl $t \geqq 0$

$$\mu_\mathfrak{v}\left(\left|\sum h_i f_i\right|^t\right) \leqq \left(2(t+2)K^2\sum h_i^2\right)^{t/2} \quad (7)$$

Für $t = 0$ ist nichts zu zeigen. Es sei also $t > 0$ und $2m$ die kleinste gerade natürliche Zahl $\geqq t$, dann ist nach der Hölderschen Ungleichung, die linke Seite von (6) $\leqq \left(\mu_\mathfrak{v}\left(\left|\sum h_i f_i\right|\right)^{2m}\right)^{t/2m} = \left(4mK^2\sum h_i^2\right)^{t/2}$ und wegen $m \geqq t/2 \geqq m-1$ ist alles gezeigt.

Hilfssatz 4: Sind weiter für alle i, alle $\mu_i(f_i^2)$ einander gleich $= L$, dann ist für $t > 0$

$$\mu_\mathfrak{v}\left(\left|\sum h_i f_i\right|^t\right) \geqq \left(tL\sum h_i^2/(3(K^2+1)(t+2))\right)^{t/2} L \quad (8)$$

Beweis: Es ist $\mu_\mathfrak{v}(|\sum h_i f_i|^2) = \sum h_i^2 \mu_i(f_i^2) = L\sum h_i^2$.

Andererseits ist

$$\mu_p(|\sum h_i f_i|^2) = \mu_p(|\sum h_i f_i|^{2t/(t+2)}|\sum h_i f_i|^{4/(t+2)})$$
$$\leqq (\mu_p(|\sum h_i f_i|^t))^{2/(t+2)} (\mu_p(|\sum h_i f_i|)^{4/t})^{t/(t+2)}$$

und alles folgt aus (5) auf den zweiten Faktor angewendet.

Wir wenden nun diese Hilfssätze auf $\lambda_n(\omega, A, f)$ an, indem wir für die h_i die n-te Zeile von A nehmen, wo wir $\to \infty$ gehen können, da $\sum |a_{nk}| \leqq ||A||$ ist. Wir erhalten dann für jedes $t > 0$

$$L(tL\alpha_n/(3K^2(t+2)))^{t/2} \leqq \mu_p(|\lambda_n(\omega, A, f)|^t) \leqq (2(t+2)K^2\alpha_n)^{t/2} \quad (9)$$

wo $K = ||f||$, $L = \mu(f^2)$, $\mu(f) = 0$, $\alpha_n = \sum a_{nk}^2$ und $\mu_p = \otimes \mu_i$ und $\mu_i = \mu$ für alle i ist.

Definition: Die Matrix A besitzt in bezug auf eine Maßfunktion μ aus V die Borelsche Eigenschaft, wenn alle Folgen ω aus P bis auf eine Menge vom Maße 0 (in bezug auf dem zu μ gehörigen Produktmaß μ_p), A, μ-gleichverteilt sind.

Satz 11: Notwendig für die Borelsche Eigenschaft einer Matrix in bezug auf ein Maß, welches nicht nur in einem einzigen Punkt konzentriert ist, ist $\lim\limits_n \alpha_n = \lim\limits_n \sum a_{nk}^2 = 0$. Die Borelsche Eigenschaft gilt[10]) für alle μ, wenn für jedes $\delta > 0$

$$\sum_{n=1}^{\infty} e^{-\delta^2/\alpha_n} < \infty \quad (10)$$

Es soll (10) die Hillsche Bedingung genannt werden.

Bemerkung: Ist μ nur in einem einzigen Punkt q konzentriert, dann besitzt A in bezug auf μ stets die Borelsche Eigenschaft, denn die Folge $\omega : x_k = q$ für alle k ist $A - \mu$ gleichverteilt und $\mu_p(\omega)$ ist 1. Besitzt X mehr als einen Punkt, so ist dieser Fall auch ein illustratives Beispiel zu Satz 12.

Beweis: Es sei zunächst angenommen, daß A, in bezug auf ein μ, dessen Träger E nicht nur aus einem Punkt besteht, die Borelsche Eigenschaft besitzt. Dann gibt es ein $f \in C(X)$ mit $\mu(f^2) > 0$ und $\mu(f) = 0$. Denn nach Voraussetzung gibt es zwei verschiedene Punkte a, b in E. Es gibt nun zwei stetige Funktionen f_1, f_2 auf X, so daß beide $\geqq 0$ und $\leqq 1$ und $f_1(a) = 1$, $f_1(b) = 0$ und $f_2(a) = 0$, $f_2(b) = 1$ ist. Es sei dann $g = f_1 - f_2$, dann leistet $f = g - \mu(g)$ das Gewünschte. Es ist $\mu(f) = 0$. Wäre nun $\mu(f^2) = 0$, dann wäre $g = \mu(g)$ im Träger E, also wäre $\mu(g)$ gleich 1 und -1, was aber nicht geht.

[10]) Für das A aus Fußnote [7]) gilt bereits der Satz, wenn $\gamma_n = 0$ $(S_n/\log\log S_n)$ wie man mittels der Methoden von R. Salem und A. Zygmund, Acta math. 91 (1954) 245—301, bes. S. 251, sofort zeigt.

Es folgt dann sofort aus der linken Seite von (9) für $t > 0$, $\lim \alpha_n = 0$.

Es sei nun (10) erfüllt. Es sei $f \in C(X)$, dann zeigen wir:

Für fast alle ω aus P ist

$$\overline{\lim_n} |\lambda_n(\omega, A, f) - \mu(f)| \leqq 4K\delta \tag{11}$$

wo $K = ||f - \mu(f)||$ ist.

Wir können o.B.d.A. $\mu(f) = 0$ annehmen. Es sei L_n die Menge in P für die $|\lambda_n(\omega, A, f)| > 4K\delta$ ist. Ihr Maß ist $\mu_p(L_n) < e^{-\delta^2/\alpha_n}$, denn setzen wir in (9) $t = 2\delta^2/\alpha_n$, so erhalten wir (o.B.d.A. $\alpha_n < 1/4\delta^2$) $(4\delta K)^t \mu_p(L_n) \leqq \mu_p(|\lambda_n|^t) \leqq (5\delta^2 K^2)^{t/2}$, also $\mu_p(L_n) \leqq (\frac{5}{16})^{t/2} \leqq e^{-\delta^2/\alpha_n}$.

Ist nun $\mathsf{U}_n = \bigcup_{k=n}^{\infty} L_k$, so ist U_n eine monoton abnehmende Mengenfolge und ihr äußeres μ_p-Maß ist $< \sum_{k=n}^{\infty} e^{-\delta^2/\alpha_k} < \eta$, wenn $n > n_0(\eta)$, so daß der Durchschnitt $\mathsf{U} = \bigcap \mathsf{U}_n$ das Maß 0 hat. Ist also ω eine Folge nicht in U, so gibt es ein n_1, so daß für alle $n > n_1$

$$|\lambda_n(\omega, A, f)| < 4\delta K$$

ist. Da nun (10) für alle $\delta > 0$ gelten soll, so folgt, daß $\lim \lambda_n(\omega, A, f) = \mu(f)$, bis auf eine Menge vom Maße 0 ist.

Wenden wir dies auf die abzählbar vielen Funktionen eines Hauptsystems H an, so erhalten wir gerade die Behauptung.

Bemerkung: Es ist (10) sicher erfüllt, wenn es eine Folge von Zahlen $q_n \geqq 0$ gibt, so daß $\sum q_n < \infty$ und $\lim_{n\to\infty} \alpha_n |\log q_n| = 0$ ist.

Einfaches Beispiel: $q_n = 1/n^2$, also $\lim_n \alpha_n \log n = 0$.

Ein anderes Beispiel: $q_n = \alpha_n^l$, wenn es ein l gibt, so daß $\sum \alpha_n^l < \infty$ ist. Dann ist $\lim \alpha_n |\log q_n| = 0$.

Ist $P_s(A)$ die Menge aller ω, welche in bezug auf A summierbar sind und $P_\mu(A)$ die Menge aller ω, welche $A - \mu$ gleichverteilt sind, so ist klar, daß $\mu_p(P_s(A)) = 1$, wenn $\mu_p(P_\mu(A)) = 1$ ist. Dies ist ja der Fall, wenn also z.B. (10) gilt. Es gilt nun aber trotzdem, daß die Mengen P_s und P_μ magere Mengen sind, d.h. Mengen erster Kategorie in P, wenn X mehr als einen Punkt enthält. Sonst ist der Satz trivialerweise falsch. Wir behaupten also (vgl. [5])

Satz 12: Besitzt X mehr als einen Punkt, dann ist $P_s(A)$ für jedes A eine Menge erster Kategorie in A, wenn A regulär ist.

Beweis: Es sei H ein Hauptsystem von X und $f_1, f_2, \ldots$ seien seine Funktionen. Dann ist nach Satz 4 eine Folge $\omega = (x_k)$ genau dann A-summierbar, wenn für alle f aus H, $\lim \lambda_n(\omega, A, f)$ existiert. Ist also für die nicht negativen ganzen Zahlen m, p, r, P_{mpr} die Menge der

Folgen ω in P, für welche $\overline{\lim_q} |\lambda_{n+q}(\omega, A, fr) - \lambda_n(\omega, A, fr)| \leqq 1/m$ für alle $n \geqq p$ gilt, so ist $P_s = \bigcap_{r=1}^{\infty} \bigcup_{m=1}^{\infty} \bigcap_{t=m}^{\infty} \bigcup_{p=0}^{\infty} P_{tpr}$. Nun ist $\lambda_n(\omega, A, f)$ stetig in ω im Raum P, es ist also P_{tpr} abgeschlossen. Nun gibt es nach Satz 9, eine Folge $\omega_0 = (x_k^p)$, welche nicht in P_s liegt, d.h. es gibt ein $\delta > 0$, ein $r = r' \geqq 1$ und zwei Folgen $(n) = (n(l))$, $(q) = (q(l))$, so daß für jedes $n = n(l)$, $q = q(l)$

$$|\lambda_{n+q}(\omega_0, A, fr) - \lambda_n(\omega_0, A, fr)| \geqq \delta$$

ist. Es genügt nun, zu zeigen, daß für dieses $r = r'$ und für genügend große t, $\bigcup_{p=0}^{\infty} P_{tpr}$ von erster Kategorie ist. Nehmen wir an, es gäbe ein $t > 2/\delta$, so daß diese Menge von zweiter Kategorie ist, d.h. nicht Vereinigungsmenge nirgends dichter Mengen von P ist. Dann muß es ein $\hat{P} = P_{tpr}$ geben, welches nicht nirgends dicht in P ist, d.h. es muß eine offene Menge $G < P$ geben, so daß $G \frown \hat{P}$ dicht in G ist, d.h. aber die Hülle von $G \frown \hat{P}$ ist in G enthalten. Nun ist $\hat{P}$ abgeschlossen, also muß diese Hülle und damit auch G in $\hat{P}$ liegen. Also gibt es ein P_{tpr}, in welcher eine offene Menge G liegt. Es muß nun G eine elementare offene Menge G^* enthalten, d.h. es gibt ein v, so daß G von der Gestalt $G_0 \prod_{i=v+1}^{\infty} X_i$ ist, wo G_0 eine offene Menge aus $\prod_{i=1}^{v} X_i$ ist (für alle i ist $Xi = X$). Wir wählen nun einen beliebigen Punkt $(x_1^*, \ldots, x_v^*)$ aus G_0, dann liegt $\omega^* = (x_1^*, \ldots, x_v^*, x_{v+1}^0, \ldots)$ in G. Es ist $\lambda_n(\omega^*, A, fr) = \lambda_n(\omega_0, A, fr) + \sum_{k=1}^{v} a_{nk}(fr(x_k^*) - fr(x_k^0))$ und es ist $|\sum_{k=1}^{v}| \leqq 2\, \|fr\| \sum_{k=1}^{v} |a_{nk}|$. Da G in P_{tpr} liegt, so ist $|\lambda_{n+q}(\omega^*, A, fr) - \lambda_n(\omega^0, A, fr)) \leqq 1/t$ für alle $n \geqq p$ und alle genügend großen q. Nun ist für $n = n(l)$, $q = q(l)$

$$|\lambda_{n+q}(\omega^*, A, fr) - \lambda_n(\omega^*, A, fr)| \geqq |\lambda_{n+q}(\omega_0, A, fr) - \lambda_n(\omega_0, A, fr)|$$
$$-2\, \|fr\| \Big(\sum_{k=1}^{v} (|a_{nk}| + |a_{n+q,\,k}|)\Big).$$

Es ist also $\overline{\lim_l} |\lambda_{n(l)+q(l)}(\omega^*, A, fr) - \lambda_{n(l)}(\omega^*, A, fr)| \geqq \delta > 2/t$ und dies ist ein Widerspruch.

Wenn die Hillsche Bedingung erfüllt ist, so ist $\mu_p(P_s(A)) = \mu_p(P_\mu(A))$. Wir zeigen nun, daß das immer gilt (vgl. [5]). Dazu bemerken wir, daß das P-Maß von $P_s(A)$ bzw. $P_\mu(A)$ nur die Werte 0 oder 1 annehmen kann, da diese Mengen homogene Mengen sind.

Satz 13: Es ist stets $\mu_p(P_s(A)) = \mu_p(P_\mu(A))$, wenn A regulär ist.

Beweis: Da $P_\mu(A) < P_s(A)$, so genügt es zu zeigen: Ist $\mu_p(P_s(A)) = 1$ dann ist auch $\mu_p(P_\mu(A)) = 1$. Es sei also für fast alle ω und alle

$f \in C(X)$, $\lim \lambda_n(\omega, A, f)$ vorhanden $= \mu_\omega(f)$. Wir legen wieder ein Hauptsystem H mit der Basis (fr) $(r = 1, 2, \ldots)$ zugrunde und setzen $\mu_\omega(fr) = \varphi_r(\omega)$. Da die λ_n in ω stetig und dem Betrage nach $\leqq \|A\|\,\|fr\|$ sind, so sind die $\varphi_r(\omega)$ integrierbar. Wenn wir zeigen können, daß für jede meßbare Menge Z aus P und jedes r, $\int\limits_Z (\varphi_r(\omega) - \mu(fr))\, d\mu_p = 0$ ist, so ist tatsächlich für fast alle ω und alle r, $\varphi_r(\omega) = \mu(fr)$ und wir sind fertig. Setzen wir $fr(x) - \mu(fr) = \psi_r(x)$, so ist zunächst $\mu(\psi_r) = 0$. Es ist $\lim\limits_n \lambda_n(\omega, A, \psi_r)$ vorhanden und gleich $\varphi_r(\omega) - \mu(fr)$. Es sei nun zunächst $Z(N)$ eine elementare Menge, d.h. also von der Gestalt $Z_N \prod\limits_{i=N+1} X_i$, wo Z_N eine offene Menge aus $\prod\limits_{i=1}^{N} X_i$ $(Xi = X)$, dann ist $\int\limits_{Z(N)} \lambda_n(\omega, A, \psi_r)\, d\mu_p = \sum a_{nk} \int\limits_{Z(N)} \psi_r(x_k)\, d\mu_p$. Nun ist für $k > N$, $\int\limits_{Z(N)} \psi_r(x_k)\, d\mu_p = \int\limits_{X_k} \psi_r(x_k)\, d\mu = 0$, also $\int\limits_{Z(N)} \lambda_n(\omega, A, \psi_r)\, d\mu_p = \sum\limits_{k=1} a_{nk} \int\limits_{Z(N)} \psi_r(x_k)\, d\mu_p$. Gehen wir also mit $n \to \infty$, so folgt $\int\limits_{Z(N)} (\varphi_r(\omega) - \mu(fr))\, d\mu_p = 0$. Es sei nun Z eine beliebige meßbare Menge in P. Dann ist $\mu_p(Z) = \inf \sum\limits_{i=l}^{\infty} \mu_p \cdot (Z(i))$, erstreckt über alle Überdeckungen von Z durch elementare Mengen $Z(i)$. Wir können also eine Überdeckung so wählen, daß $\sum\limits_i \mu_p(Z(i)) - \mu_p(Z) < \varepsilon$ ($\varepsilon > 0$ beliebig) und weiter ein m so, daß $\sum\limits_{i=m+1}^{\infty} \mu_p(Z(i)) < \varepsilon$ wird. Setzen wir $\bigcup\limits_i Z(i) = Z^*$, $\bigcup\limits_{i=1}^{m} Z(i) = Z^*_m$, $\bigcup\limits_{i=m+1}^{\infty} Z(i) = Z^{**}$, so ist $Z^* = Z^*_m \cup Z^{**}$.

Bedeutet A/B (wenn $B \subset A$) das Komplement der Menge B in A, dann ist $\mu_p(Z^*/Z) = \mu_p(Z^*) - \mu_p(Z) \leqq \sum\limits_{i=1}^{\infty} \mu_p(Z_i) - \mu_p(Z) < \varepsilon$

$$\mu_p(Z^*/Z^*_m) \leqq \mu_p(Z^{**}) \leqq \sum_{i=m+1}^{\infty} \mu_p(Zi) < \varepsilon,$$

es ist also $\left|\int\limits_Z\right| = \left|\int\limits_{Z^*} - \int\limits_{Z^*/Z}\right| \leqq \left|\int\limits_{Z^*_m} + \int\limits_{Z^*/Z^*_m}\right| + \left|\int\limits_{Z^*/Z}\right| \leqq (2\,\|A\|\,\|fr\|)\varepsilon$, da ja $\int\limits_{Z^*_m}$ nach dem vorhergehenden, Null ist. Also ist $\int\limits_Z = 0$, w. z. b. w.

Wir wollen nun Satz 11 verallgemeinern[10]):

Satz 14: Es sei (X_i) eine Folge von abgeschlossenen Mengen aus X, (μ_i) aus V mit $\mu_i(X_i) = 1$ gegeben. Weiter sei $\lim(\mu_{i+1} - \mu_i) = 0$, A regulär, und es seien $(s) = (s(l))$, $(n) = (n(l))$ zwei monoton wachsende nicht beschränkte Folgen natürlicher Zahlen, so daß für alle $n \geqq n(l)$, $\sum\limits_{k=1}^{s(l-1)} |a_{nk}| < 1/l$ und $\sum\limits_{k=s(l)+1}^{\infty} |a_{nk}| < 1/l$ für alle n mit $n(l-1) \leqq n \leqq n(l)$ ist. Es sei nun Q der Produktraum $\prod\limits_{k=1}^{\infty} Y_k$, $\mu_Q = \bigotimes\limits_k \nu_k$, wo $Y_k = X_l$, $\nu_k = \mu_l$

für $s(l-1) < k \leqq s(l)$. Es ist also Q die Menge aller Folgen $\omega = (x_k)$, wo die x_k in X_l für alle k mit $s(l-1) < k \leqq s(l)$ liegen.

Dann behaupten wir: Erfüllt A die Hillsche Bedingung (10), dann ist für (μ_Q) fast alle Folgen ω aus Q, $V(\omega, A)$ die Menge U der Häufungsmaße der Folge (μ_i).

Bemerkung: Man erhält Folgen $(s(l))$, $(n(l))$ einfach durch Rekursion, wenn man beachtet, daß ja $\lim a_{nk} = 0$ und $\sum |a_{nk}| < \infty$ ist.

Beweis: Mit genau der gleichen Überlegung wie bei Satz 11 zeigt man, daß für (μ_Q)-fast alle Folgen ω aus Q und alle r, $\lim\limits_n \lambda_n(\omega, A, fr - \nu(fr)) = 0$ ist, wo die fr ein Hauptsystem H durchlaufen. Es sei nun ein solches ω ins Auge gefaßt.

Jetzt zeigen wir: Ist μ ein Häufungsmaß der Folge (μ_l), dann ist sie auch Verteilungsmaß von ω. Es gibt nämlich eine Folge $(l) = (l(t))$, so daß $\lim\limits_{(l)} \mu_l = \mu$ ist.

Dann ist für $n = n(l)$, wo $l = l(t)$

$$\sum_{k=1}^{\infty} a_{nk} \nu_k(fr) = \Big(\sum_{k=1}^{s(l-1)} + \sum_{s(l)+1}^{\infty}\Big) + \sum_{k=s(l-1)+1}^{s(l)} a_{nk}\mu_l(fr) = O(1/l) + \mu_l(fr)\beta_n$$

$(\beta_n = \sum a_{nk})$. Für $t \to \infty$ geht die obige Summe also gegen $\mu(fr)$, also ist $\lim\limits_{(n)} \lambda_n(\omega, A, fr) = \mu(fr)$ für die Folge $n = n(l(t))$ und für jedes r. Damit ist die Behauptung schon bewiesen.

Es sei nun umgekehrt μ ein Verteilungsmaß von ω, dann zeigen wir, daß es Häufungsmaß der μ_l ist. Erst hier benützen wir, daß $\lim(\mu_{l+1} - \mu_l) = 0$ ist. Es gibt nach Voraussetzung eine Folge $(n) = (n(t))$, so daß $\lim\limits_{(n)} \lambda_n(\omega, A, fr) = \mu(fr)$ für jedes r ist. Dann ist also $\lim\limits_{(n)} \sum\limits_k a_{nk} \nu_k(fr) = \mu(fr)$. Wir bestimmen nun $l = l(t)$ so, daß $n(l(t)) \leqq n(t) < n(l(t) + 1)$ ist. Dann schätzen wir mit $l = l(t)$, $n = n(t)$

$$|\mu(fr) - \mu_l(fr)| \leqq |\mu(fr) - \sum a_{nk}\nu_k(fr)| + |\sum a_{nk}\nu_k(fr) - \mu_l(fr)|$$

ab. Es ist nun weiter

$$\begin{aligned}\sum a_{nk}\nu_k(fr) &= O(1/l) + \mu_l \sum_{k=s(l-1)}^{s(l)} a_{nk} + \mu_{l+1} \sum_{k=s(l)+1}^{s(l+1)} a_{nk} \\ &= O(1/l) + (\mu_{l+1} - \mu_l)O(1) + \mu_l\beta_n .\end{aligned}$$

Diese Summe geht also für $t \to \infty$ gegen $\mu_l(fr)$, also geht $\mu_l(fr)$ gegen $\mu(fr)$ für jedes r und damit ist alles gezeigt.

Die Bedingung (10) kann in gewissem Sinne nicht mehr verschärft werden. Ist nämlich A die Matrix von P. Erdös (in der Gestalt bei Hill): $a_{nk} = 1/[\log m]$ $(m^2 < n \leqq (m+1)^2$, $n - 2m < k \leqq n - 2m + [\log m])$ und 0 sonst, also $\lim \alpha_n \log n = 2$, dann besitzt (A, μ) für kein μ, welches nicht in einem Punkte konzentriert ist, die Borelsche Eigenschaft.

Beweis: Sind a, b zwei Punkte aus dem Träger von μ, so gibt es zueinander fremde Umgebungen dieser Punkte und für wenigstens eine dieser Umgebungen U, sagen wir von a, muß ihr Maß $\mu(U) \leqq \frac{1}{2}$ sein. Da aber andererseits $\mu(U) > 0$ ist, so haben wir für das Maß des Komplements C von U: $\frac{1}{2} \leqq \mu(C) < 1$. Weiter gibt es in U sicher eine Umgebung V von a, so daß ihre Hülle W in U liegt und $\mu(W) > 0$ ist. Ist also f die stetige Funktion mit $0 \leqq f \leqq 1$, welche auf C den Wert 1 und auf W den Wert 0 hat (C und W sind ja disjunkt), so ist stets $1 - \psi \geqq f \geqq \varphi$ (φ, ψ die charakteristischen Funktionen von C bzw. W) und $1 > 1 - \mu(W) \geqq \mu(f) \geqq \mu(C) \geqq \frac{1}{2}$. Es ist also mit $\mu = \mu(C)$, $L_m = [\log m]$, $t_m = m^2$, die Reihe $\sum \mu^{Lm}$ divergent, also strebt (es ist $t_{m+1} > L_m + t_m$) für jedes $M > N$

$$\mu_\mathfrak{p}\Big(\prod_{m=N}^{M} (1 - \varphi(x_{tm+1}) \dots \varphi(x_{tm+Lm}))\Big) = \prod_{m=N}^{M} (1 - \mu^{Lm})$$

für $M \to \infty$ gegen 0. Es gibt also für μ-fast alle Folgen ω, Folgen $(m(s)) \to \infty$, so daß $\varphi(x_{tm+v}) = 1$ ($v = 1, \dots, L_m$, $m = m(s)$). Es ist also mit $n = n(s) = m^2(s) + 2m(s)$ $(s = 1, 2, \dots)$

$$\lambda_n(f) = \sum a_{nk} f(x_k) = 1 > \mu(f)\,,$$

was zu zeigen war. Da $\lim \alpha_n = 0$, gibt es natürlich auch für fast alle ω auch Folgen (n) mit $\lim \lambda_n(f) = \mu(f)$.

§ 4. Aus Satz 14 folgt sofort[10])

Satz 15: Ist U eine abgeschlossene Menge aus $V(X)$ und erfüllt A die Hillsche Bedingung, dann gibt es Folgen ω aus X, für welche $V(\omega, A) = U$ ist, wenn es eine Folge von Verteilungsmaßen μ_i mit $\lim(\mu_{i+1} - \mu_i) = 0$ gibt, für welche die Menge der Häufungsmaße gleich U ist. Dabei können die Glieder von ω aus der Vereinigungsmenge von abgeschlossenen Mengen X_i aus X genommen werden, für welche $\mu_i(X_i) = 1$ ist. Daraus folgt[10])

Satz 16: Es sei U eine abgeschlossene konvexe Menge aus V und A erfülle die Hillsche Bedingung. Dann gibt es stets eine Folge ω aus X, ja sogar aus der Vereinigungsmenge aller Träger der Maße von U, für welche $V(\omega, A)$ gleich U ist.

Beweis: Es gibt eine abzählbare dichte Menge $\{\nu_k\}$ von Maßfunktionen in U. Dann ist die Menge der Häufungsmaße der Folge (μ'_k): $\nu_1, \nu_1, \nu_2, \nu_1, \nu_2, \nu_3, \dots$ genau U.

Wir bilden uns nun die Folge der Maße (μ_i), wo

$$\mu_i = \big(((s+1)^2 - l) \,/\, (2s+1)\big)\, \mu'_k + \big((l - s^2) \,/\, (2s+1)\big)\, \mu'_{k+1}\,,$$

wenn $s^2 \leqq l < (s+1)^2$ ist.

Diese μ_i liegen wieder in U. Weiter ist $\lim(\mu_{i+1} - \mu_i) = 0$. Es sind also alle Voraussetzungen von Satz 15 erfüllt.

Wir zeigen jetzt

Satz 17: Es sei $\omega^* = (Z_k)$ eine Folge aus einer Menge Y in X, $\omega = (x_k)$ eine Folge, welche in Y dicht liegt, dann können die Glieder von ω so umgeordnet werden, daß für die neue Folge $\omega' = (u_k)$, $W(\omega', A) = W(\omega^*, A)$, $V(\omega', A) = V(\omega^*, A)$ ist, wenn nur $\lim \alpha_n = \lim \sum a_{nk}^2 = 0$ ist.

Beweis: Es seien wieder $f_1, f_2, \ldots$ eine Basis aus einem Hauptsystem H. Nach Voraussetzung gibt es zu jedem Z_k eine Teilfolge $(x_{k(n)})$ von ω, so daß $\lim x_{k(n)} = Z_k$ ist. Es gibt also sicher ein x'_k aus ω, so daß $|fr(Z_k, -fr(x'_k)| < 1/k$ für $r = 1, 2, \ldots, k$ $(k = 1, 2, \ldots)$. Es ist also für jedes r, $\lim \sum a_{nk}(fr(x'_k) - fr(Z_k)) = 0$, da ja die $fr(x'_k) - fr(Z_k)$ in k eine Nullfolge sind und A ein Töplitzsches Summationsverfahren ist. (Aus unserer Voraussetzung folgt ja $\lim a_{nk} = 0$.) Es gibt nun zu jeder natürlichen Zahl p ein $n(p)$, so daß für alle $n > n(p)$, $\alpha_n < 1/p^4$ ist, also $|a_{nk}| < l/p^2$ für alle $n > n(p)$ und alle k ist. Dabei sei $n(p+1) > n(p)$ gewählt. Weiter gibt es natürliche Zahlen $k(p)$ so, daß für alle n mit $n(p) \leqq n < n(p+1)$, $\sum\limits_{k=k(p)+1}^{\infty} |a_{nk}| < 1/p$ ist, da $\sum |a_{nk}| < \infty$. Wir bezeichnen die Glieder in ω, welche nicht in (x'_k) vorkommen und jene x_k mit $k = k(p)$ kurz mit t_i, welche wir uns irgendwie durchnumeriert denken. Wir setzen nun $u_k = x'_k$, wenn $k \neq k(p)$, $u_k = t_p$, wenn $k = k(p)$ ist. Dann ist für jedes r

$$\sum_n = |\sum a_{nk}(fr(u_k) - fr(x'_k)| = O\left(\sum_{s=1}^{\infty} |a_{n,k(s)}|\right).$$

Zu jedem n gibt es ein $n(p)$ mit $n(p) \leqq n < n(p+1)$, also ist[11])

$$\sum_n = O\left(\sum_{s=1}^{p} + \sum_{s=p+1}^{\infty}\right) = O\left(p \cdot \frac{1}{p^2}\right) + O\left(\frac{1}{p}\right).$$

Es ist also für alle r

$$\lim \sum a_{nk}(fr(u_k) - fr(z_k)) = 0$$

und damit ist alles gezeigt.

Aus diesem Satz und Satz 15 folgt sofort[10])

Satz 18: Gelten die Voraussetzungen des Satzes 15, dann können die Glieder jeder Folge ω, welche dicht in der Vereinigungsmenge aller Träger der Maße von U ist, also insbesondere dicht in X ist, stets so umgeordnet werden, daß für die neue Folge ω^*, $V(\omega^*, A) = U$ ist.

Aus Satz 16 folgt ebenso[10])

---

[11]) Zur Beweismethode R. P. Agnew, Bull. Amer. 52 (1946) 128—132.

Satz 19: Es sei U eine abgeschlossene konvexe Menge von Verteilungsmaßen, dann kann jede Folge ω, welche dicht in der Vereinigungsmenge aller Träger der Maße von U ist, also insbesondere dicht in X ist, stets so umgeordnet werden, daß für die neue Folge ω^* die Menge $V(\omega^*, A)$ ihrer Verteilungsmaße gerade U ist, wenn nur A die Hillsche Bedingung erfüllt, wenn also insbesondere $\lim\limits_n \sum a_{nk}^2 \log n = 0(1)$ ist. Man kann fragen, ob solche Sätze gelten, wenn z. B. die Hillsche Bedingung nicht erfüllt ist, ob also insbesondere Satz 15 gilt, wenn man nur weiß, daß es zu jedem μ eine Folge ω gibt, welche A — μ-summierbar ist. Dazu zeigen wir:

Satz 20: Es sei U eine abgeschlossene Menge aus $W(X)$ und es gäbe eine Folge (μ_i) aus W, für welche die Menge der Häufungsmaße gerade U ist. Es seien weiter Mengen X_i gegeben und es gäbe nun für jedes i Folgen ω_i aus X_i, welche A — μ_i-summierbar sind. Dann gibt es eine Folge ω aus $\bigcup\limits_i X_i$, für welche $W(\omega, A) > U$ ist, wenn A regulär ist. Ist weiter $\lim(\mu_{i+1} - \mu_i) = 0$ und erfüllt A folgende Bedingung[12]): Für jede beschränkte Zahlenfolge (s_k) mit $\lim \sum a_{nk} s_k = 0$ gilt gleichmäßig in t, $\lim\limits_n \sum\limits_{k=1}^{t} a_{nk} s_k = 0$, dann kann sogar $W(\omega, A) = U$ behauptet werden.

Beweis: Es seien die (fr) Basis eines Hauptsystems H. Es gilt für alle r und i, $\lim\limits_n \lambda_n(\omega_i, A, fr) = \mu_i(fr)$. Es sei $\omega_i = (x_k^{(i)})$, dann definieren wir rekursiv, monoton wachsende nicht beschränkte Folgen natürlicher Zahlen $(s(i))$, $(n(i))$ so, daß für alle $n \geqq n(i)$

$$\sum_{k=1}^{s(i-1)} |a_{nk}| < 1/i, \quad |\lambda_n(\omega_i, A, fr) - \mu_i(fr)| < 1/i \tag{12}$$

für $j = 1, \ldots, i$ gilt, dann $s(i) > s(i-1)$ so, daß $\sum\limits_{s(i)+1}^{\infty} |a_{nk}| < 1/i$ für alle $n \leqq n(i)$ ist. Dann sei $\omega = (x_k)$ mit $x_k = x_k^{(i)}$ für $s(i-1) < k \leqq s(i)$. Es sei nun $\lim \mu_{i(t)} = \mu \, (\mu \in U)$. Dann ist für $n = n(i(t))$ und jedem r

$$|\lambda_n(\omega, A, fr) - \mu(fr)| \leqq \lambda_n(\omega, A, fr) - \lambda_n(\omega_{i(t)}, A, fr)| + |\lambda_n(\omega_{i(t)}, A, fr) \\ - \mu_{i(t)}(fr)| + |\mu_{i(t)}(fr) - \mu(fr)|.$$

Der erste Klammerausdruck ist $O\Big(\sum\limits^{s(i-1)} + \sum\limits_{s(i)+1}\Big) |a_{nk}| = O(1/i(t))$. Es ist also tatsächlich $\lim \lambda_n(\omega, A, fr) = \mu(fr)$.

Damit ist der erste Teil der Behauptung schon bewiesen. Jetzt der zweite Teil! Es ist nun $\lim(\lambda_n(\omega_i, A, fr) - \lambda_n(\omega_{i+1}, A, fr) = \mu_{i+1} - \mu_i$. Wir wählen i_0 so, daß für alle $i > i_0$, $|\mu_{i+1} - \mu_i| < 1/i$ ist und dann

[12]) Zu dieser Bedingung vgl. W. Jurkat und A. Peyerimhoff, Math. Z. 56 (1952) 152—178. Für das A in [7]) ist sie erfüllt.

zusätzlich zu (12), $n(i)$ noch so, daß für alle $n \geqq n(i)$ und alle t, also insbesondere für $s(i)$, $|\sum^{t} a_{nk}(fr(x_k^{(i)}) - fr(x_k^{(i+1)})| \leqq 2/i + \sum_{t+1}^{\infty} |a_{nk}|$. Es sei μ aus $W(\omega, A)$, dann gibt es eine Folge $(n) = (n(t))$ mit $\lim \lambda_n (\omega, A, fr) = \mu(fr)$. Wir wählen $i = i(t)$ so, daß $n(i) \leqq n(t) < n(i+1)$. Dann ist für $i = i(t)$, $\lim_{(i)} \mu_i = \mu$, also in U. Es ist nämlich für $i = i(t)$, $n = n(t)$

$$
\begin{aligned}
|\mu - \mu_i(fr)| &\leqq |\lambda_n(\omega, A, fr) - \mu(fr)| + |\lambda_n(\omega, A, fr) - \mu_i(fr)| \\
&\leqq |\lambda_n(\omega, A, fr) - \mu(fr)| + O(1/i) + |\mu_{i+1}(fr) - \mu_i(fr)| \\
&\quad + |\sum_{k=1}^{s(i)} a_{nk}(fr(x_k^{(i)}) - fr(x_k^{(i+1)}))|
\end{aligned}
$$

und daraus folgt alles.

Eingegangen am 10. 4. 1955

Zur Überdeckung der Ebene durch konvexe Scheiben

Anz. Österr. Akad. Wiss., Math.-Naturwiss. Kl. **94** (1957) 79–82

Es sei die Ebene durch konvexe Scheiben, welche einen Mittelpunkt besitzen und durch Schiebung auseinander hervorgehen, so überdeckt, daß ihre Mittelpunkte ein Gitter mit Mascheninhalt D bilden. Ist nun A die Anzahl der Scheiben S, welche mit einer willkürlich herausgegriffenen Scheibe S_0 Punkte gemeinsam haben (S_0 nicht gerechnet), so ist

$$DA \leqq 8\ F \tag{1}$$

(F = Flächeninhalt der Scheibe). Die Schranke in (1) ist scharf. Wird nämlich ein quadratisches Gitter ins Auge gefaßt und betrachtet man die Quadrate mit Flächeninhalt D, deren Mittelpunkte gerade die Gitterpunkte sind, so überdecken diese Quadrate die Ebene und an jedes Quadrat schließen sich acht Quadrate an.

Man kann (1) durch mehr aussagende Ungleichungen ersetzen. Es ist nämlich sogar

$$A \leqq 8\,q - 2\ (C - 1) \tag{2}$$

wo $q = F/D$ und $C = \left[\ \frac{1}{4}\ (\sqrt{32q+17} - 1)\right]$ ist ($[a]$ = nächstkleinere ganze Zahl an a). Es ist (2) eine Verschärfung von (1), da bekanntlich bei einer Überdeckung stets $q \geqq 1$ ist. Die Ungleichungen (1) und (2) werden aus der Abschätzung

$$A \leqq 4\,q + n \tag{3}$$

folgen, wo n die Anzahl der Geraden durch den Mittelpunkt von S_0 ist, auf welchen die Mittelpunkte der Scheiben S liegen, welche mit S_0 Punkte gemeinsam haben.

Die Abschätzung (1) wurde von mir ohne Beweis in der Festschrift zum 60. Geburtstag von G. Szegö 1955 angegeben.

Das analoge Problem im Raum ist bedeutend schwieriger. Bemerkungen hiezu am Schluß der vorliegenden Arbeit. Nun zum Beweis der Ungleichungen (1), (2), (3).

Nach Voraussetzung bilden die Mittelpunkte s der Scheiben S ein Gitter G mit Mascheninhalt D. Wir denken uns ein rechtwinkeliges Koordinatensystem so zugrunde gelegt, daß S_o den Koordinatenursprung o als Mittelpunkt besitzt. Dann möge s auch stets den Ortsvektor des Mittelpunktes der Scheibe S bedeuten und jeder Punkt von S hat die Gestalt $x+s$, wo x in S_o liegt und umgekehrt (die Scheiben gehen ja durch Schiebung auseinander hervor). Wenn nun S mit S_o einen Punkt y gemeinsam hat, dann liegt also $y-s$ und y auf S_o, also da o Mittelpunkt von S_o, auch $-(y-s)$ in S_o; also wegen der Konvexität $\frac{1}{2}(y+s-y) = \frac{1}{2}s = t$ auf S_o. Liegt umgekehrt $\frac{1}{2}s$ in S_o, dann haben S_o und S einen Punkt gemeinsam, nämlich $\frac{1}{2}s$. Die Anzahl A der Scheiben S, welche mit S_o einen Punkt gemeinsam haben, ist also genau die Anzahl der Punkte $t = \frac{1}{2}s \neq o$, welche auf S_o liegen. Die Punkte t bilden ein Gitter G_1 mit Mascheninhalt $\frac{1}{4}D = D_1$. Nun liegt mit t auch stets $-t$ in S_o. Es ist also A eine gerade Zahl $2N$. Es seien nun $l_1, l_2, \ldots l_m$ die verschiedenen Halbstrahlen durch o, auf welchen die A-Punkte t in S_o liegen, also die Halbstrahlen, auf welchen die Mittelpunkte der zugehörigen Scheiben S liegen. Die Anzahl der Punkte t auf einen Halbstrahl l_i bezeichnen wir mit k_i. Auf Grund der vorhergehenden Überlegung muß zu jedem Halbstrahl l auch der entgegengesetztgerichtete Halbstrahl, und zwar mit gleichem k auftreten; es ist also $m = 2n$ und $k_1+k_2+\ldots+k_n = N$. Es ist sicher $n > 1$, denn sonst würden die Mittelpunkte s aller Scheiben S, welche mit S_o Punkte gemeinsam haben, auf einer Geraden l liegen. Dann könnten wir an diese Scheiben zwei Stützgeraden parallel zu l legen. Bezeichnen wir den Streifen, begrenzt durch diese Stützgeraden, mit L. In diesem Streifen L liegt auch S_o. Vergrößern wir aber S_o von o aus ähnlich zu sich selbst im Verhältnis $1:1+\varepsilon$ ($\varepsilon > o$ beliebig), so erhalten wir eine Scheibe T_ε, und in dieser Scheibe liegt sicher ein Punkt $x(\varepsilon)$, welcher nicht in L liegt, also auch von keiner Scheibe in L überdeckt liegt. Nun soll doch die Menge aller Scheiben S die Ebene lückenlos überdecken. Es muß also eine Scheibe $S(\varepsilon)$ mit Mittelpunkt $s(\varepsilon)$ den Punkt $x(\varepsilon)$ überdecken. Wenn wir nun ε eine Nullfolge durchlaufen lassen, so besitzt die Menge der Punkte $x(\varepsilon)$ sicher einen Häufungspunkt $\bar{x}$, der zu S_o gehört. Die Gitterpunkte $s(\varepsilon)$ bilden aber eine diskrete Menge, also muß für unendlich viele ε der Punkt $s(\varepsilon)$ der gleiche

Punkt $\bar{s}$ sein. Dann liegt wegen der Abgeschlossenheit von S der Punkt $\bar{x}$ auf S, also hätten wir eine Scheibe S, welche mit S_o einen Punkt $\bar{x}$ gemeinsam hätte, deren Mittelpunkt $\bar{s}$ aber nicht auf l liegt. Wir denken uns die Halbstrahlen $l_1, l_2, \ldots l_m$ so numeriert, daß l_{i+1} der nächste Halbstrahl nach l_i ist, wenn o im entgegengesetzten Uhrzeigersinn durchlaufen wird ($l_{m+1} = l_1$). (Die Halbstrahlen liegen ja in einer Ebene!) Bei dieser Numerierung ist also $k_{n+1} = k_1$ usw. Die Punkte t, welche auf einem Halbstrahl l_i liegen, sind äquidistant. Ist nämlich $\bar{t}$ der Punkt auf l_i, welcher o am nächsten liegt und ist t ein weiterer Punkt auf l_i auf S_o, also $t = \lambda\bar{t}$ ($\lambda > o$), so sei $\lambda = l + \rho$ ($l = [\lambda]$, $o \leqq \rho < l$). Dann liegt auch $l\bar{t}$ in S_o und damit auch $t^* = t - l\bar{t} = \rho\bar{t}$ in S_o und auf l_i. Wäre nun $\rho > o$, so würde t^* näher an t liegen, was nicht sein soll. Es liegen also auf jedem Halbstrahl l_i die Punkte $j\bar{t}_i$ ($j = 1, 2, \ldots k_i$). Es liegen nun die Dreiecke mit den Endpunkten o, $k_i t_i$, $k_{i+1} t_{i+1}$ mit einem Flächeninhalt $\geqq \frac{1}{2} k_i k_{i+1} D/4$ in S_o, und da diese Dreiecke disjunkt sind, so ist also

$$\frac{1}{8} D \sum_{i=1}^{2n} k_i k_{i+1} = \frac{1}{4} D \sum_{i=1}^{n} k_i k_{i+1} \leqq F, \text{ also } (k_{n+1} = k_1) \sum_{i=1}^{n} k_i k_{i+1} \leqq 4q$$

Nun ist stets $\sum_{i=1}^{n} (k_i - 1)(k_{i+1} - 1) \geqq o$ (die k_i sind ja natürliche Zahlen), also $\sum_{i=1}^{n} k_i k_{i+1} \geqq 2 \sum_{i=1}^{n} k_i - n$ also

$$A \leqq 4q + n$$

Damit ist (3) bewiesen. Nun ist $n \leqq \sum_{i=1}^{n} k_i = \frac{1}{2} A$, also folgt sofort (1).

Es sei nun k das Maximum vom $k_1, \ldots, k_n$, dann ist $\sum_{i=1}^{n} k_i \geqq n - 1 + k$, also

$$\frac{A}{2} + k - 1 \leqq 4q \tag{4}$$

Nehmen wir an, es wäre $A \leqq 4C^2 - 2$ [wo C bei (2) definiert ist], dann ist, da ja $4C + 1 \leqq (32q + 17)^{1/2}$, $2C^2 - 1 \geqq 4q - C + 1$, also (2) richtig. Ist dies nicht der Fall, so muß die gerade Zahl A sogar $\geqq 4C^2$ sein. Unter den $A+1$-Punkten u, bestehend aus den A-Punkten t und o, muß es mindestens ein Paar u, u' geben, so daß $\bar{t} = \frac{1}{2} C (u - u')$ ein Gitterpunkt aus G_1 ist, denn es gibt modulo $2\,C$ nur $(2\,C)^2$ verschiedene Restklassen für die Gitterpunkte, und wir haben aber mindestens $4\,C^2+1$ solche. Mit $\bar{t}$ sind auch $1.\bar{t}, 2.\bar{t}, \ldots C\bar{t}$ Gitterpunkte und sie liegen alle in S_o, da ja $C\bar{t} = \frac{1}{2}(u - u')$ in S_o liegt. Es ist also $k \geqq C$, also folgt aus (4) wieder (2). Damit ist alles gezeigt.

Schlußbemerkung: Das analoge Problem im Raum, allgemeiner im R_n kann ganz genau so behandelt werden, abgesehen von der Bemerkung über die Anordnung der Halbstrahlen. Zieht man statt dessen einen Satz heran, welchen ich früher bewiesen habe („Über Gitterpunkte in Parallelepipeden", Journal f. d. r. u. a. Mathematik 87 [1950], Hilfssatz 1), so erhält man $2^n n!\, q \geqq A - n + 1$, und da $q \geqq 1$

$$A \leqq (2^n n! + n - 1)\, q,$$

wo $q = V/D$, V = Volumen der konvexen Körper, D = Mascheninhalt des Gitters. Für $n = 3$ erhält man so $A \leqq 50\, q$. Die Schranken sind aber nicht scharf.

Zum Hauptsatz der Theorie der Gleichverteilung

Anz. Österr. Akad. Wiss., Math.-Naturwiss. Kl. **94** (1957) 313–317

Eine Folge $\omega = (x_m)$ von reellen Zahlen heißt gleichverteilt mod 1 zum Verteilungsmaß ν, wenn für jede stetige periodische Funktion $f(x)$ mit Periode 1

$$\lim_{m \to \infty} \lambda_m (\omega, f) = \lim_{m \leftarrow \infty} \frac{1}{m} \left[f(x_1) + \ldots + f(x_m) \right] = \nu(f) = \int_G f \, d\nu (x) \qquad (1)$$

ist (G: Intervall $0 \leqq x \leqq 1$). Die Folge ω heißt schlechtweg gleichverteilt mod 1, wenn

$$\nu = \mu (f) = \int_G f (x) \, dx$$

ist. Der Hauptsatz besagt nun, daß ω gleichverteilt ist, wenn für jede natürliche Zahl h die Folge

$$\omega_h = (x_{m+h} - x_m)$$

gleichverteilt ist.

Es sei nun eine Folge $[\sigma_m(f)]$ von Verteilungsmaßen gegeben, dann sei für jedes natürliche h

$$\nu_h (f) = \int_G \ldots \int_G f(x_1 + \ldots + x_h) \, d\sigma_1 (x_1) \ldots d\sigma_h (x_h) \qquad (2)$$

Die ν_m sind wieder Verteilungsfunktionen. Ist für jedes natürliche m $D = D_m(x) = exp\ (2\pi\ imx)$, dann behaupten wir folgende Verallgemeinerung des Hauptsatzes:

Satz 1: Die Folge ω ist gleichverteilt mod 1, wenn für jedes natürliche h die Folge ω_h gleichverteilt mod 1 zum Verteilungsmaß ν_h ist und für jedes m

$$\prod_{j=1}^{\infty} \sigma_j(D_m) = 0 \tag{3}$$

ist.

Beweis: Nach der Fundamentalgleichung von van der Corput ist für jedes natürliche m und q ($1 \leqq q \leqq m$) und für beliebige komplexe Zahlen u_j ($1 \leqq j \leqq m$)

$$q^2 \Big|\sum_{j=1}^{m} u_j\Big|^2 \leqq (m+q-1)\Big(q\sum_{j=1}^{m} |u_j|^2 + 2R\Big), \tag{4}$$

wo

$$\mathrm{R} = \text{Realteil} \sum_{h=1}^{q-1}(q-h)\sum_{j=1}^{m-h}(u_j \mid u_{j+h}), \quad \left[(u \mid v) = u\overline{v}\right].$$

Wir setzen nun in (4) $u_j = D(x_j)$, dann folgt

$$|\lambda_m(\omega, D)|^2 \leqq (1+q/m)\Big[1/q + 2\sum_{h=1}^{q-1}(q-h)\,|\lambda_{m-h}(\omega_h, D)|\Big].$$

Nun soll nach Voraussetzung $\lim \lambda_m(\omega_h, D) = \nu_h(D)$ sein. Wir erhalten also in (5) für $m \to \infty$

$$\overline{\lim}\, |\lambda_m(\omega, D)|^2 \leqq 1/q + (2/q^2)\sum_{h=1}^{q-1}(q-h)\,|\nu_h(D)|. \tag{6}$$

Dabei ist nach (2), da $D(x_1 + \ldots + x_h) = D(x_1)\ldots D(x_h)$

$$\nu_h(D) = \sigma_1(D)\ldots\sigma_h(D).$$

Also ist nach (3) $\nu_h(D)$ eine Nullfolge, daher strebt die rechte Seite von (6) für $q \to \infty$ gegen 0 und damit ist alles gezeigt.

Bemerkung 1: Aus (3) folgt, daß $\nu_h(f)$ gegen $\mu(f)$ strebt. Man kann unter den gleichen Voraussetzungen wie in Satz 1 zeigen, daß mit ω auch alle Folgen (x_{km+l}) für jedes natürliche k, l gleichverteilt sind.

Satz 1 gestattet auch eine quantitative Verschärfung für die Diskrepanzen und eine Verallgemeinerung auf gewisse Mittelwerte

$$\lambda_m = \left[\gamma_1 f(x_1) + \ldots + \gamma_m f(x_m)\right] / (\gamma_1 + \ldots + \gamma_m).$$

Ein wichtiger Spezialfall von Satz 1 entsteht, wenn die σ_j die Gestalt $\int_G f p_j\, dx$ mit nicht negativem p_j haben. Die ν_h sind dann ebenfalls von der Gestalt $\int_G f t_h\, dx$, wo

$$t_h(x) = \int\limits_G p_h(z)\, t_{h-1}(x-z)\, dz = \int\limits_B \ldots \int p_1(x_1) \ldots p_h(x_h)\, dx_1 \ldots dx_h$$

$$(B\colon\ 0 \leqq x_1 + \ldots + x_h \leqq x,\ 0 \leqq x_i \leqq 1,\ 1 \leqq i \leqq h).$$

Ist $J_j = \inf\limits_G p_j$, dann ist (3) erfüllt, wenn

$$\Sigma J_j = \infty$$

Beweis: Es ist $|\sigma_j(D)| = |\int\limits_G p_j D\, dx| = |\int\limits_G (p_j - J_j) D\, dx|$

$$\leqq \int\limits_G (p_j - J_j)\, dx = 1 - J_j$$

also wegen (4) gilt (3).

Tieferliegend ist

$$p_1(x) = p_2(x) = \ldots = p(x) \tag{5}$$

in diesem Fall gilt stets (3), auch wenn $J = 0$.

Beweis: Es ist $|\sigma(D_m)| \leqq 1$. Wäre nun $|\sigma(D_m)| = 1$, also $\sigma(D_m) = \exp(2\pi i\lambda)$, so wäre

$$\int\limits_G [1 - \cos 2\pi m (x - \lambda)]\, \rho(x)\, dx = 0.$$

Es gibt nun sicher ein Intervall U, wo $\rho(x) > 0$ ist. In U müßte nun die Klammer in (4) stets 0 sein, also $\exp[2\pi i m (x - \lambda)] = 1$ sein, d. h. $D_m(x)$ konstant, was nicht geht. Es sind also $|\sigma(D_m)| < 1$, und da alle σ_j einander gleich sind, so folgt (3).

Wir haben bis jetzt den eindimensionalen Fall betrachtet, aber alles gilt auch für beliebig viele Dimensionen. Wie behandeln gleich den allgemeinen Fall von Folgen $\omega = (x_m)$ aus einer beliebigen kompakten Gruppe G mit abzählbarer Basis. Die ν sind dann Verteilungsmaße über G, μ ist das normierte Haarsche Maß. Für die $D = D_m$ sind dann die nicht trivialen irreduziblen Klassen unitärer Darstellungen von G zu nehmen. In (2) ist dann natürlich die Funktion $f(x_1.x_2 \ldots .x_h)$ zu nehmen. Satz 1 gilt dann ungeändert, wo jetzt $\omega_h = (x_{m+h}\, x_m^{-1})$. Die Bedingung (3) hängt nur von der Klasse in D ab, denn betrachten wir statt $D(x)$, $\tilde{D}(x) = U^* D(x) U$ (U unitäre Matrix unabhängig von x), so ist

$$\sigma(\tilde{D}) = U^* \sigma(D)\, U$$

und für jedes Teilprodukt haben wir

$$\widetilde{\Pi}_k = \prod_{j=1}^{k} \sigma_j(\widetilde{D}) = U^* \prod_{j=1}^{k} \sigma_j(D)\, U = U^* \Pi_k U$$

Geht nun Π_k gegen 0 (Nullmatrix), so auch $\widetilde{\Pi}_k$, denn wir haben, wenn $\widetilde{\Pi}_{ij}^{(k)}$ ein Matrixelement von $\widetilde{\Pi}_k$, stets

$$\widetilde{\Pi}_{ij}^{(k)} = \sum_{\iota, m} \Pi_{\iota m}^{(k)}\, \bar{u}_{\iota i}\, u_{mj}$$

Der Beweis von Satz 1 bleibt ungeändert, da (4) auch für Matrizen u_j gilt, dabei bedeutet jetzt

$$(u \mid v) = \operatorname{Spur} \bar{v}^{\top} u, \quad |u|^2 = (u \mid u)$$

Bemerkung 2: Aus (3) folgt wieder, daß die ν_h gegen das Haarsche Maß μ streben.

Betrachten wir jetzt die Spezialfälle (4) und (5), also $\nu_j = \int\limits_G f p_j \, d\mu$.

$$\text{Es ist } \sigma_j(D) = \int\limits_G [p_j(x) - J_j]\, D\, d\mu$$

also (D sei r zeilig)

$$\Pi_k = \int\limits_G \dots \int\limits_G [p_1(x_1) - J_1] \dots [p_k(x_k) - J_k]\, D(x_1) \dots$$

$$\dots D(x_k)\, d\mu(x_1) \dots d\mu(x_k)$$

daher $|\Pi_k| \leqq \sqrt{r} \prod_{j=1}^{k} (1 - J_j)$ und daraus folgt (3).

Für den Fall (5) muß ich annehmen, daß G zusammenhängend ist. Wir können annehmen, daß $\sigma(D) = S$ eine Dreiecksmatrix (s_{ij}) ($s_{ij} = 0$ für $i > j$) ist, denn sonst wählen wir eine unitäre Matrix U, sodaß für $\widetilde{D} = U^*DU$, $\sigma(\widetilde{D}) = U^*SU$ eine solche Matrix wird. Dann haben alle Elemente von S einen Absolutbetrag < 1. Es ist nämlich $|s_{ij}| = |\int\limits_G d_{ij}(x)\, p\, d\mu| \leqq 1$, denn es ist ja D unitär. Gäbe es nun ein s_{ij} von der Gestalt $e^{i\lambda}$, so wäre

$$\int\limits_G (1 - d_{ij} e^{-i\lambda})\, p dx = 0\,,$$

also auch

$$\int_G \left[1 - \mathrm{Real}\,(d_{ij}e^{-i\lambda})\right] p\,dx = 0\ .$$

Es gibt nun sicher eine offene Menge U, wo $\rho > 0$, dann muß wie früher der Realteil in U gleich 1 sein, also $d_{ij} = e^{i\lambda}$, also der $\sum_{k=i}^{r} |d_{ik}|^2 = 1$, $d_{ik} = 0$ für alle $k \neq j$ und natürlich auch $d_{kj} = 0$ für alle $k \neq i$.

Es würde also für alle x aus U die Darstellung $D(x)$ reduzibel sein und da G zusammenhängend, würde dies in ganz G gelten, was nicht möglich ist.

Es sind also alle Elemente in S von einem Betrag $\alpha < 1$. Dann gilt für die Elemente $s_{ij}^{(k)}$ von S^k die Abschätzung

$$|s_{ij}^{(k)}| \leqq \binom{j-i+k-1}{k-1} \alpha^k \;(i \leqq j),\quad s_{ij}^{(k)} = 0\;(j > i) \tag{7}$$

(7) ist für $k = 1$ richtig. Nun ist $S^{k+1} = S^k S$, also

1 $$s_{ij}^{(k+1)} = \sum_{\iota=1}^{r} \alpha_{i\iota}^{(k)}\, \alpha_{\iota j}$$

also $s_{ij}^{(k+1)} = 0$ für $j > i$ und für $i \leqq j$

$$s_{ij}^{(k+1)} \leqq \alpha^{k+1} \sum_{\iota=i}^{j} \binom{\iota-i+k-1}{k-1} = \alpha^{k+1} \sum_{m=0}^{j-i} \binom{m+k-1}{k-1} = \alpha^{k+1} \binom{j-i+k}{k}$$

w. ztw.

Aus (7) folgt weiter, da $1 \leqq i \leqq j \leqq r$, für $i < j$

$$|s_{ij}^{(k)}| \leqq \binom{j-i+k-1}{j-i} \alpha^k < (r+k-1)^{r-1} \alpha^k$$

2 und dies gilt auch noch für $i = j$. Da $\alpha > 1$, so folgt für h, daß $\lim_k S^k = 0$ w. zw.

Die „Bemerkung 2“ wurde für den Spezialfall (5) von Ya. T. Rivkind 1955 bewiesen (nach Math. Review 18, S. 680; die Arbeit war dem Verfasser nicht zugänglich). Bemerkung 1 wurde von P. Levy, Bull. Soc. 67 (1939), gezeigt.

Corrections

5.1 $= \sum_{\iota=1}^{r} s_{i\iota}^{(k)} s_{\iota j}$ 5.2 Da $\alpha < 1$, so folgt für $k \to \infty$, daß

Erbliche Eigenschaften in der Theorie der Gleichverteilung

Herrn Professor Dr. O. Varga zum 50. Geburtstag gewidmet

Publ. Math. 7 (1960) 181–186

§ 1. Eine Folge $\langle x_k \rangle$ reeller Zahlen $x_1, x_2, \ldots$ heißt bekanntlich gleichverteilt modulo 1, wenn für jede stetige periodische Funktion $f(x)$ mit der Periode 1

$$\lim_{n\to\infty} \frac{1}{n} \sum_{k=1}^{n} f(x_k) = \int_0^1 f(x)\,dx.$$

Dies ist nach H. WEYL genau dann der Fall, wenn für jede ganze Zahl $l \neq 0$

$$(1) \qquad \lim_{n\to\infty} \frac{1}{n} \sum_{k=1}^{n} e(l x_k) = 0 \qquad (e(x) = e^{2\pi i x}).$$

Ist $\langle \mathfrak{x}_k \rangle$ eine Folge von reellen Vektoren im s-dimensionalen Raum $R_s (s \geqq 1)$, so ist analog die Folge genau dann gleichverteilt modulo 1, wenn für jeden Gittervektor $\mathfrak{l} = (l_1, \ldots, l_s) \neq \mathfrak{o} = (o, \ldots o)$

$$(1') \qquad \lim_{n\to\infty} \frac{1}{n} \sum_{k=1}^{n} e(\mathfrak{l}\mathfrak{x}_k) = 0.$$

Daraus folgt, daß $\langle \mathfrak{x}_k \rangle$ genau dann gleichverteilt ist, wenn für jedes $\mathfrak{l} \neq \mathfrak{o}$ die Zahlenfolge $\langle \mathfrak{l}\mathfrak{x}_k \rangle$ modulo 1 gleichverteilt ist. VAN DER CORPUT[1]) hat gezeigt: Ist für jede natürliche Zahl h die Folge $\langle \mathfrak{y}_k^{(h)} \rangle = \langle \mathfrak{x}_{k+h} - \mathfrak{x}_k \rangle$ gleichverteilt, so ist auch $\langle \mathfrak{x}_k \rangle$ gleichverteilt. Der Beweis beruht, wenn wir uns auf Zahlenfolgen beschränken, auf der Fundamentalungleichung

$$(2) \qquad \left| \frac{1}{n} \sum_{k=1}^{n} e(l x_k) \right|^2 \leqq 2q^{-1} \left(1 + 2q^{-1} \sum_{h=1}^{q-1} (q-h) \left| \frac{1}{n-h} \sum_{j=1}^{n-h} e(l y_j^{(h)}) \right| \right)$$

(q natürliche Zahl beliebig mit $1 \leqq q \leqq n$).

---

[1]) *Acta Math.* **56** (1931), 373—456.

Ist $\langle \mathfrak{x}_k \rangle$ gleichverteilt, so muß $\langle \mathfrak{y}_k^{(h)} \rangle$ nicht gleichverteilt sein (Beispiel $\langle \alpha k \rangle$, α irrational).

N. M. Korobov und A. G. Postnikow[2]) haben bemerkt, daß aus der Gleichverteilung von $\langle \mathfrak{y}_k^{(h)} \rangle$ für jedes h folgt, daß sogar für jedes nichtnegative ganze s und $r (s \geqq 1)$ die Folge $\langle \mathfrak{x}_{ks+r} \rangle$ gleichverteilt ist. Dies sieht man am besten so: Es ist

$$\sum = \sum_{k=1}^{n} e(\mathfrak{l}\mathfrak{x}_{ks+r}) = \frac{1}{s} \sum_{j=1}^{s} \sum_{t=s+1}^{sn+r} e(\mathfrak{l}\mathfrak{x}_t + j(t-r)/s) \tag{3}$$

also

$$\left| \sum \right| \leqq \operatorname*{Max}_{1 \leqq j \leqq s} \left| \sum_{k=s+1}^{sn+r} e(\mathfrak{l}\mathfrak{x}_k + jk/s) \right| .$$

Die Folge $\langle z_k \rangle = \langle \mathfrak{l}\mathfrak{x}_k + jk/s \rangle$ ist gleichverteilt für jedes j, da die Folge $\langle z_k^{(h)} \rangle = \langle \mathfrak{l}\mathfrak{y}_k^{(h)} + jh/s \rangle$ gleichverteilt ist, da ja nach Voraussetzung für jedes $h \langle \mathfrak{y}_k^{(h)} \rangle$ gleichverteilt sein soll und eine additive Konstante nicht stört. Wir nennen nun eine Eigenschaft E von Folgen $\langle x_k \rangle$ erblich, wenn folgendes gilt: Besitzt für jede natürliche Zahl h die Folge $\langle y_k^{(h)} \rangle = \langle x_{k+h} - x_k \rangle$ die Eigenschaft E, dann gilt sie auch für $\langle x_k \rangle$ und alle Teilfolgen $\langle x_{ks+r} \rangle$. Diese Definition kann auch für Folgen $\langle \mathfrak{x}_k \rangle$ von Vektoren und allgemeiner für Folgen $\langle x_k \rangle$ von Gruppenelementen x_k aus einer Gruppe G formuliert werden, nur ist dann $y_k^{(h)} = x_{k+h} x_k^{-1}$. Man zeigt[3]) z. B. leicht, daß auf jeder kompakten Gruppe mit abzählbarer Basis die Gleichverteilung ebenfalls erblich ist.

Es sollen nun einige weitere Begriffe aus der Theorie der Gleichverteilung auf ihre Erblichkeit untersucht werden, wir wollen uns aber dabei auf Zahlenfolgen beschränken.

§ 2. Eine Folge $\langle x_k \rangle$ heißt gleichmäßig-gleichverteilt, kurz g-g-Folge,
1 wenn gleichmäßig in $p (p = 0, 1, 2, \ldots)$ $\lim\limits_{n \to \infty} \frac{1}{n} \sum\limits_{k=1}^{n} e(x_{k+p}) = 0$ ist. Ein Beispiel dafür ist $\langle k\alpha \rangle$ (α irrational). Diese Eigenschaft ist wieder erblich[3]). Daraus folgt, daß $\langle P(k) \rangle$ (P Polynom vom Grad $\geqq 1$, der Koeffizient der höchsten Potenz von k irrational) g-g-Folge ist. Bei den g-g-Folgen gilt, (darauf beruht die Bedeutung des Begriffes): Ist $\langle x_k \rangle$ eine g-g-Folge und $\langle x_k' \rangle$ eine Folge mit

$$\lim_{k \to \infty} (x_{k+1}' - x_k' - (x_{k+1} - x_k)) = 0$$

(dies ist sicher der Fall, wenn $\lim\limits_{k \to \infty} (x_k' - x_k) = 0$) dann ist auch $\langle x_k' \rangle$ eine g-g-Folge, also auch gleichverteilt. Die gleiche Eigenschaft haben die s-g-g-

---

2) *Dokladi Acad. Nauk SSSR* **84** (1952), 217—20.

3) *Rendiconti Circ. Mat. Palermo* **4** (2) (1955), 1—15.

Folgen. Darunter verstehe ich jene Folgen $\langle x_k\rangle$, für die[4])

2 (4) $$\lim_{n\to\infty}\overline{\lim_{H\to\infty}}\frac{1}{H}\sum_{r=0}^{H}\left|\frac{1}{n}\sum_{k=1}^{n}e(lx_{rn+k})\right|=0.$$

Jede g-g-Folge ist eine s-g-g-Folge. Es soll nun gezeigt werden, daß auch diese letztere Eigenschaft erblich ist.

Es sei also $\langle y_k^{(h)}\rangle$ für jedes h eine s-g-g-Folge und es soll nun gezeigt werden, daß dann $\langle x_{ks+r}\rangle$ ebenfalls eine s-g-g-Folge ist. Es genügt, dies für $\langle x_k\rangle$ zu zeigen.

Mit $M_{r,n}(x)=\frac{1}{n}\sum_{k=1}^{n}e(lx_{k+rn})$ folgt aus (2)

$$\frac{1}{H}\sum_{r=0}^{H}|M_{r,n}(x)|^2\leqq 2q^{-1}(1+2q^{-1}\sum_{h=1}^{q-1}(q-h)H^{-1}\sum_{r=0}^{H}|M_{r,n-h}(y^{(h)})|$$

also mit $n\to\infty$

$$\overline{\lim_{n\to\infty}}\ \overline{\lim_{H\to\infty}}\ H^{-1}\sum_{r=0}^{H}|M_{r,n}(x)|^2\leqq 2q^{-1}$$

und da q beliebig, folgt die Behauptung, da ja

$$\left(H^{-1}\sum_{r=0}^{H}|M_{r,n}|\right)^2\leqq H^{-1}\sum_{r=0}^{H}|M_{r,n}|^2.$$

§ 3. Eine Folge $\langle x_k\rangle$ heiße eine v-n-g-Folge (gleichverteilt zur Stufe v) (v = natürliche Zahl), wenn die Folge $\langle\mathfrak{x}_{k,v}\rangle=\langle(x_k,x_{k+1},\dots,x_{k+v-1})\rangle$ im R_v gleichverteilt ist. Aus (1′) folgt sofort, daß eine solche Folge eine w-n-g-Folge für jedes $w\leqq v$ ist, also auch gleichverteilt ist. Beispiel $\langle\alpha k^v\rangle$ (α irrational). Ist nämlich $\mathfrak{l}=(l_1,\dots,l_v)\neq\mathfrak{o}$, so ist

$$\sum_{j=0}^{v-1}l_{j+1}(k+j)^v=\sum_{r=0}^{v}\binom{v}{r}c_r k^{v-r}$$

wo die $c_r=\sum_{j=0}^{v-1}l_{j+1}j^r$ $(r=0,1,\dots,v-1)$ nicht alle Null sein können, also ist $\mathfrak{l}\mathfrak{x}_{k,v}$ ein Polynom von einem Grad $\geqq 1$ mit irrationalen Koeffizienten, also
3 ist $\langle\mathfrak{x}_{k,v}\rangle$ gleichverteilt. Die Eigenschaft v-n-g-Folge ist natürlich erblich, aber jetzt gilt: Ist $\langle x_k\rangle$ eine v-n-g-Folge, so ist $\langle y_k^{(h)}\rangle$ für jedes $h<v$ eine $(v-h)$-n-g-Folge. Der Beweis ist einfach: Ist $w=v-h$ und $\mathfrak{l}_w=(l_1,\dots,l_w)\neq\mathfrak{o}$, so ist zu zeigen

(5) $$\lim_{n\to\infty}\frac{1}{n}\sum_{k=1}^{n}e(\mathfrak{l}_w\mathfrak{y}_{k,w}^{(h)})=0.$$

[4]) *Arch. Math.* **6** (1955), 115—20.

Wir setzen $\mathfrak{m}_v = (0, \ldots 0, l_1, \ldots, l_w)$, $\mathfrak{n}_v = (l_1, \ldots, l_w, 0, \ldots, 0)$ dann ist mit $\mathfrak{p}_v = \mathfrak{m}_v - \mathfrak{n}_v$ die Summe in (5) gleich $\sum_{k=1}^{n} e(\mathfrak{p}_v \mathfrak{x}_{k,r})$. Es ist $\mathfrak{p}_v \neq \mathfrak{o}$, denn nehmen wir an $\mathfrak{p}_v = \mathfrak{o}$, dann wäre, wenn $h > w$ für $j > h$, $p_j = l_{j-h} = 0$ also $\mathfrak{l} = \mathfrak{o}$. Ist aber $h \leqq w$, so wäre zunächst für $j \leqq h \leqq w$, $p_j = -l_j = 0$. Jetzt wendet man vollständige Induktion an. Ist nämlich $h < j \leqq w$, so folgt aus $p_j = l_{j-h} - l_j = 0$, $l_j = 0$, da $1 \leqq j - h < j$.

§ 4. Es liegt daher nahe, Folgen zu betrachten, welche für jedes v Folgen von der Stufe v sind. Wir nennen sie normal gleichverteilt oder kurz n-g-Folgen[5]). Diese Eigenschaft ist natürlich erblich, aber jetzt ist diese Eigenschaft auch notwendig, denn aus § 3 folgt: Ist $\langle x_k \rangle$ ein n-g-Folge, so auch $\langle y_k^{(h)} \rangle$ für jedes h (es kann also z. B. $\langle \alpha k^v \rangle$ für kein v eine n-g-Folge sein) eine n-g-Folge und daraus folgt nach § 1, daß auch $\langle x_{ks+r} \rangle$ für jedes s und r eine n-g-Folge ist. Es ist ja für jedes s, r, v die Folge $\langle \mathfrak{x}_{ks+r,v} \rangle$ gleichverteilt, also für jedes s, r mit $0 \leqq r < s$, $\langle \mathfrak{x}_{ks+r,s} \rangle$ gleichverteilt (Eigenschaft B) daher

$$\lim_{n \to \infty} \frac{1}{n} \sum_{k=1}^{n} \left(e \left(\sum_{j=1}^{v} l_j x_{(k+j-1)s+r} \right) \right) = \lim_{n \to \infty} \frac{1}{n} \sum_{k=1}^{n} e(\mathfrak{m} \mathfrak{x}_{ks+r,(v-1)s+1}) = 0$$

mit $\mathfrak{m}_{(j-1)s+1} = l_j$ $(j = 1, \ldots, v)$ und 0 sonst. Wir nennen jede Folge $\langle x_k \rangle$ mit Eigenschaft B eine B-normalgleichverteilte Folge, kurz B-n-g-Folge, weiter soll jede Folge $\langle x_k \rangle$, für die für jedes s, $\langle \mathfrak{x}_{ks,s} \rangle$ gleichverteilt ist, eine P-normalgleichverteilt Folge, kurz P-n-g-Folge genannt werden.

Wir haben also gezeigt, daß jede n-g-Folge eine B-n-g-Folge[6]), also auch eine P-n-g-Folge ist. Wir zeigen nun: Jede B-n-g-Folge ist eine n-g-Folge. Dies sieht man sofort. Ist nämlich $v \geqq 1$ und $\mathfrak{l}_v = (l_1, \ldots, l_v) \neq \mathfrak{o}$, dann ist

4 $$\sum_{k=1}^{n} e(\mathfrak{l}_v \mathfrak{x}_{k,v}) = \sum_{k=1}^{v-1} + \sum_{w=0}^{v-1} \sum_{t=1}^{[(N-j)/v]} e(\mathfrak{l} \mathfrak{x}_{tv+w,v})$$

und da nach Voraussetzung $\langle \mathfrak{x}_{tv+w,v} \rangle$ gleichverteilt ist, folgt alles. Wir werden

---

[5]) *Anzeiger d. math.-naturwits. Kl. d. Öster. Akad. d. Wiss.* (1957), 94—96.

Math. Nachr. **18** (1958), 188—202.

Hier werden gleich allgemeine Mittelbildungen und belie kompakte Räume mit abzählbarer Basis betrachtet.

International Congress of Math. Edinburgh 1958. S. 30. In Zeile 17 v. u. ist „eine solche Folge" durch „k-n-g-Folge" zu ersetzen.

Wie ich dem *Zentralblatt* **72**, 38 entnehme, hat schon N. M. Korobov n-g-Folgen eingeführt und konstruiert. *Uspecchi Matem. Nauk* (NS) **4** No. 1. (29) 189—90 (1949).

[6]) Dies entspricht bekannten Sätzen aus der Theorie der normalen Zahlen, vgl. J. Niven, Irrational Numbers, New York, 1956, Ch. 8. Die hier entwickelte Methode liefert auch den einfachsten Beweis dieser Sätze.

nun zeigen, daß jede P-n-g-Folge eine n-g-Folge, also auch eine B-n-g-Folge ist. Wir werden gleich mehr zeigen[7]):

Gibt es eine monoton wachsende Folge (s_j) *natürlicher Zahlen, so daß für alle* $s=s_j$ $\langle \mathfrak{x}_{ks,s}\rangle$ *gleichverteilt ist, dann ist* $\langle x_k\rangle$ *eine n-g-Folge, also eine B-n-g-Folge.*

Beweis. Es sei wieder v beliebig, $\varepsilon>0$ und $s=s_j$ so groß gewählt, daß $v/s<\varepsilon$, dann n so groß, daß $s/n<\varepsilon$, dann ist mit $s=s_j$, $\mathfrak{l}_v=(l_1,\dots,l_v)\neq\mathfrak{o}$

$$\sum_{k=1}^{n} e(\mathfrak{l}_v\mathfrak{x}_{k,v})=\sum_{k=1}^{s-1}+\sum_{r=0}^{s-1}\sum_{t=1}^{[(n-r)/s]} e(\mathfrak{l}\mathfrak{x}_{st+r,v})=\Sigma_1+\Sigma_2.$$

Es sei $\Sigma_2=\sum\limits_{r=0}^{s-v}+\sum\limits_{r=s-v+1}^{s-1}=\Sigma_3+\Sigma_4$. Es ist $|\Sigma_4|\leqq vn/s$. In Σ_3 führen wir die Vektoren $\mathfrak{m}_{r,s}=(m_{r,1},\dots,m_{r,s})$ mit $m_{r,u}=l_{u-r}$ für $r+1\leqq u\leqq r+v\leqq s$ und $m_{r,u}=0$ für die übrigen u ein. Dann haben wir

$$\Sigma_3=\sum_{r=0}^{s-v}\left(\sum_{t=1}^{[(n-r)/s]} e(\mathfrak{m}_{r,s}\mathfrak{x}_{st,s})\right).$$

Jeder Klammerausdruck kann nach Voraussetzung für genügend großes n sicher kleiner als $\varepsilon n/s$ gemacht werden, also erhalten wir

$$\left|\frac{1}{n}\sum_{k=1}^{n} e(\mathfrak{l}_v\mathfrak{x}_{k,v})\right|\leqq s/n+\varepsilon\sum_{r=0}^{s-v}1/s+v/s<3\varepsilon.$$

Zur Rechtfertigung des Namens „Normalgleichverteilung" zeigen wir noch, daß fast alle Folgen $\omega=\langle x_k\rangle$ mit $0\leqq x_k\leqq 1$ $(k=1,2,\dots)$ normalgleichverteilt sind (fast alle in bezug auf das Torusmaß der Folgen ω). Dazu genügt es, zu zeigen, daß fast alle Folgen P-n-g-Folgen sind. Nun ist für jedes $\mathfrak{l}_s$

$$\int_0^1\cdots\int_0^1\left|\frac{1}{n}\sum_{k=1}^{n} e(\mathfrak{l}_s\mathfrak{x}_{ks,s})\right|^2 dx_1\dots dx_{ns+s-1}=1/n$$

also für fast alle Folgen ω, $\lim\limits_{m\to\infty}\frac{1}{m^2}\sum\limits_{k=1}^{m^2} e(\mathfrak{l}_s\mathfrak{x}_{ks,s})=0$ und da stets jedes n zwischen einem m^2 und $(m+1)^2$ liegt, folgt alles. Es sei noch bemerkt, daß jede Folge $\omega=\langle x_k\rangle$ dicht in $(0,1)$ so umgeordnet werden kann, daß sie eine n-g-Folge ist.[8]) Nehmen wir nämlich eine P-n-g-Folge $\langle z_k\rangle$ her, dann gibt es zu jedem k ein x_k' aus ω, so daß $|x_k'-z_k|<1/k$, da ω dicht in $(0,1)$.

---

[7]) Für Normale Zahlen von C. T. Long gezeigt.

[8]) Vgl. auch die zweite Arbeit unter [5]).

Dabei seien die x_k' so gewählt, daß sie in ω verschiedene Indizes tragen. Dann ist also $\lim_{k\to\infty}(x_k'-z_k)=0$, also ist auch

$$\lim\,(e(\mathfrak{l}_s\mathfrak{x}'_{ks,s})-e(\mathfrak{l}_s\mathfrak{z}_{ks,s}))=0 \quad \text{für jedes } s \text{ und } \mathfrak{l}_s.$$

Da $\langle z_k\rangle$ eine n-g-Folge, so ist also auch $\langle x_k'\rangle$ eine n-g-Folge. Die übrigen Glieder aus ω und jene x_k' mit k-Quadrat denken wir uns in einer Folge $\langle t_k\rangle$ angeordnet. Wir setzen nun $u_k=x_k'$, wenn k kein Quadrat und $u_k=t_k$ sonst Die so entstehende Folge $\langle u_k\rangle$ ist eine Umordnung von ω und ist eine n-g-Folge, denn es sind die Glieder in der Summe $\frac{1}{n}\left(\sum_{k=1}^{n}(e(\mathfrak{l}_s\mathfrak{u}_{ks,s})-e(\mathfrak{l}_s\mathfrak{x}'_{ks,s})\right)$ höchstens für jene k ungleich 0, für die es ein $j\leqq s$ gibt, so daß $ks+j-1$ ein Quadrat ist. Die Anzahl ist aber höchstens $O(\sqrt{n})$ also ist $\langle u_k\rangle$ eine n-g-Folge.

(Eingegangen am 11. Februar 1959.)

Corrections

182.1 wenn für jedes $l\neq 0$ gleichmäßig in p ($p=0,1,2,\cdots$) $\lim_{n\to\infty}\frac{1}{n}\sum_{k=1}^{n}e(lx_{k+p})=0$ ist.

183.2 Die Summen über r lauten richtig $\sum_{r=1}^{H}$.

183.3 Statt "Die...gilt:" besser: "Ist für jedes h $\langle y_k^{(h)}\rangle$ eine v-n-g-Folge, dann ist $\langle x_{ks+r,v}\rangle$ für alle r,s gleichverteilt.

184.4 $\sum_{k=1}^{n}(l_vx_{k,v})=\sum_{k=1}^{v-1}(l_vx_{k,v})+\sum_{w=0}^{v-1}\sum_{t=1}^{[(n-w)/v]}e(l_vx_{tv+w,v})$

185.5 eine nicht beschränkte Folge

Funktionen von beschränkter Variation in der Theorie der Gleichverteilung

Enrico Bompiani zu seinem wissenschaftlichen Jubiläum

Ann. Mat. Pura Appl., IV. Ser., **54** (1961) 325–334

Zusammenfassung. - *Ist w eine im s-dimensionalen Einheitswürfel E definierte Funktion von beschränkter Variation, $x_1, \dots, x_n$ eine Folge von Punkten in E, dann wird der Betrag $\left|\frac{1}{n}(w(x_1) + \dots + w(x_n)) - \int_E w dx\right|$ durch die Diskrepanz der Folge nach oben abgeschätzt.*

Wir betrachten Punkte x mit den Koordinaten $p_j(x)$ $(j = 1, \dots, s)$ im s-dimensionalen Einheitswürfel $\bar{E} = \bar{E}^s$: $0 \leqq \xi_j \leqq 1 (j = 1, \dots s)$ Die Projektion von
1 x auf $\bar{E}^l_{i_1, \dots, i_l}(\xi)$: $0 \leqq \xi_i \leqq 1$ $(i = i_1, \dots, i_l)$, $\xi_i = \xi_i (i \neq i_j, j = 1, \dots l)$, $l \leqq s$, nennen wir $p^{(l)}(x) = p^{(l)}_{i_1, \dots, i_l}(x)$ Es sei nun ω: $x_1, x_2, \dots$, eine unendliche Folge von Punkten im E_s, $0 \leqq \xi_i < 1$ $(i = 1, \dots s)$ ω_n die endliche Folge $x_1, \dots, x_n (n \geqq 1)$. Es sei $Q = Q^s$: $\alpha_j \leqq \xi_j < \beta_j (j = 1, \dots s)$ ein beliebiges Intervall in E^s, $n'(Q)$ die Anzahl der Punkte von ω_n in Q dann ist $\nu_n(Q, \omega) = n'/n$ die relative Häufigkeit mit welcher die Punkte von ω_n in Q auftreten. Wir sagen nun, die Folge ω ist gleichverteilt, wenn für jedes Q, $\lim \nu_n$ existiert und

$$\lim_{n \to \infty} \nu_n(Q, \omega) = \mu(Q) \tag{1}$$

ist, wo $\mu(Q)$ das Volumen von Q ist.

Es sei $w(x)$ im Riemannschen Sinne integrierbar auf E

$$\mu(w) = \int_E w(x) dx \quad (dx = dp_1, \dots, dp_s), \quad m(w, \omega_n) = \frac{1}{n} \sum_{k=1}^{n} w(x_k)$$

dann hat H. Weyl gezeigt: Ist die Folge ω gleichverteilt, dann ist für jedes w

$$\lim_{n \to \infty} m(w, \omega_n) = \mu(w) \tag{2}$$

Für $w = \chi_Q(x)$ (charakteristische Funktion von Q) erhalten wir (1) zurück. Man nennt

$$D_n(\omega) = D(\omega_n) = \sup_{Q \in E} | \nu_n(Q, \omega) - \mu(Q) |$$

die Diskrepanz von ω_n und es ist bekannt, dass (1) genau dann gilt, wenn

$$\lim_{n \to \infty} D_n(\omega) = 0.$$

Es ist nun eine wichtige Aufgabe in der Theorie der Gleichverteilung für möglichst umfassende Klassen von Funktionen w die Differenz $m(w, \omega_n) - \mu(w)$ mittels der Diskrepanz $D(\omega_n)$ nach oben abzuschätzen. Im Falle $s = 1$ wurde dies von J. F. KOKSMA (1) durchgefürt und zwar unter der Voraussetzung dass w von beschränkter Variation ist, nach dem spezielle Funktionen bereits früher von H. BEHNKE (2) behandelt wurden. Hier soll nun der mehr-dimensionale Fall behandelt werden. Dazu muss zunächst auseinandergesetzt werden, was unter Funktionen von beschränkter Variation verstanden werden soll

Es sei $Z = Z^l(\xi)$ irgendeine Zerlegung von $\bar{E}^l(\xi) = \bar{E}^l_{i_1,\dots,i_l}$ $(l \leq s)$ in achsenparallele Intervalle q: $\alpha_j \leq \xi_j \leq \beta_j$ $(j = i_1, \dots, i_l)$, $\xi_i = \xi_i$ $(i \neq i_j)$. Ist dann $f(\xi)$ eine auf $\bar{E}^l$ definierte Funktion, dann sei

$$(I) \qquad V_l(\xi;\ f) = V_l(\bar{E}^l_{i_1,\dots,i_l}(\xi);\ f) = \sup_Z \sum_{q \in Z} |\Delta^{(l)}_q f|$$

erstreckt über alle Zerlegungen von $\bar{E}^l(\xi)$. Dabei ist

$$\Delta^{(l)}_q f(q) = \sum_{e \in q} (-1)^{\varepsilon(e)} f(e)$$

woalle Eckpunkte von q durchläuft und $\varepsilon(e)$ die Anzahl der α in den Koordinaten von e ist. Ist nun $V_l(f) < \infty$ dann heisst f von beschränkter Variation im Sinne von VITALI.

Es sei nun $l = s$ Jede Zerlegung Z von $\bar{E}^s$ bedingt eine Zerlegung Z^l aller $\bar{E}^l_{i_1,\dots,i_l}$ (0) für alle l mit $1 \leq l \leq s$. Ist dann $f(x)$ auf $\bar{E}^s$ und allen $\bar{E}^l$ von beschränkter Variation, dann heisst f auf E^s von beschränkter Variation im Sinne von HARDY und KRAUSE. Diese Definition der beschränkten Variation wollen wir zugrunde legen.

Wir wollen nun zeigen (3), $w(x)$ auf den ganzen R^s periodisch mit der Periode 1 fortgesetzt,

SATZ 1. - Ist $w(x)$ von beschränkter Variation, dann ist

$$(3) \qquad |m(w) - \mu(w)| \leq V(w) D(\omega_n)$$

(1) «Mathematica B», Zutphen 11, 7-11.

(2) Vgl. KOKSMA, *Diophantische Approximationen*, Kap. 7.

(3) Vorgetragen im Institute for Advanced Study November 1959.

wo

$$V(w) = \sum_{l=1}^{s} \sum_{i_1,\dots,i_l} V(E^l_{i_1,\dots,i_l}(1);\, w) \leq \sum_{l=1}^{s} \binom{s}{l} K_l$$

wo $K_l = \text{Max } V_l(w)$ über alle E^l.
Dabei ist ω_n: $x_1, \dots, x_n$ eine ganz beliebige Folge.

Zum Beweis von (3) wollen wir zunächst einige einfache Eigenschaften von

$$\Delta f = \sum_{e \in q} (-1)^{\varepsilon(e)} f(e)$$

feststellen, wo wir $s = l$ annehmen wollen. Für beliebige nicht negative ganze Zahlen mit $m + n = s$ ist E^s das cartesische Produkt von E^m und E^n, also q das Produkt von Intervallen q_1, q_2 aus E^m und E^n. Dann ist sofort klar, dass $\Delta_q f = \Delta_{q_1}(\Delta_{q_2} f) = \Delta_{q_2}(\Delta_{q_1} f)$, $f(x) = f(x_1, x_2)$ wo $x_1 \in E^m$, $x_2 \in E^n$, denn es ist

$$\sum_{e \in q} (-1)^{\varepsilon(e)} f(e) = \sum_{e_1, e_2} (-1)^{\varepsilon_1(e_1) + \varepsilon_2(e_2)} f(e_1, e_2).$$

Es liege nun eine Zerlegung $\tilde{Z}$ des ganzen s-dimensionalen R^s in achsenparallele Intervalle q: $\alpha_j \leq \xi_j \leq \beta_j$ vor, $a(x)$, $b(x)$ Funktionen auf R^s, welche ausserhalb eines Würfels verschwinden, dann ist

$$(4) \qquad \sum_q a(0(q)) \Delta_q b = (-1)^s \sum_q b(u(q)) \Delta_q a.$$

Dabei bezeichne stets $u(q)$ den untersten «Eckpunkt» $(\alpha_1, \dots, \alpha_s)$ ebenso $0(q)$ den obersten «Eckpunkt» $(\beta_1, \dots, \beta_s)$ von q.

Für $s = 1$ ist (4) richtig, denn ist $\tilde{Z}$ die Zerlegung des R^1 in Intervalle dann ist der obere Eckpunkt β gleich dem unteren Eckpunkt α' des nachfolgenden Intervalls und es ist die linke Seite von (4) (Abelsche Umformung)

$$\sum_q a(\beta)\, (b(\beta) - b(\alpha)) = -\sum_q b(\alpha)\, (a(\beta) - a(\alpha)).$$

Jetzt nehmen wir an, dass (4) im R^{s-1} $(\xi_2, \dots, \xi_s)$ bereits richtigist, dann kann man jedes Intervall q im R^s als Produkt von Intervallen q_1 in R^{s-1}, q_2 in R^1 auffassen und die linke Seite von (4) wird

$$\sum_{q_1, q_2} a(0(q_1), 0(q_2)) \Delta_{q_1}(\Delta_{q_2} b(q_1, q_2)) = (-1)^{s-1} \sum_{q_1, q_2} \Delta_{q_2} b\, u(q_1), q_2) \Delta_{q_1} a(q_1, 0(q_2)) =$$

$$= (-1)^s \sum_{q_1, q_2} b(u(q_1), u(q_2)) \Delta_{q_2}(\Delta_{q_1} a).$$

Es liege nun eine Zerlegung Z des $\bar{E}^s$: $0 \leq \xi_j \leq 1$ $(j = 1, \ldots, s)$ in Intervalle q: $\alpha_j \leqq \xi_j \leqq \beta_j$ vor. Die Zerlegung Z bewirkt auf
2 den $\bar{E}^{l\,(1)}_{i_1,\ldots,i_l}$: $0 \leq \xi_i \leq 1$ $(i = i_j)$, $\xi_i = 1$ $(i \neq i_j)$ Zerlegungen $\bar{Z}^l$ in Intervalle $\bar{q}^l$. Die Projektionen eines Punktes x in $\bar{E}^s$ auf einen $\bar{E}^{l\,(1)}_{i_1,\ldots,i_l}$ bezeichnen wir mit $\bar{p}^{(l)}(x) = \bar{p}^{(l)}_{i_1,\ldots,i_l}(x)$. Es seien nun $a(x)$, $b(x)$ Funktionen auf $\bar{E}^s$, $b(x) = 0$ wenn eine Koordinate von x Null ist. Dann gilt

$$\sum_Z a(0(q))\Delta_q b = \sum_{l=0}^{s} (-1)^l \sum_{\bar{E}^l(1)} \sum_{\bar{q}^l \in \bar{Z}^l} \bar{b}(u(\bar{q}))\Delta^{(l)}_{\bar{q}_l} a(\bar{q}^l) \tag{5}$$

wo rechts über alle $\binom{s}{l}$ Intervalle $\bar{E}^l$ (1) summiert ird.

Beweis: Wir erweitern Z zu einer Zerlegung $\tilde{Z}$ des R^s und definieren $a(x) = b(x) = 0$ wenn x nicht in $\bar{E}^s$ dann können wir (4) anwenden. Man braucht dabei nur jene Glieder in (4) berücksichtigen, welche von q in $\tilde{Z}$ stammen, welche entweder zu Z gehören oder mit $\bar{E}$ Randpunkte gemeinsam haben. Ist q ein Quader $\alpha_j \leqq \xi_i \leqq \beta_j$ wo $0(q)$ nicht in $\bar{E}^s$, dann ist $a(0(q)) = 0$. Wenn $0(q)$ in $\bar{E}^s$, aber nicht $u(q)$ dann ist ein $\beta_j = 0$; dann ist aber $b(q) = 0$ für alle Eckpunkte von q, also ist die Summe links in (4) gleich der linken Seite von (5). Nun betrachten wir die rechte Seite in (4)! Wenn $u(q)$ nicht in $\bar{E}$ so ist $b(u(q)) = 0$. Es liege jetzt $u(q)$ auf dem Rande von $\bar{E}$ also seien die Koordinaten von $u(q)$ für ein l: $0 \leq \alpha_i < 1$ $(i = i_1, \ldots, i_l)$, und sonst $\alpha_i = 1$ $(i \neq i_j)$. Die Projektion von q auf $\bar{E}^l$ ist dann das Intervall $\bar{q}^l$: $\alpha_i \leqq \xi_i < \beta_i$ $(i = i_1, \ldots, i_l)$, $\xi_i = 1$ $(i \neq i_j)$. In $\Delta_q a(q) = \Sigma (-1)^{\varepsilon(e)} a(e)$ sind alle Glieder Null, wo e nicht auf $\bar{E}^l$ liegt. Wenn aber e auf $\bar{E}^l$ liegt, dann ist $\varepsilon(e) = s - l + \varepsilon(\bar{e})$ wenn $e(\bar{q}^l) = \bar{e}$ ein Eckpunkt von $\bar{q}^l$ ist, denn e enthält ja dann $(s-l)$-mal die Koordinate $\alpha_i = 1$. Damit ist (5) bewiesen.

Nun kann der Beweis von (3) leicht geführt werden. Wir wollen noch für $m(w, \omega_n) = m(w)$ setzen. Wir denken uns die auf E^s definierte Funktion w auf den ganzen R^s periodisch mit der Periode 1 fortgesetzt, damit ist w auch auf $\bar{E}^s$ definiert und es besitzt also w auf einen $\bar{E}^l_{i_1,\ldots,i_l}(1)$ den gleichen Wert wie auf $E^{l\,(0)}_{i_1,\ldots,i_l}$ Es sei nun Z eine Zerlegung $\bar{E}^s$ inIntervalle q. Es sei

$$G(q) = \sup_q w \qquad g(q) = \inf_q w(x).$$

Ist $\overset{*}{\chi}_q(\xi)$ die charakteristische Funktion von $\overset{*}{q}$, $\alpha_j < \xi_j \leqq \beta_j$ dann ist für alle x in $0 < \xi_j \leqq 1$

$$\sum_q \chi_q(x) = 1$$

also ist

$$l(w) = m(w) - \mu(w) = l(w \sum_q \overset{*}{\chi}_q) = \sum_q l(w \overset{*}{\chi}_q).$$

Es ist $w(x) = w_1^*(x) - w_2^*(x) - (-1)^s w(0)$, wo die w^* monotou nicht ab nehmend in jedem q und $V(w) = V(w_1^*) + V(w_2^*)$, also genügt es (3) für ein w^* zu zeigen. Es ist $G^*(q) = w^*(o(q))$, $g^*(q) = w^*(u(q))$. Es sei

$$f_1(x) = \Sigma\, w^*(o(q))\, \chi_q^*(x),\ f_2(x) = \Sigma\, w^*(u(q))\, \chi_q^*(x).$$

Es ist

$$V(f_i) \leq V(w^*),\ l(f_i) = \Sigma\, f_i(o(q)) l(\chi_q^*).$$

Nun ist

$$m(f_1) \leq m(w^*) \leq m(f_2)$$

Dabei haben wir benützt, das $m(f)$, $\mu(f)$ linear homogene, nicht negative Funktionale in f sind, und dass $m(1) = \mu(1) = 1$, also für $0 \leq f \leq 1$, $m(f) \leq 1$. Es ist also, wenn wir die analoge Überlegung für μ machen

$$\text{(6)} \qquad | l(w^*) | \leq \operatorname*{Max}_i | l(f_i) | + \sum_q (G^*(q) - g^*(q)) \mu(\chi_q^*).$$

Es sein nun $z(x;\ \xi)$ bei festen ξ die Funktion $\chi_x(\xi)$, wo χ_x die charakteristische Funktion des Intervalls $0 \leq \xi_i \leq x_i$ $(i = 1, \dots, s)$ wenn alle $x_i > 0$ und $= 0$ wenn ein $x_i = 0$. Diese Funktion ist auf E^s definiert. Es ist, wenn $q \in Z$, $\Delta_q z = \chi_q(\xi)$. Dies ist für $s = 1$ richtig, denn $\chi_\beta(\xi) - \chi_\alpha(\xi) = \chi_q(\xi)$ $(q\colon \alpha < x \leq \beta)$ allgemein folgt es durch vollständige Induktion: Es ist ja, wenn q wieder ein Produkt von $q_1 \subset E^1$, $q_2 \subset E^{s-1}$ ist, $\Delta_q z(q_1, q_2) = \Delta_{q_1}(\Delta_{q_2} \chi_{x_1, x_2}) = \Delta_{q_1} \chi_{x_1, q_2} = \chi_{q_1, q_2}$ Es ist also

$$\text{(7)} \qquad \sum_q f_i(O(q)) l(\chi_q) = \sum_q f_i(O(q)) \Delta_q(l(\chi_x)).$$

Wir können jetzt (5) mit $a = f_i$, $b = l(\chi_x)$ anwenden und erhalten für die linke Seite von (7), (es ist $l(1) = 0$)

$$\text{(8)} \qquad \sum_{l=1}^{s} (-1)^l \sum_{\bar{E}^l} \sum_{q_l \subset \bar{Z}^l} l(\chi_{u(\bar{q}_l)}) \Delta_{ql} f_i(q^l).$$

Beachten wir nun, dass $l(\chi_{u(q)}) = m(\chi_u) - \mu(\chi_u)$ dem Betrage nach $\leq D(\omega_n)$ ist, so folgt sofort, dass (8) dem Betrage nach höchstens gleich der rechten Seite von (3) ist. Lassen wir jetzt Z eine ausgezeichnete Zerlegungsfolge durchlaufen, so folgt die Behauptung; denn $\Sigma\,(\overset{*}{G}(q) - \overset{*}{g}(q))\, \mu(\chi_q)$ geht dabei gegen 0, da w im Riemannschen Sinne integrierbar ist. Ordnen wir der Folge $\omega_n\colon x_1, \dots, x_n$ die Folge $\omega^l_{n(i_1, \dots, i_l)}\colon \bar{p}^l_{i_1, \dots, i_l}(x_j)$ $(j = 1, \dots, n)$ zu mit den Diskrepanzen $D^l_{i_1, \dots, i_l}$ für alle l dann sehen wir, dass wir die rechte Seite von (3) ersetzen können, durch die Schranke

$$\text{(3')} \qquad \sum_l \sum_{i_1, \dots, i_l} D^l_{i_1, \dots, i_l} V^{(l)}_{i_1, \dots, i_l}(w).$$

Nehmen wir an $w(x)$ in E^s besitze alle stetigen Ableitungen

$$(9) \qquad \frac{\partial^l w}{\partial x_{i_1}, \dots, \partial x_{i_l}}$$

($i_1, \dots, i_l$ alle voneinander verschieden), dann folgt sofort, dass (3″) gleich

$$(10) \qquad \sum_l \sum_{i_1,\dots,i_l} D^l_{i_1,\dots,i_l} \int \cdots \int_{E^l_{i_1,\dots,i_l}} \left| \frac{\partial^l w(p^l(x))}{\partial x_{i_1}, \dots, \partial x_{i_l}} \right| dp^l(x)$$

wo $dp^l(x) = d\xi_{i_1}, \dots, d\xi_{i_l}$ ist. Es sind die Formeln (3), (3′) auch dann anwendbar, wenn w nicht die Voraussetzungen von (9) erfüllt, z. B auf $w(x_1, \dots, x_n) = | x_1, \dots, x_n |$, da $| \Delta_q w(x) | \leq (\beta_1 - \alpha_1) \dots (\beta_s - \alpha_s)$ ist.

Wir haben beim Beweis der vorher abgeleiteten Formeln nur benützt, dass $m(w)$, $\mu(w)$ nicht negative, lineare und homogene Funktionale mit $m(1) = \mu(1)$ ist, μ total stetig mit beschräukter Dichte. Setzen wir also

$$D = \sup_Q | m(\chi_Q) - \mu(\chi_Q))$$

über alle Intervalle Q in E so bleibt alles richtig.

Ein wichtiger Spezialfall: Es sei $x(t)$ eine reelle integrierbare Funktion definiert für $0 \leq t \leq T$ mit Werten in E^s, dann sei

$$m(w) = \frac{1}{T} \int_0^T w(x(t)) dt, \ \mu(w) = \int_{E^s} w(\xi) d\xi$$

und wir kommen zum Fall der C-Gleichverteilung.

Es soll nun eine Anwendung von (3) gemacht werden: Es sei B eine quadrierbare Menge in E^s, w aber nun eine Funktion von beschränkter Variation auf B, d. h. es seien nun alle

$$V_l(B, w) = \sup_\xi V(B, w, \xi), \ V(B, w, \xi) = \sup \sum_{q \in Z^l(\xi)}^{*} | \Delta_q^{(l)}(w) |$$

beschränkt, wo jetzt nur die q von $Z^l(\xi)$ berücksichtigt werden, die ganz in B liegen. Dann soll $| m(w\chi_B) - \mu(w\chi_B) |$ abgeschätzt werden. Dabei setzen wir voraus, dass B und alle seine Projectionen $B^l_{i_1,\dots,i_l}$ von jeder Geraden parallel zu den Koordinatenachsen in höchstens h Intervallen geschnitten wird. Ist B konvex, so ist dies mit $h = 1$ erfüllt.

Es sei Z_0 eine Zerlegung von E^s, in Würfeln mit Kantenlänge

$$\rho = \frac{1}{r} (r \geq 1, \text{ natürliche Zahl}).$$

Es sei A die Vereinigungsmenge der Quader, welche ganz in int B liegen, C die Vereinigungsmenge der Quader, welche einen Punkt des Randes ∂B von B enthalten. Dann ist Funktion $v(x) = w(x)\chi_A(x)$ in E^s definiert. Wir wenden nun die früheren Uberlegungen auf $v(x)$ an, Dann erhalten wir (3) mit v statt w. Die Summe in (l) für $s = l$, wo$Z$ aus Z_0 durch Verfeinerung hervorgeht, zerlegt sich in 3 Teile, entsprechend den drei Klassen, in welche die q von Z zerfallen: 1. Die q welche nicht in $A \cup C$ liegen. Die zugehörigen Summanden sind Null. 2. Die q welche in A liegen. Der Betrag der Summe über jene q ist

$$\text{(3*)} \qquad \leq \sum_{q \in Z}^{*} | \Delta_q w(q) |$$

3. Die q welche in C liegen. Ist $G = \sup_B w$, dann ist die Summe über jene q

$$2^s GN(C) + 2h \sum_{l=1}^{s} \sum_{i_1, \ldots, i_l} V_{i_1, \ldots, i_l}(B, w)$$

wo N die Anzahl der Quader Q von Z_0 in C ist, also

$$\text{(11)} \qquad | m(v) - \mu(v) | \leq (4^s GN(C) + 2h \Sigma) D$$

wo Σ der Ausdruck $\sum_{l=1}^{s} \sum_{i_1, \ldots, i_l} V_{i_1, \ldots, i_l}(B, w)$.

Wir haben nun $m(w\chi_B) - m(v) \leq m(w(\chi_B - \chi_A)) \leq Gm(\chi_C)$
Nun ist $m(\chi_C) = m(\chi_C) - \mu(\chi_C) + \mu(\chi_C) \leq N(C)D + \mu(\chi_C)$
Weiter ist $\mu(w\chi_B) - \mu(v) \leq G\mu(\chi_C)$ also ist, da $\mu(\chi_C) = N(C)\rho^s$

$$| m(w\chi_B) - \mu(w\chi_B) | \leq GN(C)(5^s D + \rho^s) + 2hD\,\Sigma.$$

Es ist also nur notwendig noch $N(C)$ abzuschätzen.

Die Würfel von Z_0 haben die Gestalt

$$\frac{h_i}{r} \leqq \xi_i < \frac{h_i+1}{r} \quad (i=1,\dots,s)$$

wo $h_1,\dots,h_s$ nicht negative ganze Zahlen sind. Wir fügen zur Menge ∂B noch die Menge aller Punkte $\xi_i = \frac{h_i}{r}$ $(i \neq i_j)$, $\xi_i = x_i (i = i_1,\dots,i_l)$ hinzu, wenn die Menge: $\xi_i = x_i (i = i_j)$, $\frac{h_i}{r} \leqq \xi_i < \frac{h_i+1}{r} (i \neq i_j)$ einen Punkt von ∂B enthält wo $0 \leqq l \leqq s$, $i_1,\dots,i_l$ beliebig ($l=0$ die leere Menge). Die so entstandene Menge bezeichnen wir mit F. Sie und jede Projektion von F wird von jeder Geraden parallel zu den Achsen in höchstens $2h$ Intervallen geschnitten, denn zu jedem Intervall, welche von diesen Geraden aus der Projektion von ∂B auf diesen Geraden ausgeschnitten werden, kommt höchstens ein Punkt hinzu. Dann ist

$$N(C) = \sum_{h_1,\dots,h_s} \chi_F(h_1\rho,\dots,h_s\rho)$$

nach einer Formel von Davenport ([4])

$$|N(C) - l^s \mu(F)| \leqq \sum_{m=0}^{s-1} (2rh)^m \mu^l_{i_1,\dots,i_l}(F).$$

Es ist $\mu(F) = 0$, $\mu^{(l)}(F) = \mu^l(B^{(l)})$) also haben wir (dabei ist $\mu^0 = 1$ definiert)

(12) $$|m(w\chi_B) - \mu(w\chi_B)| \leqq G \sum_{m=0}^{s-1} (2rh) \mu^l_{i_1,\dots,i_l}(B)(5^s D + r^{-s}) + 2hD\Sigma.$$

Wir wählen jetzt $r = [D^{-\frac{1}{s}}]$ also, da $D < 1$, so ist die rechte Seite von (12)

(13) $$|m(w\chi_B) - \int_B w\,dx| \leqq 12\, G D^{\frac{1}{s}} \sum_{m=0}^{s-1} (2h)^m \mu^{(l)}_{i_1,\dots,i_l}(B) + 2hD \sum_{l=1}^{s} \sum_{i_1,\dots,i_l} V_{i_1,\dots,i_l}(B,w)$$

([4]) Journal London 26 (1951) 179-183.

und dies wollten wir zeigen. Berücksichtigt man noch, dass ja alle $\mu^l \leq 1$ so folgt

$$(14)\qquad | m(w\chi_B) - \int_B w dx | < (12h)^s (G + \sum_l \sum_{i_1,\ldots,i_l} V_{i_1,\ldots,i_l}(B, w)) D^{\frac{1}{s}}$$

Ein wichtiger Spezialfall ist $w = \chi_B$, dann ist $\Sigma = 0$ und wir erhalten

$$(15)\qquad | m(\chi_B) - \int_B dx | < (12h)^s D^{\frac{1}{s}}$$

und diese Schranke gilt gleichmässig für alle *B*, welche mit allen ihren Projektionen von jeder achsenparallelen Geraden in höchstens *h* Intervallen geschnitten werden.

In den abgeleiteten Formeln kann *m* ein beliebiges Funktional sein, dagegen muss $\mu(f) = \int f dx$ sein.

Corrections

325.1 $\xi_i = \xi$ 328.2 den $\overline{E}^{\,l}_{i_1,\ldots,i_l}(1)$:

Zur angenäherten Berechnung mehrfacher Integrale

Professor Paul Funk zum 75. Geburtstag gewidmet

Monatsh. Math. **66** (1962) 140–151

(Eingegangen am 23. März 1961)

Es sei $f(x_1, \ldots, x_s)$ eine im Riemannschen Sinne integrierbare komplexwertige Funktion auf dem Quader

$$Q = Q^s : a_j \leqq x_j \leqq b_j \qquad (j = 1, \ldots, s)$$

Es sollen Annäherungswerte für das Integral $\mu(f) = \int_Q f(x)\, dx$ aufgestellt werden. Es kann o. B. d. A. angenommen werden, daß

$$f(x) \text{ auf } x_j = b_j \qquad (1 \leqq j \leqq s)$$

die gleichen Werte annimmt wie für $x_j = a_j$. Sonst betrachten wir z. B. die Funktion

$$f^*(x) = f^*(x_1, \ldots, x_s) = f(b_1 - |\, x_1 - b_1 \,|, \ldots, b_s - |\, x_s - b_s \,|)$$

auf $Q^* : a_j \leqq x_j \leqq 2\, b_j - a_j \qquad (j = 1, \ldots, s)$

Es ist tatsächlich z. B.

$$f^*(2\, b_1 - a_1, x_2, \ldots, x_s) = f^*(a_1, x_2, \ldots, x_s)$$

Weiter ist $\int_{Q^*} f^*\, dx = 2^s \int_Q f\, dx$, da ja auf $Q : f^* = f$ und Q^* sich aus 2^s Quadern Q zusammensetzt. O. B. d. A. kann angenommen werden, daß Q der Quader $K : 0 \leqq x_j \leqq 1$ $(j = 1, \ldots, s)$ ist, denn sonst betrachten wir $f^*(t_1, \ldots, t_s) = f(a_1 + (b_1 - a_1)t, \ldots, a_s + (b_s - a_s)\, t)$ auf K und es ist $\int_K f^*(t)\, dt = \dfrac{1}{V(Q)} \int_Q f(x)\, dx$ ($V(Q)$ Volumen von Q). Es kann also stets angenommen werden, daß $f(x)$ auf K definiert ist und f auf den Seiten $x_j = 0$ des Quaders K den gleichen Wert annimmt, wie auf den Seiten $x_j = 1$. Es kann also f auf den ganzen R^s periodisch mit der Periode 1 fortgesetzt werden.

Jede Riemannsche Summe von $\mu(f)$ liefert natürlich einen Näherungswert für $\mu(f)$. Teilen wir z. B. die Kanten von K in n gleiche Teile ($n \geqq 1$), dann zerlegt sich K in $N = n^s$ Würfel,

$$\frac{k_j}{n} \leqq x_j < \frac{k_j + 1}{n} \qquad (j = 1, \ldots, s; 0 \leqq k_j \leqq n - 1)$$

und

$$\Sigma_N = \frac{1}{n^s} \sum_{k_1, \ldots, k_s}^{n-1} f\left(\frac{k_1}{n}, \ldots, \frac{k_s}{n}\right)$$

ist eine solche Riemannsche Summe und $\lim\limits_{N \to \infty} \Sigma_N = \mu(f)$. Es ist

$$\Delta_N^1(f) = | \Sigma_N - \mu(f) |$$

der Fehler und es sei $\delta_N^1 = \sup\limits_f \Delta_N^1(f)$.

Es ist δ_N^1 für großes N sicher nicht kleiner als $O\left(\frac{1}{n}\right) = O(N^{-\frac{1}{s}})$ und das gilt auch für $\sup\limits_{f \varepsilon B} \Delta_N^1$ wenn B die Klasse der Funktionen von beschränkter Variation auf K ist.

Einfaches Beispiel: $f(x) = x_1 \ldots x_s$ auf $0 \leqq x_j < 1 \quad (j = 1, \ldots, s)$ und 0 sonst.

Dann ist

$$\Sigma_N = \frac{1}{2^s}\left(1 - \frac{1}{n}\right)^s$$

also ist

$$\Delta_N^1 = 2^{-s} \left| \left(1 - \frac{1}{n}\right)^s - 1 \right| > \frac{2^{-s}}{n}$$

Schreiben wir die $N = n^s$ Punkte $\left(\frac{k_1}{n}, \ldots, \frac{k_s}{n}\right)$ als Folge $\Omega_N : \bar{x}_1, \ldots, \bar{x}_N$ so ist also

$$\Delta_N^1 = \left| \frac{1}{N} \sum_{k=1}^{N} f(\bar{x}_k) - \mu(f) \right| \tag{1}$$

Eine obere Abschätzung für Δ_n^1 wird später gegeben.

Für $s = 1$ gibt es eine Reihe von Quadraturformeln, z. B. die Simpsonsche Formel, welche für $\mu(f)$ gute Näherungsformeln liefert, jedenfalls dann, wenn f genügend oft differenzierbar ist. Für $s > 1$ existieren solche Formeln vor allem im Falle $s = 2$ oder 3, welche für großes s

praktisch unbrauchbar werden (abgesehen davon, daß hier keine vernünftigen Fehlerschätzungen bekannt sind), da die Anzahl N der zu berechnenden Punkte bei vorgeschriebener Genauigkeit exponentiell mit s wächst (vgl. z. B. Δ_N^1). Gerade solche Fälle haben sich in der letzten Zeit als sehr wichtig erwiesen. Heute verwendet man daher eine statistische Methode, die sogenannte Monte-Carlo-Methode. Es kann nämlich $\mu(f)$ als Erwartungswert $E(f)$ von f in Bezug auf die Wahrscheinlichkeitsdichte $\equiv 1$ auf K und 0 sonst im R^s aufgefaßt werden; die Verteilung p ist also im Sinne der Wahrscheinlichkeitstheorie gleichverteilt auf K. Sind nun $X_1, \ldots, X_N$ unabhängige Zufallsvariable mit der Verteilungsfunktion p, so ist $F = \frac{1}{N} \sum_{j=1}^{N} f(X_j)$ eine statistische Annäherung an $E(f)$, denn $E(F) = E(f) = \mu(f)$ und für die Streuung $\sigma_F{}^2$ von F gilt

$$\sigma_F{}^2 = E((F - E(F))^2) = \frac{1}{N} E((f - E(f))^2) = \frac{\sigma_f{}^2}{N} \tag{2}$$

Es hat also die statistische Näherung F eine Standardabweichung $\frac{\sigma_f}{\sqrt{N}}$, also nimmt die Wahrscheinlichkeitsverteilung des Fehlers wie $N^{-1/2}$ ab. Für die numerische Berechnung müssen nun die Stichprobenvariablen $X_1, \ldots, X_N$ durch Stichprobenwerte $x_1, \ldots, x_N$ ersetzt werden. Aus (2) kann man schließen, daß es sicher solche Stichprobenwerte gibt, so daß $\Delta_N = \left| \frac{1}{N} \sum_{j=1}^{N} f(x_j) - \mu(f) \right| \leqq \frac{\sigma_f}{\sqrt{N}}$ ist, aber diese Werte hängen von f ab und sind natürlich im allgemeinen unbekannt. Es ist bekannt, daß mit Wahrscheinlichkeit 1

$$\lim_{N\to\infty} \frac{1}{N} \sum_{j=1}^{N} f(x_j) = \mu(f)$$

ist, d. h. maßtheoretisch ausgedrückt folgendes:

Ist $P = \bigotimes_{i=1}^{\infty} K_i \; (K_i = K)$ der Raum aller Folgen $\omega = (x_1, x_2, \ldots)$ mit Elementen aus K, $d\tilde{\mu}$ das Produktmaß in P zum Lebesgueschen Maß auf K, dann ist für $\tilde{\mu}$ — fast alle Folgen ω eben $\lim \frac{1}{N} \sum_{j=1}^{N} f(x_j) = \mu(f)$ und man kann aus (2) sogar leicht schließen, daß sogar für fast alle Folgen $\Delta_N = 0(N^{-1/4} \log^2 N)$ und es gilt sogar noch mehr, nämlich $0(N^{-1/2} \log\log N)$.

Aber es kann natürlich passieren, daß eine irgendwie gewählte Folge $x_1, \ldots, x_N, \ldots$ von Stichprobenwerten gerade in einer Menge von Maß 0 in Bezug auf $\tilde{\mu}$ liegt. Es wurden verschiedene Methoden entwicjekt, um Pseudo-Zufallszahlen x_j aufzustellen, von denen zu hoffen ist, daß sie in der oben erwähnten Menge von Maß 1 liegt, aber es blieb ganz offen, ob dann mit diesen Werten $\frac{1}{N} \sum_{j=1}^{N} f(x_j)$ ein Näherungswert von $\mu(f)$ ist und wie groß der Fehler ist.

N. M. Korobow[1] hat nun eine feste Folge $x_1, \ldots, x_N$ angegeben, für die der zugehörige Fehler die Größenordnung des statistischen Fehlers der Monte-Carlo-Methode hat, und zwar für alle Funktionen f, welche 2 s-mal differenzierbar sind. Genauer lautet sein Satz so:

Ist $\sum_h C_h e^{2\pi i \langle h x\rangle}$ ($\langle h x\rangle$ skalares Produkt von h und x) die Fourierreihe von f, ist $\sigma = \sum |C_h| < \infty$, sind alle Ableitungen

$$\frac{\partial^{2r} f}{\partial x_{l_1}^2 \ldots \partial x_{l_r}^2} \quad (1 \leqq r \leqq s,\ 1 \leqq l_1 < l_2 < \ldots < l_r \leqq s)$$

stetig und dem Betrage nach $\leqq C$ und ist p eine Primzahl $> s$, dann ist

$$\left| \frac{1}{p^2-1} \sum_{k=1}^{p^2-1} f\left(\frac{k}{p^2}, \frac{k^2}{p^2}, \ldots, \frac{k^s}{p^2}\right) - \mu(f) \right| \leqq \frac{(s-1)\,\sigma}{\sqrt{p}} + \frac{s\,C}{10\,p} \tag{3}$$

An sich wäre statt $\frac{k^j}{p^2}$ zu schreiben $\frac{k^j}{p^2} - \left[\frac{k^j}{p^2}\right]$, da aber f periodisch mit der Periode 1 ist, so macht dies nichts aus. Eine analoge Abschätzung gilt für die Folge

$$\Omega = \left(\frac{k}{p}, \frac{k^2}{p}, \ldots, \frac{k^s}{p}\right) \bmod 1 \tag{4}$$

Beim Beweis von (3) müssen über f also Differenzierbarkeitsvoraussetzungen gemacht werden. Ich möchte nun zunächst zeigen, daß solche Abschätzungen bis auf logarithmische Faktoren gegeben werden können, unter der alleinigen Voraussetzung, daß f von beschränkter Variation ist; es kann also f auch unstetig sein. Dies erscheint mir gerade für $s > 1$ wertvoll. Dazu benütze ich folgenden Satz[2], welchen ich unlängst bewiesen habe:

---

[1] Dokladi CCCR, **115** (1957), 1062–65.

[2] Annali di Matematica (IV) **54**, 325–334 (1961).

Hilfssatz 1: Ist $f \in B$ (Funktion von beschränkter Variation auf K, $\omega = (x_1, \ldots, x_N)$ eine beliebige Folge über K

$$D_N(\omega) = \sup_{Q \subset K} \left| \frac{1}{N} \sum_{j=1}^{N} \chi_Q(x_j) - V(Q) \right|$$

(χ charakteristische Funktion des Quaders $Q \subset K$), dann ist

$$\left| \frac{1}{N} \sum_{j=1}^{N} f(x_j) - \mu(f) \right| \leqq V(f)\, D_N(\omega) \tag{5}$$

wo $V(f)$ die Variation von f auf K ist.

Besitzt f alle gemischten Ableitungen bis zur Ordnung s und sind diese stetig, dann ist $f \in B$ und die Schranke $V(f)$ in (5) kann ersetzt werden durch

$$\sum_{l} \int_{K^l} \left| \frac{\partial^l f}{\partial x_{i_1} \ldots \partial x_{i_l}} \right| d\, x_{i_1} \ldots d\, x_{i_l} \quad (1 \leqq i_1 < i_2 < \ldots < i_l \leqq s) \tag{6}$$

wo sich die Summe über alle l-Tupel $(i_1 < i_2 < \ldots < i_l)$ für $l = 1, \ldots, s$ erstreckt und K^l der Quader $0 \leqq x_j \leqq 1$ ist, mit $j = i_k$ $(k = 1, \ldots, l)$, $x_j = 0$ $(j \neq i_k)$ ist. Für die Anwendung von (5) muß $D_N(\omega)$ (sie ist stets $\leqq 1$) die sogenannte Diskrepanz abgeschätzt werden. Hier gilt nach *Erdös-Turan-Koksma*[3]:

Ist M eine beliebige natürliche Zahl $\geqq 300$, dann ist stets

$$D_N(\omega) < 2^s \frac{300}{M} + 30^s \sum_{0 < ||h|| \leq M} |S_N(h, \omega)|\, R^{-1}(h) \tag{7}$$

wobei

$$||h|| = \text{Max}\,(|h_1|, \ldots |h_s|),\ R(h) = \prod_{j=1}^{s} \text{Max}\,(|h_j|, 1),\ (h_1, \ldots h_s) \text{ ganz},$$

$S_N(h, \omega)$ die Weylsche Summe $\frac{1}{N} \sum_{k=1}^{N} e^{2\pi i \langle h x_k \rangle}$ ist.

Nehmen wir für ω die Folge Ω aus (4) mit $N = p - 1$, dann ist

$$S_N(h, \omega) = \frac{1}{p-1} \sum_{k=1}^{p-1} e^{2\pi i (h_1 k + \ldots + h_s k^s)/p}.$$

Aus der von *H. Hasse* für Geschlecht 1 und dann von *A. Weil* allgemein bewiesenen Vermutung für die Kongruenzzetafunktion folgt[4]

---

[3] *J. F. Koksma*, Math. Centrum Amsterdam, Scriptum **5**, 51 S. (1950).

[4] Vgl. z. B. Enzyklopädie d. Math. Wiss. 2. Aufl. I 2, 29 (L.K. HUA) 1959, B. G. Teubner.

$$|S_N| < \frac{s}{\sqrt{P}}$$

also wenn wir in (7) $M = p$ nehmen,

$$D_N(\omega) < \frac{300 \cdot 2^s}{p} + \frac{(60 \log p)^s}{\sqrt{p}} < \frac{(70 \log p)^s}{\sqrt{p}}$$

und wir erhalten also (für $p \leqq 300$ ist die Ungleichung in trivialer Weise richtig)

Satz 1: Ist f von beschränkter Variation auf $K^s (s \geqq 2)$ mit Variation $V(f)$, dann ist stets

$$\Delta_p = \left| \frac{1}{p-1} \sum_{k=1}^{p-1} f\left(\frac{k}{p}, \ldots, \frac{k^s}{p}\right) - \int_K f\,dx \right| < V(f)\,\frac{(70 \log p)^s}{\sqrt{p}} \tag{8}$$

wenn p irgend eine Primzahl ist. Für $s = 2$, $f(x_1, x_2) = e^{2\pi i(x_1+x_2)}$ ist $\mu(f) = 0$, die Summe links in (8) eine Gaußsche Summe vom Betrag $\sqrt{p}$, also ist in diesem Fall $\Delta_p \geqq p^{-1/2}$, also ist (8) scharf in Bezug auf die Größenordnung $p^{-1/2}$ auch dann, wenn nur die Klasse der analytischen Funktionen in Betracht gezogen wird. Aber auch die Potenz von $\log p$ kann nicht einfach weggelassen werden. Es sei nämlich f die folgende Funktion: $f(x_1, x_2) = 1$ für $0 < x_1 < 1/2$, $0 < x_2 < \alpha$ ($0 < \alpha < 1$, α sonst beliebig), dann ist $\mu(f) = \frac{\alpha}{2}$ die Summe links in (8) die Anzahl der quadratischen Reste modulo p, welche im Intervall $> 0, \alpha <$ liegen und es ist nach *Paley*[4] bekannt, daß diese Anzahl
1 sicher nicht $o(\sqrt{p} \log \log p)$ ist, also ist sicher nicht $\Delta_p = o\left(\frac{\log \log p}{p}\right)$
Offen bleibt nur, ob $\log^s p$ nicht durch eine Funktion, welche schwächer gegen ∞ geht, ersetzt werden kann. Ist f s-mal differenzierbar, so kann $V(f)$ wieder durch (6) ersetzt werden.

Führt man die analoge Überlegung für die Folge Ω_N^1 aus (1) durch, so erhält man sehr leicht, wenn man $D_N(\omega)$ direkt abschätzt.

Satz 2: Für alle $f \in B$ ist stets

$$\Delta_N{}^1 = \left| \frac{1}{N} \sum_{k_1, \ldots, k_s}^{n-1} f\left(\frac{k_1}{n}, \ldots, \frac{k_s}{n}\right) - \mu(f) \right| \leqq \frac{s\,V(f)}{N^{1/s}} \tag{9}$$

L. C. Hsu[5] und *L. W. Lin* haben vorgeschlagen, mehrfache Integrale durch einfache Integrale zu approximieren und diese durch geläufige

[5] Acta math. Acad. Sci. Hungar. **9**, 279–90 (1958).

Quadratformeln zu berechnen. Sie geben Abschätzungen für die Differenz (A beliebige Zahl > 0, ganz)

$$| \mu(f) - \frac{1}{T} \int_0^T f(A^{q-1} t, A^{q-2} t, \ldots, A^{q-s} t) \, dt \tag{10}$$

wo $T = A^{r+s-q}$, wenn f r-mal differenzierbar ist und q beliebige natürliche Zahl ist. Die Abschätzung hängt von den Fourierkoeffizienten von f und A ab. Unter Verwendung der Simpsonschen Formel einer Ordnung N für das einfache Integral in (10) erhalten sie Näherungswerte für $\mu(f)$, welche von der Größenordnung $N^{-\varepsilon_r}$ sind, wo $\varepsilon_r = \dfrac{4r}{5r+4s-4}$ also $\lim\limits_{r \to \infty} \varepsilon_r = \dfrac{4}{5}$ und sicher $> \dfrac{4}{7}$ für $r \geqq 2s$, also unter sehr starken Voraussetzungen nur maximal einen um $\dfrac{8}{5}$ größeren Exponenten als in (8), trotzdem ist die Methode sehr nützlich.

Benützt man das kontinuierliche Analogon zu (5)

$$| \frac{1}{T} \int_0^T f(x(\tau)) \, d\tau - \mu(f) | \leqq V(f) \, D_s(x(\tau)) \tag{5'}$$

wo $x(\tau) = (x_1(\tau), \ldots, x_s(\tau))$ integrierbare Vektorfunktion auf K

$$D_s = \sup_{Q \subset k} | \frac{1}{T} \int_0^T \chi_Q(x(\tau)) \, d\tau - V(Q) |$$

und nimmt man $x(\tau) = \tau g$, wo $g = (A^{s-1}, A^{s-2}, \ldots, A^0)$, dann erhält man für $f \in B$

$$| \frac{1}{A} \int_0^A f(\tau g) \, d\tau - \mu(f) | < V(f) \frac{(70 \log A)^s}{A} \tag{11}$$

Es ist nämlich

$$D_s < 2^s \frac{300}{M} + 30^s \sum_{0 < ||h|| \leq M} R^{-1}(h) \, | \frac{1}{A} \int_0^A e^{2\pi i \tau \langle hg \rangle} \, d\tau | \tag{7'}$$

Nun ist für $M = A - 1$ für alle ganzen $h \neq 0$ mit $||h|| < A$ sicher

$$| \langle h g \rangle | = | h_1 A^{s-1} + \ldots + h_s | \geqq 1$$

also ist das Integral in (7′) sicher $\leqq (2\pi A)^{-1}$ und damit folgt die Abschätzung (11). Es sollen nun Folgen ω angegeben werden, für welche die Fehlerabschätzung Δ_N an Genauigkeit alle bisher angegebenen

Formeln übertrifft. Es sei wieder p eine Primzahl, dann nennen wir einen Gitterpunkt $g = (g_1, \ldots, g_s)$, $(1 \leqq g_j \leqq p - 1,\ j = 1, \ldots, s)$ einen *guten* Gitterpunkt modulo p, wenn für alle Gitterpunkte $h = (h_1, \ldots, h_s)$ mit $0 < \| h \| \leqq \frac{p-1}{2}$ und $\langle h g\rangle \equiv 0 \pmod p$

$$R(h) \geqq p\ (8 \log p)^{-s} \tag{12}$$

Wir werden in Hilfssatz 2 zeigen, daß es stets gute Gitterpunkte g gibt. Sie können stets in endlich vielen Schritten bestimmt werden, und zwar sind höchstens p^{2s} elementare Rechenoperationen auszuführen.

Hilfssatz 2: Es gibt stets gute Gitterpunkte g mod p.

Beweis: Wir bilden die Summe

$$\Sigma = \frac{1}{(p-1)^s} \sum_{g \bmod p}' \sum_{\substack{0<\|h\|<p/2 \\ <hg> \equiv 0 \pmod p}} R^{-1}(h)$$

wo stets $g_j \not\equiv 0 \pmod p$ für alle j sei. Mit g_j durchläuft auch $h_j g_j$ alle Zahlen relativ prim zu p, wenn $h_j \neq 0$, also ist die Summe gleich

$$\frac{1}{(p-1)^s} \sum_{0<\|h\|<p/2} R^{-1}(h) \sum_{\substack{g \bmod p \\ <e(h)g> \equiv 0 \pmod p}}' 1$$

wo $e(h)$ als Komponenten die Zahlen 0 oder 1 hat, je nachdem die entsprechenden Komponenten von h Null sind oder ungleich Null sind. Nun hat eine solche Kongruenz $g_1{}^* + \ldots + g_l{}^* \equiv 0 \pmod p$, $l \leqq s$ höchstens p^{s-1} Lösungen mod p, also wird $\Sigma < \frac{2^s}{p} \sum_{0<\|h\|<p/2} R^{-1}(h) < \frac{(8 \log p)^s}{p}$.

Es gibt also ein g, so daß

$$\Sigma_2 = \sum_{\substack{0<\|h\|<p/2 \\ <hg> \equiv 0 \pmod p}} R^{-1}(h) < \frac{(8 \log p)^s}{p} \tag{13}$$

also gilt die Schranke rechts in (13) für alle Glieder in Σ_2 und dies ist gerade die Behauptung.

Satz 3: Ist p Primzahl, g ein guter Gitterpunkt modulo p, dann ist für jedes $f \in B(K)$

$$\left| \frac{1}{p} \sum_{k=0}^{p-1} f\left(\frac{kg}{p}\right) - \mu(f) \right| < V(f) \frac{(1000 \log p)^{2s}}{p} \tag{14}$$

Bemerkung 1: Diese Abschätzung für den Fehler ist, abgesehen von den Exponenten in der Potenz von log p die *genaueste* Abschätzung, welche überhaupt möglich ist. Aus einem Satz von *K. F. Roth*[6] folgt, daß für jede beliebige Folge ω stets $D_N \geqq 4^{-s} \dfrac{(\log N)^{(s-1)/2}}{N}$ ist.

Daraus folgt: Zu jeder Folge $\omega = (x_1, \ldots, x_N)$ gibt es stets Funktionen $f \in B(K)$, so daß

$$\left| \frac{1}{N} \sum_{k=1}^{N} f(x_k) - \mu(f) \right| \geqq V(f)\, 4^{-s} \frac{(\log N)^{(s-1)/2}}{N} \tag{15}$$

Dies gilt auch dann, wenn das arithmetische Mittel links in (15) mit Gewichten λ_j versehen wird, also $\sum\limits_{k=1}^{N} \lambda_k f(x_k)$ mit $\sum\limits_{k=1}^{N} \lambda_k = 1$ betrachtet wird, wie dies z. B. bei der Simpsonschen Formel der Fall ist.

Bemerkung 2: Es ist bei der Konstruktion der guten Gitterpunkte g mod p nicht notwendig anzunehmen, daß p Primzahl ist. Es kann z. B., was gerade für die Praxis wichtig ist, p eine Potenz von 2 sein. Auf diese Dinge und auf eine genauere Berechnung von g soll an anderer Stelle eingegangen werden. Ebenso soll gezeigt werden, daß man den Exponenten in der Potenz von log p in (14) noch verkleinern kann. Für $s = 2$ kann man auch andere Folgen $x_1, \ldots, x_N$ angeben, welche das gleiche leisten wie in (14). Eine solche Folge, welche in der Theorie der Gleichverteilung schon an anderer Stelle aufgetreten ist, ist folgende: Ist $n \geqq 1$ ganz, dann sei $x_1, \ldots, x_N (N = 2^n)$ die Folge der Punkte

$$\left(\frac{t_1}{2} + \ldots + \frac{t_n}{2^n}, \frac{t_n}{2^n} + \ldots + \frac{t_1}{2^n} \right)$$

wie die t_j unabhängig voneinander die Werte 0 oder 1 annehmen. Für diese Folge ist bekannt, daß $D_N \leqq 2 \dfrac{\log N}{N}$ ist.

Nun zum Beweis von Satz 3: Betrachten wir die zugehörige Weylsche Summe

$$S_p(h, \omega) = \frac{1}{p} \sum_{k=0}^{p-1} e^{\frac{2\pi i k}{p} \langle h\, g \rangle} \qquad (0 < \| h \| < p/2),$$

so ist ihr Wert $= 0$, wenn $\langle h\, g \rangle \not\equiv 0 \pmod p$ und sonst 1, also wird die Summe rechts in (7) mit $M = \dfrac{p-1}{2}$ sicher

$$< \frac{2^s\, 600}{p} + 30^s \sum_{0 < \| h \| < p/2,\ \langle h\, g \rangle \equiv 0 \bmod p} R^{-1}(h) \tag{16}$$

[6] Mathematika 1, 73–79 (1954); verallgemeinert für $s > 2$ von *J. H. Halton*, Numerische Mathematik 2, 84–90 (1960).

Ist nun in $h\,h_{l-1}$ die 1. Komponente $\neq 0$, so ist also

$$h_{l-1}\,g_{l-1} + h_l\,g_l + \ldots + h_s\,g_s \equiv 0 \pmod p$$

oder (da $(g_{l-1}, p) = 1$)

$$h_{l-1} \equiv h_l\,g_l' + \ldots + h_s\,g_s' \pmod p,$$

also

$$h_{l-1} \equiv \langle h^{(l)}\,g^{(l)}\rangle \bmod p$$

wo

$$g^{(l)} = (g_l', \ldots g_s'),\; h^{(l)} = (h_l, \ldots, h_s)$$

Nach (12) ist

$$R(h^{(l)})\,|\,h_{l-1}\,| \geqq (8\log p)^{-s}\,p \tag{17}$$

Die Summe in (16) rechts ist $\sum\limits_{l=2}^{s} \Sigma^{(l)}, \Sigma^{(l)} = \sum\limits_{0<\|h^l\|<p/2} R^{-1}(h^{(l)})\,|\,h_{l-1}^{-1}\,|$.

Betrachten wir eine solche l-Summe z. B. für $l = 2$ (setzen $h^{(2)} = k, g^{(2)} = a$ $h_1 = h(k)$). Wir zerlegen den Summationsbereich $0 < \|k\| < p/2$ in Bereiche $P_r : 2^{r_j-1} \leqq \mathrm{Max}\,(|\,k_j\,|, 1) < 2^{r_i}$ ($j = 2, \ldots s$; ist $k_j = 0$, dann sei $r_j = 1$; $r_2, \ldots, r_s$ laufen von 1 bis $\log_2 p/2$. In P_r ist $2^{r_2+\cdots+r_s} \geqq$ $R(h) > 2^{r_2+\cdots+r_s-s+1}$ also

$$\Sigma^{(2)} < 2^s \sum_{r_2\ldots r_s=1}^{\log p} 2^{-(r_2+\cdots+r_s)} \sum_{k\subset P_r} h^{-1}(k).$$

Es ist nach (17) in P_r stets

$$|\,h(k)\,| \geqq p\,2^{-(r_2+\cdots+r_s)}\,(4\log p)^{-s} = C_r.$$

Zu jedem $h(k)$ gibt es also eine natürliche Zahl v, so daß

$$v\,C_r \leqq |\,h(k)\,| < (v+1)\,C_r \qquad (v = 1, \ldots, [C_r^{-1}]).$$

In jedem solchen v-Intervall können höchstens zwei solche $h(k)$ ($k \in P_r$) liegen. Gäbe es z. B. drei solche $h(k^1), h(k^2), h(k^3)$, dann müßten zwei, z. B. $h(k^1), h(k^2)$ das gleiche Vorzeichen haben, und es wäre dann $|\,h(k^1 - k^2)\,| < C_r$ mit

$$\|\,k^1 - k^2\,\| < 2^{r_2+\cdots+r_s-s+1}$$

was (17) widerspricht. Es ist also

$$\sum_{k\subset P_r} < 2 \sum_v \frac{1}{v\,C_r}, \text{ also } p\,\Sigma^{(2)} < 2^{s+2}(8\log p)^s \sum_{r_2\ldots r_s}^{\log p} (r_2 + \ldots + r_s) < 20^s(\log p)^{2s}.$$

Dies gilt nun für alle $\Sigma^{(l)}$ also ist die Summe in (16) rechts $< 30^s\,\dfrac{(\log p)^{2s}}{p}$, also eingesetzt in (16) liefert Satz 3.

Bemerkung: Aus (13) entnehmen wir: Es gibt stets einen guten Gitterpunkt g, so daß für jedes $f \in B(K)$

$$\left| \frac{1}{p} \sum_{k=0}^{p-1} f\left(\frac{g\,k}{p}\right) - \mu(f) \right| \leqq V(f) \frac{(80 \log p)^s}{p} \tag{18}$$

Es ist nur notwendig, in (16) die Abschätzung (13) einzusetzen. Wenn wir voraussetzen, daß f differenzierbar ist bis zu einer Ordnung $m\,s$ ($m \geqq 2$, natürliche Zahl,) dann können wir (14) verschärfen.

Satz 4: Ist p Primzahl, g ein guter Gitterpunkt modulo p, besitzt f alle stetigen Ableitungen

$$\frac{\partial^{mr} f}{\partial x_{l_1}^m \ldots \partial x_{l_r}^m} \quad (1 \leqq r \leqq s,\, l_1 < l_2 < \ldots < l_r;\, m \geqq 2)$$

und sind sie alle von einem Betrag $\leqq K_m$, dann ist

$$\left| \frac{1}{p} \sum_{k=0}^{p-1} f\left(\frac{k\,g}{p}\right) - \mu(f) \right| \leqq \frac{K_m}{p^m} (30 \log p)^{(m+1)\,s-1} \tag{19}$$

Beweis: Unter diesen Voraussetzungen läßt sich f in eine absolutkonvergente Fourierreihe $\Sigma\, C_h\, e^{2\pi i \langle h\,x \rangle}$ entwickeln, und die linke Seite von (17) ist gleich

$$\sum_{h \neq 0} C_h\, S_p(h, \omega) = \sum_{\langle h\,g \rangle \equiv 0 \,(\mathrm{mod}\, p),\, h \neq 0} C_h$$

Weiter ist $|C_h| \leqq K_m((2\,\pi)^s\, R(h))^{-m}$, also genügt es

$$\Sigma^* = \sum_{\langle h\,g \rangle \equiv 0 \,(\mathrm{mod}\, p),\, h \neq 0} R^{-m}(h)$$

abzuschätzen. Diese Summe

$$= \sum_{0 < \|h\| < p/2}^{*} + \sum_{\|h\| > p/2}^{**} = \Sigma_1 + \Sigma_2$$

Da in der zweiten Summe sicher in $R(h)$ ein Faktor die Gestalt

2 $l_p + H$ ($l \neq 0$, $|H| \leqq \frac{p-1}{2}$) hat, also $\geqq \frac{|l|}{2}\, p$ hat, so ist

$$\Sigma_2 \leqq \left(\frac{2}{p}\right)^m \sum_{0 < \|h^*\|} R^{-m}(h^*) < \left(\frac{2}{p}\right)^m \left(\left(\sum_{n=1}^{\infty} \frac{1}{n^m} + 1 \right)^s - 1 \right) < \left(\frac{6\,s}{p}\right)^m.$$

Bei der ersten Summe gehen wir wie in Satz 3 vor, also

$$\Sigma_1 < \sum_{l=2}^{s} \sum_{0 < \|h^{(l)}\| < p/2} (R(h^{(l)})\, h_{l-1})^{-m}$$

Eine l-Summe ist wieder

$$< 2^{ms} \sum_{r_1, \ldots, r_s} 2^{-m(r_1 + \cdots + r_s)} \sum_{k \subset P_r} h^{-m}(k) < 2^{ms} . 2 p^{-m} . (8 \log p)^{ms} \sum_{r_1, \ldots, r_s} \sum_{v} v^{-m} <$$
$$< 8 (20 \log p)^{ms} (\log p)^{s-1} p^{-m}$$

und damit ist alles gezeigt.

Bemerkung: Aus (13) entnehmen wir wieder, daß es stets einen guten Gitterpunkt g gibt, so daß sogar für die linke Seite von (18) die schärfere Abschätzung

$$\frac{K_m}{p^m} 20^s (\log p)^{ms}$$

gilt.

Bemerkung 2: Satz 4 gilt für $m > 1$, (m beliebig), wenn die partiellen Ableitungen nach Liouville definiert werden.

Corrections

145.1 $\Delta_p = o(\frac{\log \log p}{\sqrt{p}})$ 150.2 $lp + H$

Rhythmische Folgen auf kompakten Gruppen I

Sitzungsber., Abt. II, Österr. Akad. Wiss., Math.-Naturwiss. Kl. **171** (1963). 67–74

(Vorgelegt in der Sitzung vom 14. Dezember 1961)

Es sei G eine kompakte Gruppe, $\omega = \{x(j)\}$ eine Folge von Elementen $x(j)$ auf G, wo j alle ganzen Zahlen durchläuft. Es ist eine wichtige Aufgabe, die Verteilung der Elemente dieser Folge in G zu untersuchen [1]. Wir wollen dazu die Theorie der rhythmischen Folgen entwickeln, welche für den Fall der Torusgruppe zuerst von van der Corput [2] eingeführt wurden.

§ 1. Wir nennen die endliche Teilfolge: $x(l)$, $x(l+1)$, ..., $x(l+N-1)$ von ω die Welle $\omega(l, N)$ mit Anfang l und Länge N. Unter $\omega(s)$ verstehen wir die Welle $\omega(-s, 2s+1)$, also die Menge aller Elemente $x(k)$ mit $|k| \leqq s$, unter $\omega_r(s)$ die Welle $\omega(r-s, 2s+1)$, also die Menge der $x(k)$ mit $|k-r| \leqq s$. Jede Welle kann in eine solche symmetrische Welle $\omega(s)$ eingebettet werden. Wir wollen im weiteren unter Umgebungen U stets Umgebungen des Einheitselementes e verstehen. Dann definieren wir:

Def. 1. Die Welle $\omega(l, N)$ wiederholt sich bis auf U, wenn es eine Welle $\omega(m, N)$ gibt, so daß

$$x(m+k)\,x^{-1}(l+k) \in U, \qquad 0 \leqq k < N. \tag{1}$$

Nun kann die Definition der rhythmischen Folgen gegeben werden.

Def. 2. Eine Folge ω heißt rhythmisch kurz rh-Folge, wenn es zu jedem U und zu jeder Welle $\omega(l, N)$ eine natürliche Zahl $L = L(U, \omega(l, N))$ gibt, so daß sich in jedem Intervall der Länge L die Welle $\omega(l, N)$ bis auf U wiederholt, d. h. zu jedem r gibt es ein m, mit $r \leqq m + N - 1 \leqq r + L$, so daß (1) gilt.

Man sieht sofort

Satz 1: Ist ω rhythmisch, M eine offene Menge in G und liegt in M wenigstens ein Element von ω, dann unendlich viele und es gibt eine natürliche Zahl L, so daß es zu jedem r ein m mit $r \leqq m \leqq r + L$ gibt, so daß $x_m \in$ M.

Beweis: Es sei $x_s \in M$. Es gibt, da M offen, eine Umgebung U, so daß $U x_s$ in M liegt. Da ω rhythmisch ist, muß nach Definition zur Welle $\omega\,(s, 1)$ ein $L\,(U, \omega\,(s, 1))$ geben, so daß es zu jedem r ein m mit $r \leqq m \leqq r + L$ gibt, so daß $x_m\, x_s^{-1} \in U$, also $x_m \in U x_s \in M$ w. z. b. w.

Daraus geht die Bedeutung der rh-Folge schon hervor. Es gibt stets solche Folgen, z. B. $\omega = \{g\}$, wo g beliebig aus G, weiter ist jede periodische Folge, d. h. eine solche Folge, für die es ein p gibt, so daß $x\,(j + p) = x\,(j)$ für alle j eine solche Folge. Mit $\omega = \{x\,(j)\}$ ist auch, mit $g \in G$, stets $\{g\, x\,(j)\}$ eine rh-Folge. In der Definition der rh-Folgen kann man sich auf symmetrische Umgebungen $\hat{U}$ beschränken, also auf solche $\hat{U}$, so daß für alle $x \in G$, $x^{-1} \hat{U} x = \hat{U}$ ist[1] und $\hat{U} = \hat{U}^{-1}$. Für kompakte Gruppen gibt es zu jedem U stets $\hat{U}$ mit $\hat{U} \subset U$.

In der Definition der rh-Folgen kann $L\,(U, \omega\,(l, N))$ unabhängig vom Anfang l gewählt werden, also $L = L\,(U, N)$ genommen werden, d. h. genauer

Satz 2: Ist ω eine rh-Folge, dann gibt es zu jedem U und jedem N ein $\hat{L}$, so daß es zu jedem l, r ein m mit $r \leqq m \leqq m + N - 1 \leqq r + L$ gibt, so daß

$$x\,(m + k)\, x^{-1} (l + k) \in U \qquad 0 \leqq k < N.$$

Beweis: Es sei U_1 eine symmetrische Umgebung, so daß $U_1^3 \subset U$. Da G kompakt, gibt es endlich viele Elemente $z_1 = e, \ldots z_s$ aus G, so daß die $U_i = U_1\, z_i$ $(i = 1, \ldots s)$ die Gruppe G überdecken, dabei kann o. B. d. A. $s > N$ angenommen werden. Wir sagen N Elemente $v_1, \ldots v_N$ gehören zur Klasse $K = \{j_1, \ldots j_N\}$, wenn $v_i \in U_1\, z_{j\,(i)}$, $(i = 1 \ldots N)$. Es gibt höchstens N^s Klassen. Jede Welle $\omega\,(l, N)$ liegt in mindestens einer dieser Klassen. Wir wählen aus jeder Klasse K eine solche Welle ω $(\hat{l}\,(K), N)$ irgendwie aus, wenn sie überhaupt eine Welle enthält, und setzen $\sup\limits_K L\,(U_1, \omega\,(\hat{l}\,(K), N)) = \hat{L}$. Es sei nun $\omega\,(l, N)$ eine beliebige

[1] Vgl. A. Weil, L'Integration dans les Groupes Topologiques(1953) S. 86.

Welle, welche in einer Klasse K liege und $\omega(\hat{l}(K), N)$ die ausgezeichnete Welle aus K, dann ist für $0 \leqq k < N$

$$x(\hat{l}(K)+k)\,x^{-1}(l+k) \in U_1 U_1^{-1} \subset U_1^2. \tag{2}$$

Da $\omega\, rh$, so gibt es zu jedem r ein m mit $r \leqq m \leqq m+N-1 \leqq r+L$ $(U\omega_1, (\hat{l}(K), N)) \leqq r+\hat{L}$, so daß für $0 \leqq k < N$, also

$$x(m+k)\,x^{-1}(l(K)+k) \in U_1, \tag{3}$$

also nach (2), (3)

$$x(m+k)\,x^{-1}(l+k) \in U_1^3 \subset U$$

w. z. b. w.

Es sei $f(x)$ eine stetige Funktion auf G mit Wertebereich in G, dann gilt

Satz 3: Ist $\omega = \{x(j)\}$ eine rh-Folge, dann ist dies auch die Folge $\omega^* = \{f(x(j))\}$.

Beweis: Bei jedem U gibt es ein U_1, so daß für alle x und alle $x_1, \ldots x_N$ mit $x\,x_j^{-1} \in U_1$ folgt $f(x)\,f^{-1}(x_j) \subset U$. Ist nun $\omega(l, N) = \{x(l+j)\}$ eine Welle, so gibt es, da ω rh-Folge ist, zu jedem r ein m mit $r \leqq m \leqq m+N-1 \leqq r+L$, so daß $x(m+j)\,x^{-1}(l+j) \in U_1$, also ist dann $f(x(m+j))\,f^{-1}(x(l+j)) \subset U$ für $0 \leqq j < N$.

Beispiel: $f(x) = x^s$, x^{-1}, also mit $\omega = \{x(j)\}$ ist auch $\{x^s(j)\}$, $\{x^{-1}(j)\}$ rh-Folge.

Bemerkung: Aus dem Beweis geht hervor, daß f nur in den Elementen (j) stetig zu sein braucht. Es sei $\omega_k = \{x_k(j)\}$ für jedes k eine Folge, welche gleichmäßig in k gegen $\{x(j)\}$ konvergiert, dann gilt

Satz 4: Ist für jedes k die Folge $\omega_k = \{x_k(j)\}$ eine rh-Folge, dann auch $\omega = \{x(j)\}$.

Beweis: Es sei U_1 eine symmetrische Umgebung mit $U_1^3 \subset U$. Wegen der gleichmäßigen Konvergenz gibt es ein n, so daß für alle j

$$x_n(j)\,x^{-1}(j) \in U_1.$$

Weiter gibt es in jeder Welle $\omega^{(n)}(l, N)$ ein $L(U_1, \omega^{(n)})$ so daß es zu jedem r ein m mit $r \leqq m \leqq m+N-1 \leqq r+L$ gibt, so daß

$$x_n(m+j)\,x_n^{-1}(l+j) \in U_1,$$

also

$$x(m+j)\,x^{-1}(l+j) \in U_1^3 \subset U.$$

§ 2: Ist $v \in G$, ist ω eine Folge $\{x(j)\}$, dann bedeute ωv die Folge $\{x(j)v\}$, ebenso $\omega(l, N)v$ die Folge $x(l)v, \ldots, x(l+N-1)v$.

Def. 3: $S(\omega)$ sei die Menge aller $v \in G$, so daß es zu jedem s und jedem symmetrischen U ein r gibt, so daß

$$x(r+j)(x(j)v)^{-1} \in U \qquad |j| \leqq s.$$

Es liegt stets das Einheitselement $e \in S$.

Wir zeigen weiter:

Mit v, v_1 liegt auch $v v_1$ in S.

Beweis: Wir wählen ein symmetrisches U_1, so daß $U_1^2 \subset U$. Nach Definition gibt es ein r, so daß

$$x(r+j)(x(j)v)^{-1} \in U_1, |j| \leqq s. \tag{4}$$

Es wurde nun s_1 so groß gewählt, daß die Welle $\omega(s_1)$ alle Elemente von $\omega_r(s)$ enthält. Es gibt dann nach Definition ein r_1, so daß

$$x(r_1+k)(x(k)v_1)^{-1} \in U_1, \qquad |k| \leqq s_1. \tag{5}$$

Da unter den k in (5) alle $r+j$ aus (4) vorkommen, so folgt aus (4) und (5) durch Multiplikation

$$x(r_1+r+j)(x(j)v v_1)^{-1} \in U_1^2 \subset U \qquad |j| \leqq s.$$

Es ist weiter S abgeschlossen.

Beweis: Es sei v ein Berührungspunkt von S, U wieder eine beliebige Umgebung von e und U_1 symmetrisch mit $U_1^2 \subset U$. Es gibt dann ein $v_1 \in S$, so daß $v_1 v^{-1} \in U_1$. Nach Definition gibt es zu u_1 und jedem s ein r_1, so daß

$$x(r_1+j)(x(j)v_1)^{-1} \in U_1 \qquad |j| \leqq S,$$

also

$$x(r_1+j)(x(j)v)^{-1} \in U_1^2 \subset U.$$

Da G kompakt, so folgt daraus, daß S mit v auch v^{-1} enthalten muß, es ist also $S(\omega)$ eine abgeschlossene Untergruppe von G.

Hilfssatz 1: Ist U_1 symmetrisch, dann gibt es stets eine natürliche Zahl $s^* = s(U_1)$ und ein $U_1^* \subset U_1$, so daß für jedes $s \geqq s^*$ folgendes gilt: Gibt es ein v und ein r mit

$$x(r+j)(x(j)v)^{-1} \in U_1{}^*$$

so hat $v\,U_1$ mit S stets einen nicht leeren Durchschnitt.

Beweis: Angenommen, dies wäre nicht richtig, dann gäbe es zu jedem s und jedem $U_1{}^*$ stets ein v' so daß $v'\,U_1 \cap S = \phi$ und ein r, so daß,

$$x(r+j)(x(j)v')^{-1} \in U_1{}^* \qquad |j| \leqq s. \tag{6}$$

Da G kompakt, so besitzen die v' einen Berührungspunkt v. Ist nun $U_1{}^*$ so gewählt, daß $(U_1{}^*)^2 \subset U_1$, so gibt es also ein v', so daß $v \in U_1{}^* v'$, und es wäre, (6) angewendet auf $U^*{}_1$

$$x(r+j)(x(j)v)^{-1} \subset (U^*{}_1)^2 \subset U_1 \quad |j| \leqq s \text{ für alle } U^*{}_1 \text{ und } s, \text{ d. h.}$$

$v \in S$, was der Definition von v widerspricht. Da S abgeschlossen, gibt es bei beliebigen symmetrischen U_3 endlich viele Elemente $v_1, \ldots, v_a$ so daß S von $U_3\,v_1, \ldots, U_3\,v_a$ überdeckt wird.

Hilfssatz 2: Zu jedem v_i gibt es ein k, so daß für jedes U_3 $v_i\,v_k \in U_3{}^2$ liegt.

Beweis: Zunächst liegen alle $v_i{}^l$ $(l = 1, 2, 3, \ldots)$ in S, und daraus folgt, daß eine Potenz $v_i{}^h \in U_3$ liegt; es ist ja S kompakt. Es liegt auch $v_i{}^{h-1}$ in S, also $v_i{}^{h-1} \in U_3\,v_k$, also $v_i\,v_k \in v_i\,U_3{}^{-1}\,v_i{}^{h-1} \in U_3{}^2$.

Def. 4: Ist $\omega = \{x(j)\}$, dann sei $\Delta\omega$ die Folge $\{x(j+1)\,x^{-1}(j)\} = \{z(j)\}$.

Dann gilt der

Hauptsatz: Die Folge ω ist genau dann rhythmisch, wenn dies für die Folge $\Delta\omega$ gilt.

Beweis 1: Ist ω eine rh-Folge, so gibt es doch zu jedem U und U_1 mit $U_1{}^2 \subset U$ und jedem l, N ein L, so daß es zu jedem r ein m mit $r \leqq m \leqq m + N \leqq L$,

so daß

$$x(m+k)\,x^{-1}(l+k) \in U_1 \qquad 0 \leqq k \leqq N \tag{7}$$

also auch $x(m+k+1)\,x^{-1}(l+k+1) \in U_1 \qquad 0 \leqq k \leqq N-1$,

also $\quad .\,z(m+k)\,z^{-1}(l+k) \in U_1\,U_1{}^{-1} \subset U \qquad 0 \leqq k < N$,

also $\Delta\omega$ rh-Folge.

Beweis 2: Es sei jetzt $\Delta\omega$ eine *rh*-Folge, dann soll also gezeigt werden, daß auch ω eine rh-Folge ist. Da jede Welle in eine symmetrische Welle $\omega(l)$ eingebettet werden kann, genügt es, solche Wellen zu betrachten. Es sei U vorgegeben, dann es gibt zu den v_i aus Hilfssatz 2 mit $U_3^8 \subset U$ nach Definition von S stets ganze Zahlen r_i, so daß, wenn $U_3^2 = U_1$,

$$x(r_i + j)\,(x(j)\,v_i)^{-1} \in U_1 \qquad |j| \leqq l,\ 1 \leqq i \leqq a. \tag{8}$$

Dabei sei U_3 symmetrisch.
Wir wählen nun s so groß, daß $\omega(s)$ alle Glieder $x(r_i + j)$ aus (8) enthält und noch $s \geqq s^*(U_1)$ nach Hilfssatz 1. Da $\Delta\omega$ rh-Folge, gibt es zu $\Delta\omega(s)$ und zu jeder Umgebung, also auch zu U_2, wo $U_2^{2s+1} \subset U_1^*$ aus Hilfssatz 1, ein $L_2^{(i)} = L^{(i)}(U_2, \Delta\omega(s))$, so daß es zu jedem r ein m mit $r - r_i \leqq m \leqq m + 2s + 1 \leqq r - r_i + L_2 \leqq r - r_i + \overline{L}$, wo $\overline{L} = \sup L_2^{(i)}$ gibt, so daß

$$z(m + t)\,z^{-1}(t) \in U_2 \qquad |t| \leqq s \tag{9}$$

und dies gilt daher auch für alle $t = r_i + j$ aus (8). Es ist nun stets für jedes t aus (9) $x(t)\,x^{-1}(-s) = z(t)\,z(t-1)\ldots z(-s)$

$$x(m + t)\,x^{-1}(m - s) = z(m + t)\,z(m + t - 1)\ldots z(m - s).$$

Nun ist $z(m + t) \in U_2\, z(t)$ also, da U_2 symmetrisch und $|t| \leqq s$

$$x(m + t)\,x^{-1}(m - s) \subset U_2^{2s+1}\, z(t)\ldots z(-s) \subset U_1^*\, x(t)\,x^{-1}(-s)$$

also, wenn wir $x(m - s)\,x^{-1}(s) = w$ setzen, so ist

$$x(m + t)\,(x(t)\,w)^{-1} \in U_1^* \qquad |t| \leqq s.$$

Da $s \geqq s^*$, so hat nach Hilfssatz 1 $w\,U_1$ mit S sicher einen nichtleeren Durchschnitt. Es liegt also $w \in S\,U_1^{-1}$. Da S von Umgebungen $v_i U_3$ überdeckt wird, so gibt es ein v_i, so daß $w \in v_i U_3\, U_1^{-1} \subset v_i\, U_3^3$ also

$$x(m + t) \in x(t)\,v_i\,U_3^4 \qquad |t| \leqq s$$

Nach Hilfssatz 2 gibt es zu v_i ein v_k, so daß $v_i v_k \in U_3^2$ also

$$x(m + t)\,v_k \in x(t)\,U_3^6 \text{ für } |t| \leqq s.$$

Nun wenden wir dies auf $t = r_k + j$ für $|j| \leqq l$ an: $x(m + r_k + j)\,v_k \in x(r_k + j)\,U_3^6$, andererseits nach (8) $x(r_k + j) \in x(j)\,v_k\,U_3^2$, also

$$x(m + r_k + j)\,x^{-1}(j) \in U_3^8 \subset U \qquad |j| \leqq l$$

und $r \leqq r - r_k + r_k \leqq m + r_k \leqq m + r_k + 2l + 1 \leqq L_2 + r - r_k + $
$+ r_k \leqq \overline{L}_2 + r$

also ω rh-Folge w. z. b. w. Daraus folgt

Satz: Ist $g \in G$ und $\varphi(j)$ ein ganzzahliges Polynom, dann ist $\{g^{\varphi(j)}\}$ rhythmisch.

Besitzt $\varphi(j)$ kein Absolutglied, dann gibt es ein j, nämlich $j = 0$ so daß für jedes U $g^{\varphi(j)} \in U$, also folgt 3 aus Satz 1

Satz: Es gibt unendlich viele j, so daß $g^{\varphi(j)} \in U$ für jedes g, U und Polynom $\varphi(j)$ ohne Absolutglied.

§ 4. Um die Stellung der rh-Folgen in der Menge aller Folgen zu charakterisieren, führen wir folgende Definition ein:

Def. 5: Ist $\omega = \{x(j)\}$, $\omega' = \{y(j)\}$, dann heißt $\omega' < \omega$, wenn es zu jedem l, N, U ein m gibt, so daß $y(l+j)\, x^{-1}(l+m+j) \in U$ für $0 \leqq j < N$.... Ist $\omega' < \omega$, $\omega < \omega'$, dann sagen wir ω und ω' sind gleich reich: $\omega \sim \omega'$. Ist $\omega' < \omega$, aber nicht $\omega \sim \omega'$, dann heißt ω' ärmer als ω. Es ist leicht einzusehen: Ist $\omega'' < \omega'$, $\omega' < \omega$, dann ist $\omega'' < \omega$. Ist $\omega'' = \{z(j)\}$ und $U_1{}^2 \in U$, dann ist nämlich bei passenden m, m_1 $z(l+j)\, y^{-1}(l+m_1+j) \in U_1$, $y(l+m_1+j)\, x^{-1}(l + m_1 + m + j) \in U_1$, $0 \leqq j < N$, also $z(l+j)\, x^{-1}(l+m+m_1+j) \in U_1{}^2 \subset U$.

Daraus folgt aus $\omega'' \sim \omega'$, $\omega' \sim \omega$ sofort $\omega'' \sim \omega$. Weiter gilt folgender

Satz 5.: Ist $\omega' < \omega$ und ist M eine beliebige offene Menge, dann ist die Anzahl der Elemente von ω, welche in M liegen, mindestens so groß wie jene von ω'.

Bew.: Sind $y(j_1), \ldots, y(j_s)$ Elemente von ω' in M, dann gibt es Umgebungen $U_1, \ldots, U_s$, so daß $U_i\, y(j_i) \subset M$ $(i = 1, \ldots, s)$, also gilt dies auch für den Durchschnitt U aller dieser U_i. Ist nun $N \geqq \operatorname{Max} j_i$, so folgt aus Def. 5 mit $l = 0$, daß es ein m gibt, so daß

$$x(m + j_i) \in U\, y(j_i) \subset M,$$

also es liegen mindestens s Lösungen in M. Es soll nun gezeigt werden;

Satz 6.: Ist ω eine rh-Folge und $\omega' < \omega$, dann ist $\omega' \sim \omega$ und ω' ist wieder eine rh-Folge.

Bew.: Da $\omega = \{x(j)\}$ eine *rh*-Folge, so gibt es zu l, N, U ein U_1 mit $U_1^2 \subset U, L = L(l, N, U_1)$, so daß es zu jedem r_1 ein m mit $r_1 \leqq m_1 \leqq \leqq m + N \leqq r_1 + L$, so daß

$$x(m_1 + j)\, x^{-1}(l + j) \in U_1 \quad 0 \leqq j < N. \tag{10}$$

Weiter gibt es ein m, sodaß für alle k mit $0 \leqq k \leqq L + N$

$$y(k + l)\, x^{-1}(m + l + k) \in U_1 \tag{11}$$

Wir nehmen jetzt $r_1 = m + l$, dann ist $l + m \leqq m_1 \leqq l + m + L$, also, $0 \leqq m_2 = m_1 - m - l \leqq L$, also $m = m_1 - m_2 - l$, und für alle $k = m_2 + j$, $0 \leqq j < N$ ist nach (11) $y(m_2 + l + j)\, x^{-1}(m_1 + j) \in U_1$, also wegen (10) $x(l + j)\, y^{-1}(m_2 + l + j) \in U_1^2 \subset U \quad 0 \leqq j < N$

Jetzt zeigen wir, daß ω' auch rhythmisch ist. Da $\omega' < \omega$, so gibt es ein m_1, so daß, wenn $U_1^3 \subset U$, $y(l + j)\, x^{-1}(l + m_1 + j) \in U_1$ $0 \leqq j < N$. Da $\{x(j)\}$ eine *rh*-Folge, so gibt es zu l, m, N, U_1 ein $L = L(l, m, N, U_1)$, so daß es zu jedem r_1 ein m mit $r_1 \leqq m \leqq m + N \leqq r_1 + L$ gibt mit

$$x(l + m_1 + j)\, x^{-1}(m + j) \in U_1 \quad 0 \leqq j < N. \tag{13}$$

Es sei nun r beliebig, so gibt es wieder, da $\omega' < \omega$, ein m_2, so daß

$$y(r + k)\, x^{-1}(r + m_2 + k) \in U_1 \quad 0 \leqq k \leqq N + L \tag{14}$$

Wir nehmen nun $r_1 = r + m_2$, dann ist, wenn wir $m_3 = m - m_2$ setzen

$$r \leqq m_3 \leqq m_3 + N \leqq r + L,$$

und wenn wir die k aus (14) von der Gestalt $m_3 - r + j$, $0 \leqq j < N$ nehmen, dann ist nach (14)

$$y(m_3 + j)\, x^{-1}(m + j) \in U_1. \tag{14'}$$

Aus (12), (13), (14') folgt $y(m_3 + j)\, y^{-1}(l + j) \in U_1^3 \subset U$ für $0 \leqq j < N$, d. h. wenn wir (14) beachten, wo r beliebig war, folgt, daß ω' ebenfalls eine *rh*-Folge ist.

Literaturverzeichnis

[1] Vgl. z. B. E. Hlawka, Math. Ann. 125 (1952) 183–207. Rend. Circ. Mat. 4 (2) (1955) 1–14.

[2] Acta Mathem. 59 (1932) 209–328.

[3] Vgl. loc. cit. in der ersten Arbeit des Verfassers S. 195, Satz 7.

Zur Geometrie der Zahlen

Jahrbuch der Akademie der Wissenschaften zu Göttingen, 1963, pp. 23–29

Die Geometrie der Zahlen kann mit Recht als ein Werk von Göttinger Mathematikern bezeichnet werden, denn es stammt schon der Name dieses Gebietes von Hermann Minkowski, der von 1902—1909 in Göttingen gewirkt hat. Wir können ziemlich genau die Entstehung dieses Gebietes fixieren. Der Ausgangspunkt war die Theorie der quadratischen Formen, die für den Fall von zwei Variablen von Lagrange, Legendre und vor allem im 5. Kapitel der Disquisitiones arithmeticae von Gauß entwickelt wurde. Der deutsche Mathematiker Seeber dehnte diese Theorie 1831 auf drei Variable aus. Diese Arbeit wurde von Gauß im gleichen Jahr in den Göttinger Gelehrten Anzeigen besprochen und geometrisch gedeutet. Die verwickelten Beweise von Seeber veranlaßten Dirichlet 1850, diese von Gauß nur angedeuteten geometrischen Einkleidungen zur Grundlage der Theorie zu machen, und er erhielt dadurch eine außerordentlich durchsichtige Darstellung der Seeberschen Theorie. Im gleichen Jahr und in der gleichen Zeitschrift, nämlich im Crelleschen Journal, erschien der Inhalt eines Briefes des französischen Mathematikers Hermite an Jacobi, in dem Hermite bereits quadratische Formen in beliebiger Variablenzahl untersucht. Minkowski gelang es nun, in einer Arbeit 1891 diese rein arithmetischen Resultate von Hermite in außerordentlich einfacher Weise geometrisch herzuleiten[1].

Der grundlegende Begriff, den Minkowski verwendet, ist der Begriff des Gitters, der auch schon von Gauß und Dirichlet verwendet wurde. Denken wir uns ein Quadrat vom Flächeninhalt 1 in der Ebene und legen wir an jede der 4 Seiten ein vollkommen gleiches Quadrat, an die noch freien Seiten dieser Quadrate wieder ein gleiches. Denken wir uns dieses Verfahren unbegrenzt fortgesetzt, dann ist die Menge aller Eckpunkte dieser Quadrate genau das, was man ein Gitter nennt, genauer ein quadratisches Gitter. Diese Quadrate erfüllen die Ebene lückenlos, das heißt, jeder Punkt der Ebene liegt in einem dieser Quadrate, und abgesehen von den Randpunkten dieser Quadrate liegt

[1] Crellesches Journal, Bd. 107, S. 278—297. Wieder abgedruckt in den Gesammelten Abhandlungen von H. Minkowski, B. G. Teubner, Leipzig und Berlin 1911, Bd. 1, S. 243—260.

jeder Punkt in genau einem Quadrat. Man kann in dieser Überlegung das Quadrat durch ein Parallelogramm vom Flächeninhalt 1 ersetzen und erhält so ein Parallelgitter. Das gleiche, was hier in der Ebene geschehen ist, kann genausogut im Raum gemacht werden, indem man Würfel oder Parallelepipede nimmt. Denken wir uns nun auf jedem Gitterpunkt eine Kreisscheibe mit gleich großem Radius gelegt, so daß der Mittelpunkt der Kreisscheibe mit dem Gitterpunkt zur Deckung kommt; das so entstehende Gebilde wollen wir ein Kreisgitter nennen. Wir wollen nun annehmen, daß diese Kreisscheiben nicht übereinandergreifen. Es ist anschaulich leicht zu sehen, daß dann der Flächeninhalt der Kreisscheiben kleiner oder höchstens gleich sein kann dem Flächeninhalt des Gitterparallelogrammes, das das Gitter erzeugt, also gleich 1. Dieses ganz einfache Prinzip, das für beliebige Dimensionen gilt, ermöglichte es Minkowski, die Sätze von Hermite herzuleiten, und sogar noch in verschärfter Form. Minkowski erkannte, daß beim Beweis dieses Satzes nur verwendet wird, daß der Kreis die Eigenschaft hat, einen Mittelpunkt zu besitzen und konvex zu sein, d. h. wenn man zwei Punkte des Kreises verbindet, gehört die Verbindungsstrecke ganz dem Kreis an. Dieser Satz gilt daher nicht nur für den Kreis, sondern für beliebige konvexe Figuren mit Mittelpunkt. In dieser Fassung heißt der Satz der *Minkowskische Fundamentalsatz*[2]. Minkowski gelang es sofort, mit Hilfe dieser Methode einen berühmten Satz über die Diskriminante algebraischer Zahlkörper zu zeigen[3] [4].

[2] Tatsächlich wird der Minkowskische Fundamentalsatz in folgenden Formen angewendet: Ist das Volumen eines konvexen Körpers K mit Mittelpunkt im s-dimensionalen Raum $\geqq 2^s$ und ist der Mittelpunkt von K ein Gitterpunkt eines Gitters G mit Determinante 1, so enthält K noch mindestens ein weiteres Paar von Gitterpunkten von G. Aber auch die erste Formulierung gestattet vielseitige Anwendungen, und zwar bei dem Problem der dichtesten Ausfüllung des Raumes durch konvexe Körper, die nicht übereinandergreifen. Vergleiche z.B. die Arbeit des Verfassers: Ausfüllung und Überdeckung konvexer Körper durch konvexe Körper, Monatshefte für Mathematik, Bd. 53, S. 81—131. Weiter H. Groemer, Monatshefte für Mathematik, Bd. 65 (1961), S. 74—81, Archiv der Mathematik 12 (1961), S. 78—80, 477—480; Mathematische Zeitschrift 81 (1963), S. 260—278. Vgl. auch C. A. Rogers, Packing and Covering Cambridge-press (im Druck).

[3] Minkowski selbst hat dies ausführlich dargestellt in seinen Werken „Geometrie der Zahlen" (1896) und in seinen „Diophantischen Approximationen", Leipzig 1907. Es sei auch hingewiesen auf die Gedächtnisrede auf H. Minkowski von Hilbert (1909), welche wohl eine der schönsten Würdigungen H. Minkowskis darstellt.

[4] Vgl. weiter: O. H. Keller, Geometrie der Zahlen, Enzyklopädie der Math. Wissenschaften, Bd. 12, H. 11, Teil 3; I. W. S. Cassels, An Introduction to the geometry of numbers, Springer-Verlag, Berlin-Göttingen-Heidelberg (1959).

Der Minkowskische Fundamentalsatz gestattet vielseitige Anwendungen nicht nur auf die Zahlentheorie, sondern auch auf die Funktionentheorie.

Es gibt außer dem klassischen Beweis von Minkowski selbst verschiedene andere Beweise arithmetischer oder analytischer Natur[5]. Den Flächeninhalt bzw. das Volumen der Körper heranzuziehen, lag natürlich bei dem Beispiel der quadratischen Formen nahe, da das Volumen eng mit der Diskriminante der quadratischen Formen zusammenhängt. Nun ist aber rein geometrisch gesehen das Volumen nur *eine* der geometrischen Größen, und man könnte natürlich erwarten, daß man durch Heranziehung weiterer Eigenschaften der betreffenden Körper schärfere Sätze erhält. Dies ist auch bei der Kugel, wie z.B. Blichfeldt gezeigt hat, und auch noch für andere konvexe Körper, wie sie z.B. von van der Corput und Schaake[6] betrachtet wurden, der Fall, doch scheinen diese Eigenschaften nicht in einfacher geometrischer Weise interpretiert werden zu können.

Davenport und van der Corput[7] haben 1945 eine Verschärfung des Minkowskischen Satzes in der Ebene unter Heranziehung der Krümmung gefunden. Für höhere Dimensionen wurden diese Unter-

---

Minkowski selbst wurde durch seinen Satz veranlaßt, seine berühmten Untersuchungen über die Geometrie der konvexen Körper anzustellen. Siehe Gesammelte Abhandlungen, Bd. 2.

[5] Einer der wichtigsten Beweise ist wohl der Beweis von Blichfeldt 1914; vergleiche dazu J. F. Koksma, „Diophantische Approximationen" (Ergebnisse der Mathematik und ihrer Grenzgebiete, Bd. 4, Springer-Verlag 1935, Kap. 2).

Der kürzeste Beweis ist jener von Mordell 1934 (Compositio math. Bd. 1, S. 248—253). Ist K ein Parallelepiped, so erhält man als Spezialfall den berühmten Minkowskischen Linearformensatz. Auch für diesen speziellen Fall wurden zahlreiche Beweise gegeben, ein analytischer Beweis ist jener von C. L. Siegel; Math. Annalen 87 (1922), S. 36—38. Die Idee dieses Beweises gestattet, eine Theorie der trigonometrischen Interpolation in mehreren Variablen zu entwickeln, die ich ohne Beweis 1962 beim Symposium über Gleichverteilung in Breukelen und 1963 in der Göttinger Mathematischen Gesellschaft näher ausgeführt habe.

Der Blichfeldtsche Beweis ist von gruppentheoretischem Charakter und gestattet eine Anwendung der Minkowskischen Prinzipien auf diskontinuierliche Gruppen. Vergleiche C. L. Siegel, Ann. of Math. 44 (1943), S. 674—685; 46 (1945), S. 708—718; E. Hlawka, Math. Annalen 125 (1952), S. 183—207. Jetzt auch Bourbaki. Intégration Chap. VII, S. 109—110.

[6] Acta Arith. 2 (1937), S. 152—160, vgl. E. Hlawka, Sitzungsberichte der Akademie der Wissenschaften Wien 154 (1945), S. 50—58; 156 (1948), S. 247—254; Rankin, Ann. of Math. 48 (1947), S. 1062—1081; 50 (1940), S. 691—704; Nederland Akad. 51 (1948), S. 274—281.

[7] Nederland Akad. Wet. 49 (1946), S. 701—707.

suchungen in zwei Wiener Dissertationen fortgesetzt[8]. Trotzdem muß gesagt werden, daß die Ergebnisse in vieler Weise unbefriedigend sind, da sie vor allem metrische Invarianten enthalten. Der Sachlage angepaßt wäre es, affine Eigenschaften der konvexen Kurven und Flächen, wie sie z.B. in dem Buch von Blaschke und Reidemeister[9] entwickelt sind, heranzuziehen. Für den Fall der Ebene hat man in jüngster Zeit eine solche Formulierung gegeben[10], aber über höhere Dimensionen wissen wir nichts. Es gibt allerdings eine Verschärfung des Minkowskischen Fundamentalsatzes, die von Siegel[11] stammt, in der mehrfache Integrale auftreten, die als Verallgemeinerung der Frauenhoferschen Beugungsintegrale aufgefaßt werden können. Ich habe diese Integrale im Jahre 1945 in der Arbeit „Über Integrale auf konvexen Körpern" mit Hilfe der Methode der stationären Phase[12] untersucht und gefunden, daß in ihren asymptotischen Entwicklungen einfache geometrische Größen auftreten. Diese Untersuchungen wurden in letzter Zeit von C. S. Herz[13] wieder-aufgenommen.

Man kann natürlich auch die Frage aufwerfen, ob die Beschränkung auf konvexe Körper notwendig ist, denn tatsächlich hat Minkowski seinen Satz auf nichtkonvexe Körper angewendet, ja sogar auf Körper von unendlichen Volumen, und zwar in dieser Weise, daß er diesem Körper konvexe Körper einschrieb. Man kann heute rückblickend feststellen, daß analoge Sätze für nicht konvexe Körper nicht gelten[14]. Das ist bereits der Fall, wenn man zu den sogenannten Sternkörpern übergeht. In dieser Hinsicht kann wirklich die Beschränkung auf konvexe Körper als sehr glücklich bezeichnet werden. Dank zahlreicher Publikationen des Mathematikers Mahler (siehe Anm. 14) wissen wir heute mehr über die Theorie der Sternkörper, die in Bei-

---

[8] Demmelbauer (1950) und K. Schmidt (1958), Eine Verschärfung des Satzes von Minkowski.

[9] Vorlesungen über Differentialgeometrie II, Grundlagen der Mathematischen Wissenschaften VII, Springer-Verlag (1923).

[10] H. Groemer, Archiv der Mathematik 10 (1959), S. 62—63.

[11] Acta math., Bd. 65 (1935), S. 309—329, Verallgemeinerungen und Anwendungen (siehe Anm. 12).

[12] Monatshefte für Mathematik, Bd. 54 (1950), S. 1—36, 81—99; Bd. 55, S. 105—137; Hamburger Abhandlungen 18 (1952), S. 53—69. Anwendung auf die Zetafunktion konvexer Körper: Monatshefte für Mathematik 54.

[13] Ann. of Math. 75 (1962), S. 81—92; Amer. Journal of Math. 84 (1962), S. 126—133. Dazu Littman, Bull. Amer. Soc. 69 (1963) 766—70.

[14] E. Hlawka, Grundbegriff der Geometrie der Zahlen, Jrb. DMV 57 (1954), S. 37—55; Cassels (s. Anm. 4).

spielen bereits von Minkowski betrachtet wurden. Ausgangspunkt für diese neue Entwicklung waren die Untersuchungen von Mordell und Davenport. Zusammenfassend kann heute gesagt werden, daß wir eine ausgebildete Theorie bei den sogenannten automorphen Sternkörpern besitzen, daß aber jeder besondere Körper eine spezielle Untersuchung erforderlich macht. Davenport[15] hat gezeigt, daß die Kreismethode von Hardy-Littlewood, welche zum erstenmal in den Göttinger Abhandlungen 1912 veröffentlicht wurde, hier von Nutzen ist.

Minkowski hat wiederholt einen Satz formuliert, der das Gegenstück zu seinem Fundamentalsatz darstellt und der die Existenz von Gittern von extremaler Eigenschaft in bezug auf die betreffenden konvexen Körper aussagt, ja er spricht ihn sogar für Sternkörper aus. Wenn wir zu dem Beispiel der Kreisscheiben zurückgehen, kann er so ausgesprochen werden: Ist der Flächeninhalt einer jeden Kreisscheibe $3/4$, so gibt es ein Gitter, in welchem die Scheiben nicht übereinandergreifen[16]. Er hat diesen Satz selbst nur für die Kugel bewiesen in seiner letzten arithmetischen Arbeit: Diskontinuitätsbereich für arithmetische Äquivalenz, 1905[17]. Dieser Satz verlangt, aus der unendlichen Menge aller Gitter solche mit einer besonderen Eigenschaft herauszugreifen. Da es sehr schwer ist, gerade durch Zufall ein solches Gitter zu finden, liegt es nahe, folgenden Vergleich zu ziehen: Man soll in einem Teich Fische, z.B. mit bestimmter Länge, finden. Bei einem einmaligen Fang wird es nur ein Zufall sein, einen solchen Fisch zu fangen. Man wird daher eine Menge Fische fangen und hoffen, daß sich ein solcher Fisch dabei findet. Man spricht in der Wahrscheinlichkeitsrechnung dabei von einer Stichprobe. Man bildet bei der Stichprobe den Mittelwert, hier in bezug auf die Länge, und zieht daraus einen Schluß auf die Stichprobe selbst. Das war der Gedankengang, den ich 1942 angewendet habe[18], und zwar war für mich der Ausgangspunkt die Theorie der binären quadratischen Formen. Damit gelang es mir, diese Behauptung von Minkowski zu beweisen und auch einen analogen Satz für allgemeinere Mengen.

---

[15] Mathematica 3 (1956), S. 81—101; Proc. London Math. Soc. (3) 8 (1958), S. 109—126, mit Birch, Mathematica 5 (1954), S. 8—12, mit Ridout, Proc. London Math. (3) 9 (1959), S. 544—555.

[16] Genauer lautet diese Behauptung folgendermaßen: Ist der Inhalt V eines Sternkörpers K mit Mittelpunkt im $R^s < 1 + 2^{-s} + 3^{-s} + \ldots$, so gibt es stets ein Gitter von Det. 1, so daß der Mittelpunkt des Körpers ein Gitterpunkt ist, aber der Körper sonst keine weiteren Gitterpunkte enthält.

[17] Z.B. Minkowski, Gesammelte Abhandlungen, Bd. 2, S. 53—100.

[18] Math. Zeitschrift 49 (1944), S. 285—312.

Man kann diese Verfahren in verschiedener Weise durchführen, und es wurden auch einfachere Beweise für diesen Satz von Rogers und anderen Mathematikern gegeben[19]. Hervorheben möchte ich noch die Arbeit von Siegel[20], der bei der Mittelbildung die volle Theorie der quadratischen Formen heranzieht. Es liegt die Frage nahe, ob sich dieser Satz noch weiterverfolgen läßt. Wenn man wieder das Bild der Stichprobe zugrunde legt, so würde dies bedeuten, daß man vom Mittelwert zu den sogenannten höheren Momenten übergeht. Einen Anhalt zur Anwendung dieser Methode gibt eine Formel in der erwähnten Arbeit von Siegel. Wenn man diese Formel formal ansetzt, so erhält man eine unendliche Reihe von mehrfachen Integralen sehr komplizierter Bauart, die noch dazu nicht konvergent ist. Es erschien lange Zeit als hoffnungslos, auf diesem Weg weiterzukommen. Rogers in London und W. Schmidt in Wien haben ab 1955 gezeigt, daß man durch passende Umordnung der Glieder dieser Reihe erreichen kann, daß die Reihe konvergent wird, und es ist ihnen gelungen, durch passende Wahl der hier auftretenden Parameter Verschärfungen dieses Satzes zu erzielen[21]. Bedeutsamer als die dabei erzielten Verschärfungen erscheinen mir die dabei verwendeten Methoden, die es ermöglichen, jedenfalls für den Fall der Ebene, und zwar für konvexe und nicht konvexe Bereiche, das Problem, das ich schon vorher bei der Besprechung des Fundamentalsatzes erwähnt habe, vollständig zu lösen.

Soweit zu sehen ist, scheint nun ein gewisser Abschluß erreicht zu sein, und es dürften wohl neue Ideen erforderlich sein, um hier weiterzukommen[22]. Ich möchte noch zum Abschluß darauf hinweisen, daß die Überlegungen, die mich zu meinem Beweis geführt haben, sich auch in ganz anderer Richtung als anwendbar zeigten, nämlich auf die

---

[19] C. A. Rogers, Ann. of Math. 48 (1947), S. 994—1002; J. W. S. Cassels, Proc. Cambridge 49 (1952), S. 165—166. Dort weitere Literaturangaben. Siehe auch G. H. Hardy-E. M. Wright, An introduction to the theory of numbers, Oxford.

[20] Ann. of Math. 46 (1945), S. 340—347. — In diesem Zusammenhang: A. Weil, Summa Brasiliensis Math. 1 (1946), S. 21—39. Ein integralgeometrischer Beweis von L. A. Santalo in: Ann. of Math. 51 (1953), S. 739—755.

[21] Von den Arbeiten seien einige angeführt. C. A. Rogers, Acta Math. 94 (1955), S. 249—287; Phil. Trans. Roy. Soc. London A (2) 48 (1955), S. 225—251; Proc. London Math. Soc. (3) 6 (1956), S. 305—320; (3) 8 (1958), S. 447—465; Mathematica 4 (1957), S. 1—6; W. Schmidt, Monatshefte für Mathematik 59 (1955), S. 274—304; 60 (1956), S. 1—10, 110—113; Acta Math. 102 (1959), S. 159—224; Proc. Amer. Math. Soc. 9 (1958), S. 390—403.

[22] Eine Arbeit von W. Schmidt, Illinois Jour. Math. 7 (1963), S. 18—23, enthält einiges Neues in dieser Richtung.

Aufstellung von Quadraturformeln hochdimensionaler Integrale, wie man sie gerade in den Anwendungen benötigt[23]. Sie werden zur Berechnung mehrfacher Integrale benützt, wie eine Arbeit aus dem Recheninstitut Stuttgart bereits zeigt.

Man kann noch die Frage erheben, ob ein analoger Satz gilt, wenn man nicht alle Gitter betrachtet, sondern z.B. nur jene Gitter, die aus einem festen Gitter durch eine Ähnlichkeitstransformation hervorgehen. Nach einem schwächeren Resultat von mir[24] hat dies Rogers[25] für die konvexen Körper nachgewiesen, aber diese Überlegungen sind, soviel ich sehe, nicht verallgemeinerungsfähig und es bleibt noch die Frage offen, für welche linearen Gruppen ein solcher Satz gilt[26].

[23] Monatshefte für Mathematik 66 (1962).

[24] Ann. math. 50 (1949), S. 314—317; er beruht auf einer Methode von C. L. Siegel. Vgl. dazu H. Davenport, Acta arith. Bd. 2 (1937), S. 262—265, ebenso E. Hlawka, Crellesches Journal 187 (1950), S. 246—252.

[25] Ann. of Math. 53 (1951), S. 531—540.

[26] Ich hatte für die orthogonale Gruppe 1942 einen analogen Satz aufgestellt (nicht publiziert). Für den Fall der Ebene hat D. B. Sawyer, Quarterly Jour. of Math. 13 (1962), S. 221—228, in dieser Hinsicht einen sehr scharfen Satz gefunden.

Discrepancy and uniform distribution of sequences*

Compos. Math. **16** (1964) 83–91

First I want to repeat some definitions: If we have a point x in R^s, we call his coordinates $p_j(x)$ $(j=1, \ldots s)$. Two points x, x' are congruent modulo 1, if $x-x'$ has integral coordinates. Now if we are only interested in properties modulo 1, we can always suppose that x is in the unit cube $E=E^s: 0 \leqq p_j < 1$ $(j=1, \ldots s)$ in other words we consider only the fractional part $\{x\}$ of x. Now we consider first infinite sequence $\omega: x_1, x_2, \ldots$ of points or more general $\omega: x_{11}; x_{21}, x_{22}; \ldots; x_N, \ldots x_{NN}; \ldots$ all lying in E^s. Now we consider an intervall $Q: \alpha_j \leqq p_j < \beta_j$ $(j=1, \ldots s)$ in E^s. Let $\chi_Q(x)$ be the characteristic function of Q. Then $N(Q, \omega) = \sum_{k=1}^{N} \chi_Q(x_k)$ is the number of points x_n of the given sequence with $n \leqq N$ lying in Q. If we have a double sequence then we define $N(Q, \omega) = \sum_{k=1}^{N} \chi_Q(x_{Nk})$. The sequence ω is called uniformly distributed modulo 1 if for each interval Q of E^s, $\lim_{N\to\infty} N(Q, \omega)/N$ exists and is equal to the volume of $Q: \mu(Q) = \int_E \chi_Q(x)dx$. This means that [1]

$$(1) \qquad \lim_{N\to\infty} \lambda_N(Q, \omega) = \lim_{N\to\infty} \frac{\sum_{k=1}^{N} \chi_Q(x_k)}{N} = \int_E \chi_Q(x)dx.$$

Sometimes it is better to write it in another form. Let Q be an arbitrary interval $\alpha_j \leqq p_j < \beta_j$ in R^s, not necessarily lying in the unit cube but with $\beta_j - \alpha_j < 1$ $(j=1, \ldots s)$ and counting now the number $F(Q, \omega, N) = \sum_{k=1}^{N} \chi'_Q(x_k)$, where χ'_Q is the function χ_Q periodically extended by translating O to all lattice points t, we see immediately that 1) $F(Q, \omega) = \sum_{k=1}^{N} \chi_Q(x_k)$ if Q is in the unit cube 2) F is periodic with period 1 and the sequence ω is uniformly distributed if and only if

$$(1') \qquad \lim_{N\to\infty} \frac{F(Q, \omega, N)}{N} = \mu(Q)$$

for each interval Q.

*) Nijenrode lecture.

In this definition it is not necessary to assume that the x_n are in the unit cube. Now it follows immediately from the definition of the Riemann integral for any function $f(x)$ (real or complex) defined in the unit cube or what is the same for any periodic function with period 1 and Riemann integrable.

$$(2) \qquad \lim_{N\to\infty} \lambda_N(f, \omega) = \lim_{N\to\infty} \frac{1}{N} \sum_{k=1}^{N} f(x_k) = \int_E f(x)dx = \mu(f)$$

and conversely. If we have any set M which has a content in the sense of Jordan we have setting $f(x) = \chi_M(x)$

$$(3) \qquad \lim_{N\to\infty} \frac{N(M, \omega)}{N} = \mu(M)$$

where $N(M, \omega)$ is the number of points $x_1, \ldots x_N$ of the sequence in M. All the formulas (1), (2) und (3) are equivalent. If we take [1] $f(x) = e^{2\pi i\langle hx\rangle}$, where h is an integral vector and $S_N(h, \omega) = 1/N \sum_{k=1}^{N} e^{2\pi i\langle hx_k\rangle}$ then we have

$$(4) \qquad \lim_{N\to\infty} S_N(h, \omega) = 0$$

(Criterion of Weyl), if only if ω is uniformly distributed 1.

It was I. Schoenberg, who generalized formula (4) in the one-dimensional case. He considered an arbitrary continuous distribution function $z(x)$ and defined: a sequence ω is asymptotically distributed modulo 1, with the distribution function $z(x)$, if $\lim \lambda_N(Q, \omega) = \int_E \chi_Q(x)dz(x)$ for any Q in E^s. If $z(x)$ has a density $\rho(x)$ we have $dz(x) = \rho(x)dx$. This definition was generalized to more than one dimension. It is now not necessary to speak more about this, because other members of this symposium will speak about these things. Now we will go back to (1), that means $\lim_N (\lambda_N(Q, \omega) - \mu(Q)) = 0$. Now we define: $N(\lambda_N(Q, \omega) - \mu(Q))$ is called the remainder term and we can ask, if we can say more about the sequence ω than $\lambda_N(Q) - \mu(Q) = o(1)$. Now we define the discrepancy: In the classical sense it is $D_N(\omega) = \sup_Q |\lambda_N(Q) - \mu(Q)|$ about all intervals Q in E^s. It has been shown by H. Weyl that for each uniformly distributed sequence ω, $\lim_N D_N(\omega) = 0$, that means that (1) is uniform in Q. It is obvious that $D_N \leqq 1$. Trivial is also that $ND_N \geqq 1$ and this was very improved by the important work of van Aardenne-Ehrenfest

[1] $\langle hx\rangle = \sum_{j=1}^{s} p_j(h) p_j(x)$.

and Roth; Cassels will speak about this. We have many results [2] in the one-dimensional case for special sequences so for the sequence $\{\alpha n\}$, α an irrational number. The results are depending on the finer arithmetical properties of α. If for instance
1 there is a number γ such that $n^{\gamma}\{n\alpha\} \geqq C$ for all n, then $ND_N = O((\log N))$ if $\gamma = 1$, $O(N^{1-(1/\gamma)+\varepsilon})$ if $\gamma > 1$ ($\varepsilon > 0$); on the other hand $ND_N = \Omega(\log N)$ for all α. Well known is the work of Vinogradov [3] about polynomials $f(n)$. In the linear case there are elementary methods using continued fractions, but the easiest way is to work with Weyl sums [4]. The form of this connection is the formula of Erdös-Turan-Koksma: for any positive integer[2] M

$$D_N \leqq 30^s \left(\frac{1}{M} + \sum_{0<||h||\leqq M} R^{-1}(h)|S_N(\omega, h)|\right)$$

$$||h|| = \text{Max}\,(|h_1|, \ldots |h_s|),$$

$$R(h) = \prod_{j=1}^{s} \text{Max}(|h_j|, 1).$$

Now I want to say some remarks on the definition of D. Some times is used another definition [5]: $D' = \sup_Q |(F(Q, N, \omega)/N) - \mu(Q)|$. It is easy to see that $D \leqq D' \leqq 2^s D$. One can also define the discrepancy D^* with the intervals $Q : 0 \leqq p_i < \beta_i < 1$. It is
2 clear that $D^* \leqq D \leqq 2^s D^*$. If we have an asymptotic distribution ρ then we define

$$D_N(\rho; \omega) = \sup_Q \left| \lambda_N(Q) - \int_E \rho(x)\chi_Q(x)dx \right|. \tag{5}$$

If we know something about D, can we say something about the difference

$$\lambda_N(f) - \mu(f)$$

where f is a Riemann integrable function? For the one-dimensional case Koksma 1942 [6] has shown the following result: if f is of bounded variation, $V(f)$ the total variation then

$$|\lambda_N(f) - \mu(f)| \leqq D_N(\omega)V(f) \tag{6}$$

generalizing results of Behnke and his own Satz 1 chap. 9 § 98. I have generalized [7] it to more dimensions. The variation $V(f)$ is to take in the sense of Hardy and Krause. For $f = P_r(x)$ the Bernouilli polynomial of degree r you get estimates for the sums

[2] $p_j(h) = h_j$.

$$(8)\qquad N(\lambda_N(P_r, \omega)-\mu(P_r)) = \sum_{k=1}^{N} P_r(x_k) = R_r^*$$

the most studied case [8] is $R_1 = R_1^* = \sum_{k=1}^{N} (\{x_k\}-\frac{1}{2})$.

I want to make another application in the one-dimensional case: consider the function $f(x) = \chi_\gamma(1/x-[1/x])$ where $\chi_\gamma(x)$ is the characteristic function of the interval $0 \leqq x < \gamma$. This function is not of bounded variation, the integral $\int_E f(x)dx = \int_0^\gamma \tau(x)dx$ with $\tau(x) = \sum_{n=1}^{\infty} 1/(n+x)^2$.

We consider therefore for a natural number l $f_l(x) = f(x)$ for $1/l \leqq x < 1$, $f_l(x) = 0$ for $0 \leqq x < 1/l$. The variation of $f_l(x)$

3 is $2l+1$, therefore we have

4
$$\begin{aligned}
|\lambda_N(f)-\mu(f)| &= \left|\frac{1}{N}\sum_{k=1}^{N} \chi_\gamma\left(\frac{1}{x_k}-\left[\frac{1}{x_k}\right]\right) - \int_0^1 \chi_\gamma \tau(x)dx\right| \\
&= \left|\frac{1}{N}\sum_{x_k<1/l} \chi_\gamma\left(\frac{1}{x_k}-\left[\frac{1}{x_k}\right]\right) - \int_0^{1/l} + \frac{1}{N}\sum_{x_k>1/l} \chi_\gamma - \int_{1/l}^1\right| \\
&\leqq |\lambda_N(f_l)-\mu(f_l)| + \left|\frac{1}{N}\sum_{x_k<1/l} 1\right| + \frac{1}{l} \\
&\leqq |\lambda_N(f_l)-\mu(f_l)| + \left|\frac{1}{N}\sum_{x_k\leqq 1/l} - \int_0^{1/l} dx\right| + \frac{2}{l} \\
&\leqq (2l+2)D_N(\omega) + \frac{2}{l}\,.
\end{aligned}$$

We take $l \sim D_N^{-\frac{1}{2}}$ and we get for each interval Q

$$\left|\frac{1}{N}\sum_{k=1}^{N} \chi_Q\left(\frac{1}{x_k}-\left[\frac{1}{x_k}\right]\right) - \int_0^1 \chi_Q(x)\tau(x)dx\right| \leqq 6\sqrt{D}$$

and therefore we have for the discrepancy D^+ of the sequence ω^+: $(1/x_k-[1/x_k])$

5
$$(7)\qquad D_N(\tau;\, \omega^+) \leqq 6\sqrt{D_N(\omega)}.$$

If ω is uniformly distributed we get the result of I. Schoenberg [9], that ω^+ is asymptotically distributed with the distribution density $\tau(x)$. I have generalized this to more than one-dimension. If $x = (p_1, \ldots p_s)$ is a point in R_s then I define $x^+ : p_j(x^+) = 1/p_j(x)-[1/p_j(x)]$ $(j=1, \ldots s)$. I consider with the sequence ω: $(x_1, x_2, \ldots)$ the sequence $\omega^+ : (x_1^+, x_2^+, \ldots)$ then I have [10]

$$(7')\qquad D_N(\rho;\, \omega^+) \leqq 6^s D_N^{1/(s+1)}(\omega)$$

$(\rho(x) = \tau(p_1) \ldots \tau(p_s))$. Another application of (6) in the more dimensional case is the following: if x is a point with coordinates $p_1 \dots p_s$, then let be $z^1 \leqq z^2 \dots \leqq z^s$ these numbers in ascending order and we set $f^{(j)}(x) = z^{(i)}$, for instance $f^{(1)}(x) = \mathrm{Min}\,(p_1, \dots p_s)$. If we have a sequence $\omega = (x_k)$ then we consider the one dimensional sequences $\omega^{(j)} = (f^{(j)}(x_k))$, $(j = 1, \dots s)$ and the s dimensional sequences $\tilde{\omega} = (f^{(1)}(x_k), \dots f^{(s)}(x_k))$; it is easy to show and well known [11] in the theory of order statistics that if the sequence ω is uniformly distributed, then the $\omega^{(j)}$ are asymptotically distributed with a density $\rho_i(x) = s\binom{s-1}{i-1}x^{i-1}(1-x)^{s-i}$ and $\tilde{\omega}$ with the density $\tilde{\rho} = s!$ in $p_1 \leqq p_2 < \dots \leqq p_s < 1$, 0 otherwise in E^s. We can easily show with the help of (6) that

$$D_N(\rho_j;\, \omega^{(j)}) \leqq C_s D_N(\omega),\; D_N(\tilde{\rho};\, \tilde{\omega}) \leqq 6^s D_N(\omega) \tag{8}$$

(C_s absolute constant).

To show it for $j = 1$ take in (6), where $\chi_\gamma(x)$ is defined as above,

$$f(x) = \sum_{k=1}^{s} \sum_{(i_1, \dots, i_k)} \prod_{j=1}^{k} \chi_\gamma(p_{i_j}) \prod_{i \neq i_j} (1-\chi_\gamma(p_i)) = \chi_\gamma(f^{(1)}(x)).$$

A weaker result than (6) we get [12] for all Riemann integrable functions f. If Z is a partition of the unit cube E in intervals, with Norm $l(Z)$, $\sigma(f, Z)$ the mean oscillation of f for the partition Z, we have

$$|\lambda_N(f) - \mu(f)| \leqq 6^s \textstyle\sum (f, D^{1/s}) \tag{9}$$

where $\sum (f, k) = \sup_{l(Z) \leqq k} \sigma(f, Z)$.

If we have a set M in E with the property that any line parallel to the coordinate axes intersects M in a set of points which, if not empty, consists of at most h intervals and the same is true for any of the m dimensional regions ($m \leqq s$) obtained by projecting M on one of the coordinate spaces defined by equating a selection of $s-m$ of the coordinates to zero then we get [13].

$$|\lambda_N(M, \omega) - \mu(M)| \leqq 20^s D_N^{1/s}(\omega) \tag{10}$$

This is true if M is convex; therefore if we consider a set $\{K\}$ of convex bodies in E we get

$$D_N(\{K\}, \omega) = \sup_{\{K\}} |\lambda_N(K, \omega) - \mu(K)| \leqq 20^s D_N^{1/s}(\omega) \tag{11}$$

We can say $D_N(\{K\}, \omega)$ is the discrepancy of ω for this set $\{K\}$. Bergström [14] 1935 has considered such a discrepancy but I do not know, for what sets of bodies. Now I want to estimate the discrepancy in the classical sense from above with the help of D_N

$(\{K\}, \omega))$. This seems to be a difficult problem. Take for example the set $\{S\}$ of all spheres. Then I can only show [15]

$$D_N(\omega) \leqq C(S)/\log D_N^{-1}(\{S\}, \omega)) \tag{12}$$

where the constant $C(S)$ is only depending on s. The difficulty stems from the fact that the covering and packing with spheres has a density different of one.

With the sequence $\omega = (x_k)$ consider the sequences $\omega^{(h)} = (x_{k+h} - x_k)$ $(h = 1, 2, \ldots)$. The fundamental theorem of van der Corput says that ω is uniformly distributed, if for each h, $\omega^{(h)}$ is uniformly distributed. Is there a connection between the discrepancies $D_N(\omega)$ and $D_N(\omega^{(h)})$? In the one-dimensional case this question was first considered by Vinogradov and then by van der Corput and Pisot [16] 1939, the final result was given by Cassels 1953 [17]. I generalized [18] the result to more dimension and we
6 have for any natural number q with $1 \leqq q \leqq s$

$$D_N(\omega) \leqq C(s)\sqrt{B}\left(1+\log^s \frac{1}{B}\right) \tag{13}$$

where

$$B = \left(\frac{1}{q} + \frac{1}{s}\right)\left(1+2\sum_{h=1}^{q-1} D_{N-h}(\omega^{(h)})\right) \tag{14}$$

If all $\omega(h)$ are uniformly distributed, then $D_{N-h}(\omega^{(h)}) \to 0$ for $N \to \infty$ therefore

7 $$\overline{\lim_N}\, D_N(\omega) \ll \frac{1}{\sqrt{q}}\left(l+\log^s \frac{1}{q}\right)$$

and therefore $\lim D_N(\omega) = 0$, if we go with $q \to \infty$.

An important result of H. Weyl is the following: Consider the one dimensional sequences $\omega(h) = (\langle x_k h \rangle)$, h integral vector, then the sequence ω is uniformly distributed if and only if the sequences $\omega(h)$ are for each h uniformly distributed. Now what is the connection between the discrepancies $D_N(\omega(h))$ and $D_N(\omega)$? This question was first considered by W. J. Coles [19], a pupil of Cassels, for the two dimensional case. I proved the following general result: For any natural M, we have

$$D_N(\omega) < C(s)\left(\frac{1}{M} + \sum_{0<\|h\|\leqq M} R^{-1}(h) D_N(\omega^{(h)})\right). \tag{15}$$

We can suppose that the h are primitive, that means that the coordinates of h are relative prime. Let us give an example. We have for the discrepancy of $\omega: \{\alpha k^g\}$ after Vinogradov (α irrational, g natural number $\geqq 1$): For any τ with $0<\tau<1$

$$(16) \qquad D_N(\omega) = 0(N^{\varepsilon}(N^{-\tau}+m^{-1}(\alpha))^{2/\gamma} \qquad (\gamma = 2^g)$$

where $m(\alpha) = \text{Min}\, q$, with $\{\alpha q\} < N^{-g+\tau}$.

Now we consider the s-dimensional sequences $\omega: (\alpha_1 k^g, \alpha_2 k^g, \ldots, \alpha_s k^g)$ where the $\alpha_1, \ldots \alpha_s$ are linear independent, then we get for the discrepancy of ω with the help of the above formula and the result of Vinogradov

8 $$(17) \quad D_N(\omega) = 0\left(N^{\varepsilon}\left(N^{-\tau}+\sum_{l=1}^{s}\sum_{(i_1,\ldots i_l)} m^{1/l}(\alpha_{i_1},\ldots \alpha_{i_l})\right)\right)^{2/\gamma}$$

where $m(\alpha_{i_1}\ldots\alpha_{i_l}) = \min(q_1\ldots q_l)$, so that

$$(18) \qquad \{\alpha_{i_1}q_1+\ldots+\alpha_{i_l}q_l\} < N^{-g+\tau}.$$

For $s = 2$ this is an old result of Kusmin and Skopin [20] 1934, who work directly with Weyl-sums. Let us make some remarks about uniform distribution in a compact space. The difficulty is the definition of discrepancy. I would propose to define the discrepancy in the following way:

$$D_N = \sup_K \left|\frac{1}{N}\sum_{k=1}^{N}\chi_K(x_k) - \mu(K)\right|$$

about all convex bodies K. This definition depends on the definition of a convex set in a compact space. This is easy, if in this space exists a Riemann metric. A special case namely the spheres $S^m: p_1^2+\ldots+p_m^2 = \langle p, p\rangle = 1$ and for the special sequence $\{g/\sqrt{n}\}$, where g are lattice points, with $\langle g, g\rangle = n$ (n natural numbers) is already known. For $m \geqq 4$, this sequence is uniformly distributed on S^m, with

$$\mu(K) = \frac{\text{surface of } K}{\text{surface of } S}$$

(K a Jordan measurable set on S^m), and we have for the discrepancy in the above sense

$$D_N = O\left(\frac{(\log\log n)^{m-2}}{(\log n)^{m/2-1}}\right)$$

($N = r(n)$ the number of solutions of $\langle g, g\rangle = n$). This was shown by Malyshew [21] but natural expressed in other terms.

NOTES

[1] Compare for these definitions the report of J. Cigler and G. Helmberg "Neuere Entwicklungen der Theorie der Gleichverteilung", Jahresberichte der DMV 64(1961) S. 1–50. (short C & H).
There is a continuous analogue studied especially by L. Kuipers and B. Meulenbeld, the so called C-uniform distribution: Let $x(t)$ be a real-valued measurable vector function in s dimension, defined on the interval $0 \leqq t < \infty$, then $x(t)$ mod 1 is C uniformly distributed, if for any Q in E^s:

$$\lim_{T\to\infty} \frac{1}{T}\int_0^T \chi_Q(\{x(t)\})dt = \mu(Q)$$

(compare § 11 of the article cited above).

[2] See § 6 of C. & H. and J. F. Koksma Diophantische Approximationen Ergebn. Math. u. Grenzgeb. Bd. 4, Heft 4, Berlin 1936.
If a is a vector in E^s and if there exist an η and a $C > 0$, that $R(h)^\eta \{\langle ha\rangle\} \geqq C$ for all lattice points $h \neq 0$, then $D_N = O(N^{-1/\eta} \log^s N)$. If $a = (\alpha_1, \ldots \alpha_s)$, where $1, \alpha_1 \ldots \alpha_s$ linear independent algebraic numbers in a field K of degree $s+1$, then $\eta = s$. It is unknown if there exist a with $\eta = 1$ (Littlewood problem). For almost all a $\eta = 1+\varepsilon(\varepsilon > 0)$. This result can be improved (W. Schmidt).

[3] I. M. Vinogradow,
The method of trigonometric sums in the theory of numbers.
See also the report of Loo-Keng Hua in Enzykl. math. Wiss. 12, Heft 13, I 1950.

[4] Easier are to find metrical theorems:
We have sequences (x_n) depending on a parameter θ and it is asked for an estimate for the Discrepancy D for all θ, except a set of measure zero. There are important results of Koksma, Cassels, Erdös compare C & H § 9. There exist metrical Theorems where instead the Lebesguemeasure is considered Hausdorffmeasure or the Capacity of the set (R. Salem). In the proof of the theorems is of great use a theorem of J. F. Koksma and Gál.
For the C-uniform distribution metrical theorems seems not to be published. In this case it seems interesting to study vector functions $x(t)$ depending not only on a parameter a but on a continuous function and to consider the Discrepancy for all θ except a set Wienermeasure zero. Some results in this direction have been found by the author (unpublished).

[5] Compare J. W. S. Cassels, An introduction to Diophantine Approximation, Cambridge Univ. Press. 1957 and the paper of J. G. van der Corput, et Ch. Pisot, Proc. Kon. Ned. Akad. v. Wet. 42, 476–486 (1939) and for the continuous case E. Hlawka, Ann. Math. pure and appl. IV. Ser. 49, 311–326 (1960).

[6] Mathematica, Zutphen B 11, 7–11 (1942).

[7] Ann. Mat. pure appl. IV, Ser. 54, 325–333 (1961).

[8] Compare the report of Koksma 2.

[9] Mathem. Z. 28, 171–199 (1928).

[10] Monatshefte für Math. 65 (1961).

[11] Compare for instance L. Schmetterer.
Einführung in die Mathematische Statistik, Springer-Verlag, Wien (1956) Kap. 7. For (8) see E. Hlawka, Math. Ann. 150, 259–267 (1963).

[12] A proof was given at the Jahresversammlung der DMV, Bonn 1962. If we define $\Delta_i f = f(\ldots, p_i+h_i, \ldots)-f(\ldots, p_i, \ldots)$ (other coordinates fixed) and if we suppose that there exists a constant C and α with $0 < \alpha < 1$ such that $|\Delta_{i_1,\ldots i_r} f| \leqq (\alpha+1/2)^r C|h_{i_1}\ldots h_{i_r}|^\alpha$ (the so called class H) for all $1 \leqq r \leqq s$, $i_1 < i_2 < \ldots < i_r$ then $|\lambda_N(f)-\mu(f)| \leqq 6^s CD_N^\alpha(\omega)$. J. F. Koksma told me at this symposium that he has for $s = 1$ and for continuous functions some results of the form (9).

[13] Compare [7].

[14] Medd. mat. Sem. Univ. Lund 2 (1935) 97 S. Kunigl. Fysiografiska Sätslkapets Lund För handlinger 6 (1936) S. 113–131.

[15] Using methods of the geometry of numbers.

[16] Compare the paper of these authors cited in [5].

[17] Math. Ann. 126, 108–118 (1953).

[18] Math. Zeitschrift 77, 273–84 (1961).

[19] Proc. Cambridge 53, 781–89 (1957).

[20] H 3 B A H CCCP (7) 1934 S. 547–60.

[21] Compare for instance the literature in the paper of Pommerenke in Acta Arithmetica V and A. V. Malysev Dokladi Akad. CCCR 133 (1960) 1294–97.

(Oblatum 29-5-63).

Corrections

85.1	$n^\gamma \min(\{n\alpha\}, 1-\{n\alpha\}) \geq C$	86.5	$\leq 12\sqrt{D_N(\omega)}$
85.2	distribution with density ρ	88.6	with $1 \leq q \leq N$
86.3	is $\leq 2l-1$, therefore	88.7	$\ll \frac{1}{\sqrt{q}}(1+\log^s q)$
86.4	$-\int_0^1 f(x)\,dx$	89.8	$\sum_{(i_1,\cdots,i_l)} m^{-1/l}(\alpha_{i_1},\cdots,\alpha_{i_l})$

Uniform distribution modulo 1 and numerical analysis*

Compos. Math. **16** (1964), 92–105

One of the oldest applications of uniform distribution modulo 1 seems to be a theorem of Fejer and Kalmar, going back to a theorem of C. Runge about Lagrange interpolation: Let C be a closed bounded pointset of the finite z-plane, consisting of a finite number of mutually exterior Jordan regions. Let the points $z_1^{(n)}, \ldots z_{n+1}^{(n)}$ $(n = 1, 2, \ldots)$ lie on the boundary B of C; a necessary and sufficient condition that the sequence of polynomials $p_n(z)$ of degree n found by interpolation to an arbitrary function $f(z)$ analytic on C in the points $z_1^{(n)}, \ldots, z_{n+1}^{(n)}$ converges uniformly to $f(z)$ is, that the transforms of the $z_1^{(n)}, \ldots z_{n+1}^{(n)}$, be uniformly distributed on the unit circle. The definition of the transform w of the point z is the following: Let K be the exterior of C and let $w = \varphi(z)$ map K on to the exterior of $|w| = 1$, so that the points at infinity in the two planes correspond to each other. I do not want to go further in the details of this theorem, because there is the excellent exposition in the wellknown book of Walsh [1]. I want to speak now about the calculation of multiple integrals.

Let us suppose, that a function f is defined on the unit cube $E = E^s$

$$(1) \qquad 0 \leqq p_1 < 1, \ldots 0 \leqq p_s < 1 \qquad (s \geqq 1)$$

periodic with period one and integrable in the Riemann-sense. This shall be at first the only condition about f. The classical formulas in numerical calculation of integrals are working with sequences

$$(2) \qquad \left(\frac{k_1}{n}, \ldots, \frac{k_s}{n}\right) \qquad 0 \leqq k_j \leqq n-1 \qquad (n \geqq 1).$$

The number of the points is $T = n^s$ and if we consider a Riemann-sum

* Nijenrode lecture.

$$\sum_T = \frac{1}{n^s} \sum_{k_1,\ldots,k_s} f\left(\frac{k_1}{n}, \ldots, \frac{k_s}{n}\right)$$

then we have [2] if f is of bounded variation,

1 $$|\Delta_T| = |\sum_T - \mu(f)| = o(T^{-1/s}), \qquad \mu(f) = \int_E f dx_1 \ldots dx_s$$

and we can easily find functions of bounded variation, for which

$$\Delta_T \geqq CT^{-1/s}.$$

If you want, that $\Delta_T < \varepsilon$ for a given ε, then you must take $T \sim (1/\varepsilon)^s$ that is for large s beyond the realm of computers. The situation is the same, if you work for instance with Simpson-formulas in more dimensions. Therefore is often used the Monte Carlo-method [3]. Here one uses a sequence of independent uniformly distributed random variables $x_1, \ldots, x_T$ in the unit cube and if you consider

$$F = \frac{1}{T} \sum_{j=1}^{T} f(x_j)$$

then the expected value $E(F)$ is $\mu(f)$ and for the variance σ_F^2 you have σ_f^2/T, therefore the probable error in mean is $T^{-\frac{1}{2}}$. This looks very nice, but you have to replace the random variables by a sample of pseudo random numbers and then the problem remains, if the error Δ_T is also of this order. This is not so, and this has been shown by numerical examples by Ph. Davis and Rabinovits [4] and also in a later paper by N. M. Korobow [9]. It seems, that there are many mathematicians, who where suspicious about the Monte Carlo-method, but it is not very easy to find this in the literature. I mention further a paper of J. Bass and J. Guilloud [5] and a lecture of J. M. Hammersley [6]. There seems to be also a paper of Richtmyer (1951) in the Los Alomos reports, in which is not worked with a Monte-Carlo-method, but with the theory of uniform distribution. Also a paper of G. Peck (1953) seems to belong to the ideas of the paper of Richtmyer. Further I want to mention a new paper of Richtmyer, Devany, Metropolis [7] where it is said, that it would be very important to find new methods for calculating multiple integrals. Perhaps it was un-
2 known to these authors the work of N. M. Korobow [7] and mine[2].

Korobow always works with functions f of class E_s^α:

$$f = \sum_h C_h e^{2\pi i \langle hx \rangle}$$

where the Fourier coefficients

$$C_h = 0(R(h)^{-\alpha}),\ R(h) = R(h_1, \ldots h_s) = \prod_{j=1}^{s} \mathrm{Max}(|h_j|, 1)$$

with $\alpha > 1$, so that the Fourier series of f is absolute convergent. In a paper [8] of 1957 he showed, that for functions of the class E_s^2 and the sequence $(k/p^2, k^2/p^2, \ldots k^s/p^2) \bmod 1$

$$\text{(3)} \qquad \begin{gathered} p \text{ prime},\ k \leqq p^2-1,\ T = p^2 \\ \Delta_T = 0(p^{-\frac{1}{2}}), \end{gathered}$$

the error is the same as in the Monte Carlo-method. In lectures, given in Princeton at the Columbia University 1959 and at the II. Hungarian Mat. Congress in Budapest 1960, I have shown, [2] that the discrepancy of *this* sequence is

$$\text{(4)} \qquad \leqq (70 \log p)^s/\sqrt{p}$$

and then you have for functions of bounded variation

$$\text{(5)} \qquad \Delta_T \leqq V(f)\,\frac{(70 \log p)^s}{\sqrt{p}}$$

using[1] the formula I (6).

You get the same result with the sequence,

$$\text{(6)} \qquad \omega_1(s, T): \left(\frac{k}{p}, \frac{k^2}{p}, \ldots, \frac{k^s}{p}\right), \qquad k \leqq p-1, \quad T = p$$

applying the theorem of Hasse and A. Weil about the zeros of the Riemann congruence zeta-function. It can be shown, that in this formula the power of $\log p$ cannot be cancelled. But there remains the possibility that the formula is true with a power of $\log \log p$. Korobow and I independently found sequences of the form $(\beta > 0)$

$$\text{(7)} \qquad \omega_2(s, T): \left(\frac{kg_1}{p}, \ldots, \frac{kg_s}{p}\right) \bmod 1, \qquad k \leqq p-1, \quad T = p$$

for which the error is of the order $0\left((\log p)^{s\beta}/p\right)$ therefore better than the Monte-Carlo-method. Korobow showed this only for the class E_s^α and his work was continued by I. N. Sobol, Solodov, Bachalov and I. Sarygin [9]. I showed it for functions of bounded variations and the results are more general and more precise [2]. More than $(\log p)^{(s-1)/2}/p$ is not possible after the theorem of

[1]) The lecture "Discrepancy and uniform distribution" is cited with I. (This Vol. p. 88).

K. Roth. Another sequence with the same properties as the sequence of Korobow and me, was found by Halton, [10] generalizing a sequence of van der Corput on the suggestion of Hammersley, but only calculating the discrepancy of the sequence. Sometimes people are astonished to hear, that it is possible to find sequences which give a smaller error than the Monte-Carlo-method for large T. But they forget the following thing: It follows from the theorem of large numbers, that for almost all sequences the error term is of the order $0(p^{-\frac{1}{2}} \log\log p)$. This means for all sequences except a set of measure zero, depending on f. But in this set of measure zero, there may exist sequences with larger errors, but also with smaller errors and my theorem and also the theorem of Korobow show, that there exist sequences, that for large classes of f the error terms constructed with these sequences are smaller than in the Monte-Carlo-method [11]. Further it must be said, that it is impossible, that an error term is independent of s for all functions of bounded variation.

This follows from the theorem of Roth.

Now I want to go into details. My method is very simple and was inspired by some methods in geometry of numbers. I call a lattice point $g = (g_1, \ldots, g_s)$ a good lattice point modulo p (p prime) if for all lattice points $h = (h_1, \ldots, h_s)$ with $0 < ||h|| \leq (p-1)/2$ and

$$\langle hg \rangle \equiv h_1 g_1 + \ldots + h_s g_s = 0 \pmod p$$
$$R(h) \geqq p\,(8 \log p)^{-s}.$$

Such good lattice points exist. I will show this for $s = 2$. We can use $g_1 = 1$ and consider all solutions g of

$$h_1 + h_2 g \equiv 0 \pmod p, \quad 0 < \operatorname{Max}(|h_1|, |h_2|) \leqq \frac{p-1}{2},$$
$$|h_1 h_2| \leqq p/(8 \log p)$$

For fixed h_1 we have $p/(8 h_1 \log p)$ possibilities for g, therefore for all h_1, h_2 the total number of g is

$$\frac{p}{8 \log p} \sum \frac{1}{h_1} < \frac{p}{4}$$

therefore exists a g_2 such that for all h_1, h_2 with $h_1 + h_2 g_2 \equiv 0 \pmod p$, we have

$$|h_1 \cdot h_2| \geqq \frac{p}{8 \log p}.$$

If we have already found $g_1 = 1, g_2, \ldots g_{s-1}$ we consider again the congruence

$$h_1 + h_2 g_2 + \ldots + h_{s-1} g_{s-1} + h_s g \equiv 0 \ (\mathrm{mod}\, p),\ R(h) < \frac{p}{(8 \log p)^{s-1}}$$

we see that the total number of all this g with the same method as before is

$$\frac{p}{(8 \log p)^{s-1}} \sum_h \frac{1}{R(h)} < \frac{p}{4}$$

and so it is shown, how we can find successively $g_1, g_2, \ldots g_s$.

The g_j exists also, if p is not a prime, but instead of the constant 8 you have to take another constant. Especially important for the numerical applications is the case that p is a power of two. Now it can be shown, that the discrepancy of the sequence (7) with the numbers g_i is smaller than

(8) $$< \frac{(1000 \log p)^{2s}}{p}$$

and it can be shown, that we can choose these numbers g so, that the discrepancy is smaller than

(8′) $$< \frac{(1000 \log p)^{s}}{p}.$$

Now I want to say some words about the Corput-Halton sequence. Let R be a natural number, then we have for any natural number n

$$n = n_0 + n_1 R + \ldots.$$

We define

3 $$\varphi_R(n) = \frac{n_0}{R} + \frac{n_1}{R} + \ldots,$$

then the sequence is

(9) $$\omega_3(s, T) : (\varphi_{R_1}(n), \ldots \varphi_{R_s}(n)) \qquad n = 1, 2, \ldots T$$

where $R_1, \ldots R_s$ have the greatest common divisor 1. For example $R_1, \ldots R_s$ can be the first s primes. It can be shown, that

$$D(\omega_3(s, T)) < \frac{(\log T)^s}{T} \prod_{i=1}^{s} \left(\frac{3R_i - 2}{\log R_i} \right),\ T > \mathrm{Max}\,(R_1, \ldots R_s).$$

This was shown by van der Corput for $s = 2$. You can get even a better sequence

$$(9') \qquad \left(\frac{n}{T}, \varphi_{R_1}(n), \ldots \varphi_{R_{s-1}}(n)\right)$$

with

$$D < \frac{(\log T)^{s-1}}{T} C_s \qquad (C_s \text{ constant}).$$

It is not necessary to work with prime numbers, because for large s there would be necessary a large table of prime numbers [12]. I would propose to work with the Fermat-numbers, because after a well known theorem, the Fermat-numbers are relative prime.

Now suppose, that we know more about f.

Let us suppose, that all derivatives for fixed m $(m \geqq 2)$

$$\frac{\partial^{mr} f}{\partial x_{l_1}^m \ldots \partial x_{l_r}^m} \qquad (1 \leqq r \leqq s, l_1 < l_2 < \ldots < l_r)$$

exist are periodic and continuous in the unit cube. Then we can show [2] [9] that for $\omega_2(s, T)$

$$(10) \qquad |\lambda_T(f) - \mu(f)| < \frac{C\,(30 \log p)^{(m+1)s-1}}{p^m}.$$

There are two possibilities to show this [13]. The first method is, to develop f in a Fourier series. But it seems better to work with a generalization of the Euler summation formula to get smaller constancy in the error. We have

$$(11) \quad f(x) = \mu(f) + \sum_{r=1}^{m} \frac{P_r(x)}{r!}\left(f^{(r-1)}(1) - f^{(r-1)}(0)\right) - \int_0^1 \frac{P_m(x-t)}{m!} f^{(m)}(t)dt$$

in one variable, therefore for periodic function with continuous derivatives till the order m

$$(12) \qquad f(x) = \mu(f) - \int_0^1 \frac{P_m(x-t)}{m!} f^{(m)}(t)dt$$

and for more dimension

$$f(x) - \mu(f) = \sum_l \int_{E^l_{l_1,\ldots l_r}} \frac{\partial^{mr} f(t)}{\partial x_{l_1}^m \ldots \partial x_{l_r}^m} P_{m_1,\ldots m_r}(x-t)dt_{l_1} \ldots dt_{l_r}$$

$$P_m(x) = P_{m_1,\ldots m_r}(x) = P_m(x_{l_1}) \ldots P_m(x_{l_r})$$

and therefore

$$|\lambda_T(f) - \mu(f)| \leqq \sum_l \int_{E_l} \left|\frac{\partial^{mr} f}{\partial x_{l_1}^m \ldots dx_{l_r}^m}\right| \left|\frac{1}{T}\sum_{n=1}^{T} P_m(x_n)\right|$$

and now we have

$$\frac{1}{T}\sum_{n=1}^{T} P_m(x_n) = \sum \frac{1}{R(h)^m}\frac{1}{T}\sum_{n=1}^{T} e^{2\pi i\langle hx_n\rangle}$$

and estimating this sum we get the above result.

Now let us suppose that the derivatives are continuous in the unit cube, but not periodic with period one. To fix the ideas I suppose that $m = 2l$ and $s = 1$. We form $\varphi(x) = \frac{1}{2}(f(x)+f(1-x))$ then we have, because $P_\nu(x)+P_\nu(1-x) = 0$ for ν odd, $(\Delta^k f = f^{(k)}(1)-f^{(k)}(0))$

$$\varphi(x) = \mu(f)+\sum_{\nu=1}^{l}\frac{P_{2\nu}}{(2\nu)!}\Delta^{2\nu-1}f-\int_0^1\frac{P_{2l}(x-t)}{(2l)!}\varphi^{(2l)}(t)dt$$

then I apply the operator

$$U^k\varphi = \left(\varphi\left(\frac{x}{2}\right)+\varphi(\tfrac{1}{2}+x)-\frac{1}{2^{k-1}}\varphi(x)\right)\Big/\left(1-\frac{1}{2^{k-1}}\right)$$

and form

$$\varphi_2 = U^2\varphi,\ \varphi_4 = U^4\varphi_2,\ \ldots\ \varphi_{2l} = U^{2l}\varphi_2(l-1)$$

and we have

(14) $$\varphi_{2l}(x) = \mu(f)-C_{2l}\int_0^1 P_{2l}(x-t)\varphi_{2l}^{(2l)}(t)dt$$

more general $(0 \leqq k \leqq l,\ \varphi_0 = \varphi)$

$$\varphi_{2k}(x) = \mu(f)-\sum_{\nu=k+1}^{l} C_k\frac{P_{2k}}{(2k)!}\Delta^{(2k-1)}(f)-\int_0^1\frac{P_{2l}(x-t)}{(2l)!}\varphi^{(2l)}(t)dt.$$

This can be easily shown by complete induction in k. For $k = 0$ it is clear and we have

$$\varphi_{2k+2} = U^{2k+2}\varphi_{2k} = \mu(f)+\sum_{\nu=k+1}^{l} U^{2k+2}P_{2\nu}(x)\Delta^{2\nu-1}f-\int_0^1 U^{2k+2}P_{2l}\varphi^{(2l)}$$

now

$$U^{2k+2}P_{2k}(x) = P_{2k+2}\left(\frac{x}{2}\right)+P_{2k+2}(\tfrac{1}{2}+x)-\frac{1}{2^{2k+1}}P_{2k+2}(x) = 0.$$

We can easily generalize it to more variables, if we apply the operators to all variables and then we came back to formula (15) where we have now instead of f the function φ as (6) shows, with $\mu(\varphi) = \mu(f)$.

We can also work with the operators

$$U_p^k\varphi = \left(\sum_{h=0}^{p-1}\varphi\left(\frac{x}{p}+\frac{h}{p}\right)-\frac{1}{p^{k-1}}\varphi(x)\right)\Big/\left(1-\frac{1}{p^{k-1}}\right)$$

or better

$$S_p^k \varphi = \left(\sum_{h=0}^{k-1} \varphi\left(x+\frac{h}{p}\right) - \frac{1}{p^{k-1}}\varphi(\{px\})\right) \Big/ \left(1-\frac{1}{p^{k-1}}\right)$$

if we work with the sequence (7).

Now let us suppose we want to calculate a multiple integral about a simplex [14]. Now we can apply the formula (9) of I. We get a much better result if we apply the second formula I (8) and we take for ω a good sequence (7) or (9) and we have for the simplex $\tilde{S}: 0 \leqq p_1 \leqq p_2 \ldots \leqq p_s < 1$ if $f(x)$ is of bounded variation on $\tilde{S}$

$$(15) \qquad \left|\lambda_T(f, \tilde{\omega}) - \frac{\tilde{\mu}(f)}{\tilde{\mu}(\tilde{S})}\right| \leqq C_s \frac{(30 \log T)^s}{T} V_{\tilde{S}}(f)$$

where $\tilde{\mu}(f) = \int_{\tilde{S}} f dx$ and $\mu(\tilde{S}) = 1/s!$ is the volume of $\tilde{S}$.

With the help of one affine mapping we can therefore construct to any simplex S a good sequence ω (T, S), so that

$$(15') \qquad \left|\lambda_T(f, \omega) - \frac{\mu(f)}{\mu(S)}\right| < C^s \frac{(\log T)^s}{T} V_s(f)$$

where $\mu(f)$ is the integral of f about S, $V_s(f)$ the variation of f on S, $\mu(S)$ the volume of S. It follows immediately that there must exist a sequence ω defined on any complex S with the same formula (15′) and it is then easy, to see, how to construct such ω, when S is a differentiable compact manifold, because it is a wellknown theorem in topology, that S can be covered by a finite complex. There exist many good sequences ω on S, but it is essentially for the numerical work to find sequences, which can easily be computed, e.g. when S is a sphere, but about these things I will speak on another occasion, but compare the sequence at the end of I.

Now we will consider some applications.

We consider an integral equation [16]

$$(16) \qquad \varphi(x) - \lambda \int_0^1 K(x, y)\varphi(y)dy = f(x).$$

We suppose that $K(x, y)$ is defined on the unit interval $E = E^2$: $0 \leqq x < 1, 0 \leqq y < 1$ and that it is periodic with period one and that it has a finite variation $V(K)$ in E^2. Further we set $\sigma(K) = \sup_{E^2} |K(x, y)|$, $M = \text{Max}\,(\sigma(K), V(K))$. We suppose further that $f(x)$ is defined on E^1: $0 \leqq x < 1$, periodic and of bounded Variation $V(f)$, $\sigma(f) = \sup|f|$, $A = \text{Max}\,(\sigma(f), V(f))$. The solution of the equation is given by the Neumann-series

$$\varphi(x) = f(x) + \sum_{n=1}^{\infty} \lambda^n \int_0^1 K^{(n)}(x, y) f(y) dy \tag{17}$$

where

$$K^{(n)}(x, y) = \int_0^1 \dots \int_0^1 K(x, x_2) K(x_2, x_3) \dots K(x_n, y) dx_2 \dots dx_n. \tag{18}$$

The series is convergent, if $|\lambda| M < 1$.

We set $\Pi^{(n)}(x_1, \dots x_n) = K(x_1, x_2) K(x_2, x_3) \dots K(x_{n-1}, x_n)$ defined on the unit cube $E^n: 0 \leqq x_1 \leqq 1, 0 \leqq x_n \leqq 1$. It is easy to show, that for the variation we have $V(\Pi^{(n)}) \leqq 2(6M)^{n-1}$ and for the variation of $F(\xi, x_1, \dots x_n) = \Pi^{(n)}(\xi, x_1, \dots x_n) f(x_n)$, we have $V(F) \leqq 2A(6M)^n$.

With these definitions, the Neumann-series has the form

$$\varphi(x) = f(x) + \sum_{m=1}^{\infty} \lambda^m \int_{E^m} F(x, x_1, \dots x_m) dx_1 \dots dx_m$$

and if s is a natural number

$$\varphi(x) = f(x) + \sum_{m=1}^{s} \lambda^m \int_{E^m} F(x, x_1, \dots, x_m) dx_1 \dots dx_m + R_1,$$

$$|R_1| \leqq A \frac{(|\lambda| M)^{s+1}}{1 - |\lambda| M}.$$

Now we apply our quadrature formula for $\int_{E^m} F\ (m \leqq s)$ and choose a sequence $\omega(m, T) = (y_{k1}, \dots y_{km})\ (k = 1, \dots T)$ with discrepancy $D(\omega(m, T))$ (m is the dimension), then we have

$$\begin{aligned} \varphi(x) = f(x) + \frac{1}{T} \sum_{k=1}^{T} \sum_{m=1}^{s} \lambda^m K(x, y_{k1}) \\ K(y_{k1}, y_{k2}) \dots K(y_{k,m-1}, y_{km}) f(y_{km}) + R \end{aligned} \tag{19}$$

where

$$|R| = |R(s, T, \omega)| < A\left(2 \sum_{m=1}^{s} (6M|\lambda|)^m D(\omega(m, T)) + (|\lambda| M)^{s+1} \Big/ (1 - |\lambda| M)\right)$$

If we consider a sequence $\omega(s, T)$ and we take for the sequence $\omega(m, T)$ $(m \leqq s)$ the sequence of points, whose coordinates are the first m coordinates of the points of $\omega(s, T)$, we have $D(\omega(m, T)) \leqq D(\omega(s, T))$ therefore we get the simpler form of the remainder term, $B = \text{Max}\,(M, 1)$.

$$|R| < A(12(6B)^s D(\omega(s, T)) + (|\lambda| M)^{s+1} / (1 - |\lambda| M)). \tag{20}$$

If we take one of the good sequences, discussed before, and we take

$$s = \left[\frac{1}{4}\frac{\log T}{\log\log T}\right],$$

we have for $\varepsilon > 0$, $u = 4\log(\varepsilon(1-|\lambda|M))/\log|\lambda|M$

$$\text{(21)} \qquad |R| < 2A\varepsilon \text{ for } T > \operatorname{Max}(\exp(u\log^2 u), \; \varepsilon^{-2}, \exp(1000\,B)) = T_0(\varepsilon).$$

If we take the sequence $(k/p, \dots k^s/p) \bmod 1$ $T = p-1$, p prime, we have

$$\text{(22)} \qquad R < 2A\sqrt{\varepsilon}$$

with the same s and we have, when we go with T to infinity

$$\text{(23)} \qquad \varphi(x) = f(x) + \lim_{T\to\infty} \frac{1}{T} \sum_{k=1}^{T} \sum_{m=1}^{s(T)} K(x, y_{k1}) \; K(y_{k1}, y_{k2}) \dots K(y_{k,m-1}, y_{km}) f(y_{km}).$$

Now let us consider the homogeneous integral equation

$$\text{(24)} \qquad \varphi(x) = \lambda \int_0^1 K(x, y)\varphi(y)dy$$

with $K(x, y) = K(y, x)$ and let us suppose, that $K(x, y)$ is always > 0, then we have after the wellknown theorem of Jentsch for the Eigenvalues $\lambda_1, \lambda_2, \dots$ of this equation $0 < \lambda_1 < |\lambda_2| \leqq |\lambda_3| \leqq \dots$. This means that the lowest Eigenvalue is simple and positive. We can always suppose, that $\lambda_1 \geqq 1$. The Eigenfunctions to these Eigenvalues let us denote $\varphi_1, \varphi_2 \dots$. Now we want to calculate the lowest Eigenvalue λ_1. After Kellogg, we have $\lambda_1 = \lim_{s\to\infty} I_{s-1}/I_s$ where $I_s = \int_0^1\int_0^1 K^{(s-1)}_{(x,y)}$ and more precise:

$$\left|\lambda_1 - \frac{I_{s-1}}{I_s}\right| < c\left(\frac{\lambda_1}{\lambda_2}\right)^{s-1}.$$

we have

$$I_s = \int_{E^s} \Pi^{(s)}(x_1, \dots x_s)dx_1 \dots dx_s.$$

If we apply again our quadrature formula, we get a result of the Type

$$\text{(25)} \qquad \lambda_1 = \frac{\sum_{k=1}^{T} K(y_{k1}, y_{k2}) \dots K(y_{k,s-2}, y_{k,s-1})}{\sum_{k=1}^{T} K(y_{k1}, y_{k2}) \dots K(y_{k,s-1}, y_{k,s})} + R$$

and we get

$$\lambda_1 = \lim_{T\to\infty} \rule{4cm}{0.4pt}$$

Our method can also be applied for linear equations

$$\varphi(i) - \frac{\lambda}{U}\sum_{k=1}^{i} K(i, k)\varphi(k) = f(i) \qquad 1 \leqq i \leqq U \tag{28}$$

because it is only a special case of an integral equation, with the kernel of bounded variation. But it is of interest to write explicitly the formulas for this case. The Neumann series is in this case

$$\varphi(i) = f(i) + \sum_{m=1}^{\infty} \lambda^m \frac{1}{U}\sum_{i=1}^{U} K^{(m)}(i, k)f(k) \tag{17'}$$

where

$$K^{(n)}(i, k) = \frac{1}{U^n}\sum_{i_2,\ldots i_n=1}^{U} K(i, i_2)K(i_2, i_3)\ldots K(i_n, k)$$

$$\begin{aligned} V(K) = &\sum_{i,k=1}^{U} |K(i,k)-K(i-1,k)-K(i,k-1)+K(i-1,k-1)| \\ &+\sum_{i=1}^{U}|K(i,1)-K(i-1,1)| + \sum_{k=1}^{U}|K(1,k)-K(1,k-1)|, \sigma(K) \\ &= \sup|K(i,k)|, V(f) = \sum_{i=1}^{U}|f(i)-f(i-1)|. \end{aligned} \tag{18'}$$

The Neumann series is convergent for $|\lambda|M < 1$. If we write the system of linear equations in the usual form

$$\varphi(i)-\mu\sum_{k=1}^{U} K(i, k)\varphi(k) = f(i) \text{ with } |\mu|M < U.$$

If we set

$$\tilde{\varphi}(x) = \varphi(i), \text{ if } \frac{i-1}{U} \leqq x < \frac{i}{U} \qquad (i = 1,\ldots U)$$

$$\tilde{K}(x,y) = K(i,k) \text{ if } \frac{i-1}{U} \leqq x < \frac{i}{U}, \frac{k-1}{U} \leqq y < \frac{k}{U} \ (i,k = 1,\ldots U)$$

$$\tilde{f}(x) = f(i) \text{ if } \frac{i-1}{U} \leqq x < \frac{i}{U} \qquad (i = 1,\ldots U)$$

then we have

$$\int_0^1 \tilde{K}(x, y)\tilde{\varphi}(y)dy = \sum_{k=1}^{U}\int_{k-1/U}^{k/U} \tilde{K}(x, y)\tilde{\varphi}(y)dy = \frac{1}{U}\sum_{k=1}^{U} K(i, k)\varphi(k)$$

and the system of linear equations has the form

$$\tilde{\varphi}(x)-\lambda\int_0^1 \tilde{K}(x, y)\tilde{\varphi}(y)dy = \tilde{f}(x).$$

It is $\tilde{K}(x, y) = K([xU]+1, [yU]+1)$ and we have

$$\varphi(i) = f(i)+\frac{1}{T}\sum_{k=1}^{T}\sum_{m=1}^{S}\lambda^m K(i, [y_{k1}U]+1)$$
$$K([y_{k1}U]+1, [y_{k2}U]+1)\ldots f([y_{km}U]+1)+R$$

where

$$|R| < 2A\varepsilon$$

for $T > T_0$ after (21).

Now let us suppose we want to calculate $\varphi(i)$ with an error smaller than δ, then we must take

$$T > T_0(\delta/2A)= \rho,$$

for instance $T = [\rho]+1$. Now we calculate the number of multiplications that are necessary to calculate a fixed $\varphi(i)$ with this error term. The number of multiplications is

$$\sum_{k=1}^{T}\sum_{m=1}^{S}(m+1) < 2Ts^2 = 0\left(T\left(\frac{\log T}{\log\log T}\right)^2\right)$$

independent of U! This is not so surprising as it looks and well-known in the Monte Carlo-method. The Gausz elimination method needs $\sim U^3$ multiplications, the usual iterations method KU^2 multiplications where K is depending on ρ. The above method is only useful when U is very large, so that $Ts^2(T) < KU^2$.

The method can also be applied to matrix inversion: If the problem is to solve $\mathfrak{A}\mathfrak{H} = \mathfrak{E}$, where $\mathfrak{A}$ is a given quadratic matrix of s^2 elements then we can get this equation in the form

$$\mathfrak{H} = \mathfrak{A}^{-1} = \mathfrak{B}+(\mathfrak{E}+\mathfrak{F}+\mathfrak{F}^2+\ldots)\mathfrak{F}\mathfrak{B},$$

where $\mathfrak{B}$ is an estimate of $\mathfrak{A}^{-1}$ and $\mathfrak{F} = \mathfrak{E}-\mathfrak{B}\mathfrak{A}$ and it can be assumed that $||\mathfrak{F}|| < 1$. Then we can work as before.

In general we have for any function $F(i_1, \ldots, i_m)$ defined for all m-tuples of integers $1 \leq i_1 \leq U, \ldots, 1 \leq i_m \leq U$ and for any sequences $\omega = (y_k)$

$$1 \leq k \leq T, y_k = (y_{k1}, \ldots y_{km})$$

after I 6

$$\left|\frac{1}{T}\sum_{k=1}^{T} F([Uy_{k1}], [Uy_{k2}], \ldots [Uy_{km}]) - \frac{1}{U^m}\sum_{i_1, \ldots i_m=1}^{U} F(i_1, \ldots i_m)\right| \leq D(\omega)V(F).$$

$V(F)$ is analogously defined as $V(K)$ for $m = 2$.

This follows from I(6), if we define for $(x_1, \ldots x_m)$ in E^m

$$f(x_1, \ldots x_m) = F([x_1 U], \ldots [x_m U]).$$

Final remarks: The methods developed above can be applied for calculating the eigen values of matrices and differential equations. I have applied them also for interpolation formulas in more than one variables.

NOTES

[1] Compare J. L. Walsh, Interpolation and Approximation (1935) Amer. Math. Soc. Coll. Publ. 20. With the theory of discrepancy this theorem can be improved.

[2] Compare E. Hlawka, Monatshefte für Mathematik 66, 140–151 (1962) Satz 2. If f has continuous derivatives there are results of L. C. Hsu and others, compare the paper of Hsu, Numerische Mathematik 3 (1961) 69–73.

[3] Compare for instance Symposium on Monte Carlo Methods, editor H. A. Meyer, John Wiley & Sons (1954).

[4] Math. Tables Aids Comput. 10 (1956) 1–8.

[5] Chiffres, Revue Assoc. France. Calcul 1, (1958) 149–155.

4 [6] Proc. Nat. Acad. Sci. 86 (1960) 844–874.

[7] Numer. Math. 4 (1961) 68–84. There are cited the papers of Richtmyer and G. Peck.

5 [8] Dokladi CCCR 115 (1937) 1062–65.

[9] Compare the paper of Korobow Trudi MHAH, LX 196–210 and the literature cited there.

[10[Numer. Math. 2 (1960) S. 84.

[11] There is an interesting note of I. M. Sobol Dokladi Acad. 114 (1957) 7067–69 but without details. It is unknown to me, whether there exists a further paper on this line of thoughts.

[12] Some numerical work has been done with this sequence for $s = 10$ at the Math. Inst. of the technical university of Stuttgart, by P. Ross & K. Arnold (Österr. Akad. Wiss.; in print).

[13] It is unknown if this is true for the Halton-sequence.
If f is analytic, then the method of Ph-Davis, Numer. Approx. Proc. Sympos. Math. Res-Center Madison April 21–23 (1958) 45–49 can be generalized to more dimension.

[14] E. Stiefel und H. Rutishauser, C. R. Acad. Sci. Paris 252 (1961) 1899–1900 ZAMM 41 (1961); E. Stiefel Einführung in die numerische Mathematik (1961), B. G. Teubner Stuttgart.

[15] There are some results of P. C. Hammer, O. J. Marlowe, A. H. Stroud, compare the report of J. Albrecht a. L. Collatz, ZAMM 38 (1958) 1–15.

[16] E. Hlawka, Sitzungsberichte Österr. Akad. Wissenschaften, Bd. 171 (1962), E. Hlawka & K. Kreiter, ebenda (1963) (in print).
N. M. Korobow has applied also his results for inhomogeneous integral equations, see [9].

[17] For numerical work it will be useful to combine it with other methods for instance with over relaxation methods.

Corrections

93.1	$= O(T^{-1/s})$,	104.4	[6] Ann.N.Y.Acad.Sci.
93.2	N.M.Korobow [8]	104.5	(1957)
96.3	$= \frac{n_0}{R} + \frac{n_1}{R^2} + \cdots$,		

Interpolation analytischer Funktionen auf dem Einheitskreis

Abhandlungen aus Zahlentheorie und Analysis (E. Landau Gedenkband), VEB Deutscher Verlag der Wissenschaften, Berlin, 1968, pp. 99–118

Es sei S die abgeschlossene Kreisscheibe vom Radius 1 in der komplexen z-Ebene. Es sei weiter f analytisch auf S, d. h., sie ist analytisch auf einer größeren Kreisscheibe $|z| \leqq R\,(R > 1)$. Auf dem Einheitskreis $|z| = 1$ sei nun eine unendliche Folge $\zeta_1, \zeta_2, \ldots$ von Punkten gegeben. Nun betrachte man die ersten N Glieder dieser Folge, und es sei $L_N = L_N(f, z)$ das Polynom in z vom Grad $N - 1$, welches durch Interpolation in den Werten von f in den Punkten $\zeta_1, \ldots, \zeta_N$ gefunden wird. Sind $\zeta_1, \ldots, \zeta_N$ alle voneinander verschieden, so ist bekanntlich

$$L_N(f, z) = \sum_{h=1}^{N} f(\zeta_h) \frac{(z - \zeta_1) \cdots (z - \zeta_{h-1})(z - \zeta_{h+1}) \cdots (z - \zeta_N)}{(\zeta_h - \zeta_1) \cdots (\zeta_h - \zeta_{h-1})(\zeta_h - \zeta_{h+1}) \cdots (\zeta_h - \zeta_N)} \tag{1}$$

(mit der üblichen Konvention, daß für $h = 1$ das Glied ζ_{h-1} und für $h = N$ jenes mit ζ_{h+1} wegzulassen ist). Setzen wir $\varrho_N(f, (\zeta_i)) = \sup_{|z|=1} |f(z) - L_N(f)|$, so erhebt sich die Frage, wann $\lim_{N\to\infty} \varrho_N = 0$ gilt.

Setzen wir $\zeta_j = e^{2\pi i \varphi_j}$, so liegen die φ_j im Einheitsintervall $I: 0 \leqq \varphi < 1$.

Dann gilt folgender Satz, der von L. Fejer [1] 1918 (ergänzt durch L. Kalmár 1926) herrührt (aufbauend auf C. Runge 1908, Theorie und Praxis der Reihen):

Es ist $\lim \varrho_N = 0$ *dann und nur dann für alle* f *(analytisch in* S*), wenn die Folge der* φ_j *gleichverteilt* mod 1 *ist. Ist* f *analytisch für* $|z| \leqq R\,(R > 1)$*, dann gilt in diesem Falle*

$$\overline{\lim_{N\to\infty}}\, \varrho_N^{1/N} \leqq \frac{1}{R}. \tag{2}$$

Dabei heißt eine Folge φ_j von reellen Zahlen φ nach H. Weyl (1916) gleichverteilt, wenn folgendes gilt:

Es sei $N'(t)$ die Anzahl der $\varphi_j (j = 1, \ldots, N)$ im Intervall I_t : $\langle 0, t \rangle$ mit $0 \leqq t \leqq 1$. Gilt dann für jedes t

$$\lim_{N\to\infty} \frac{N'(t)}{N} = t,$$

dann ist die Folge gleichverteilt.

Wir können dies auch gleich so formulieren: Ist $\chi_t(\varphi)$ die charakteristische Funktion von I_t, so ist $N'(t) = \sum_{h=1}^{N} \chi_t(\varphi_h)$, und es sei $\Delta_N(t) = \frac{1}{N} \sum_{h=1}^{N} \chi_t(\varphi_h) - t$; dann besagt eben die Definition der Gleichverteilung, daß $\lim_{N\to\infty} |\Delta_N(t)| = 0$ für alle t gilt.

Wir wollen nun die Abschätzung (2) vertiefen und die feineren Eigenschaften der Folge (φ_j) ins Spiel bringen. Wir führen dazu den Begriff der Diskrepanz D_N nach van der Corput ein: Es ist $D_N = \sup |\Delta_N(t)|$. Schon von H. Weyl wurde gezeigt, daß die Bedingung $\lim_{N\to\infty} D_N = 0$ notwendig und hinreichend dafür ist, daß eine Folge gleichverteilt ist. Man kennt für viele Folgen Abschätzungen für D_N und auch die Größenordnung, z. B. für die Folge $(h - 1)\,\alpha$ ($h = 1, 2, \ldots$; α irrational). Wir werden z. B. zeigen:[1]) Es ist für beliebige $\varphi_1, \ldots, \varphi_N$

$$\varrho_N^{1/N}(f) \leqq \frac{1}{R} C_1^{1/N} \cdot C_2^{2D_N}(R); \tag{3}$$

dabei sind C_1, C_2 Größen, die von f und R abhängen. Für die genauere Formulierung sei auf (22) verwiesen. Andererseits gibt es eine analytische Funktion f in $|z| \leqq R$, so daß

$$\varrho_N^{1/N}(f) \geqq \frac{1}{R} C_1^{1/N} C_2^{D_N^2}(R) \tag{4}$$

ist [siehe dazu (23)].

Wir werden uns bei der Ableitung dieser Abschätzungen auf die bekannte Gleichung

$$f(z) - L_N(f, z) = R_N(f, z),$$

$$R_N(f, z) = \frac{1}{2\pi i} \int_{|\zeta| = R} \frac{f(\zeta)}{\zeta - z} \frac{\omega_N(z)}{\omega_N(\zeta)} d\zeta \tag{5}$$

stützen, wobei $\omega_N(z) = (z - \zeta_1) \cdots (z - \zeta_N)$ ist.

Wir wollen an den einfachen Beweis erinnern:

Es ist $\frac{1}{\zeta - z} = \frac{1}{\zeta - \zeta_j} + \frac{z - \zeta_j}{\zeta - \zeta_j} \frac{1}{\zeta - z}$, und beginnend bei $j = 1$ folgt

$$\frac{1}{\zeta - z} = \sum_{j=0}^{N-1} \frac{\omega_i(z)}{\omega_j(\zeta)} \frac{1}{\zeta - \zeta_{j+1}} + \frac{\omega_N(z)}{\omega_N(\zeta)} \frac{1}{\zeta - z}.$$

---

[1]) Dieser Satz wurde ohne Beweis vom Verfasser angegeben in Nuffic International Summer Sessions in Science 1962 Breukelen, S. 104.

Multiplikation mit $f(\zeta)$ und Integration längs $|\zeta| = R$ liefert die Behauptung. Sie liefert gleichzeitig die Darstellung von L_N, wenn nicht alle $\zeta_1, \ldots, \zeta_N$ verschieden sind.

Wie das Restglied zeigt, läuft alles auf die Abschätzung von $|\omega_N(z)|$ nach oben und unten hinaus.

Wir wollen noch ein weiteres Interpolationsproblem besprechen: Es sei jetzt $f(z)$ analytisch in einem Kreisring $r \leqq |z| \leqq R$ mit $1 > r > 0$, $R > 1$, und es liegen wieder die Stellen $\zeta_1, \zeta_2, \ldots$ vor und es mögen die Interpolationspolynome L_N jetzt für ungerades $N = 2M + 1$ für f betrachtet werden. Nehmen wir an, es liege in der Form (1) vor, dann ist für

$$z = e^{2\pi i\vartheta}, \quad z - \zeta_h = e^{2\pi i\vartheta} - e^{2\pi i\varphi_h} = 2ie^{i\pi(\vartheta+\varphi_h)} \sin \pi(\vartheta - \varphi_h),$$

also ist

$$(z - \zeta_1) \cdots (z - \zeta_{h-1})(z - \zeta_{h+1}) \cdots (z - \zeta_{2M+1}) = z^M \prod_{j \neq h} 2i \sin \pi(\vartheta - \varphi_j)\, e^{i\pi\varphi_j},$$

also

$$L_{2M+1}(f) = \sum_{h=1}^{2M+1} f(\zeta_h) \frac{z^M}{\zeta_h^M} \prod_{j \neq h} \frac{\sin \pi(\vartheta - \varphi_j)}{\sin \pi(\varphi_h - \varphi_j)}.$$

Ersetzen wir f durch $z^M f(z)$, dann ist

$$\overline{L_M}(f) = \frac{1}{z^M} L_{M+1}(z^M f) = \sum_{h=1}^{2M+1} f(\zeta_h) \prod_{j \neq h} \frac{\sin \pi(\vartheta - \varphi_j)}{\sin \pi(\varphi_h - \varphi_j)}.$$

Ist die Funktion $f(z)$ auf dem Einheitskreis reell, so steht links die bekannte trigonometrische Interpolationsformel. In den praktischen Rechnungen wählt man $\varphi_j = \frac{1}{n}$ $(j = 1, \ldots, N)$, und dieser Fall ist auch theoretisch ausführlich untersucht worden. Es haben aber schon Euler und insbesondere U. J. Leverrier 1843 vorgeschlagen, auch hier an den Stellen $\varphi_j = (j - 1)\alpha$ (α irrational) zu approximieren.

Dieses Verfahren ist dann in numerischer Hinsicht von G. J. Hoüel 1865 und J. F. Encke 1860 weiter ausgestaltet worden. (Vgl. dazu H. Burkhardt, Jahresberichte X 2, 1 (1908) S. 237ff.). Es ist allerdings bemerkenswert, daß das verwendete α tatsächlich nicht irrational ist. Es nimmt z. B. Hoüel $4\alpha = \frac{8461}{180 \cdot 60}$. Aber er sagt, da $8461 = 180 \cdot 47 + 1$ eine Primzahl ist, verhält sich α so, als wäre es irrational. Tatsächlich ist die Kettenbruchentwicklung von $4\alpha = [1, 3, 1, 1, 1, 1, 1, 1, 2, 2, 1, 1, 1, 1, 4]$ also sehr lang. Von A. Wintner 1932 wurde die Frage nach dem Fehler $\varrho_M(f) = \sup_{|z|=1} |\overline{L_M}(f) - f|$ ($\overline{L_M}$ ist kein Polynom in z) für den Fall $\varphi_j = (j - 1)\alpha$ (α irrational) aufgeworfen und von G. Pólya 1934 beantwortet. Allgemein gilt wieder: Ist die Folge (φ_j) gleichverteilt, dann gilt

$$\lim_{M\to\infty} \bar{\varrho}_M^{1/N}(f) \leqq \operatorname{Max}\left(\frac{1}{R}, r\right), \tag{6}$$

also tatsächlich $\lim \bar{\varrho}_M = 0$, und diese Bedingung ist auch notwendig. Auch hier werden wir Restabschätzungen für jedes N geben, und es gilt auch in diesem Fall

$$\bar{\varrho}_M \leqq C_1 \left(R^{-(M+1)} C_2^{2D_N}(R) + r^M C_3^{2D_N}(r)\right). \tag{7}$$

Die Abschätzung stützt sich auf die Gleichung

$$f(z) - \overline{L_M}(f) = \overline{R_N},$$

$$\overline{R_N} = \frac{1}{2\pi i}\int_{|\zeta|=R} \frac{z^M f(\zeta)}{\zeta^M(\zeta - z)} \frac{\omega_N(z)}{\omega_N(\zeta)}\, d\zeta - \frac{1}{2\pi i}\int_{|\zeta|=r} \frac{z^M f(\zeta)}{z^M(\zeta - z)} \frac{\omega_N(z)}{\omega_N(\zeta)}\, d\zeta. \tag{6'}$$

Der Gedankengang der vorliegenden Arbeit beruht darauf, zunächst $\omega_N(z)$ für $|z| \neq 1$ nach oben und unten abzuschätzen nach einer Methode, welche in der Theorie der Gleichverteilung oft verwendet wird und die auf H. Behnke und J. F. Koksma zurück geht. Dann benutzen wir diese Abschätzungen, um $\omega_N(z)$ für $|z| = 1$ abzuschätzen.

Dabei ergibt sich Gelegenheit, einen Satz von Erdös-Turán auf einfache Art herzuleiten. Es sei bemerkt, daß es eine wichtige Aufgabe ist (auf die Erdös auch hingewiesen hat), möglichst scharfe Abschätzungen für $\omega_N(z)$ zu geben. Hier liegen noch viele Probleme offen. Dann leiten wir die zitierten Sätze her und besprechen einige Beispiele. Der weitere Teil der Arbeit beschäftigt sich mit folgender Fragestellung: Wenn eine Folge (φ_i) gleichverteilt ist, dann ist sie auch überall dicht im Einheitsintervall bzw. die Interpolationsstellen ζ_j sind dicht auf dem Einheitskreis. Die Umkehrung gilt jedoch nicht. Wir zeigen aber, daß die Folgen dann so umgeordnet werden können, daß für große N die Diskrepanz $D_N = O\left(\frac{\log N}{N}\right)$ ist. Andererseits läßt sie sich auch so umordnen, daß $D_N \geqq \frac{1}{2}$ wird. Die Güte der Interpolation hängt also von der Reihenfolge der Interpolationsschritte ab, was bei endlich vielen Stellen natürlich nicht der Fall ist. Dies dürfte auch bei numerischen Anwendungen von Bedeutung sein. In diesem Zusammenhang geben wir noch eine Methode an, um D_N bei gegebenen Stellen $\varphi_1, \ldots, \varphi_N$ in endlich vielen Schritten beliebig genau berechnen zu können.

§ 1

Wir schicken zunächst einen Hilfssatz vorweg:

Hilfssatz 1. *Es sei $f(\varphi)$ periodisch mit der Periode* 1 *und stetig differenzierbar; dann ist für jede endliche Folge von reellen Zahlen $\varphi_1, \ldots, \varphi_N$*

$$\left|\frac{1}{N}\sum_{h=1}^{N} f(\varphi_h) - \int_0^1 f(\varphi)\,d\varphi\right| \leqq D_N \int_0^1 |f'(\varphi)|\,d\varphi \tag{8}$$

mit

$$D_N = \sup_t \left|\frac{1}{N}\sum_{h=1}^{N} \chi_t(\varphi_h) - t\right|.$$

Beweis. Es ist

$$f(\varphi) = f(1) - \int_\varphi^1 f'(t)\,dt.$$

Nun ist

$$\int_\varphi^1 f'(t)\,dt = \int_0^1 f'(t)\,\chi_t(\varphi)\,dt,$$

also

$$f(\varphi) = f(1) - \int_0^1 f'(t)\,\chi_t(\varphi)\,dt,$$

d. h.

$$\frac{1}{N}\sum_{h=1}^{N} f(\varphi_h) - \int_0^1 f(\varphi)\,d\varphi = -\int_0^1 f'(t)\left(\frac{1}{N}\sum_{h=1}^{N} \chi_t(\varphi_h) - t\right)dt,$$

und daraus folgt bereits die Behauptung.

1 Wir setzen nun im weiteren stets $\zeta_h = e^{2\pi i\varphi}$ $(h = 1, \ldots, N)$ und für jedes komplexe z

2 $\omega_N(z) = \prod_{h=1}^{N} (z - {}_h)$; dann gilt

Hilfssatz 2. *Für jedes $|z| > 1$ ist*

$$|z|\left(\frac{|z|-1}{|z|+1}\right)^{2D_N} \leqq |\omega_N(z)|^{1/N} \leqq |z|\left(\frac{|z|+1}{|z|-1}\right)^{2D_N}. \tag{9}$$

Beweis. Wir setzen $\zeta = \frac{1}{z} = r\,e^{2\pi i\vartheta}$ $(r < 1)$ – es ist dann $|\zeta| < 1$ – und nehmen in Hilfssatz 1

$$f(\varphi) = \ln|1 - \zeta\,e^{2\pi i\varphi}| = \frac{1}{2}\ln\left(1 + r^2 - 2r\cos 2\pi(\varphi + \vartheta)\right).$$

Dann ist

$$f'(\varphi) = \frac{2r\pi\sin 2\pi(\varphi + \vartheta)}{1 + r^2 - 2r\cos 2\pi(\varphi + \vartheta)}.$$

Nun ist $f(\varphi)$ eine periodische Funktion, wir können daher bei der Abschätzung von $\int_0^1 f'(\varphi)\,d\varphi$ annehmen, daß $\vartheta = 0$ ist. Dann ist $f'(\varphi) \geqq 0$ im φ-Intervall $\left[0, \frac{1}{2}\right]$, $f(\varphi) \leqq 0$ im Intervall $\left[\frac{1}{2}, 1\right]$. Es ist also

$$\int_0^1 f'(\varphi)\,d\varphi = 2\int_0^{1/2} f'(\varphi)\,d\varphi = 2\left(f\left(\frac{1}{2}\right) - f(0)\right) = 2\ln\frac{1+r}{1-r}$$

und daher wegen $\int_0^1 f(\varphi)\,d\varphi = 0$

$$\left|\frac{1}{N}\sum_{h=1}^{N} \ln|1-\zeta\zeta_h|\right| \leqq 2D\ln\frac{1+r}{1-r}$$

oder

$$-2D\ln\frac{1+r}{1-r} \leqq \frac{1}{N}\sum_{h=1}^{N}\ln|(1-\zeta\zeta_h)| \leqq 2D\ln\frac{1+r}{1-r},$$

also

$$\left(\frac{1-r}{1+r}\right)^{2D} \leqq \left|\prod_h (1-\zeta\zeta_h)\right|^{1/N} \leqq \left(\frac{1+r}{1-r}\right)^{2D}, \tag{10}$$

und daraus folgt die Behauptung.

Bemerkung. Wegen $|\zeta_h| = 1$ ist trivialerweise

$$|z| - 1 \leqq |\omega_N(z)|^{1/N} \leqq 1 + |z|.$$

Die Abschätzungen von Hilfssatz 2 sind also nur für kleine D_N interessant.

Hilfssatz 3. *Für* $|z| < 1$ *ist*

$$\left(\frac{1-|z|}{1+|z|}\right)^{D} \leqq |\omega_N(z)|^{1/N} \leqq \left(\frac{1+|z|}{1-|z|}\right)^{2D}. \tag{11}$$

Beweis. Die Folge $\varphi_1' = 1 - \varphi_1, \ldots, \varphi_N' = 1 - \varphi_N$ besitzt ebenfalls die Diskrepanz D_N. Setzen wir $\zeta_h' = e^{2\pi i(1-\varphi_h)} = \zeta_h^{-1}$ und wenden (10) mit $\zeta = z$ und den ζ_h' an, so folgt die Behauptung.

Für $|z| = 1$ sind die Abschätzungen von $|\omega_N(z)|$ wertlos. Nun ist $\omega_N(z)$ eine analytische Funktion, also ist $1 \leqq \operatorname{Max}_{|z|=1} |\omega_N(z)| \leqq \operatorname{Max}_{|z|=R} |\omega_N(z)|$, wenn $R > 1$ ist. Wählen wir $R_1 = 2D + \sqrt{4D^2 + 1}$ und wenden (9) an, so erhalten wir

$$|\omega_N(z)|^{1/N} \leqq R_1\left(\frac{R_1+1}{R_1-1}\right)^{2D} = f(D) = R_1\left(\frac{1+\sqrt{4D^2+1}}{2D}\right)^{2D} = \beta(D). \tag{12}$$

Wir können aber eine Verschärfung angeben:

Hilfssatz 4. *Für* $|z| = 1$ *ist*

$$|\omega_N(z)|^{1/N} \leqq \gamma(D), \tag{13}$$

$$\gamma(D) = \sqrt{R_2}\left(\frac{1 + \sqrt{16D^2 + 1}}{4D}\right)^{2D}, \quad R_2 = 4D + \sqrt{16D^2 + 1}. \tag{14}$$

Bemerkung. Es ist tatsächlich $\gamma < \beta$, denn es ist

$$\frac{1}{4} + \sqrt{D^2 + \frac{1}{16}} < \frac{1}{2} + \sqrt{D^2 + \frac{1}{4}}$$

und

$$R_2 = 4D + \sqrt{16D^2 + 1} \leqq 4D\sqrt{4D^2 + 1} + 8D^2 + 1 = R_1^2.$$

Um Hilfssatz 4 zu beweisen, genügt es zu zeigen: Für $|z| = 1$ ist

$$|\omega_N(z)|^{1/N} \leqq \sqrt{R}\left(\frac{R+1}{R-1}\right)^{2D}, \quad R > 1. \tag{15}$$

Wir geben für (15) zwei Beweise. Zunächst einen algebraischen Beweis, dessen Idee von I. Schur stammt: Ist $\zeta = Re^{i\vartheta}$ $(R > 1)$, dann ist für jedes $\zeta_h = e^{2\pi i\varphi_h}$

$$|\zeta - \zeta_h|^2 = R^2 + 1 - 2R\cos 2\pi(\vartheta - \varphi_h) = R\left(R + \frac{1}{R} - 2\cos 2\pi(\vartheta - \varphi_h)\right)$$

$$\geqq R(2 - 2\cos 2\pi(\vartheta - \varphi_h)) = R|e^{2\pi i\vartheta} - e^{2\pi i\varphi_h}|^2,$$

also

$$\prod_h |\zeta - \zeta_h| \geqq R^{N/2}\prod_{h=1}^{N}|e^{2\pi i\vartheta} - \zeta_h|, \tag{16}$$

und da ϑ beliebig ist, gilt für $z = e^{2\pi i\vartheta}$

$$|\omega_N(z)|^{1/N} \leqq R^{-(1/2)}|\omega_N(Rz)|^{1/N}.$$

Wendet man (9) auf $\omega_N(Rz)$ an, so folgt (15).

Der zweite Beweis benutzt den Hadamardschen Dreikreisesatz, angewendet auf die Kreise $|z| = \varrho$ mit $\varrho = \frac{1}{R}$, 1, R. Setzen wir $\operatorname{Max}_{|z|=1} |\omega_N(z)|^{1/N} = M$, dann erhalten wir

$$M^{2\log R} \leqq \left(\frac{R+1}{R-1}\right)^{2D\log R}\left(R\left(\frac{R+1}{R-1}\right)^{2D}\right)^{\log R},$$

also

$$M^2 \leqq R\left(\frac{R+1}{R-1}\right)^{4D}.$$

Wir wollen nun eine einfache Abschätzung für γ geben: Es ist

$$1 + \sqrt{1 + 16D^2} \leqq 2(1 + 2D) \leqq 2e^{2D},$$

also

$$\frac{\gamma}{\sqrt{R_2}} < \left(\frac{1 + 2D}{2D}\right)^{2D} < \frac{e^{4D^2}}{D^D} < e^{4D(\ln(1/D)+1)},$$

da $D \leqq 1$ ist, und somit

$$\gamma < (1 + 4D)\, e^{4D(\ln(1/D)+1)};$$

und wegen $1 + 4D < e^{4D}$ ist dann

$$|\omega_N(z)|^{1/N} < e^{8D(\ln(1/D)+1)} \quad \text{für alle } |z| = 1. \tag{17}$$

Andererseits gilt stets $\operatorname{Max} |\omega_N(z)| \geqq 1$, d. h., es existieren z, für die der Ausdruck links in (17) sicher $\geqq 1$ ist.

Es gilt aber nach einem Satz von Erdös–Turán schärfer für $M = \operatorname*{Max}_{|z|=1} |\omega_N(z)|^{1/N}$: Es existiert eine absolute Konstante $\alpha > 0$, so daß

$$M \geqq e^{\alpha D^2} \tag{18}$$

ist.

Wir wollen für diesen Satz einen einfachen Beweis geben und setzen $\prod_{h=1}^{N} (1 - z\zeta_h) = P(z)$. Dann ist, da $P(z)$ analytisch ist,

$$\operatorname*{Max}_{|z|=r} |P(z)| \leqq M^N \quad \text{für} \quad r < 1.$$

$\left(\text{Es ist } \operatorname*{Max}_{|z|=1} |P(z)| = \operatorname*{Max}_{|z|=1} |\omega_N(z)|\right)$, also ist für alle z mit $|z| = r < 1$

$$\frac{1}{N} \log |P(z)| = \frac{1}{N} \sum_{h=1}^{N} \log |1 - z\zeta_h| \leqq \log M.$$

Setzen wir $z = re^{i\vartheta}$ und entwickeln den Logarithmus in seine Potenzreihe, so ist also für alle ϑ

$$\sum_{h=1}^{\infty} \frac{r^h}{h} (R_h \cos 2\pi h\vartheta + I_h \sin 2\pi h\vartheta) \leqq \log M \tag{19}$$

mit

$$R_h = \operatorname{Re} S_h,\ I_h = \operatorname{Im} S_h,\ S_h = \frac{1}{N} \sum_{h=1}^{N} e^{2\pi i \varphi_h}.$$

Dies ist eine absolut und gleichmäßig konvergente Fourierreihe in ϑ. Multiplizieren wir (19) für ein natürliches l mit der nichtnegativen Funktion $1 + \cos 2\pi l\vartheta$ und integrieren nach ϑ von 0 bis 1, so bleibt in (19) nur das Glied mit $h = l$ stehen, und wir erhalten

$$\frac{1}{2}\,\frac{r^l}{l}\,R_l \leqq \log M.$$

(Wir könnten uns auch auf die Parsevalsche Gleichung berufen.) Nehmen wir $1 - \cos 2\pi l\vartheta$, so erhalten wir analog

$$-\frac{1}{2}\,\frac{r^l}{l}\,R_l \leqq \log M, \quad \text{d. h.} \quad r^l\,\frac{|R_l|}{l} \leqq 2\log M.$$

Diese Ungleichung gilt für jedes $r < 1$, also ist

$$|R_l| \leqq 2l \log M.$$

Nehmen wir $1 + \sin 2\pi l\vartheta$ bzw. $1 - \sin 2\pi l\vartheta$, so erhalten wir analog

$$|I_l| \leqq 2l \log M,$$

also

$$|S_l| \leqq 2\sqrt{2}\,l \log M.$$

Nun gilt nach Erdös und Turán (vgl. [4])

$$D_N \leqq 1000\left(\frac{1}{K} + \sum_{0<|l|\leqq K} \frac{|S_l|}{l}\right) \quad \text{für jedes } K > 1;$$

wir erhalten somit

$$D_N \leqq 10^3\left(\frac{1}{K} + 4\sqrt{2}\,K \log M\right).$$

Nun ist trivialerweise $M \leqq 2$. Nehmen wir also

$$K = \left[\sqrt{\frac{4}{\log M}}\right] > 2, \quad \text{so ist} \quad K > \sqrt{\frac{1}{\log M}}$$

und

$$D_N \leqq 10^3\left(\sqrt{\log M} + 8\sqrt{2}\sqrt{\log M}\right) = 10^3\left(1 + 4\sqrt{2}\right)\sqrt{\log M}$$
$$< 2 \cdot 10^4 \sqrt{\log M}.$$

Daraus folgt (18) mit einem $\alpha = \frac{1}{4}10^{-8}$.

Es folgt sofort

Hilfssatz 5. *Für jedes $R > 1$ existiert stets ein z mit $|z| = R$, so daß*

$$|\omega_N(z)|^{1/N} \geqq \sqrt{R}\,e^{\alpha D^2} \tag{20}$$

ist.

Beweis. Dies folgt aus (16) und (18).

Hilfssatz 6. *Für jedes $R > 1$ existiert stets ein z mit $|z| = R$, so daß*

$$|\omega_N(z)|^{1/N} < R \tag{21}$$

gilt.

Beweis. Es ist

$$\frac{1}{N} \log |\omega_N(z)| = \log R + \frac{1}{N} \sum_{h=1}^{N} \log \left| 1 - \frac{1}{z} \zeta_h \right|.$$

Nun ist

$$\int_0^{2\pi} \log \left| 1 - \frac{e^{-i\vartheta}}{R} \zeta_h \right| d\zeta = 0,$$

also ist für $z = Re^{i\vartheta}$

$$\int_0^{2\pi} \frac{1}{N} \lg |\omega_N(R\, e^{i\vartheta})|\, d\vartheta = \log R,$$

und daraus folgt die Behauptung.

Jetzt können wir leicht den folgenden Satz beweisen.

3 **Satz 1.** *Es sei $f(z)$ analytisch für $|z| \leqq R\,(R < 1)$, und $L_N(z,f)$ sei das zugehörige Interpolationspolynom in den Stellen $\zeta_h = e^{2\pi i \varphi_h}$ $(h = 1, \ldots, N)$; dann ist*

$$\sup_{|z|=1} |f(z) - L_N(z,f)| \leqq \frac{K}{R-1} \left(\gamma(D) \frac{1}{R} \left(\frac{R+1}{R-1} \right)^{2D} \right)^N \tag{22}$$

mit

$$\gamma_N < e^{8D(\ln(1/D)+1)}, \quad K = \frac{1}{2\pi} \int_0^{2\pi} |f(R\, e^{i\varphi})|\, d\varphi.$$

Für $D \to 0$ geht der Ausdruck in der Klammer (22) gegen $\frac{1}{R}$.

Beweis. Nach (5) ist

$$f(z) - L_N(z) = \frac{1}{2\pi i} \int_{|\zeta|=R} \frac{f(\zeta)}{\zeta - z} \frac{\omega_N(z)}{\omega_N(\zeta)} d\zeta.$$

Wendet man für $|z| = 1$ und $|\zeta| = R$ (13), (17) und die linke Abschätzung in (9) an, so folgt die Behauptung.

Wir zeigen nun

Satz 2. *Ist $R > 1$, dann existiert ein ζ mit $|\zeta| = R$, so daß für $f_0(z) = \frac{1}{z - \zeta}$*

$$\sup_{|z|=1} |f_0(z) - L_N(z, f_0)| \geqq \frac{1}{R+1} \left(\frac{e^{\alpha D^2}}{R} \right)^N \tag{23}$$

gilt.

Beweis. Es ist $f_0(z) - L_N(z, f_0) = \dfrac{1}{z-\zeta}\,\dfrac{\omega_N(z)}{\omega_N(\zeta)}$.

Nun gibt es nach (21) ein ζ mit $|\omega_N(\zeta)| < R^N$. Weiter ist $|z - \zeta| \leqq R + 1$, also folgt aus (18) die Behauptung.

Zusatz. Für $|z| = \sigma \geqq 1$ gilt

$$\sup_{|z|=\sigma} |f_0(z) - L_N(z, f_0)| \geqq \frac{1}{R+\sigma}\left(\frac{\sqrt{\sigma}}{R}\, e^{\alpha D^2}\right)^N. \tag{24}$$

Dies folgt sofort aus (20), also für $\sigma = R^2$ sicher Divergenz.

Analog folgt

Satz 3. *Ist $f(z)$ analytisch in $R_1 \leqq |z| \leqq R_2$ $(R_2 > 1, 0 < R_1 < 1)$, ist $N = 2M + 1$, $\overline{L_M}(f) = \dfrac{1}{z^M} L_n(z^M f)$, wobei L_N das Interpolationspolynom von $z^M f$ in den ζ_h $(h = 1, \dots, N)$ ist, dann ist*

$$\overline{\varrho_M} = \sup_{|z|=1} |f(z) - \overline{L_M}(f)| \leqq S$$

mit

4 $$S = \gamma^N(D)\left[\frac{K_2}{R_2 - 1}\,\frac{1}{R_2^{M+1}}\left(\frac{R_2+1}{R_2-1}\right)^{2ND} + \frac{K_1}{1-R_1}\,R_1^M\left(\frac{1+R_1}{1-R_1}\right)^{2ND}\right]$$

und

$$K_r = \frac{1}{2\pi}\int_0^{2\pi} |f(R_i e^{i\varphi})|\, d\varphi \quad (r = 1, 2).$$

Dies folgt aus (6'), (9), (10), (13).

§ 2

Wir wollen nun annehmen, daß α eine reelle Zahl ist und $\varphi_h = (h-1)\alpha$ $(h = 1, \dots, N)$, also $\zeta_h = e^{2\pi i (h-1)\alpha}$. Es werde α in einen Kettenbruch entwickelt, der endlich oder unendlich sein kann: $\alpha = [a_0, a_1, a_2, \dots]$, und es seien $\dfrac{p_0}{q_0}, \dfrac{p_1}{q_1}, \dots$ die Näherungsbrüche. Es sei $i(N)$ durch $q_i \leqq N < q_{i+1}$ definiert, dann gilt (vgl. [4])

$$D_N < 4\left(\frac{1}{q_i} + \frac{i + a_1 + \cdots + a_i}{N}\right). \tag{25}$$

Ist insbesondere $N = q_i$, dann ist $D_{q_i} \leqq \dfrac{2}{N}$. In diesem Fall ist also

$$\gamma_N < e^{16(\log N + 2)/N} \leqq e^{c(\log N/N)}$$

und

$$\varrho_N(f) < \frac{K}{R-1}\,\frac{e^{c_1 \lg N}}{R^N}\left(\frac{R+1}{R-1}\right)^4. \tag{22'}$$

In diesem Fall könnte man noch schärfer abschätzen.

Es werde nun $A = \operatorname*{Max}_{h=1,\dots,i} a_h$ gesetzt. Ist $A \neq 1$, dann gilt bekanntlich $\sum_{h=1}^{i} a_h < \frac{2A}{\log A}\log q_i + 4A$; ist $A = 1$, dann ist diese Summe $\leqq \frac{2}{\log 2}\log q_i$. Weiter gibt es zu jedem i ein N, so daß $q_i < N < q_{i+1}$ und

$$D_N > K\frac{(a_1 + \cdots + a_{i+1})}{N} \tag{26}$$

ist. Es sei nun α eine positive Irrationalzahl, dann kann man aus (26) folgern, daß es unendlich viele N gibt, so daß

$$D_N > \frac{1}{720}\,\frac{\log N}{N}$$

gilt, also wegen (23)

$$\varrho_N(f_0) \geqq \frac{1}{R+1}\,\frac{e^{c_1(\log^2 N/N)}}{R^N}. \tag{23'}$$

Sind insbesondere die a_i beschränkt, also $a_i \leqq K$ für alle i, dann ist dies richtig für die N_i, definiert durch

$$N_i = q_{4i+1} + N_{i-1}, \quad N_1 = 95 \quad (i = 2, \dots),$$

und dann ist

$$D_{N_i} \geqq \frac{\log N_i}{8(A+1)^6}.$$

Andererseits ist für jedes N

$$D_N \leqq 4A\,\frac{\log N}{N}, \tag{25'}$$

also wegen (22)

$$\varrho_N(f) < \frac{K}{R-1}\,\frac{e^{c\log^2 N}}{R^N}\left(\frac{R+1}{R-1}\right)^{8A\log N}. \tag{22''}$$

Dies gilt insbesondere für quadratische Irrationalitäten.

Gilt

$$q_{i+1} < c q_i^r \quad \text{mit} \quad c > 0, \quad r > 1, \tag{26'}$$

dann ist

$$D_N \leqq CN^{-(1/r)} \quad \text{und} \quad D_N \geqq C_1 N^{-(1/r)} \tag{27}$$

für unendlich viele N.

Diese Abschätzung kann insbesondere angewendet werden, wenn α eine algebraische Zahl ist mit einem Grad $\geqq 3$ mit $r = 1 + \varepsilon\,(\varepsilon < 0)$; dann ist

$$D_N \leqq CN^{-1/(1+\varepsilon)}. \tag{27'}$$

Nach K. F. Roth ist nämlich stets

$$\left|\alpha - \frac{p}{q}\right| \geqq \frac{c}{q^{2+\varepsilon}}$$

Somit gilt für die i-ten Näherungsbrüche von α

$$\frac{1}{q_i q_{i+1}} > \left|\alpha - \frac{p_i}{q_i}\right| \geqq \frac{c}{q_i^{2+\varepsilon}}, \quad \text{also} \quad \frac{1}{q_{i+1}} \geqq \frac{c}{q_i^{1+\varepsilon}}.$$

Nehmen wir $\alpha = \pi$, dann ist nach K. Mahler [5]

$$\left|\pi - \frac{p_i}{q_i}\right| > \frac{1}{q_i^{42}},$$

also

$$q_{i+1} < q_i^{41},$$

also gilt (26') mit $r = 41$.

Interessant ist der Fall $\alpha = \dfrac{1}{\pi}$, dann gilt ebenfalls

$$\left|\frac{1}{\pi} - \frac{p}{q}\right| > \frac{c_1}{q^{42}};$$

denn gäbe es p, q mit $\left|\dfrac{1}{\pi} - \dfrac{p}{q}\right| < \dfrac{c_1}{q^{42}}$ (o.B.d.A. $p < q$), dann wäre

$$\left|\frac{q}{p} - \pi\right| < \frac{\pi c_1}{p q^{41}} < \frac{\pi c_1}{p^{42}}.$$

Dann sind die ζ_h von der Gestalt e^{2ih} $(h = 0, 1, \ldots, N-1)$, und die Diskrepanz ist $\leqq CN^{-(1/41)}$.

Betrachtet man die Menge aller Folgen $(\varphi_1, \varphi_2, \ldots)$ mit dem zugehörigen Haarschen Maß, dann ist für fast alle Folgen für großes N nach Cassels

$$D_N < (1+\varepsilon)\sqrt{\frac{\log\log N}{N}}, \tag{27''}$$

und es gibt Folgen von natürlichen Zahlen N, so daß

$$D_N > (1-\varepsilon)\sqrt{\frac{\log\log N}{N}} \tag{27'''}$$

ist; also ist für unendlich viele N

$$\varrho_N(f) < \frac{K}{R-1}\,\frac{e^{16(1+\varepsilon)\sqrt{N\log\log N}\,\log N}}{R^N}\left(\frac{R+1}{R-1}\right)^{(1+\varepsilon)2\sqrt{N\,\mathrm{lg}\,\mathrm{lg}\,N}} \tag{28}$$

und

$$\varrho_N(f_0) > \frac{1}{R+1}\,\frac{e^{\alpha(1-\varepsilon)^2\log\log N}}{R^N}. \tag{28'}$$

Es ist daher für fast alle Folgen $\overline{\lim\limits_{N\to\infty}}\, R^N\varrho_N(f_0) = \infty$.

Bemerkung. Für jede Folge gibt es nach K. F. Roth unendlich viele N, so daß $D_N > c\dfrac{\sqrt{\log N}}{N}$ gilt. Für alle N ist $ND_N \geqq \dfrac{1}{2}$ [vgl. (35)]

Aus (28) folgt, daß für fast alle Folgen $\lim\limits_{N\to\infty}\varrho_N(f) = 0$ ist. Dieses Resultat ist schon lange bekannt und kann direkt sehr leicht hergeleitet werden (vgl. auch J. H. Curtiss [5]).

Aus (27'), (27''), (17), (18) folgt für $|z| = 1$ für fast alle Folgen

$$|\omega_N(z)| \leqq e^{O(\sqrt{N}\sqrt{\log\log N}\log N)}$$

und

$$\max_{|z|=1}|\omega_N(z)| \geqq e^{\Omega(\log\log^2 N)}, \text{ also } \overline{\lim_{N\to\infty}}\,|\omega_N(z)| = \infty.$$

§ 3

Will man D_N für eine gegebene endliche Folge $\varphi_1, \ldots, \varphi_N$ abschätzen, so kann man so vorgehen: Es war $D_N = \sup\limits_t \dfrac{1}{N}\left|\sum\limits_{h=1}^{N}\chi_t(\varphi_h) - t\right|$, wobei das Supremum erstreckt wird über alle t mit $0 \leqq t < 1$. Es sei m eine natürliche Zahl $\geqq 1$ und man betrachte die m Intervalle j_l: $0 \leqq x < \dfrac{l}{m}$ $(l = 0, \ldots, m)$ und setze

$$\Delta_{l,m} = \left|\frac{1}{N}\sum_{h=1}^{N}\chi_{l/m}(\varphi_h) - \frac{l}{m}\right|.$$

Dabei ist $\chi(t)$ die charakteristische Funktion des Intervalls $0 \leqq \varphi < t$. Es ist $\Delta_{l,m} = \left|\frac{N'(j_l)}{N} - \lambda(j_l)\right|$, wobei $N'(j_l)$ die Anzahl der φ_h im Intervall j_l und $\lambda(j_l)$ die Länge des Intervalls j_l ist. Es kann also $\Delta_{l,m}$ für jedes l in endlich vielen Schritten bestimmt werden. Wir setzen nun $\operatorname{Max}_l \Delta_{l,m} = \Delta_m$. Dann ist

$$\Delta_m \leqq D_N \leqq \Delta_m + \frac{1}{m}. \tag{29}$$

Der Beweis von (29) ist, was die linke Seite betrifft, trivial. Um die rechte Seite zu beweisen nehmen wir ein Intervall $0 \leqq \varphi \leqq t$ und setzen $l = [tm]$. Es ist $0 \leqq l \leqq tm < l + 1$, und da $t \leqq 1$ ist, gilt $l \leqq m$. Damit ist

$$\frac{l}{m} \leqq t < \frac{l+1}{m}.$$

Es ist

$$\sum_h \chi_{l/m}(\varphi_h) \leqq \sum_h \chi_t(\varphi_h) \leqq \sum \chi_{(l+1)/m}(\varphi_h).$$

Daher ist

$$-\Delta_{l,m} - \frac{1}{m} \leqq \frac{1}{N}\sum \chi_t(\varphi_h) - t \leqq \Delta_{l+1,m} + \frac{1}{m},$$

also

$$\left|\frac{1}{N}\sum \chi_t(\varphi_h) - 1 - \Delta_m\right| \leqq \frac{1}{m},$$

und daraus folgt die Behauptung.

Für die Anwendung wird man $m \geqq N$ wählen müssen.

Es soll noch kurz der Fall behandelt werden, daß die φ_h nur bis auf einen Fehler $< \varepsilon$ bekannt sind; es gilt folgender Satz, der an sich bekannt ist:

Hilfssatz 7. *Sind* $(\varphi_1, \ldots, \varphi_N)$ *und* $(\psi_1, \ldots, \psi_N)$ *zwei Folgen und ist*

$$|\varphi_h - \psi_h| < \varepsilon \quad (h = 1, \ldots, N),$$

dann ist

$$|D - D^*| < \varepsilon. \tag{30}$$

Dabei wurde $D = D_N((\varphi_h))$, $D^* = D_N((\psi_h))$ *gesetzt.*

Beweis. Es sei $\langle 0, \alpha)$ zunächst ein Intervall I_α mit $\varepsilon < \alpha < 1 - \varepsilon$. Es sei $N_2(\alpha - \varepsilon)$ die Anzahl der ψ_h mit $0 \leqq \psi_h < \alpha - \varepsilon$. Da $0 \leqq \varphi_h < \psi_h + \varepsilon$ ist, gilt für die zugehörigen φ_h: $0 \leqq \varphi_h < \alpha$, d. h., wenn $N_1(\alpha)$ die Anzahl aller φ_h und I_α ist, daß $N_1(\alpha) \geqq N_2(\alpha - \varepsilon)$ gilt. Analog folgt aus $0 \leqq \varphi_h < \alpha$ die Ungleichung $0 \leqq \psi_h < \alpha + \varepsilon$, damit also

$$N_2(\alpha + \varepsilon) \geqq N_1(\alpha).$$

Nun folgt aus der Definition von D^*

$$N_2(\alpha \pm \varepsilon) = N(\alpha \pm \varepsilon) + \vartheta N \cdot D^* \quad \text{mit} \quad |\vartheta| \leqq 1,$$

d. h., wir erhalten

$$N(\varepsilon + D^*) \geqq N_1(\alpha) - N\alpha \geqq -N(\varepsilon + D^*)$$

oder

$$|N_1(\alpha) - N\alpha| \leqq N(\varepsilon + D^*).$$

Diese Ungleichung ist trivialerweise richtig, wenn α nicht der vorher eingeführten Bedingung genügt, da stets $N_1 \leqq N$ ist. Daraus folgt $D \leqq \varepsilon + D^*$. Vertauschen wir D mit D^*, so folgt (30). Wir wollen noch folgende Verallgemeinerung von Hilfssatz 7 beweisen:

Hilfssatz 8. *Es seien* $(\varphi_1, \ldots, \varphi_N)$, $(\psi_1, \ldots, \psi_N)$ *zwei Folgen mit*

$$|\varphi_h - \psi_h| < \varepsilon_h \quad (h = 1, \ldots, N).$$

Weiter sei $S(\varepsilon)$ *für* $\varepsilon > 0$ *die Anzahl der* $\varepsilon_h > \varepsilon$. *Dann gilt für die Diskrepanz* D, D^*

$$|D - D^*| < \varepsilon + \frac{S(\varepsilon)}{N}. \tag{31}$$

Wenn alle ε_h *gleich* ε *sind, so erhalten wir wieder* (30).

Beweis. Wir betrachten wie im Beweis von (30) Intervalle $I : \langle 0, \alpha)$ mit $\varepsilon < \alpha < 1 - \varepsilon$, und es sei $N_2(\alpha - \varepsilon)$ die Anzahl der ψ_h mit $0 \leqq \psi_h < \alpha - \varepsilon$. Für die zugehörigen φ_h mit $0 \leqq \varphi_h < \psi_h + \varepsilon_h$ gilt sicher, wenn $\varepsilon_h \leqq \varepsilon$ ist, daß sie in $\langle 0, \alpha)$ liegen. Ist also $N_1(\alpha)$ die Anzahl aller φ_h in I_α, dann ist $N_1(\alpha) \geqq N_2(\alpha - \varepsilon) - S$. Analog folgt $N_1(\alpha) < N_2(\alpha + \varepsilon) + S$. Schließt man nun weiter wie beim Beweis von (30), so ergibt sich

$$|N_1(\alpha) - N\alpha|\, \varepsilon \leqq N(\varepsilon + D^*) + S$$

und daraus folgt (31).

Daraus leiten wir nun folgenden Satz her:

Satz 4. *Es sei* (φ_j) *eine unendliche Folge mit der Folge der Diskrepanzen* D_N $(N = 1, \ldots)$. *Es sei* (ψ_j) *eine weitere unendliche Folge, welche im Einheitsintervall überall dicht liegt. Weiter sei* $\sigma(t) \geqq 1$ *monoton wachsend im stärkeren Sinne mit* $\lim_{t \to \infty} \sigma(t) = \infty$. *Dann können wir diese Folge so umordnen, daß für die neue Folge* $\vartheta_1, \vartheta_2, \ldots$ *die zugehörige*

Diskrepanz D_N^* für jedes N die Ungleichung

$$|D_N^* - D_N| \leqq \frac{\sigma(N)}{N} \tag{32}$$

erfüllt.

Beweis. Da die Folge (ψ_h) dicht ist im Einheitsintervall, gibt es sicher in ihr Elemente $\psi_h' = \psi_{j_h}$ $(h = 1, 2, \ldots)$, so daß

$$|\psi_h' - \varphi_h| < \frac{1}{\tau(h)} = \varepsilon_h$$

gilt, wobei $\tau(t)$ die inverse Funktion zu σ ist.

Ist $\varepsilon > 0$, so ist die Anzahl $S(\varepsilon)$ der $\varepsilon_h \geqq \varepsilon$, also der $\tau(h) \leqq \frac{1}{\varepsilon}$, sicher $\leqq \sigma\left(\frac{1}{\varepsilon}\right)$. Es gilt also nach (31) für die Diskrepanz D_N' der Folge (ψ_h') mit $\varepsilon = \frac{1}{N}$

$$|D_N' - D_N| \leqq \frac{1}{N} + \frac{\sigma(N)}{N} \leqq \frac{2\sigma(N)}{N}. \tag{33}$$

Wir setzen nun $\vartheta_h = \psi_h' = \psi_{i_h}$, wenn h nicht von der Gestalt $[\tau(l)]$ ist. Alle übrigen Glieder der Folge (ψ_h) denken wir uns irgendwie durchnumeriert: $\lambda_1, \lambda_2, \ldots$ Setzen wir $\vartheta_h = \lambda_l$ für $h = [\tau(l)]$, dann ist für jedes α

$$\left|\frac{1}{N}\sum_{h=1}^{N} (\chi_\alpha(\vartheta_h) - \chi_\alpha(\psi_h'))\right| \leqq \frac{1}{N} + \frac{\sigma(N)}{N};$$

denn die Anzahl der $h = [\tau(l)] < N$ ist höchstens $\sigma(N)$. Es ist also

$$|D_N^* - D_N'| \leqq \frac{1 + \sigma(N)}{N},$$

d. h.

$$|D_N^* - D_N| < \frac{4\sigma(N)}{N}.$$

Wählt man statt σ die Funktion $\frac{\sigma}{4}$, so folgt die Behauptung.

Bemerkung. Aus (32) folgt: Die Folge (ψ_j) kann so umgeordnet werden, daß für die Diskrepanz der neuen Folge

$$|D_N^* - D_N| \leqq D_N\sigma(N) \tag{34}$$

gilt.

Beweis. Es ist stets für jede Folge (φ_j)

$$D_N \geqq \frac{1}{2N}. \tag{35}$$

Es sei φ_j irgendein Glied mit $j \leqq N$. Dann betrachten wir die Intervalle $\langle 0, \varphi_j - \varepsilon)$ und $\langle 0, \varphi_j + \varepsilon)$ mit $\varepsilon < \frac{1}{2N}$. (Der Fall, daß $\varphi_j = 0$ $(j \leqq N)$ ist, erledigt sich analog.) Dann ist

$$N'(\varphi_j + \varepsilon) - N'(\varphi_j - \varepsilon) \geqq 1,$$

also

$$2ND \geqq |N'(\varphi_j + \varepsilon) - N(\varphi_j + \varepsilon) - N'(\varphi_j - \varepsilon) - N(\varphi_j - \varepsilon)| \geqq 1 - 2N\varepsilon,$$

und daraus folgt sofort (35).

Es erhebt sich die Frage, ob man erreichen kann, daß $c_1 D_N \leqq D_N^* \leqq c_2 D_N$ ist, wobei c_1, c_2 absolute Konstante sind, und zwar für alle N. (Für ein festes N ist dies leicht zu zeigen.)

Nehmen wir einerseits z. B. für die Folge (φ_j) eine Folge mit (25'), andererseits die Folge 0, 0, ..., für welche die Diskrepanzen D_N alle 1 sind, so folgt: Jede Folge (ψ_j), welche im Einheitsintervall überall dicht liegt, also auch jede gleichverteilte Folge, kann einerseits so umgeordnet werden, daß die neue Folge (ϑ_j') eine Diskrepanz $D_N' = O\left(\frac{\log N}{N}\right)$ besitzt (also gleichverteilt ist), andererseits so, daß für die Folge (ϑ_j'') die zugehörigen Diskrepanzen $D_N'' \geqq \frac{1}{2}$ für alle N sind, also (ϑ_j'') nicht gleichverteilt ist.

Bemerkung. Die Tatsache, daß jede in E dichte Folge so umgeordnet werden kann, daß sie dort gleichverteilt ist, stammt bereits von J. von Neumann. Aus (22) entnimmt man: Ist $f(z)$ analytisch in $|z| \leqq R$ $(R > 1)$ und ist die unendliche Folge $\varphi_1, \varphi_2, \ldots$ gleichverteilt, also $\lim\limits_{n \to \infty} D_N = 0$, dann gilt eine Abschätzung

$$\varrho_N(f, \varphi_h) \leqq \frac{C^{o(N)}}{R^N} \tag{36}$$

wobei C von R abhängt.

Daraus folgt natürlich der bekannte Satz, welcher in der Einleitung zitiert wurde: Es ist $\lim\limits_{N \to \infty} \varrho_N(f, \varphi_h)$ genau dann und nur dann $= 0$ für alle in $|z| \leqq 1$ analytischen Funktionen, wenn die Folge (φ_h) gleichverteilt ist. Dies folgt sofort aus (22') und (23'). Ist nämlich nicht $\lim D_N = 0$, dann gibt es eine Teilfolge $(D_{N'})$ mit $\lim D_{N'} = \varepsilon > 0$, und dann kann (23) angewendet werden mit $R \overset{\varepsilon}{=} e^{\alpha\varepsilon^2/4} > 1$ mit $f_0(N) = \dfrac{1}{z - \zeta(N)}$, wo $|\omega_N(\zeta(N))| \leqq R^N$ mit passendem $\zeta(N)$ mit $|\zeta(N)| = R$. Es existiert eine Teilfolge (N_1) von N mit $\lim \zeta(N_1) = \zeta$, für genügend großes N_1 ist also

$$|\zeta(N_1) - \zeta| < (R - 1)^2.$$

Nun ist für jedes j wegen $|\zeta(N) - \zeta_j| \geqq R - 1$

$$|\zeta - \zeta_j| \leqq |\zeta - \zeta(N_1)| + |\zeta(N_1) - \zeta_j| \leqq |\zeta(N_1) - \zeta_j|\, R,$$

also

$$\omega_{N_1}(\zeta) \leqq \omega_{N_1}(\zeta(N_1))\, R^{N_1} \leqq R^{2N_1}.$$

Somit ist für $f_0(\zeta) = \dfrac{1}{z - \zeta}$

$$\varrho_{N_1}(f_0, \varphi_h) \geqq \frac{1}{R+1} \left(e^{\alpha\left(D_{N_1}{}^2 - \frac{\varepsilon^2}{2}\right)}\right)^{N_1}, \tag{23''}$$

und es ist dann sogar $\lim\limits_{N_1 \to \infty} \varrho_{N_1} = \infty$.

Aus Satz 4 erkennt man aber, daß es bei Interpolation mittels einer unendlichen Folge auf die Reihenfolge ganz wesentlich ankommt, im Gegensatz zu endlichen Folgen. Wesentlich ist natürlich, daß die Folge φ_h überall dicht in E, also die Folge $(e^{2\pi i\varphi_h})$ dicht auf dem Einheitskreis sein muß, wenn die Interpolationspolynome für alle in $|z| \leqq R$ $(R > 1)$ analytischen Funktionen $f(z)$ gegen diese Funktion konvergieren soll. Ist die Folge ζ_h gleichverteilt, dann ist dies stets der Fall. Man kann aber, wie aus der Schlußbemerkung von § 3 hervorgeht, (φ_j) so umordnen, daß für die neue Folge (ϑ_j)

$$\varrho_N(f, \vartheta_h) \leqq \frac{e^{8\log^2 N}}{R^N} \cdot \left(\frac{R+1}{R-1}\right)^{8\log N} \frac{C}{R-1}$$

ist. Man kann sie aber auch in eine „schlechte" Folge (ϑ_h'') umordnen, daß es eine analytische Funktion f in $|z| \leqq R$ gibt mit

$$\varrho_N(f_0, \vartheta_h'') \geqq \frac{C_1^N}{R^N} \cdot \frac{1}{R+1},$$

wobei C_1 eine absolute Konstante ist. (Man kann $C_1 = e^\beta$, $\beta = \dfrac{1}{4}\alpha$, $\alpha = \dfrac{1}{4} 10^{-8}$ wählen.)

Wenn also $1 < R < C$ ist, so ist das Verfahren divergent.

Literatur

[1] J. L. Walsh, Interpolation and approximation by rational functions in the complex domain, Amer. Math. Soc. Providence 1956 (2. Ed).

[2] Abschätzungen ohne Verwendung der Diskrepanz bei O. Kis, Hung. Acta Math. 7 (1956), 173—200.

[3] Annals of Math. **51** (1950), 105—119. Ein funktionentheoretischer Beweis bei T. Ganelius, Arkiv för Matematik 3 (1958), Nr. 1, 1—50.

[4] Vgl. etwa Cigler-Helmberg, Jahresbericht DMV **64** (1962), 1—50.

[5] J. H. Curtiss, Fourth Berkeley Symposium on Mathematical Statistics and Probability 1961, University of California Press, Berkeley and Los Angeles, Vol. II, pp. 79—93.

Corrections

103.1 $\zeta_h = e^{2\pi i\varphi_h}$ $(h = 1, \cdots, N)$

103.2 $\omega_N(z) = \prod_{h=1}^{N} (z - \zeta_h);$

108.3 $(R > 1)$

109.4 $S = \gamma(D)[\cdots]$

Ein metrischer Satz in der Theorie der C-Gleichverteilung

Herrn Professor N. Hofreiter zum 65. Geburtstag gewidmet

Monatsh. Math. **74** (1970) 108–118

(*Eingegangen am 8. Mai 1969*)

Es sei f eine reellwertige Funktion, definiert auf $I^+: 0 \leq t$ und dort integrierbar, dann heißt sie C-gleichverteilt modulo 1, wenn für jedes Intervall $J: \langle \alpha, \beta \langle \subset \langle 0, 1 \rangle$

$$\lim_{T \to \infty} \frac{1}{T} \int_0^T \chi(\{f(t)\}; \alpha, \beta)\, dt = \beta - \alpha, \tag{1}$$

wo $\chi(.\,; \alpha, \beta)$ die charakteristische Funktion des Intervalls $\alpha \leq x < \beta$ und $\{f(t)\}$ der Wert von $f(t)$ modulo 1 ist. Es ist dieser Begriff das „kontinuierliche" Analogon[1] zum Begriff der gleichverteilten Folge $(f(n))$ modulo 1, wo für jedes Intervall $\langle \alpha, \beta \langle$ verlangt wird:

$$\lim_{N \to \infty} \frac{1}{N} \sum_{n=1}^{N} \chi(\{f(n)\}; \alpha, \beta) = \beta - \alpha. \tag{1'}$$

Wir wollen gleich eine Verallgemeinerung durchführen: Es sei a eine reellwertige Funktion, definiert auf $I_2^+: T > 0,\ t \geq 0$, integrabel in bezug auf t mit $\int_0^\infty a(T, t)\, dt = 1$ für alle $T > 0$. (Wir wollen zusätzlich noch zur Vereinfachung verlangen, daß a nicht negativ ist.) Dann heißt f A-gleichverteilt, wenn für jedes $J \subset \langle 0, 1 \rangle$

$$\lim_{T \to \infty} \int_0^\infty a(T, t)\, \chi(\{f(t)\}; \alpha, \beta)\, dt = \beta - \alpha. \tag{2}$$

---

[1] Vgl. dazu den Bericht von J. Cigler und G. Helmberg, Neuere Entwicklungen der Theorie der Gleichverteilung, Jahresberichte der Deutschen Mathematiker-Vereinigung **64** (1961) S. 1–50, insbesondere § 11. Weiter auch die Dissertation von P. J. Holewijn, Contributions to the theory of asymptotic distribution modulo 1 (1965) S. 1–49 Uitgeverij Waltmann, Delft.

Ein wichtiger Spezialfall ist die sogenannte B-Gleichverteilung, wo b nicht negativ und integrabel auf I^+, $\beta(b, T) = \int_0^T b(t)\,dt$ für $T > 0$ und $a(T, t) = b(t)/\beta(T)$ für $0 \leq t \leq T$ und sonst 0.

Für $b = 1$ für alle t, erhalten wir (1). Das diskrete Analogon zu (2) ist die A-Gleichverteilung für Folgen $(f(n))$. Es sei a definiert für alle natürlichen Zahlen N und n und es sei $\sum_{n=1}^{\infty} a(N, n) = 1$, dann ist $(f(n))$ A-gleichverteilt, wenn

$$\lim_{N\to\infty} \sum_{n=1}^{\infty} a(N, n)\, \chi(\{f(n)\}; \alpha, \beta) = \beta - \alpha \tag{2'}$$

für alle J.

Es gilt folgendes notwendiges und hinreichendes Kriterium für A-Gleichverteilung (sog. Weylsches Kriterium): f ist genau dann A-gleichverteilt, wenn für jede natürliche Zahl $k \neq 0$

$$\lim_{T\to\infty} \int_0^{\infty} a(T, t)\, e^{\pi i k f(t)}\, dt = 0. \tag{3}$$

(Im diskreten Fall ist das Integral durch die analoge Summe zu ersetzen.) Dieses Kriterium gestattet in vielen Fällen festzustellen, ob eine Funktion (bzw. eine Folge) A-gleichverteilt ist. Im diskreten Falle gilt aber sogar folgendes: Es sei P der Raum aller Folgen $(f(n))$ mit $0 \leq f \leq 1$ und μ_p das Produktmaß zum Lebesgueschen Maß auf $\langle 0, 1\rangle$. Dann sind fast alle Folgen gleichverteilt im Sinne von (1'), ja es sind fast alle Folgen A-gleichverteilt, wenn für jedes $\delta > 0$ (sog. Hillsche Bedingung)

$$\sum_{n=1}^{\infty} e^{-\delta^2/\alpha_N} < \infty, \tag{4}$$

wo $\alpha_N = \sum_{n=1}^{\infty} a^2(N, n)$. Aus (4) folgt $\lim_{N\to\infty} \alpha_N = 0$.

Wir wollen nun in dieser Arbeit ein kontinuierliches Gegenstück zu den zitierten Sätzen aufstellen. Dazu betrachten wir den Raum C_0 aller stetigen Funktionen f auf I^+ mit $f(0) = 0$ mit dem zugehörigen Wiener-Maß μ_w (die zugehörigen Integrale wollen wir ebenfalls mit μ_w bezeichnen). Es ist bekanntlich $\mu_w(C) = 1$. Wir werden zeigen[2]:

---

[2] Dieser Satz wurde von mir o. B. in Nuffic Int. Summer Sessions 1962 S. 90 angegeben und mit Beweis im Privatissimum vorgetragen. Ein metrischer Satz in anderer Richtung bei P. J. HOLEWIJN (vgl. Fußnote 1).

Satz 1: *Fast alle $f \in C_0$ sind C-gleichverteilt.*

Wir werden auch ein Analogon zur Hillschen Bedingung aufstellen, müssen allerdings einige einschränkende Voraussetzungen machen, die vielleicht überflüssig sind:

1) Für jede stetige Funktion φ auf I^+ sei $\psi(\varphi, T) = \int_0^\infty a(T, t)\, e^{i\varphi(t)}\, dt$ gleichmäßig stetig für genügend große T, d. h. für alle $T > T_1$.

2) $\alpha(T) = \int_0^\infty a^2(T, t)\, dt$ existiere, sei > 0 für große T und es sei $\lim_{T\to\infty} \alpha(T) = 0$.

3) Für jedes $\delta > 0$ sei

$$\int_0^\infty e^{-\delta^2/\alpha(T)}\, dT < \infty. \tag{5}$$

Wir wollen (5) auch die Hillsche Bedingung nennen,

Dann gilt

Satz 2: *Fast alle $f \in C_0$ sind A-gleichverteilt, wenn die Voraussetzungen* 1, 2, 3 *erfüllt sind.*

Es ist Satz 1 eine Folgerung aus Satz 2, denn für $a(T, t) = \frac{1}{T}$ $(0 \le t \le T)$ und 0 sonst ist $\psi(\varphi, T) = \frac{1}{T} \int_0^T e^{i\varphi(t)}\, dt$, und es ist

$$|\,\psi(T) - \psi(T')\,| \le 2(T - T') \text{ für alle } T, T' \ge T_1 \ge 1,$$

denn es ist

$$\psi(T) - \psi(T') = \frac{1}{T} \int_{T'}^{T} e^{i\varphi(t)}\, dt + \left(\frac{1}{T} - \frac{1}{T'}\right) \int_0^{T'} e^{i\varphi(t)}\, dt,$$

also $|\,\psi(T) - \psi(T')\,| \le \dfrac{2\,|\,T - T'\,|}{T} \le |\,T - T'\,|$ für $T \ge 1$.

Voraussetzung 2 ist ebenfalls erfüllt, da $\alpha(T) = \frac{1}{T}$, und ebenso Voraussetzung 3, denn es ist

$$\int_0^\infty e^{-\delta^2/\alpha(t)}\, dT = \int_0^\infty e^{-\delta^2 T}\, dT.$$

Ein anderes Beispiel wäre $a(T, t) = \frac{1}{t+1} \frac{1}{lg(T+1)}$ für $0 \leq t \leq T$ und 0 sonst.

Bemerkung: Ist $\alpha(T)$ gleichmäßig stetig in T für $T \geq T'$, dann folgt die Vor. 2 aus der Vor. 3.

Beweis: Es sei $\delta > 0$, dann existiert nach Voraussetzung ein $\eta(\delta)$, so daß für alle $T, T' > T_1$

$$| \alpha(T) - \alpha(T') | < \delta^2$$

für $| T - T' | < \eta(\delta)$. Man wähle eine natürliche Zahl n so groß, daß $\frac{1}{2^n} < \eta(\delta)$. Nun ist $\int\limits_0^\infty e^{-\delta^2/\alpha(T)}\, dT = \sum\limits_{r=0}^{\infty} \int\limits_{\frac{r}{2^n}}^{\frac{r+1}{2^n}} e^{-\delta^2/\alpha(T)}\, dT < \infty$; also existiert ein $r_0(\eta)$ so, daß $\int\limits_{\frac{r}{2^n}}^{\frac{r+1}{2^n}} e^{-\delta^2/\alpha(T)}\, dT < \frac{1}{e\, 2^n}$ ist für alle $r > r_0(n)$.

Setzen wir jetzt $T_2 = \operatorname{Max}\left(\frac{r_0}{2^n}, T_1+1\right)$, dann ist für alle $T > 2T_2$ sicher $\alpha(T) < 5\delta^2$. Nehmen wir nämlich an, es gäbe ein $T > 2T_2$, so daß $\alpha(T) \geq 5\delta^2$, dann ist für alle T' mit $| T - T' | < \eta(\delta)$ sicher

$$\alpha(T') \geq 4\delta^2.$$

Es sei nun $r = [2^n T]$, dann gilt dies sicher für alle T' im Intervall $\langle \frac{r}{2^n}, \frac{r+1}{2^n} \rangle$, denn dann ist, da ja $r \leq 2^n T < r+1$,

$\frac{r}{2^n} - T \leq T' - T < \frac{r+1}{2^n} - T$, also $| T' - T | < \frac{1}{2^n} < \eta(\delta)$, und es ist $T' \geq \frac{r}{2^n} \geq T - \frac{1}{2^n} > T - 1 > T_1$. Dann ist aber für diese r

$$\int\limits_{\frac{r}{2^n}}^{\frac{r+1}{2^n}} e^{-\delta^2/\alpha(T)}\, dT > \frac{1}{2^n}\, e^{-\frac{1}{4}}$$ und dies ist ein Widerspruch.

Da δ beliebig ist, folgt die Behauptung.

Für den Beweis von Satz 2 benötigen wir einige Hilfssätze:

Hilfssatz 1: Es sei h eine nichtnegative quadratisch integrierbare Funktion auf $\langle 0, T \rangle$, $\beta > 0$; dann ist

$$\int_0^T h(x)\, e^{-\beta x}\, dx \int_0^x h(y)\, e^{\beta y}\, dy \leq \frac{1}{\beta} \int_0^T h^2(x)\, dx. \tag{6}$$

Beweis: Wir können o. B. d. A. annehmen, daß h beschränkt ist. Wir setzen $u(x) = \int_0^x h(y)\, e^{\beta y}\, dy$ für alle x in $\langle 0, T\rangle$. Es ist u fast überall differenzierbar, $u(x) = O(x)$ für $x \to 0$.

Es genügt zu zeigen:

$$\int_0^T e^{-2\beta x}\,(u'^2 - \beta u u')\, dx \geq 0, \tag{6'}$$

denn es ist $u'(x) = h(x)\, e^{\beta x}$. Für $T = 0$ ist (6′) richtig. Es sei $T > 0$ und ε sei beliebig gewählt mit $0 < \varepsilon < T$. Nun ist für ein x mit $\varepsilon \leq x \leq T$, für welches u' existiert,

$$u'^2 - \beta u u' - \left((u' - u)\,(\beta + \frac{1}{x})\right)^2 = u u'(\beta + \frac{2}{x}) - u^2(\beta + \frac{1}{x})^2,$$

also[3]

$$e^{-2x\beta}(u'^2 - \beta u u') = e^{-2x\beta}\left(u' - u(\beta + \frac{1}{x})\right)^2 + \frac{d}{dx}\left(e^{-2x\beta}\frac{u^2}{2}(\beta + \frac{2}{x})\right),$$

also

$$\int_\varepsilon^T e^{-2x}(u'^2 - \beta u u')\, dx = \int_\varepsilon^T e^{-2x\beta}(u' - u(\beta + \frac{1}{x}))^2\, dx + e^{-2+\beta}\frac{u^2(T)}{2}(\beta + \frac{2}{T}) + R,$$

wo $R = -e^{-2\varepsilon\beta}\,\dfrac{u^2(\varepsilon)}{2}\,(\beta + \dfrac{2}{\varepsilon})$. Gehen wir nun mit $\varepsilon \to 0$, so folgt die Behauptung.

Hilfssatz 2: Es seien Φ_1, Φ_2 nichtnegative integrierbare Funktionen auf dem Intervall J: $\langle 0, T\rangle$, m eine natürliche Zahl ≥ 1. Es sei weiter Δ_{2m} das $2m$-dimensionale Simplex $0 \leq \tau_1 < \tau_2 < \ldots < \tau_{2m} \leq T$, und es sei $\Phi_k = \Phi_1$, wenn k ungerade, und $= \Phi_2$, wenn k gerade. Dann ist

$$J_{2m} = \int_{\Delta_{2m}} \Phi_1(\tau_1) \ldots \Phi_{2m}(\tau_{2m})\, d\tau_1 \ldots d\tau_{2m} \leq \frac{1}{m!}\left(\int_0^T \Phi_2(x)\, dx \int_0^x \Phi_1(y) dy\right)^m. \tag{7}$$

[3] Man kommt auf diese Identität, wenn man (6′) als Variationsproblem auffaßt und die Weierstraß-Hilbertsche Formel anschreibt (vgl. z. B. G. H. Hardy, J. E. Littlewood, G. Pólya, Inequalities, 1952, Cambridge University Press, insbesondere Ch. VII).

Wir führen den Beweis durch vollständige Induktion. Wir wollen noch $n = 2m$ setzen und unter $\Delta_n(\xi)$ für $0 \leq \xi \leq T$ das Simplex $0 \leq \tau_1 < \tau_2 < \ldots < \tau_n \leq \xi$ verstehen. Weiters setzen wir

$$\Phi_1(\tau_1)\ldots\Phi_n(\tau_n) = \psi_n(\tau_1\ldots\tau_n),\ J_n(\xi) = \int\limits_{\Delta_n(\xi)} \psi_n(\tau_1\ldots\tau_n)\, d\tau_1\ldots d\tau_n.$$

Dann ist also zu zeigen: *für alle* ξ *in* $\langle 0, T\rangle$ *ist* $J_n(\xi) \leq \dfrac{1}{m!} J_2^m(\xi)$.

Nehmen wir an, für $n = 2m$ sei es bereits erwiesen. Es ist

$$J_{n+2}(\xi) = \int\limits_0^\xi \Phi_2(\tau_{n+2})\, d\tau_{n+2} \int\limits_0^{\tau_{n+2}} \Phi_1(\tau_{n+1})\, d\tau_{n+1}\, J_n(\tau_{n+1});$$

also nach Induktionsvoraussetzung ist

$$J_{n+2}(\xi) \leq \frac{1}{m!} \int\limits_0^\xi \Phi_2(\eta)\, d\eta \int\limits_0^\eta \Phi_1(\xi)\, J_2^m(\xi)\, d\xi. \tag{8}$$

Nun ist J_2 fast überall differenzierbar und $J_2'(\eta) = \Phi_2(\eta)\, F(\eta) \geq 0$, wo $F(\eta) = \int\limits_0^\eta \Phi_1(\xi)\, d\xi$. Dann ist

$$\int\limits_0^\eta \Phi_1(\xi)\, J_2^m(\xi)\, d\xi = F(\eta)\, J_2^m(\eta) - m \int\limits_0^\eta F(\eta)\, J_2^{m+1}(\eta)\, J_2'(\eta)\, d\eta,$$

also $\leq F(\eta)\, J_2^m(\eta)$, also ist

$$J_{n+2}(\xi) \leq \frac{1}{m!} \int\limits_0^\xi \Phi_2(\eta)\, F(\eta)\, J_2^m(\eta)\, d\eta = \frac{1}{m!} \int\limits_0^\xi J_2^m(\eta)\, J_2'(\eta)\, d\eta$$

und dies ist die Behauptung für $n + 2$.

Daraus folgt sofort

Hilfssatz 3: Es mögen die gleichen Voraussetzungen wie in Hilfssatz 2 gelten, es sei nun aber $\Phi_1(\tau) = h(\tau)\, e^{+\beta\tau}$, $\Phi_2(\tau) = h(\tau)\, e^{-\beta\tau}$, *wo* h, β *die Voraussetzungen von Hilfssatz 1 erfüllen; dann ist*

$$\int\limits_{\Delta_{2m}} \Phi_1(\tau_1)\ldots\Phi_{2m}(\tau_{2m})\, d\tau_1\ldots d\tau_{2m} \leq \frac{1}{m!\,\beta^m} \Big(\int\limits_0^T h^2(x)\, dx\Big)^m. \tag{9}$$

Dies folgt sofort aus (7) und (6).

Hilfssatz 4: Es sei $P(x \mid y; t) = (2\pi t)^{-\frac{1}{2}}\, e^{-\frac{1}{2}(y-x)^2/t}$, m, k *ganz* $\neq 0$, $m \geq 1$. *Seien* $\tau_1\ldots\tau_{2m}$ *reelle Zahlen mit* $0 < \tau_1 < \ldots < \tau_{2m}$, $\xi_1\ldots\xi_{2m}$ *reelle Zahlen,*

$$Q(\xi_1 \ldots \xi_{2m}; \tau_1 \ldots \tau_{2m}) =$$
$$= P(0 \mid \xi_1; \tau_1) P(\xi_1 | \xi_2; \tau_2 - \tau_1) \ldots P(\xi_{2m-1} | \xi_{2m}; \tau_{2m} - \tau_{2m-1}).$$

Sind dann $\varepsilon_1 \ldots \varepsilon_{2m}$ *Zahlen von der Gestalt* ± 1, k *ganz, dann gilt für*

$$J_{2m} = \int_{R_{2m}} Q(\xi_1 \ldots \xi_{2m}; \tau_1 \ldots \tau_{2m}) \, e^{2\pi i k(\varepsilon_1 \xi_1 + \ldots + \varepsilon_{2m} \xi_{2m})} \, d\xi_1 \ldots d\xi_{2m} \tag{10}$$

die Abschätzung

$$J_{2m} \leq \prod_{j=1}^{m} e^{-2k^2 \pi^2 (\tau_{2j} - \tau_{2j-1})}. \tag{11}$$

Beweis. Wir machen in (10) die Transformation von Determinante 1

$$\xi_1 = \eta_1, \ \xi_2 = \eta_1 + \eta_2, \ldots \ \xi_{2m} = \eta_1 + \ldots + \eta_{2m}.$$

Es ist $\sum_{i=1}^{2m} \varepsilon_i \, \xi_i = \sum_{i=1}^{2m} \delta_i \, \eta_i$, wo $\delta_i = \sum_{j=i}^{2m} \varepsilon_j$. Es ist δ_i ganz und zwar gerade, wenn i ungerade, und ungerade, wenn i gerade. Es ist also $\delta_i^2 > 0$, wenn i ungerade, $\delta_i^2 \geq 1$, wenn i gerade, da die Anzahl der Summanden ± 1 in δ_i ja $2\,m - i + 1$ ist. Dann wird mit $\tau_0 := 0$

$$J_{2m} = \prod_{j=1}^{2m} \int_{-\infty}^{\infty} \frac{1}{\sqrt{2\pi(\tau_j - \tau_{j-1})}} \, e^{-\frac{\eta_j^2}{2(\tau_j - \tau_{j-1})} + 2\pi i \eta_j \delta_j} \, d\eta_j =$$

$$= \prod_{j=1}^{2m} e^{-2\pi^2 k^2 \delta_j^2 (\tau_j - \tau_{j-1})} = \prod_{j \text{ gerade}} \cdot \prod_{j \text{ ungerade}}$$

$$\leq \prod_{j \text{ gerade}} = \prod_{i=1}^{m} e^{-2\pi^2 k^2 (\tau_{2i} - \tau_{2i-1})}.$$

Daraus folgt sofort

Hilfssatz 5: Es sei mit den Bezeichnungen von Hilfssatz 4

$$\psi(\xi_1 \ldots \xi_{2m}; k) = \prod_{i=1}^{2m} \cos 2\pi k \xi_i \ \textit{oder} \ \prod_{i=1}^{2m} \sin 2\pi k \xi_i; \tag{12}$$

dann gilt für das Integral

$$J'_{2m}(\tau_1 \ldots \tau_{2m}) = \int_{R_{2m}} Q(\xi_1, \ldots, \xi_{2m}, \tau_1, \ldots, \tau_{2m}) \, \psi(\xi_1 \ldots \xi_{2m}, k) \, d\xi_1 \ldots \xi_{2m} \tag{13}$$

die gleiche Abschätzung (11).

Beweis: Es ist ja $2^{2m}\,\psi = \prod_{j=1}^{2m} (e^{2\pi i k \xi_j} + e^{-2\pi i k \xi_j})$ (wenn wir den Fall des Kosinus betrachten) ausmultipliziert eine Summe von 2^m Funktionen $e^{2\pi i k(\varepsilon_1 \xi_1 + \cdots + \varepsilon_{2m} \xi_{2m})}$, und aus (11) folgt alles.

Es sei nun

$$I_m = \mu_w(|\int_0^T e^{2\pi i k f(\tau)}\, h(\tau)\, d\tau\,|^{2m}), \tag{14}$$

wo $f \in C_0$ und h die Voraussetzungen von Hilfssatz 1 erfüllt. Wir trennen $J = \int_0^T e^{2\pi i k g(\tau)}\, h(\tau)\, d\tau$ in Real- und Imaginärteil $J_1 + iJ_2$; dann ist $|J|^{2m} = (J_1^2 + J_2^2)^m \leq 2^{m-1}(J_1^{2m} + J_2^{2m})$, also $I_m \leq 2^{m-1}(I_m^1 + I_m^2)$, wo $I_m^l = \mu_w(J_l^{2m})$ $(l = 1, 2)$. Setzen wir $\Phi(\tau) = h(\tau) \cos k f(\tau)$ bzw. $h(\tau) \sin k f(\tau)$, dann ist z. B.

$$J_1^{2m} = \int_0^T \ldots \int_0^T \Phi(\tau_1) \ldots \Phi(\tau_{2m})\, d\tau_1 \ldots d\tau_{2m} = (2m)! \int_{\Delta_{2m}} \Phi(\tau_1) \ldots \Phi(\tau_{2m}) d\tau_1 \ldots d\tau_{2m},$$

wo Δ_{2m} in Hilfssatz 2 definiert ist.

Nun ist[4]

$$\mu_w(\int_{\Delta_{2m}} \Phi(\tau_1) \ldots \Phi(\tau_{2m}) d\tau_1 \ldots d\tau_{2m}) = \int_{\Delta_{2m}} h(\tau_1) \ldots h(\tau_{2m})\, J'_{2m}(\tau_1 \ldots \tau_{2m})\, d\tau_1 . d\tau_{2m},$$

wo J'_{2m} durch (13) definiert ist; also ist nach (11)

$$I'_m \leq (2m)! \int_{\Delta_{2m}} \prod_{i=1}^{2m} h(\tau_i) \prod_{i=1}^{m} e^{-2k^2\pi^2(\tau_{2i} - \tau_{2i-1})}\, d\tau_1 \ldots d\tau_{2m},$$

also nach (9) mit $\beta = 2k^2\pi^2$

$$I_m^1 \leq \frac{(2m)!}{m!\,\beta^m} \left(\int_0^T h^2(x)\, dx\right)^m.$$

Die gleiche Abschätzung gilt für I_m^2 und wir haben also

Hilfssatz 6: Es ist für ganzzahlige k und $m \geq 1$

$$I_m = \mu_w(|\int_0^T e^{2\pi i k f(\tau)}\, h(\tau)\, d\tau\,|^{2m}) \leq \frac{(2m)!}{m!} \left(\frac{1}{k^2\pi^2} \int_0^T h^2(x)\, dx\right)^m; \tag{15}$$

also ist auch sicher

---

[4] Zu den hier benützten Sätzen über das Wiener-Integral vgl. z. B. M. Kac, Probability and related topics in physical sciences (1959), Interscience Publishers, London-New York, insbesondere Ch. 4.

$$I_m \leq (2m \int_0^T h^2 dx)^m. \tag{15'}$$

Daraus folgt sofort

Hilfssatz 7: Es sei k ganz und ungleich 0, *s eine positive reelle Zahl, σ die nächst größere ganze Zahl an* $\frac{s}{2}$, *dann ist*

$$\mu_w(|\int_0^T e^{2\pi i k f(\tau)} h(\tau)\, d\tau|^s) \leq (2\,\sigma \int_0^T h^2(\tau)\, d\tau)^{\frac{s}{2}}. \tag{16}$$

Beweis: Es ist nach der Hölderschen Ungleichung und da auch $\sigma \geq 1$ und nach (15')

$$\mu_w(|\ldots|^s) \leq \mu_w(|\ldots|^{2\sigma})^{\frac{s}{2\sigma}} \leq (2\sigma)^{\frac{s}{2\sigma}} (\int_0^T h^2(\tau)\, d\tau)^{\frac{s}{2}}.$$

Es ist $\sigma \leq \left[\frac{s}{2}\right] + 1$, also $2\sigma \leq s + 2$.

Wir wollen nun voraussetzen, daß die Integrale $\int_0^\infty h(\tau)\, d\tau$ und $\int_0^\infty h^2(\tau) d\tau$ existieren, dann können wir in (16) mit $T \to \infty$ gehen und erhalten

Hilfssatz 8: Es ist für jede ganze Zahl $h \neq 0$ *und positive Zahl s*

$$\mu_w(|\int_0^\infty e^{2\pi i k f(\tau)} h(\tau)\, d\tau|^s) \leq (s+2)^{\frac{s}{2}} (\int_0^\infty h^2(\tau)\, d\tau)^{\frac{s}{2}}. \tag{17}$$

Jetzt kann der Beweis von Satz 2 leicht geführt werden.

Wir müssen nach dem Weylschen Kriterium zeigen, daß für fast alle $f \in C_0$ und alle natürlichen Zahlen k die Behauptung (3) zutrifft. Es sei zunächst k fest. Es sei $\delta > 0$ beliebig. Nach Voraussetzung 2 ist $\lim_{T\to\infty} \alpha(T) = 0$, also ist für $T > T_0(\delta)$ sicher $\delta^2 > 10\alpha(T)$.

Wir setzen nun in (16) $h(\tau) = a(T, \tau)$ und $s(\delta, T) = \dfrac{2\delta^2}{\alpha(T)} > 2$, dann erhalten wir

$$\mu_w(|(3\delta)^{-1} \int_0^\infty a(T, \tau)\, e^{2\pi i k f(\tau)}\, d\tau|^{s(\delta, T)}) \leq e^{-\delta^2/\alpha(T)}, \tag{18}$$

denn es ist die linke Seite von (16), noch durch $(3\delta)^s$ dividiert,

$$\leq \left(\left(\frac{s+2}{9\delta^2}\right) \alpha(T)\right)^{\frac{s}{2}} \leq \left(\frac{11}{45}\right)^{\frac{s}{2}} \leq e^{-\delta^2/\alpha(T)}.$$

Es sei nun n eine natürliche Zahl und $\gamma_{r,n} = \frac{r}{2^n}$ $(r = 0, 1, 2 \ldots)$.

Wir wollen noch den Ausdruck in μ_w links in (18) mit $\Phi(T, f)$ bezeichnen. Dann ist (wir lassen im Augenblick den Index n weg) nach (18)

$$\mu_w\Big(\int_{\gamma_r}^{\gamma_{r+1}} \Phi(T, f)\, dT\Big) \leq \int_{\gamma_r}^{\gamma_{r+1}} e^{-\delta^2/\alpha(T)}\, dT. \tag{19}$$

Es sei nun $\mu_w(r, n)$ das Wienersche Maß der Menge $L_{r,n}$ aller $f \in C_0$, so daß

$$\int_{\gamma_r}^{\gamma_{r+1}} \Phi(T, f)\, dT \geq 2^n. \tag{20}$$

Dann folgt aus (19) $\mu(L_{r,n}) \leq \int_{\gamma_r}^{\gamma_{r+1}} e^{-\delta^2/\alpha(T)}\, dT$. Nun ist

$$\sum_{r=0}^{\infty} \int_{\gamma_r}^{\gamma_{r+1}} e^{-\delta^2/\alpha(T)}\, dT = \int_0^{\infty} e^{-\delta^2/\alpha(T)}\, dT$$

konvergent. Es gibt also zu jedem $\eta > 0$ ein $r_0(n, \delta, \eta)$ so, daß für $s \geq r_0$ $\int_{\gamma_s}^{\infty} e^{-\delta^2/\alpha(T)}\, dT < \eta$, also ist $\mu(U_{s,n}) < \eta$, wo $U_{s,n} = \bigcup_{r=s}^{\infty} L_{r,n}$. Es besitzt also, da $(U_{s,n})$ eine in s monoton abnehmende Mengenfolge ist, $U_n = \bigcap_s U_{s,n}$ das Maß 0, also besitzt $U = \bigcup_{n=1}^{\infty} U_n$ ebenfalls das Wienersche Maß 0. Ist also $f \in C_0$ aber nicht in U, dann gilt für alle n und alle $r > r_1(n, \delta)$

$$\int_{\frac{r}{2^n}}^{\frac{r+1}{2^n}} \Phi(T, f)\, dT \leq \frac{1}{2^n}. \tag{21}$$

Wir behaupten nun: Für jedes solche f ist für $T > T_4(\delta)$

$$|\,\psi(T, f)\,| = \Big|\int_0^{\infty} a(T, \tau)\, e^{2\pi k f(\tau)}\, d\tau\,\Big| < 10\,\delta. \tag{22}$$

Nach Voraussetzung 2 ist das Integral $\psi(T)$ gleichmäßig stetig auf T für $T > T_1$. Es existiert also ein $\eta(\delta)$ so. daß für alle $T, T' \geq T_1$ mit $|\,T - T'\,| < \eta$, $|\,\psi(T) - \psi(T')\,| < \delta$. Dies gilt also auch sicher für alle $T \geq 2\,\mathrm{Max}\,(T_1, 1) = T_2$. Man wähle nun $n(\delta)$ so groß, daß $\frac{1}{2^n} < \eta$.

Dann gilt für alle $r > r_1(n(\delta), \delta)$ sicher (21). Es sei $T_3 = r_1/2^n$ und $T_4 = 2 \operatorname{Max}(T_2, T_3)$. Nehmen wir nun an, (22) sei für ein $T' > T_4$ falsch. Dann ist für alle T mit $|T - T'| < \delta$ auch $|\psi(T)| \geq 9\delta$.

Es sei nun $r = [2^n T']$. Es ist $r \geq 2^n T' - 1 \geq 2r_1 - 1 \geq r_1$. Betrachten wir die T mit $\frac{r}{2^n} \leq T < \frac{r+1}{2^n}$. Es ist $|T' - T| \leq \frac{1}{2^n} < \eta$, also gilt auch für diese T $|\psi(T)| \geq 9\delta$. Dann ist nach (18)

$$\Phi(T) = |\psi(T)/3\delta|^{s(\delta, T)} \geq 3^s \geq 3$$

für alle diese T. Dann ist aber auch $2^n \int\limits_{\frac{r}{2^n}}^{\frac{r+1}{2^n}} \Phi(T)\, dT \geq 9$. Dies ist aber ein Widerspruch zu (21).

Bis jetzt war δ fest, und es gilt also für alle $f \in C_0$ bis auf eine Menge $U(\delta)$ vom Maße 0 die Behauptung (22). Wir lassen jetzt δ eine Nullfolge (δ_n) durchlaufen, dann gilt für alle $f \in C_0$ bis auf die Menge $V = \bigcup\limits_{n=1}^{\infty} U(\delta_n)$, welche das Maß 0 hat, daß $\lim\limits_{T \to \infty} \int\limits_0^T a(T, \tau)\, e^{2\pi i k f(\tau)}\, d\tau = 0$. Nun war auch k fest, es hängt also V von k ab. Nun ist aber auch $V = \bigcup\limits_{k=1}^{\infty} V(k)$ vom Maß 0, also gilt für alle $f \notin V$ das Weylsche Kriterium und dies ist die Behauptung.

Bemerkung. Für den Beweis von Satz 1 hätte schon $m = 1$ und der Hilfssatz 1 genügt, der in diesem Fall direkt nachzurechnen ist.

Discrepancy and Riemann Integration

Studies in Pure Mathematics (Papers presented to Richard Rado),
Academic Press, London and New York, 1971, pp. 121–129

Let E^s be the unit cube $0 \leqslant \xi_i < 1$ $(i = 1, \ldots, s)$, $\bar{E}^s$ the closed cube $0 \leqslant \xi_i \leqslant 1$ $(i = 1, \ldots, s)$. If $\omega = (x_i)$, $i \in N$ is a sequence in E^s, then ω is called† u.d.(mod 1), if for all intervals $J \subset E^s$

$$\lim_{N \to \infty} \mu_N(\chi(J)) = \mu(\chi(J)). \tag{1}$$

where

$$\mu_N(\chi(J)) = \frac{1}{N} \sum_{i=1}^{N} \chi(J; x_i), \quad \mu(\chi(J)) = \int_{E^s} \chi(J; x)\, dx = \operatorname{Vol}(J)$$

($\chi(J)$ is the indicator function of J). More generally, the sequence ω is called u.d.(mod 1) with density ρ, if we have, instead of (1),

$$\lim \mu_N(\chi(J)) = \int_{E^s} \chi \rho\, dx = \mu(\chi; \rho). \tag{1'}$$

It is supposed, that ρ is Riemann-integrable on $\bar{E}^s$ with $\int \rho\, dx = 1$ and non-negative.

It is well known that as a consequence of (1) for any periodic, Riemann-integrable function f of period one

$$\lim \mu_N(f) = \lim \frac{1}{N} \sum_{i=1}^{N} f(x_i) = \int_{E^s} f\, dx = \mu(f). \tag{2}$$

In the case (1′) we have instead of $\mu(f)$ the integral $\int f \rho\, dx = \mu(f, \rho)$. The discrepancy $D_N(\omega; \rho)$ of the sequence ω is defined as

$$\sup_J |\mu_N(\chi(J)) - \mu(\chi(J); \rho)|. \tag{3}$$

† Uniformly distributed.

If $\rho \equiv 1$ (the classical case), we write $D_N(\omega)$. Now we ask: is it possible to estimate

$$D_N(f, \omega, \rho) = |\mu_N(f) - \mu(f, \rho)| \tag{4}$$

from above with the help of D_N? If f is of bounded variation $V(f)$ in the sense of Hardy–Krause on $\bar{E}^s$, then it is known, that

$$D_N(f, \omega, \rho) \leqslant V(f)\, D_N(\omega; \rho).$$

Now we will give an estimate of $D_N(f)$ for general f. The result is simplest in the case $\rho \equiv 1$.

THEOREM 1†

$$D_N(f, \omega) \leqslant (1 + 2^{2s-1}) \sum (f, [D^{-1/s}]^{-1}) \tag{5}$$

Here

$$\sum(f, k) = \sup_{n(p) \leqslant k} \sigma(f; p),$$

where we consider all partitions p of $\bar{E}^s$ in intervals with norm $n(p) \leqslant k$ and $\sigma(f; p)$ is the mean oscillation of f for the partition p. The norm $n(p)$ is the largest length of all the edges of the intervals in p. (Usually the norm $n'(p)$ is the largest diameter of the intervals in p but we have $n(p) \leqslant n'(p) \leqslant \sqrt{s}\, n(p)$). For arbitrary ρ see (14).

If ω is u.d.(mod 1), then $D_N \to 0$; f is integrable; therefore the right side of (5) tends to zero, as it should do.

For the proof of (5), we will first consider the case $s = 1$, because here the details of the proof are much simpler. Let $p(L)$ be the partition of $E^1 = E$ in the intervals $J(z)$:

$$\frac{z-1}{L} \leqslant \xi < \frac{z}{L}$$

(L natural number; $z = 1, \ldots, L$), therefore $n(p(L)) = L^{-1}$. Further let

$$G(z) = \sup_{J(z)} f, \qquad g(z) = \inf_{J(z)} f$$

and

$$\Phi_2 = \sum_{z=1}^{L} G(z)\chi(J(z)), \quad \Phi_1 = \sum_{z=1}^{L} g(z)\chi(J(z)),$$

† The result was announced in 'Asymptotic Distribution Modulo 1'. Nuffic International summer session in Breukelen 1962 (P. Nordhoff N.V. Groningen) p. 87, formula (9).

then we have

$$\Phi_1 \leqslant f \leqslant \Phi_2 \quad \text{and} \quad \mu(\Phi_1) \leqslant \mu(f) \leqslant \mu(\Phi_2),$$
$$\mu_N(\Phi_1) \leqslant \mu_N(f) \leqslant \mu_N(\Phi_2);$$

further

$$\mu(\Phi_2) - \mu(\Phi_1) = \sigma(f; p(L)),$$

if $\rho \equiv 1$, which we suppose. It is now clear that

$$\mu(f) - \mu_N(f) \leqslant \sigma(f) + \mu(\Phi_1) - \mu_N(\Phi_1)$$

and

$$\mu(f) - \mu_N(f) \geqslant -\sigma(f) + \mu(\Phi_2) - \mu_N(\Phi_2);$$

therefore

$$D_N(f) \leqslant \max_i D_N(\Phi_i) + \sigma(f; p(L)). \tag{6}$$

It is enough to estimate $D_N(\Phi_2)$. We have

$$\mu_N(\Phi_2) = \frac{1}{N} \sum_{i=1}^{N} \Phi_2(x_i) = \sum_z G(z) \frac{1}{N} \sum_{i=1}^{N} \chi(J(z); x_i).$$

Now we set for $z = 1, \ldots, L$.

$$S(z) = \frac{z}{L} - \frac{1}{N} \sum_{i=1}^{N} \sum_{r=1}^{z} \chi(J(r); x_i).$$

We define $S(z) = 0$ for $z \leqslant 0$ and $z > L$ (z integer). Now $\sum_{r=1}^{z} \chi(J(r))$ is the indicator function of the interval $J'(z) : 0 \leqslant \xi < z/L$; therefore

$$S(z) = \frac{z}{L} - \frac{1}{N} \sum_{i=1}^{N} \chi(J'(z); x_i).$$

For $z = L$, $S(z) = 0$, because $J'(L) = E$.

We have $S(z) - S(z-1) = \dfrac{1}{L} - \dfrac{1}{N} \sum_{i=1}^{N} \chi(J(z); x_i)$; therefore

$$\begin{aligned} S = \mu(\Phi_2) - \mu_N(\Phi_2) &= \sum_{z=1}^{L} G(z) \left(\frac{1}{L} - \frac{1}{N} \sum_{i=1}^{N} \chi(J(z); x_i) \right) \\ &= \sum_{z=1}^{L} G(z) \left(S(z) - S(z-1) \right) \\ &= G(L)\, S(L) + \sum_{z=1}^{L-1} \left(G(z) - G(z+1) \right) S(z). \end{aligned}$$

After the definition of D_N, we have $|S(z)| \leqslant D_N$; therefore

$$|S| \leqslant D_N \sum_{z=1}^{L-1} |G(z+1) - G(z)|. \tag{7}$$

Now we set $L = 2M$ (M natural number) and we have

$$\sum_{z=1}^{L-1} = \sum_{l=1}^{M} |G(2l) - G(2l-1)| + \sum_{l=1}^{M-1} |G(2l+1) - G(2l)| = \Sigma_1 + \Sigma_2.$$

Now consider in p the union of $J(2l-1)$, $J(2l)$ ($l = 1, \ldots, M$), which is the interval $J_1''(l)$:

$$\frac{2(l-1)}{L} \leqslant \xi < \frac{2l}{L}.$$

Then we get a partition p_1'' of $\bar{E}$ with norm $n(p_1'') = 2/L$. If

$$\sup_{J_1''(l)} f = G_1''(l), \qquad \inf_{J_1''(l)} f = g_1''(l),$$

then we have

$$G(2l) - G(2l-1) \leqslant G''(l) - g''(l);$$

therefore

$$\Sigma_1 \leqslant M\,\sigma\left(p_1''; \frac{1}{M}\right).$$

If we consider in p the union of $J(2l)$, $J(2l+1)$ ($l = 1, \ldots, M-1$), which is the interval $J_2''(l)$:

$$\frac{2l-1}{L} \leqslant \xi < \frac{2l+1}{L},$$

then the intervals $J_2''(l)$ ($l = 1, \ldots M-1$), with the intervals $J(1)$ and $J(l)$, define a partition p_2'' of $\bar{E}$, with norm $2/L$ (all intervals of p', have the length $2/L$, except the first and the last, which have the length $1/L$). We have in Σ_2

$$G(2l+1) - G(2l) \leqslant G_2''(l) - g_2''(l)$$

(G'', g'' defined similarly as above) and therefore

$$\Sigma_2 \leqslant M\sigma\left(p_2''; \frac{1}{M}\right)$$

and

$$|S| \leqslant M\,D_N\left(\sigma\left(p_1''; \frac{1}{M}\right) + \sigma\left(p_2''; \frac{1}{M}\right)\right); \tag{8}$$

therefore

$$D_N(f) \leqslant (2\,M\,D_N + 1)\sum\left(f, \frac{1}{M}\right).$$

Now we take $M = [D_N^{-1}]$ and we have the result for $s = 1$.

Now let us consider the case for general ρ. We have only to define

$$S(z) = \int \rho\chi(J'(z); x)\, dx - \frac{1}{N} \sum_{i=1}^{N} \chi(J'(z); x_i).$$

Then we get (8) again. Further we have to consider

$$\mu(\Phi_2) - \mu(\Phi_1) = \sum_{z=1}^{L} (G(z) - g(z)) \int \chi(J(z); x)\, dx.$$

We define

$$K = L \sup_z \int \chi(J(z); x)\, dx, \quad \text{then} \quad \mu(\Phi_2) - \mu(\Phi_1) \leqslant K\sigma\left(p(L); \frac{1}{L}\right)$$

and we have with $L = 2[D_N^{-1}]$

$$D_N(f, \omega, \rho) \leqslant (2 + K) \sum (f, [D_N^{-1}]^{-1}). \tag{5'}$$

Now let us consider the multidimensional case. At first we make some formal remarks. Let F be a real-valued function defined on Z^s and let us assume that $F = 0$ outside some bounded set in Z^s. By e we denote a point in Z^s, whose coordinates e_i have only the values 0 or 1. There exist 2^s such points.

Now we define the following functions on Z^s:

$$\Delta F = \sum_e F(z + e)\,(-1)^{\sigma(e)}, \qquad \Delta' F = \sum_e F(z - e)\,(-1)^{\sigma(e)}$$

$(\sigma(e) = e_1 + \ldots + e_s)$. It is clear that Δ, Δ' are linear operators. If $F_1, \ldots, F_s$ are real-valued functions on Z and $F = F_1 \times \ldots \times F_s$, then

$$\Delta F = \prod_{j=1}^{s} (F_j(z_j) - F_j(z_j + 1)), \qquad \Delta' F = \prod_j (F_j(z_j) - F_j(z_j - 1))$$

because

$$\Delta F = \sum_e \prod_{j=1}^{s} F_j(z_j + e_j)\,(-1)^{e_j} = \prod_{j=1}^{s} \sum_{e_j=0,1} F_j(z_j + e_j)\,(-1)^{e_j}.$$

We have the Abel transformation

$$\sum_z F(z)\,\Delta H(z) = \sum_z H(z)\,\Delta' F(z) \tag{9}$$

because

$$\sum_z F(z)\,\Delta H(z) = \sum_e (-1)^{\sigma(e)} \sum_z F(z) H(z + e) = \sum_e (-1)^{\sigma(e)} \sum_z F(z - e) H(z).$$

Now let us consider again a partition $p(L)$ of E^s in the intervals $J_s(z)$:

$$\frac{z_i-1}{L} \leqslant \xi_i < \frac{z_i}{L} \qquad (i=1,\dots,s) \quad (1 \leqslant z_i \leqslant L),$$

where $z=(z_1,\dots,z_s)$. The set of $z=(z_1,\dots,z_s)$ with $1 \leqslant z_i \leqslant L$ we denote $Z(L)$. The norm $n(p(L))$ is L^{-1}. By $J'_s(z)$ we denote the interval

$$0 \leqslant \xi_i < \frac{z_i}{L} \qquad (i=1,\dots,s).$$

We define $\chi(J_s(z))=0$ if $z \notin Z(L)$ and similarly for $J'(z)$. Now we have $\chi(J_s(z);\xi)=\chi(J_1(z_1);\xi_1)\dots\chi(J_1(z_s);\xi_s)$ because $\xi \in J_s$ if and only if

$$\frac{z_i-1}{L} \leqslant \xi_i < \frac{z_i}{L}.$$

The same is true for J'. Now for each $\xi \in E^s$ we have with $F(z)=\chi(J_s(z);\xi)$

$$\Delta' \chi(J'_s(z);\xi)=\chi(J_s(z);\xi) \tag{10}$$

because

$$\Delta' \chi(J'_s(z);\xi)=\prod_{i=1}^{s} \chi(J'_1(z_i);\xi_i)-\chi(J'_1(z_i-1);\xi_i)=\prod_{i=1}^{s} \chi(J_1(z_i);\xi_i).$$

This point has now been settled, and we can use the same methods as in the case $s=1$.

We set

$$G(z)=\sup_{J(z)} f, \qquad g(z)=\inf_{J(z)} f, \qquad (G(z)=g(z)=0 \text{ if } z \notin Z(L))$$

$$\Phi_1=\sum_z g(z)\chi(J(z)), \qquad \Phi_2=\sum_z G(z)\chi(J(z))$$

and we have again (6). Now

$$S=\mu(\Phi_2)-\mu_N(\Phi_2)=\sum_z G(z)\Big(\mu(J(z))-\mu_N(J(z))\Big).$$

Because of (10) we have $\mu(J(z))=\Delta'\mu(J'(z))$, $\mu_N(J(z))=\Delta'\mu_N(J'(z))$; therefore by (9),

$$S=\sum_z G(z)\Big(\Delta'\big(\mu(J'(z))-\mu_N(J'(z))\big)\Big)=\sum_z \Delta G(z)\big(\mu(J'(z))-\mu_N(J'(z))\big). \tag{11}$$

On the right-hand side of (11) it is enough to consider only the z with $z \in Z(L)$.

For $z_L = (L, \ldots, L)$ we have $\mu(J'(z)) = \mu_N(J'(z)) = 1$; therefore it is enough to consider the $z \in Z'(L) = Z(L) - \{z_L\}$. We have therefore

$$|S| \leqslant D_N \sum_{z \in Z'(L)} |\Delta G(z)| = D_N \Sigma'.$$

Now we set $L = 2M$ and we divide $Z'(L)$ in 2^s classes $Z''(L)$, where in each class $z \equiv z' \pmod 2$; this means $z_i \equiv z_i' \pmod 2$ $(i = 1, \ldots, s)$. Therefore we have

$$\Sigma' = \sum_{Z''} \sum_{z \in Z''}{}' .$$

We write

$$\sum_{z \in Z''}{}' = \Sigma''.$$

Now we have

$$\Sigma'' = \sum_{z \in Z'(L) \cap Z''} \left| \sum_e (-1)^{\sigma(e)} G(z+e) \right| . \tag{12}$$

It is impossible for $z, z' \in Z''$ that $z + e = z' + e'$ because then $e \equiv e' \pmod 2$, which is only the case for $e = e'$, except $z = z'$, $e = e'$. In

$$\Sigma'''(z) = \sum_e (-1)^{\sigma(e)} G(z+e)$$

we have only to consider the e for which $z + e$ belongs to $Z(L)$. This e we denote by d, the whole set by D and their number by $A(z)$. It is clear that $0 \in D$, because $z \in Z'(L) \subset Z(L)$. We have further

$$\Sigma'''(z) = \sum_{d, \sigma(d) \equiv 0(2)} G(z+d) - \sum_{d, \sigma(d) \equiv 1(2)} G(z+d).$$

We have

$$\sum_{\sigma(d) \equiv 0(2)} 1 = \sum_{\sigma(d) \equiv 1(2)} 1,$$

because

$$\sum_d (-1)^{\sigma(d)} = \prod_{i=1}^{s} \sum_{d_i} (-1)^{d_i}.$$

Now there exists always a $d \neq (0, \ldots, 0)$ in D, because of $z \in Z'(L)$ there exists a coordinate of z, let us say z_j, with $z_j < L$ and then the point $e = (e_1, \ldots, e_s)$ with $e_j = 1$ and $e_i = 0$ for $i \neq j$ is such a d. Therefore for $i = j$ we have

$$\sum_{d_j} (-1)^{d_j} = 1 + (-1) = 0.$$

Therefore $A(z)$ is even.

Now we consider $J''(z) = \bigcup_{d \in D} J(z+d) = \bigcup_{e} J(z+e) \cap \bar{E}^s$.

$J''(z)$ is the intersection of the cube

$$\frac{z_j - 1}{L} \leqslant \xi_j < \frac{z_j + 1}{L} \qquad (j = 1, \ldots, s)$$

with E^s. We have $\mu(J''(z)) = A(z)L^{-s}$. We set

$$G''(z) = \sup_{J''(z)} f, \qquad g''(z) = \inf_{J''(z)} f;$$

then we have

$$|\Sigma'''(z)| \leqslant \frac{A(z)}{2}(G''(z) - g''(z)); \tag{13}$$

therefore

$$\sum{}'' \leqslant \tfrac{1}{2} \sum_{z \in Z''} A(z)(G''(z) - g''(z)) = \frac{L^s}{2} \sum_{z \in Z'(L) \cap Z''} (G''(z) - g''(z))\, \mu(J''(z));$$

therefore

$$\sum{}'' \leqslant \frac{L^s}{2} \sum_{z \in Z''}{}' (G''(z) - g''(z))\, \mu(J''(z))$$

where we consider all z, for which $J''(z) \cap E^s \neq \emptyset$. The set of all these $J''(z)$ defines a partition p'' of E^s, where the edges of $J''(z)$ have only the value L^{-1} or $2L^{-1}$. The value of L^{-1} is only possible if $J''(z) \not\subset E^s$. Now let us denote $\mathscr{P}$ the set of all p of the above form and let

$$\Sigma^*\left(f, \frac{1}{M}\right) = \sup_{p \in \mathscr{P}} \sigma(p, n(p)),$$

then we have

$$\Sigma'' \leqslant \tfrac{1}{2} L^s\, \Sigma^*(f).$$

Now we have 2^s such sums Σ'' and therefore

$$\Sigma' \leqslant 2^{s-1} L^s\, \Sigma^*(f, M^{-1})$$

and further

$$|S| \leqslant 2^{s-1} L^s D_N\, \Sigma^*(f)$$

and for $\rho = 1$

$$'D_N(f) \leqslant (2^{2s-1} M^s D_N + 1) \sum{}^*\left(f, \frac{1}{M}\right). \tag{14}$$

We take $M = [D_N^{-1}]^{1/s}$ and we have the result, even in a somewhat stronger form.

In the general case, defining again

$$K = \sup_{J'(z)} \frac{\int \rho(x)\, \chi(J'(z); x)\, dx}{\mu\big(\chi(J'(z))\big)},$$

we get

$$D_N(f; \omega, \rho) \leqslant (2^{2s-1} + K)\, \angle \left(f, \frac{1}{M}\right) \tag{15}$$

with the same M as above.

A final remark: If we assume that f is continuous we get with the help of (15) with the same M

$$D_N(f) \leqslant (2^{2s-1} + 1)\, \omega \left(\frac{1}{M}\right) \tag{14'}$$

where ω is the modulus of continuity of f. We can also take (for $s = 1$ this has been done by Koksma) $M = [N^s]$; then we get

$$D_N(f) \leqslant (2^{2s} + 1)\, N\, D_N\, \omega\left([N^{1/s}]^{-1}\right); \tag{14''}$$

but (14″) is not as generally applicable as (14).

Über eine Methode von E. Hecke in der Theorie der Gleichverteilung

Herrn Professor C. L. Siegel zum 75. Geburtstag gewidmet in Dankbarkeit und Verehrung

Acta Arith. **24** (1973) 11–31

E. Hecke [3] zeigte 1921 folgenden Satz: Ist α irrational, dann ist die Potenzreihe $\sum_{n=1}^{\infty} \{n\alpha\} z^n$ über den Einheitskreis nicht fortsetzbar. Dabei bedeutet $\{x\} = x - [x]$. Er benützte dabei die Theorie der Gleichverteilung. Er ging von folgender Bemerkung aus: Ist für eine reelle Zahlenfolge (c_k) der Limes $\lim_{N\to\infty} \frac{1}{N} \sum_{k=1}^{N} c_k = c$ vorhanden, dann ist $\lim_{r\to 1-0} (1-r) \sum_{k=1}^{\infty} c_k r^k$ ebenfalls vorhanden und $= c$. Da die Folge $\{n\alpha\}$ gleichverteilt ist, so ist für jede im Riemannschen Sinne integrierbare, periodische Funktion f mit der Periode 1

$$\lim_{N\to\infty} \frac{1}{N} \sum_{k=1}^{N} f(k\alpha) = \int_0^1 f(x)\,dx. \tag{1}$$

Nimmt man jetzt $z_q = re^{2\pi i q\alpha}$ (q natürliche Zahl, $0 < r < 1$) und wendet (1) auf die Funktion $f(x) = \{x\} e^{2\pi i q x}$ an, so erhält man, da für $q \neq 0$ ja $\int_0^1 f\,dx = (2\pi i q)^{-1}$, daß

$$\lim_{r\to 1} (1-r) \sum_{n=1}^{\infty} \{n\alpha\} e^{2\pi i q n\alpha} r^n = (2\pi i q)^{-1}. \tag{2}$$

Da die Folge $(e^{2\pi i q\alpha})$ auf $|z| = 1$ dicht liegt, so folgt aus (2) die Behauptung von Hecke. Er zeigte darüber hinaus: Sind α, β linearunabhängig über Z, so ist

$$\lim_{r\to 1} (1-r) \sum_{n=1}^{\infty} \{n\alpha\} e^{2\pi i n\beta} r^n = 0. \tag{3}$$

Dies folgt daraus, daß unter dieser Voraussetzung die zweidimensionale Folge $(n\alpha, n\beta)$ gleichverteilt mod 1 ist und für jede im Riemannschen Sinne integrierbare Funktion f, periodisch mit der Periode 1, in zwei Variabeln

$$\lim_{N\to\infty} \frac{1}{N} \sum_{k=1}^{N} f(k\alpha, k\beta) = \int_0^1\int_0^1 f(x_1, x_2)\,dx_1\,dx_2.$$

Die Funktion $f(x_1, x_2) = \{x_1\}e^{2\pi i x_2}$ liefert dann die Behauptung. Die Methode von Hecke wurde von vielen Autoren verallgemeinert, ich hebe nur W. Schwarz [8] und F. W. Caroll und J. H. B. Kemperman [1] hervor. So zeigen die letztgenannten Autoren: Für jede Funktion Φ (unter den gleichen Voraussetzungen wie oben für f) gilt für jedes ganzzahlige $q \geqslant 1$

$$(3') \qquad \lim_{r\to 1} (1-r) \sum_{n=1}^{\infty} \Phi(\alpha n) e^{2\pi i q n \alpha} r^n = \hat{\Phi}$$

wo $\hat{\Phi} = \int_0^1 \Phi(x) e^{2\pi i q x}\,dx$. Wenn $\hat{\Phi} \neq 0$, so wird die Stelle $e^{2\pi i q \alpha}$ eine Stelle starker Singularität der Funktion $\sum \Phi(\alpha n) z^n$ genannt. Wir stellen nun die Frage: Kann man z.B. in (2) bzw. (3') mehr aussagen d.h. kann man über die Differenz

$$\delta(\Phi, r, q) = (1-r) \sum \Phi(\alpha n) e^{2\pi i q n \alpha} - \hat{\Phi}$$

in Abhängigkeit von r eine Abschätzung stärker als (1) herleiten? Diese Abhängigkeit wird von den feineren Eigenschaften von α und Φ abhängen. So werden wir zeigen: Besitzt die Kettenbruchentwicklung von α beschränkte Teilnenner und ist Φ von beschränkter Variation, so ist

$$|\delta(r)| \leqslant K(\Phi)\,(1-r) \log \frac{1}{1-r}$$

wo K nur von Φ und q abhängt. Besitzt Φ höhere Ableitungen, dann gilt sogar

$$|\delta(r)| \leqslant K(\Phi)\,(1-r).$$

Da es die gleiche Mühe macht und um auch (3) behandeln zu können behandeln wir gleich Potenzreihen $\sum_{n=0}^{\infty} \Phi(\alpha_1 n, \ldots, \alpha_s n) z^n$, wo $\alpha_1, \ldots, \alpha_s$ linear unabhängig über Z sind.

Wir gehen sogar weiter, indem wir eine beliebige Folge $\omega = (x_0, x_1, \ldots)$ im R^s zu Grunde legen. Dann betrachten wir $\sum_{n=0}^{\infty} f(x_n) r^n$, wo wieder f die gleichen Voraussetzungen wie oben erfüllt, aber jetzt auf R^s definiert ist. Von Interesse ist für uns der Fall, daß $f(x) = \chi_J(x_1 - [x_1], \ldots, x_s - [x_s])$, wo χ_J die charakteristische Funktion eines Teilintervalls J des s-dimensionalen Einheitswürfels I^s ist. Ist nämlich $D_N(\omega)$ die Diskrepanz von ω welche definiert ist als

$$\sup_J \left| \frac{1}{N+1} \sum_{k=0}^{N} \chi_J(\{x_k\}) - \operatorname{Vol} J \right|$$

erstreckt über alle J mit $0 = (0, \ldots, 0)$ als linken unteren Eckpunkt ($\operatorname{Vol} J$ = Volumen von J), dann sei

$$D_N^*(\omega) = \sup_{m \geqslant N} D_m(\omega).$$

Weiter führen wir folgenden Ausdruck ein, welchen ich Abelsche Diskrepanz von ω nennen möchte $(0 < r < 1)$

$$D_A(\omega, r) = \sup_J \left|(1-r)\sum_{k=0}^{\infty} \chi_J(\{x_k\}) r^k - \operatorname{Vol} J\right|$$

und dazu noch $D_A^*(\omega, r) = \sup_{r' \geqslant r} D_A(\omega, r')$.

Wir zeigen zunächst (Formel (10))

$$D_A(\omega, r) \leqslant D_A^*(\omega, r) \leqslant 4 D_M^*(\omega)$$

wo $M = [(1-r)^{-1/2}]$. Dann zeigen wir als „Taubersche Abschätzung"

$$D_N(\omega) \leqslant D_N^*(\omega) \leqslant 6^8 \left(\log D_A^{-1}(\omega, e^{-\frac{\log(N+1)}{N+1}})\right)^{-1/2} \quad \text{(Formel (38))}.$$

Es läßt sich leicht zeigen, daß der Exponent $1/2$ in der obigen Abschätzung durch 1 ersetzt werden kann, was aber in der Arbeit nicht weiter ausgeführt wird. Wie weit sich diese Abschätzungen verbessern lassen wäre ein interessantes Problem. Es ist bekannt, daß $D_N(\omega) \geqslant (N+1)^{-1}$, ja nach W. Schmidt gibt es unendlich viele N_i so daß bei passendem $c(\omega)$ gilt

$$D_{N_i}(\omega) \geqslant c \frac{\log(N_i+1)}{N_i+1}.$$

Für die oben eingeführte Abelsche Diskrepanz läßt sich leicht zeigen, daß $D_A(\omega, r) \geqslant 1 - r$.

Es ist zu vermuten, daß unendlich viele r_i mit $\lim_{i\to\infty} r_i = 1$ existieren, so daß

$$D_A(\omega, r_i) \geqslant c(1-r_i) \log \frac{1}{1-r_i}.$$

Dies wird insbesondere für die Folge $\omega = (n\alpha)$ gelten, wenn α irrational ist. Dieser Spezialfall läßt sich vermutlich einfacher beweisen, wenn man die Methode von Ostrowski (vgl. Koksma [5]) zum Beweis von

$$D_N((n\alpha)) = \Omega\left(\frac{\log N}{N}\right)$$

auf den vorliegenden Fall überträgt.

§ 1. Es sei f eine im Riemannschen Sinne integrierbare Funktion in s Variablen, periodisch mit der Periode 1, weiter sei $\omega = (x_0, x_1, \ldots)$ eine Folge im R^s. Wir betrachten nun die Reihe $\sum_{n=0}^{\infty} f(x_n) r^n$. Wir setzen

$\lambda(f) = \int\limits_{I^s} f(x)dx$, wo I^s der Würfel $0 \leqslant \xi_i < 1$ $(i = 1, \ldots, s)$. Wir wollen nun das Verhalten von

$$\delta(r, f, \omega) = \left|(1-r) \sum_{n=0}^{\infty} f(x_n) r^n - \lambda(f)\right|$$

untersuchen. Wir setzen $c_n = f(x_n) - \lambda(f)$, $C_n = c_0 + \ldots + c_n$, dann ist für ganzzahliges $M \geqslant 0$

$$\sum_{n=0}^{\infty} c_n r^n = \sum_{n=0}^{M-1} c_n r^n - C_{M-1} r^M + (1-r) \sum_{n=M}^{\infty} C_n r^n \tag{4}$$

$(\sum\limits_{n=0}^{-1} = C_{-1} = 0)$. Wir setzen weiter

$$\Delta_n(f) = \frac{1}{n+1} \sum_{k=0}^{n} f(x_k) - \lambda,$$

so ist $C_n = (n+1)\Delta_n(f)$. Da f integrierbar ist, so existiert ein K, so daß $|f| \leqslant K$. Wir erhalten aus (4) für $0 \leqslant r < 1$

$$\delta(r, f, \omega)(1-r)^{-1} \leqslant 2MK + M|\Delta_M(f)| + (1-r) \sum_{n=M}^{\infty} (n+1)|\Delta_n(f)| r^n \tag{5}$$

insbesondere für $M = 0$

$$\delta(r, f, \omega) \leqslant (1-r)^2 \sum_{n=0}^{\infty} (n+1)|\Delta_n(f)| r^n. \tag{5'}$$

Es sei jetzt vorausgesetzt, daß f von beschränkter Variation $V(f)$ im Sinne von Hardy–Krause ist (für $s = 1$ ist es die übliche Variation), dann gilt [4]

$$|\Delta_n(f)| \leqslant D_n(\omega) V(f). \tag{6}$$

Wir führen wie in der Einleitung die modifizierte Diskrepanz $D_n^* = \sup\limits_{m \geqslant n} D_m$ ein. Es ist die Folge (D_n^*) monoton abnehmend,

$$\sum_{n=M}^{\infty} (n+1) D_n r^n \leqslant D_M^* \sum_{n=M}^{\infty} (n+1) r^n < D_M^* (1-r)^{-2}$$

erhalten. Wir aus (5)

$$\delta(r, f, \omega)(1-r)^{-1} \leqslant 2MK + V(f)\big(M + (1-r)^{-1}\big) D_M^*. \tag{7}$$

Für $M = 0$ erhalten wir aus (5')

$$\delta(r, f, \omega)(1-r)^{-1} \leqslant V(f)(1-r) \sum_{n=0}^{\infty} (n+1) D_n r^n. \tag{8}$$

Wir nehmen nun in (7) $M = [(1-r)^{-1/2}]$. Es ist $M \geqslant 1$ und wir erhalten aus (7)

$$\delta(r, f, \omega) \leqslant 2K(1-r)^{1/2} + V(f)\left((1-r)^{1/2}+1\right)D_M^*.$$

Nun ist $D_M^* \geqslant D_M \geqslant (M+1)^{-1}$, also

$$\delta \leqslant 2\left(K+V(f)\right)D_M^*(\omega). \tag{9}$$

Es sei nun J ein Teilintervall von I^s, $f(x) = \chi_J(x_1-[x_1], \ldots, x_s-[x_s])$ wo χ_J die charakteristische Funktion von J ist. Dann ist $\lambda(f)$ das Volumen $\mu(J)$ von J, $V(f) = K = 1$ und wir erhalten, wenn wir

$$D_A(\omega, r) = \sup_J \left|(1-r)\sum_{n=0}^{\infty} \chi_J(x_n) - \mu(J)\right|$$

die Abelsche Diskrepanz von ω, einführen

$$D_A(\omega, r) \leqslant 4D_{M(r)}^*(\omega) \tag{10}$$

wo $M(r) = [(1-r)^{-1/2}]$. Dabei wollen wir auch in D_A nur die J betrachten, welche 0 als linken unteren Eckpunkt haben (obwohl dies nicht notwendig wäre). Führen wir noch die modifizierte Abelsche Diskrepanz $D_A^*(\omega, r) = \sup\limits_{r' \geqslant r} D_A(\omega, r')$ ein, so folgern wir leicht aus (10), daß auch

$$D_A^*(\omega, r) \leqslant 4D_M^*(\omega) \tag{10'}$$

gilt.

Man kann unter Umständen schärfere Abschätzungen als (9) herleiten. Nehmen wir an, es gelte eine Abschätzung

$$D_n(\omega) \leqslant \frac{1}{n+1}\sum_{k=0}^{n} \frac{1}{h(k)}$$

wo h eine monoton wachsende Funktion ist. Dann folgt aus (8)

$$\delta(r, f, \omega) \leqslant V(f)\,(1-r)^2 \sum_{n=0}^{\infty} r^n \sum_{k=0}^{n} \frac{1}{h(k)}.$$

Nun ist

$$\sum_{n=0}^{\infty} r^n \sum_{k=0}^{n} h^{-1}(k) = (1-r)^{-1} \sum_{n=0}^{\infty} h^{-1}(n) r^n,$$

also erhalten wir

$$\delta(r, f, \omega) \leqslant V(f)\,(1-r) \sum_{n=0}^{\infty} \frac{r^n}{h(n)}. \tag{11}$$

Wenn man über f stärkere Voraussetzungen macht, kann man ebenfalls (9) verschärfen. Nehmen wir zuerst $s = 1$ und nehmen an, daß f stetige Ableitungen bis zu einer Ordnung m besitzt.

Nach der Eulerschen Summenformel gilt

$$f(x) = \lambda(f) + \sum_{r=1}^{m} \frac{1}{r!} P_r(x)\left(f^{(r-1)}(1) - f^{(r-1)}(0)\right) - \frac{1}{m!}\int_0^1 P_m(x-t) f^{(m)}\,dt.$$

Dabei ist P_r das r-te Bernoullische Polynom. Da f periodisch ist, so haben wir, wenn wir noch $P_0 = 1$ setzen

$$f(x) = \lambda(f) - \frac{1}{m!}\int_0^1 P_m(x-t) f^{(m)}(t)\,dt = \sum_{k=0,1} \frac{(-1)^k}{(mk)!}\int_0^1 P_{mk}(x-t) f^{(mk)}(t)\,dt.$$

Ist jetzt $s > 1$ dann wenden wir diese Summenformel s-mal an. Dabei nehmen wir an, daß f stetige Ableitungen bis zur Ordnung ms besitzt. Wir erhalten dann, wenn wir $k = (k_1, \ldots, k_s)$,

$$P_{mk}(x-t) = P_{mk_1}(x_1 - t_1) \ldots P_{mk_s}(x_s - t_s), \qquad (mk)! = (mk_1)! \ldots (mk_s)!,$$

$|k| = k_1 + \ldots + k_s$ setzen

$$f(x) = \sum_k \frac{(-1)^{|k|}}{(mk)!} \int_{I^s} P_{mk}(x-t) \frac{\partial^{m|k|} f}{\partial t_1^{k_1} \ldots \partial t_s^{k_s}}\, dt_1 \ldots dt_s.$$

Dabei erstreckt sich die Summe über alle Gitterpunkte k mit Koordinaten 0 oder 1. Es ist dann

$$\left|\frac{1}{n+1}\sum_{j=0}^{n} f(x_j) - \lambda(f)\right| \leqslant \sum_{k \neq 0} \int_I \left|\frac{\partial^{m|k|} f}{\partial t_1^{k_1} \ldots \partial t_s^{k_s}}\right| \left|\frac{1}{n+1}\sum_{j=0}^{n} P_{mk}(x_j - t)\right| dt_1 \ldots dt_s.$$

Es sei nun L eine obere Schranke für alle Integrale

$$\int_I \left|\frac{\partial^{m|k|} f}{\partial t_1^{k_1} \ldots \partial t_s^{k_s}}\right|$$

dann ist

$$\left|\frac{1}{n+1}\sum_{j=1}^{n} f(x_j) - \lambda(f)\right| \leqslant 2^s L \sum_{k \neq 0} \sup_t \left|\frac{1}{n+1}\sum_{j=0}^{n} P_{mk}(x_j - t)\right|.$$

Nun besitzt $P_m(x-t)$ für $s = 1$ und $m > 1$ die absolut konvergente Fourierentwicklung $\sum\limits_{h \neq 0} h^{-m} e^{2\pi i h x}$, also ist

$$|\Delta_n(f)| \leqslant 2^s L \sum_{h \neq 0} R^{-m}(h) \left|\frac{1}{n+1}\sum_{j=0}^{n} e^{2\pi i \langle h x_j \rangle}\right|. \tag{12}$$

Dabei ist

$$R(h) = \prod_{i=1}^{s} \operatorname{Max}(|h_i|, 1), \qquad h = (h_1, \ldots, h_s), \qquad \langle hx \rangle = h_1 x_1 + \ldots + h_s x_s.$$

§ 2. Wir wählen nun $s = 1$, für (x_n) die Folge (αn), wo α irrational ist und setzen für ganzzahliges q und für eine im Riemannschen Sinne integrierbare Funktion Φ, periodisch mit der Periode 1

$$f(x) = \Phi(x) e^{2\pi i q x}.$$

Setzen wir weiter von Φ beschränkte Variation $V(\Phi)$ voraus, so ist auch f von beschränkter Variation und es ist

$$V(f) \leqslant V(\Phi) + 2\pi C q \tag{13}$$

wo C eine obere Schranke von Φ in I ist. Es ist ja $V(e^{2\pi i q}) = 2\pi q$. Wir benötigen nun eine Abschätzung für D_n. Wir wollen eine Abschätzung geben welche die klassischen Abschätzungen von Ostrowski und Behnke umfaßt (vgl. [5]).

Wir nennen (nach einer modifizierten Definition von S. Lang [6]) α vom Kotypus γ, wo γ eine Funktion auf N ist, wenn für alle natürlichen n ganze Zahlen p, q existieren mit g.g.T. $(p, q) = 1$, so daß

$$\left|\alpha - \frac{p}{q}\right| < \frac{1}{q^2} \quad \text{und} \quad \gamma(n) \leqslant q \leqslant n$$

ist.

Leichter nachzuprüfen ist der Typus φ (φ wieder auf N definiert) von α. Er ist so definiert: Für alle ganzzahligen p, q wo $q \geqslant 1$ ist

$$\left|\alpha - \frac{p}{q}\right| \geqslant q^{-1}\varphi^{-1}(q).$$

Ist φ monoton wachsend im strengen Sinn, so ist α vom Kotypus $\overline{\varphi}^1$ ($\overline{\varphi}^1$: inverse Funktion von φ). Der Beweis ist leicht: Nach Dirichlet existieren für alle $n \geqslant 1$ ganze Zahlen p, q mit g.g.T. $(p, q) = 1$ so daß

$$\left|\alpha - \frac{p}{q}\right| \leqslant \frac{1}{qn} \quad \text{und} \quad 1 \leqslant q \leqslant n.$$

Es ist also $n \leqslant \varphi(q)$, also $\overline{\varphi}^1(n) \leqslant q \leqslant n$. Wir werden nun zeigen: Ist α vom Kotypus γ, wo γ monoton wachsend auf N, dann ist

$$D_n(\omega) \leqslant \frac{1}{n+1}\left(\frac{5}{2}\sum_{k=1}^{n}\frac{1}{\gamma(k)} + 1\right). \tag{14}$$

Definieren wir noch $\gamma(0) = 1$, so ist

$$D_n(\omega) < \frac{5}{2(n+1)}\sum_{k=0}^{n}\frac{1}{\gamma(k)}. \tag{14'}$$

Beweis. Es ist für jedes Teilintervall J: $0 \leqslant x < \gamma$ ja

$$\chi_J(x) = \gamma + \{x + 1 - \gamma\} - \{x\}.$$

Es ist also mit $1-\gamma = \delta$

$$\frac{1}{n+1}\sum_{k=0}^{n}\chi_J(x_k)-\gamma = \frac{1}{n+1}\sum_{k=0}^{n}\{x_n+\delta\}-\frac{1}{2}-\left(\frac{1}{n+1}\sum_{k=0}^{n}\{x_k\}-\frac{1}{2}\right)$$

also

$$D_n(\omega)\leqslant\frac{2}{n+1}\sup_{0\leqslant\delta<1}\left|\sum_{k=0}^{n}\{x_n+\delta\}-\frac{n+1}{2}\right|.$$

Nach Voraussetzung existieren ganze p, q mit $\alpha = \frac{p}{q}+\frac{\vartheta_1}{q^2}$, $|\vartheta_1|<1$, g.g.T. $(p,q)=1$ und $\gamma(n)\leqslant q\leqslant n$.

Wir betrachten nun

$$\Sigma = \sum_{k=n+1-q}^{n}\{\alpha n+\delta\} = \sum_{r=0}^{q-1}\{\alpha(n-q+1)+\delta+r\alpha\} = \sum_{r=0}^{q-1}\{r\alpha+\delta_2\}$$

wo $\delta_2 = \alpha(n-q+1)+\delta$ gesetzt ist. Es ist $\delta_2\geqslant 0$. Weiter ist

$$\Sigma = \sum_{r=0}^{q-1}\left\{\frac{rp}{q}+\frac{\Psi(r)}{q}\right\},$$

wo $\Psi(r) = \frac{r\delta_1}{q}+q\delta_2$. Es ist

$$q\delta_2-1\leqslant q\delta_2-\frac{r}{q}\leqslant\Psi(r)\leqslant q\delta_2+\frac{q-1}{q}<q\delta_2+1.$$

Setzen wir also $q\delta_2-1 = c$, so ist $c<\Psi(r)<c+2$. Nun ist

$$\Sigma = \sum_{r=0}^{q-1}\left(\frac{rp+[c]}{q}+\frac{\Psi(r)-[c]}{q}\right).$$

Setzen wir $\Psi_1(r) = \Psi(r)-[c]$ so ist also mit $\lambda = c-[c]$ sicher $\lambda<\Psi_1(r)<\lambda+2$ mit $0\leqslant\lambda<1$. Nun durchläuft mit r auch $rp+[c] = a(r)$ alle Zahlen von 0 bis $q-1$ modulo q, da g.g.T. $(p,q)=1$. Es ist r durch a eindeutig bestimmt, also $r = r(a)$.

Setzen wir $\Psi_1(r(a)) = \Psi_2(a)$ so ist

$$\Sigma = \sum_{a=0}^{q-1}\left\{\frac{a}{q}+\frac{\Psi_2(a)}{q}\right\} = \sum_{a=0}^{q-1}\frac{a}{q}+\sum_{a=0}^{q-1}\delta_3(a)$$

wo

$$\delta_3(a) = \left\{\frac{a}{q}+\frac{\Psi_2(a)}{q}\right\}-\frac{a}{q} = \frac{\Psi_2(a)}{q}-\left[\frac{a}{q}+\frac{\Psi_2(a)}{q}\right].$$

Wir unterscheiden nun zwei Fälle:

1) $0\leqslant a\leqslant q-3$. Dann ist

$$\frac{\lambda}{q}\leqslant\frac{a}{q}+\frac{\Psi_2(a)}{q}<\frac{q-3}{q}+\frac{\lambda+2}{q}<1$$

also ist

$$\delta_3(a) = \frac{\Psi_2(a)}{q} \quad \text{und} \quad \frac{\lambda}{q} \leqslant \delta_3(a) < \frac{\lambda+2}{q}.$$

2) $a = q-2$ oder $q-1$. Dann ist

$$0 \leqslant \frac{q-2}{2} \leqslant \frac{a}{q} + \frac{\Psi_2(a)}{q} < \frac{q-1}{a} + \frac{\lambda+2}{q} < 1 + \frac{2}{q} < 2.$$

Es ist also $0 \leqslant [\ ] \leqslant 1$, also

$$\frac{\lambda}{q} - 1 < \delta_3(a) < \frac{\lambda+2}{q}.$$

Wir erhalten also insgesamt

$$\sum_{a=0}^{q-1} \delta_3(a) = \sum_{a=0}^{q-3} + \sum_{a=q-2}^{q-1} < \frac{\lambda+2}{q} q = \lambda + 2 < 3$$

andererseits

$$\sum_{a=0}^{q-1} \delta_3(a) \geqslant (q-2)\frac{\lambda}{q} + 2\left(\frac{\lambda}{q} - 1\right) = q\frac{\lambda}{q} - 2 \geqslant -2.$$

Es ist also

$$\Sigma = \frac{q-1}{2} + \Sigma\delta_3(a) < \frac{q}{2} + \frac{5}{2}$$

also

$$\left|\Sigma - \frac{q}{2}\right| < \frac{5}{2}.$$

Wir zeigen nun durch vollständige Induktion nach n

$$(15) \qquad \left|\sum_{k=0}^{n} \{ak + \delta\} - \frac{n+1}{2}\right| \leqslant \frac{5}{2} \sum_{k=1}^{n} \frac{1}{\gamma(k)} + 1.$$

Für $n = 0$ ist (15) richtig. Nehmen wir an, daß (15) für alle $n' < n$ richtig ist. Es existieren ganze p, q mit $\left|a - \frac{p}{q}\right| < \frac{1}{q^2}$ mit $\gamma(n) \leqslant q \leqslant n$, dann ist

1

$$\begin{aligned}\left|\sum_{k=0}^{n} \{ak + \delta\} - \frac{n+1}{2}\right| &= \left|\sum_{k=0}^{n-q} - \tfrac{1}{2}(n-q+1) + \sum_{k=n-q+1}^{n} - \frac{q}{2}\right| \\ &\leqslant \left|\sum_{k=0}^{n-q} - \tfrac{1}{2}(n-q+1)\right| + \left|\sum_{k=n-q+1}^{n} - \frac{q}{2}\right| \\ &\leqslant \frac{5}{2} \sum_{k=1}^{n-q} \frac{1}{\gamma(k)} + 1 + \frac{5}{2}\end{aligned}$$

Nun ist

$$\sum_{k=n-q+1}^{n} \frac{1}{\gamma(k)} \geqslant \frac{q}{\gamma(n)} \geqslant 1,$$

also

2

$$\left| \sum_{k=n-q+1}^{n} - \frac{q}{2} \right| \leqslant \frac{5}{2} \sum_{k=n-q+1}^{n} \frac{1}{\gamma(k)}$$

und daraus folgt die Behauptung.

Ist $\gamma(n) \geqslant 1$ für alle $n \geqslant n_0$ so gilt auch

$$D_n^* \leqslant \frac{5}{2(n+1)} \sum_{k=0}^{n} \frac{1}{\gamma(k)}. \tag{16}$$

Es ist nämlich

$$D_{m+1} \leqslant \frac{5}{2(m+2)} \sum_{k=0}^{m+1} \frac{1}{\gamma(k)} \leqslant \frac{5}{2(m+1)} \sum_{k=0}^{m} \frac{1}{\gamma(k)} \quad \text{für} \quad m > n_0$$

da γ monoton wachsend.

Ist γ für alle $t \geqslant 1$ definiert, $\gamma(1) \geqslant 1$ und γ monoton wachsend, dann ist

$$\sum_{k=0}^{n} \frac{1}{\gamma(k)} \leqslant 1 + \frac{1}{\gamma(1)} + \int_1^n \frac{1}{\gamma(t)},$$

also

$$D_n^* \leqslant \frac{5}{2(n+1)} \left(1 + \frac{1}{\gamma(1)} + \int_1^n \frac{dt}{\gamma(t)}\right). \tag{16'}$$

Betrachten wir einige Beispiele:

1) Es habe α beschränkte Teilnenner in ihrer Kettenbruchentwicklung, dann existiert bekanntlich ein $c(\alpha)$, so daß für alle p, q sicher

$$\left| \alpha - \frac{p}{q} \right| \geqslant \frac{1}{c(\alpha) q^2}.$$

Es ist also α vom linearen Typus $\varphi(t) = c(\alpha) t$, also vom Kotypus $\gamma(t) = \frac{1}{c(\alpha)} t$. Wir können o.B.d.A. $c \geqslant 1$ annehmen, dann ist also

$$D_n \leqslant D_n^* \leqslant \frac{5c(\alpha)}{2(n+1)} \left(1 + \sum_{k=1}^{n} \frac{1}{k}\right) = O\left(\frac{\log n}{n}\right).$$

Dann folgt aus (9) und (12) für

$$\delta(r, \Phi, q) = \left|(1-r)\sum \Phi(an)e^{2\pi iqna}r^n - \hat{\Phi}_q\right|$$

$$\delta(r, \Phi, q) \leqslant c_1\big((C+1)q+V(\Phi)\big)\,(1-r)^{1/2}\log\frac{1}{1-r},$$

wo c_1 nur von a abhängt. Dabei ist $\hat{\Phi}_q$ die Fouriertransformierte $\int\limits_0^1 \Phi(x)e^{2\pi iqx}dx$ von Φ. Benützen wir (8), so erhalten wir für die rechte Seite von (8) mit $h(n) = \frac{2}{5}\gamma(n) = \frac{2}{5c(a)n}$ die Abschätzung

$$\big(V(\Phi)+Cq\big)\,(1-r)\left(1+\frac{2}{5c}\sum_{n=1}^{\infty}\frac{r^n}{n}\right),$$

also

3 (17) $$\big(V(\Phi)+Cq\big)\,(1-r)\left(1+\frac{2}{5c}\log(1-r)\right).$$

Wählen wir mit Hecke $\Phi(x) = \{x\}$ so ist für $q \neq 0$, $\hat{\Phi} = (2\pi iq)^{-1}$, $V(\Phi) = C = 1$ und wir erhalten

4 (18) $$\left|(1-r)\Sigma\{an\}e^{2\pi iaqn}r^n - \frac{1}{2\pi iq}\right| \leqslant (1+q)\,(1-r)\left(1+\frac{2}{5c}\log(1-r)\right).$$

Nehmen wir jetzt an, es existiere ein $k > 0$, so daß für alle p, q $\left|a - \frac{p}{q}\right| \geqslant \frac{1}{cq^{k+1}}$ bei passendem c. Wir sagen a ist vom Typus k. Dies ist sicher der Fall wenn a vom Typus Iη (vgl. [5]) mit $k = \eta+\varepsilon$ $(\varepsilon > 0)$. Dann ist $\varphi(t) = ct^k$, $\gamma(t) = \frac{1}{c}t^{1/k}$ und

(19) $$D_n = O(n^{-1/k}).$$

Wir haben nun $\sum n^{-1/k}r^n$ zu betrachten und wir benützen, daß

$$\lim_{r\to 1}\,(1-r)^{1-1/k}\sum n^{-1/k}r^n = K > 0$$

ist. Wir erhalten also

(20) $$\delta(r, \Phi, r) \leqslant c_2\big((C+1)q+V(\Phi)\big)\,(1-r)^{1/k}.$$

Es sei jetzt a vom Typus IIη d.h. für alle p, q ist $\left|a - \frac{p}{q}\right| > \frac{1}{q}e^{-cq^{\eta+\varepsilon}}$ bei passendem c. Dann ist $\varphi(t) = e^{ct^{\eta+\varepsilon}}$ und $\gamma(t) = \left(\frac{\log t}{c}\right)^{1/(\eta+\varepsilon)}$. Wir erhalten für die linke Seite von (20) dann eine Abschätzung

$$C_3\big((C+1)q+V(\Phi)\big)(1-r)\left(1+\sum_{n=1}^{\infty}(\log n)^{-1/(\eta+\varepsilon)}r^n\right).$$

Dabei hängen C_2 und C_3 nur von α ab.

Wir können für Φ statt der Funktion von Hecke auch noch andere interessante Spezialfälle betrachten.

1. Es sei $\Phi = \chi_J$ (Indikatorfunktion eines Intervalls J: $\alpha \leqslant x < \beta$ von I). Dann ist

$$\hat{\Phi} = \frac{1}{2\pi i q}(e^{2\pi i q\beta} - e^{2\pi i q\alpha}) \quad \text{wenn} \quad q \neq 0.$$

Es ist $\hat{\Phi} \neq 0$, wenn $\lambda(J) = \beta - \alpha$ irrational.

2. Ein damit zusammenhängendes Beispiel ist folgende Funktion:

$$\Phi(x) = -1 \text{ für } 0 \leqslant x < \tfrac{1}{2}, \quad \Phi(x) = +1 \text{ für } \tfrac{1}{2} \leqslant x < 1.$$

Dann ist

$$\hat{\Phi} = 2\,\frac{1-e^{\pi i q}}{2\pi i q} \neq 0 \quad \text{wenn } q \text{ ungerade.}$$

Nach F. W. Caroll und J. H. B. Kemperman [1] sind interessante Beispiel mit $\hat{\Phi} \neq 0$:

3. $\Phi(x) = \sum_{j=0}^{\nu} b_j x^{\gamma_j}$ $(0 = \gamma_0 < \gamma_1 < \ldots < \gamma_\nu)$. Es ist $C = V(\Phi) = \sum_{j=0}^{\nu} |b_j|$.

4. $\Phi(x) = e^{iA\sin 2\pi(x+B)}$ $(A \neq 0)$. Es ist $C = 1$, $V(\Phi) = A$.

Setzen wir jetzt voraus, daß Φ bis zur Ordnung m stetig differenzierbar ist. Dann wenden wir (5′) und (12) mit $s = 1$ an. Nach H. Behnke (vgl. [5], S. 203) gilt für Irrationalzahlen α vom Typus Iη (vgl. auch (34), (35)) für

$$\Sigma_m^n(\alpha) = \sum_{h\neq 0}\frac{1}{|h|^m}\left|\sum_{k=0}^{n} e^{2\pi i n h\alpha}\right| \quad \text{wenn} \quad m \geqslant 1$$

(21) $$\Sigma_m^n(\alpha) = O(n^{1-\frac{m}{\eta}+\varepsilon} + n^{\varepsilon})$$

und für $m > \eta$

(22) $$\Sigma_m^n(\alpha) = O(1).$$

Wir beachten noch daß

$$\lim_{r\to 1}(1-r)^{2-\frac{m}{\eta}+\varepsilon}\sum n^{1-\frac{m}{\eta}+\varepsilon}r^n = c_1 > 0, \quad \lim(1-r)^{1+\varepsilon}\sum n^{\varepsilon}r^n = c_2$$

dann erhalten wir aus (21)

$$\delta(r, \Phi, q) \leqslant c_4 L(\Phi)\left((1-r)^{\frac{m}{\eta}-s}+(1-r)^{1-s}\right). \tag{23}$$

Wenn $m > \eta$, dann erhalten wir aus (22)

$$\delta(r, \Phi, q) \leqslant 2(1-r)L(\Phi). \tag{24}$$

Besitzt α beschränkte Kettenbruchnenner dann ist $\eta = 1$, also gilt (24) für $m \geqslant 2$. Es ist (24) eine Verschärfung von (17). Es ist $\hat{\Phi} \neq 0$ für großes q nach W. Schwarz [8] wenn Φ differenzierbar bis zur Ordnung $m+2$, aber $\Phi^{(m+1)}(1) \neq \Phi^{(m+1)}(0)$.

Wenden wir uns nun dem mehrdimensionalen Fall zu. Es sei jetzt $\alpha_1, \ldots, \alpha_s$ Zahlen, welche über Z linear unabhängig sind und $\beta = q_1\alpha_1 + \ldots + q_s\alpha_s$. Wir wollen jetzt das Verhalten von $(1-r)\sum \Phi(\alpha_1 n, \ldots$ $\ldots, \alpha_s n)z^n - \hat{\Phi}$ untersuchen, wo $z = e^{2\pi i\beta}$ und

$$\hat{\Phi} = \int_I \Phi(x)e^{2\pi i(q_1x_1+\ldots+q_sx_s)}\,dx_1\ldots dx_s.$$

Dazu benötigen wir eine Abschätzung der Diskrepanz der Folge (αn) $(\alpha = (\alpha_1, \ldots, \alpha_s))$. Hier ist im allgemeinen wenig bekannt. Wir wollen voraussetzen, daß eine Funktion φ definiert auf R^+ existiert, so daß $\varphi(t)/t$ monoton wachsend ist und für alle Gitterpunkte $h = (h_1, \ldots, h_s) \neq 0$

$$(\langle h\alpha\rangle) \geqslant \varphi^{-1}(|h|). \tag{25}$$

Dabei ist $\langle h\alpha\rangle = h_1\alpha_1 + \ldots + h_s\alpha_s$, $|h| = \mathrm{Max}(|h_1|, \ldots, |h_s|)$, $(\gamma) = \inf\limits_n |\gamma - n|$ (erstreckt über alle ganzen Zahlen n). Es ist auch φ monoton wachsend. Ein wichtiger Spezialfall ist $\varphi(t) = ct^k$ für $k \geqslant 1$ insbesonder $\varphi(t) = ct^s$. Der letztere Fall tritt z.B. sicher ein, wenn $\alpha_1, \ldots, \alpha_s$ ganze Zahlen eines reellen algebraischen Zahlkörpers K vom Grade $s+1$ sind, so daß $\alpha_1 \ldots \alpha_s$ linear unabhängig über Z sind.

Nach einer Formel von P. Erdös–P. Turán J. F. Koksma gilt für jede Folge $\omega = (n\alpha)$ und jede natürliche Zahl $M > 1$

$$D_n(\omega) < 40^s\left(\frac{1}{M} + \sum_{0<|h|\leqslant M} R^{-1}(h)\,|S_n(h, \omega)|\right) \tag{26}$$

wo

$$S_n(h, \omega) = \frac{1}{n+1}\sum_{k=0}^{n} e^{2\pi ik\langle h\alpha\rangle} \leqslant 2\left(n(\langle h\alpha\rangle)\right)^{-1}.$$

Wir zerlegen den Summationsbereich $0 < |h| \leqslant M$ in Bereiche

$$P_r\colon\ 2^{r_j-1} \leqslant \mathrm{Max}(|h_i|, 1) < 2^{r_j} \quad (j = 1, \ldots, s).$$

Ist $h_j = 0$, dann sei $r_j = 1$. Es habe h_j für festes j in P_r das gleiche Vorzeichen. Die $r_1, \ldots, r_j$ laufen von 1 bis $[\log_2 M]$. In P_r ist

$$2^{r_1+\cdots+r_s-s} \leqslant R(h) < 2^{r_1+\cdots+r_s},$$

also wird

$$\Sigma = \sum R^{-1}(h)\,|S_n| \leqslant 2n^{-1} \sum_{r_1,\ldots,r_s} 2^{-(r_1+\cdots+r_s)} \sum_{h\in P_r} (\langle ha\rangle)^{-1}.$$

Wir können in einem festen P o.B.d.A. $\mathrm{Max}(r_1, \ldots, r_s) = r_1$ annehmen. Dann ist in P_r

$$(\langle ha\rangle) \geqslant \varphi^{-1}(|h|) \geqslant \varphi^{-1}(2^{r_1}) = C_r.$$

Zu jedem $h \in P_r$ gibt es eine natürliche Zahl v, so daß

$$vC_r \leqslant (\langle ha\rangle) < (v+1)C_r \qquad (v = 1, \ldots, [C_r^{-1}]).$$

Zu gegebenen v kann es höchstens zwei Gitterpunkte $h \in P_r$ geben, denn gäbe es drei solche, dann müßte für zwei solche h, h' ja $\langle ha\rangle$ – nächste ganze Zahl an $\langle ha\rangle$ das gleiche Vorzeichen haben und es wäre $\big(\langle (h-h')a\rangle\big) < C_r$ und $0 < |h-h'| < 2^{r_1-1}$ in Widerspruch zu (25). Wir haben also

$$\sum_{h\in P_r} (\langle ha\rangle)^{-1} \leqslant \frac{2}{C_r} \sum \frac{1}{v} < \frac{4}{C_r}(1 + |\log C_r|)$$

also wird

$$\Sigma < 8n^{-1} \sum_{r_1,\ldots,r_s} 2^{-(r_1+\cdots+r_s)} \varphi(2^{r_1}) \big(1 + |\log\varphi(2^{r_1})|\big)$$

also

$$\Sigma < 8n^{-1} \sum_{r_1} \frac{\varphi(2^{r_1})}{2^{r_1}} \big(1 + |\log\varphi(2^{r_1})|\big) \tag{27}$$

also

$$\Sigma < 8n^{-1} \frac{\varphi(M)}{M} \big(1 + |\log\varphi(M)| \log M\big).$$

Es ist also

$$D_n < 50^s \left(\frac{1}{M} + \frac{\varphi(M)}{Mn} \big(1 + |\log\varphi(M)|\big) \log M\right).$$

Es sei nun γ die inverse Funktion zu φ, dann nehmen wir $M = [\gamma(n)]$ und erhalten

$$D_n < 50^s \left(\frac{1}{\gamma(n)} \big(1 + (1 + \log n) \log\gamma(n)\big)\right)$$

also

$$D_n = O\left(\frac{\log n \log \gamma(n)}{\gamma(n)}\right). \tag{28}$$

Ist z.B. $\varphi(t) = ct^k$ (wir sagen wieder α ist vom Typus k) dann ist $\gamma = \frac{1}{c}\, t^{1/k}$ und wir erhalten

$$D_n = O(n^{-1/k}\log^2 n). \tag{29}$$

Wir wollen gleich für $m > 1$

$$\Sigma_m^n(\alpha) = \sum_{n \neq 0} R^{-m}(h) \left|\sum_{k=0}^{n} e^{2\pi i k\langle h\alpha\rangle}\right|$$

abschätzen und zwar für den Fall, daß α vom Typus k ist. Es ist für natürliches $M \geqslant 1$

$$\frac{1}{n+1}\Sigma_m^n(\alpha) = \sum_{|h|\leqslant M} + \sum_{|h|>M}.$$

Dann kommt, wenn man wie oben vorgeht, statt (27)

$$\sum_{|h|\leqslant M} \leqslant \frac{1}{n}\sum_{r_1} \varphi(2^{r_1})2^{-mr_1}\bigl(1+|\log\varphi(2^{r_1})|\bigr) = \frac{1}{n}\sum_{r_1} 2^{r_1(k-m)}(1+r_1 k) \tag{30}$$

also wenn $k \geqslant m$

$$\sum_{|h|\leqslant M} \leqslant \frac{M^{k-m}}{n}(1+\log^2 M).$$

Wir nehmen $M = [n^{1/k}]$, dann ist also

$$\sum_{|h|\leqslant M} = n^{-\frac{m}{k}+\varepsilon}. \tag{31}$$

5 Wir schätzen nun $\sum_{|k|>M}$ ab und beachten, daß für $M_1 = [n^{m/(m-1)^k}]$

$$\sum_{|h|\geqslant M_1} R^{-m}(h)\,|S_n(\alpha)| \leqslant \sum_{|h|\geqslant M_1} R^{-m}(h) = O(M_1^{-m+1})$$

also

$$\sum_{|h|\geqslant M_1} = O(n^{-m/k}). \tag{32}$$

Wir haben nun $\sum_{M\leqslant|h|\leqslant M_1} R^{-m}(h)\,|S_n(\alpha)|$ abzuschätzen. Wir betrachten zuerst die Glieder in dieser Summe für die $(\langle h\alpha\rangle)n \geqslant |h|$ ist. Für sie ist

6 $$R^{-m}(h)\,|S_n(\alpha)| \leqslant R^{-m}(h)\,(n\langle h\alpha\rangle)^{-1} \leqslant R^{-m}(h)\,|h|.$$

Es ist also die Summe dieser Glieder $\leqslant \sum_{M\leqslant|h|} |h|^{-m-1} = O(M^{-m}) = O(M^{-m/k})$. (Es ist ja $R(h) \geqslant |h|$.) Wir müssen also nur mehr die Glieder mit $(\langle h\alpha\rangle)n \leqslant |h|$ betrachten. Die Summe dieser Glieder ist

(33)
$$\leqslant \sum_{|h|\leqslant M_1, n(\langle h\alpha\rangle)\leqslant|h|} R^{-m}(h) \leqslant \sum_{\mathrm{Max}(|h_2|,\ldots,|h_s|)\leqslant M_1} \frac{1}{(h_2\ldots h_s)^m} \sum_{\substack{|h_1|\leqslant M \\ (\langle h_1\alpha_1+\ldots+h_s\alpha_s\rangle)\leqslant M_1/n}} h_1^{-m}.$$

Dabei ist ein h_j durch 1 zu ersetzen wenn $h_j = 0$. Die innere Summe in (33) ist nach (19) mit $h_2\alpha_2 + \ldots + h_s\alpha_s = \varrho$ sicher

$$\leqslant M_1^{-m} \sum_{\langle h_1\alpha_1+\varrho\rangle\leqslant M_1/n,\, |h_1|\leqslant M_1} 1 = M_1^{-m}\left(\frac{M_1^2}{n} + O(M_1^{1-\frac{1}{k}})\right)$$

also

$$\leqslant M_1^{2-m} n^{-1} + O(M_1^{1-m-\frac{1}{k}}) = O(n^{-\frac{m}{k}-\frac{m}{(m-1)k^2}})$$

also erhalten wir insgesamt $O(n^{-m/k})$.

Es ist also, wenn $k \geqslant m$

(34)
$$\Sigma_m^n(\alpha) = O(n^{-m/k+\varepsilon}).$$

Wenn $m > k$ dann folgt sofort aus (30)

(35)
$$\Sigma_m^n(\alpha) = O(1).$$

Wir erhalten nach (8) und (29) mit $x_n = \alpha n$ im R^s und

$$\delta(f, r) = \left|(1-r)\sum f(\alpha n) r^n - \lambda(f)\right|$$

$$\delta(f, r) \leqslant C_1 V(f)\left(\sum n^{1-\frac{1}{k}+\varepsilon} r^n\right)(1-r)^2$$

also

(36)
$$\delta(f, r) \leqslant C_2 V(f)\,(1-r)^{1/k-\varepsilon}$$

wo C_1, C_2 nur von α abhängen, wenn α vom Typus k ist. Nehmen wir $s = 2$, $f(x_1, x_2) = \Phi(x_1) e^{2\pi i x_2}$, dann ist $\lambda = 0$ und wir erhalten

$$\left|(1-r)\sum \Phi(\alpha, n) e^{2\pi i \alpha_2 n} r^n\right| \leqslant C_2 V(\Phi)\,(1-r)^{1/k-\varepsilon}.$$

Wenn $s > 1$ dann erhalten wir

$$\left|(1-r)\sum \Phi(\alpha_1 n, \ldots, \alpha_s n) z^n - \hat{\Phi}\right| \leqslant C_3\big(V(\Phi) + q(C+1)\big)(1-r)^{1/k-\varepsilon}$$

wenn $|\Phi| \leqslant C$, $z = re^{2\pi i\beta}$, $\beta = a_1 q_1 + \ldots + a_s q_s$, $q = (q_1, \ldots, q_s)$. Weiter erhalten wir für $a = (a_1, \ldots, a_s)$, $a_{s+1} = \beta$, wenn $(a_1, \ldots, a_s, \beta)$ vom Typus k ist

$$\left|(1-r)\sum \Phi(a_1 n, \ldots, a_s n) e^{2\pi i n\beta} r^n\right| \leqslant C_4 V(\Phi)(1-r)^{1/k-\varepsilon}.$$

Die letzteren Resultate lassen sich mittels (34) bzw. (35) verschärfen, wenn Φ stetige Ableitungen bis zur Ordnung sm besitzt.

§ 3. Wir wollen nun die in § 1 eingeführte Diskrepanz $D_A(\omega, r)$ etwas näher untersuchen. Wir stellen zunächst fest: Es ist

$$D_A(\omega, r) \geqslant 1 - r. \tag{37}$$

Es sei J ein Teilintervall von I mit einer Länge $< \varepsilon$ welche das Glied x_0 der Folge enthält. Dann ist $D_A(\omega, r) \geqslant 1 - r - \varepsilon$ also da ε beliebig $D_A(\omega, r) \geqslant 1 - r$. Wir führen wieder auch

$$D_A^*(\omega, r) = \sup_{r' \geqslant r} D_A(\omega, r')$$

ein. Es ist stets $D_A^*(\omega, r') \leqslant D_A^*(\omega, r)$ für $r' \geqslant r$. Wir haben schon festgestellt, daß

$$D_A^*(\omega, r) \leqslant 4 D_{M(r)}^*(\omega, r)$$

mit $M(r) = [(1-r)^{-1/2}]$. Wir wollen nun D_A^* mittels D^* nach unten abschätzen.

Wir wollen folgendes zeigen: Es existiert eine absolute Konstante C z.B. 6^8, so daß

$$D_N^*(\omega) \leqslant C\left(\log D_A^{-1}\left(e^{-\frac{\log(N+1)}{N+1}}\right)\right)^{-1/2}. \tag{38}$$

Man könnte mittels der neueren Untersuchungen von Korevaar und Ganelius (vgl. [2]) den Exponenten $-1/2$ in (38) durch -1 ersetzen. Wir begnügen uns mit (38) da der Beweis, wobei wir eine Methode von Postnikov [7] benützen, am einfachsten wird. Wie schon in der Einleitung bemerkt, wäre eine wesentliche Verschärfung von (38) wünschenswert. Es sei nun J ein Teilintervall von I^s. Nach Voraussetzung ist, wenn wir für $D_A^*(\omega, r)$ kurz $D^*(r)$ schreiben

7 $$(1-r)\sum \chi_J(x_n) r^n = J + \delta D^*(r).$$

Dabei bedeuten im folgenden alle δ Zahlen mit $|\delta| \leqslant 1$. Es ist allgemein für jedes $k \geqslant 0$

$$(1-r^{k+1})\sum \chi_J(x_n) r^{n(k+1)} = J + \delta D^*(r^{k+1}).$$

Nun ist

$$\frac{1-r}{1-r^{k+1}} - \frac{1}{k+1} \leqslant 1 - r$$

also

$$(1-r)\sum \chi_J(x_n)r^{n(k+1)} = \frac{1-r}{1-r^{k+1}}(1-r^{k+1})\sum \chi_J(x^n)r^{n(k+1)}$$
$$= \frac{J}{k+1} + 2\delta D^*(r^{k+1}) + \delta(1-r)$$

also nach (37)

$$(1-r)\sum \chi_J(x_n)r^{n(k+1)} = \frac{J}{k+1} + 3\delta D^*(r^{k+1}).$$

Es sei nun $P_g(x)$ ein Polynom $b_0+b_1x+\ldots+b_gx^g$ vom Grade g. Dann haben wir, da D^*monoton abnimmt

$$(39)\qquad (1-r)\sum \chi_J(x_n)P_g(r^n)r^n = J\int_0^1 P_g(x)\,dx + 3\delta D^*(r^{g+1})\sum_{k=0}^{g}|b_k|$$

da

$$\int_0^1 P_g(x)\,dx = \sum_{k=0}^{g}\frac{b_k}{k+1}.$$

Es sei nun f eine stetige Funktion auf $\langle 0,1\rangle = \bar{I}$. Dann existiert, wenn $\sup|f| = M(f)$ auf $\bar{I}$ zu jedem g ein Polynom P_g so daß für alle $x \in \bar{I}$ stets

$$|f(x)-P_g(x)| \leqslant 12\omega\left(\frac{1}{2g};f\right)$$

und für die Koeffizienten b_j von P_g gilt $\sum_{k=0}^{g} \leqslant 2\cdot 6^g M$. Dabei ist $\omega(\delta,f)$ der Stetigkeitsmodul von f. Es ist

$$(1-r)\sum \chi_J(x_n)f(r^n)r^n$$
$$= (1-r)\sum \chi_J P_g(r^n)r^n + (1-r)\sum\big(f(r^n)-P_g(r^n)\big)\chi_J(x_n)r^n.$$

Da

$$|f-P| \leqslant 12\omega\left(\frac{1}{2g}\right)$$

so ist der zweite Term im obigen Ausdruck $\leqslant 12\omega\left(\frac{1}{2g},f\right)$. Weiter ist

$$\int_0^1 P_g(x)\,dx = \int_0^1 f(x)\,dx + \int_0^1 \big(P_g(x)-f(x)\big)\,dx$$

also erhalten wir

$$(40)\quad (1-r)\sum \chi_J(x_n)f(r^n)r^n = J\int_0^1 f(x)\,dx + 24\delta\omega\left(\frac{1}{2g}\right) + \delta M(f)6^{g+1}D^*(r^{g+1}).$$

Wir nehmen nun in (40) $r = e^{-1/(N+1)}$ und bemerken daß

$$(N+1)^{-1} \geqslant 1-e^{-1/(N+1)} \geqslant (N+1)^{-1} - \tfrac{1}{2}(N+1)^{-2} \geqslant \tfrac{1}{2}(N+1)^{-1}.$$

Weiter sei

$$\alpha = 1+\frac{1}{\sqrt{g}},\qquad \beta = 1-\frac{1}{\sqrt{g}}$$

und

$$f_1(x) = \begin{cases} 0 & \text{für } 0 \leqslant x \leqslant e^{-\alpha}, \\ \dfrac{e}{e^{-1}-e^{-\alpha}}(x-e^{-\alpha}) & \text{für } e^{-\alpha} \leqslant x \leqslant e^{-1}, \\ \dfrac{1}{x} & \text{für } e^{-1} \leqslant x \leqslant 1 \end{cases}$$

und

$$f_2(x) = \begin{cases} 0 & \text{für } 0 \leqslant x \leqslant e^{-1}, \\ \dfrac{e^{\beta}}{e^{-\beta}-e^{-1}}(x-e^{-1}) & \text{für } e^{-1} \leqslant x \leqslant e^{-\beta}, \\ \dfrac{1}{x} & \text{für } e^{-\beta} \leqslant x \leqslant 1. \end{cases}$$

Es ist $M(f_1)$ und $M(f_2) \leqslant e$,

$$\int_0^1 f_1(x)\,dx = 1+\frac{1-e^{1-\alpha}}{2},\qquad \int_0^1 f_2(x)\,dx = 1-\frac{e^{-\beta}-e^{-1}}{2}e^{\beta},$$

$$\omega(\varrho, f_1) = \frac{e}{e^{-1}-e^{-\alpha}}\varrho,\qquad \omega(\varrho, f_2) = \frac{e^{\beta}}{e^{-\beta}-e^{-1}}\varrho.$$

Weiter bemerken wir, daß

$$1-e^{1-\alpha} \leqslant \alpha-1,\qquad (1-e^{1-\alpha})^{-1} \leqslant \frac{3}{\alpha-1},\qquad 1-e^{\beta-1} \leqslant 1-\beta,$$

$$e^{\beta}(e^{-\beta}-e^{-1})^{-1} \leqslant e(e^{-\beta}-e^{-1})^{-1} \leqslant 4e(1-\beta)^{-1}.$$

Wir berücksichtigen daß

$$(1-e^{-1/(N+1)})\sum_{n=0}^{N}\chi_J(x_n) \geqslant \frac{1}{N+1}\sum_{n=0}^{N}\chi_J(x_n) - \frac{1}{2(N+1)},$$

dann erhalten wir aus (40) mit $f = f_1$

$$\frac{1}{N+1}\sum_{n=0}^{N}\chi_J(x_n) \leqslant J + \frac{1}{2(N+1)} + \frac{\alpha-1}{2} + \frac{36e^2}{g(\alpha-1)} + e\cdot 6^{g+1} D^*(r^{g+1}).$$

Weiters liefert (40) mit $f = f_2$ auf der linken Seite

$$(1-e^{-1/N})\left(\sum_{n\leqslant\beta N}\chi_J(x_n) + \sum_{\beta N\leqslant n\leqslant N}\chi_J(x_n)\frac{e^{\beta}}{e^{-\beta}-e^{-1}}(e^{-n/(N+1)}-e^{-1})e^{-n/(N+1)}\right)$$

also

$$\frac{1}{N+1}\sum_{n=0}^{N}\chi_J(x_n) \geqslant J - \frac{1-\beta}{2} - \frac{36e}{g(1-\beta)} - e\cdot 6^{g+1} D^*(r^{g+1})$$

also erhalten wir

$$\left|\frac{1}{N+1}\sum_{n=0}^{N}\chi_J(x_n) - J\right| \leqslant \frac{1}{2(N+1)} + \frac{500}{\sqrt{g}} + 3\cdot 6^{g+1} D^*(r^{g+1}).$$

Das gilt für jedes J, also haben wir eine Abschätzung für $D_N(\omega)$. Nun ist $D_N(\omega) \geqslant \frac{1}{N+1}$, also erhalten wir

$$D_N(\omega) \leqslant \frac{10^3}{\sqrt{g}} + 6^{g+2} D^*(r^{g+1}). \tag{41}$$

Wir setzen nun zunächst voraus, daß $N \geqslant g$ und $D^*(\varrho) \leqslant 6^{-8}$ wo $\varrho = e^{-\frac{\log(N+1)}{N+1}}$. Dann nehmen wir für

$$g = \left[\frac{1}{2\log 6}\log D^{-1}(\varrho)\right] - 1.$$

Es ist $g \geqslant 3$. Es ist weiter nach (37)

$$D^*(\varrho) \geqslant 1 - \varrho \geqslant \frac{1}{2}\frac{\log(N+1)}{N+1},$$

also

$$\log\frac{1}{D^*(\varrho)} \leqslant \log 2 + \log(N+1) - \log\log(N+1) \leqslant \log(N+1).$$

Es ist also $g+1 < \log(N+1)$. Dann ist

$$r^{g+1} \geqslant e^{-\frac{\log(N+1)}{N+1}} = \varrho, \quad \text{also } D^*(r^{g+1}) \leqslant D^*(\varrho).$$

Es wird also $6^{g+1} D^*(r^{g+1}) \leqslant \sqrt{D^*(\varrho)}$. Nun ist $D^* \leqslant \log^{-1} \dfrac{1}{D^*}$ also erhalten wir

$$D_N(\omega) \leqslant 6^8 (\log D^{*-1}(\varrho))^{-1/2}. \tag{42}$$

Ist $D^*(\varrho) \geqslant 6^{-8}$ so ist (42) trivial, ebenso für $N \leqslant g$. Da die rechte Seite von (41) für $N \geqslant 2$ monoton abnimmt, so folgt (38).

Literaturverzeichnis

[1] F. W. Caroll and J. H. B. Kemperman, *Noncontinuable analytic functions* Duke Math. Journal 32 (1963), S. 65–83.

[2] T. H. Ganelius, *Tauberian Remainder Theorems* (Lecture Notes in Mathematics Nr. 232), Berlin – Göttingen – Heidelberg 1971.

[3] E. Hecke, *Über analytische Funktionen und die Verteilung von Zahlen mod. Eins*, Abh. Math. Sem. Univ. Hamburg 1 (1921), S. 54–71.

[4] E. Hlawka, *Funktionen von beschränkter Variation in der Theorie der Gleichverteilung*, Ann. Mat. Pura ed Appl. IV 54 (1961), S. 325–333.

[5] J. F. Koksma, *Diophantische Approximationen*, 1936.

[6] S. Lang, *Introduction to Diophantine Approximation*, Reading, Mass., 1966.

[7] A. G. Postnikov, *The remainder term in the Tauberian theorem of Hardy and Littlewood* (Russian), Dokl. Akad. Nauk SSSR (N.S.) 77 (1951), S. 193–196.

[8] W. Schwarz, *Irrationale Potenzreihen*, Arch. Math. 13 (1962), S. 228–240.

MATHEMATISCHES INSTITUT DER UNIVERSITÄT WIEN

Eingegangen 26. 4. 1972 (277)

Corrections

19.1 $= \left| \sum_{k=0}^{n-q} \{\alpha k + \delta\} - \frac{1}{2}(n-q+1) + \sum_{k=n-q+1}^{n} \{\alpha k + \delta\} - \frac{q}{2} \right|$

$\leq \left| \sum_{k=0}^{n-q} \{\alpha k + \delta\} - \frac{1}{2}(n-q+1) \right| + \left| \sum_{k=n-q+1}^{n} \{\alpha k + \delta\} - \frac{q}{2} \right|$

20.2 $\left| \sum_{k=n-q+1}^{n} \{\alpha k + \delta\} - \frac{q}{2} \right| \leq$

21.3 $(1 - \frac{2}{5c} \log(1-r))$

21.4 wie 21.3

25.5 $M_1 = [n^{m/(m-1)k}]$

25.6 $\leq R^{-m}(h)|h|^{-1}$

27.7 In dieser und den folgenden Formeln (S.28, 29, 30) ist J durch Vol J zu ersetzen.

Mathematische Modelle der kinetischen Gastheorie

Rheinisch-Westfälische Akademie der Wissenschaften,
Natur-, Ingenieur- und Wirtschaftswissenschaften, Vorträge N 240,
Westdeutscher Verlag, Opladen, 1974, 15 pp.

In der klassischen kinetischen Gastheorie, entwickelt von Clausius, Maxwell, Boltzmann und Gibbs, betrachten wir das System Σ, das wir zunächst physikalisch beschreiben wollen. Es sei Σ ein hochverdünntes Gas von N Molekülen, die in einem Kasten K vom Volumen V eingeschlossen sind. Die Temperatur sei so hoch und die Dichte sei so gering angenommen, daß jedes Molekül als ein klassisches Teilchen mit wohldefiniertem Ort und Impuls betrachtet werden kann. Weiterhin nehmen wir an, daß die Moleküle stets als unterscheidbar angesehen werden können. Wir treiben also klassische Physik. Die Wände des Kastens seien Flächen, die auf ein auftreffendes Gasmolekül nur so wirken, daß sie es elastisch reflektieren. Weiter nehmen wir an, daß das Gas so hoch verdünnt ist, daß die Moleküle in keiner Weise aufeinander wirken; außerdem sollen keine Kräfte von außen wirken. Wir machen noch die schwerwiegende Einschränkung, daß zwischen den Molekülen auch keine Stöße auftreten. Diese Einschränkung machen wir nur gezwungenermaßen. Es ist ja gerade ein Vorzug der klassischen Boltzmannschen Theorie, Stöße zu berücksichtigen; allerdings handelt es sich in dieser Theorie nur um Zweierstöße. Die Möglichkeit, daß drei oder mehr Moleküle gleichzeitig stoßen, wird vernachlässigt. Die Theorie führt dann zur berühmten Boltzmannschen Transportgleichung. Erst in letzter Zeit wurden von physikalischer Seite Dreier- und Viererstöße berücksichtigt und die Boltzmannsche Gleichung in dieser Hinsicht erweitert. Es stellte sich aber heraus, daß dann in der Gleichung divergente Terme auftreten (E. G. D. Cohen und I. R. Dorfmann). Aber auch die Boltzmannsche Theorie birgt große mathematische Schwierigkeiten in sich.

Ein erster Ansatz wurde von dem russischen Mathematiker Sinaĭ 1963 angekündigt und eine mehrere hundert Seiten umfassende Arbeit in Aussicht gestellt.

Wir machen noch die weitere Annahme, daß der Kasten, der das Gas einschließt, ein Würfel von einer Länge $\frac{1}{2}L$ ist. Nur in diesem Fall bietet sich bis jetzt die Möglichkeit, die Bewegung der Teilchen im Würfel genauer zu verfolgen. Bereits der Fall, daß das Gas z. B. in einem Tetraeder eingeschlossen ist, bietet Schwierigkeiten, die bis heute nicht überwunden werden konnten. Dies ist sogar der Fall, wenn wir das unphysikalische Beispiel betrachten, daß sich die Moleküle stets in einer Ebene bewegen und in einem Dreieck eingesperrt sind und an den Seiten des Dreiecks elastisch reflektiert werden. Mit diesem Fall hat sich schon Poincaré beschäftigt. In letzter Zeit wurde das spezielle Problem des Dreiecks in Angriff genommen; es liegen aber, soviel ich weiß, noch keine abschließenden Resultate vor.

Was sagt nun die kinetische Gastheorie aus? Die Aussage lautet: Wenn die Anzahl der Teilchen sehr groß ist, dann nimmt das System Σ, bei beliebiger Ausgangslage der Teilchen, im Kasten nach einer genügend langen Zeit eine Gleichgewichtslage ein, und zwar so, daß die Moleküle im Kasten gleich dicht verteilt sind. Man meint dabei, daß in jedem Gebiet G in K vom Volumen $V(G)$ die Anzahl der Teilchen angenähert $N\frac{V(G)}{V(K)}$ ist. Es ist klar, daß über Σ noch zusätzliche Annahmen getroffen werden müssen, und zwar vor allem über die Geschwindigkeiten $v_1, \ldots, v_N$ der N Moleküle. Denn nehmen wir an, daß die Gasmoleküle zur Zeit $t = 0$ sich an einem und demselben Ort befinden und alle die gleiche Geschwindigkeit parallel zu einer der Kanten des Kastens haben, dann ist klar, daß sich die Moleküle nicht im ganzen Kasten ausbreiten können.

§ 1

Die kinetische Gastheorie ist ein Spezialfall der Theorie dynamischer Systeme. Ein mechanisches System Σ wird bestimmt durch $2n$ Funktionen $q_1, \ldots, q_n, p_1, \ldots, p_n$ der Zeit t, die verallgemeinerten Koordinaten und Impulse des Systems, definiert auf einer Mannigfaltigkeit X, und sie sind Lösungen eines Systems von Differentialgleichungen

$$\frac{dq_j}{dt} = \frac{\partial H}{\partial p_j}, \quad \frac{dp_j}{dt} = -\frac{\partial H}{\partial q_j} \qquad (j = 1, \ldots, N).$$

Dabei ist H Funktion der p und q und heißt Hamilton-Funktion des Systems. Man betrachtet gleich allgemeine Systeme von gewöhnlichen Differentialgleichungen

$$\frac{dx_j}{dt} = f_j(x_1, \ldots, x_k) \quad \text{für} \quad j = 1, \ldots, k$$

auf einer Mannigfaltigkeit X im R^k. Ist x_0 ein beliebiger Anfangspunkt, so gibt es dazu eine Lösung $x(t) = (x_1(t) \ldots x_k(t))$. Sie definiert eine Kurve (die Geschichte des Systems vom Anfang x_0) auf X. Damit wird auf X eine einparametrige Gruppe von Transformationen S_t des Phasenraumes X

definiert, wo die Transformation S_t jeden Punkt x_0 zum Punkt $x(t)$ entlang seiner Trajektorie bewegt. In diesem Zusammenhang erwähnen wir gleich den Satz von Poincaré, der aussagt: Für fast alle x_0 auf X und für jede Funktion f auf X kommt $f(x(t))$ seinem Wert $f(x_0)$ bis auf einen beliebig kleinen Fehler unendlich oft beliebig nahe, wenn das System in einem kompakten Teilbereich des Phasenraumes X verbleibt (Poincaré-zykel). Dies wurde vor allem von Zermelo gegen die Theorie von Boltzmann eingewendet, die versucht, die Irreversibilität thermodynamischer Systeme mechanisch zu erklären. Da, wie sich Zermelo ausdrückt, es sehr unwahrscheinlich ist, daß sich die Systeme der Natur gerade in der Ausnahmemenge vom Maße Null des Poincaréschen Satzes aufhalten, ist dies ein schwerwiegender Einwand. Boltzmann selber erwiderte darauf, daß die Widerkehrzeit bei einem Gas von 10^{18} Teilchen ebenfalls von der Größenordnung 10^{18} sec, daher physikalisch uninteressant ist. Wir erwähnen in diesem Zusammenhang den Loschmidtschen Einwand, daß die Gesetze der Mechanik invariant sind in bezug auf die Umkehrung der Zeit. Das steht daher auch im Widerspruch zur Existenz irreversibler Prozesse.

Zur Diskussion dieses Einwandes verweisen wir auf den Artikel von J. Progogine in „The Boltzmann Equations, 1973, Springer-Verlag, S. 40". Kehren wir zur allgemeinen Theorie zurück. In der sogenannten Ergodentheorie betrachtet man noch allgemeinere Systeme $(X, (S_t))$, wo X eine (differenzierbare) Mannigfaltigkeit und die Gruppe (S_t) eine Gruppe von Diffeomorphismen von X ist. Man ordnet nun diesem System $(X, (S_t))$ zwei Räume zu, nämlich 1) den Funktionenraum X'' der Funktionen $f(x)$ auf den Phasenraum X. Die Gruppe S_t erzeugt auf X'' die Gruppe S_t'', definiert durch $(S_t'' f)(x) = f(S_t x)$. 2) Man betrachtet den Raum X^{**} aller Wahrscheinlichkeitsmaße auf X. Die Gruppe (S_t) erzeugt die Gruppe (S_t^{**}), definiert durch

$$(S_t^{**} \lambda)(A) = \lambda(S_t A)$$

für alle $A \subset X$, für die $S_t A$ λ-meßbar ist für alle t und alle $\lambda \in X^{**}$.

Ein Maß $\lambda \in X^{**}$ heißt invariant für die Gruppe (S_t), wenn $S_t^{**} \lambda = \lambda$ für alle t.

Man nennt (X, λ, S_t) ein klassisches dynamisches System.

Wird das System durch eine Hamiltonfunktion H erzeugt, dann haben wir ein natürliches invariantes Maß auf den Mannigfaltigkeiten von konstanter Energie, nämlich die mikrokanonische Verteilung μ. Nach dem bekannten Liouvillschen Satz ist sie auf jeder Energiefläche E definiert durch

$$d\mu = \frac{d\sigma}{\|\text{grad}\, H\|},$$

wo $\| \; \|$ = Länge von grad H und σ das Volumenelement auf E ist, induziert durch das Volumenelement des R^{2n}.

Ein klassisches System (X, λ, S_t) heißt ergodisch, wenn für jede integrierbare Funktion f gilt

$$\lim_{T\to\infty} \frac{1}{2T} \int_{-T}^{T} f(S_t x_0)\, dt = \int_X f d\lambda$$

für fast alle x_0.

Dieser Begriff der Ergodizität wurde in anderer Form von Boltzmann eingeführt.

Ein ergodisches System ist z. B. das System auf dem Torus X, definiert durch $S_t(x_1, \ldots, x_n) = (x_1 + \omega_1 t \pmod 1) \ldots, x_n + \omega_n t \pmod 1))$, wenn
1 die $\omega_1, \ldots, \omega_n$ linear unabhängig über Q sind und λ das Lebesquesche Maß ist. Als Hauptproblem ist natürlich nachzuweisen, daß ein mechanisches System ergodisch ist. Wenn dies nicht der Fall ist, wird es erzwungen. Man schließt oft so: Es sei Σ ein System, erzeugt von einer Hamiltonfunktion $H_0(q, p)$, und es sei nicht ergodisch. Dann wird durch eine Störung, welche beliebig klein gemacht werden kann, das gestörte System ergodisch. Genauer formuliert: Man betrachtet ein System, erzeugt von einer Hamilton-Funktion $H(p, q, \lambda)$, welche noch von einem Parameter λ abhängt, mit $H(p, q, 0) = H_0(p, q)$, für kleines λ und nimmt an, daß es ergodisch ist. Es wurde aber von Kolmogorow–Arnold–Moser gezeigt, daß es nichtergodische Systeme H_0 (wo H_0 separierbar ist) gibt und trotzdem die gestörten Systeme nicht ergodisch sind. Wie wir schon angedeutet haben, hat Sinaĭ angekündigt, daß das System, gebildet von Kugeln, eingeschlossen in einem Würfel, die an den Wänden elastisch reflektiert werden und deren Zusammenstöße untereinander ebenfalls elastisch sind, ein ergodisches System bildet (auf eine genauere Formulierung sei verzichtet). Der Beweis ist aber bis heute nicht erbracht, und es bleiben noch viele Probleme offen.

§ 2

Wir wollen zu unserem System Σ zurückkehren, welches wir in der Einleitung beschrieben haben. Wir wollen zunächst die Lage $x(t)$ eines Teilchens zur Zeit t im Würfel von der Kantenlänge $\frac{1}{2} L$ bestimmen. Um die Sache einfach zu gestalten, wollen wir zunächst annehmen, daß wir statt eines Würfels ein eindimensionales Intervall $0 \leqq x \leqq \frac{1}{2} L$ haben. Nach

Voraussetzung ist $x(0) = u$ die Anfangslage und v die Geschwindigkeit des Teilchens. Wir führen noch die Funktion $x'(t) = u + vt$ ein. Es beschreibt $x'(t)$ die Lage des Teilchens zur Zeit t, wenn es nicht im Intervall eingesperrt wäre. Solange $0 \leqq x' \leqq \frac{1}{2} L$ gilt, ist $x(t) = x'(t)$. Am Intervallende $\frac{1}{2} L$ wird aber das Teilchen reflektiert, also $x(t) = L - x'(t)$, solange $\frac{L}{2} \leqq x'(t) \leqq L$ gilt. Dann landet das Teilchen am anderen Intervallende 0, wird dort reflektiert und hat dann wieder dieselbe Richtung wie x'. Allgemein zeigt man durch vollständige Induktion, daß $x(t) = (k+1) L - x'(t)$, sofern $(k + \frac{1}{2}) L \leqq x'(t) \leqq (k+1) L$, und $x(t) = x'(t) - kL$, sofern $kL \leqq x'(t) \leqq (k + \frac{1}{2}) L$ gilt. Wir erhalten also $x(t) = x'(t) - \left[\frac{x'(t)}{L}\right] L$, sofern $0 \leqq \frac{x'(t)}{L} - \left[\frac{x'(t)}{L}\right] \leqq \frac{1}{2}$, und $x(t) = \left(\left[\frac{x'(t)}{L}\right] + 1\right) L - x'(t)$, sofern $\frac{1}{2} \leqq \frac{x'(t)}{L} - \left[\frac{x'(t)}{L}\right] < 1$ ist.

Analog erhalten wir im dreidimensionalen Fall des Würfels bzw. gleich im s-dimensionalen Fall des Intervalls $0 \leqq \xi_i \leqq \frac{1}{2} L$ $(i = 1, \ldots, s)$ die Lage $x(t) = (x^1(t), \ldots, x^s(t))$, wenn $x(0) = (u^1, u^2, \ldots, u^s) = u$ und die Anfangsgeschwindigkeit $v = (v^1, \ldots, v^s)$ ist, wenn wir auf jede Koordinate $x^j(t)$ die vorhergehende Formel des eindimensionalen Falles anwenden.

Um die weiteren Überlegungen zu vereinfachen, nehmen wir wieder $s = 1$. Wir stellen die Frage, wann sich das Teilchen zur Zeit t in einem Teilintervall $J = [a, b]$ von $[0, \frac{1}{2} L]$ aufhält. Es sei χ_J die charakteristische Funktion von J, d. h. $\chi_J(\xi) = 1$, wenn $\xi \in J$ $\chi_J(\xi) = 0$, wenn $\xi \notin J$, dann lautet die Anwort

$$\chi_J(x(t)) = \chi_J\left(x'(t) - \left[\frac{x'(t)}{L}\right] L\right) + \chi_{J'}\left(x'(t) - \left[\frac{x'(t)}{L}\right] L\right),$$

wo J' das Intervall ist, das aus J durch Spiegelung am Punkt $\frac{L}{2}$ entsteht und welches Teilintervall von $\left[\frac{L}{2}, L\right]$ ist. Dies sieht man sofort so ein:

1. Fall: $0 \leqq \frac{x'(t)}{L} - \left[\frac{x'(t)}{L}\right] < \frac{1}{2}$. Dann ist, wie oben gezeigt,

$$x(t) = x' - L\left[\frac{x'}{L}\right] \leqq \frac{L}{2}, \text{ also } \chi_J(x(t)) = \chi_J\left(x'(t) - \left[\frac{x'(t)}{L}\right] L\right)$$

und $\chi_{J'}\left(x'(t) - \left[\frac{x'(t)}{L}\right] L\right) = 0$. Also ist die Formel richtig.

Im 2. Fall: $\frac{1}{2} \leqq \frac{x'(t)}{L} - \left[\frac{x'(t)}{L}\right] < 1$ ist aber $x(t) = \left(1 + \left[\frac{x'(t)}{L}\right]\right)L - x'(t)$, also $\chi_J(x(t)) = \chi_{J'}\left(x'(t) - \left[\frac{x'(t)}{L}\right]L\right)$ und $\chi_J\left(x'(t) - \left[\frac{x'(t)}{L}\right]L\right) = 0$.

Wir betrachten nun das doppelte Intervall $[0, L]$ und J, J' als Teilintervalle von $[0, L]$. Wir vereinfachen jetzt die Formel, indem wir $L = 1$ setzen, und erhalten $\chi_J(x(t)) = \chi_J(x'(t)) + \chi_{J'}(x'(t))$. Dabei haben wir noch die charakteristische Funktion auf der ganzen Zahlengeraden periodisch mit der Periode 1 fortgesetzt, d. h. es ist $\chi_J(\xi) = \chi_J(\xi')$, wenn $\xi' - \xi$ eine ganze Zahl ist.

Betrachten wir jetzt N Massenpunkte mit den Anfangslagen $u_1, \ldots, u_N$ und den Anfangsgeschwindigkeiten $v_1, \ldots, v_N$, setzen $x'_k = u_k + v_k t$, dann ist

$$\frac{1}{N}\sum_{k=1}^{N}\chi_J(x_k(t)) - 2J = \frac{1}{N}\sum_{k=1}^{N}(\chi_J(x'_k(t)) - J) + \frac{1}{N}\sum_{k=1}^{N}(\chi_{J'}(x'_k(t)) - J).$$

Im s-dimensionalen Fall erhalten wir für $L = 1$, also für den Würfel $K: 0 \leqq \xi_i \leqq \frac{1}{2}$ und für ein Teilintervall J dieses Würfels analog

$$\frac{1}{N}\sum_{k=1}^{N}\chi_J(x_k(t)) - 2^s J = \frac{1}{N}\sum_{r=1}^{2^s}\left(\sum_{k=1}^{N}\chi_{J_r}(x'_k(t)) - J\right),$$

wo die 2^s-Intervalle J_r durch Spiegelung an den Seitenwänden von K aus J hervorgehen. Es ist χ wieder die charakteristische Funktion der zugehörigen Intervalle, periodisch mit der Periode 1 auf dem ganzen R^s fortgesetzt.

§ 3

Wir wollen nun das Verhalten von

$$\Delta_J(t) = \frac{1}{N}\sum_{k=1}^{N}\chi_J(x_k(t)) - 2^s V(J)$$

der lokalen Diskrepanz in J untersuchen. Trivialerweise ist $|\Delta_J(t)| \leqq 1$ für alle t. Nach der kinetischen Gastheorie soll nun bei beliebiger Anfangslage von Σ für großes N und t diese lokale Diskrepanz klein sein.

Wir setzen noch $\sup\limits_{J \in K} |\Delta_J(t)| = D_N(t)$, die Diskrepanz von Σ. Nach der kinetischen Gastheorie soll auch für großes t und N $D_N(t)$ klein sein, also

Δ_J klein sein, gleichmäßig in J, unabhängig von der Anfangslage. Dies kann nicht der Fall sein, wie schon Zermelo für allgemeine mechanische Systeme bemerkt hat. Für unser einfaches System Σ kann man dies leicht sehen. Nehmen wir wieder $s = 1$ und alle $u_k = \frac{1}{4}$ (die Teilchen sind also zur Zeit $t = 0$ im Mittelpunkt von K). Weiter sei $J: |\xi - \frac{1}{4}| < \varepsilon < \frac{1}{4}$. Dann gibt es für beliebiges ganzzahliges q sicher ein $\bar{t}$ und ganze Zahlen $g_1, \ldots, g_n$, so daß $|v_k \bar{t} - g_k| < \varepsilon < \frac{1}{4}$. Es ist also $|x_k(\bar{t}) - \frac{1}{4}| < \varepsilon$ und $q \leqq \bar{t} \leqq q\, \varepsilon^{-N}$. Es liegen also alle $x_k(\bar{t})$ in J, daher ist $\Delta_J = 1 - 2\varepsilon$, und $D_N(\bar{t}) \geqq 1 - 2\varepsilon$.

Es kommt also Σ nach einer gewissen Zeit seiner Anfangslage beliebig nahe.

Egervary und P. Turan haben 1951 in Studia Mathematica 12 (170–180) für $s = 3$ und sehr speziellen Annahmen über die Geschwindigkeiten, nämlich

$$v_k = ((N+h)^2,\ (N+h)^2\sqrt{2},\ (N+h)^2\sqrt{3}) \qquad (h = 1, \ldots, N)$$

die Idee entwickelt, daß man $D_N(t)$ nicht für alle t betrachtet, sondern nur für viele t, d. h., daß man eine gewisse Teilmenge der Zeit, in der man das System betrachtet, vernachlässigt. Wir werden das später noch präzisieren. Diesem Standpunkt wollen wir uns anschließen, aber wesentlich allgemeinere Annahmen für die Geschwindigkeiten machen.

Im Geschwindigkeitsraum sei Ω_N die konvexe Hülle der $v_1, \ldots, v_N$, die wir uns vom Nullpunkt aus abgetragen denken. Es ist also Ω_N die kleinste konvexe Menge, die $v_1, \ldots, v_N$ enthält. Den Durchmesser von Ω_N bezeichnen wir mit $2c$. Ist $s = 1$, dann ist Ω_N ein Intervall der Länge $2c$. Wir wollen noch annehmen, was unwesentlich ist, daß Ω den Nullpunkt als Mittelpunkt besitzt. Dies ist sicher der Fall, wenn mit jeder Geschwindigkeit auch die entgegengesetzte Geschwindigkeit unter den $v_1, \ldots, v_N$ vorkommt. Wir wollen c noch die Maximalgeschwindigkeit nennen.

Ist nun r eine Richtung von o aus, dann verstehen wir unter einer Schicht S_r von Ω_N in der Richtung r die Menge, die aus Ω_N durch zwei Ebenen senkrecht zu r ausgeschnitten wird. Den Abstand der beiden Ebenen nennen wir die Höhe $h(S_r)$ von S_r.

Die geometrische Annahme, die wir nun machen, ist die, daß ein $\tau(\Omega)$ existiert, wo $\tau(\Omega) = \tau(\lambda\Omega)$, so daß für alle S_r gilt

$$\frac{V(S_r)}{V(\Omega)} \leqq \frac{\tau(\Omega)}{c}\, h(S_r), \tag{A}$$

wo V Volumen bedeutet. (Es bedeutet $\lambda\Omega$ die konvexe Menge, die durch Streckung mit dem Verhältnis λ aus Ω vom Nullpunkt aus entsteht.) Be-

sitzt Ω innere Punkte, dann existiert immer ein solches τ. Denn dann existiert eine konvexe Distanzfunktion $F(x)$ mit $F(tx) = |t|\, F(x)$, so daß Ω die Gestalt $F(x) \leqq \mu$ hat. Es sei Ω_0 der sogenannte Eichkörper
2 $F(x) \leqq 1$. Ist d der Durchmesser von Ω_0, so ist $\mu d = 2c$. Weiter existieren Konstanten C_1, C_2, so daß $C_1 |x| \leqq F(x) \leqq C_2 |x|$. Es ist bei passendem x_0

$$V(S_r)/V(\Omega) = \int\limits_{\substack{|r(x-x_0)| \leq \frac{h}{2} \\ F(x) \leq \mu}} dx \Big/ \int\limits_{F(x) \leq \mu} dx = \int\limits_{\substack{\left|r\left(x-\frac{x_0}{\mu}\right)\right| \leq \frac{h}{2\mu} \\ F(x) \leq 1}} dx / V(\Omega_0)$$

$$\leqq \int\limits_{\substack{\left|r\left(x-\frac{x_0}{\mu}\right)\right| \leq \frac{h}{2\mu} \\ |x| \leq \frac{1}{c_1}}} dx / V(\Omega_0) \qquad \leqq \frac{h}{\mu} V^{-1}(\Omega_0) = \frac{h}{c} \frac{2}{d} V^{-1}(\Omega_0).$$

Es ist also $\tau(\Omega) = \dfrac{2}{d} V^{-1}(\Omega_0)$.

Wir setzen nun

$$\check{D}_N = \sup_{r, S_r} \left| \frac{N(S_r)}{N} - \frac{V(S_r)}{V(\Omega)} \right|, \tag{0}$$

wo $N(S_r)$ die Anzahl der $v_1, \ldots, v_N$ in S_r ist und sich das Supremum über alle Richtungen r und alle Schichten S_r erstreckt. Wir nennen $\check{D}_N$ die Diskrepanz der Geschwindigkeiten von Σ. Es ist $\check{D}_N \leqq 1$. Man kann zeigen, daß für fast alle Systeme $(v_1, \ldots, v_N)$ in Ω_N stets $\check{D}_N \leqq (1 + \varepsilon) \times \dfrac{\log \log N}{\sqrt{N}}$ ist $(\varepsilon > 0)$.

§ 4

Wir wollen nun für $t \geqq 0$

3 $$M(T, \Sigma) = \int\limits_{-T}^{T} D_N^2(t)\, dt$$

untersuchen. Es stellt sich folgende Abschätzung heraus:

$$\varrho(T) = \frac{M(T)}{T} \leqq K_3 \operatorname{Min}\left(1, \frac{\tau(\Omega)}{cT} + \check{D}_N (\log^{2s} \check{D}_N + 1)\right), \tag{1}$$

wo K_3 eine absolute Konstante ist, z. B. $K_3 = 100^s$.

Wir wollen den Beweis kurz skizzieren und benützen eine Formel von Erdös–Turan–Koksma: Es sei $K \geqq 1$ eine natürliche Zahl, dann ist

$$D_N \leqq K_2 \left(\frac{1}{K} + \sum_{0 \leqq \|h\| \leqq K} R^{-1}(h)\, |S(h)| \right).$$

Dabei ist K_2 eine absolute Konstante, $S(h) = \frac{1}{N} \sum_{k=1}^{N} e^{2\pi i h x'_k(t)}$, h Gitterpunkt in R^s, $\|h\| = \operatorname{Max}(|p_1(h)|, \ldots, |p_s(h)|)$ ($p_j(h)$: $j$te Komponente
4 von h), $R(h) = \prod_{j=1}^{s} \operatorname{Max}(1, |p_j(h)|)$ $hx = \sum_{j=1}^{s} p_j(h)\, p_j(x)$.

Es ist nun

$$\frac{M(T)}{T} \leqq \frac{2}{T} \int_{-2T}^{2T} \left(1 - \frac{|t|}{2T}\right) D_N^2\, dt \leqq \left(\sqrt{\frac{1}{K}} + \sum_{0 < \|h\| \leqq K} R^{-1}(h) \sqrt{M} \right)^2$$

wo

$$M_1 \leqq \frac{1}{N} \sum_{i,k=1}^{N} \frac{\sin^2 2\pi T h (v_k - v_i)}{(TN\pi h (v_k - v_i))^2}.$$

Es ist die innere Summe bei festem i gleich

$$\sum_k = \sum_{r=0} \sum_{\substack{k \\ \frac{r}{T} \leqq |h(v_k - v_i)| < \frac{r+1}{T}}} \frac{\sin^2 2\pi T h (v_k - v_i)}{(T N\pi h (v_k - v_i))^2}$$

$$\leqq \frac{1}{N} \sum_{\substack{h \\ |h(v_k - v_i)| < \frac{1}{T}}} 1 + \sum_{r=1}^{\infty} \frac{1}{r^2} \frac{1}{N} \sum_{\substack{k \\ \frac{r}{T} \leqq |h(v_k - v_i)| < \frac{r+1}{T}}} 1 .$$

Wenn wir (A) anwenden, so ist nach Definition von $\check{D}_N$, und *nur* das verwenden wir,

$$\frac{1}{N} \sum_{\substack{k \\ \frac{r}{T} \leqq |h(v_k - v_i)| < \frac{r+1}{T}}} 1 \leqq 2 \frac{\tau(\Omega)}{c} \frac{1}{|h|\, T} + \check{D}_N. \qquad (*)$$

Dann wird

$$\sum_k \leqq \frac{\tau(\Omega)}{cT} \frac{1}{|h|} + \check{D}_N,$$

und wir erhalten nach kurzer Rechnung die Behauptung, wenn man $K = [\check{D}_N^{-1}]$ nimmt.

Bemerkung: Das Resultat bleibt also richtig, wenn die $v_1, \ldots, v_N$ alle auf dem Rand von Ω_N liegen und $\check{D}_N$ durch (*) definiert wird.

Man kann zeigen, daß die Abschätzung von $M(T)$ nicht so schlecht ist. Wenn $u_1 = \ldots = u_N = 0$, so ist $\frac{M(T)}{T} \geqq K_1\left(\frac{1}{cT} - \check{D}_N\right)$ bei passendem K_1. Am einfachsten wird die Abschätzung für $T_0 = \frac{\tau(\Omega)}{c\check{D}_N}$, denn dann ist

$$\varrho(T_0) = \frac{M(T_0)}{T_0} \leqq K_3 \operatorname{Min}(1, \check{D}_N(1 + \log^s \check{D}_N^{-1})).$$

T_0 hat eine einfache physikalische Bedeutung, es ist nämlich im wesentlichen die Relaxationszeit. Wenn L nicht 1 ist, dann ist

$$T_0 = \frac{L\tau(\Omega)}{c\check{D}_N}.$$

Es sei nun $\alpha > 0$ eine reelle Zahl und $\mathfrak{W}_\alpha^\pm(T)$ die Menge aller t im Intervall $\langle 0, \pm T\rangle$, wo $D_N(t) \geqq \alpha$. Wir wollen $\mathfrak{W}_\alpha^\pm(T)$ kurz die α-Ausnahmemenge nennen. Mit $\mu(\mathfrak{W}_\alpha)$ bezeichnen wir das Lebesguesche Maß von $\mathfrak{W}_\alpha$. Dann ist unmittelbar klar, daß

$$\frac{\mu(\mathfrak{W}_\alpha)}{T} \leqq \frac{\varrho(T)}{\alpha^2} = \frac{K_3}{\alpha^2}\left(\frac{\tau(\Omega)}{cT} + \check{D}_N(\log^{2s}\check{D}_N^{-1} + 1)\right) \tag{2}$$

ist.

Insbesondere ist für T_0

$$\mu(\mathfrak{W}_\alpha^\pm(T_0)) \leqq K_3 \frac{\tau(\Omega)}{c\alpha^2}(1 + \log^{2s}\check{D}_N^{-1}). \tag{2'}$$

Nehmen wir $\check{D}_N = 0(N^{-1/2} \log\log N)$, $\tau(\Omega) = 1$, $N = c^3$, $\alpha = c^{-1/3}$, dann ist $D_N(t) < c^{-1/3}$ im Intervall $[0, \sqrt{c}]$ ausgenommen einer Menge vom Maße $0(c^{-1/3}\log^6 c)$.

Ist z. B. $c = 10^4$, dann ist T_0 ungefähr $1\frac{1}{2}$ Stunden und $\mu(\mathfrak{W}_\alpha)$ ungefähr 0,1 Minuten.

Man kann $\mu(\mathfrak{W}_\alpha)$ auch nach unten abschätzen.

Es ist $\int\limits_{\mathfrak{W}_\alpha} D_N^2 = M(T) - \int\limits_{C(\mathfrak{W}_\alpha)} D_N^2 \geqq M(T) - \alpha^2 T$,

also

$$\mu(\mathfrak{W}_\alpha) \geqq M(T) - 2T\alpha^2.$$

Die Abschätzung ist nur brauchbar für kleine α.

Wir haben $\mathfrak{W}_\alpha^+$ und $\mathfrak{W}_\alpha^-$ gleichzeitig betrachtet, d. h., es ist auch in der Vergangenheit $D_N(T) < \alpha$ in $< -T, 0 >$, ausgenommen die Ausnahmemenge $\mathfrak{W}_\alpha^-$. Wir haben also zeitliche Symmetrie vor uns, wie es sein muß. Wir können mit Schrödinger sagen: Ist $t = 0$ ein Endzustand, dann ist anzunehmen, daß das System nicht aus einem noch abnormeren Zustand hervorgegangen ist, und zwar mit großer Wahrscheinlichkeit. Wir wollen nun weitere Folgerungen aus (1) ziehen.

Es sei f eine Funktion von beschränkter Variation $V(f)$ im Würfel K, dann ist

$$\left|\frac{1}{N}\sum_{k=1}^{N} f(x_k(t)) - 2^s \int_K f\,dx\right| \leqq V(f)\,D_N. \tag{3}$$

Diese Formel wurde für $s = 1$ von Koksma in Verallgemeinerung einer Formel von H. Behnke und im mehrdimensionalen Fall vom Verfasser aufgestellt. Es gilt also, daß für alle t in $\langle 0, T\rangle$ ausgenommen die Menge $\mathfrak{W}_\alpha$

$$\left|\frac{1}{N}\sum_{k=1}^{N} f(x_k(t)) - 2^s \int_K f\,dx\right| \leqq \alpha V(f). \tag{3'}$$

Dabei ist $\mathfrak{W}_\alpha$ von f unabhängig. Nehmen wir z. B. $s = 1, f = x$, dann ist $\int_K f\,dx = \frac{1}{8}$, und für den Schwerpunkt $S(t) = \frac{1}{N}\sum_{k=1}^{N} x_k(t)$ der N Teilchen gilt $|s(t) - \frac{1}{4}| < \alpha$ für alle t in $\langle 0, T\rangle$, ausgenommen in $\mathfrak{W}_\alpha$ (es ist $\frac{1}{4}$ der Mittelpunkt von K).

Man kann weiter folgendes zeigen: Es sei B eine konvexe Menge in K, also z. B. eine Kugel, dann gilt

$$\left|\frac{1}{N}\sum_{k=1}^{N} \chi_B(x_k(t)) - 2^s\, V(B)\right| \leqq 12^s \alpha^{\frac{1}{s}}$$

für alle t in $\langle 0, T\rangle$, wieder ausgenommen in $\mathfrak{W}_\alpha$, wo $\mathfrak{W}_\alpha$ unabhängig von B ist. Es ist χ_B die charakteristische Funktion von B, also $N_B = \sum_{k=1}^{N} \chi_B(x_k(t))$ die Anzahl der Teilchen B zur Zeit t. Betrachten wir den Ausdruck

$$H(B, t) = \frac{N_B}{2^s\, NV(B)} \log \frac{N_B}{2^s\, NV(B)}$$

und nennen ihn die lokale Entropie von Σ in B zur Zeit t. Dann zeigt man leicht, daß für alle t in $\langle 0, T\rangle$, ausgenommen in $\mathfrak{W}_\alpha$,

$$|H(B, t)| < \frac{\alpha^{\frac{1}{s}}}{2^s\, V(B)}\left(1 + \frac{\alpha^{\frac{1}{s}}}{2^s\, V(B)}\right)$$

gilt. Wenn also $H(B, 0) \neq 0$ und α genügend klein, dann ist für alle t in $\langle 0, T \rangle$, ausgenommen $\mathfrak{W}_\alpha$,

$$|H(B, t)| < H(B, 0).$$

Das heißt, die Entropie hat, ausgenommen in $\mathfrak{W}_\alpha$, im Vergleich zum Anfangszustand $t = 0$ abgenommen. Das gleiche gilt für $\langle -T, 0 \rangle$, wo jetzt 0 als Endzustand erscheint.

Wir beschränken uns jetzt auf $s = 1$ und $\Omega = \langle -c, c \rangle$ und führen $p_J(t) = \sum_{k=1}^{N} \chi_J(x_k(t))\, v_k^2$ ein und nennen p_J den Druck von Σ in J. Dann kann man zeigen, daß in $\langle 0, T \rangle$

$$\left| p_J(t) - \frac{2\, V(J)\, N c^2}{3} \right| < \gamma N c^2,$$

wo $\gamma \leqq K_4 (\alpha + \check{D}_N)$, ausgenommen wieder eine Ausnahmemenge $\mathfrak{W}_\alpha$.

Definieren wir als Dichte $\varrho(J, t) = \frac{1}{N} \sum_{k=1}^{N} \chi_J(x_k(t))$ in J und als Temperatur $T(J, t)$ von Σ in J:

$$T(J, t) = p(J, t) / k V(J) \qquad (k = \text{Boltzmannkonstante}),$$

dann ist $\frac{p}{kT} - \varrho = 0(\gamma)$ in $\langle 0, T \rangle$, ausgenommen in $\mathfrak{W}_\alpha$.

Dies können wir als Zustandsgleichung von Σ auffassen. Es werden also die Gleichungen des idealen Gases durch Ungleichungen ersetzt.

§ 5

Wir haben bisher nur ein System Σ betrachtet.

Fassen wir nun A Systeme $\Sigma_1, \ldots, \Sigma_A$ ins Auge, alle eingesperrt im Kasten K von der Länge $\frac{1}{2}$. Die Anzahl der Moleküle $X_j = (x_{j1}, \ldots, x_{jN})$ seien in jedem System Σ_j gleich N, aber die Anfangslagen seien $U_j = (u_{j1}, \ldots, u_{jN})$ und die Geschwindigkeiten $V_j = (v_{j1}, \ldots, v_{jN})$ für $1 \leqq j \leqq A$.

Wir betrachten nun $\mathfrak{S}(A) = \Sigma_1 \times \cdots \times \Sigma_A$, eingesperrt im Kasten K^N. Wir betrachten also X_j $(1 \leqq j \leqq A)$ als A Moleküle im K^N mit den Anfangslagen U_j und den Geschwindigkeiten V_j im Geschwindigkeitsraum R^{sN}.

Es sei $\Omega(A)$ wieder die konvexe Hülle von $V_1, \ldots, V_A$, und wir führen wie früher eine Diskrepanz $\check{\mathfrak{D}}(A)$ der Geschwindigkeiten ein. Wir können die früheren Überlegungen jetzt auf $\mathfrak{S}(A)$ eingesperrt in K^N anwenden.

Es sei nun m eine ganze Zahl mit $0 \leqq m \leqq N$.

Wir fragen nun: Für wie viele Systeme Σ_j ist für ein Intervall J von K

$$\sum_{k=1}^{N} \chi_J(x_{j,k}(t)) = m$$

zur Zeit t? Es sei $A_m(t)$ die Anzahl dieser Systeme.

Es ist, wenn $\exp \xi = e^{2\pi i \xi}$,

$$\frac{A_m(t)}{A} = \frac{1}{A} \sum_{j=1}^{A} \int_0^1 \exp\left(\beta\left(\sum_{k=1}^{N} \chi_J(x_{j,k}(t) - m)\right)\right) dt,$$

denn es ist für ganzes s doch $\int_0^1 \exp(\beta(s-m))\, d\beta = 1$, wenn $s = m$ und 0 sonst.

5 Nun ist $e(\beta)\,\chi_J(x) = (e(\beta) - 1)\,\chi_J(x) - 1$, also

$$f_\beta(x_1, \ldots, x_N) = e\left(\beta \sum_{k=1}^{N} \chi_J(x_k(t))\right) = \prod_{k=1}^{N} ((e(\beta) - 1)\,\chi_J(x_k(t)) - 1).$$

Wir wenden nun (3′) auf die Funktion f_β an, die alle damaligen Voraussetzungen erfüllt, die Variation ist $\leqq (\pi\beta)^N$, und haben für alle t in $[0, T]$, ausgenommen eine Menge $\mathfrak{W}_\alpha^-$ vom Maße $\varrho_A T/\alpha^2$, wo ϱ_A genauso wie in (1) definiert ist, nun angewendet auf $\mathfrak{S}(A)$,

$$\left|\frac{1}{A} \sum_{j=1} f_\beta(x_{j1}(t), \ldots, x_{jN}(t) - 2^{Ns} \int_{K^n} f_\beta(x_1, \ldots, x_N)\, dx_1 \ldots dx_N)\right| < \alpha(\pi\beta)^N.$$

Integriert man über β, so erhält man

$$\left|\frac{A_m(t)}{A} - 2^{Ns} \binom{N}{m} \left(\frac{1}{2^s} - V(J)\right)^{N-m} V(J)^m\right| < \alpha\pi^N \quad \text{für alle } t \text{ in } [0, T],$$

ausgenommen $\mathfrak{W}_\alpha^-$. Für $m = 0$ erhält man

$$\left|\frac{A_0}{A} - (1 - 2^s\, V(J))^N\right| < \alpha\pi^N.$$

Es sei $A_\eta(t)$ die Anzahl der Systeme, für die ein $\eta > 0$ existiert mit

$$\left|\frac{1}{N} \sum_{k=1}^{N} \chi_J(x_{j,k}(t)) - 2^s\, V(J)\right| < \eta \sqrt{\frac{pq}{N}},$$

wo $p = 2^s V(J)$, $q = 1 - 2^s V(J)$, so erhält man

$$\left| \frac{A_\eta(t)}{A} - \frac{2}{\sqrt{2\pi}} \int_0^\eta e^{-\frac{\xi^2}{2}} d\xi \right| < \Delta_J$$

für alle t in $[0, T]$, ausgenommen $\mathfrak{W}_\alpha^-$. Dabei ist

$$\Delta_J = \Delta_1 + \Delta_2,$$

wo $\Delta_1 = 0((Npq)^{-1/2})$, $\Delta_2 = 0(\alpha \pi^N (pN + \eta\sqrt{Npq})/A)$, wenn $\eta < \sqrt{\frac{Np}{q}}$, $Npq \leqq 25$.

Damit sind wir zu einer statistischen Betrachtung der Gesamtheit der Systeme Σ_j übergegangen.

Wir möchten zusammenfassend noch folgende Bemerkung machen:

Wir kennen viele Beispiele dafür, wie kausale Gesetze aus Gesetzen der Wahrscheinlichkeit folgen, z. B. die Formeln der Versicherungsmathematik oder die Gesetze der Thermodynamik als Folgerungen der statistischen Mechanik. Hier haben wir gezeigt, wie die Wahrscheinlichkeitsgesetze aus kausalen Gesetzen folgen. Es sind, mit Lipschutz-Yevick zu sprechen, Kausalität und Wahrscheinlichkeit nur Abstraktionen, zwei Gesichter ein und derselben Realität.

Literatur

Zur Ergodentheorie verweisen wir auf das Buch:

V. I. Arnold und A. Avez, Problèmes ergodiques de la Mécanique Classique, Gauthier Villard.

Zu den Untersuchungen von Ja. G. Sinaĭ vgl.:

Ja. G. Sinaĭ, Uspeki mat. Nauk 25, Nr. 2 (152) 141–192 (1970), welche das ebene Billiard behandelt.

E. Hlawka, Symposia Mathematica IV (Istituto Nazionale di Alta Matematica (1970) 81–97, und

Sitzungsberichte der Österr. Akademie der Wissenschaften 174 (1965) 287–307 und 178 (1969) 1–12.

M. Lipschutz-Yevick, Probability and Determinism, American Journal of Physics 25 (1957) 570–580,

gibt noch weitere interessante Beispiele zu dem in dieser Arbeit vertretenen Gesichtspunkt.

(Auf diese Arbeit von Lipschutz-Yevick hat mich P. Turan aufmerksam gemacht.)

Summary

In this paper an ideal gas is considered, where all particles are in a cube. No assumptions are made about the initial positions of the particles, but it is assumed that there are no collisions. It is shown that under general conditions on the initial velocities of the particles they are uniformly distributed in all convex domains of the cube during the whole time except from a "small" portion of it. In the classical kinetic gas theory this is a postulate.

Résumé

Dans cet article il est consideré un gaz ideal, dont tout les particles se trouvent dans un cube. Il n'y a pas d'assomptions concernant les positions initiales des particles, mais il est supposé qu'il n'y a pas des collisions. Il est démontré que sous conditions générales concernant les vélocités initiales des particles ils sont uniformement distribués dans toutes les domaines convexes du cube durant tout le temps sauf une «petite» partie de lui. Dans la théorie de gaz classique c'est un postulat.

Corrections

8.1 wenn $1, \omega_1, \cdots, \omega_n$ linear unabhängig über $\mathcal{Q}$ sind

12.2 so ist $cd = 2\mu$.

12.3 $M(T) = M(T, \Sigma) = \int_{-T}^{T} \check{D}_N^2(t)\,dt$

13.4 $, |p_j(h)|)$ und $hx =$

17.5 $e(\beta)$ steht für $\exp(\beta)$.

Anwendung zahlentheoretischer Methoden auf Probleme der numerischen Mathematik I

Sitzungsber., Abt. II, Österr. Akad. Wiss., Math.-Naturwiss. Kl. **184** (1975), 217–225

(Vorgelegt in der Sitzung der math.-nat. Klasse am 11. April 1975)

Wir wollen in einer Reihe von Arbeiten zeigen, wie zahlentheoretische Methoden, insbesondere die Theorie der Gleichverteilung, auf Probleme der numerischen Mathematik angewendet werden können. In der vorliegenden 1. Arbeit wird gezeigt, wie das Maximum einer stetigen positiven Funktion durch zahlentheoretische Methoden bestimmt werden kann.

Die Abfassung dieser Note wurde angeregt durch eine Unterhaltung mit Herrn Prof. Dr. Zinterhof (Universität Salzburg) am 19. März 1975 anläßlich eines gemeinsamen Besuches der IIASA in Laxenburg, dem ich für die Anregung herzlich danke.

Wir setzen in dieser Arbeit voraus, daß die zu untersuchende Funktion f differenzierbar ist. In einer Fortsetzung soll einerseits diese Voraussetzung abgeschwächt werden, anderseits sollen Suchmethoden zur Bestimmung von Extrema mit Hilfe der Gleichverteilung aufgestellt werden.

Für die Durchsicht der Arbeit möchte ich Frau Dr. Binder und Herrn Dr. Schoissengeier (Universität Wien) danken.

Wir nehmen zunächst an, es soll das Maximum μ einer stetigen Funktion f auf $\bar{I} = < 0,1 >$ bestimmt werden. Wir wollen noch annehmen, daß $f(0) = f(1)$ ist. Wir wollen dabei noch annehmen, daß sie differenzierbar auf dem offenen Intervall $I = >0,1<$ ist.

Es sei x_0 eine Stelle des Maximums von f auf $\bar{I}$. Es ist für jede natürliche Zahl p

$$\int_0^1 f^p(x)\,dx \leqslant f^p(x_0) = \mu^p. \tag{1}$$

Andererseits gilt folgendes: Es ist für jedes x in $\bar{I}$

$$|f(x) - f(x_0)| = \left|\int_x^{x_0} |f'(t)\,dt\right| \leqslant \int_x^{x_0} |f'(t)|\,dt \leqslant M\,|x - x_0|,$$

wo $M = \sup\limits_{t \in \hat{I}} |f'(t)|$ ist.

Es ist also

$$f(x) \geqslant f(x_0) - M\,|x - x_0|. \tag{2}$$

Wir wählen x so nahe an x_0, daß $|x - x_0| \leqslant \frac{f(x_0)}{M}$ ist. Es liegt also x im Intervall $A = <0,1> \cap < x_0 - \frac{f(x_0)}{M},\ x_0 + \frac{f(x_0)}{M} >$.

Es ist

$$\int_0^1 f^p(x)\,dx \geqslant \int_A f^p(x)\,dx \geqslant \int_A (f(x_0) - M\,|x - x_0|)^p\,dx.$$

Es sei nun B ein Teilintervall von A der Länge ξ, das x_0 enthält, wo $\xi \leqslant \frac{f(x_0)}{M}$ und auch $\xi < 1$ ist. Es ist

$$\int_I f^p\,dx \geqslant \int_B (f(x_0) - M\,|x - x_0|)^p\,dx \geqslant (f(x_0) - M\,\xi)^p \int_B dx \geqslant$$
$$\geqslant (f(x_0) - M\,\xi)^p\,\xi.$$

Es ist natürlich $0 < \xi < 1$, da wir $f(x_0) > 0$ annehmen. Weiters ist $\xi \leqslant \frac{f(x_0)}{M}$.

Es ist also

$$0 \leqslant f(x_0) - \left(\int f^p\right)^{\frac{1}{p}} \leqslant f(x_0) - (f(x_0) - M\,\xi)\,\xi^{\frac{1}{p}}. \tag{3}$$

Es ist $\xi^{\frac{1}{p}} \leqslant \dfrac{1}{1+\frac{1}{p}\log\frac{1}{\xi}}$, denn es ist $\xi^{\frac{1}{p}} = e^{\frac{1}{p}\log\xi} = e^{-\frac{1}{p}\log\frac{1}{\xi}}$.

Nun ist $e^{\alpha} \geqslant 1 + \alpha$ für $\alpha \geqslant 0$, also $e^{\frac{1}{p}\log\xi} \geqslant \dfrac{1}{1+\frac{1}{p}\log\frac{1}{\xi}}$, also ist

$$f(x_0) - (f(x_0) - M\,\xi)\,\xi^{\frac{1}{p}} \leqslant f(x_0) - \frac{f(x_0) - M\xi}{1+\frac{1}{p}\log\frac{1}{\xi}} \leqq$$

$$\leqq f(x_0)\frac{\frac{1}{p}\log\frac{1}{\xi}}{1+\frac{1}{p}\log\frac{1}{\xi}} + M\xi \leqslant \frac{f(x_0)}{p}\log\frac{1}{\xi} + M\xi,$$

also mit passendem ϑ_1, wo $|\vartheta_1| < 1$

$$f(x_0) = \left(\int f^p\right)^{\frac{1}{p}} + \left(M\xi + \frac{f(x_0)}{p}\log\frac{1}{\xi}\right)\vartheta_1. \tag{4}$$

Nun sei (x_n) eine Folge mit Diskrepanz D_N, dann ist ($|\vartheta_2| < 1$)

$$\int f^p = \frac{1}{N}\sum_{n=1}^{N} f^p(x_n) + \vartheta_2 D_N p \int_0^1 |f'|\,f^{p-1}\,dx.$$

Nun ist $\int\limits_0^1 |f'\,f^{p-1}|\,\mathrm{d}x \leqslant M\int\limits_0^1 f^{p-1}\,\mathrm{d}x \leqslant M\,f^{p-1}(x_0)$, also ist

$$\left(\int f^p\right)^{\frac{1}{p}} = \left(\frac{1}{N}\sum_{n=1}^{N} f^p(x_n)\right)^{\frac{1}{p}} + D_N^{\frac{1}{p}}\,p^{\frac{1}{p}}\,f^{1-\frac{1}{p}}(x_0)\,M^{\frac{1}{p}}\,\vartheta_2,$$

also

$$f(x_0) = \left(\frac{1}{N}\sum_{n=1}^{N} f^p(x_n)\right)^{\frac{1}{p}} + \left(M\xi + \frac{f(x_0)}{p}\log\frac{1}{\xi}\right)\vartheta_1 + D_N^{\frac{1}{p}}\,p^{\frac{1}{p}}\,f^{1-\frac{1}{p}}(x_0)\,M^{\frac{1}{p}}\,\vartheta_2. \tag{5}$$

Wir nehmen nun

$$p = \operatorname{Max}\left(\left[\frac{lg\,\frac{1}{D_N}}{lg\,lg\,\frac{1}{D_N}}\right] + 1, 2\right) \text{ und } \xi = \frac{1}{p} \leqslant \frac{lg\,lg\,\frac{1}{D_N}}{lg\,\frac{1}{D_N}}. \tag{*}$$

Dann erhalten wir

1 $$0 \leqslant f(x_0) - \left(\frac{1}{N}\sum_{n=1}^{N} f^p(x_n)\right)^{\frac{1}{p}} +$$

$$+ \vartheta_3 \frac{lg\,lg\,\frac{1}{D_N}}{lg\,\frac{1}{D_N}}\left(M + f(x_0)\,\frac{lg\,lg\,\frac{1}{D_N}}{lg\,lg\,\frac{1}{D_N}}\right) + \frac{3}{lg\,\frac{1}{D_N}} f(x_0)^{1-\frac{1}{p}} M^{\frac{1}{p}}. \tag{6}$$

Die Schranke hängt noch von $f(x_0)$ ab.
Es ist stets

$$f(x_0) \leqslant \int_0^1 f\,\mathrm{d}x + M = K. \tag{7}$$

Beweis: Ist $f(x_0) \leqslant M$, so ist es richtig, wenn nicht, so ist

$$f(x_0) > M, \text{ also für alle } x \;\; f(x) \geqslant f(x_0) - M\,(x - x_0),$$

also

$$\int_0^1 f(x)\,\mathrm{d}x \geqslant f(x_0) - M \int_I (x - x_0)\,\mathrm{d}x \geqslant f(x_0) - M,$$

also, wenn wir $J = \int_0^1 f\,\mathrm{d}x$ setzen, so ist

$$f(x_0) \leqslant J + M \text{ Schranke für } f(x_0).$$

Man kann J abschätzen durch $J = \frac{1}{N}\sum_{n=1}^{N} f(x_n) + \vartheta_4 D_N \int_0^1 |f'(x)|\,\mathrm{d}x$,

also

$$f(x_0) \leqslant M + \frac{1}{N}\sum_{n=1}^{N} f(x_n) + M\,D_N.$$

Wenn $f(x_0) \geqslant M$, dann ist $\dfrac{M}{f(x_0)} \leqslant 1$, also haben wir

$$f(x_o) - \left(\frac{1}{N}\sum_{n=1}^{N} f^p(x_n)\right)^{\frac{1}{p}} \leqq \frac{\left(lg\, lg \frac{1}{D_N}\right)^2}{lg \frac{1}{D_N}}(M+K) + \frac{3K}{lg \frac{1}{D_N}} \leqq$$

2

$$\leqq \left(\frac{1}{N}\sum_{n=1}^{N} f^p(x_n)\right)^{\frac{1}{p}} + \frac{\left(lg\, lg \frac{1}{D_N}\right)^2}{lg \frac{1}{D_N}}(M+4K)$$

Wenn $f(x_0) \leqslant M$, dann haben wir

$$f(x_o) \leqq \left(\frac{1}{N}\sum_{u=1}^{N} f^p(x_n)\right)^{\frac{1}{p}} + 2M\frac{\left(lg\, lg \frac{1}{D_N}\right)^2}{lg \frac{1}{D_N}} + \frac{3}{lg \frac{1}{D_N}} M^{1-\frac{1}{p}} M^{\frac{1}{p}},$$

$$\text{also} \qquad \mu \leqslant \left(\frac{1}{N}\sum_{n=1}^{N} f^p(x_n)\right)^{\frac{1}{p}} + 5M\frac{\left(lg\, lg \frac{1}{D_N}\right)^2}{lg \frac{1}{D_N}}. \tag{8}$$

Wir haben also Satz 1:

$$|\mu - \mu_p| \leqq 5\, M\, V \quad \text{wo} \quad \mu_p = \left(\frac{1}{N}\sum_{n=1}^{N} f^p(x_n)\right)^{\frac{1}{p}}$$

mit p aus (*) und

$$V = \left(\log\log \frac{1}{D_N}\right)^2 \Big/ \log \frac{1}{D_N}.$$

Jetzt betrachten wir den s-dimensionalen Fall: Es ist wieder

$$\int_{I^s} f^p(x)\, \mathrm{d}x \leqslant f^p(x_0) = \mu^p.$$

Andererseits ist

$$f(x) - f(x_0) = -\int_0^1 \frac{df}{dt}(x_0 + t(x - x_0))\, \mathrm{d}t = \int_0^1 \operatorname{grad} f(x - x_0)\, \mathrm{d}t,$$

also

$$|f(x) - f(x_0)| \leqslant |x - x_0| \int_0^1 |\mathrm{grad}\, f|\, dt,$$

also, wenn $M_1 = \sup |\mathrm{grad}\, f|$,

$$|f(x) - f(x_0)| \leqslant s\, M_1 \|x - x_0\|, \text{ wobei } \|y\| = \mathrm{Max}\,(|y_1|, \ldots, |y_s|),$$

also ist

$$f(x) \geqslant f(x_0) - s\, M_1 \|x - x_0\|. \tag{9}$$

Es sei nun $A = I^s \cap$ Würfel mit Mittelpunkt x_0 und Kantenlänge $\dfrac{f(x_0)}{s M_1}$.

Es ist, wenn $B: \|x - x_0\| < \xi$ in A liegt,

$$\int_{I^s} f^p\, dx \geqslant \int_B (f(x_0) - s\, M_1 \|x - x_0\|)^p\, dx \geqslant (f(x_0) - s\, M_1\, \xi)^p\, \xi^s.$$

Es ist also

$$f(x_0) - (\textstyle\int f^p)^{\frac{1}{p}} \leqslant f(x_0) - (f(x_0) - s\, M_1\, \xi)^s\, \xi^{\frac{s}{p}}. \tag{10}$$

Es ist wieder $\quad \xi^{\frac{s}{p}} = e^{-\frac{s}{p} lg \frac{1}{\xi}} \leqslant \dfrac{1}{1 + \frac{s}{p}\, lg \frac{1}{\xi}}$,

also

$$f(x_0) - (\textstyle\int f^p)^{\frac{1}{p}} \leqslant f(x_0) - \dfrac{(f(x_0) - sM\xi)}{1 + \frac{s}{p}\, lg \frac{1}{\xi}} = \vartheta_5 \left(\dfrac{s f(x_0)\, lg \frac{1}{\xi}}{p} + sM\xi \right). \tag{11}$$

Nun ist

$$\int f^p = \frac{1}{N} \sum_{n=1}^{N} f^p(x_n) + \vartheta_6 D_N \left(\int_{I^s} \left| \frac{\partial^s f^p}{\partial x_1 \ldots \partial x_s} \right| dx_1 \ldots dx_s + R \right), \tag{12}$$

wo R Glieder enthüllt, in welchen nur Ableitungen von f vorkommen, deren Ordnung $< s$ ist (vgl. das Buch von Kuipers und Niederreiter, Uniform distribution of sequences, Wiley 1974, S. 153).

Nun ist

$$\frac{\partial^s f^p}{\partial x_1 \dots \partial x_s} = p(p-1)\dots(p-s+1) f^{p-s} \frac{\partial f}{\partial x_1} \dots \frac{\partial f}{\partial x_s} +$$

$$+ p(p-1)\dots(p-s+2) f^{p-s+1} \left(\frac{\partial^2 f}{\partial x_1 \partial x_2} \frac{\partial f}{\partial x_3} \dots \right) +$$

$$+ \dots + p f^{p-1} \frac{\partial^s f}{\partial x_1 \dots \partial x_s}.$$

Wir wollen nun voraussetzen, daß $\operatorname{Max}\left(\left| \frac{\partial f}{\partial x_1} \right|, \dots, \left| \frac{\partial^s f}{\partial x_1 \dots \partial x_s} \right| \right) \leqslant M$,

dann ist

$$\left| \frac{\partial^s f^p}{\partial x_1 \dots \partial x_s} \right| \leqslant p^s [f^{p-s}(x_0) M^s + f^{p-s+1}(x_0) M^{s-1} \tbinom{s}{2} +$$

$$+ f^{p-s+2}(x_0) M^{s-2} \tbinom{s}{3} + \dots + f^{p-s+j}(x_0) M^{s-j} \tbinom{s}{j+1} +$$

$$+ \dots + f^{p-1}(x_0) M].$$

Es ist wieder

$$f(x_0) \leqslant s M + \int_{I^s} |f| \, \mathrm{d}x = K. \tag{13}$$

Beweis: Ist $f(x_0) < M s$, dann ist (13) richtig.

Es sei $f(x_0) > M s$, dann ist, wenn $p_k(x)$ die k-te Koordinate von x ist,

$$f(x) \geqslant f(x_0) - M \sum_k p_k(x - x_0) \geqslant f(x_0) - M s,$$

also ist $\int_{I^s} f \geqslant f(x_0) - M s$ und damit ist (13) bewiesen.

Es sei nun $f(x) > M s \geqslant M$, dann ist

$$\left| \frac{\partial^s f^p}{\partial x_1 \dots \partial x_s} \right| \leqslant p^s f^p(x_0) \left[\left(\frac{M}{f(x_0)} \right)^s + \binom{s}{2} \left(\frac{M}{f(x_0)} \right)^{s-1} + \dots \right.$$

$$\left. \dots + \frac{M}{f(x_0)} \right] \leqslant p^s (2^s - 1) K^p.$$

Wenn $f(x_0) \leqslant M s$, dann ist $f^{p-s+j} \leqslant s^{p-s+j} M^{p-s+j}$, also ist

$$\left|\frac{\partial^s f^p}{\partial x_1 \ldots \partial x_s}\right| \leqslant p^s M^p \left[s^{p-s} + \binom{s}{2} s^{p-s+1} + \ldots + s^{p-1}\right] \leqslant$$
$$\leqslant p^s s^{p-1} (2^s - 1) M^p,$$

also stets

$$\left|\frac{\partial^s f^p}{\partial x^1 \ldots \partial x_s}\right| \leqslant p^s s^p (2^s - 1) K^p. \tag{14}$$

Es ist also

$$\left(\int f^p\right)^{\frac{1}{p}} \leqslant \left(\frac{1}{N} \sum_{n=1}^{N} f^p(x_n)\right)^{\frac{1}{p}} + D_N^{\frac{1}{p}} K p^{\frac{s}{p}} s (2^s - 1)^{\frac{1}{p}}, \tag{15}$$

also

$$0 \leqslant \left(f(x_0) - \left(\frac{1}{N} \sum_{n=1}^{N} f^p(x_n)\right)^{\frac{1}{p}}\right) \leqslant s \left[\frac{K \lg \frac{1}{\xi}}{p} + M\xi + D_N^{\frac{1}{p}} K p^{\frac{s}{p}} (2^s - 1)^{\frac{1}{p}}\right]$$
$$\leqslant sK \left[\frac{\lg \frac{1}{\xi}}{p} + \xi + D_N^{\frac{1}{p}} e^{s \frac{\log p}{p}} (2^s - 1)^{\frac{1}{p}}\right].$$

Wir nehmen wieder $\xi = \frac{1}{p} \mid p = \operatorname{Max}\left(\left[\frac{\lg \frac{1}{D_N}}{\lg \lg \frac{1}{D_N}}\right], s\right)$,

dann ist

$$\left(f(x_0) - \left(\frac{1}{N} \sum_{n=1}^{N} f^p(x_n)\right)^{\frac{1}{p}}\right) \leqslant sK \frac{\left(\lg \lg \frac{1}{D_N}\right)^2}{\lg \frac{1}{D_N}} \left(3 + e^{s \frac{\log p}{p}} \left(2^s - 1\right)^{\frac{1}{p}}\right),$$

also

$$\mu \leqslant sK \frac{\left(\lg \lg \frac{1}{D_N}\right)^2}{\lg \frac{1}{D_N}} (3 + 3s) + \mu_p. \tag{16}$$

Es muß aber sein $\xi \leqslant \mathrm{Min}\left(\frac{1}{p}, 1, \frac{f(x_0)}{sM}\right)$. Es ist wieder

$$\mu_p = \left(\frac{1}{N}\sum_{n=1}^{N} f^p(x_n)\right)^{\frac{1}{p}}.$$

$$\text{Es muß also } \frac{lg\, lg \frac{1}{D_N}}{lg \frac{1}{D_N}} \leqslant \frac{f(x_0)}{sM} \text{ sein.}$$

Ist aber $\frac{f(x_0)}{sM} \leqslant \frac{lg\, lg \frac{1}{D_N}}{lg \frac{1}{D_N}}$, so gilt (16) auch, denn dann ist

$$\mu = f(x_0) \leqslant \frac{Ms\, lg\, lg \frac{1}{D_N}}{lg \frac{1}{D_N}}.$$

Wir haben also Satz 2:

$$|\mu - \mu_p| \leqslant R$$

wo

$$R = 3s(1+s)KU^2$$

$$U = \log\log \frac{1}{D_N} \Big/ \log \frac{1}{D_N}$$

$$p = \mathrm{Max}\left(\left[\frac{1}{U}\right], s\right)$$

und

$$\mu_p = \left(\frac{1}{N}\sum_{n=1}^{N} f^p(x_n)\right)^{\frac{1}{p}} \text{ ist.}$$

Literatur

Wilde-Beightler: Foundations of Optimation, S. 236ff. Prentice-Hall 1976.

Corrections

220.1 $0 \le f(x_0) - \left(\frac{1}{N}\sum_{n=1}^{N} f^p(x_n)\right)^{\frac{1}{p}} \le$

$\le \vartheta_3 \frac{lg\, lg \frac{1}{D_N}}{lg \frac{1}{D_N}}\left(M + f(x_0)\, lg\, lg \frac{1}{D_N}\right) + \frac{3}{lg \frac{1}{D_N}} f(x_0)^{1-\frac{1}{p}} M^{\frac{1}{p}}.$

221.2 $\le \frac{(lg\, lg \frac{1}{D_N})^2}{lg \frac{1}{D_N}}(M + 4K)$

Über die Gleichverteilung gewisser Folgen, welche mit den Nullstellen der Zetafunktion zusammenhängen

Professor C. L. Siegel zum 80. Geburtstag in Verehrung gewidmet

Sitzungsber., Abt. II, Österr. Akad. Wiss., Math.-Naturwiss. Kl. **184** (1975). 459–471

(Vorgelegt in der Sitzung der math.-nat. Klasse am 7. November 1975 durch das w. M. E. Hlawka)

Es sei $(\rho_n) = (\beta_n + i\,\gamma_n)$ die Folge der Nullstellen der Riemannschen Zetafunktion, dann hat Rademacher [1] bemerkt, daß unter Annahme der Riemannschen Vermutung (kurz mit R. V. zitiert) für jedes positive z die Folge $\frac{1}{2\,\pi}\,\gamma_n \log z$ modulo 1 gleichverteilt ist. Besonders wichtig ist der Fall, daß z von der Gestalt q^k ist, wo q eine Primzahl und k eine natürliche Zahl ist. Der Beweis dieser Bemerkung stützt sich auf das Resultat von E. Landau [2] 1911, daß für $x > 1$ und $x = q^k$

$$\Sigma^*\, x^{\rho_n} = -\frac{T}{2\,\pi} \log q + 0\,(\log T) \tag{1}$$

ist, und wenn x keine Potenz einer Primzahl ist

$$\Sigma^*\, x^{\rho_n} = 0\,(\log T), \tag{2}$$

gilt Σ^* bedeute stets die Summation $\sum\limits_{0<\gamma_n\leq T}$.

Dabei hängt die 0-Abschätzung von x ab. Wenn $0 < x < 1$ ist und wenn x von der Gestalt q^{-k} ($k \geqslant 1$, q prim) ist, so ist das erste Glied in (1) mit x zu multiplizieren.

Bekanntlich ist [3] $N(T) = \sum_{0<\gamma_n\leq T} 1 = (2\pi)^{-1} T \log T + 0(T)$.

Wir wollen nun als erstes bemerken, daß die Bemerkung von Rademacher richtig bleibt, wenn wir *nicht* voraussetzen, daß R. V. gilt. Dazu benützen wir das Resultat von A. Selberg (vgl. [3], S. 208, Theorem 924), daß $\sum^* \left|\beta_n - \frac{1}{2}\right| = 0(T)$ ist.

Dann folgt aus (1) und (2)

$$\frac{1}{N(T)} \sum{}^* x^{\beta_n + i\gamma_n} = 0\left(\frac{1}{\log T}\right).$$

Andererseits ist

$$\left|\sum{}^* x^{(\frac{1}{2} + i\gamma_n} - x^{\beta_n + i\gamma_n)}\right| \leqslant \sum{}^* \left|x^{\frac{1}{2}} - x^{\beta_n}\right|.$$

Nun ist, da ja stets $|\beta_n| < 1$,

$$|x^{\frac{1}{2}} - x^{\beta}| \leqslant x^{\beta} \log x \left|\beta - \frac{1}{2}\right| \leqslant x \log x \left|\beta - \frac{1}{2}\right| \tag{3}$$

also

$$\sum{}^* |x^{\frac{1}{2}} - x^{\beta_n}| = 0\left(\sum{}^* \left|\beta_n - \frac{1}{2}\right|\right) = 0(T),$$

also ist

$$\frac{1}{N(T)} \sum{}^* x^{\frac{1}{2} + i\gamma_n} = 0\left(\frac{1}{\log T}\right)$$

und

$$\frac{1}{N(T)} \sum{}^* x^{i\gamma_n} = 0\left(\frac{1}{\log T}\right).$$

Es sei nun z reell und > 1. Setzt man $x = z^h$ für $h \geqslant 1$, so folgt nach dem Weylschen Kriterium

(wo $e(\alpha) = e(2\pi i \alpha)$, $e(x) = e^x$),

daß für die Folge $\xi(z) = (\xi_n(z))$

$$\xi_n(z) = \frac{1}{2\pi}\gamma_n \log z \quad \text{mit} \quad 0 < \gamma_n \leqslant T$$

$$\frac{1}{N(T)}\Sigma^* e(h\xi_n) = 0\left(\frac{1}{\log T}\right)$$

ist, also ist die Folge $\xi(z)$ gleichverteilt modulo 1.

Es scheint noch nicht bemerkt worden zu sein, daß auch die s-dimensionale Folge

$$\xi(z) = (\xi_n^1, \ldots, \xi_n^s), \quad \text{wo} \quad \xi_n^j = \frac{1}{2\pi}\gamma_n \log z_j$$

und $z_1, \ldots, z_s$ multiplikativ unabhängig sind, ebenfalls gleichverteilt modulo 1 ist.

Wir brauchen für ganze Zahlen $h_1, \ldots h_s$, wo $h = (h_1, \ldots, h_s) \neq \neq (0 \ldots 0)$ ist, nur

$$x = z_1^{h1} \ldots z_s^{hs}$$

zu nehmen. Es ist sicher $x \neq 1$, denn das besagt ja gerade die multiplikative Unabhängigkeit von $z_1, \ldots, z_s$. Es ist keine Beschränkung anzunehmen, daß $x > 1$ ist, und wir erhalten wieder, daß

$$\frac{1}{N(T)}\Sigma^* e(< h\xi >) = 0\left(\frac{1}{\log T}\right)$$

ist. (Es ist $< h\,\xi > = \sum_{j=1}^{s} h_j\,\xi^j$).

Ein Beispiel wäre $z_j = q_j$, wo $q_1, \ldots, q_s$ verschiedene Primzahlen sind.

Es sei noch bemerkt, daß wir statt dem Selbergschen Resultat auch das schwächere Resultat von Littlewood (vgl. [3], S. 204), nämlich

$$\Sigma^* \left|\beta_n - \frac{1}{2}\right| = 0\,(T \log\log T),$$

hätten nehmen können.

Wir wollen nun die Diskrepanz D_T der Folge $\xi(z)$ abschätzen, und zwar beschränken wir uns zunächst auf den Fall, daß z eine natürliche Zahl und $s = 1$ ist.

Wir werden zeigen, daß unter Annahme von R. V.

$$D_T(\xi(z)) \leqslant C \frac{\log z}{\log T} \tag{4}$$

ist, wo C die absolute Konstante ist. Allgemein gilt nur

$$D_T = 0\left(\frac{\log z}{\log\log T}\right).$$

Dabei ist

$$D_T(\xi) = \sup_J \left| \frac{1}{N(T)} \Sigma^* i_J(\xi_n) - \lambda(J) \right|$$

wo J ein Teilintervall von $I = \langle 0, 1 \langle$ ist, i_J seine Indikatorfunktion und $\lambda(J)$ seine Länge ist (im s-dimensionalen Fall ist J Teilintervall von I^s und $\lambda(J)$ sein s-dimensionales Volumen). Wir stützen uns bei der Berechnung von D_T auf die Formel [4] von Erdös-Turan-Koksma

$$D_T(\xi) \leqslant C_s\left(\frac{1}{M} + \sum_{\|h\| \leq M} R^{-1}(h) \, | \, \hat{\Sigma}(h) \, | \right), \tag{5}$$

wo

$$\hat{\Sigma}(h) = \frac{1}{N(T)} \Sigma^* e(\langle h\, \xi_n \rangle) \tag{6}$$

und für den Gittervektor $h = (h_1 \ldots h_s)$

$$R(h) = \prod_{j=1}^{s} \operatorname{Max}(|h_j|, 1)$$

und $\|h\| = \operatorname{Max}(|h_j|)$ ist. Dabei ist $M \geqslant 1$ eine noch wählbare natürliche Zahl.

Wir müssen nun die Abschätzungen (1) und (2) bei Landau genauer verfolgen. Landau geht dabei von folgender Formel aus:

Es sei $\sum_{n=1}^{\infty} \frac{a_n}{n^s}$ für $\sigma = 2$ absolut konvergent, dann gilt für die zugehörige Funktion f

$$\left| \int_{2+\gamma i}^{2+Ti} x^s f(s)\, ds - 2\pi i h(x, T) \right| \leqslant P(x), \tag{7}$$

wo $h(x, T) = 0$, wenn $x > 0$ und x nicht ganz ist, aber $h(x, T) =$

1 $= \frac{1}{2\pi}(T - \gamma)\, a^x$ ist, wenn $x > 0$ und x ganz ist. Dabei ist

$$P(x) = 2x^2 \sum_{n=1}^{\infty}{}^{*} \frac{|a_n|}{n^2 \left| \log \frac{x}{n} \right|} \tag{8}$$

und die Summation in (8) bezieht sich auf alle $n \neq x$ (γ ist eine Zahl, welche $< \inf \gamma_n$ ($\gamma_n > 0$) und zwischen 0 und 2 liegt).

Es wird nun (7) auf

$$f(s) = \frac{\zeta'(s)}{\zeta(s)} = \sum_{p,\, m} \frac{\log p}{p^{ms}}$$

angewendet. Es ist also

$$P_1(x) = 2\, x^2\, P_2(x) \tag{8'}$$

und

$$P_2(x) = \sum_{p^m \neq x} p^{-2m} \left| \log \frac{x}{p^m} \right|^{-1} \log p \tag{9}$$

abzuschätzen. Dabei wollen wir nun x als *natürliche* Zahl annehmen. Wir zerlegen P_2 in vier Summen

$$P_2 = \Sigma_1 + \Sigma_2 + \Sigma_3 + \Sigma_4, \tag{10}$$

wo

$$\Sigma_1 = \sum_{p^m > 2x} \leqslant C_1 \sum_{p^m \geq 2x} p^{-2m} \leqslant C_1 \sum_{n > x2} \frac{\log n}{n^2} \leqslant \frac{C_2}{x} \log x$$

ist, wo $C_1 = \frac{1}{\log 2}$ ist.

Weiters ist

$$\Sigma_4 = \sum_{p^m < \frac{x}{2}} \leqslant C_1 \sum_{p^m < \frac{x}{2}} \frac{\log p}{p^{2m}} \leqslant C_1 \sum_{n=1}^{\infty} \frac{\log n}{n^2} = C_2.$$

Es sei $\Sigma_2 = \sum\limits_{x < p^m < 2x}$. Setzen wir $\frac{p^m}{x} = 1 + \alpha$, so ist also $0 < \alpha < 1$.

Nun ist aber $1 + \alpha \geqslant e^{\frac{\alpha}{2}}$. Es ist also

$$\log \frac{p^m}{x} = \log (1 + \alpha) \geqslant \frac{\alpha}{2},$$

also

2 $$\Sigma_2 \leqslant 2 \sum_{x < p^m < 2x} \log p^{-2m} \left(\frac{p^m}{x} - 1\right)^{-1} = 2x \sum_{x < p^m < 2x} p^{-2m} (p^m - x)^{-1} \log p.$$

Da nun x eine natürliche Zahl ist, so ist $p^m - x \geqslant 1$ und wir haben

3 $$\Sigma_2 \leqslant 2x \sum_{x < p^m < 2x} \log p^{-2m} \leqslant C x \sum_{x < p < 2x} \frac{\log p}{p^2} \leqslant C_4.$$

Analog ist

4 $$\Sigma_3 = \sum_{\frac{x}{2} < p^m < x} \leqslant x \sum_{\frac{x}{2} < p^m < x} \log p^{-m} (x - p^m)^{-1} \leqslant C_4.$$

Wir haben also insgesamt

$$P_2(x) \leqslant C_5,$$

also ist

$$P_1(x) \leqslant 2 C_5 x^2. \tag{11}$$

Nach (7) ist bei uns

$$h(x, T) = -\frac{1}{2\pi} (T - \gamma) \log q,$$

falls $x = q^k$, wo q prim ist und 0 sonst. Es ist also

$$\int_{2+\gamma i}^{2+Ti} x^s \frac{\zeta'(s)}{\zeta(s)} \, ds = 2\pi i \, h(x, T) + 0(x^2),$$

wo jetzt 0 nur absolute Konstanten enthält.

Jetzt wird der Cauchysche Integralsatz auf das Integral

$$\int x^s \frac{\zeta'(s)}{\zeta(s)} \, ds$$

erstreckt über das Rechteck $2 + \gamma i$, $2 + T i$, $\alpha + T i$, $\alpha + \gamma i$ angewendet, wo $\alpha < 0$ ist. Weiters ist $T \geqslant 2$ und von allen γ verschieden angenommen.

Es kommt dann nach dem Residuensatz

$$J = \int_{2+\gamma i}^{2+Ti} = \int_{2+\gamma i}^{\alpha+\gamma i} + \int_{\alpha+\gamma i}^{\alpha+Ti} + \int_{\alpha+Ti}^{2+Ti} + 2\pi i \sum{}^{*} x^{\rho_n}.$$

Nun zeigt man, daß die drei Integrale rechts für $\alpha \to -\infty$ einen Limes besitzen, wenn $x > 11$. Es ist also

$$J = -\int_{-\infty+\gamma i}^{2+\gamma i} + \int_{-\infty+Ti}^{2+Ti} + 2\pi i \sum{}^{*} x^{\rho_n}.$$

Es ist also

$$\sum{}^{*} x^{\rho_n} = h(\gamma, T) - \frac{1}{2\pi i} \int_{-\infty+Ti}^{2+Ti} x^s \frac{\zeta'(s)}{\zeta(s)} \, ds + 0(x^2).$$

Es ist

$$\int_{-\infty+Ti}^{2+Ti} = \int_{-\infty+Ti}^{-1+Ti} + \int_{-1+Ti}^{2+Ti}.$$

Es ist

$$\left| \int_{-\infty+Ti}^{-1+Ti} x^s \frac{\zeta'(s)}{\zeta(s)} \, ds \right| \leqslant C_8 \frac{\log T}{x}.$$

Weiters ist

$$\int_{-1+Ti}^{2+Ti} x^s \frac{\zeta'(s)}{\zeta(s)} \, ds - \int_{-1+Ti}^{2+Ti} x^s \, ds \sum_{|\gamma_n - T| < 1} \frac{1}{s - \rho_n} = 0(x^2 \log T).$$

Wird nun in jedem Integral $J_n = \int\limits_{-1+Ti}^{2+Ti} \frac{x^s}{s - \rho_n} ds$ der gerade Weg durch einen Halbkreis über ihm nach oben oder unten ersetzt, je nachdem, ob $T - 1 < \gamma_n < T$ oder $T < \gamma_n < T + 1$ ist, so erhält man für jedes $J_n = 0\,(x^2)$. Da wir im ganzen $0\,(\log T)$ Glieder haben, erhalten wir endgültig

$$\sum{}^* x^{\rho_n} = - T\,\psi(x) + 0\,(x^2 \log T), \tag{12}$$

wo

$$\psi(x) = + \frac{1}{2\pi} \log q,$$

wenn $x = q^k$, und $\psi(x) = 0$, wenn $x \neq q^k$ ist.

Prof. P. Turan hat bemerkt, daß man überall x^2 durch $x \log^2 x$ ersetzen kann, wenn man längs der Geraden $Rs = 1 + \frac{1}{\log x}$ integriert. Ich danke Prof. Turan für seine freundlichen Bemerkungen.

Dabei ist x als natürliche Zahl vorausgesetzt.

Benützen wir (3) und das Selbergsche Resultat, dann erhalten wir

$$\sum{}^* x^{\frac{1}{2} + i\gamma_n} = - T\,\psi(x) + 0\,(x^2\,T),$$

also

$$\sum{}^* x^{i\gamma_n} = - T \frac{\psi(x)}{x^{\frac{1}{2}}} + 0\,(x^{\frac{3}{2}}\,T). \tag{13}$$

Wir setzen nun $x = z^h$, wo h eine natürliche Zahl ist, dann erhalten wir

$$\hat{\Sigma}(h) = - T \frac{\psi(z^h)}{N(T)\,z^{\frac{h}{2}}} + 0\left(\frac{z^{\frac{3h}{2}}}{\log T}\right) \tag{14}$$

bzw., wenn R. V. gilt,

$$\hat{\Sigma}(h) = - T \frac{\psi(z^h)}{N(T)\,z^{\frac{h}{2}}} + 0\left(\frac{z^{\frac{3h}{2}}}{T}\right). \tag{14'}$$

Nehmen wir also den Fall $s = 1$, dann erhalten wir

$$D_T(\xi(z)) \leqslant C_1\left(\frac{1}{M} + 0\left(\sum_{h=1}^{M} \frac{z^{\frac{3h}{2}}}{h \log T}\right)\right).$$

Nun ist

$$\sum_{h=1}^{M} \frac{z^{\frac{3h}{2}}}{h} = 0\left(\frac{z^{\frac{3M}{2}}}{M}\right),$$

also ist, wenn wir $M = \left[\frac{\log\log T}{\log z}\right] + 1$ nehmen,

$$D(\xi) = 0\left(\frac{\log z}{\log\log T}\right). \tag{15}$$

Wenn aber (14′) gilt, so nehmen wir $M = \left[\frac{\log T}{\log z}\right] + 1$ und erhalten (4).

Dabei haben wir berücksichtigt, daß, wenn

$$z = q^k, \quad \text{ja} \quad T\,\psi(z^h) \,/\, (N(T)\, z^{\frac{h}{2}}) = 0\,(\log^{-1} T) \quad \text{ist.}$$

Wir betrachten nun den mehrdimensionalen Fall, wo jetzt wieder $z_1, \ldots z_s$ natürliche Zahlen sein sollen, welche multiplikativ unabhängig sind. Wir setzen für $h = (h_1 \ldots h_s) \neq 0$

$x(h) = z_1^{h_1} \ldots z_s^{h_s}$, wo wir annehmen können, daß $x > 1$ ist.

Wir erhalten dann analog

$$\hat{\Sigma}(h) = T\,\frac{\psi(x(h))}{N(T)\, x^{\frac{1}{2}}(h)} + 0\left(\frac{x(h)^{\frac{3h}{2}}}{\log T}\right), \tag{16}$$

wo im 0-Glied $\log T$ durch T zu ersetzen ist, wenn R. V. gilt. Wendet man (5) an, so erhält man

$$D_T(\xi) = 0\left(\frac{\log(z_1 \ldots z_s)}{\log\log T}\right) \tag{17}$$

bzw., wenn R. V. gilt,

$$D_T(\xi) = 0\left(\frac{\log(z_1 \ldots z_s)}{\log T}\right). \tag{17'}$$

Die erzielten Resultate erscheinen sehr schlecht, sind es aber nicht, wie wir gleich sehen werden, jedenfalls wenn R. V. gilt. Unter dieser Annahme gilt nun für $z = q^m$

$$D_T(\xi(q^m)) \geqslant K(T),$$

wo

$$K(T) = \psi(q)\, q^{-\frac{m}{2}} \log^{-1} T - 0\,(q^{\frac{3m}{2}}\, T^{-1}). \tag{18}$$

Wir gehen dazu von der Gleichung

$$e(x) = 1 - 2\pi i \int_x^1 e(t)\,\mathrm{d}t \tag{19}$$

aus.

Ist i_t die Indikatorfunktion des Intervalls $< 0, t <$, dann schreibt sich (19) so:

$$e(x) = 1 - 2\pi i \int_0^1 e(t)\, i_t(x)\,\mathrm{d}t.$$

Daraus folgt, wenn wir von dieser Gleichung den Realteil nehmen,

$$\frac{1}{N(t)} \sum{}^* \cos 2\pi \xi_n - \int_0^1 \cos 2\pi x\,\mathrm{d}x = 2\pi \int_0^1 \sin 2\pi t\, J(t)\,\mathrm{d}t, \tag{20}$$

wo

$$J(t) = \frac{1}{N(t)} \sum{}^* i_t(\xi_n) - t.$$

Es ist also, wenn wir (14') benützen,

$$\int_0^1 \sin 2\pi t\, J(t)\,\mathrm{d}t = K(T). \tag{20'}$$

Nun ist

$$\int_{1/2}^1 \sin 2\pi t\, J(t)\,\mathrm{d}t = -\int_0^{1/2} \sin 2\pi t\, J\left(\frac{1}{2} + t\right)\mathrm{d}t,$$

also erhalten wir

$$K(t) = \int_1^{1/2} \sin 2\pi t\, \Delta(t)\, dt, \tag{21}$$

wo

$$\Delta(t) = \frac{1}{N(t)} \Sigma^* i\left(< t, \frac{1}{2} + t >, \xi_n\right) - \frac{1}{2}$$

und $i\,(< t, \frac{1}{2} + t <, *)$ die Indikatorfunktion des Intervalls $A(t) = \\ = < t, \frac{1}{2} + t >$ von der Länge $\frac{1}{2}$ ist.

Aus (20) folgt, daß es ein t aus $< 0, 1 <$ gibt, so daß

$$\Delta(t) \geqslant K(T)\,\pi \tag{22}$$

ist. Es ist also damit (18) bewiesen.

Betrachten wir für dieses t zu $A(t) = < t, \frac{1}{2} + t >$ das Komplement $B(t) = < 0, t > \cup < \frac{1}{2} + t, 1 <$, welches ebenfalls die Länge $\frac{1}{2}$, so ist, wenn wir

$$\lambda(A(t)) = \frac{1}{N(t)} \Sigma^* i\,(A(t), \xi_n)$$

und

$$\lambda(B(t)) = \frac{1}{N(t)} \Sigma^* i\,(B(t), \xi_n)$$

setzen trivialerweise $\lambda(A(t)) + \lambda(B(t)) = 1$, also folgt aus (22)

$$\lambda(A(t)) - \lambda(B(t)) \geqslant 2K\pi,$$

also liegen im Intervall $A(t)$ mindestens $C(q)\,T$ mehr Glieder der Folge ζ als im Komplement $B(t)$, wenn T genügend groß ist, z. B. $\geqslant q^{4m}$ ist. Man kann auch die Folge $\xi(q^m)$ mit einer Folge $\xi(z)$, wo z nicht von der Gestalt q^m ist, vergleichen. Setzen wir kurz $x_n = \xi_n(q^m)$ und $y_n = \xi_n(z)$, dann folgt aus (20)

$$\frac{1}{N(t)} \Sigma^* (\cos 2\pi x_n - \cos 2\pi y_n) = 2\pi \int_0^1 \sin 2\pi t\, \widehat{J}(t)\, dt,$$

wo

$$\widehat{J}(t) = \frac{1}{N(t)} \Sigma^* (i_t(x_n) - i_t(y_n))$$

ist. Nach (21) erhalten wir

$$\widehat{K}(T) = 2\pi \int_0^{1/2} \sin 2\pi t \, \widehat{\Delta}(t) \, dt,$$

wo

$$\widehat{\Delta}(t) = \frac{1}{N(t)} \Sigma^* (i(A(t), x_n) - i(A(t), y_n))$$

ist, und

$$\widehat{K}(t) = \psi(q) q^{-\frac{m}{2}} \log^{-1} T - 0(T^{-1})$$

ist, wo der 0-Term von q und z abhängt.

Es gibt daher wieder ein t, so daß im Intervall $A(t)$ von der Länge $\frac{1}{2}$ um $C(q, z)$ T mehr Glieder der Folge $\xi(q^m)$ liegen als Glieder der Folge $\xi(z)$. Dabei hängt C von q und z ab.

W. Schmidt hat bemerkt, daß man analog $\xi(q^m)$ mit $\xi(q_1{}^l)$ vergleichen kann, wenn $q^{-\frac{m}{2}} \log q > q_1{}^{-\frac{m}{2}} \log q_1$.

Wir wollen noch drei Bemerkungen anschließen:

1. Rademacher vermutet, daß (22) für $t = \frac{1}{4}$ richtig ist.

2. Analoge, allerdings schwächere Abschätzungen wie (4) und (15) gelten, wenn z eine ganze algebraische Zahl ist.

3. Die Überlegungen lassen sich auf L-Reihen wie auf die Dedekindsche Zetafunktion übertragen. Die Ausnahmestellung von q^m wird durch $N(q^m)$ übernommen, wo q ein Primideal aus dem entsprechenden Zahlkörper ist. Interessant wäre es, dies auf andere Zetafunktionen, wie z. B. auf die $\Sigma \frac{\tau(n)}{n^s}$ auszudehnen, wo τ die bekannte Ramanujansche Funktion ist.

Literatur

[1] Collected papers, Vol. II, S. 455.
[2] Math. Annalen **71** (1912), 548–564.
[3] Titchmarsh: The theory of the Riemann Zetafunktion, Oxford, Univ. Press 1951.
[4] Kuipers-Niederreiter: Uniform distribution of sequences, Academic Press (1974).

Corrections

463.1 $= -\frac{1}{2\pi}(T - \gamma)\Lambda(x)$ ist,

464.2 $\leq 2 \sum_{x<p^m<2x} p^{-2m}(\frac{p^m}{x} - 1)^{-1} \log p =$

464.3 $\leq 2x \sum_{x<p^m<2x} p^{-2m} \log p \leq$

464.4 $\leq x \sum_{\frac{x}{2}<p^m<x} p^{-2m}(x - p^m)^{-1} \log p \leq$

469.5 $K(T) = \int_0^{1/2} \sin 2\pi t\Delta(t)\, dt,$

90 Jahre Geometrie der Zahlen*

Jahrbuch Überblicke Mathematik 1980, Bibliographisches Institut, Mannheim, 1980, pp. 9–41

Die Geometrie der Zahlen wendet geometrische Methoden auf die Zahlentheorie, die Lehre der ganzen Zahlen, an. Wenn man nicht auf Lagrange zurückgehen will, so sind Gauß und Dirichlet die ersten, die mit geometrischen Methoden Probleme der Zahlentheorie, insbesondere die Theorie der quadratischen Formen, behandelt haben. Systematisch wurde dieses Gebiet von Minkowski entwickelt. Die erste Arbeit ist im Journal f. reine u. angewandte Mathematik, Band 107, 1891 erschienen.

Minkowski hat seine Untersuchungen in dem berühmten Buch „Geometrie der Zahlen" (1896) zusammengefaßt. Die Geometrie der Zahlen wurde dann von H. F. Blichfeldt (1873–1945), L. J. Mordell (1888–1972), C. L. Siegel (geb. 31. 12. 1896), K. Mahler (1903 geb.), H. Davenport (1907–1969), J. W. Cassels, C. A. Rogers (1920 geb.) entscheidend beeinflußt. Das Buch „Diophantische Approximationen" (1936 erschienen) von J. F. Koksma (1904–1964), dessen zweites Kapitel der Geometrie der Zahlen gewidmet ist, war die Keimzelle vieler Arbeiten in der Geometrie der Zahlen.

Die Schule in Manchester um Mordell und die Schule in Wien um Furtwängler (1869–1940), weitergeführt von N. Hofreiter (geb. 1904) und dem Verfasser haben das Gebiet systematisch gepflegt. Von der Wiener Schule sei nur W. Schmidt (geb. 1933) hervorgehoben.

Die stürmische Entwicklung dieses Gebietes in den 40- und 50-ziger Jahren endet ungefähr um 1960. Das Lehrbuch von Lekkerkerker: „Geometry of Numbers" (1969) stellt das Erzielte zusammen. Die Entwicklung ging aber in stilleren Bahnen weiter. Es ist hier die russische und australische Schule hervorzuheben. Eine Frucht der Geometrie der Zahlen ist die diskrete Geometrie, die vor allem durch die Schule um Fejes-Toth (geb. 1915) besonders gefördert wurde.** Das Interesse an der Geometrie der Zahlen ist jetzt wieder im Ansteigen begriffen, da es gelungen ist, wichtige Probleme zu lösen bzw. weitere Fortschritte auch bei Problemen, die noch nicht gelöst sind, zu erzielen. Diese Fortschritte wurden einerseits durch Vertiefung schon vorhandener Methoden der Geometrie der Zahlen, andererseits durch Heran-

---

* Die Anregung zu diesem Artikel verdanke ich Professor J. J. Burckhardt (Zürich) und Professor R. Liedl (Innsbruck).

** A. Florian hat hier wichtige Beiträge geleistet.

ziehung von Methoden der analytischen Zahlentheorie, insbesondere der Kreismethode von Hardy (1877–1947) und Littlewood (1885–1977) erzielt.

§ 1. Minkowskis Geometrie der Zahlen I

Im Februar 1896 erschien die 1. Lieferung des Buches „Geometrie der Zahlen" von Hermann Minkowski (1864–1909). Minkowski war damals Professor in Königsberg, nahm im Oktober 1896 in Zürich seine Vorlesungen auf und wirkte dort bis 1902. Danach folgte er einem Ruf nach Göttingen. Die 2. Lieferung des Werkes „Geometrie der Zahlen" erschien erst 1910, nach seinem Tod, aber nicht in der Form, in der sie Minkowski geplant hatte, sondern sie stellt im wesentlichen den Schluß des 5. Kapitels dar, wurde in seinem Nachlaß gefunden und wurde von Hilbert und Speiser (1885–1956) herausgegeben. Über die Gründe, warum sich das Erscheinen der 2. Lieferung verzögert hat, kann man nur vermuten. Im Vorwort der Herausgeber wird nur angeführt, daß sich einige unerwartete Schwierigkeiten eingestellt hätten. Ich werde mir erlauben, später noch auf diesen Punkt zurückzukommen (vgl. § 8).

Wie kam nun Minkowski zur Begründung seiner Theorie der Geometrie der Zahlen? Seit frühester Jugend hat er sich schon mit der Theorie der quadratischen Formen beschäftigt und hier fundamentale Resultate erzielt. Er schließt direkt an die Zahlentheoretischen Briefe von Hermite an Jacobi (1804–1851) an, die 1850 im Journal für reine und angewandte Mathematik, Band 40 veröffentlicht wurden. Die geometrische Deutung, die Dirichlet (1805–1859) der Reduktionstheorie der ternären, quadratischen Formen gegeben hatte, erschien in ausführlicher Darstellung ebenfalls im gleichen Band des Journals. Minkowski verknüpft nun im gleichen Journal, Band 107, 1891 die beiden Methoden, indem er noch in sehr vorsichtiger Weise den Begriff eines n-dimensionalen Gitters Γ im n-dimensionalen Zahlenraum $\mathbb{R}^n$ eingeführt. Damals hatte der $\mathbb{R}^n$ für $n > 3$ noch lange nicht sein Bürgerrecht in der Mathematik erworben.

Was versteht man nun unter einem Gitter Γ im $\mathbb{R}^n$? Die einfachste, obwohl nicht invariante Definition ist folgende:

Sie ist eine Menge aller Punkte $g = (g_1 \ldots g_n)$ von der Gestalt

$$\begin{aligned} g_1 &= b_{11}k_1 + \ldots + b_{1n}k_n \\ &\vdots \\ g_n &= b_{n1}k_1 + \ldots + b_{nn}k_n \end{aligned} \tag{1}$$

oder in Matrizenform geschrieben $g = Bk$ wo B die Matrix der b_{ij} ist und $k = (k_1, \ldots k_n)$

ist. Dabei sind b_{ij} reell und die Determinante Det $B \neq 0$. Die $k_1, \ldots k_n$

durchlaufen dabei alle ganzen Zahlen. Wenn B die Einheitsmatrix E ist, dann erhalten wir das sogenannte Würfelgitter Γ_0 (nützlich ist auch die Bemerkung, daß jedes Γ affines Bild des Würfelgitters Γ_0 ist). Durch das Gitter Γ ist B nicht eindeutig bestimmt. Zwei Matrizen A und B erzeugen genau dann das gleiche Gitter, wenn $B = AU$, wo U eine ganzzahlige, unimodulare Matrix ist.

Wir können (1) noch in der Form schreiben:

$$(1') \quad g = k_1 b_1 + \ldots + k_n b_n$$

wo b_i der Gitterpunkt mit den Koordinaten $b_i = (b_{1i}, \ldots b_{ni})$ ist. Man nennt die n Vektoren $(b_1 \ldots b_n)$ eine Basis (genauer eine Gitterbasis) des Gitters Γ. Man kann also das Gitter Γ dahin charakterisieren, daß es in dem von den $(b_1 \ldots b_n)$ aufgespannten Koordinatensystem im $\mathbb{R}^n$ die Menge aller Punkte mit ganzzahligen Koordinaten ist. Unter der Determinante des Gitters $d(\Gamma)$ versteht man den Betrag $|\mathrm{Det}\, B|$. Er ist unabhängig vom gewählten B. Geometrisch ist $d(\Gamma)$ das Volumen des Quaders F, der von n Vektoren einer Gitterbasis aufgespannt wird.

Aus der Gruppeneigenschaft des Gitters Γ folgt leicht folgende, wichtige Eigenschaft: Die Determinante von n Gitterpunkten eines n-dimensionalen Gitters ist stets ein ganzzahliges Vielfaches der Determinante des Gitters $d(\Gamma)$. Es ist keine Einschränkung der Allgemeinheit, wenn man, weil es oft zweckmäßig ist, annimmt, daß die Determinate des Gitters gleich 1 ist. Denn ist dies nicht der Fall, so betrachtet man das zu ihm ähnliche

Gitter mit der Matrix $\dfrac{A}{\sqrt[n]{d(\Gamma)}}$, wenn A eine erzeugende Matrix von Γ ist.

Weiters sehen wir sofort: es enthält die Diskretheit von Γ (vgl. § 4 (1)):

Ist L die Kantenlänge eines Würfels $\mathbf{W}$ mit dem Mittelpunkt im Nullpunkt. so ist die Anzahl $N(\Gamma, L)$ der Gitterpunkte von Γ in W asymptotisch $L^n d^{-1}(\Gamma)$, das heißt

$$(2) \quad \lim_{L \to \infty} \frac{N(\Gamma, L)}{L^n} = d^{-1}(\Gamma)$$

Nach dieser Einleitung können wir den Satz, den Minkowski 1891 im oben zitierten Journal für angewandte Mathematik aufgestellt hat, folgendermaßen aussprechen:

Jedes Ellipsoid mit dem Mittelpunkt in $0 = (0, 0 \ldots 0)$ und mit einem Volumen $V > 2^n d(\Gamma)$ enthält mindestens einen Gitterpunkt $\neq 0$, aus jedem Gitter mit der Determinante d. Minkowski führt dafür folgenden Beweis:

Mit Hilfe einer affinen Abbildung kann man den Fall eines Ellipsoids auf den Fall der Kugel $\mathbf{K}$ zurückführen, deren Volumen V ist. Es sei ρ der Kugelradius. Wir schlagen um jeden Gitterpunkt g von Γ eine Kugel $\mathbf{S}(g)$ mit dem Radius $\frac{1}{2}\rho$ und dem Volumen $V(\mathbf{S}) = \frac{1}{2^n} V(\mathbf{K})$. Wir betrachten nun nur die Gitterpunkte von Γ, die in einem Würfel **W** mit dem Mittelpunkt in 0 und der Kantenlänge L liegen. Die Anzahl dieser Gitterpunkte ist, wie wir schon bemerkt haben, asymptotisch $L^n \mathrm{d}^{-1}(\Gamma)$. Daher ist das Gesamtvolumen der Kugeln $\mathbf{S}(g)$, wobei $g \in \mathbf{W}$, asymptotisch $L^n V(\mathbf{S})\mathrm{d}^{-1}(\Gamma)$. Die betrachteten Kugeln $\mathbf{S}(g)$ liegen sicher alle im Würfel mit Mittelpunkt 0 und der Kantenlänge $L + \rho$. Da nun $V(\mathbf{S}) > \mathrm{d}(\Gamma)$ ist, so ist für großes $L: V > \left(1 + \frac{\rho}{L}\right)^n \mathrm{d}(\Gamma)$ also $V \cdot L^n > (L + \rho)^n \mathrm{d}(\Gamma)$. Das heißt, es können nicht alle Kugeln $\mathbf{S}(g)$ paarweise disjunkt sein; daher gibt es zwei verschiedene Kugeln $\mathbf{S}(g_1)$, $\mathbf{S}(g_2)$, so daß der Abstand $|g_2 - g_1| < \rho$ ist. Hier benützen wir die Dreiecksungleichung. Der Gitterpunkt $g = g_2 - g_1$ ist also $\neq 0$ und leistet das Gewünschte. Minkowski hat nun erkannt, daß bei dieser Überlegung nur die Eigenschaften des Abstandes $f(x) = d(0, x)$ des Punktes x vom Nullpunkt, benützt werden.

1) $f(\lambda x) = |\lambda| f(x)$ also insbesondere
$f(-x) = f(x)$ und
2) $f(x + y) \leqslant f(x) + f(y)$.

Jede solche Funktion f nennt nun Minkowski eine Distanzfunktion. Bei der Kugel ist eben

$$f(x) = \sqrt{x_1^2 + x_2^2 + \ldots + x_n^2}.$$

Die sogenannte Würfelmetrik

$$f(x) = \max(|x_1| \ldots |x_n|)$$

bzw. die L^p-Metrik

$$f(x) = (|x_1|^p + \ldots + |x_n^p|)^{1/p}$$

haben auch diese Eigenschaften.

Ersetzt man nun die Kugel durch den Körper $f(x) \leqslant \rho$, so erhält man eine Menge $\mathbf{K}(\rho)$, die 0 als Mittelpunkt besitzt, das heißt mit x liegt auch stets $-x$ in $\mathbf{K}$. Es ist $\mathbf{K}$ konvex, das heißt mit je 2 Punkten x und y in $\mathbf{K}$, liegt auch die Verbindungsstrecke $\overline{xy}$, das heißt die Menge aller Punkte $\lambda x + (1 - \lambda)y$, $(0 \leqslant \lambda \leqslant 1)$ in $\mathbf{K}$. Alle Mengen $\mathbf{K}(\rho)$ gehen durch Ähnlichkeitstransformation aus $\mathbf{K}(1)$, dem sogenannten Eichkörper, hervor. Im Fall der euklidischen Metrik ist es die Einheitskugel. Unter sehr allgemeinen Voraussetzungen über f kann man zeigen, daß $\mathbf{K}(\rho)$ stets ein Volumen im Jordan'schen Sinne besitzt, und ihr Volumen $V(\rho) = J\rho^n$ ist, wobei J das Volumen des Eichkörpers ist.

Minkowski zeigt dies alles ausführlich in seiner Geometrie der Zahlen und dazu auch die Umkehrung, nämlich, daß es zu jeder konvexen Menge $\mathbf{K}$ mit dem Mittelpunkt 0, wenn sie kompakt ist und innere Punkte besitzt, eine Distanzfunktion gibt, deren Eichkörper $\mathbf{K}$ ist. Nach diesen allgemeinen Überlegungen kann Minkowski, in Verallgemeinerung seines Satzes über das Ellipsoid, folgenden Satz, als Fundamentalsatz von Minkowski bezeichnet, aussprechen:

Jeder konvexe Körper mit dem Mittelpunkt in 0 und Volumen $V \geqslant 2^n d(\Gamma)$ enthält mindestens einen Gitterpunkt $\neq 0$, aus jedem Gitter Γ. Ist $V > 2^n d(\Gamma)$, so liegt der Gitterpunkt sogar im Inneren von **K**. Der Grenzfall $V = 2^n d(\Gamma)$ entsteht durch Grenzübergang aus dem Fall $V > 2^n d(\Gamma)$. Wendet man diesen Satz auf den Fall des Würfels mit dem Mittelpunkt 0 an, so erhält man den berühmten Linearformensatz von Minkowski:

Es seien n Linearformen

$$\begin{aligned} &L_1(x_1, \ldots x_n) = a_{11}x_1 + \ldots + a_{1n}x_n \\ &L_2(x_1, \ldots x_n) = a_{21}x_1 + \ldots + a_{2n}x_n \\ &\ldots\ldots\ldots\ldots\ldots\ldots\ldots\ldots \\ &L_n(x_1, \ldots x_n) = a_{n1}x_1 + \ldots + a_{nn}x_n \end{aligned} \tag{3}$$

mit Determinate $\Delta \neq 0$, und n positive reelle Zahlen $\lambda_1, \ldots \lambda_n$ gegeben, so daß

$$\lambda_1, \ldots \lambda_n \geqslant |\Delta| \tag{4}$$

ist, dann gibt es ganze Zahlen $k_1, \ldots k_n$, die nicht zugleich alle 0 sind, so daß

$$\begin{aligned} &|L_1(k_1, \ldots k_n)| \leqslant \lambda_1 \\ &|L_2(k_1, \ldots k_n)| \leqslant \lambda_2 \\ &\ldots\ldots\ldots\ldots\ldots\ldots \\ &|L_n(k_1, \ldots k_n)| \leqslant \lambda_n \end{aligned} \tag{5}$$

ist.

Daraus folgert Minkowski in einfacher Weise, daß wir immer ganze Zahlen $k_1, \ldots k_n$ finden können, die nicht zugleich 0 sind, so daß

$$|L_1 L_2 \ldots L_n| < |\Delta| \tag{6}$$

ist.

Wichtig ist in (6), daß hier das $<$-Zeichen gilt. Minkowski zeigt dies dadurch, daß er zeigt, daß in (5) das $=$-Zeichen in den n Ungleichungen, bis auf eine Ungleichung weggelassen werden kann. Eine Beweisvariante von J. von Neumann (1903–1957) für (6) geht folgendermaßen vor:

Der Wertvorrat von $L_1(k_1, \ldots k_n)$, wenn $(k_1, \ldots k_n)$ alle ganzzahligen n-Tupel durchläuft, ist abzählbar. Es gibt daher ein λ_1 verschieden von allen $L_1(k_1, \ldots k_n)$. Man wählt $\lambda_2, \ldots \lambda_n$ so, daß (4) gilt. Dann folgt direkt aus (5) Ungleichung (6).

Minkowski zeigt, daß ein $c_n < 1$ existiert, so daß für passendes $(k_1, \ldots k_n) \neq (0, \ldots 0)$

$$|L_1, \ldots L_n| \leqslant c_n |\Delta| \tag{7}$$

ja, daß sogar

(8) $\lim_{n\to\infty} c_n = 0.$

ist.

Er zeigt dies mit

(9) $c_n = \frac{n!}{n^n}$

indem er für den Körper $|x_1| + \ldots + |x_n| \leqslant \rho$ mit Volumen $\frac{2^n\rho^n}{n!}$ seinen Fundamentalsatz anwendet (man nimmt natürlich stets ρ so, daß $J\rho^n = 2^n d(\Gamma)$ ist (J Volumen des Eichkörpers), also das kleinste ρ, welches der Fundamentalsatz noch zuläßt), und dann die arithmetisch-geometrische Ungleichung

$$|x_1, \ldots x_n| \leqslant \left(\frac{|x_1| + \ldots + |x_n|}{n}\right)^n$$

anwendet. Für die Anwendungen auf die algebraische Zahlentheorie muß man auch den Fall betrachten, daß in (7) $2s$ konjugiert komplexe Linearformen auftreten. Minkowski betrachtet dann ($r = n - 2s$) den Körper $|x_1| + \ldots + |x_r| + \sqrt{2}\,(\sqrt{y_1^2 + z_1^2} + \ldots. + \sqrt{y_s^2 + z_s^2}) \leqslant \rho$ mit dem Volumen $V(\rho) = \left(\frac{\pi}{4}\right)^s \frac{2^n}{n!} \cdot \rho^n$. Dann folgt

$$|x_1, \ldots x_r (y_1^2 + z_1^2) \ldots (y_s^2 + z_s^2)| \leqslant 2^s \left(\frac{\rho}{n}\right)^n.$$

Man betrachte nun, wenn $L_1, \ldots L_r$ die reellen Linearformen und $L_{r+1}, \overline{L_{r+1}}, \ldots L_{r+s}, \overline{L_{r+s}}$ paarweise konjugiert komplexe Linearformen sind, die n reellen Linearformen $L_1 \ldots L_r, M_{r+1} = \frac{1}{\sqrt{2}}(L_{r+1} + \overline{L_{r+1}}), K_{r+1} = \frac{1}{\sqrt{2}\,i}(L_{r+1} - \overline{L_{r+1}}), \ldots M_{r+s} = \frac{1}{\sqrt{2}}(L_{r+s} + \overline{L_{r+s}}), K_{r+s} = \frac{1}{\sqrt{2}\,i}(L_{r+1} + \overline{L_{r+1}})$, welche dem Betrage nach die gleichen Determinante $|\Delta|$ haben. Man erhält dann

(7′) $|L_1, \ldots L_n| \leqslant c_n^* |\Delta|$

wo

(9′) $c_n^* = \left(\frac{4}{\pi}\right)^s \frac{n!}{n^n}$

ist. Daraus folgert Minkowski den berühmten Satz, daß die Diskriminante $D_n(K)$ eines algebraischen Zahlkörpers vom Grad n für $n > 1$ stets > 1 ist, also mindestens durch eine Primzahl teilbar ist. Wenn wir uns auf $s = 0$ beschränken, so zeigt Minkowski sogar

$$(10) \quad \sqrt[n]{D_n} \geqslant e^2 + 0(1) = 7{,}389 + 0(1).$$

Mit Hilfe der Geometrie der Zahlen wurde von C. A. Rogers, Acta Math. 82 (1950) sogar $D_n^{1/n} > 32{,}561 + 0(1)$ erreicht. Dieses Resultat wurde von Mullholland J. London Math. 35 (1960) 241–50 auf den allgemeinen Fall verallgemeinert. Ich möchte an einem Satz von Bombieri, Riv. Un. Parma 8 (1957) zeigen, wie (7) bzw. (7') angewendet wird.

Es sei $f(x) = a_0x^n + a_1x^{n-1} + \ldots + a_n = a_0(x - \alpha_1) \ldots (x - \alpha_n)$ ein Polynom über $\mathbf{Z}$. Wir können o.B.d.A. $a_0 = 1$ annehmen, denn sonst betrachten wir $a_0^{n-1}f$. Wir betrachten nun die n Linearformen

$$L_j(x) = x_1 + \alpha_j x_2 + \ldots + \alpha_j^{n-1} x_n.$$

Da die Determinante $\Delta = \mathrm{Det}(1, \alpha_j, \ldots \alpha_j^{n-1})$ ist, so ist $|\Delta|^2 = D$ die Diskriminante von f.

Nach (7) bzw. (7') existieren ganze Zahlen $(k_1, \ldots k_n) \neq (0, \ldots 0)$, so daß

$$A = |L_1, \ldots L_n| \leqslant c_n \sqrt{|D|}\,.$$

Da $\alpha_1, \ldots \alpha_n$ ganze algebraische Zahlen sind, so ist A eine ganze rationale Zahl. Ist also $\sqrt{|D|} < \dfrac{1}{c_n}$, so muß $A = 0$ sein, d.h. ein $L_j(k_1, \ldots k_n) = k_1 + \alpha_j k_2 + \ldots + \alpha_j^{n-1} k_n$ ist 0, d.h. α_j genügt einer Gleichung $(n - 1)$-ten Grades, also ist f *reduzibel*. Ist also f *irreduzibel*, so muß

$$\sqrt{|D|} \geqslant \frac{1}{c_n} \quad \text{bzw.} \quad \frac{1}{c_n^*}$$

sein. Man sieht, wie man zu den Resultaten von Minkowski kommt. Viele prominente Zahlentheoretiker (so z.B. H. Hasse (1898 geb.) waren der Ansicht, daß die Geometrie der Zahlen doch ein zu schwaches Werkzeug für die Zahlentheorie ist, da, nach einer Vermutung von E. Artin (1898–1962) und Ph. Furtwängler ein c_n in (7) existieren sollte, so daß sogar $\lim_{n\to\infty} \sqrt[n]{c_n} = 0$ ist. Es haben aber 1964 Golod und Shafarewich (Iz. Akad. Nauk SSSR (1964) 261–72) (vgl. auch P. Roquette S. 231–49 in Algebraic Number Theory, J. W. S. Cassels und A. Fröhlich ed. Academics Press, London 1967) gezeigt, daß diese Vermutung falsch ist. Sie zeigten z.B. die Existenz eines total reellen Zahlkörpers von beliebig hohem Grad, so daß $D_n^{1/n} \leqslant 5123{,}1\ldots$ ist. Es ist H. M. Stark (The analytic Theory of algebraic numbers, Bulletin Vol. 81 (1975) 961–72) gelungen, ohne Geometrie der Zahlen, allerdings mit starken analytischen Hilfsmitteln mit Hilfe der Dedekindschen Zetafunktion noch stärkere Abschätzungen für $D_n^{1/n}$ zu zeigen. Diese Untersuchungen wurden von Odlyzko, Acta Arith. 29 (1976) 275–97 fortgesetzt. So gelingt es ihm sogar $D_n^{1/n} \geqslant 55$ zu zeigen. Martinet, Ann. Fourier 29 (1979) 159–170 und G. Poitou, Sem. Pisot (1976) haben dies weiter verfolgt.

Minkowski hat seine Methoden auch auf die Approximation von Irrationalzahlen angewendet. Er betrachtet z.B. den Körper Max $(|x_1|+|x_n|, \ldots$ $\ldots |x_{n-1}|+|x_n|) \leqslant \rho$ vom Volumen $V = \dfrac{2^n \rho^n}{n}$. Durch Anwendung der arithmetisch-geometrischen Ungleichung folgt für $j = 1, \ldots n-1$

$$(11) \quad |x_j|^{n-1}|x_n| \leqslant (n-1)\left(\frac{|x_j|+|x_n|}{n}\right)^n.$$

Setzen wir $n - 1 = m$ und seien $\alpha_1, \ldots, \alpha_m$ Irrationalzahlen, dann sind $L_j = t(\alpha_j x_n - x_j)$ $(j = 1, \ldots m)$, $L_n = \dfrac{x_n}{t^{n-1}}\, n$ Linearformen von Determinante 1. Dabei ist t ein Parameter, der noch gewählt werden kann. Mit Hilfe des Minkowskischen Fundamentalsatzes folgert man nun, wenn man $x_j = p_j$ $(j = 1, \ldots m)$, $x_n = q$ setzt: Es gibt unendlich viele Brüche $\dfrac{p_i}{q}$ $(j = 1, \ldots m)$, so daß

$$(12) \quad \left|\alpha_i - \frac{p_j}{q}\right| < \frac{m}{m+1}\,\frac{1}{q^{1+1/m}}$$

ist. Dies ist eine Verschärfung gegenüber dem Dirichletschen Approximationssatz, der statt $\dfrac{m}{m+1}$ nur 1 hat. Später ist es gelungen (12) noch weiter zu verschärfen. Als eine Anwendung aus späterer Zeit sei ein zahlengeometrischer Beweis des Satzes von Lagrange (1736–1813), daß sich jede natürliche Zahl m als Summe von vier Quadraten darstellen läßt, gegeben. Es kann o.B. der Allgemeinheit m als quadratfrei angenommen werden. Es gibt ganze Zahlen a, b, so daß $a^2 + b^2 + 1 \equiv 0 \pmod m$ ist. Die Existenz solcher Zahlen zeigt man für jede Primzahl p direkt. Dann wendet man den chinesischen Restsatz an. Man betrachtet die vierdimensionale Kugel $x_1^2 + x_2^2 + x_3^2 + x_4^2 \leqslant \rho^2$ mit dem Volumen $\dfrac{\pi^2}{2}\rho^4$. Man wendet nun den Minkowskischen Fundamentalsatz auf das Gitter, gegeben durch die vier Linearformen $L_1 = x_1, L_2 = x_2, L_3 = ax_1 + bx_2 + mx_3$, $L_4 = -bx_1 + ax_2 + mx_4$ mit Determinante m^2 an. Es folgt dann also die Existenz von ganzen Zahlen $k_1, \ldots k_4$ (nicht alle 0), so daß

$$A = k_1^2 + k_2^2 + (ak_1 + bk_2 + mk_3)^2 + (-bk_1 + ak_2 + mk_4)^2 \leqslant$$

$\frac{4\sqrt{2}}{\pi} m < 2m$. Es ist A eine natürliche Zahl. Weiters ist $A \equiv 0 \pmod{m}$,

also muß $A = m$ sein, damit ist der Satz von Lagrange gezeigt.

Diese Beispiele zeigen, daß man in der Geometrie der Zahlen ein machtvolles Instrument besitzt.

Weitere Anwendungen stammen von L. Redei (1950) vgl. E. Hlawka M. A. 125 (1952) 183–207.

Wir wollen nun kurz berichten, wie das Buch „Geometrie der Zahlen" aufgenommen wurde. C. H. Hermite (1822–1901) hat sich über die 1. Lieferung enthusiastisch geäußert. Von R. Fricke (1861–1930) liegt in den Fortschritten der Mathematik eine ausführliche Besprechung vor. Ich möchte aber auch die Überlieferung bemühen. Der Physiker Sommerfeld (1868–1951) sagte einmal zu mir, daß viele Mathematiker nach dem Erscheinen dieses Buches Minkowski über Hilbert (1862–1944) gestellt haben. Er sagte ungefähr, Minkowski habe mehr Ideen gehabt, aber Hilbert wäre fleißiger gewesen. Wenn man die Biographie von Blumenthal (1876–1944) über Hilbert in den gesammelten Werken von Hilbert liest, so findet man ungefähr das gleiche Urteil und das kann nicht ohne Zustimmung von Hilbert erfolgt sein. Heute würde man allerdings dieses Urteil nicht so formulieren.

Felix Klein (1849–1925), der in seinen Vorlesungen über Zahlentheorie, die von Sommerfeld herausgegeben wurden, Gitter in der Ebene verwendet, faßte das Werk von Minkowski als die Fortsetzung seines eigenen Werkes auf, wie man in seinen Vorlesungen über die Mathematik des 19. Jhdt.'s nachlesen kann. Dies trifft aber in keiner Weise zu. Die Puristen unter den Zahlentheoretikern, haben die Geometrie der Zahlen als der Analysis angehörig gefunden und sich bemüht, dieses Werkzeug möglichst zu eliminieren. So wurden für den Minkowski'schen Linearformsatz verschiedene, elementare Beweise gegeben. Am bekanntesten ist der Beweis von A. Hurwitz (1859–1919), der ihn zunächst für ganzzahlige Koeffizienten mit Hilfe des Schubfachprinzipes beweist. Daraus folgt unmittelbar der Fall der rationalen Koeffizienten und durch Grenzübergang kann der allgemeine Fall erledigt werden. Das Werk von Minkowski ist nicht nur für die Zahlentheorie von Bedeutung, sondern darüber hinaus hat gerade die allgemeine Theorie der konvexen Körper Anlaß zur Entwicklung der metrischen Räume und der Funktionalanalysis überhaupt gegeben. So wurde Hilbert zur Begründung der Minkowskischen Geometrie geführt. F. Hausdorff (1868–1942) und E. Helly (1884–1943) schließen direkt an Minkowski an. Hervorgehoben sei die Arbeit „Über Systeme Linearer Gleichungen mit unendlich vielen Unbekannten", Monatsh. Math. 31 (1921) 60–91; über Helly vgl. den Bericht der Rheinisch-westfälischen Hochschule Aachen v. P. L. Butzer ed. 1979. Durch den Fundamentalsatz ist die Leistung von Minkowski nicht erschöpft, er hat diesen Satz durch die Betrachtung der sukzessiven Minima bedeutend vertieft. Dies soll im nachfolgenden Paragraph besprochen werden.

§ 2. Minkowski: Geometrie der Zahlen

Es soll nun auf die 2. Lieferung des Werkes „Geometrie der Zahlen" eingegangen werden. Es sei f die Distanzfunktion eines konvexen Körpers K im $\mathbb{R}^n$. Dann führt Minkowski die n Minima $M_1, M_2, \ldots M_n$ ein: Es ist

$$
\begin{aligned}
&M_1 = f(m_1) = \min_{g \neq 0} f(g)\\
(1)\quad &M_2 = f(m_2) = \min_{g} f(g) \quad \text{l.u.a. von } m_1\\
&M_3 = f(m_3) = \min_{g} f(g) \quad \text{l.u.a. von } m_1, m_2\\
&\cdots\cdots\cdots\cdots\cdots\cdots\cdots\cdots\\
&M_n = f(m_n) = \min_{g} f(g) \quad \text{l.u.a. von } m_1, m_2, \ldots, m_{n-1}
\end{aligned}
$$

Es gilt $0 \leqslant M_1 \leqslant M_2 \leqslant \ldots \leqslant M_n$.

Diese Definition scheint zunächst von den Gitterpunkten $m_1, m_2, \ldots, m_n$ abzuhängen. Dies ist aber nicht der Fall, denn wir können z.B. M_k so charakterisieren: Ist $q \leqslant M$, so enthält der Körper $f(x) \leqslant q$ im Inneren weniger als k linear unabhängige Gitterpunkte, während er für $q > M_k$ mindestens k linear unabhängige Gitterpunkte im Inneren enthält. Die $m_1, \ldots m_n$, die linear unabhängig sind, sind nicht eindeutig bestimmt und bilden ein System von sogenannten Minimalpunkten. Sie erzeugen im allgemeinen aber nicht das Gitter Γ. Für den Fall des Kreises, bzw. der Kugel wurden diese Begriffe in der Reduktionstheorie der quadratischen Formen, von Gauß und Dirichlet eingeführt. Hier bilden die Minimalpunkte noch eine Gitterbasis, aber schon für den Fall der 4-dimensionalen Kugel, wurde von Julia (1893–1978) 1912 bemerkt, obwohl dies wahrscheinlich schon vorher bekannt war, daß die 4 Minimalpunkte keine Gitterbasis bilden.

Minkowski studiert in seinem Buch, Diophantische Approximationen ganz allgemein für beliebige Eichkörper, den 3-dimensionalen Fall.

Der 4-dimensionale Fall wurde unter anderem in verschiedenen Dissertationen unter Furtwängler, Hofreiter und Hlawka untersucht. Das abschließende Resultat findet sich in der Dissertation von E. Brunngraber (1944) bzw. K. H. Wolff (1952).

Der 5-dimensionale Fall wurde ebenfalls für Spezialfälle, vor allem für den der Kugel, untersucht. Die Fallentscheidungen werden aber immer umfangreicher, so daß sich kein praktischer Nutzen bisher gezeigt hat.

Vorausgreifend sei bemerkt, daß K. Mahler 1938 und unabhängig davon H. Weyl 1939 in einfachster Weise gezeigt haben, wie man aus einer Minimalbasis $m_1, m_2, \ldots, m_n$ eine Gitterbasis $b_1, \ldots, b_n$ konstruieren kann. Dies erfolgt unter Benutzung der von Minkowski entwickelten Methode der Adaption des Gitters. Der Gedankengang ist kurz folgender:
Es seien allgemein $a_1, a_2, \ldots a_n$. l.u.a. Gitterpunkte aus Γ. Man wählt nun aus der endlichen Menge aller Gitterpunkte $\lambda_1 a_1$ $(0 < \lambda_1 \leqslant 1)$ denjenigen $b_1 = \lambda_{11} a_1$, für welchen λ_1 den kleinsten Wert hat. Dann wählt man unter den endlich vielen Gitterpunkten $\lambda_1 a_1 + \lambda_2 a_2$ mit $0 \leqslant \lambda_1 < \lambda_{11}, 0 < \lambda_2 \leqslant 1$ jenen

$$b_2 = \lambda_{21} a_1 + \lambda_{22} a_2$$

für welchen λ_2 am kleinsten ist. Dann wählt man unter allen Gitterpunkten $\lambda_1 a_1 + \lambda_2 a_2 + \lambda_3 a_3$ $\quad 0 \leqslant \lambda_1 < \lambda_{11}, 0 \leqslant \lambda_2 < \lambda_{22}, 0 < \lambda_3 \leqslant 1$ jenen $b_3 = \lambda_{31} a_1 + \lambda_{32} a_2 + \lambda_{33} a_3$ aus, für den λ_3 am kleinsten ist usw.; die $b_1 b_2, \ldots b_n$ bilden dann eine Basis von Γ.

Diesen Prozeß (in dessen Schilderung wir H. Weyl, Die Idee der Riemannschen Fläche (2. Aufl. 1923)) wörtlich, abgesehen von der Bezeichnung, gefolgt sind, findet auch in anderen Gebieten der Mathematik wichtige Anwendung. Die Überlegung zeigt, daß jede diskrete Gruppe im $\mathbb{R}^n$, welche n l.u.a. Vektoren enthält, ein Gitter ist (invariante Definition des Gitters).

Man erhält, wenn $a_k = m_k$ $(k = 1, \ldots n)$: $f(b_k) \leqslant \delta_k M_k$, wo δ_k unter einer festen Schranke liegt.

Es ist $\delta_1 = 1$ und für δ_k, wenn $k \geqslant 2$ ist, kann man $(3/2)^{k-2}$ nehmen, Es ist wirklich erstaunlich, daß Minkowski das nicht gesehen hat. H. Weyl vergleicht Minkowski mit Moses, der das gelobte Land nur von der Ferne sah.

Der Minkowski'sche Fundamentalsatz liefert sofort das Resultat:

(2) $\quad JM_1^n \leqslant 2^n \, \mathrm{d}(\Gamma)$.

In diesem 2. Teil vertieft nun Minkowski diese 1. Ungleichung in die sogenannte 2. Fundamentalungleichung

$$\frac{2^n}{n!} \, \mathrm{d}(\Gamma) \leqslant JM_1 \ldots M_n \leqslant 2^n \mathrm{d}(\Gamma),$$

Die linke Seite folgt sofort daraus, daß die linear unabhängigen Punkte $\dfrac{m_1}{M_1}, \ldots \dfrac{m_n}{M_n}$ im Eichkörper $\mathbf{E} : f(x) \leqslant 1$ liegen, also auch der von ihnen aufgespannte Oktaeder in $\mathbf{E}$ liegt. Sein Volumen ist $\geqslant \dfrac{2^n d}{n! \, M_1 \ldots M_n}$ und es ist $\leqslant J$, dem Volumen des Eichkörpers. Die rechte Seite der Unlgeichung aber ist sehr tiefliegend, vom Fall der Kugel abgesehen, wo die 2. Ungleichung auf die 1. Ungleichung zurückgeführt werden kann. Für diesen Fall soll eine kurze Skizze gegeben werden (dabei nehmen wir $d(\Gamma) = 1$). Wir können o.B.d.A. (sonst passende Drehung) annehmen, daß

$$\begin{aligned} m_1 &= (m_{11}, 0, \ldots 0) \\ m_2 &= (m_{21}, m_{22}, \ldots 0) \\ &\ldots\ldots\ldots\ldots\ldots \\ m_n &= (m_{n1}, m_{n2}, \ldots m_{nn}) \end{aligned}$$

ist.
Es ist $|\mathrm{Det}(m_1, \ldots m_n)| \geqslant 1$ und es ist

$$M_j = |m_j| = \sqrt{m_{1j}^2 + \ldots m_{jj}^2} \geqslant |m_{jj}|.$$

Wir betrachten das Ellipsoid

$$\mathbf{E}^* : \frac{x_1^2}{M_1^2} + \ldots + \frac{x_n^2}{M_n^2} < 1.$$

Es kann kein Gitterpunkt $g \neq 0$ in $\mathbf{E}^*$ liegen.

Beweis: Wir können stets schreiben:
$g = a_1 m_1 + \ldots + a_k m_k$, bei passendem k, wo $a_k \neq 0$. $(1 \leqslant k \leqslant n)$. Es ist $g = (g_1, \ldots g_k, 0, \ldots 0)$, wo $g_k \neq 0$ ist. Es ist also $|g| \geqslant M_k$, daher ist

$$\frac{g_1^2}{M_1^2} + \ldots + \frac{g_k^2}{M_k^2} \geqslant \frac{|g|^2}{M_k^2} \geqslant 1.$$

Das Volumen des Ellipsoids ist gleich $(M_1, \ldots M_n)J$. Nach dem Fundamentalsatz von Minkowski ist daher $M_1 \ldots M_n J \leqslant 2^n$. Minkowski studiert nun ausführlich den Fall des Gleichheitszeichens. Dabei entwickelt er die geometrische Theorie der konvexen Körper, die man heute die Theorie von Brunn-Minkowski nennt (vgl. Leichtweiss, Konvexe Mengen, im Druck).

§ 3. Minkowski: Diophantische Approximationen

Es soll nun das 2. Buch von Minkowski: „Diophantische Approximationen", das 1907 mit dem Untertitel „Mathematische Vorlesungen an der Universität Göttingen II" erschienen und H. Weber (1842–1912) gewidmet ist, besprochen werden. Es ist aus einer Vorlesung, die Minkowski im Wintersemester 1903/04 gehalten hat, hervorgegangen und von A. Axer ausgearbeitet worden. Dieses Buch, das reich an Figuren ist (es enthält 82 Zeichnungen) im Gegensatz zur „Geometrie der Zahlen", gilt oft nur als Einführung in die Geometrie der Zahlen (so in der Besprechung des Buches von Fueter (1880–1950) in den Fortschritten der Mathematik), da es sich nur auf den zwei- und dreidimensionalen Fall beschränkt, wenn man vom letzten Kapitel absieht. Dies stimmt aber höchstens teilweise. Daher soll der Inhalt dieses Buches kurz besprochen werden.

Das 1. Kapitel behandelt das Schubfachprinzip und bringt einen Beweis des Linearformensatzes für 2 und 3 Dimensionen von Hilbert, der für die Denkweise von Hilbert außerordentlich charakteristisch ist. Ausgehend von einem Spezialfall wird durch Kontinuitätsmethoden (heute würde man topologische Methoden sagen) der allgemeine Fall erreicht.

Das 2. Kapitel: Zahlengitter in 2 Dimensionen bespricht außer den Potenzsummen $|x|^p + |y|^p \leqslant \rho^p$ (vgl. § 7(6)) den inhomogenen Linearformensatz (vgl. § 6(5)).

Das 3. Kapitel: Zahlengitter in drei Dimensionen behandelt die kritischen Gitter des Würfels (vgl. § 7(6)) und den elliptischen Zylinder ($|x| \leqslant \rho$, $y^2 + z^2 \leqslant 2\rho^2$), Oktaeder ($|x| + |y| + |z| \leqslant \rho$) und den Doppelkegel ($|x| + \sqrt{2(y^2 + z^2)} \leqslant \rho$). Diese Körper werden dann im 4. Kapitel zur Theorie der algebraischen Zahlen, auf kubische Zahlkörper beschränkt, benötigt. Im 3. Kapitel wird noch weiters die dichteste Lagerung von konvexen Körpern, (vgl. § 9), insbesondere die dichteste Lagerung von Kugeln, behandelt. Die Methode der Adaption von Gittern wird hier ebenfalls geometrisch gedeutet. Besonders bemerkenswert sind im 3. und 4. Kapitel die verwendeten Methoden, die man heute als tópologische Methoden bezeichnet, die von K. Mahler entwickelt wurden (vgl. § 7)) und hier schon in intuitiver und spezieller Form auftreten.

Auf das 4. und 5. Kapitel: Zur Theorie der Ideale, das die Geometrie der Zahlen auf kubische Körper anwendet, soll hier nicht weiter eingegangen werden.

Das 6. Kapitel: Annäherung komplexer Größen durch Zahlen des Körpers der 3. und 4. Einheitswurzeln, soll etwas ausführlicher besprochen werden. Es wurde in der Vorlesung 1903/04 nicht vorgetragen, sondern ist die Wiedergabe eines unveröffentlichten Manuskriptes von Minkowski durch Axer. Zunächst wird das Zahlengitter von 4 Dimensionen behandelt, aber dann wird hier, modern ausgedrückt, Geometrie der Zahlen nicht mehr in ganzen, rationalen Zahlen, sondern in ganzen Zahlen aus dem Gaußschen Zahlkörper $\mathbb{Q}(i)$ bzw. aus $\mathbb{Q}(i\sqrt{3})$ betrieben. Es wird z.B. folgendes gezeigt: Es sei $L_1(x, y) = ax + by$, $L_2(x, y) = cx + dy$, wobei die Koeffizienten *komplex* sind und die Determinante $D = ad - bc \neq 0$ ist, dann gibt es ganze Zahlen x, y aus $\mathbb{Q}(i)$, so daß

$$\operatorname{Max}(|L_1(x, y)|,\ |L_2(x, y)|) \leqslant \sqrt{\frac{\sqrt{2}}{3-\sqrt{3}}|D|} = 1{,}05610..\sqrt{|D|}$$

und die auftretende Schranke kann nicht mehr verbessert werden.

Der vorgeführte Beweis ist sehr kompliziert und ist wohl nie ganz in allen Einzelheiten nachgerechnet worden. Direkte Beweise haben E. Hlawka (Monatshefte 49 (1941) 321–26) und Cassels, W. Ledermann und K. Mahler (Phil. Trans. Roy. Soc., London Ser. 1,234 (1951) 585–626) gegeben. Eine allgemeine Theorie der Geometrie der Zahlen über algebraische Zahlkörper wurde von H. Weyl (1885–1955), Trans. 48, 1940) begonnen, und von K. Rogers und Swinnerton-Dyer (Trans. 88, 1958) systematisch aufgebaut. Eine Geometrie der Zahlen über p-adische Körper wurde von K. Mahler (J. d. M. V. 44) 1934 begonnen und in den Annals of Math. 42, 1941 weiter entwickelt. Hervorgehoben sei die Arbeit von C. Mc Feat in der Dissertationes Math. 47 (1971), die eine Geometrie der Zahlen über den Adelen entwickelt.

§ 4. Das Prinzip von Blichfeldt

Einen neuen Zugang zum Fundamentalgesetz von Minkowski hat H. F. Blichfeldt in seiner Arbeit: A new principle in the geometry of numbers, with some applications (Trans. Amer. Math. Soc., Bd. 15 (1914) 227–235) gefunden. Bevor wir darauf eingehen, seien einige Daten über diesen bedeutenden Mathematiker angeführt. Er wurde am 9.1.1873 in Dänemark geboren, wanderte in jungen Jahren nach USA aus, wo er verschiedene handwerkliche Berufe ergriff. Er studierte dann in Stanford und dissertierte bei S. Lie in Leipzig 1898. Er war die meiste Zeit in Stanford tätig, seit 1927 war er Head of Department. Für seine Verdienste wurde ihm 1938 vom dänischen König der Dannebrogorden verliehen. Er starb am 16.2. 1945. Seine Arbeiten, deren Anzahl nur 22 betrug, beschäftigten sich mit Gruppen- und Zahlentheorie. Auf diesen Gebieten hat er bedeutende Arbeiten veröffentlicht und Richard Brauer (1901–77) rechnete ihn zu den größten Mathematikern. Nach G. Kowalewski (1876–1950) soll er schweigsam wie Gauß gewesen sein.

Es soll jetzt auf die Grundidee von Blichfeldt eingegangen werden. Dabei benützen wir eine Darstellung, die in der Idee auf G. Birkhoff (1884–1944) und R. Remak (1888–1940), (Math. Zeitschrift 26, (1927) 694–699) zurückgeht. Remak hat wertvolle Beiträge zur Geometrie der Zahlen geleistet, er hat in der wichtigen Arbeit in Crelle 167 (1931) die Geometrie der Zahlen noch auf die Theorie der algebraischen Zahlentheorie angewendet (vgl. auch Compositio 10 (1952), und Comp. 12 (1955) und C. L. Siegel, Nachrichten Akadem., Göttingen (1969) 71–80 (Ges. Abl. IV, 66–81)). Er galt als schwieriger Kollege und Sonderling. Ich verweise auf das Buch von H. Behnke: Ein Leben an deutschen Universitäten im Wandel der Zeit (Vandenhoeck & Ruprecht) 39 ff.

Es sei Γ ein Gitter mit der Basis $b_1, b_2 \ldots b_n$. Es läßt sich dann jeder Punkt x eindeutig in der Form schreiben:

$$x = \xi_1 b_1 + \xi_2 b_2 + \ldots + \xi_n b_n$$

Wir setzen nun $\xi_j = [\xi_j] + \lambda_j$ wo $0 \leqslant \lambda_j < 1$ ist. Dadurch ist λ_j eindeutig bestimmt. Es ist nun

$g = [\xi_1]b_1 + \ldots + [\xi_n]b_n$ Gitterpunkt von Γ und $y = \lambda_1 b_1 + \ldots \lambda_n b_n$.
ein Punkt des Quaders F, der von $b_1, b_2, \ldots b_n$ aufgespannt wird.

Sein Volumen $V(\mathsf{F})$ ist wie schon vorher bemerkt gleich $\mathrm{d}(\Gamma)$. Daher läßt sich jeder Punkt x eindeutig in der Form

$$x = g + y \; (g \in \Gamma, y \in \mathsf{F})$$

darstellen. Es ist gruppentheoretisch gesprochen F ein Fundamentalbereich von Γ. Wir können das auch so ausdrücken. Es ist

$$\bigcup_{g \in \Gamma} (\mathsf{F} + g) = \mathbb{R}^n \quad \text{und es ist stets}$$

$$\mathsf{F} + g \cap \mathsf{F} + g_1 = \emptyset \quad \text{wenn} \quad g \neq g_1 \quad \text{wo } g, g_1 \in \Gamma \text{ sind.}$$

Daraus folgt sofort:

Für jede integrierbare Funktion φ (die außerhalb eines „großen" Würfels verschwindet) gilt die Gleichung

$$(1) \qquad \int_{\mathbb{R}^n} \varphi(x)\,\mathrm{d}x = \sum_{g \in \Gamma} \int_{\mathsf{F}+g} \varphi(x)\,\mathrm{d}x$$

($\mathrm{d}x$ ist stets das Volumselement $\mathrm{d}x_1 \ldots \mathrm{d}x_n$)

Wir nehmen nun an, es gibt eine Konstante $c > 0$, so daß

$$\int_{\mathbb{R}^n} \varphi(x)\mathrm{d}x > c\,V(\mathsf{F}),$$

so folgt daraus, daß es ein $x \in \mathsf{F}$ geben muß, so daß

$$(2) \qquad \sum_{g \in \Gamma} \varphi(x + g) > c.$$

Denn wäre für alle x aus F: $\sum_{g \in \Gamma} \varphi(x + g) \leqslant c$, so folgt durch Integration über F aus (1), daß

$\int_{\mathbb{R}^n} \varphi(x)\,\mathrm{d}x \leqslant c\;V(\mathsf{F})$ wäre. Das ist ein Widerspruch. Ist φ eine ganzzahlige Funktion, so folgt aus (2), daß ein x existiert, so daß

$$(3) \qquad \sum_{g} \varphi(x + g) \geqslant [c] + 1 \quad \text{ist}$$

Das ist gerade das Blichfeldt'sche Prinzip, insbesondere wenn φ die Indikatorfunktion einer Menge S ist.

Das bedeutet geometrisch folgendes:

Ist das Volumen $V(\mathsf{S}) > c\,V(\mathsf{F})$, dann kann man stets die Menge S so verschieben, d.h. es gibt eine Translation x, so daß $\mathsf{S} + x$ mindestens

$[c]+1$ Gitterpunkte aus Γ enthält. Es soll nun kurz gezeigt werden, wie daraus der Minkowskische Fundamentalsatz folgt.

Es sei $\mathbf{K}$ eine konvexe Menge mit Mittelpunkt 0 und $V(\mathbf{K}) > 2^n\, d(\Gamma)$. Wir nehmen nun in (3) $\varphi(x) = i_{\mathbf{K}}(2x)$, i Indikatorfunktion von $\mathbf{K}$. Es ist

$$\int_{\mathbb{R}^n} i_{\mathbf{K}}(2x)\, dx = \frac{V(\mathbf{K})}{2^n} > d(\Gamma) = V(\mathbf{F})$$

Mit $c = 1$ folgt aus (3), daß es ein x und zwei verschiedene Gitterpunkte g_1, g_2 gibt, so daß

$$2(x+g_1) \quad \text{und} \quad 2(x+g_2) \text{ in } \mathbf{K} \text{ liegen.}$$

Da 0 Mittelpunkt von $\mathbf{K}$ ist, so liegt auch

$-2(x+g_2)$ in $\mathbf{K}$ und da $\mathbf{K}$ konvex ist, liegt auch der Punkt

$$\frac{1}{2}\left(2(x+g_1) - 2(x+g_2)\right) = g_1 - g_2 = g \text{ in } \mathbf{K}.$$

Es ist $g \neq 0$ und damit ist der Minkowskische Satz erneut bewiesen.

Analog zeigt man die Verallgemeinerung des Minkowskischen Satzes, die von van der Corput (1890–1978) stammt.

Ist $V(\mathbf{K}) \geqslant 2^n l\, d(\Gamma)$, l natürliche Zahl, so enthält der konvexe Körper $\mathbf{K}$, l verschiedene Gitterpunktpaare ungleich Null. Man braucht in (3) nur $c = l$ zu nehmen.

Es sei noch bemerkt, daß man in der Geometrie der Zahlen, eine l-Theorie entwickelt hat, in der man Bedingungen sucht, daß eine Menge l von Null verschieden Gitterpunkte enthält. Auf diese Theorie wollen wir der Einfachheit halber nicht eingehen und uns stets auf den Fall $l = 1$ beschränken. Blichfeldt zeigte in seiner vorher zitierten großen Arbeit und auch in späteren Arbeiten, wie man mit Hilfe seines Prinzips für spezielle konvexe Körper den Minkowskischen Fundamentalsatz verschärfen kann. Es soll im Falle der Kugel gezeigt werden

(es ist dies nicht das schärfste bekannte Resultat):

Ist $V(\mathbf{K}) > 2^{\frac{n}{2}}\left(\frac{n}{2}+1\right) d(\Gamma)$, so liegt im Inneren der Kugel ein Gitterpunkt

$g \neq 0 \in \Gamma$. Dies ist für $n > 2$ eine Verschärfung des Minkowskischen Satzes. Man nimmt in (3)

$$\varphi(x) = \operatorname{Max}\left\{0, 1 - \frac{2|x|^2}{\rho^2}\right\}$$

ρ = Radius von $\mathbf{K}$.

Es ist

$$\int_{\mathbb{R}^n} \varphi(x)\,dx = \frac{V(K)}{2^{\frac{n}{2}}\left(\frac{n}{2}+1\right)} > d(\Gamma)$$

Es folgt aus (3), daß ein x existiert, so daß

$$\sum_{g\in\Gamma} \varphi(x+g) > 1 \quad \text{ist, d.h. es ist}$$

$$\sum_{|x+g|<\frac{\rho}{\sqrt{2}}} \left(1-\frac{2(x+g)^2}{\rho^2}\right) > 1$$

Denken wir uns die Punkte $x+g$, die (*) erfüllen, der Reihe nach durchnumeriert

$$x_1 = x+g_1, \ldots x_s = x+g_s, \quad \text{so folgt aus (*)}$$

$$\sum_{i=1}^{s} |x_i|^2 < \frac{\rho^2}{2}(s-1)$$

Nun gilt, wie man leicht zeigt,

$$\sum_{i\neq j} |x_i - x_j|^2 \leqslant 2s \sum_{i=1}^{s} |x_i|^2. \text{ Es ist also}$$

$$\sum_{i\neq j} |g_i - g_j|^2 = \sum_{i\neq j} |x_i - x_j|^2 < \rho^2 s(s-1)$$

Es gibt daher ein i und ein j, $i \neq j$, so daß $0 < |g_i - g_j| < \rho$ ist. Der Punkt $g = g_i - g_j$ leistet das Gewünschte.

Das allgemeine Problem, wie sich der Minkowskische Fundamentalsatz für konvexe Körper, wenn der Körper genügend glatt ist, verschärfen läßt, ist vor allem für hohe Dimensionen noch ungelöst. Im Fall der Ebene haben Davenport und van der Corput 1946 (Proceedings Amsterdam 49, 701–707) unter Heranziehung der Krümmung ein scharfes Resultat erzielt.

In der Dissertation von E. Demmelbauer (1955) wurde dies auf den 3-dimensionalen Raum und in der Dissertation von Karl Schmidt (1958) auf den n-dimensionalen Raum verallgemeinert.

H. Groemer hat im Fall der Ebene, durch Heranziehung des Affinumfanges ein scharfes Resultat erhalten.

Bemerkung:
Das Blichfeldt'sche Prinzip läßt sich stark verallgemeinern. Es kann auf den Fall übertragen werden, daß $\mathbb{R}^n$ eine lokalkompakte Gruppe, Γ eine diskrete Untergruppe mit kompaktem Fundamentalbereich ist. Vgl. z.B. die Arbeit des Verfassers, Zur Theorie des Figurengitters Math. Ann. 125 (1952) 183–207.

§ 5. Eine Formel von Siegel

Es sei wieder φ eine integrierbare Funktion wie in § 4. Wir bilden uns die Funktion

$$(1) \quad \Phi(x) = \sum_{g \in \Gamma} \varphi(x+g)$$

Es ist nach § 4(1):

$$(2) \quad \int_F \Phi(x)\,dx = \int_{\mathbb{R}^n} \varphi(x)\,dx$$

Wir bemerken nun, daß Φ eine periodische Funktion mit Γ als Periodengitter ist; d.h. es ist $\Phi(x+g_1) = \Phi(x)$, wenn $g_1 \in \Gamma$, für alle x. Es liegt daher nahe zur Funktion Φ die zugehörige Fourierreihe

$$(3) \quad \sum_l c_l\, e(-\langle l x\rangle)$$

zu betrachten, wo $e(\alpha) = e^{2\pi i\alpha}$ und $\langle lx\rangle$ das skalare Produkt von l und x $l_1x_1 + \ldots + l_nx_n$ ist. Dabei erstreckt sich die Summe über alle Gitterpunkte l des polaren Gitters Γ^* zu Γ. (Zur Definition des polaren Gitters vergleiche die Einleitung des folgenden § 6). Es ist

$$c_l = \int_F \Phi(x)\, e(\langle lx\rangle)\,dx$$

Man rechnet sofort nach, daß

$$(4) \quad c_l = \int_{\mathbb{R}^n} \varphi(x)\, e(\langle lx\rangle)\,dx$$

Die Fourierreihe muß nicht konvergieren, aber es gilt die Parseval'sche Gleichung

$$(5) \quad \int_F |\Phi(x)|^2\,dx = \sum_l |c_l|^2$$

Die linke Seite von (5) lautet entwickelt

$$(6) \quad \int_F \sum_g \varphi(x+g) \sum_h \varphi(x+h)\,dx = \sum_k \int_{\mathbb{R}^n} \varphi(x)\varphi(x+k)\,dx$$

daher erhalten wir die grundlegende Formel

$$(7)\quad \sum_g \int_{\mathbb{R}^n} \bar{\varphi}(x)\,\varphi(x+g)\,dx = \sum_l |c_l|^2$$

Es sei jetzt $\mathbf{K}$ ein konvexer Körper mit Mittelpunkt 0, $i_{\mathbf{K}}$ die Indikatorfunktion von $\mathbf{K}$. Dann sei wie schon in § 4, $\varphi(x) = i_{\mathbf{K}}(2x)$. Es ist

$$(8)\quad \begin{aligned} c_l &= \frac{1}{2^n}\, b_l \text{ wo} \\ b_l &= \int_{\mathbb{R}^n} i_{\mathbf{K}}(x)\, e\, \frac{1}{2}\,(\langle lx\rangle)\,dx = \int_{\mathbf{K}} e\left(\frac{1}{2}\,(\langle lx\rangle)\right)dx \end{aligned}$$

ist.

Andererseits ist

$$(9)\quad \begin{aligned} \int_{\mathbb{R}^n} \varphi(x)\varphi(x+g)\,dx &= \frac{1}{2^n}\int_{\mathbb{R}^n} i_{\mathbf{K}}(x)\,i_{\mathbf{K}}(x+2g)\,dx = \\ &= \frac{1}{2^n}\, V(\mathbf{K} \cap \mathbf{K} + 2g) \end{aligned}$$

Wir erhalten also folgende Formel

$$(10)\quad V(\mathbf{K}) + \sum_{g \neq 0} V(\mathbf{K} \cap \mathbf{K} + 2g) = \frac{1}{2^n}\,(V^2(\mathbf{K}) + \sum_{l \neq 0} |b_l|^2)$$

Wir wollen dies die Formel von C. L. Siegel (1897 geb.) nennen. Es ist $V(\mathbf{K} \cap \mathbf{K} + 2g)$ genau dann $\neq 0$, wenn $g \in \mathbf{K}$; wenn also kein $g \neq 0$ aus Γ in $\mathbf{K}$ liegt, so ist

$$(11)\quad V(\mathbf{K}) \geqslant \frac{1}{2^n}\,(V^2(\mathbf{K}) + \sum_{b \neq 0} |b_l|^2).$$

Diese Formel von Siegel stellt eine Verschärfung des Satzes von Minkowski dar. Sie wurde von ihm 1935 (Acta Mathematica 65, 307–323) aufgestellt. nachdem schon Siegel für den Fall des Würfels auf andere Art eine analoge Formel in den Math. Annalen 87 (1922) 36–38 aufgestellt hatte. Diese letztere Arbeit schloß an eine Reihenentwicklung von Diskriminanten algebraischer Zahlkörper in den Göttinger Sitzungsberichten 1922 (S. 17–24) an.

Diese Formeln wurden verschiedentlich verallgemeinert, ich verweise auf die Arbeiten von R. Rado (Journal London, Bd. 21 (1946)), Cassels, (Journal London 22 (1947)) Bombieri (Bull. Italiano (1950)), E. Hlawka, (Math. Ann. 125 (1952)). Ramanathan in Acta Arithmetica 5 (1956) und Deuring, im Landau Gedächtnisbd. (1968) 11, haben (10) zur Aufstellung von Reihenentwicklungen der Diskriminanten von algebraischen Zahlkörpern bzw. Algebren benützt.*

Eine Schwierigkeit der Anwendung dieser Formel besteht darin, daß die Integrale b_l schwer auszuwerten sind. Mit diesen Integralen, die eine Verallgemeinerung der Frauenhofer'schen Beugungsintegrale sind, hat sich der Verfasser in mehreren Arbeiten beschäftigt, so z.B. in den Monatsheften 54 (1950).

§ 6. Der Übertragungssatz von Mahler

Dieser Satz von Mahler bezieht sich auf die von Minkowski entwickelte Theorie der konvexen Körper. Man kann jedem konvexen Körper $\mathbf{K}$ einen polaren oder dualen konvexen Körper $\mathbf{K}^*$ zuordnen: Ist f die Distanzfunktion von $\mathbf{K}$, so ist die Distanzfunktion H von $\mathbf{K}^*$ definiert durch:

$$(1) \qquad H(y) = \sup_x \frac{|\langle x, y\rangle|}{f(x)}$$

Es ist also stets

$$f(x)H(y) \geqslant |\langle x, y\rangle| .$$

Dann kann man jedem Gitter Γ ein polares Gitter Γ^* zuordnen. Es sei Γ nach § 1 Formel (1) durch die Matrix B gegeben. Dann ist dem polaren Gitter Γ^* die Matrix A zugeordnet, wo $B^T A = E$ ist (B^T transponierte Matrix zu B). Abstrakt ausgedrückt: Ist g ein Gitterpunkt aus Γ und g_1^* ein Gitterpunkt aus Γ^*, so ist das Skalarprodukt $\langle g, g_1^*\rangle$ stets eine ganze Zahl.

Es seien nun $M_1, M_2, \ldots M_n$ die Minima von f in bezug auf Γ und $M_1^*, \ldots$ $\ldots, M_n^*$ die Minima von H in bezug auf Γ^*, dann sagt der sogenannte Übertragungssatz von K. Mahler (1939) aus, daß

$$M_j \approx \frac{1}{M_{n-j+1}^*} ,$$

das heißt genauer, es ist

$$(2) \qquad 1 \leqslant M_j M_{n-j+1}^* \leqslant c_n, \quad (\text{z.B. } c_n = (n!)^2).$$

Der Beweis ist ganz einfach.

Es seien $m_1, m_2, \ldots m_n$ ein System von Minimalvektoren zu f und m_1^*, $m_2^*, \ldots m_n^*$ eines von H. Betrachten wir $m_1, \ldots m_j$ und $m_1^*, \ldots m_{n-j+1}^*$, so kann das erste System auf dem zweiten System nicht senkrecht stehen, sonst hätten wir $n+1$ linear unabhängige Vektoren. Es existiert also ein

---

* **Eine neue Arbeit stammt von Bela Uhrin**

$m_k (k \leqslant j)$ und ein $m_l^* (l \leqslant n - j + 1)$, sodaß $|\langle m_k, m_l^* \rangle| \geqslant 1 > 0$ ist, d.h. es ist

$$1 \leqslant f(m_k)\, H(m_l^*) \leqslant f(m_j)\, H(m_{n-j+1}^*) = M_j M_{n-j+1}^*$$

nun ist (2. Fundamentalgleichung)

$$JM_1 \ldots M_n \leqslant 2^n\, d\,(\Gamma)$$
$$J^* M_1^* \ldots M_n^* \leqslant 2^n\, d\,(\Gamma^*)^{-1}$$

also $JJ^* M_1 M_n^* M_2 M_{n-1}^* \ldots \leqslant 4^n$,

daraus folgt

$$M_j M_{n-j+1}^* < \frac{4^n}{JJ^*}.$$

1 Es ist ein geometrischer Satz, daß $JJ^* \geqslant \frac{4^n}{n!}$ ist, und daraus folgt unsere Behauptung. Eine Anwendung davon soll nun besprochen werden, und zwar soll (2) auf das inhomogene Problem der Geometrie der Zahlen angewendet werden.

Es ist festzustellen, wann

$$\bigcup_{g \in \Gamma} (\mathbf{K} + g) = \mathbb{R}^n \text{ ist.}$$

f sei die Distanzfunktion von $\mathbf{K}$ und x_0 ein beliebiges Produkt des $\mathbb{R}^n$. Es seien $m_1, \ldots m_n$ wieder ein System von Minimalpunkten. Es gilt, weil sie linear unabhängig sind:

$$x_0 = \xi_1 m_1 + \ldots + \xi_n m_n.$$

Wir bestimmen ganze Zahlen k_j so, daß $|\xi_j - k_j| \leqslant \frac{1}{2}$ ist und betrachten den Gitterpunkt $g = k_1 m_1 + \ldots + k_n m_n$. Es ist

$$\begin{aligned} f(x_0 - g) &\leqslant |\xi_1 - k_1| f(m_1) + \ldots + |\xi_n - k_n| f(m_n) \leqslant \\ &\leqslant \frac{1}{2}\,(M_1 + \ldots + M_n) \leqslant \frac{nM_n}{2} \end{aligned} \tag{3}$$

Benützt man die zweite Minkowskische Fundamentalgleichung, so kann man M_n durch M_1 abschätzen und erhält

$$f(x_0 - g) \leqslant \frac{2^n nJ}{M_1^{n-1}} = t. \tag{4}$$

Schätzt man mit dem Mahler'schen Übertragungssatz M_n durch das erste Minimum des dualen Körpers ab, so erhält man

$$(4') \quad f(x_0 - g) \leqslant \frac{(n!)^2 n}{M_1^*} = t^*$$

Der Körper $f(x) \leqslant t$ (bzw. t^*) und seine Translate überdecken also den $\mathbb{R}^n$. Man kann folgende Frage stellen:

Wie sieht dies aus, wenn der Körper nicht konvex ist? Ein Spezialfall wurde bereits von Minkowski behandelt und zwar handelt es sich hier um das berühmte Problem des inhomogenen Linearformsatzes. In diesem Problem wird die Frage gestellt, wie muß bei gegebenem Gitter Γ die Konstante c_n gewählt werden, damit der Körper

$$(5) \quad |x_1 \ldots x_n| \leqslant \frac{1}{c_n} \, \mathrm{d}(\Gamma)$$

und seine Translate den $\mathbb{R}^n$ überdecken. Die Minkowskische Vermutung besagt, daß dieses c_n existiert und das das größte $c_n = 2^n$ ist. Man muß dabei beachten, daß nicht einmal die Existenz von c_n selbstverständlich ist. Minkowski selbst bewies diese Vermutung für $n = 2$ in seinen diophantischen Approximationen. Für $n = 3$ wurde es von Remak bewiesen. Die Existenz von c_n wurde zuerst von Cebotarew (1894–1947) 1934 und nach einer anderen Methode von Siegel in einem Brief an Mordell (1937)
2 bewiesen. Čebotarew zeigte dies mit der guten Schranke $c_n = 2^{\frac{n}{2}}$. Für $n = 4$ wurde die Minkowskische Vermutung von F. J. Dyson bewiesen. Für $n = 5$ wurde es 1972 von Skubenko bewiesen.

Mit dieser Vermutung und der Verbesserung der Schranke haben sich viele Mathematiker beschäftigt, so z.B. Mordell, Davenport, Macbeath und P. Gruber. Die beste Schranke, die man bisher gefunden hat (1977), ist:

$$c_n = 2^{\frac{n}{2}} \cdot \left(\frac{n}{lg^2 n} \right)^{\frac{1}{3}} \epsilon_n \,,$$

wo $\lim\limits_{n \to \infty} \epsilon_n = e^{-2} = 0{,}13\ldots$ ist.

Es wurde gesprächsweise die Vermutung geäußert, daß für großes n die Minkowskische Vermutung zu modifizieren ist.

§ 7. Die topologische Theorie der Geometrie der Zahlen

Mahler hat in einer Reihe von großen Arbeiten, (beginnend 1946 in Proc. Roy. Soc. London A 187, 151–187) eine allgemeine Theorie entwickelt, deren Begriffsbildungen sich als fundamental erwiesen haben. Man kann den Ursprung dieser Begriffsbildung für den 2- und 3-dimensionalen Raum auf Minkowski selbst zurückführen, wo in anschaulicher Weise nicht nur ein Gitter betrachtet wird, sondern gleich eine ganze Menge. Man könnte diese Mahlersche Theorie als die topologische Theorie der Geometrie der Zahlen bezeichnen.

Wir betrachten jetzt die Menge aller Gitter im $\mathbb{R}^n$. Es soll nun der grundlegende Begriff der konvergenten Folge von Gittern erklärt werden.

Definition 1: Eine Folge $\Gamma_1, \Gamma_2, \ldots$ von Gittern heißt konvergent gegen ein Grenzgitter Γ, wenn folgendes gilt:

1) Zu jedem Gitter g von Γ gibt es eine konvergente Folge von Gitterpunkten g_r aus $\Gamma_r (r = 1, 2, \ldots)$, sodaß $\lim\limits_{r \to \infty} g_r = g$.

2) Jeder Häufungspunkt jeder Folge von Punkten g_r aus Γ_r gehört zu Γ.

Aus dieser Definition folgt sofort, daß eine konvergente Folge nur ein Grenzgitter haben kann.

Es gilt nun der wichtige

Satz 1: *Damit eine Folge von Gittern $\Gamma_1, \Gamma_2, \ldots$ konvergent ist gegen ein Grenzgitter Γ, ist notwendig und hinreichend, daß Γ eine Gitterbasis $b_1, \ldots b_n$ und die Γ_r Gitterbasen $b_1(r), \ldots b_n(r)$ besitzen, so daß*

$$\lim_{r \to \infty} b_i(r) = b_i (i = 1, \ldots, n)$$

ist.

Aus dem Satz folgt sofort, daß jedes Gitter Γ Grenzgitter von konvergenten Folgen von Gittern ist. Ist nämlich $a_1, \ldots a_n$ Basis von Γ, dann leistet die Folge $\{\Gamma_r\}$ der Gitter mit den Basen $a_i\left(1 + \frac{1}{r}\right)$ das Gewünschte.

Folgende Definition ist naheliegend: Ein Gitter Γ heißt Häufungsgitter einer unendlichen Menge $\mathbf{M}$ von Gittern, wenn es in $\mathbf{M}$ eine unendliche konvergente Folge mit Grenzgitter Γ gibt.

Weiters gilt folgende

Definition 2: Eine Menge $\mathbf{M}$ von Gittern heißt beschränkt, wenn es zwei positive Konstante c_1, c_2 unabhängig von $\mathbf{M}$ gibt, sodaß für alle Γ aus $\mathbf{M}$

1) $d(\Gamma) \leqslant c_1$
2) $|g| \geqslant c_2$ für alle $g \neq 0$ aus Γ gilt.

Dann gilt

Satz 2: *Jede konvergente Folge $\{\Gamma\}$ ist beschränkt.*

Selbstverständlich ist nicht jede beschränkte Folge konvergent, aber es gilt der wichtige K. Mahler'sche *Auswahlsatz:*

Jede beschränkte unendliche Folge $\{\Gamma_r\}$ von Gittern Γ_r enthält eine konvergente unendliche Teilfolge. Es sei $\mathbf{S}$ eine Menge. Wir betrachten nun alle Gitter Γ, die keinen Gitterpunkt $\neq 0$ in $\mathbf{S}$ haben. Solche Gitter heißen *zulässige* Gitter in bezug auf **S**. Nicht jede Menge braucht zulässige Gitter besitzen, ein Beispiel dafür ist der ganze $\mathbb{R}^n$. Solche Mengen, die keine zulässigen Gitter besitzen, heißen vom unendlichen Typus, die anderen Mengen vom endlichen Typus. Jede beschränkte Menge ist vom endlichen Typus. Es ist ein schwieriges Problem, festzustellen, von welchem Typus die Menge ist. So ist es eine alte Vermutung, daß der Körper $|x_1^2 \pm \ldots \pm x_n^2| < 1$ für $n \geqslant 5$ von unendlichem Typus ist, wenn auch negative Vorzeichen vorkommen. Für großes n hat hier Davenport (siehe § 10) durch Anwendung der Kreismethode von Hardy und Littlewood einen Durchbruch erzielt. Mit Hilfe der gleichen Methode ist es vor kurzem W. Schmidt gelungen, das gleiche für den Körper $|x_1^r + \ldots + x_n^r| < 1$ zu zeigen, wenn r ungerade und n sehr groß ist.

Wenn **S** vom endlichen Typus ist, dann betrachten wir das Infimum der Determinante $d(\Gamma)$ aller zulässigen Gitter von **S**. Dieses Infimum heißt die *Minimaldiskrimante* $\Delta(\mathbf{S})$ von **S**. Ist $\Delta(\mathbf{S}) = 0$, dann heißt **S** vom Nulltypus (Beispiel: $\mathbf{S} = \{0\}$). Man erweitert die Definition von Δ, indem man für Mengen vom unendlichen Typus $\Delta(\mathbf{S}) = \infty$ festlegt.

Aus der obigen Definition folgt sofort, daß jedes Gitter Γ mit $d(\Gamma) < \Delta(\mathbf{S})$ einen Gitterpunkt $g \neq 0$ in **S** hat. Es ist daher nur notwendig, $\Delta(\mathbf{S})$ zu bestimmen und noch weiter alle Gitter Γ zu betrachten, für die $d(\Gamma) = \Delta(\mathbf{S})$ ist, denn für diese Gitter ist es offen, ob sie zulässig sind oder nicht. Selbstverständlich sind folgende zwei Tatsachen: Ist $\mathbf{S}_1$ Teilmenge von $\mathbf{S}_2$, so ist $\Delta(\mathbf{S}_1) \leqslant \Delta(\mathbf{S}_2)$ und $\Delta(\lambda\mathbf{S}) = |\lambda|^n \Delta(\mathbf{S})$. Mahler selbst entwickelt in seinen Arbeiten diese obigen Begriffsbildungen nur für das Innere $\mathbf{S}^0$. Die zugehörige Minimaldeterminante $\Delta(\mathbf{S}^0)$ heißt die Minimaldeterminante vom **S** im schwachen Sinn. (Es ist noch zu erwähnen, daß Mahler nur Sternbereiche betrachtet. **S** ist Sternmenge, wenn für alle $x \in \mathbf{S}$ und alle λ mit $0 \leqslant \lambda \leqslant 1$ auch $\lambda x \in \mathbf{S}$.* Besitzt **S** ein Volumen, so ist $\mathrm{Vol}(\lambda\mathbf{S}) = \lambda^n \mathrm{Vol}(\mathbf{S})$. Dabei ist $\lambda\mathbf{S}$ die Menge aller $\lambda x (x \in \mathbf{S})$). Später hat es sich als zweckmäßig erwiesen, allgemeinere Mengen zugrunde zu legen.

Nun kommt folgende, weitere grundlegende

Definition 3: Ein Gitter Γ heißt *kritisches* Gitter von **S**, wenn $d(\Gamma) = \Delta(\mathbf{S})$ ist und es eine konvergente Folge zulässiger Gitter von **S** gibt, deren Grenzgitter Γ ist. Ein Gitter Γ heißt schwaches kritisches Gitter von **S**, wenn es kritisches Gitter von $\mathbf{S}^0$ ist. Für offene Mengen und, was wichtiger ist, für Sternbereiche fallen die Begriffe kritisches und schwach kritisches Gitter zusammen. Man kann in diesem Fall zeigen: Ein Gitter Γ ist in bezug auf einen Sternbereich genau dann kritisch, wenn

1) $d(\Gamma) = \Delta(\mathbf{S})$ ist und

2) wenn Γ zulässig in bezug auf das Innere $\mathbf{S}^0$ ist.

Man kann folgenden **Satz 4** zeigen:

---

* Eine Sternmenge ist *Sternbereich*, wenn O innerer Punkt ist.

Eine beschränkte Menge besitzt genau dann ein kritisches Gitter, wenn $\Delta(\mathbf{S}) > 0$ *ist.* Die Bestimmung von $\Delta(\mathbf{S})$ und darüber hinaus, die Bestimmung der kritischen Gitter ist eine sehr schwierige Aufgabe. Schon Minkowski hat sich mit diesem Problem für konvexe Körper beschäftigt, so z.B. für den ebenen Bereich $|x_1|^p + |x_2|^p < 1$ für $p \geqslant 1$. Er spricht in seinem Buch „Diophantische Approximationen" eine Vermutung über die Minimaldiskriminante in Abhängigkeit von p aus. Diese Vermutung hat sich nicht für alle Werte von p als richtig erwiesen. Sie wurde von der russischen Schule, insbesondere um Malyšew studiert. So ist die Minkowski'sche Vermutung richtig, für alle $p \geqslant 5$ (1975–1977).

Für Sternbereiche, die nicht konvex sind, werden die Minimaldiskriminanten von Mordell, Davenport und ihrer Schule bestimmt. Es handelt sich hier immer um 2- und 3-dimensionale Probleme. Besonders hervorzuheben ist die Bestimmung der Minimaldiskriminante für die Bereiche

$$\mathbf{S}_1\colon |x_1x_2x_3| < 1 \text{ mit } \Delta(\mathbf{S}_1) = 7, \mathbf{S}_2 = \frac{1}{2}|x_1|(x_2^2 + x_3^2) \leqslant 1 \quad \text{mit } \Delta(\mathbf{S}_2) = \sqrt{23}.$$

Zu bemerken ist noch,daß zuerst ein kritisches Gitter zu bestimmen versucht wird, Man benützt dazu die Eigenschaft der kritischen Gitter, extreme Gitter zu sein. Ein Gitter Γ heißt extrem in bezug auf $\mathbf{S}$, wenn es eine Folge von zulässigen Gittern Γ_r gibt, die Γ als Grenzgitter haben und wenn für jede solche Folge bis auf endlich viele $d(\Gamma_r) \geqslant d(\Gamma)$ ist. Die kritischen Gitter sind jene extremen Gitter mit kleinster Determinante. Man versucht nun, Eigenschaften von extremen Gittern zu finden. Man sieht sofort, daß kein Gitterpunkt $g \neq 0$ eines extremen Gitters im Inneren von $\mathbf{S}$ liegen kann. Man kann nun zeigen: Ist $\mathbf{S}$ beschränkt, dann besitzt jedes extreme Gitter n linear unabhängig Gitterpunkte auf dem Rand von $\mathbf{S}$. Gelingt es, diese Gitterpunkte auf dem Rand von $\mathbf{S}$ zu finden, dann kann man eine Abschätzung für $d(\Gamma)$ finden und bilden sie eine Gitterbasis, dann hat man bereits den Wert der Determinante eines extremen Gitters. Können die Determinante der Menge aller extremen Gitter nur endlich viele Werte annehmen, so liefert der kleinste Wert die Minimaldiskriminante. Diese Methode wurde schon implizit von Minkowski benützt, ist aber für höher dimensionale Fälle fast nicht brauchbar. Für konvexe Körper kann man diese Methode in jedem Fall in endlich vielen Schritten durchführen, aber nur theoretisch. Ein wichtiger, vorbereitender Schritt ist das Problem, das schon früher erwähnt wurde (§ 2), wann nämlich Minimalpunkte eines konvexen Körpers das ganze Gitter erzeugen. Für alle Dimensionen kennt man die Minimaldiskriminante des Würfels und auch alle kritischen Gitter. Es sind dies alle Gitter, abgesehen von Permutationen, von der Gestalt:

$$\begin{aligned} &L_1(x_1, \ldots, x_n) = x_1 \\ (6)\quad &L_2(x_1, \ldots, x_n) = a_{21}x_1 + x_2 \\ &\ldots\ldots\ldots\ldots\ldots\ldots\ldots\ldots \end{aligned}$$

$$L_n(x_1, \ldots, x_n) = a_{n1}x_1 + a_{n2}x_2 + \ldots + x_n.$$

Minkowski hat dies vermutet. G. Hajos (1912–1972) hat dies bewiesen (1940). Für die Kugel kennt man die Minimaldiskriminante bis zu 8 Dimensionen. Mit der Bestimmung von kritischen bzw. extremen Gitter beschäftigt sich vor allem die australische Schule. Diese extremen Gitter gestatten auch wichtige Anwendungen, z.B. in der Theorie des Codes.

Für den höher dimensionalen Fall haben sich andere Methoden, so die Maßtheorie der Gitter bzw. analytische Methoden als kräftiger erwiesen. Es sei (f Distanzfunktion)

$$\zeta(s, \Gamma, f) = \sum_{g \in \Gamma - 0} \frac{1}{f(g)^s}.$$

Die Reihe ist konvergent für $s > n$. Sie wurde von Minkowski, G. Herglotz (1881–1953) und auch von E. Hlawka (Monatsh. 54 (1950) 100–107) studiert. R. Rankin (1915 geb.) hat für $n = 2$ als Erster im Fall des Kreises gezeigt, daß ζ für großes s ein Extrem annimmt, wenn Γ kritisch ist. Der Fall der Kugel wurde von der russischen Schule um Delone (1890 geb.) untersucht. Es hat sich herausgestellt, daß die Funktion für großes s ein lokales Minimum hat, wenn Γ ein extremes Gitter ist. Es sei zur allgemeinen Theorie noch folgendes bemerkt: Fast alle Körper, für die man die Minimaldiskriminante bestimmt hat, sind Sternkörper mit Automorphismen, d.h. sie gestatten lineare Transformationen, die den Nullpunkt fest lassen. Daraus folgen, wie Mahler und später Davenport und Rogers gezeigt haben, schon Aussagen über die Minimaldiskriminante. Man versucht auch, echte Teiler $\mathbf{S}_1$ von $\mathbf{S}$ zu finden. Dabei heißt $\mathbf{S}_1$ echter Teiler von $\mathbf{S}$, wenn $\mathbf{S}_1$ echte Teilmenge von $\mathbf{S}$ ist und $\Delta(\mathbf{S}_1) = \Delta(\mathbf{S})$ ist. Besitzt $\mathbf{S}$ keine echten Teiler, so heißt $\mathbf{S}$ irreduzibel. Man kann sich daher auf irreduzible Mengen beschränken, wenn $\mathbf{S}$ eine Sternmenge ist, wie C. A. Rogers gezeigt hat.

§ 8. Die Maßtheorie der Geometrie der Zahlen

Im allgemeinen muß man sich auf Abschätzungen für $\Delta(\mathbf{S})$ beschränken. Ist $\mathbf{S}$ konvex, so liefert der Minkowski'sche Fundamentalsatz sofort, wenn V das Volumen von $\mathbf{S}$ ist, die untere Abschätzung $\Delta(\mathbf{S}) \geqslant \frac{V}{2^n}$. Für allgemeine Sternbereiche gibt es, wie Mahler gezeigt hat, keine solche Abschätzung nach unten. Dagegen gibt es eine allgemeine Abschätzung für symmetrische Sternbereiche nach oben, wie schon Minkowski vermutet hat, und sie lautet sogar schärfer:

$\Delta(\mathbf{S}) \leqslant \frac{V}{2\zeta(n)}$ wo ζ die Riemann'sche Zetafunktion ist. Minkowski konnte dies für die Kugel beweisen, doch ist der Beweis nicht verallgemeinerungsfähig. Dies dürfte der Grund sein, warum Minkowski den zweiten Teil seiner Geometrie der Zahlen nicht veröffentlicht hat. Die obige Vermutung konnte ich 1943 allgemein beweisen, ich konnte sogar zeigen, daß für beliebige Mengen **S**, wenn sie nur quadrierbar sind, $\Delta(\mathbf{S}) \leqslant V$ ist. Das heißt anders ausgesprochen: Ist **S** eine beliebige beschränkte Menge und ist sein Volumen < 1, so gibt es ein in bezug auf **S** zulässiges Gitter Γ mit $d(\Gamma) = 1$. Dies folgt aus folgendem Deformationssatz, den ich damals aufgestellt habe: Es sei φ eine beliebige, quadrierbare Menge, die außerhalb eines Würfels verschwindet. Dann gibt es zu jedem $\epsilon > 0$ ein Gitter Γ mit $d(\Gamma) = 1$, sodaß

$$(1) \qquad \sum_{g \neq 0} \varphi(g) < \int_{\mathbb{R}^n} \varphi(x)\, dx + \epsilon$$

Nimmt man für φ die Indikatorfunktion von **S**, so ist dann

$$(2) \qquad \sum_{g \neq 0} \varphi(g) \leqslant V(\mathbf{S}) + \epsilon$$

Ist $V(\mathbf{S}) < 1$, so ist bei passender Wahl von ϵ, $\sum_{\varphi \neq 0} \varphi(g) < 1$, also $\varphi(g) = 0$ für alle $g \neq 0$ aus Γ.

Ist **S** symmetrisch, so funktioniert die gleiche Schlußweise, wenn $V(\mathbf{S}) < 2$ ist.

Bei Sternbereichen kommt die ζ-Funktion dadurch hinein, daß man in der obigen Schlußweise nur primitive Gitterpunkte betrachtet. Dieser Deformationssatz* wurde von Siegel und anderen Mathematikern vertieft und vor allem von C. A. Rogers und W. Schmidt zu einer Maßtheorie in der Menge A aller Gitter Γ mit $d(\Gamma) = 1$ ausgebaut. Man kann bei der Definition dieses Maßes nach Siegel die Minkowski'sche Reduktionstheorie der quadratischen Formen benützen. Diese Methode wurde von Macbeath und Rogers vereinfacht. Wie Rogers bemerkt hat, ist es noch einfacher und damit schließt er an den Beweis meines Deformationssatzes an, ein Integral auf A so zu definieren:

$$\int_{\mathrm{A}} \rho(\Gamma)\, d\Gamma = \lim_{\omega \to 0} \int_0^1 \ldots \int_0^1 \rho\,(\Gamma(\vartheta_1, \ldots, \vartheta_{n-1}, \omega))\, d\vartheta_1, \ldots d\vartheta_{n-1},$$

wo das Gitter $\Gamma(\vartheta_1, \ldots, \vartheta_{n-1}, \omega)$ erzeugt wird durch die Vektoren

$$(*) \qquad \begin{array}{l} (\omega, 0, \ldots, 0, \vartheta_1 \omega^{-n+1}) \\ (0, \omega, \ldots, 0, \vartheta_2 \omega^{-n+1}) \\ \ldots\ldots\ldots\ldots\ldots\ldots \\ (0, \ldots, 0, \omega, \vartheta_{n-1}\, \omega^{-n+1}) \\ (0, \ldots, 0, \omega^{-n+1}) \end{array}$$

* Von R. de Buda wurde (1) auf Codierung angewendet

Dann gilt in dieser Sprechweise

(2′) $\int \Phi(\Gamma, \mathbf{S})\, d\Gamma = V(\mathbf{S})$

wo

(3) $$\Phi(\Gamma, \mathbf{S}) = \sum_{g \neq 0, g \in \Gamma} \varphi(g)$$

ist (*Siegelschen Mittelwertsatz*). Wird (2′) auf Adele verallgemeinert, so erhält man nach L. Verner (L'enseignement 23 (1977)) die Siegel'sche Maßformel in der Theorie der quadratischen Formen.

Man kann erwarten, daß man schärfere Resultate erzielt, wenn man für Φ andere Funktionen wählt. Es sei nun $\mathbf{A}(\mathbf{S})$ die Menge der zulässigen Gitter in bezug auf $\mathbf{S}$ und $\alpha(\Gamma, \mathbf{S})$ die Indikatorfunktion dieser Menge. Es ist $\alpha(\Gamma, \mathbf{S}) = \prod_{g \neq 0} (1 - i_s(g))$, wo i_s die Indikatorfunktion von $\mathbf{S}$ ist. Es ist $m(\mathbf{A}(\mathbf{S})) = \int \alpha(\Gamma, \mathbf{S})\, d\Gamma$ das Maß der zulässigen Gitter von $\mathbf{S}$ mit $d(\Gamma) = 1$. Ist dieses Maß > 0, dann existiert ein zulässiges Gitter $\Gamma(\mathbf{S})$ mit $d(\Gamma) = 1$. Daher ist die Minimaldeterminante des Körpers $\Delta(\mathbf{S}) \leqslant 1$.

Wenn $\mathbf{S}$ ein Sternbereich ist, kann man sogar $\Delta(\mathbf{S})$ bestimmen. Es gilt nämlich

(4) $$\Delta(\mathbf{S}) = \frac{1}{\lambda_0^n} \text{ wo } \lambda_0 = \min \lambda \{m(\mathbf{A}(\lambda \mathbf{S})) = 0\},$$

Die Durchführung dieses Programms ist sehr mühsam und erfordert ungeheuere Rechnungen. $m(\mathbf{A}(\mathbf{S}))$ wird durch eine unendliche Reihe dargestellt. Die Glieder dieser Reihe sind mehrfache Integrale. Für den Fall der Ebene konnte diese Summe für gewisse Bereiche ausgewertet werden. Wir wollen kurz einige Resultate, die erhalten wurden, angeben. Es ist $m(\mathbf{A}(\mathbf{S})) = e^{-V}(1 - R)$, wobei $|R| < V^{n-1} n^{-n+1} e^{V+n}(1+\epsilon) + \epsilon$ für $V \leqslant n-1, n \geqslant n_0(\epsilon)$.

Ist $Q(\mathbf{S}) \equiv V(\mathbf{S})\Delta(\mathbf{S})^{-1}$, so folgt daraus $Q(\mathbf{S}) \geqslant nr - 2$, wobei $r \sim 0{,}278$ ist. Man erhält auf andere Weise ein besseres Resultat, nämlich: $Q(\mathbf{S}) \geqslant 16/15$ für $n = 2$ (Schmidt 1963). Allgemein: $Q(\mathbf{S}) \geqslant ns - c_2$, wobei $s \sim 0{,}34$ ist.

§ 9. Ausfüllung und Überdeckung

Wir wollen diese Ergebnisse auch geometrisch deuten, damit kommen wir zurück auf den ursprünglichen Beweis von Minkowski. Es sei eine Menge S gegeben, weiters eine Menge von Vektoren $a_1, a_2, \ldots$ und wir bilden die Translate $S + a_i$. Die Menge der Translate bildet eine *Ausfüllung*, wenn verschiedene Translate stets disjunkt sind, und sie bilden eine *Überdek-*

kung, wenn die Vereinigungsmenge aller dieser Translate gleich dem ganzen Raum $\mathbb{R}^n$ ist. Ist $\mathbf{K}_\rho$ eine Kugel mit dem Mittelpunkt Null und Radius ρ, so heißt

$$\lim_{\rho\to\infty} \frac{V(\bigcup_i (\mathbf{S}+a_i)\cap \mathbf{K}_\rho)}{V(\mathbf{K}_\rho)}$$

wenn er existiert, die Dichte der Ausfüllung bzw. Überdeckung. Bildet die Menge der a_i ein Gitter Γ, so spricht man von einer gitterförmigen Ausfüllung bzw. Überdeckung.

Es folgt leicht, wenn **S** konvex ist, daß die Dichte bei einer solchen Ausfüllung existiert (aus den Ausführungen in § 1) und gleich $V(\mathbf{S})/\mathrm{d}(\Gamma)$ ist. Man nennt das Supremum der Dichte aller Ausfüllungen durch **S** und seine Translate, die Dichte $\delta(\mathbf{S})$ der dichtesten Ausfüllung. Betrachtet man nur gitterförmige Ausfüllungen durch **S** und seine Translate, so nennt man das zugehörige Supremum $\delta_L(\mathbf{S})$ die Dichte der dichtesten, gitterförmigen Ausfüllung. Es ist

$$\delta_L(\mathbf{S}) = \frac{V(\mathbf{S})}{2^n\Delta(\mathbf{S})}.$$

Damit gewinnt die Größe der Minimaldiskriminante eine geometrische Bedeutung. Nehmen wir den Fall der Kugel, so erhalten wir für δ_L für $n \leqslant 8$ folgende kleine Liste:

n	2	3	4	5	6	7	8
δ_L	0,9	0,74	0,61	0,46	0,379	0,2952	0,253

Man kann zeigen, daß die Dichte δ_L bei der Kugel $n \to \infty$ gegen Null geht*. Es sei noch eine weitere Bemerkung hinzugefügt: Es ist klar, daß stets $\delta_L(\mathbf{S}) \leqslant \delta(\mathbf{S}) \leqslant 1$ ist. Im Fall der Ebene hat C. A. Rogers gezeigt, daß diese beiden Größen einander gleich sind. Wie es für höhere Dimensionen, besonders für den 3-dimensionalen Raum steht, ist offen. Mit der Ausfüllung des 3-dimensionalen Raumes durch Kugeln hat sich bereits Newton (1643–1727)** beschäftigt. Die Definitionen, die wir im Falle der Ausfüllungen eingeführt haben, können wir auch für die Überdeckungen durchführen, nur ist bei den analogen Größen δ^* bzw. δ^*_L, das Supremum durch das Infinum der Dichten zu ersetzen. Wir geben für den Fall der Kugel wieder eine kleine Liste:

---

* Nach § 4 (Fall der Kugel **K**) ist ja $\delta_L(\mathbf{K}) \leqslant 2^{-\frac{n}{2}}\left(\frac{n}{2}+1\right)$

** Vor ihm J. Kepler (1571–1630) in der Arbeit „Strena seu de Nive Sexangula" (Neujahrsgabe(n) oder vom sechseckigen Schnee) 1611 (Diesen Hinweis verdanke ich Dezsö Nagy (Budapest))

n	2	3	4	5	
δ_L^*	1,209	1,4635	1,765	2,42	...

Man kann zeigen, daß die Dichte der dünnsten Überdeckung durch Kugeln für $n \to \infty$ ebenfalls $\to \infty$ geht. Im Fall der Ebene gilt wieder $\delta^* = \delta_L^*$. Mit Hilfe der Maßtheorie der Geometrie der Zahlen hat man für die Dichte von Überdeckungen ebenfalls gute Resultate erzielt.

§ 10. Die analytische Methode von H. Davenport

Zum Schluß soll nun skizziert werden, wie Davenport (1957) gezeigt hat, daß der Körper $|x_1^2 + \ldots + x_r^2 - (x_{r+1}^2 +, \ldots + x_n^2)| < 1$ für $n \geq 21$ von unendlichem Typus ist.

das heißt, daß für jede indefinite Form Q und zu jedem $\epsilon > 0$ stets ganze Zahlen $x_1 \ldots x_n$ existieren,* sodaß $|Q(x_1 \ldots x_n)| < \epsilon$ ist. Der Beweis wird indirekt geführt, indem man annimmt, daß stets

$$(*) \quad |Q(x_1 \ldots x_n)| \geq \epsilon$$

ist. Es sei $(c_1 \ldots c_n)$* eine Lösung von $Q(c_1 \ldots c_n) = 0$ und P eine große Zahl. Man bildet nun $P_j = \left[\frac{2}{3} c_j P\right]$ $(j = 1 \ldots n)$ und die Summe

$$S(\alpha) = \sum_{x_1=P_1}^{2P_1} \ldots \sum_{x_1=P_n}^{2P_n} e^{2\pi i\alpha(Q(x_1 \ldots x_n)|\epsilon)}$$

Man geht von folgender Idee aus:

Es ist mit

$$K(\alpha) = \frac{\sin^2 \pi\alpha}{(\pi\alpha)^2}$$

$$(*) \quad \int_{-\infty}^{\infty} e^{2\pi i\xi\alpha} K(\alpha)\, d\alpha = 0 \quad \text{für} \quad |\xi| \geq 1.$$

Aus (*) folgt:

$$\int_{-\infty}^{\infty} e^{2\pi i(Q(x)|\epsilon)} K(\alpha)\, d\alpha = 0,$$

also auch

$$\int_{-\infty}^{\infty} S(\alpha) K(\alpha)\, d\alpha = 0.$$

* Nicht alle 0.

Es wird gezeigt, daß diese Behauptung falsch ist, daß nämlich tatsächlich $\int > P^{n-2}$ gilt. Dabei wird allerdings ein stärkerer Kern K_n statt K benützt. Das Integral wird in folgende vier Integrale zerlegt

$$\int = I_1 + I_2 + I_3 + I_4$$

Dabei ist $I_4 = \int\limits_{|\alpha| > P^4} \frac{1}{}$.

Es ist

$$(1) \quad I_4 < P^{\frac{3}{4}n}$$

Weiters wird gezeigt, daß

3 $$I_1 = \int\limits_{|\alpha| < p-3/4} > p^{n-2}$$

Das wird so gezeigt, daß $S(\alpha)$ durch das Integral

$$I(\alpha) = \int \ldots \int e^{2\pi i(Q(x_1 \ \ldots x_n)|\epsilon)} \, dx_1 \ldots dx_n$$

approximiert wird, das die Größenordnung P^{n-2} hat.

Weiters ist

4 $$I_1 = \int\limits_{|\alpha| < p - \frac{3}{4}} > p^{n-2}$$

$$I_2 = \int\limits_{\frac{1}{p^{\frac{3}{4}}} < |\alpha| < \frac{1}{\sqrt{P}}} < P^{n-2-\frac{1}{4}}$$

Nun betrachten wir

$$I_3 = \int\limits_{\frac{1}{\sqrt{P}} < |\alpha| < P^{\frac{1}{4}}}$$

Hier liegt die Hauptschwierigkeit. In der ersten Arbeit nimmt Davenport $n > 100$ an und kann zeigen, daß

$$I_3 < P^{n-2-\frac{1}{18}} (\log P)^n$$

ist. Damit ist der Widerspruch gegeben. Die Behandlung von I_3 war der entscheidende Durchbruch, während die Behandlung der anderen Integrale eine bekannte Technik von Hardy-Littlewood benützt. Zu I_3 möchte ich noch folgendes bemerken: Es wird, wie in der Theorie des Waring'schen Problems üblich,

$$|S(\alpha)^2| = \sum_{x,y} e(\alpha(Q(x) - Q(y)) = \sum_{x,z} e(\alpha(Q(x+z) - Q(x))$$

wo $Q(x+z) - Q(x)$ Bilinearform in x und z ist, betrachtet. Hier wird nun die Geometrie der Zahlen benützt, indem folgendes gilt:

Ist K eine Zahl, sodaß $M_l \leqslant K < M_{l+1}$, so ist die Anzahl der Gitterpunkte $f(g) \leqslant K$ kleiner als

$$\frac{K^l}{M_1 \ldots M_l} .$$

Diese Methode hat Davenport immer weiter verfeinert, sodaß er die Schranke 21 erreicht hat. Man vermutet, daß es schon ab $n = 5$ gilt. Diese Davenport'sche Methode wurde neuerdings von W. Schmidt* auf den Fall:

$$|x_1^r + \ldots + x_n^r| < 1$$

für ungerades r und großes n angewendet, wie dies schon früher gesagt wurde.

Literatur

Davenport, H., Collected papers, Academic Press 1970
Minkowski, Gesammelte Abhandlungen I, II, Teubner 1911, herausgegeben von D. Hilbert unter Mitwirkung von A. Speiser und H. Weyl
Siegel, C. A., Gesammelte Abhandlungen I, II, III, Springer 1966, Band IV (1979)

Lehrbücher

Cassels J. W. S., An Introduction to the Geometry of Numbers, Springer 1959
Florian A., Diskrete Geometrie (in Vorbereitung)
Lekkerkerker G. C., Geometry of Numbers, North-Holland 1969
Mahler K., The Geometry of Numbers (1956) Boulder Lectures
Minkowski H., Geometrie der Zahlen, Teubner 1896, Nachdruck 1910, 1925, Nachdruck New York–Chelsea 1953
Minkowski H., Diophantische Approximationen, Teubner 1907, Nachdruck 1927, Chelsea 1957, Physica Verlag 1961
Rogers C. A., Packing and Covering, Cambridge Tracts 54 (1964)

---

* Diophantine Inequalities for Forms of odd Degree (1979) (im Druck)

Berichte (eine Auswahl)

Gruber P., Geometry of numbers, 1978. Proc. Symp. Siegen
Hlawka E., Zur Geometrie der Zahlen, Jahrb. Akad. Wiss., Göttingen 1963, 23–29
Keller O. H., Geometrie der Zahlen, Enz. Math. Wiss. I, 2, 27, Teubner 1954
Koksma J. F., Diophantische Approximationen, Ergebnisberichte Springer 1936, Nachdruck Chelsea 1950, Springer 1974
Saaty T. L. and *Alexander* J. M., Optimization and the Geometry of Numbers; Packing and Covering, Siam Review 17 (1975), 475–519.

Corrections

29.1 $JJ^* \geq \frac{4^n}{(n!)^2}$ 39.3 $\int\limits_{|\alpha|<p^{-3/4}}$ 39.4 wie 39.3

30.2 $c_n = 2^{\frac{n}{2}}$

Approximation von Irrationalzahlen und pythagoräische Tripel*

Bonner Mathematische Schriften **121** (1980) 1–32

Es ist mir eine besondere Freude, hier sprechen zu dürfen.** Zuerst möchte ich aber Herrn Prof. Peschl die Grüße des Mathematischen Instituts der Universität Wien und der Österr. Akademie der Wissenschaften überbringen. Das Symposium ist einem Meister der Analysis, der komplexen Analysis gewidmet, aber Prof. Peschl hat sich stets auch für Zahlentheorie interessiert und Arbeiten aus diesem Gebiet verfaßt und angeregt. Ich hebe nur seine Arbeit über die Fermatsche Kongruenz $x^p + y^p + z^p \equiv 0(\bmod p^2)$ hervor, welche in den Memoirs de l'Académie de Sciences Toulouse VI (1965) erschienen ist. Daher erlaube ich mir, über ein Thema aus diesem Gebiet zu sprechen. Es lautet: „Approximation von Irrationalzahlen und phytagoräische Tripel". Zunächst möchte ich etwas über die Approximation von Irrationalzahlen sagen. Bekanntlich läßt sich jede irrationale Zahl α durch rationale Zahlen beliebig genau approximieren, da die rationalen Zahlen auf der Zahlengeraden dicht liegen. Das heißt, zu jedem $\varepsilon > 0$ gibt es mindestens eine rationale Zahl $\frac{p}{q}$, so daß

$$\left|\alpha - \frac{p}{q}\right| < \varepsilon$$

ist. Wir bestimmen diese $\frac{p}{q}$ am einfachsten so, daß wir für eine beliebige, natürliche Zahl n αn betrachten, das heißt ein Vielfaches von α. Nehmen wir jetzt $p = [\alpha n]$, wo [ ] die nächstkleinere ganze Zahl bedeutet und $q = n$, so ist

$$\left|\alpha - \frac{p}{q}\right| < \frac{1}{n}.$$

Ist $n > \frac{1}{\varepsilon}$, so ist die Aufgabe gelöst. Man kann dies noch etwas verbessern, indem man die nächste ganze Zahl von αn nimmt. Es ist wohl bekannt, wie *Dirichlet* diese Approximation verbessert hat. Er hat gezeigt, daß es immer ein p und q gibt, so daß

$$\left|\alpha - \frac{p}{q}\right| < \frac{1}{q^2}$$

ist. Es gibt sogar unendlich viele solche $\frac{p}{q}$.

---

* Gegenüber dem Original unter Mitarbeit von Herrn Dr. Turnwald leicht korrigierte Fassung

** Vortrag am 9. Dez. 1977 in Bonn, anläßlich des 70. Geburtstages von Prof. E. Peschl.

Man kann dies auf verschiedene Arten zeigen, am einfachsten mit dem Dirichletschen Schubfachprinzip. Man betrachtet wieder die Vielfachen von α, aber jetzt nicht nur eines, sondern gleich mehrere: $0\alpha, 1\alpha, \ldots, N\alpha$. Wir betrachten nun $x_n = n\alpha - [n\alpha]$. Diese Zahlen liegen alle im Intervall $0 \leq \xi < 1$. Wir teilen dieses Intervall in N gleiche Teile von der Länge $\frac{1}{N}$. Dann muß eines dieser Intervalle von den $N+1$ Zahlen x_n mindestens 2 solche Zahlen enthalten. Daraus folgt sofort: es gibt ein $\frac{p}{q}$, so daß

$$\left|\alpha - \frac{p}{q}\right| < \frac{1}{qN}$$

mit $1 \leq p \leq N$ ist. Das ist der berühmte Dirichletsche Approximationssatz. Daraus folgt wieder die obige Behauptung, d.h. es gibt unendlich viele (p, q), so daß

$$\left|\alpha - \frac{p}{q}\right| < \frac{1}{q^2}. \tag{1}$$

Es ist vielleicht nicht sehr bekannt, daß man diesen Satz verschärfen kann, indem man die Ungleichung $1 \leq q \leq N$ durch $1 \leq q \leq N-1$ ersetzen kann. Man braucht dazu nur x_N, welches früher $\alpha N - [\alpha N]$ war, durch $x_N = 1$ zu ersetzen; also das halboffene Intervall $[0,1)$ durch $[0,1]$ ersetzen.

Es erhebt sich die Frage, ob man dies noch weiter verbessern kann. *Khintchine* zeigte, daß es jedenfalls mit $\frac{N}{2}$ nicht immer geht. Aber wie sieht es dazwischen aus? Es sei $c(\alpha)$ eine Zahl, mit der Eigenschaft, daß es zu jeder natürlichen Zahl N ganze Zahlen p und q gibt, mit der Eigenschaft, daß

$$1 \leq q \leq Nc(\alpha) \quad \text{und} \quad \left|\alpha - \frac{p}{q}\right| < \frac{1}{Nq}$$

ist. Es sei $\varepsilon > 0$. Es wurde 1968 von *Davenport* und *W.Schmidt* gezeigt, daß es zu fast allen α eine Folge N_j gibt, so daß

$$1 \leq q \leq N_j(1-\varepsilon), \quad \left|\alpha - \frac{p}{q}\right| < \frac{1}{N_j q}$$

nicht lösbar ist. Es ist aber vollkommen unbekannt, ob man z.B. N durch $N - \sqrt{N}$ ersetzen kann.

Wir haben uns jetzt mit der Approximation nach oben beschäftigt. Was kann man über die Abschätzung nach unten sagen? Im allgemeinen nichts, denn man kann zeigen: Ist $\phi > 0$ eine beliebige Funktion von q, so gibt es immer ein α, ja sogar unendlich viele, so daß

$$\left|\alpha - \frac{p}{q}\right| < \frac{1}{q\phi(q)}$$

gilt. Man setzt einfach

$$\alpha = \sum_{n=1}^{\infty} \frac{1}{g_1 \cdots g_n}$$

wo die (g_n) eine monoton wachsende Folge bilden. Man wählt dann die g_n so, daß $g_{N+1} \geq \phi(g_1 \dots g_N) + 1$ ist. Ist α algebraisch vom Grade g, so sagt der Satz von *Liouville* aus, daß für alle

$$\frac{p}{q} : \left|\alpha - \frac{p}{q}\right| > \frac{c}{q^g}$$

ist, wo c von α abhängt.

Man sieht also, daß im Fall $g = 2$, für den Fall quadratischer Irrationalitäten, sich der Exponent 2 in (1) nicht verschärfen läßt. Ist $g > 2$, so hat *K. Roth* bekanntlich gezeigt, daß man im Satz von *Liouville* g durch $2 + \varepsilon$ ersetzen kann. Die vorher zitierten Sätze haben zu einer Typeneinteilung der irrationalen Zahlen geführt, die im Ursprung wohl auf *H. Behnke* zurückgeht, aber erst von *Serge Lang* allgemein formuliert wurde.

Ich will aber diese Einteilung etwas modifizieren. Es sei f eine positivwertige Funktion auf N, dann heißt eine Zahl α vom Typus f (oder besser vom Typus $\leq f$) wenn für alle Brüche

$$\frac{p}{q} : \left|\alpha - \frac{p}{q}\right| \geq \frac{1}{qf(q)}$$

gilt. Insbesondere sagt man, α ist vom Potenztypus μ, wenn für jedes $\varepsilon > 0$, $f(q) = q^{\mu-1+\varepsilon}c(\varepsilon)$ genommen werden kann. α heißt schlecht approximierbar, wenn man $\mu = 2$ und $\varepsilon = 0$ nehmen kann.

Die quadratischen Irrationalitäten sind schlecht approximierbar. Es heißt α von einem Typus $\geq f$, wenn es unendlich viele $\frac{p}{q}$ mit

$$\left|\alpha - \frac{p}{q}\right| \leq \frac{1}{qf(q)}$$

gibt. Im besonderen betrachtet man den Fall des Potenztypus $f(q) = q^{\mu-1+\varepsilon}c(\varepsilon)$. Ist α vom Typus $\geq f$ und vom Typus $\leq f$, dann heißt α vom Typus f im scharfen Sinne. Aus den obigen Ausführungen folgt, daß man den Approximationssatz im wesentlichen nicht mehr verschärfen kann, was die Potenz q^2 betrifft. Es ist aber eine andere Verschärfung möglich. Z.B. sagt der Satz von *Hurwitz*, daß es zu jeder irrationalen Zahl α unendlich viele $\frac{p}{q}$ gibt, so daß

$$\left|\alpha - \frac{p}{q}\right| < \frac{1}{\sqrt{5}q^2}$$

ist, und daß $\sqrt{5}$ durch keine größere Zahl ersetzt werden kann. Wir haben bis jetzt erst eine irrationale Zahl α betrachtet. *Dirichlet* und *Kronecker* haben gleich ein ganzes System $\alpha_1, \alpha_2, \dots, \alpha_n$ betrachtet und Dirichlet hat mit dem gleichen Schubfachprinzip gezeigt:

Sind $\alpha_1, ..., \alpha_n$ beliebige reelle Zahlen, so gibt es zu jedem N eine natürliche Zahl q mit $1 \leq q \leq N^n$, so daß für geeignete ganze Zahlen $p_1, \ldots, p_n$

$$\left|\alpha_j - \frac{p_j}{q}\right| < \frac{1}{qN} \leq q^{-(1+\frac{1}{n})}$$

ist.

Gleichzeitig hat er gezeigt: Es gibt ganze Zahlen $q_1, \ldots, q_n, p$ so daß

$$|q_1\alpha_1 + \cdots + q_n\alpha_n - p| \leq \max(|q_1|, \ldots, |q_n|)^{-n}$$

gilt (und nicht alle q_i gleich Null sind).

Man kann auch hier die Frage stellen, ob man N^n durch eine kleinere Zahl ersetzen kann. Dies ist tatsächlich der Fall. Mit dieser Frage hat sich 1956 erstmals *Raisbeck* (Candian Journal 2, 283-88), beschäftigt und auf meine Anregung hat *Kaindl* (Sitzungsberichte der Österr. Akademie d. Wissenschaften, 25. Juni 1976, G. Kaindl: "Simultane diophantische Approximationen") folgendes gezeigt: Man kann N^n durch $N^n - N^{n-1} \ldots - N - 1$ ersetzen. Diese Schranken können für n=2,3,4 nicht mehr verbessert werden. Man kann auch hier wieder z.B. den Begriff des Potenztypus einführen. Man sagt $\alpha_1, ..., \alpha_n$ sind vom Potenztypus $\leq \mu$, wenn für alle $\varepsilon > 0$ und $(h_1, \ldots, h_n, p) \neq (0, \ldots, 0, 0)$

$$R(h)^{\mu+\varepsilon} |h_1\alpha_1 + \ldots + h_n\alpha_n - p| \geq c(\varepsilon) > 0$$

ist, wo

$$R(h) = R(h_1, \ldots, h_n) = \prod_{j=1}^{n} \max(|h_j|, 1)$$

ist. Sind also alle $h_1, \ldots, h_n \neq 0$, so ist $R(h) = |h_1 \cdots h_n|$. Sind $1, \alpha_1 \ldots \alpha_n$ linear unabhängige algebraische Zahlen, so hat *W. Schmidt* gezeigt, daß das System $(\alpha_1, ..., \alpha_n)$ vom Potenztypus ≤ 1 ist.

Kehren wir zum eindimensionalen Fall $n = 1$ zurück, so ist klar, daß die Folge der $\frac{p}{q}$ irgendwie von α abhängen wird. Besonders interessant ist, was man über den Nenner q sagen kann. Wir haben am Anfang gesehen, daß wir, wenn wir statt $\frac{1}{q^2}$ nur $\frac{1}{q}$ als Approximationsfehler verlangen, den Nenner beliebig vorschreiben können, allerdings kann der Zähler nicht beliebig gewählt werden. Die erste Frage, die man stellen kann, und die 1941 tatsächlich *Duffin* und *Schaeffer* gestellt haben, ist folgende:

Es sei ω eine monoton wachsende Folge natürlicher Zahlen, weiters sei ε_j eine beliebige Folge positiver Zahlen. Unter welchen Bedingungen gibt es dann unendlich viele ganze Zahlen p_k, so daß

$$\left|\alpha - \frac{p_k}{q_k}\right| < \frac{\varepsilon_k}{q_k}$$

ist, wobei noch g.g.T $(p_k, q_k) = 1$ und die q_k aus der gegebenen Folge sein sollen.

Der Spezialfall $\varepsilon_j = \frac{\varepsilon}{q_j}$ wurde 1970 von *Erdös* im Journal of Number Theory, Band 2, in einer tiefliegenden Arbeit erledigt.

Man kann nun die Frage stellen, ob man nicht an die Näherungsbrüche $\frac{p}{q}$ andere, z. B. zahlentheoretische Bedingungen stellen kann. Gehen wir vom folgenden klassischen Problem aus: Betrachten wir ein rechtwinkeliges Dreieck so ist $\cos\gamma = \frac{a}{c}$ und $\sin\gamma = \frac{b}{c}$ und es ist im allgemeinen $\frac{a}{c}$ bzw. $\frac{b}{c}$ irrational.

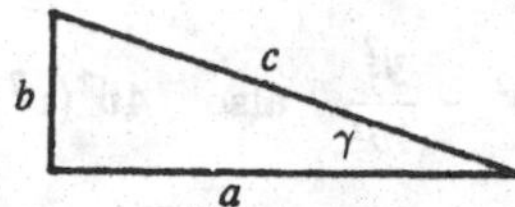

Diese Tatsache hat ja zur ersten Grundlagenkrise der Mathematik geführt. Es gibt aber auch anständige Dreiecke, nämlich die pythagoräischen Dreiecke, deren Seiten x, y, z ein pythagoräisches Tripel bilden. d.h. daß x, y, z natürliche Zahlen mit $x^2 + y^2 = z^2$ sind. Anders ausgesprochen, es ist $(\frac{x}{z})^2 + (\frac{y}{z})^2 = 1$ d.h. es ist der Punkt $(\frac{x}{z}, \frac{y}{z})$ ein Punkt mit rationalen Koordinaten auf dem Einheitskreis $\xi^2 + \eta^2 = 1$.

Es gibt bekanntlich unendlich viele solcher Tripel, z.B. sind

$$x = q(v^2 - u^2), \quad y = 2uvq, \quad z = (u^2 + v^2)q$$

solche Tripel und man zeigt in der Zahlentheorie, daß sich bis auf die Reihenfolge von x, y jedes pythagoräische Tripel in dieser Gestalt darstellen läßt, wo g.g.T $(u, v) = 1$ und u oder v gerade ist. Wir stellen uns nun folgende Frage: Kann jedes beliebige, rechtwinkelige Dreieck approximiert werden?

Sind mehrere rechtwinkelige Dreiecke gleichzeitig gegeben, kann man sie dann gleichzeitig durch pythagoräische Dreiecke approximieren? Zunächst ist nicht klar, was unter gleichzeitig zu verstehen ist. Man kommt auf diese Frage, wenn man die Quadratwurzelschnecke ins Auge faßt. Sie besteht aus den rechtwinkeligen Dreiecken mit den Längen: $(1, 2, \sqrt{2})$; $(\sqrt{2}, 1, \sqrt{3})$, $(\sqrt{3}, 1, \sqrt{4})$, $(\sqrt{4}, 1, \sqrt{5})$, $(\sqrt{5}, 1, \sqrt{6}), \ldots$, allgemein: $(\sqrt{n}, 1, \sqrt{n+1})$.

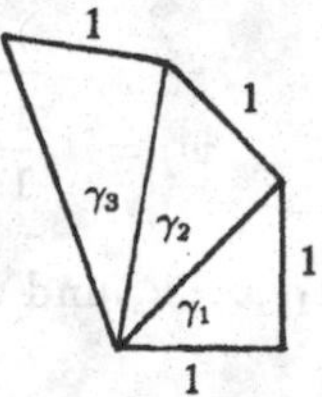

Arithmetisch gesprochen, handelt es sich um die Approximation von

$$\sin\gamma_j = \frac{1}{\sqrt{j+1}} \qquad j = 1, 2, 3, \ldots$$

Wir wollen nun folgendes zeigen: Sind $\alpha_1, \ldots, \alpha_s$ positive Zahlen, alle < 1, dann gibt es zu jedem genügend großen $N > 1$ positive ganze Zahlen $u_1, u_2, \ldots, u_s, v$, so daß $1 \leq v \leq N^s$, alle $u_j < v$ sind und

$$\left|\alpha_j - \frac{v^2 - u_j^2}{v^2 + u_j^2}\right| < \frac{2}{Nv}, \quad \left|\sqrt{1 - \alpha_j^2} - \frac{2vu_j}{v^2 + u_j^2}\right| < \frac{2}{Nv} \tag{2}$$

ist.

Es sind $x_j = v^2 - u_j^2$, $y_j = 2u_jv$, $z_j = v^2 + u_j^2$ pythagoräische Tripel und es hat stets $x_j + z_j$ denselben Wert $2v^2$ für $j = 1, ..., s$. In diesem Sinne soll hier die simultane Approximation verstanden werden, indem die Summe von Zähler und Nenner immer gleich bleibt. Es ist mir nicht bekannt, daß eine solche simultane Approximation bisher betrachtet wurde. Es ist

$$x_j = v^2 - \frac{y_j^2}{4v^2} \quad \text{also} \quad 4v^2(v^2 - x_j) = y_j^2$$

d.h. die Punkte (x_j, y_j) liegen auf einer Parabel mit Brennpunkt in O. Wir benützen hier eine Bemerkung, die *Stiefel* der Mathematikerin *O. Taussky* mitgeteilt hat. ("From Pythagoras Theorem via Sums of Square to Celestical Mechanics", Blakeslee Lecture, Smith College, May 8, 1975, Olga Taussky- Todd (Californian Institute of Technology, 1-13)).

Es ist $z_j \leq 2v^2$, daher erhalten wir

$$\left|\alpha_j - \frac{x_j}{z_j}\right| \leq \frac{2\sqrt{2}}{N\sqrt{z_j}} \leq \frac{2}{z_j^{\frac{1}{2}(1+\frac{1}{s})}}. \tag{3}$$

Daraus folgt: Ist $\varepsilon > 0$, so gibt es $\frac{x_j}{z_j}$ mit

$$\left|\alpha_j - \frac{x_j}{z_j}\right| < \varepsilon.$$

Für $j = 1$ wurde dies in qualitativer Form von Sierpinski* gezeigt. Wiederentdeckt von 2 Physikern 1976 (*Deakin* und *Troup*, Stevinus epitaph: "Thermodynamics meets Number Theory", Intern. J. Math.Ed. Sc.Tech.7, (1976); S 271-276).

Wir setzen

$$w_j = \sqrt{\frac{1 - \alpha_j}{1 + \alpha_j}}$$

und wählen ganze Zahlen $u_1, \ldots, u_s$ und v mit $1 \leq v \leq N^s$, so daß

$$|w_j - \frac{u_j}{v}| < \frac{1}{NV}$$

für $j = 1, \ldots, s$. Für $N \geq 1/\min(1 - w_j, w_j)$ folgt $0 < u_j < v$. Wir betrachten nun die Funktion

$$f(w) = \frac{1 - w^2}{1 + w^2}.$$

---

* Sierpinski: Pythagoräische Tripel (polnisch), 1954, Pánstwowe Wydawnictwo Nankowe (englische Ausgabe 1962), § 13, S. 75. 1962

Wegen $f(w_j) = \alpha_j$ und

$$|f'(w)| = \left|\frac{-4w}{(1+w^2)^2}\right| \leq \frac{2}{1+w^2} < 2$$

folgt aus dem Mittelwertsatz

$$\left|\alpha_j - \frac{v^2 - u_j^2}{v^2 + u_j^2}\right| = \left|f(w_j) - f(\frac{u_j}{v})\right| \leq 2|w_j - \frac{u_j}{v}| = \frac{2}{Nv}.$$

Entsprechend erhält man mit Hilfe der Funktion $g(w) = \frac{2w}{1+w^2}$

$$\left|\sqrt{1-\alpha_j^2} - \frac{2vu_j}{v^2+u_j^2}\right| = \left|g(w_j) - g(\frac{u_j}{v}\right| \leq 2\left|w_j - \frac{u_j}{v}\right| < \frac{2}{Nv}$$

wegen

$$|g'(w)| = \left|\frac{2(1-w^2)}{(1+w^2)^2}\right| \leq 2.$$

Wir wollen nun zeigen, daß dieser Approximationssatz, was den Exponenten der Potenz betrifft, nicht verschärft werden kann.

Wir nehmen für α eine quadratische Irrationalität, am einfachsten $\alpha = \frac{1}{\sqrt{2}}$. Dann ist

$$\sqrt{\frac{1-\alpha}{1+\alpha}} = \sqrt{2} - 1.$$

Es gilt

$$|\sqrt{2} - 1 - \frac{u}{v}| \geq \frac{c_1}{v^2}$$

für passendes c_1. Wegen $|f'(w)| \geq |w|$ für $|w| \leq 1$ folgt aus dem Mittelwertsatz

$$\left|\frac{1}{\sqrt{2}} - \frac{v^2-u^2}{v^2+u^2}\right| = \left|f(\sqrt{2}-1) - f\left(\frac{u}{v}\right)\right| \geq \frac{c_1}{v^2}\bar{w}$$

für ein $\bar{w}$ zwischen $\frac{u}{v}$ und $\sqrt{2}-1$. Wegen $f(\sqrt{2}-1) \neq f(0)$ gilt daher für passendes $c > 0$

$$\left|\frac{1}{\sqrt{2}} - \frac{v^2-u^2}{v^2+u^2}\right| \geq \frac{c}{v^2}.$$

Wir bemerken noch folgendes: Es seien (p_n, q_n) Lösungen der Pellschen Gleichung $p_n^2 - 2q_n^2 = \pm 1$ Dann ist

$$\left|\sqrt{2} - \frac{p_n}{q_n}\right| < (\sqrt{2}q_n^2)^{-1}.$$

Man kann also $u_n = p_n - q_n$, $v_n = q_n$ nehmen. Analoges kann man im Mehrdimensionalen zeigen.* Es gilt nicht nur für quadratische Irrationalitäten, sondern auch allgemein für algebraische Zahlen (es ist dann nur v^2 durch $v^{2+\varepsilon}$ zu ersetzen)

$$\left|\alpha - \frac{v^2 - u^2}{v^2 + u^2}\right| \geq \frac{c}{v^{2+\varepsilon}}.$$

Wir wollen jetzt die Ungleichung (2) anders schreiben. Wir beschränken uns auf den Fall s=1 und setzen $\alpha = \cos\gamma$, dann ist $\sqrt{1-\alpha^2} = \sin\gamma$.

Weiters bestimmen wir uns einen Winkel β mit $0 < \beta < \frac{\pi}{2}$ durch $\cos\beta = \frac{x}{z}$, $\sin\beta = \frac{y}{z}$. Dann erhalten wir

$$|\cos\beta - \cos\gamma| \leq \frac{2}{Nv} \quad \text{und} \quad |\sin\beta - \sin\gamma| \leq \frac{2}{Nv},$$

oder, komplex zusammengefaßt

$$\left|e^{i\gamma} - e^{i\beta}\right| \leq \frac{2\sqrt{2}}{Nv}.$$

Daraus folgt: Für jede natürliche Zahl m gilt :

$$\left|e^{im\beta} - e^{im\gamma}\right| \leq (2^m - 1)\frac{2\sqrt{2}}{Nv}, \tag{4}$$

wenn nur $c = \frac{2\sqrt{2}}{Nv} < 1$ ist. Der Beweis ist sehr einfach. Wir setzen $e^{i\gamma} = e^{i\beta} + c\vartheta$, wo $|\vartheta| \leq 1$ ist und erheben zur m-ten Potenz.

Wir haben

$$e^{i\beta} = \frac{x+iy}{z}.$$

Daraus folgt

$$\frac{x_m + iy_m}{z_m} = e^{im\beta} = \left(\frac{x+iy}{z}\right)^m \tag{I}$$

dabei ist

$$x_m + iy_m = (x+iy)^m, \quad u_m = z^m. \tag{II}$$

Das Bemerkenswerte ist, daß (x_m, y_m, z_m) wieder ein pythagoräisches Tripel bilden. Wir nennen es das m-te Tripel zum Tripel (x, y, z). Es ist natürlich

$$x_m^2 + y_m^2 = (x^2 + y^2)^m.$$

Diese Formel findet sich schon bei *Vieta*, "Genesis triangolarum", der sie aus der Formel

$$(x^2 + y^2)(x_1^2 + y_1^2) = x_2^2 + y_2^2$$

---

* Herr W. Schmidt hat mich am 12.01.1979 auf eine Arbeit von Thomas (1925) aufmerksam gemacht. Er selbst hat eine Arbeit angekündigt.

mit $x_2 = xx_1 - yy_1$, $y_2 = xy_1 + x_1y$ herleitet (*Glushkov*, Historia math. 4, 127-136 (1977)).

Es ist dies nach Vieta eine Methode, aus zwei rechtwinkeligen Dreiecken ein drittes rechtwinkeliges Dreieck herzustellen. So benutzt Vieta schon die Moivresche Formel, welche erst 1707 aufgestellt wurde, wenn auch in reeller Form. (Ich verweise weiter auf den Artikel von *Bachmakova* und *Slavutin* in Arch. History Exact Sci. 16, 289-306 (1977) und auf die dort zitierte Abhandlung „Zetetica" von Vieta).

Aus (II) folgen die Rekursionsformeln

$$\begin{aligned} x_{m+1} &= xx_m - yy_m \\ y_{m+1} &= yx_m + xy_m. \end{aligned} \tag{III}$$

Wir wollen gleich zeigen: Ist $x \equiv 0 (\bmod 3), y \equiv 0 (\bmod 4)$, so ist für alle $l \geq 1$

$$x_{2l-1} \equiv 0 (\bmod 3), \quad y_{2l} \equiv 0 (\bmod 12). \tag{IV}$$

Dabei setzen wir $x_1 = x$, $y_1 = y$, $z_1 = z$. Wir sehen sofort, daß $y_2 = 2xy \equiv 24 \equiv 0(12)$ ist. Wir wenden vollständige Induktion nach l an. Es ist

$$x_{2l+1} = xx_{2l} - yy_{2l}.$$

Da $x \equiv 0(\bmod 3)$, $y_{2l} \equiv 0(\bmod 3)$, so folgt $x_{2l+1} \equiv 0(\bmod 3)$. Es ist $y_{2l+2} = yx_{2l+1} + xy_{2l+1}$. Da $x_{2l+1} \equiv 0(\bmod 3)$, $y \equiv 0(\bmod 4)$,$xy_{2l+1} = xyx_{2l} + x^2y_{2l} \equiv 0(\bmod 12)$, so folgt $y_{2l+2} \equiv 0(\bmod 12)$.

Wir nehmen als Beispiel das Seilspannerdreieck $x = 3$, $y = 4$, $z = 5$ und erhalten folgende kleine Tabelle von Paaren

$$(x_m, y_m) : (3,4), (-7, 2^3.3), (-9.13, 4.11), (-17.31, -3.2^4.7).$$

Wir können (I) noch etwas anders schreiben. (Vgl. auch Taussky-Todd, Sums of Squares, Amer. Math. Monthly 77 (1970), 805-30). Es sei $x = v^2 - u^2$, $y = 2uv$. Dann ist $x + iy = (v + iu)^2$, $z = |v + iu|^2$ also $x_m + iy_m = (v + iu)^{2m}$, $z_m = (v + iu)^{2m}$. Es ist dann

$$e^{im\beta} = \left(\frac{v + iu}{|v + iu|}\right)^{2m}.$$

Kehren wir zu (4) zurück. Es hängt β in (4) natürlich von γ ab. Wir lassen nun β fest und betrachten die Vielfachen $2h\beta$, wobei β zu einem pythagoräischen Tripel (x, y, z) gehört. Es ist bekannt, daß $\beta/\pi = \sigma$ irrational ist. Man kann einen Beweis dafür dem Buch von *Meschkowski* ("Ungelöste und unlösbare Probleme in der Geometrie", 1960, vgl. auch Anhang Nr. 5) entnehmen. Ein wichtiger Teil des Beweises findet sich auch in der vorliegenden Arbeit bei der Behandlung von G.

Daraus folgt, daß die Vielfachen von σ mod 1 überall dicht liegen, d.h. die Punkte $(e^{2\pi ih\sigma}) = (e^{2ih\beta})$ liegen auf dem Einheitskreis überall dicht.

Aus der Theorie der Gleichverteilung (vgl. Kuipers-Niederreiter, "Uniform distribution of sequences", John Wiley, 1974) folgt, daß diese Punkte $e^{2ih\beta}$, welche das pythagoräische Tripel (x_{2h}, y_{2h}, z_{2h}) liefern, sogar gleichverteilt sind.

Daraus folgt: Ist γ ein beliebiger Winkel, so existiert bei festem β immer ein h, so daß

$$|e^{i\gamma} - e^{2ih\beta}| < \varepsilon$$

ist. Wir wollen nun eine Schranke für dieses h bestimmen. Dazu berechnen wir jetzt die Diskrepanz D_N der Folge $\omega = (h\sigma)$.

Es gilt nach *Erdős* und *Turan* (vgl. Kuipers-Niederreiter)

$$D_N \leq 6 \quad \frac{1}{M} + \sum_{h=1}^{M} \frac{1}{h} |W(h) \tag{5}$$

wo

$$W(h) = \frac{1}{N} \sum_{k=1}^{N} e^{2\pi i k h \sigma}$$

ist. Nun ist $\sigma = \frac{\beta}{\pi}$, also ist

$$NW(h) = \sum_{k=1}^{N} e^{ikh\beta} = \frac{e^{2iNh\beta} - 1}{e^{2ih\beta} - 1} e^{2ih\beta}.$$

Es ist also

$$N\,|W| \leq \frac{1}{|\sin h\beta|}.$$

Nun ist

$$e^{ih\beta} = (\frac{x + iy}{z})^h,$$

also

$$\sin h\beta = \sum_{l=0}^{\frac{h-1}{2}} (-1)^l \binom{h}{2l+1} \cos^{h-(2l+1)} \beta \sin^{2l+1} \beta = \frac{G}{z^h}$$

wo

$$G = \sum_{l=0}^{\frac{h-1}{2}} (-1)^l \binom{h}{2l+1} x^{h-(2l+1)} y^{2l+1}$$

ist.

Es ist $|G| \leq 1$, denn sonst wäre $\sin h\beta = \sin \pi h \sigma = 0$, also σ rational. Man kann dem Ausdruck für G, wenn h ungerade ist, leicht ansehen, daß er $\neq 0$ sein muß . Es ist nämlich $G = yG_1$, wo $G_1 = hx^{h-1} + yG_2$. Wenn $v \equiv 1(2)$ und $u \equiv 0(2)$ ist, ist hx ungerade und yG_2 gerade.

Es ist also

$$|\sin h\beta| \geq z^{-h} = (u^2 + v^2)^{-h}. \tag{6}$$

Aus dieser Abschätzung folgt, daß $\min_g |h\sigma - g| \geq C_1 e^{-hc_2}$ ist mit $c_2 = \log z$. Es ist also σ vom Exponentialtypus. Dabei heiß t eine Irrationalzahl σ von diesem Typus, wenn für alle $h \neq 0$ und alle $g : |h\sigma - g| \geq C_1 e^{-hc_2}$ ist.

Wenn wir unsere Überlegungen fortsetzen, so erhalten wir

$$N\,|W(h)| \leq z^h. \tag{7}$$

Es ist also

$$D_N \leq 6 \quad \frac{1}{M} + \frac{1}{N}\sum_{h=1}^{M}\frac{z^h}{h} \quad \leq 6\left(\frac{1}{M} + \frac{2}{N}\frac{z^M}{M}\right).$$

Wir nehmen nun $M = [\frac{\log N}{\log z}]$ und erhalten, wenn $N \geq z^{10}$

$$D_N < (20\log z)/\log N \tag{8}$$

für Nyz^{10} ist dies ebenfalls richtig, da $D_N \leq 1$ gilt. Wir können dies so ausdrücken: Es sei $2\pi\rho \leq \alpha \leq 2\pi\sigma$ ein Bogen auf dem Einheitskreis, dann gilt für die Anzahl N' der Punkte $P_h = e^{2ih\beta} = e^{2\pi ih\sigma}$, die in diesem Bogen liegen, die Abschätzung

$$\left|\frac{N'}{N} - (\sigma - \rho)\right| < \frac{20\log z}{\log N}. \tag{9}$$

Ist also für eine natürliche Zahl L

$$\sigma - \rho \geq \frac{L}{N} + \frac{20\log z}{\log N} \tag{10}$$

so liegen in diesem Bogen mindestens $L + 1$ Punkte $e^{2ih\beta}$ mit $1 \leq h \leq N$. Nehmen wir den Spezialfall $\rho = \frac{\gamma}{2\pi}$, $\sigma = \rho + \varepsilon$, $L \geq O$, so gilt

$$|e^{i\gamma} - e^{2ih\beta}| < 2\pi\varepsilon$$

mit $1 \leq h \leq e^{\frac{20\log z}{\varepsilon}}$, wenn wir $N \geq e^{\frac{20\log z}{\varepsilon}}$ nehmen. Nehmen wir $\varepsilon = \frac{20\log z}{\log N}$, dann folgt mit geeignetem g:

$$\left|\frac{\gamma}{2\pi} - h\sigma - g\right| \leq 20\frac{\log z}{\log N} \tag{11}$$

wo $1 \leq h \leq N$ ist. Wir haben hier ein inhomogenes Gegenstück zu dem Dirichletschen Approximationssatz.

Da $h \leq N$ ist, folgt aus (10)

$$\left|\frac{\gamma}{2\pi} - h\sigma - g\right| \leq \frac{20\log z}{\log h}.$$

Diesen Satz können wir aber viel schärfer aus dem Satz von Tschebyschew herleiten, der besagt, daß sogar

$$|\gamma - 2h\beta - g| \leq \frac{2}{h}$$

ist.

Das h läßt sich hier allerdings nicht lokalisieren, wie wir das vorher konnten.

Betrachten wir die allgemeine Situation, wie dies in (10) angenommen ist. Diese $L+1$ Punkte $P_n = e^{2ih\beta}$, die in dem Bogen liegen, haben eine besondere Eigenschaft.

Es ist der Abstand von je zwei Punkten

$$d(P_h, P_k) = \left|e^{2ih\beta} - e^{2ik\beta}\right| = 2.\,|\sin(h-k)\beta|$$

also stets rational und, wie man sofort sieht ist $z^N d(P_h, P_k)$ sogar ganzzahlig. Da man L beliebig vorgeben kann, wenn man nur N genügend groß nimmt, liegen auf dem Kreisbogen $(2\pi\rho, 2\pi\sigma)$ mit dem Radius z^N mindestens $L+1$ Punkte mit ganzzahligem Abstand. Man kann

$$N = \max\left([\frac{2L}{\sigma - \rho}] + 1, [\exp(\frac{40 \log z}{\sigma - \rho})] + 1\right) \tag{12}$$

nehmen.

Es sei daran erinnert, daß nach einem Satz von *Amery* und *Erdös* folgendes gilt: Ist M eine unendliche Menge in der Ebene, so daß stets $d(P_i, P_j) \in N$, wenn $P_i, P_j \in M$, so liegt M auf einer Geraden.

Es gibt dagegen stets eine Menge M von Kardinalzahlen $\geq L$ auf einem Kreis, so daß $d(P_i, P_j) \in N$. Dabei kann der Radius $\leq z^N$ mit N aus (12) genommen werden. Dies stellt eine Verschärfung eines Satzes von *Choquet* und *Krewerceras* (1945) dar. Wir können nun bei festem k, z.B. $k = 0$, nach der mittleren Entfernung

$$\frac{1}{N}\sum_{h=1}^{N} d(P_h, P_o) = \frac{2}{N}\sum_{h=1}^{N} |\sin h\beta|$$

fragen. Nun ist ja die Folge $(h\sigma - [h\sigma])$ gleichverteilt. Dann gilt für jede periodische, im Riemannschen Sinne integrierbare Funktion f

$$\lim_{N\to\infty} \frac{1}{N}\sum_{h=1}^{N} f(h\sigma) = \int_0^1 f(t)dt. \tag{13}$$

Mit $f(t) = |\sin \pi t|$ erhalten wir

$$\lim_{N\to\infty} \frac{1}{N}\sum_{h=1}^{N} d(P_h, P_0) = \frac{4}{\pi} \tag{14}$$

Man kann mit Hilfe von (8) sogar zeigen, daß

$$\left|\frac{1}{N}\sum_{h=1}^{N} d(P_h, P_0) - \frac{4}{\pi}\right| \leq \frac{80\log z}{\log N} \tag{15}$$

ist. Es gilt nämlich folgendes (vgl. Kuipers-Niederreiter): Ist f in (13) nicht nur integrierbar, sondern sogar von beschränkter Variation $V(f)$, so gilt

$$\left|\frac{1}{N}\sum_{h=1}^{N} f(h\sigma) - \int_0^1 f(t)dt\right| \leq D_N V(f). \tag{13'}$$

Ist f stetig differenzierbar, so ist $V(f)$ bekanntlich $\int_0^1 |f'(t)|\, dt$.
Es ist nach (I)

$$\sin h\beta = \frac{y_h}{z_h}.$$

Wir erhalten also aus (14):

$$\lim_{N\to\infty} \frac{1}{N}\sum_{h=1}^{N} d(P_h, P_o) = 2\lim_{N\to\infty}\frac{1}{N}\sum_{h=1}^{N}\left|\frac{y_h}{z_h}\right| = \frac{4}{\pi}.$$

Wir erhalten sogar aus (15)

$$\left|\frac{1}{N}\sum_{h=1}^{N}\left|\frac{y_h}{z_h}\right| - \frac{2}{\pi}\right| \leq \frac{40\log z}{\log N}. \tag{15'}$$

Wir können also $\frac{2}{\pi}$ als Grenzwert von rationalen Zahlen darstellen, die mit Hilfe eines festen pythagoräischen Tripels gebildet werden. Man sieht, wie man einer so alten Sache neue Aspekte abgewinnen kann. Ein weiteres Beispiel ist folgende Formel

$$\lim_{N\to\infty}\prod_{h=1}^{N}\left|\frac{z_{2h}}{x_{2h}}\right|^{\frac{1}{N}\left|\frac{12x_{2h}}{y_{2h}}\right|} = e^{\pi}. \tag{15''}$$

Man beachte, daß nach (IV) $y_{2h} \equiv 0 (\mathrm{mod}\, 12)$ ist. Es ist dies ein Gegenstück zu einer Formel von Gauß (Werke III, S. 377, VII, S. 14. - Brent, Math. Comp. 30 (1976)).

Wir nehmen in (13')

$$f(t) = (\log|\cos\pi t|)\left|\frac{\cos\pi t}{\sin\pi t}\right|.$$

Dabei sei an den Stellen $t = \frac{1}{2}$ und $\frac{3}{2}$ $f(t) = 0$. Ebenso sei $f(0) = f(1) = 0$. Dann ist f sogar von beschränkter Variation. Es ist

$$\int_0^1 f(t)dt = -\frac{\pi}{12},$$

denn es ist

$$\int_0^1 (\log|\cos\pi t|\,|\cot\pi t|\,dt = \frac{2}{\pi}\int_0^{\frac{\pi}{2}} (\log\cos t)\cot t\,dt =$$

$$= \frac{1}{\pi}\int_0^1 \frac{1}{x}\log(1-x^2)dx = \frac{1}{\pi}(-\frac{\pi}{12}) = -\frac{\pi}{12}.$$

Außerdem sieht man leicht, daß $V(F) \leq 2$ ist.

Wir erhalten also aus (13'), wobei wir σ durch 2σ ersetzen (z ist durch z^2 zu ersetzen)

$$\left|\frac{1}{N}\sum_{h=1}^{N}\log|\cos 2\pi h\sigma||\cot 2\pi h\sigma| + \frac{\pi}{12}\right| \leq \frac{80\log z}{\log N} = \frac{\Delta_N}{12}.$$

Daraus folgt durch Entlogarithmieren

$$e^{\pi-\Delta_N} \leq \prod_{h=1}^{N}|\cos 2\pi h\sigma|^{-\frac{12}{N}|\cot 2\pi h\sigma|} \leq e^{\pi+\Delta_N}.$$

Daraus folgt

$$\left|\prod_{h=1}^{N}|\cos 2\pi h\sigma|^{-\frac{12}{N}|\cot 2\pi h\sigma|} - e^{\pi}\right| \leq e^{\pi}(e^{\Delta_N}-1).$$

Daraus folgt die Behauptung sogar in der Abschätzung

$$\left|\prod_{h=1}^{N}\left|\frac{z_{2h}}{x_{2n}}\right|^{\frac{1}{N}\left|\frac{12x_{2h}}{y_{2n}}\right|} - e^{\pi}\right| \leq e^{\pi}\left(z^{\frac{960}{\log N}}-1\right). \qquad (15''')$$

Wir haben bisher nur rechtwinkelige Dreiecke betrachtet. Man kann nun verallgemeinerte pythagoräische Dreiecke betrachten, mit der Gleichung:

$$x^2+y^2-2xy\cos\rho = z^2 \qquad (16)$$

wobei $\cos\rho = \frac{p}{q}$ rational ist.

Die Punkte $(\frac{x}{z}, \frac{y}{z})$ liegen auf der Ellipse

$$\xi^2+\eta^2-2\xi\eta\cos\rho = 1.$$

Interessant ist der Fall $\cos\rho = \frac{1}{2}$. Es gilt wieder

$$\frac{x}{z} = \frac{v^2-u^2}{v^2+u^2-2uv\cos\rho}, \quad \frac{y}{z} = \frac{2v(v\cos\rho-u)}{v^2+u^2-2uv\cos\rho}$$

Man kann jetzt fast alles übertragen, die Rechnungen werden aber bedeutend komplizierter. Ich möchte nur einige qualitative Resultate anführen (vgl. aber den Anhang):

Ist γ ein beliebiger Winkel und $\varepsilon > 0$, so gibt es stets ganze Zahlen (x, y, z), welche (16) erfüllen, also ein verallgemeinertes pythagoräisches Tripel bilden, so daß

$$\left|\frac{\sin\gamma}{\sin\rho} - \frac{x}{z}\right| < \varepsilon \qquad \left|\frac{\sin(\gamma+\rho)}{\sin\rho} - \frac{y}{z}\right| < \varepsilon$$

ist. Wir setzen nun

$$\sin\beta_1 = \frac{x}{z}\sin\rho \quad , \quad \sin(\beta_1+\rho) = \frac{y}{z}\sin\rho$$

dann ist

$$e^{i\beta_1} = \frac{y - xe^{-i\rho}}{z}.$$

Es liegt

$$\cos\beta_1 = \frac{y - x\cos\rho}{z}$$

in Q. Es ist weiter

$$\sin\beta_1 = \frac{x}{z}\sin\rho = \frac{x}{z}\sqrt{1 - p^2/q^2}.$$

Wenn $p = 0$, $q = 1$ ist, erhalten wir wieder die gewöhnlichen pythagoräischen Tripel.

Wir setzen nun $\beta_1 = \pi\gamma$ und

$$\beta = \beta_1 + \frac{\pi}{2} = \pi(\gamma + \frac{1}{2}) = \pi\sigma.$$

Im pythagoräischen Fall ist σ und damit γ stets irrational. Dies braucht im allgemeinen Fall nicht so zu sein. Es seien u und v ganze Zahlen. Es sei nun

$$\cos\rho = \frac{p}{q} = \frac{u^2 + v^2 - 4uv}{2(u^2 + v^2 - uv)},$$

$$\frac{x}{z} = \frac{u^2 + v^2 - uv}{v^2 - u^2} \quad , \quad \frac{y}{z} = \frac{v(v - 2u)}{v^2 - u^2};$$

bzw.

$$\frac{x}{z} = \frac{u^2 + v^2 - uv}{v^2 - u^2} \quad , \quad \frac{y}{z} = \frac{u(u - 2v)}{v^2 - u^2};$$

dann ist γ rational und zwar ist im ersten Fall $\gamma = \frac{1}{3}$ und im zweiten Fall $\gamma = \frac{2}{3}$. Ein dritter Fall liegt für

$$\cos\rho = \frac{p}{q} = \frac{u^2 - v^2}{u^2 + v^2}$$

und

$$\frac{x}{z} = \frac{u^2+v^2}{2uv}, \quad \frac{y}{z} = \frac{u^2-v^2}{2uv}$$

vor. Es ist dann hier $\gamma = \frac{1}{2}$.

Man kann nun zeigen (Anhang 4), daß dies aber die einzigen Fälle sind. In jedem anderen Fall ist γ und damit σ irrational.

Es liegen dann wieder die Punkte

$$e^{2\pi i h\sigma} = e^{2ih\beta} = (-1)^h e^{2ih\beta_1}$$

auf dem Einheitskreis überall dicht und sogar gleichverteilt.

Es gilt

$$e^{im\beta_1} = \frac{y_m - x_m \cos\rho + i x_m \sin\rho}{z_m},$$

wobei die Rekursionsformeln

$$\frac{x_{m+1}}{z_{m+1}} = \frac{x y_m + x_m y - 2 x x_m \cos\rho}{z z_m}$$

und

$$\frac{y_{m+1}}{z_{m+1}} = \frac{y y_m - x x_m}{z z_m}$$

bestehen. Wir können damit alle Formeln, welche wir für $\cos\rho = 0$ erhalten haben, auf den allgemeinen Fall übertragen.

Man kann sogar eine beliebige positiv definite quadratische Form Q betrachten und die rationalen Punkte auf der Ellipse $Q(\xi,\eta) = 1$ untersuchen, wenn auf dieser Ellipse überhaupt ein solcher Punkt liegt. Eine andere Möglichkeit wäre, die rationalen Punkte auf S_3 und S_7 (allgemein auf der S_r) zu untersuchen. Man kann auch die rationalen Punkte auf der Lemniskate, allgemein auf Kurven vom Geschlecht 0, in dieser Hinsicht untersuchen. Es sei auch auf Anhang 3 verwiesen, wo das allgemeine Schema herausgestellt wird.

Wir wollen die irrationalen Zahlen β von einem anderen Winkel aus betrachten. Es kommt, wenn h und k alle ganzen Zahlen durchlaufen, $h\beta + k$ jedem Wert beliebig nahe, da ja $h\beta$ mod 1 gleichverteilt ist. Es würde naheliegend sein, die Reihe

$$\sum_{h\neq 0,k} \frac{1}{|h\beta + k|}$$

zu betrachten. Diese Reihe ist aber divergent. Wir wollen nun

$$T(j,t) = \sum_{h\neq 0,k} \frac{1}{|h|^{2t}(h\beta+k)^j}$$

betrachten, wo j und t natürliche Zahlen sind. Man kann nun zeigen, (vgl. eine Arbeit des Verfassers in der Acta Arithmetica (Turan-Gedächtnisband)), daß die Reihe für $j < \frac{2t}{\mu}$ konvergent ist, wenn β vom Typus μ ist. Weiters ist

$$\Gamma(j)T(j,t) = \lim_{w\to 0} \sum_{l=1}^{\infty} l^{j-1} e^{-lw} P_{2t}(l\beta)$$

wo P_{2t} (abgesehen von einem Faktor) die Bernoullische Funktion der Ordnung $2t$ ist.

Es gilt allgemein (auch wenn j keine natürliche Zahl ist) für $w > 0$ und $1 < j < \frac{2t}{\mu}$

$$\sum_{l=1}^{\infty} l^{j-1} e^{-lw} P_{2t}(l\beta) = \Gamma(j) \sum_{h\neq 0,k} \frac{1}{|h|^{2t}} \frac{1}{(w+i(h\beta+k))^j}.$$

Ist nun $j > 0$ und $j < \frac{2t}{\mu}$, dann gilt für $0 < w$, $0 < w_0$:

$$\sum_{l=1}^{\infty} l^{j-1}(e^{-lw} - e^{-lw_0}) P_{2t}(l\beta) = \Gamma(j) S,$$

wo

$$S = \sum_{h\neq 0,k} \frac{1}{|h|^{2t}} \left((w + 1(h\beta+k))^{-j} - (w_0 + i(h\beta+k))^{-j}\right)$$

ist. Eine andere Formel, gültig für $t > \mu$ ist

$$\sum_{h\neq 0,k} \frac{1}{h^{2t}} \left(\frac{\sin w\pi(h\beta+k)^2}{w\pi(h\beta+k)}\right)^2 = \sum_{|l|\leq w} (w - |l|) P_{2t}(l\beta).$$

Man kann die obigen Formeln zum Studium der Zetafunktionen

$$Z(r,s,p) = \sum_{l=1}^{\infty} \frac{P_r(l\beta)}{ls}$$

die von *Hecke, Behnke* und *Hardy-Littlewood* eingeführt wurden, benützen.

Anhang

1.) Das Beispiel auf Seite 7 läßt sich verallgemeinern. Es sei

$$\alpha = \frac{1}{\sqrt{k^2+1}}, \quad \beta = \frac{1}{\sqrt{k^2+1}}, \quad w = (\sqrt{k^2+1} - 1)/k$$

Wir betrachten die Pellsche Gleichung $p^2 - Dq^2 = \pm 1$ mit $D = k^2 + 1$. Die Grundlösung ist $p_0 = k$, $q_0 = 1$. Die allgemeine Lösung (p_n, q_n) erhält man aus

$$p_n + q_n\sqrt{k^2+1} = (k + \sqrt{k^2+1})^n.$$

Ist (p,q) eine beliebige Lösung der Pellschen Gleichung, so folgt aus der Pellschen Gleichung

$$p - q\sqrt{D} = \pm(p + q\sqrt{D})^{-1}$$

also ist, da $D = k^2 + 1$

$$\left|\sqrt{k^2+1} - \frac{p}{q}\right| < \frac{1}{q^2}.$$

Daraus folgt

$$\left|\frac{\sqrt{k^2+1} - 1}{k} - \frac{p-q}{kq}\right| < \frac{1}{kq^2}.$$

Wir setzen $u = p - q$, $v = kq$, $x = v^2 - u^2$, $y = 2uv$, $z = v^2 + u^2$, dann ist nicht nur $x^2 + y^2 = z^2$, sondern auch $y - kx = \pm k$ und wir haben für

$$\frac{1}{\sqrt{k^2+1}}$$

die Näherungsbrüche

$$\frac{x}{z} = ((qk)^2 - (p-q)^2)/((qk)^2 + (p-q)^2).$$

Es sei noch bemerkt, daß $x + z = 2(qk)^2$ ist.

2.) Wir bringen noch zwei weitere Beispiele zu Seite 14. Es sei

$$f_2(t) = \tan \pi t \quad \text{in} \quad 0 \le t \le \frac{1}{4} \quad \text{und} \quad \frac{3}{4} \le 1 \quad \text{und}$$

$$f_2(t) = \cot \pi t \quad \text{in} \quad \frac{1}{4} \le t \le \frac{3}{4}.$$

Es ist

$$\int_0^1 f_2(t)dt = \frac{2}{\pi}\int_0^1 \frac{d\xi}{1+\xi} = \frac{2}{\pi}\log 2.$$

Wir erhalten daher

$$\lim_{N\to\infty} \frac{1}{N}\sum_{n=1}^{N} \min\left(\left|\frac{x_n}{y_n}\right|, \left|\frac{y_n}{x_n}\right|\right) = \frac{2}{\pi}\log 2.$$

Ein weiteres Beispiel erhalten wir durch die Funktion (wobei k eine reelle Zahl mit $0 < k < 1$ ist)

$$f_3(t) = \frac{k\,|\cos 2\pi t|}{\sqrt{1 - k^2\cos^2 2\pi t}}.$$

Es ist

$$\int_0^1 f_3(t)dt = \frac{2}{\pi}\int_0^1 \frac{k\xi\, d\xi}{\sqrt{(1-k^2\xi^2)(1-\xi^2)}}.$$

Nun ist

$$\int \frac{k\xi d\xi}{\sqrt{(1-k^2\xi^2)(1-\xi^2)}} = -\left(\log(k\sqrt{1-\xi^2}+\sqrt{1-k^2\xi^2})\right)$$

also ist

$$\int_0^1 f_3(t)dt = \frac{1}{\pi}\log\frac{1+k}{1-k}.$$

Setzen wir

$$A_n = \frac{k\frac{x_{2n}}{z_{2n}}}{\sqrt{1-k^2(\frac{x_{2n}}{z_{2n}})^2}}$$

so erhalten wir

$$\lim_{N\to\infty}\frac{1}{N}\sum_{n=1}^{N} A_n = \frac{1}{\pi}\log\frac{1+k}{1-k}.$$

Daraus folgt

$$\lim_{N\to\infty}\left(\frac{1+k}{1-k}\right)^{N/\sum_{n=1}^{N}A_n} = e^{\pi}.$$

Für $k=\frac{1}{2}$ erhalten wir

$$\lim_{N\to\infty} 3^{B_N} = e^{\pi} \quad \text{mit} \quad \frac{1}{B_N} = \frac{1}{N}\sum_{n=1}^{N}\frac{x_{2n}}{\sqrt{(4z_{2n}^2-x_{2n}^2)}}.$$

3.) Wir wollen die Behauptung auf Seite 16 beweisen. Es war

$$x = q(v^2-u^2), \quad y = 2vq(v\cos\rho - u), \quad z = q(u^2+v^2-2uv\cos\rho)$$

mit $\cos\rho = \frac{p}{q}$. Wir setzen

$$f(w) = \frac{1-w^2}{1+w^2-2w\cos\rho} \qquad g(w) = \frac{-2(w-\cos\rho)}{1+w^2-2w\cos\rho}.$$

Es ist

$$f'(w) = 2\frac{2w-(1+w^2)\cos\rho}{(1+w^2-2w\cos\rho)^2} \qquad g'(w) = \frac{-2(\sin^2\rho-(w-\cos\rho)^2)}{1+w^2-2w\cos\rho}.$$

Es ist also für $0\le w<1$

$$|f'(w)| \le \frac{8}{\sin^4\rho}, \quad |g'(w)| \le \frac{8}{\sin^4\rho}.$$

Es ist also mit $w_o = \frac{u}{v}$

$$|f(w)-f(w_o)| \le \frac{8}{\sin^4\rho}|w-w_0|, \quad |g(w)-g(w_o)| \le \frac{8}{\sin^4\rho}|w-w_o|.$$

Es ist

$$f(w_0) = \frac{x}{z} \quad , \quad g(w_o) = \frac{y}{z}.$$

Es ist weiter, wenn α eine Zahl ist mit

$$\alpha^2 < \frac{1}{\sin^2 \rho} \quad w = \frac{\alpha \cos \rho - \sqrt{1 - \alpha^2 \sin^2 \rho}}{1 + \alpha}.$$

Es ist dann $f(w) = \alpha$ und

$$g(w) = \alpha \cos \rho + \sqrt{1 - \alpha^2 \sin^2 \rho} = \beta.$$

Wir haben also, wenn $\left|w - \frac{u}{v}\right| \leq \frac{1}{Nv}$ ist

$$\left|\alpha - \frac{x}{z}\right| \leq \frac{8}{Nv \sin^4 \rho}$$

und

$$\left|\beta - \frac{y}{z}\right| \leq \frac{8}{Nv \sin^4 \rho}.$$

Wir wollen nun den allgemeinen Sachverhalt herausarbeiten. Es sei eine rationale Mannigfaltigkeit X_m in R^s durch die Gleichungen ($n = s - m$):

$$\phi_1(x_1, ..., x_s) = 0, \ldots, \phi_n(x_1, ..., x_s) = 0$$

gegeben. Wir betrachten nun jenen Teil $\overline{X_m}$ (wenn vorhanden), der durch eine Parameterdarstellung

$$x_1 = \frac{P_1(\tau_1, ..., \tau_m)}{Q(\tau_1, ..., \tau_m)} \quad , ..., \quad x_s = \frac{P_s(\tau_1, ..., \tau_m)}{Q(\tau_1, ..., \tau_m)}$$

gegeben ist. Dabei seien $P_1, ..., P_s, Q$ homogene, ganzzahlige Polynome vom Grad L. Weiter sollen reelle Zahlen c_o, c_1 vorhanden sein, so daß für alle $\tau_1, ..., \tau_m$

$$c_1(\tau_1^2 + ... + \tau_m^2)^{\frac{L}{2}} \geq Q(\tau_1, ..., \tau_m) \geq c_o(\tau_1^2 + ... + \tau_m^2)^{\frac{L}{2}}$$

ist. Weiters gebe es noch auf $\overline{X_m}$ $m - 1$ Funktionen $\psi_2, ..., \psi_m$, so daß

$$t_2 = \frac{\tau_2}{\tau_1} = \psi_2(x_1, ..., x_s), ..., t_m = \frac{\tau_m}{\tau_1} = \psi_m(x_1, .., x_s)$$

ist.

Es sei nun $(\alpha_1, ..., \alpha_s)$ ein Punkt auf $\overline{X_m}$. Es gibt dann $m - 1$ Zahlen $\overline{t_2}, ..., \overline{t_m}$, so daß für $j = 2, ..., m$

$$\overline{t_j} = \psi_j(\alpha_1, ..., \alpha_s)$$

gilt. Es ist dann

$$\alpha_j = \rho_j(1, \overline{t_2}, ..., \overline{t_m}).$$

Dabei haben wir

$$\rho_j = \frac{P_j}{Q}$$

gesetzt. Es gibt dann nach dem Dirichletschen Approximationssatz bei vorgegebenem N ganze Zahlen $n_1, n_2, ..., n_m$, so daß $(j = 2, ..., m)$

$$\left|\overline{t_j} - \frac{n_j}{n_1}\right| \leq \frac{1}{Nn_1}$$

mit $1 \leq n_1 \leq N^{m-1}$ ist.

Es sind dann auch die $s+1$ Zahlen $\xi_j = P_j(n_1, ..., n_m)$, $\quad \eta = Q(n_1, ..., n_m)$ ganze Zahlen und es ist

$$\rho_j(1, \frac{n_2}{n_1}, ..., \frac{n_m}{n_1}) = \frac{\xi_j}{\eta}.$$

Es ist

$$\alpha_j - \frac{\xi_j}{\eta} = \rho_j(1, \overline{t_2}, \ldots, \overline{t_m}) - \rho_j(1, \frac{n_2}{n_1}, \ldots, \frac{n_m}{n_1}) = \sum_{k=2}^{m} c_{j,k}(\overline{t_k} - \frac{n_k}{n_1}),$$

wo die Ableitungen

$$c_{j,k} = \frac{\partial \rho_j}{\partial \tau_k}$$

an einer Zwischenstelle $1, \hat{t}_2, ..., \hat{t}_m$ zu nehmen sind. Es ist

$$Q^2 \frac{\partial \rho_j o}{\partial \tau_k} = \frac{\partial P_j}{\partial \tau_k} Q - \frac{\partial Q}{\partial \tau_k} P_j.$$

Es ist nun nach Voraussetzung $Q(1, t_2, ..., t_k) \geq c_o$. Es ist also, wenn M eine obere Schranke für die Summe der Beträge der Koeffizienten von

$$\frac{\partial P_j}{\partial \tau_k} Q - \frac{\partial Q}{\partial \tau_k} P_j$$

(für beliebige k, j) ist

$$\max \left|\frac{\partial \rho_j}{\partial \tau_k}\right| \leq \frac{M}{c_0} \max\left(1, |\hat{t}_s|, \ldots, |\hat{t}_k|\right)^{2l-1}.$$

Wir erhalten also für $j = 1, \ldots, s$

$$\left|\alpha_j - \frac{\xi_j}{\eta}\right| \leq \frac{C}{Nn_1}$$

mit

$$C = (m-2)c_0^{-2} \max\left(1, |\hat{t}_s|, \dots |\hat{t}_k|\right)^{2l-1}.$$

Es ist nun andererseits

$$\eta = Q(n_1, ..., n_m) = n_1^L Q(1, \frac{n_2}{n_1}, ..., \frac{n_m}{n_1}).$$

Da

$$Q(1, \frac{n_2}{n_1}, ..., \frac{n_m}{n_1}) \leq c_1 \quad 1 + \left|\frac{n_2}{n_1}\right|^2 + \cdots + \left|\frac{n_m}{n_1}\right|^2 \ ^{\frac{L}{2}}$$

ist, so folgt für großes N, da ja

$$\left|\frac{n_j}{n_1}\right| \leq \frac{1}{Nm} + |\overline{t_j}|$$

gilt, daß

$$\eta \leq c_1 n_1^L A(\overline{t_2}, ..., \overline{t_m})$$

ist, wo A eine Schranke ist, welche von $\overline{t_2}, ..., \overline{t_m}$ alleine abhängt. Liegen die $\alpha_1, ..., \alpha_s$ in einem beschränkten Bereich von $\overline{X_m}$, so kann man auch eine obere Schranke für $A(\overline{t_2}, ..., \overline{t_m})$ finden, welche nur von diesem Bereich abhängt. Es ist dann

$$\left|\alpha_j - \frac{\xi_j}{\eta}\right| \leq \frac{B}{N\eta^{\frac{1}{L}}}$$

wo $B = C(c_1 A)^{\frac{1}{L}}$ ist.

Betrachten wir einige Beispiele. Es sei S_r die Sphäre $x_o^2 + ... + x_r^2 = 1$ und es sei $\overline{S_r} = S_r - \{x_0 = 1\}$.

Dann haben wir

$$x_0 = \frac{\tau_1^2 + ... + \tau_r^2 - \tau_o^2}{\tau_1^2 + ... + \tau_r^2 + \tau_o^2}, \quad x_j = \frac{2\tau_o\tau_j}{\tau_1^2 + ... + \tau_r^2 + \tau_o^2} \quad (j = 1, \dots, r)$$

und es ist

$$\frac{\tau_j}{\tau_o} = \frac{x_j}{1 - x_o}.$$

Ein weiteres Beispiel ist die Mannigfaltigkeit von *Veronese*, gegeben durch die Gleichungen

$$(x_1x_2)^2 + (x_1x_3)^2 + (x_2x_3)^2 = 2x_1x_2x_3 \quad \text{(Römerfläche)}$$

$$x_4x_2^2 = x_5x_3^2, \quad x_1x_3 = 2x_2x_4.$$

Es sei $\overline{V}$ der Teilbereich mit $x_4 \neq 0$. Dann ist, $\tau^2 = \tau_1^2 + \tau_2^2 + \tau_3^2$ gesetzt,

$$x_1 = \frac{2\tau_1\tau_3}{\tau^2}, \quad x_2 = \frac{2\tau_2\tau_3}{\tau^2}, \quad x_3 = \frac{2\tau_1\tau_2}{\tau^2}, \quad x_4 = \frac{\tau_1^2}{\tau^2}, \quad x_5 = \frac{\tau_3^2}{\tau^2}.$$

Es ist

$$\frac{x_1}{x_4} = \frac{2\tau_3}{\tau_1} \quad , \quad \frac{x_3}{x_4} = \frac{2\tau_2}{\tau_1} \quad \text{bzw.} \quad \frac{\tau_3}{\tau_2} = \frac{x_1}{x_3} \quad , \quad \frac{\tau_1}{\tau_2} = \frac{x_1}{x_2}.$$

Setzen wir

$$X_1 = \frac{1}{2}x_1, \quad X_2 = \frac{1}{2}x_2, \quad X_3 = \frac{1}{2}x_3, \quad X_4 = x_4 - x_5$$

dann erhalten wir das Modell der projektiven Ebene (vgl. *Hilbert-Cohn-Vossen*, "Anschauliche Geometrie", Julius Springer Verlag, Berlin 1932).

$$(X_1X_2)^2 + (X_2X_3)^2 + (X_1X_3)^2 = X_1X_2X_3$$

$$X_1(X_3^2 - X_2^2) = X_2X_3X_4.$$

Ein Beispiel einer Kurve vom Geschlecht O ist die Lemniskate X_1:

$$(x_1^2 + x_2^2)^2 - a^2(x_1^2 - x_2^2) = 0.$$

Für den Fall $\overline{X_1} = X_1 - \{(0,0)\}$ erhalten wir die Parameterdarstellung

$$x_1 = \frac{(\tau_1^2 + \tau_2^2)a\tau_1\tau_2}{(\tau_1^4 + \tau_2^4)}, \quad x_2 = \frac{(\tau_1^2 - \tau_2^2)a\tau_1\tau_2}{(\tau_1^4 + \tau_2^4)},$$

mit

$$a\frac{\tau_2}{\tau_1} = \frac{x_1 + x_2^2}{x_1 + x_2}.$$

4.) Wir wollen den Beweis der Behauptung auf Seite 16 nachtragen. Es gilt zunächst der Hilfssatz: Ist für ein $\beta_1 = \pi\gamma \quad \cos\beta_1 \in Q$ so ist entweder γ irrational oder $\gamma = 1, \frac{1}{3}, \frac{2}{3}$ oder $\frac{1}{2}$ modulo 1. Den Beweis für diesen Satz führen wir in Anhang 5.

Wir betrachten zunächst die Fälle $\gamma = \frac{1}{3}, \frac{2}{3}$. Es ist dann $\cos\beta_1 = \pm\frac{1}{2}$, $\sin\beta_1 = \frac{\sqrt{3}}{2}$. Da

$$\cos\beta_1 = \frac{y}{z} - \frac{p}{q}\frac{x}{z}, \quad \sin\beta_1 = \frac{x}{z}\sqrt{1 - \frac{p^2}{q^2}}$$

so erhalten wir die Gleichungen (wir nehmen jetzt den Fall $\cos\beta_1 = \frac{1}{2}$).

$$\frac{x}{z} = \frac{1}{2}q\sqrt{\frac{3}{q^2 - p^2}} \quad , \quad \frac{y}{z} = \frac{1}{2}\left(1 + p\sqrt{\frac{3}{q^2 - p^2}}\right).$$

Da x und z ganzzahlig sein sollen, so muß die Quadratwurzel rational sein, also von der Gestalt $\frac{d}{c}$ mit g.g.T $(c, d) = 1$. Wir haben also

$$\sqrt{q^2 - p^2} = \frac{c}{d}\sqrt{3}.$$

Es ist also $d^2(q^2 - p^2) = 3c^2$, also ist die linke Seite dieser Gleichung durch 3 teilbar. Es kann nicht d durch 3 teilbar sein, denn sonst wäre auch c durch 3 teilbar, d.h. es gibt eine ganze Zahl k mit $q^2 - p^2 = 3k$. Dann ist aber $\sqrt{k} = \frac{c}{d}$, also muß k ein Quadrat r^2 sein. Wir haben also die Gleichung $q^2 - p^2 = 3r^2$.

Setzen wir $\xi = \frac{q}{r}$, $\eta = \frac{p}{r}$, so haben wir die rationalen Lösungen der Gleichung: $\xi^2 - \eta^2 = 3$ zu finden. Da (2,1) eine spezielle Lösung ist, so können wir ansetzen: $\xi = e + ut$, $\eta = 1 + vt$ und erhalten, wenn wir in geläufiger Weise mit der Hyperbel $\xi^2 - \eta^2 = 3$ schneiden,

$$t = 2\frac{2u - v}{v^2 - u^2},$$

also

$$\xi = 2\frac{u^2 + v^2 - uv}{v^2 - u^2} \quad , \quad \eta = \frac{4uv - (u^2 + v^2)}{v^2 - u^2}$$

also erhalten wir, (L passende rationale Zahl)

$$q = 2L(u^2 + v^2 - uv), \quad p = L(u^2 + v^2 - 4uv), \quad r = L(v^2 - u^2).$$

Weiters ist

$$\frac{x}{z} = \frac{u^2 + y^2 - uv}{v^2 - u^2} \quad , \quad \frac{y}{z} = \frac{y(y - 2u)}{y^2 - u^2}$$

und

$$\cos\rho = \frac{p}{q} = \frac{u^2 + v^2 - 4uv}{2(u^2 + v^2 - uv)}.$$

Im Fall $\cos\beta_1 = -\frac{1}{2}$ erhalten wir

$$\frac{y}{z} = \frac{u(u - 2v)}{v^2 - u^2}.$$

Ist $\gamma = 1$, so ist $\cos\beta_1 = -1$, $\sin\beta_1 = 0$ also wäre $q = \pm p$, was nicht geht (außer im Fall $\cos\rho = \pm 1$, also $\sin\rho = 0$, was uns nicht interessiert). Es bleibt noch der Fall $\gamma = \frac{1}{2}$. Dann ist $\cos\beta_1 = 0$, also muß $x = kq$, $y = kp$ mit passendem k sein. Es ist dann

$$\sin\beta_1 = \frac{x}{z}\sqrt{1 - p^2/q^2} = 1, \quad \text{also ist} \quad z = \pm k\sqrt{q^2 - p^2}.$$

Es muß dann $\sqrt{q^2 - p^2}$ rational sein. Es gibt dann ein ganzzahliges r so daß $q^2 - p^2 = r^2$, also ist $q^2 = r^2 + p^2$. Dann bilden aber (r, p, q) ein gewöhnliches pythagoräisches Tripel. Es gibt dann ganze u, v so daß

$$r = 2uvL, \quad p = L(u^2 - v^2), \quad q = L(u^2 + v^2)$$

für passendes rationales L. Es ist also

$$\frac{x}{z} = \frac{u^2 + v^2}{2uv} \quad , \quad \frac{y}{z} = \frac{u^2 - v^2}{2uv}$$

und
$$\cos\rho_1 = \frac{p}{q} = \frac{u^2-v^2}{u^2+v^2}.$$

5.) Wir wollen den Beweis des Hilfssatzes aus 4.) nachtragen. Wir folgen der Beweisidee von *Hadwiger* (Elemente der Mathematik 1 (1946), 98-100). Wir nehmen an, daß $\gamma = \frac{a}{b}$ ist mit $(a,b) = 1$ und $b > 0$, und setzen $Z = e^{\frac{i\pi a}{b}}$. Es soll $\cos\frac{\pi a}{b} = U$ in Q sein. Dann ist
$$\sin\frac{\pi a}{b} = \pm\sqrt{1-U^2} = V.$$
Es ist also $Z = U + iV$. Dann ist $Z_m = Z^m = A_m + iB_m V$, wo A_m, B_m rationale Zahlen sind. Es ist $Z_{m+2b} = Z_m$ d.h. es ist $A_{m+2b} = A_m$, $B_{m+2b} = B_m$, d.h. die A_m und B_m nehmen nur endlich viele verschiedene Werte an. Sie besitzen daher alle einen gemeinsamen Nenner d, d.h. wir können setzen
$$A_m = \frac{C_m}{d},\quad B_m = \frac{D_m}{d}$$
wo C_m, D_m ganze Zahlen sind. Weiters bemerken wir folgendes: Es ist für jedes ganze $k \geq 0$ $(1-Z_m)^k$ eine endliche ganzzahlige Linearkombination von $1, Z_m, \ldots, Z_{mk}$, läßt sich also ebenfalls mit ganzzahligen X_{mk}, Y_{mk} in der Form schreiben
$$(1-Z_m)^k = \frac{X_{mk} + iY_{mk}V}{d}.$$
Wir nehmen zunächst an (1. Fall): Es ist b ungerade. Dann gibt es ein m, so daß $ma \equiv 1(2b)$ lösbar ist. Es ist (mit $\beta_1 = \pi\gamma$)
$$|1-Z_m| = \left|2\sin\frac{m\beta_1}{2}\right| = 2\left|\sin\frac{\pi}{2b}\right|,$$
da ja $|\sin\pi x|$ die Periode 1 hat. Wir erhalten also
$$|1-Z_m|^k = \left|2\sin\frac{\pi}{2b}\right|^k = \frac{1}{d}\,|X_{mk} + iY_{mk}V|.$$

Wenn nun $b > 3$ ist, so ist $\sin\frac{\pi}{2b} < \sin\frac{\pi}{6} = \frac{1}{2}$. Für $k \to \infty$ geht also $|1-Z_m|^k \to 0$. Andererseits ist $|X_{mk} + iY_{mk}V| \geq |V|$ und damit ist ein Widerspruch erzielt. Es muß also, wenn a ungerade, ist $b \leq 3$ sein. Im 2. Fall, a gerade: $a = 2a_1$, ein Fall, den gerade Hadwiger behandelt, betrachtet man die Kongruenz $ma_1 \equiv 1(\bmod b)$ und erhält
$$|1-Z_m| = \left|2\sin\frac{m\beta_1}{2}\right| = 2\left|\sin\frac{\pi}{b}\right|.$$

Wenn $b > 6$ ist, erhält man wieder den Widerspruch. Man braucht also nur noch den Fall $1 \leq b \leq 6$ untersuchen. Die Fälle $b = 4, 5, 6$ fallen sofort aus, da ja in diesen Fällen $\cos\beta_1$ irrational ist. Damit ist der Hilfssatz vollständig bewiesen.

Gleichverteilung auf Produkten von Sphären

J. Reine Angew. Math. **330** (1982), 1–43

Die bisherige Theorie der Gleichverteilung beschäftigt sich mit der Gleichverteilung auf dem Torus, dem Produkt von Kreisen vom Radius 1, andererseits mit der Gleichverteilung auf kompakten Räumen. Hier soll nun die Theorie der Gleichverteilung auf Produkten von Sphären entwickelt werden. [Für den Fall der Sphäre S von m Dimensionen selbst und für die Verteilung der Punkte $\left(\frac{x}{\sqrt{n}}\right)$ auf der Sphäre, wo $x=(x_1,\ldots,x_{m+1})$ ganzzahlige Lösungen der Gleichung $x^2=n$ sind, liegen Resultate von A. W. Malyschew ([9], [10]) und C. Pommerenke [11] vor.] In der vorliegenden Arbeit soll nun eine allgemeine Theorie entwickelt werden. Im § 1 benützen wir ein Gegenstück zum Landauschen Integral, das auch Pommerenke benützt, und approximieren stetige Funktionen durch Produkte von harmonischen Polynomen. Im § 2 entwickeln wir diese Polynome in Kugelfunktionen mit Benützung der Formel von Funk-Hecke. Dabei verwenden wir das Buch von Claus Müller, Spherical Harmonics [10]. In § 3 leiten wir mit Hilfe der Greenschen Funktion auf der Sphäre, angeregt [durch das Buch von Courant-Hilbert, Methoden der Mathematischen Physik I (5. Kapitel), [4], und] durch die bedeutende Arbeit von W. Freeden „Über eine Klasse von Integralformeln der mathematischen Geodäsie" [6] Gegenstücke zur sogenannten Formel von Koksma-Hlawka her und zwar gleich für kompakte geschlossene differenzierbare Mannigfaltigkeiten. In § 4 wird dies für die Sphäre bzw. für Produkte von Sphären genauer ausgeführt, und es werden die zugehörigen Diskrepanzen studiert. In § 5 liegen statt den Kugelfunktionen die Funktionen $(cx)^k$, wo c isotope Vektoren sind, für die Aufstellung eines Kriteriums für die Gleichverteilung auf Sphären nach Hecke und Pommerenke [11] zugrunde. In einem Anhang studieren wir die Transformation von Box und Muller [3], die gleichverteilte Folgen auf dem Torus in gleichverteilte Folgen im Raum mit der Normalverteilungsdichte überführt. Damit gelingt es, gleichverteilte Folgen auf der Sphäre durch gleichverteilte Folgen auf dem Torus zu erzeugen. Die auftretenden Abschätzungen sind sehr grob und können sicher durch Verwendung stärkerer Methoden verbessert werden.

§ 1. Es liege die Sphäre S^m von der Dimension m vor, d.h. jeder Punkt x des S^m genügt der Gleichung $x^2=x_1^2+\cdots+x_{m+1}^2=1$. Wir betrachten nun mit Pommerenke (loc. cit) für ganzzahlige $L\geqq 1$

$$W_L(x)=\int(xy+1)^L\,dV(y)\,. \tag{1}$$

Dabei sei $dV(y)$ das Oberflächenelement (im R^{m+1}) von $S=S^m$, also das Volumselement von S^m. Weiters sei xy das Skalarprodukt von x und y. Die Integration erstreckt sich über die ganze Sphäre.

Hilfssatz 1. *Es ist W_L unabhängig von x und es gilt*

1 (2) $$W_L \geqq 2^L (\sqrt{e}\, V^{-1}(K_m) L^{\frac{m}{2}})^{-1}.$$

Dabei ist $V_m = V(K_m)$ das Volumen der m-dimensionalen Kugel.

Beweis. Durch eine Drehung D können wir erreichen, daß $x = De_1$ ist, wobei $e_1 = (1, 0, \ldots, 0)$ ist. Dabei geht y in den Punkt w über, d.h. es ist $y = Dw$. Es wird dann, da ja dV gegenüber Drehungen invariant ist, $dV(y) = dV(w)$. Daher wird

$$W_L = \int (1+w_1)^L dV(w),$$

da ja $xy = ew$ ist. Somit ist W_L unabhängig von x. Weiters ist, da $w^2 = 1$, $1 + w_1 \geqq 0$ ist

(3) $$W_L \geqq \int' (w_1+1)^L dV(w),$$

wobei sich die Integration über den Teil T der Sphäre

(4) $$1 - \frac{1}{L} \leqq w_1 \leqq 1$$

erstreckt. Es ist also

(5) $$W_L \geqq \left(2 - \frac{1}{L}\right)^L V(T).$$

Dabei sei $V(T)$ das Volumen von T. Es ist

2 (6) $$\left(2 - \frac{1}{L}\right)^L = 2^L \left(1 - \frac{1}{2L}\right)^L \geqq 2^L e^{-\frac{1}{2}}.$$

Es gilt ganz allgemein für das Volumen $V(T(\eta))$ des Bereiches $T(\eta)$

(7) $$1 - \eta \leqq w_1 \leqq 1$$

die Abschätzung

(8) $$V(T(\eta)) \geqq V(K_m)\, \eta^{\frac{m}{2}}.$$

Es enthält ja die Menge $T(\eta)$, da $w_2^2 + \cdots + w_{m+1}^2 = 1 - w_1^2$, die m-dimensionale Kugel $K: w_2^2 + \cdots + w_{m+1}^2 \leqq \eta$ mit dem Radius $\sqrt{\eta}$, da dann $1 - w_1^2 \leqq \eta$ ist. Damit ist (8) schon gezeigt.

Hilfssatz 2. *Es sei $\|x, y\|$ der sphärische Abstand der Punkte x, y auf S^m. Weiters sei $C(x, \delta)$ die Menge aller Punkte auf S^m mit $\|x, y\| \leqq \delta$. Dann ist*

3 (9) $$J(L, \delta) = \int' (1+xy)^L dV(y) \leqq e^{-\frac{L\delta^2}{16}} V(S^m),$$

wobei sich die Integration über alle y, die nicht in $C(x, \delta)$ liegen, erstreckt, d.h. es ist stets

(10) $$\|x, y\| \geqq \delta.$$

Beweis. Wir üben wie in Hilfssatz 1 eine Drehung D aus, so daß $x=De_1$, $y=Dw$ ist. Dann ist

$$J(\delta)=\int'(1+w_1)^L\,dV(w).$$

Es ist $\|x,y\|=\|e_1,w\|$. Wir bezeichnen nun mit $|e_1,w|$ den euklidischen Abstand von e_1 und w in R^{m+1}. Es gilt nun

$$|e_1,w|\leqq\|e_1,w\|\leqq 2|e_1,w|. \tag{11}$$

Dies sieht man auch leicht analytisch ein. Setzen wir nämlich

$$w_1=\cos\vartheta,\quad w_2^2+\cdots+w_{m+1}^2=\sin^2\vartheta,$$

so ist $\|e_1,w\|=\vartheta$ mit $0\leqq\vartheta\leqq\pi$. Weiters ist

$$|e_1,w|=\sqrt{(1-\cos\vartheta)^2+\sin^2\vartheta}=2\sin\frac{\vartheta}{2}=\sqrt{2(1-w_1)}.$$

Nun ist bekanntlich für $0\leqq\alpha\leqq\frac{\pi}{2}$ stets $\sin\alpha\geqq\frac{2}{\pi}\cdot\alpha$. Mit $\alpha=\frac{1}{2}\vartheta$ erhalten wir daher

$$|e_1,w|=2\sin\frac{\vartheta}{2}\leqq\vartheta\leqq\frac{\pi}{2}\left(2\sin\frac{\vartheta}{2}\right)<2|e_1,w|.$$

Aus (11) folgt daher, daß

$$J^L\leqq J_2=\int'(1+w_1)^L\,dV(w) \tag{12}$$

ist, wo sich in J_2 der Integrationsbereich auf alle w mit $|e_1,w|\geqq\frac{\delta}{2}$ erstreckt, also über alle w mit $w_1\leqq 1-\frac{1}{8}\delta^2$. Daher ist

$$J\leqq\left(2-\frac{\delta^2}{8}\right)^L V(S^m).$$

Nun ist

$$\left(1-\frac{\delta^2}{16}\right)^L\leqq e^{-\frac{L\delta^2}{16}}.$$

Damit ist (9) bewiesen.

Es sei nun f eine stetige Funktion auf S^m, dann definieren wir als das Landau-Pommerenkesche Integral

$$T_L(f)(x)=\int f(y)\,K(xy,L)\,dV(y), \tag{13}$$

wobei

$$K(z,L)=(1+z)^L\,W_L^{-1} \tag{14}$$

ist. Es ist der Kern $K(xy,L)$ auf S nicht negativ. Weiter sei ω_f der Stetigkeitsmodul der Funktion f auf S, d.h. es gilt

$$|f(x)-f(y)|\leqq\omega_f(\delta), \tag{15}$$

wenn die sphärische Distanz $\|x, y\| \leqq \delta$ ist. Weiters bezeichnen wir mit

$$A(f) = \sup |f(x) - f(y)|, \tag{16}$$

wo x, y alle Punkte auf S durchlaufen, die Abweichung von f auf S. Da sich x, y auf S stets durch einen Großkreisbogen mit einer Länge $\leqq \pi$ verbinden lassen, so folgt aus der Definition von (15), daß

$$A(f) \leqq \left(\left[\frac{\pi}{\delta}\right] + 1\right) \omega_f(\delta) \tag{17}$$

ist. Wir behaupten nun folgenden

Satz 1. *Es ist stets für alle x auf S^m*

$$|f(x) - T_L(f)(x)| \leqq R(L, \delta), \tag{18}$$

wo

$$R(L, \delta) = \omega_f(\delta) + c_m A(f) e^{-\frac{L\delta^2}{16}} L^{\frac{m}{2}} \tag{19}$$

und

4 $$c_m = \sqrt{e}\, V^{-1}(K^m)\, V(S^m). \tag{20}$$

Nehmen wir

$$\delta = 4 \sqrt{\frac{m+1}{2}} \sqrt{\frac{\lg L}{L}}, \tag{21}$$

dann ist nach (17)

5 $$R(L) \leqq \omega_f\left(4m \sqrt{\frac{\lg L}{L}}\right)(1 + c_m). \tag{22}$$

Es ist $\lim\limits_{m \to \infty} c_m = 0$, also

6 $$R(L) \leqq k_m \omega_f\left(\sqrt{\frac{\lg L}{L}}\right). \tag{22'}$$

Erfüllt f eine Lipschitzbedingung $|f(x) - f(y)| \leqq K \|x, y\|$, so erhalten wir

$$R(L) \leqq K k_m \sqrt{\frac{\lg L}{L}}. \tag{22''}$$

Dabei ist

7 $$k_m \leqq 5 \sqrt{e}\, V^{-1}(K^m)\, V(S^m). \tag{23}$$

Wir gehen nun zum Beweis des Satzes 1 über. Es ist nach (14) und (1)

$$\int K(xy, L)\, dV(y) = 1.$$

Wenn wir

$$U(x) = f(x) - T_L(f)(x) \tag{24}$$

setzen, so erhalten wir

$$U(x) = J_1(x) + J_2(x), \tag{25}$$

wobei

$$J_1(x) = \int (f(x) - f(y))\, K(xy, L)\, \iota(C(x, \delta, y))\, dy \tag{26}$$

und

$$J_2(x) = \int (f(x) - f(y))\, K(xy)\, (1 - \iota(C(x, \delta, y)))\, dy \tag{27}$$

ist. Dabei ist $C(x, \delta)$ wie vorher die Kappe $\|xy\| \leqq \delta$ und ι ihre Indikatorfunktion. Da $K \geqq 0$ ist, so ist nach Definition (14)

$$|J_1| \leqq \omega_f(\delta). \tag{28}$$

Nach Definition 16 ist

$$|J_2(x)| \leqq A(F) \int K(xy, L)\, (1 - \iota(C(x, \delta, y)))\, dy. \tag{29}$$

Nun ist $1 - \iota(C(x, \delta, y))$ die Indikatorfunktion der Menge $\|x, y\| \geqq \vartheta$. Nun ist aber nach (2) bzw. (9)

8 $$K(xy, L) \leqq \frac{\sqrt{e}}{2^L} V^{-1}(K^m)\, V(S^m)\, e^{-\frac{L\delta^2}{16}}\, L^{\frac{m}{2}}. \tag{30}$$

Damit ist Satz 1 schon bewiesen.

Wir wollen jetzt diesen Satz verallgemeinern. Wir betrachten t Sphären $S_1, \ldots, S_t$ mit den Dimensionen $m_1, \ldots, m_t$ und das Produkt $\Pi = S_1 \times \cdots \times S_t$. Ist x ein Punkt auf Π, so bezeichnen wir seine Projektion auf die j-te Sphäre mit $p_j(x)$. Es ist also $x = (p_1, \ldots, p_t)$. Weiters sei f eine stetige Funktion auf Π. Wir ordnen dann f die Funktion

$$T_L(f)(x) = \int f(y)\, K(xy, L)\, d\hat{\Pi}(y) \tag{31}$$

zu. Dabei sei $d\hat{\Pi}(y) = dV_1(q_1) \cdots dV_t(q_t)$ das Volumselement von Π. Es sei

$$dV_j(q_j)$$

das Volumselement der j-ten Sphäre. Weiters sei mit $L = (L_1, \ldots, L_t)$ (L_j natürliche Zahlen)

$$K(xy, L) = K_1(p_1 q_1, L_1) \cdots K_t(p_t q_t, L_t). \tag{32}$$

Dabei ist $p_j q_j$ das Skalarprodukt $p_j(x)\, q_j(y)$ auf der Sphäre S_j. Es ist wieder K nicht negativ und $\int K(xy, L)\, d\hat{\Pi}(y) = 1$.

Wir bezeichnen mit $C(x, \delta)$ $(\delta = (\delta_1, \ldots, \delta_t))$ das Produkt der Kappen

$$C(p_1(x), \delta_1) \times \cdots \times C(p_t(x), \delta_t). \tag{33}$$

Das Komplement von $C(x, \delta)$ ist die Vereinigung aller Mengen $\hat{C}_j = S_1 \times \cdots \times S_t$, wo an der j-ten Stelle die Sphäre S_j durch $S_j - C(p_j, \delta_j)$ zu ersetzen ist. Weiters setzen wir noch

$$\|x, y\| = \operatorname{Max}(\|x_1, y_1\|, \ldots, \|x_t, y_t\|). \tag{34}$$

Wir behaupten dann folgenden

Satz 2. *Es ist für alle x auf Π*

(35) $$|f(x)-T_L(f)(x)| \leqq R(\delta, L),$$

wo

(36) $$R(\delta, L) = R_1(\delta) + R_2(\delta, L).$$

Dabei ist

(37) $$R_1(\delta) \leqq \omega_f(\delta),$$

wobei $\omega_f(\delta) = \sup |f(x)-f(y)|$ *über* $\|x_1, y_1\| \leqq \delta_1, \ldots, \|x_t, y_t\| \leqq \delta_t$ *der Stetigkeitsmodul von f auf Π ist. Weiters ist*

(38) $$R_2 \leqq A(f) \sum_{j=1}^{t} c(m_j) L_j^{\frac{m_j}{2}} e^{-\frac{L_j \delta_j^2}{16}},$$

wobei c(m) in (20) definiert ist und A(f) die Abweichung von f auf Π ist. Es gilt wieder

(39) $$A(f) \leqq \operatorname{Max}\left(\left[\frac{\pi}{\delta_j}\right]+1\right)\omega_f(\delta).$$

Wählt man

9 (40) $$\delta_j = 4\sqrt{m_j+1} \cdot \sqrt{\frac{\lg L_j}{L_j}},$$

so erhält man, wenn z.B. $L_1 = \cdots = L_t = L$, $m_1 = m_2 = \cdots = m_t = m$ ist,

10 (41) $$R(L) \leqq t k_m \omega_f\left(\sqrt{\frac{\lg L}{L}} \cdots \sqrt{\frac{\lg L}{L}}\right).$$

Der Beweis erfolgt genau so, wie bei Satz 1. Wir setzen wieder $U(x) = f(x) - T_L(f)(x)$. Dann ist $U(x) = J_1(x) + J_2(x)$, und es gilt wieder (28). Bei J_2 ist jetzt zu beachten, daß $1 - \iota(C(x, \delta), y) = \sum_{j=1}^{t} \iota(\hat{C}_j(y, \delta), y)$ gilt. Jedes Integral $\int K(xy, L)\, \iota(\hat{C}_j)\, d\Pi(y)$ ist gleich

$$\int K_j(p_j(x)\, p_j(y), L_j)\, \iota(\hat{C}_j, y)\, dV_j(w),$$

und es folgt aus (30) die Behauptung.

Es sei nun eine Folge $\omega_N = (x_1, \ldots, x_N)$ auf Π gegeben. Wir betrachten nun für jede im Jordanschen Sinne integrierbare Funktion f

11 (42) $$\mu_N(f) = \sum_{n=1}^{N} f(x_n)$$

und

(43) $$\mu(f) = \frac{1}{V(\Pi)} \int f(x)\, d\hat{\Pi}(x) = \int f(x)\, d\Pi.$$

Dabei erstreckt sich die Integration, wie immer, wenn nichts anderes gesagt wird, auf ganz Π. Wir setzen weiters

(44) $$\varepsilon_N(f) = \varepsilon_N(f, \omega_N) = \mu_N(f) - \mu(f).$$

Wir wollen nun eine unendliche Folge $\omega=(x_1, x_2, \ldots)$ gleichverteilt nennen, wenn für jede stetige Funktion auf Π

$$\lim_{N\to\infty} \varepsilon_N(f, \omega_N)=0 \tag{45}$$

ist. Ist f stetig auf Π, dann können wir $\varepsilon_N(f)$ durch $\varepsilon_N(T_L(f))$ abschätzen. Dann haben wir, wie sofort aus Satz 2 folgt,

12 $$|\varepsilon_N(f, \omega_N)| \leqq k_m \omega_f\left(\sqrt{\frac{\lg L}{L}}\right)+|\varepsilon_N(T_L)|, \tag{46}$$

ja sogar allgemein

$$|\varepsilon_N(f, \omega_N)| \leqq 2R(L, \delta)+|\varepsilon_N(T_L)|. \tag{46'}$$

Wir wollen aber noch allgemeinere Funktionen, nicht nur stetige Funktionen zulassen, sondern auch unstetige Funktionen betrachten. Es sei E eine im Jordanschen Sinne meßbare Menge auf Π und es sei $\iota(E)$ die Indikatorfunktion von E. Weiters sei g eine nicht negative, beschränkte stetige Funktion auf E. Dann betrachten wir die Funktion

$$f=g\,\iota(E). \tag{47}$$

Auch hier wollen wir f durch Polynome T approximieren. Zu diesem Zweck betrachten wir die Menge $E_2(\delta)$ (kurz mit E_2 bezeichnet) aller Kappen $C(x, \delta)$, wo x in E liegt. Es wird $\Pi-E_2$ von der Menge $\bigcup_{j=1}^{t} \hat{C}_j$ überdeckt, wobei wir $\hat{C}_j$ bereits vorher definiert haben. Weiters sei $E_1(\delta)$ die Menge aller y aus E, so daß für alle x'' aus $\Pi-E$, sicher $\|y, x''\| \geqq \delta$ ist, d.h. für mindestens ein j $(1\leqq j\leqq t)$ ist $\|y_j, x_j''\| \geqq \delta$.

Es ist $E_3(\delta)=E_2(\delta)-E_1(\delta)$ die Menge aller y aus $E_2(\delta)$, für welche es ein x' aus E gibt, so daß $\|x', y\| \leqq \delta$ und für alle $x'' \in \Pi-E$ stets $\|x'', y\| \geqq \delta$ ist. Es ist also $E_3(\delta)$ die Menge aller Punkte y aus Π, die vom Rand ∂E von E einen Abstand $\leqq \delta$ haben. Verbindet man nämlich x', x'' durch einen kürzesten Bogen (d.h. verbindet man jedes x_j' mit x_j'' durch einen Großkreisbogen), so muß es auf dem Bogen einen Punkt p geben, der Randpunkt von E ist und für den $\|p, y\| \leqq \delta$ ist, da ja die Kappen konvexe Mengen auf Π sind. Da E quadrierbar ist, so ist sicher

$$\lim_{\delta\to 0} V(E_3(\delta))=0.$$

Wir nennen E *normal*, wenn eine positive Zahl C existiert, so daß für alle δ

$$V(E_3(\delta)) \leqq C\delta \tag{48}$$

gilt. Jede Kappe ist normal, jede konvexe Menge auf Π ist normal. Wir führen nun die Funktion $I_L(E, \delta)$ durch

$$I_L(E_3, x)=T_L(\iota(E_3, x)) \tag{49}$$

ein, wo natürlich

$$T_L(\iota(E_3, x))=\int K(xy, L)\,\iota(E_3, y)\,d\hat{\Pi}(y) \tag{49'}$$

ist. Wir nennen I die Randfunktion von E. Es ist $I_L(x) \geqq 0$ für alle x. Weiters ist

$$\int I_L(x)\,dx=\int \iota(E_3, y)\,d\hat{\Pi}(y)=V(E_3). \tag{50}$$

Wir betrachten nun die Funktion f, die durch (47) auf ganz Π definiert ist. Wir setzen

(51) $$M(f)=\sup f=\sup g=M(g).$$

Weiters setzen wir

(52) $$T_L(f)(x)=\int K(xy,L)f(y)\,d\hat{\Pi}(y)$$

und wir betrachten

(53) $$U_L(f)(x)=f(x)-T_L(f)(x).$$

Wir behaupten nun

Satz 3. *Es ist für alle x auf Π, alle σ mit $\sigma\leqq\delta$ ($M(g)=M$ gesetzt)*

(54) $$U(x)\leqq M\,I(x)+R(\sigma,\delta,L)$$

und

(54′) $$U(x)\geqq -M\,I(x)-R(\sigma,\delta,L).$$

Dabei ist

(55) $$R(\sigma,\delta)=R_1(\sigma)+R_2(\delta),$$

wobei

13 (56) $$R_1(\sigma)=\omega g(\sigma)+A(f)\,e^{-\frac{L\delta^2}{16}}L^{\frac{m}{2}}$$

und

14 (56′) $$R_2(\delta)=M(g)\,e^{-\frac{L\delta^2}{16}}L^{\frac{m}{2}}$$

15 *ist. Nehmen wir* $\delta=\sigma=\sqrt{\dfrac{\lg L}{L}}$, *so setzen wir*

(57) $$R(\sigma,\delta,L,f)=R(L,f).$$

Bevor wir Satz 3 beweisen, wollen wir die Voraussetzung fallen lassen, daß $g\geqq 0$ auf E ist. Dazu brauchen wir nur in Satz 3 die Funktion g durch die nicht negative Funktion $M(g)+g$ zu ersetzen. Es ändert sich dann U nicht, aber $M(g)$ ist in (56′) durch $2M(g)$ zu ersetzen und wir erhalten Satz 3′.

Es sei g eine stetige beschränkte Funktion auf E, dann gilt

(58) $$U(x)\leqq 2M\,I(x)+R(L)$$

und

(58′) $$U(x)\geqq -2M\,I(x)-R(L).$$

Dabei ist

16 (59) $$R(L)=\omega_g\left(\sqrt{\frac{\lg L}{L}}\right)+\left(A(g)+2M(g)\right)\sqrt{\frac{\lg L}{L}}.$$

Ist g auf E konstant, dann ist $f=\iota(E)$ die Indikatorfunktion von E, und es wird in diesem Fall, wenn E konvex ist,

$$(59') \qquad R(L, \iota(E)) = O\left(\sqrt{\frac{\lg L}{L}}\right).$$

Wir beweisen nun Satz 3. Zuerst beweisen wir (54). Wenn x nicht in E liegt, so ist $f(x)=0$ und (54) ist trivialerweise richtig, da ja stets $T_L \geqq 0$ ist. Wir nehmen an, daß x in E liegt, dann ist $f(x)=g(x)$. Nun gilt (trivial)

$$1 = \int K(xy, L)\, \iota_2(y)\, d\hat{\Pi}(y) + \int K(xy, L)\, \iota_2'(y)\, d\hat{\Pi}(y),$$

wo ι_2 die Indikatorfunktion von E_2 und ι_2' die Indikatorfunktion des Komplements von E_2 in Π ist. Damit ist (54) äquivalent mit der Ungleichung

$$(60) \qquad 0 \leqq \int (f(y)-f(x))\, \iota_2(y)\, K(xy)\, d\hat{\Pi}(y) - f(x) \int \iota_2'(y)\, K(xy)\, d\hat{\Pi}(y) + R_4,$$

wo

$$(61) \qquad R_4 = R(\sigma, \delta, L) + M I(x)$$

ist. Nun ist rechts das 1. Integral gleich

$$(62) \qquad \int A(x, y)\, \iota(E, y)\, d\hat{\Pi}(y) + \int A(x, y)\, (\iota_2(y) - \iota(y))\, d\hat{\Pi}(y),$$

wobei wir $A(x, y) = (f(y)-f(x))\, K(xy, L)$ gesetzt haben. Das 2. Integral in (62) ist, da in $E_2 - E$ ja $f(y)=0$ ist und x in E ist, $f(x)=g(x)$ und $E_2 - E \subseteqq E_2 - E_1$ ist,

$$\geqq -g(x) \int K(xy, L)\, \iota_3(y)\, dy \geqq -M I(x).$$

Das 1. Integral ist aber, da in E ja $f=g$ ist,

$$\int (g(y)-g(x))\, \iota(E, y)\, K(xy, L)\, d\hat{\Pi}(y),$$

und wir können die Beweismethode von Satz 2 anwenden.

Man sieht, daß das Integral $\leqq R_1(\sigma)$ ist. Im 2. Integral in (60) ist ja nach Definition von E_2, da x in E liegt, $\|xy\| \geqq \delta$. Nach der gleichen Ungleichung wie in Satz 2 folgt, daß dieses Integral $\geqq -R_2(\delta)$ ist. Damit ist (54) bewiesen.

Jetzt zeigen wir (54′). Es ist ($\iota_1 = \iota(E_1)$, $\iota_1' = 1 - \iota_1$)

$$1 = \int K(xy, L)\, \iota_1(y)\, d\hat{\Pi}(y) + \int K(xy, L)\, \iota_1'(y)\, d\hat{\Pi}(y).$$

Daher ist (54′) äquivalent zu

$$(62') \qquad f(x) + R_1 \geqq \int K(xy, L)\, g(y)\, \iota_1(y)\, d\hat{\Pi}(y).$$

Das rechte Integral ist gleich

$$(63) \qquad \int K g \iota_1\, d\hat{\Pi}(y) + \int K g (1-\iota_1)\, d\hat{\Pi}(y).$$

Das 2. Integral in (63) ist $\leqq M I(x)$. Es ist also nur mehr zu zeigen, daß

$$(64) \qquad f(x) + R \geqq \int K g(y)\, \iota_1\, d\hat{\Pi}(y).$$

ist. Ist x nicht in E, dann ist, da y in E_1 liegt, $\|xy\| \geqq \delta$, daher $\int K g(y)\, \iota_1(y)\, d\hat{\Pi}(y) \leqq R_2(\delta)$, daher ist (64) richtig. Liegt x in E, dann ist $f(x)=g(x)$ und um (64) zu zeigen, ist zu zeigen, daß $R_1 + \int K(g(x)-g(y))\, \iota_1(y)\, d\hat{\Pi}(y) + \int K g(x)\, \iota_1'(y)\, d\hat{\Pi}(y) \geqq 0$ ist. Links ist das 1. Integral $\leqq R_1(\sigma)$. Das 2. Integral ist $\geqq 0$. Damit ist (54′) bewiesen.

Setzen wir $f_2 = T_L(f) + 2MI + R$ und $f_1 = T_L(f) - 2MI - R$, so ist

$$f_2 - f_1 = 2(2MI + R).$$

Dann ist $f_1 \leqq f \leqq f_2$. Daher ist auch $\mu_N(f_1) \leqq \mu_N(f) \leqq \mu_N(f_2)$, $\mu(f_1) \leqq \mu(f) \leqq \mu(f_2)$.
17 Daher ist $\varepsilon_N(f) \leqq \varepsilon_N(f_2) + 2\mu(f_2 - f_1)$,

18 $$\varepsilon_N(f) \geqq \varepsilon_N(f_1) - 2\mu(f_2 - f_1).$$

Daher haben wir

Satz 4. *Es sei g eine stetige beschränkte Funktion auf der quadrierbaren Menge E, mit dem Stetigkeitsmodul ω_g und der oberen Schranke $M(g)$ für $|g|$. Es sei $f = g\iota(E)$, wo ι die Indikatorfunktion von E ist. Weiters sei $E_3(\delta)$ die Menge aller y aus Π mit $\|\partial E, y\| \leqq \delta$.*

19 *Dann ist mit* $\delta = \sqrt{\frac{\lg L}{L}}$

20 (65) $$|\varepsilon_N(f, \omega)| \leqq 2\left(|\varepsilon_N(T_L(f))| + M\varepsilon_N\left(T_L(\iota(E_3))\right)\right) + R.$$

Dabei ist

(65′) $$R(L) = R_1 + R_2,$$

21 (65″) $$R_1 = M(g)\, V(E_3),$$

22 (65‴) $$R_2 = k_m\left(\omega_g\left(\sqrt{\frac{\lg L}{L}}\right) + M(g)\sqrt{\frac{\lg L}{L}}\right).$$

Ist insbesondere E normal, dann ist $V(E_3) \leqq C\sqrt{\frac{\lg L}{L}}$ und wenn E eine Kappe ist, so sind auch E_1 und E_2 Kappen und $E_3 = E_2 - E_1$ ist ein Kappenring von der Breite 2δ. Ist weiters $g = \iota(E)$, so ist

(66) $$|\varepsilon_N(\iota(E), \omega)| \leqq 2\left(|\varepsilon_N(T_L(\iota(E)))| + |\varepsilon_N(T_L(\iota(E_2 - E_1)))|\right) + R,$$

wobei

(66′) $$R(L) \leqq k_m\sqrt{\frac{\lg L}{L}}$$

wird. Ein anderes wichtiges Beispiel erhalten wir so: Es sei h eine Funktion auf S^m, die überall stetig differenzierbar ist, außer im Punkte q, wo sie unendlich wird, von der Ordnung $\|x, q\|^{-\alpha}$ ($\alpha \leqq m-2$, $m \geqq 3$; im Falle $m = 2$ sei die Ordnung $\lg\|x, q\|$). Es sei nun C die Kappe (q, ρ) und E sei das Komplement $\hat{C}$ zu C. Dann sei $g_\rho = h\iota(E)$, dabei sei $\rho > \sqrt{\frac{\lg L}{L}}$. Dann ist

$$M(g_\rho) = O(\rho^{-\alpha}), \quad \omega_{g_\rho} = O(\rho^{-\alpha-1})\sqrt{\frac{\lg L}{L}}, \quad \int g_\rho\, d\Pi(y) = \int h\, d\Pi(y) + O(\rho^{m-\alpha}).$$

Es ist daher

23 (67) $$|\varepsilon(f_\rho, \omega_N)| \leqq 2\,|\varepsilon_N(T_L(f_\rho))| + 2\,|\varepsilon_N(T_L(\iota(E)))| + O\left(\rho^{-\alpha}\sqrt{\frac{\lg L}{L}} + \rho^{-\alpha-1}\sqrt{\frac{\lg L}{L}}\right).$$

Wir wählen L so, daß $\rho^{-\alpha-1} L^{-1/2} = \rho^{m-\alpha}$, dann erhalten wir

24 (67') $$|\mu_N(h\iota(\hat{C}, q)) - \int h d\hat{\Pi}| \leqq \varepsilon_N(T_L(f_\rho)) + \varepsilon_N(T_L(\iota(E_3))) + O(\rho^{m-\alpha} \lg \rho).$$

§ 2. Es sei $\Pi = S_1 \times \cdots \times S_t$, wo $\dim S_j = m_j$, $L = (L_1, \ldots, L_t)$,

$$T_L(x_1, \ldots, x_t) = \int_\Pi f(y) K(xy, L) d\hat{\Pi}.$$

Dabei ist $f(y) = f(y_1, \ldots, y_t)$, $d\hat{\Pi} = dV(y_1) \cdots dV_t(y_t)$. Wir entwickeln T_L in eine Reihe nach Kugelfunktionen H_{kj}. Da T_L in jeder Variablen x_j vom Grad L_j ist, gilt

(1) $$T_L(x) = \sum_{k,j} c_{kj} H_{kj}(x).$$

Dabei bedeutet $k = (k_1, \ldots, k_t)$, wo $0 \leqq k_s \leqq L_s$, $j = (j_1, \ldots, j_t)$, wo $1 \leqq j_s \leqq Z(m_s, k_s) = Z_{(s)}$,

(2) $$c_{kj} = c_{k_1 j_1} \cdots c_{k_t j_t}, \quad H_{k,j}(x) = H_{k_1 j_1}(x_1) \cdots H_{k_t j_t}(x_t).$$

$Z(m, k)$ ist die Anzahl der linear unabhängigen Kugelfunktionen H_{kj} vom Grade k und

25 von der Größenordnung k^{m-2}. Da die H_{kj} auf Π orthogonal sind, d.h.

$$\int H_{kj}(x) H_{k'j'}(x) d\hat{\Pi} = V(\Pi)$$

genau dann, wenn $k = k'$, $j = j'$ und 0 sonst, so folgt $\left(d\Pi = \frac{1}{V(\Pi)} d(\hat{\Pi})\right)$:

$$c_{kj} = \int_\Pi T_L(x) H_{k,j}(x) d\Pi = \int_\Pi f(y) d\Pi(y) \int_\Pi H_{kj}(x) K(xy, L) d\Pi.$$

Es ist $\left(d\mu(p_r) = \frac{1}{V(S_r)} dV(p_r)\right)$

$$\int K(xy) H_{kj}(x) d\Pi(y) = \prod_{r=1}^{t} \int K_r(p_r \cdot q_r) H_{k(r)j(r)}(p_r) d\mu(p_r).$$

Wir haben also ($S_r = S$, $L_r = L$, $\dim S_r = m$ gesetzt) zu berechnen:

$$\frac{1}{W_L} \int (1 + pq)^L H_{kj}(p) d\mu(p).$$

Da der Integrand nur von pq abhängt, so ist nach dem Satz von Funk-Hecke (vgl. C. Müller [10])

26 $$\frac{1}{W_L} \int (1 + pq)^L H_{kr}(p) d\mu(p) = L_k(L) H_{kr}(\mu),$$

27 wobei $L_k = \frac{V(S)}{W_L} J(L, k)$ mit

(3) $$J(L, k) = \int_{-1}^{1} (1 + z)^L P_k(z) (1 - z^2)^{\frac{m}{2} - 1} dz$$

ist. Nach dem gleichen Satz ist dann weiters $W_L = V(S) J(L)$ mit

$$J(L) = \int_{-1}^{1} (1 + z)^L (1 - z^2)^{\frac{m}{2} - 1} dz.$$

Da nun $|P_k(z)| \leqq 1$ ist, so ist

$$|J(L, k)| \leqq J(L), \tag{4}$$

daher

$$|L_k| \leqq 1. \tag{5}$$

Dies gilt nicht nur für S, sondern auch für Π. Führen wir die Fourierkoeffizienten von f

$$b_{k,r}(f) = \int_{\Pi} f(y)\, H_{kr}(y)\, d\Pi(y) \tag{6}$$

ein, so erhalten wir

$$c_{k,r} = L_k(L)\, b_{kr}(f), \tag{7}$$

und es ist

$$|c_{k,r}| \leqq |b_{kr}(f)|. \tag{8}$$

Wir beachten noch, daß $\int_{\Pi} T_L(f)(x)\, d\Pi(x) = \int_{\Pi} f(x)\, d\Pi$. Weiters ist nach (1) und (7) für $t=1$ $T_L(x) = \sum_k L_k \sum_j \int f(y)\, H_{kj}(y)\, H_{kj}(x)\, d\Pi(y)$. Nun ist nach dem Additionstheorem (vgl. C. Müller loc. cit) $\sum_j H_{kj}(x) H_{kj}(y) = \dfrac{Z(m,k)\, P_k(xy)}{V(S^m)}$. So haben wir auch $T_L(x) = \sum_k L_k \int f(y)\, P_k(xy) \dfrac{Z(m,k)}{V(S^m)}\, d\Pi$. Dabei bedeutet, wie oben gesagt, $Z(m,k)$ die Anzahl der linear unabhängigen Kugelfunktionen vom Grade k. Es ist (für $t=1$)

$$Z(m,k) = \frac{(2k+m-1)\,\Gamma(k+m-1)}{\Gamma(k+1)\,\Gamma(m)}.$$

(Bei C. Müller [10] steht statt Z der Buchstabe N.) Bei der dortigen Bezeichnung ist $q = m+1$, k ist durch n zu ersetzen. Es ist $Z(m,k)$ von der Größenordnung k^{m-1}. Ist Π ein Produkt von t Faktoren S_m, dann ist $k = (k_1, \ldots, k_t)$ und

$$Z(m,k) = Z(m,k_1) \cdots Z(m,k_t).$$

Es sei nun $\omega_N = (x_1, \ldots, x_N)$ eine Folge auf Π und wir haben

$$\varepsilon_N(T_L(f)) = \frac{1}{N} \sum_{n=1}^{N} T_L(x_n) - \int T_L\, d\Pi; \tag{8'}$$

dann ist

$$\varepsilon_N(T_L) = \int f(y) \left(\frac{1}{N} \sum_{n=1}^{N} K(x_n y) - \int K(xy)\, d\Pi(x) \right) d\Pi(y) \tag{9}$$

abzuschätzen. Es ist zu beachten, daß in dieser Darstellung f durch $f - \mu(f)$ ersetzt werden kann. Da $|f(y) - \mu(f)| \leqq A(f)$ ist, gilt

$$|\varepsilon_N(T_L(f))| \leqq A(f) \int \left| \frac{1}{N} \sum_{n=1}^{N} K(x_n y) - \int K(xy)\, d\Pi(x) \right| d\Pi(y). \tag{9'}$$

Andererseits haben wir nach (1) $\varepsilon_N(T_L(f)) = \sum_{k,j} c_{kj}\varepsilon_N(H_{kj})$, also ist

28 $$\eta_N(T_L(f)) = |\varepsilon_N(T_L(H))| \leqq \sum{}' |b_{kj}||\varepsilon_N(H_{kj})|,$$

wobei $k=(k_1,\ldots,k_t) \neq (0,\ldots,0)$ ist. Es ist ja $H_0=1$. Weiter folgt aus (9) folgende, triviale Abschätzung $\eta_N(T_L) \leqq A(f)$. Wir haben also

$$|\varepsilon_N(T_L(f))| \leqq \sum_{k,j}{}' |b_{kj}||\varepsilon_N(H_{kj})|. \tag{10}$$

Nach der Schwarzschen Ungleichung folgt aus (10)

$$|\varepsilon_N(T_L(f))| \leqq \sqrt{\sum_{k,j} |b_{kj}|^2}\,\sqrt{\sum_{k,j} \varepsilon_N^2(H_{k,j})}. \tag{11}$$

Und nach der Vollständigkeitsrelation ist $\sum' |b_{k,j}|^2 = \int f^2 - (\int f)^2 = \int (f-\mu(f))^2 \leqq A^2(f)$.

Wir setzen $\eta_N(H_{kj}) = W_{kj}$ und $W_k = \sqrt{\dfrac{\sum (W_{kj})^2}{Z(m,k)}} \leqq \operatorname{Max} |W_{kj}|$; dann erhalten wir

$|\varepsilon_N(T_L)| \leqq A(f) \operatorname{Min}\left(1, \sum_{k=1}^{L} Z(m,k)\, W_k\right)$. Nach § 1 (22) bzw. (41) erhalten wir (vgl. § 1 (17), (39))

29 $$|\varepsilon_N(f)| \leqq c_m\, \omega_f\left(\sqrt{\frac{\lg L}{L}}\right)\left(1+\sqrt{L}\operatorname{Min}\left(1, \sum_{k=1}^{L} k^{m-1} W_k\right)\right), \tag{12}$$

wo $k^{m-1} = k_1^{m-1} \cdots k_t^{m-1}$ bedeutet. Liegt der allgemeine Fall § 1, Satz 4 vor, so erhalten wir folgendes: Da $\int \iota^2(E_3, y)\, d\Pi(y) = V\left(E_3, \sqrt{\dfrac{\lg L}{L}}\right) = V(L)$, ist nach § 1, (65), (65′), (65″)

30 $$|\varepsilon_N(f)| \leqq (M(g)) \operatorname{Min}\left(1, \sum_{k=1}^{L} k^{m-1} W_k\right) + R(L), \tag{12'}$$

$$\leqq M(g)\,(\operatorname{Min}(1, \sum |b_{kj}(f)|\,|W_{kj}|) + R(L), \tag{12''}$$

wobei $R(L)$ in (65′) definiert ist.

Ist also g eine Lipschitzfunktion auf E und E normal, so erhalten wir

31 $$|\varepsilon_N(f)| \leqq \left(M(g) + \sqrt{\frac{\lg L}{L}}\right) \operatorname{Min}\left(1, \sum_{k=1}^{L} k^{m-1} W_k\right) + \left(M(g)+C\right)\sqrt{\frac{\lg L}{L}}.$$

Ist g konstant und 1 auf E, so erhalten wir

$$|\varepsilon_N(\iota(E))| \leqq c\left(\sqrt{\frac{\lg L}{L}} + \operatorname{Min}\left(1, \sum_{k=1}^{L} k^{m-1} W_k\right)\right),$$

wo c von E abhängt. Ist E eine Kappe $C(p,\rho)$, so ist c von ρ unabhängig. Wir können aber schärfere Resultate erhalten, wenn wir (11) beibehalten. Dann ist mit

32 $|b_k| = \sqrt{\sum_j |b_{kj}|^2 / Z(m)}$, $\varepsilon_N(\iota(E))| \leqq \sum_{kj}^{L} k^{m-1} |b_k|\, W_k + \sqrt{\dfrac{\lg L}{L}}$. Wir werden später zeigen, daß bei einer Kappe $C(p,\rho)$, $|b_{kj}| \leqq \dfrac{|H_{kj}(p)|}{k^{\frac{m-1}{2}}}$ ist (§ 4, Bemerkung 5.).

Daraus folgt

$$(13) \qquad |\varepsilon_N(\iota(E))| \leqq c_m \left(\sqrt{\frac{\lg L}{L}} + \sum_{k=1}^{L} k^{\frac{m-1}{2}} W_k \right).$$

Zur Abschätzung (9′) ist noch folgendes zu bemerken: Es sei Π in ein Produkt $\Pi_1 \Pi_2$ zerlegt. Man kann also jedes x aus Π in der Form (x', x'') schreiben, wo x' in Π_1, x'' in Π_2 liegt. Es ist dann $K(xy, L) = K_1(x'y', L)\, K_2(x''y'', L)$. Weiters sei

$$\omega_N = \omega_{N(1)} \times \omega_{N(2)}$$

selbst Produkt einer Folge ω_1 von $N(1)$ Elementen auf Π_1 und einer solchen Folge auf Π_2 von $N(2)$ Elementen. Dann ist

$$\frac{1}{N} \sum_{n=1}^{N} K(x_n y, L) = \frac{1}{N(1)} \sum_{n(1)=1}^{N(1)} K_1(x'_{n(1)} y', L) \frac{1}{N(2)} \sum_{n(2)=1}^{N(2)} K_2(x''_{n(2)} y'', L).$$

Es sei darüber hinaus auch f ein Produkt $f_1 f_2$, wo f_1 auf Π_1, f_2 auf Π_2 definiert ist. Dann ist $\int f(y)\, K(xy, L)\, d\Pi(y) = \int f(y')\, K_1(x'y', L)\, d\Pi_1(y) \int f(y'')\, K_2(x''y'', L)\, d\Pi_2(y'')$. Wir setzen $\sup |f(y_1)| = M(f_1)$, $\sup |f(y_2)| = M(f_2)$. Es ist dann

$$(9') \qquad \mu_N(K, y) - \mu(K, y) = \big(\mu_{N(1)}(K_1, y') - \mu(K_1, y')\big)\, \mu_{N(2)}(K_2, y'') + \big(\mu_{N(2)}(K_2, y'')\big) - \mu(K_2, y'')\, \mu(K_1, y'),$$

also nach (9′)

33 $$(9'') \qquad |\mu_N(T_L(f))| \leqq |\varepsilon_{N(1)}(T_L(f_1))| M(f_2) + |\varepsilon_{N_2}(T_L(f_2))| M(f_1).$$

Es ist ja $|\int \mu_{N(2)}(K_2, y'')\, f_2(y'')\, d\Pi(y'')| \leqq M(f_2) \frac{1}{N(2)} \sum \int K_2(x''_{n(2)} y'')\, d\Pi_2 = M(f_2)$, da ja $\int K_2(x''y'', L)\, d\Pi_2(y'')$ von x'' unabhängig und gleich 1 ist. Analog zeigt man den 2. Teil rechts in (9″).

Ist ω_N ein Produkt $\omega_1 \times \cdots \times \omega_t$, wo ω_j eine Folge von S_j ist, so erhalten wir, wenn $f = f_1 \cdots f_t$, $M^t = M(f_1) \cdots M(f_t)$ ist,

$$(9''') \qquad |\varepsilon_N(T_L(f))| \leqq M^t \sum_{s=1}^{t} |\varepsilon_{N(s)}| (T_L(f_s))\, M^{-1}(f_s).$$

Im besonderen erhalten wir, wenn wir auf (9″) die Abschätzung (9), (11) auf (12) anwenden

$$(9'''') \qquad |\varepsilon_N(T_L(f))| \leqq M^{t-1} \sum_{s=1}^{t} \sum_{k,j} |b_{kj}(f_s)|\, |W_{kj}|.$$

§ 3. Wir wollen nun die Greensche Funktion G auf der Sphäre ins Spiel bringen. Zunächst wollen wir gleich den allgemeinen Fall so ins Auge fassen, daß wir statt der Sphäre eine zusammenhängende, kompakte, geschlossene differenzierbare Mannigfaltigkeit X haben, die die topologische Struktur einer Sphäre von der Dimension m besitzt. Es sei Δ der Beltramioperator auf X, der bekanntlich (in lokalen Koordinaten x_i) durch

$$(1) \qquad \Delta U = \frac{1}{\sqrt{g}} \sum_{i,k=1}^{m} \frac{\partial}{\partial x_k} \left(g^{ik} \sqrt{g} \frac{\partial U}{\partial x_i} \right)$$

definiert ist. Dabei sei auf X eine Riemannsche Metrik $ds^2 = \sum g_{ik}\, dx_i\, dx_k$ gegeben. Dabei mögen $g = \mathrm{Det}(g^{ik})$ und die g^{ik} die übliche Bedeutung haben. Es kommt auf die

Gestalt von ΔU nicht an, es kann Δ ein beliebiger, selbstadjungierter Differentialoperator sein, der auf ganz X definiert ist. Dies ist beim Beltramioperator, wie er eben definiert ist, nicht der Fall, aber es ist wohl bekannt, wie man den Ausdruck auf ganz X definieren kann, wenn er in jeder Karte von X definiert ist. Das Wesentliche ist, daß auf jedem Bereich B auf X, der durch eine zweimal stetige differenzierbare Hyperfläche ∂B begrenzt ist, die 2. Greensche Formel

$$\int_B (v\Delta u - u\Delta v)\, dV = \int_{\partial B} \left(v\frac{\partial u}{\partial n} - u\frac{\partial v}{\partial n}\right) do$$

gilt. Dabei ist dV das Volumselement auf X und do das Oberflächenelement von ∂B. Es ist die Normalableitung $\frac{\partial}{\partial n}$ durch

$$\frac{\partial}{\partial n} = \sum_{l,k=1}^{m} g^{lk} \frac{D_k}{\sqrt{\sum g^{ik} D_i D_k}} \frac{\partial}{\partial x_l}$$

definiert, wobei wiederum die D_j (nach Feller, [5]) so definiert sind: Ist ∂B in Parameterdarstellung gegeben durch

$$x_1 = \varphi_1(\lambda_1, \ldots, \lambda_{m-1}), \ldots, x_m = \varphi_m(\lambda_1, \ldots, \lambda_{m-1}),$$

dann sei $D_i\,(i = 1, \ldots, m)$ die Funktionaldeterminante

$$D_j = (-1)^{j-1} \frac{\partial(x_1, \ldots, \hat{x}_j, \ldots, x_m)}{\partial(\lambda_1, \ldots, \lambda_{m-1})}.$$

(Das Zeichen $\hat{x}$ deutet an, daß das x wegzulassen ist.) Das Oberflächenelement dc ist dann durch

$$do = \sqrt{g} \sqrt{\sum_{l,k=1}^{m} g^{lk} D_l D_k}\, d\lambda_1 \cdots d\lambda_{m-1}$$

gegeben. Man kann zeigen, daß die Definitionen von $\frac{\partial}{\partial n}$ und do von der Parameterdarstellung unabhängig sind und auch unabhängig von der Karte $x_1, \ldots, x_m$.

Es heißt nun eine Funktion $G(p, q)$ auf $X - \{q\}$ Greensche Funktion, wenn sie die Gleichung

$$\Delta G = 1 \tag{2}$$

erfüllt und in q eine Singularität von der Gestalt $\big(V = V(X)$ Volumen von $X\big)$

$$\frac{V}{(m-2)\,\big(r(p,q)\big)^{m-2} E(q)} \tag{3}$$

bzw. (für $m=2$) $\frac{V \log v(p,q)}{E(q)}$ besitzt. Dabei soll $r(p,q)$ die geodätische Distanz von p und q sein. $\big(E(q)$ wird noch definiert.$\big)$ Eine solche Funktion, die übrigens nicht eindeutig bestimmt ist, existiert. Man kann ja zu G eine beliebige Lösung von $\Delta U = 0$ hinzufügen. Man kann auch eine additive Konstante $c(q)$ hineinfügen, so daß

$$\int G(p,q)\, dV(p) = 0. \tag{4}$$

Wir wollen diese Funktion G Greensche Funktion von X nennen.

Wir wollen noch folgendes annehmen: Die geodätischen Kugeln $K(Q,\rho): r(P,Q)\leqq\rho$ mit dem Mittelpunkt Q und Radius ρ sollen für kleine e geschlossen und für $e=0$ nur aus Q bestehen. Wir führen r und $\varphi_1,\ldots,\varphi_{m-1}$ als geodätische Polarkoordinaten ein. Dann ist das Linienelement ds von X von der Gestalt

$$ds^2=a_0\,dr^2+\sum_{i,j=1}^{m-1} a_{ij}\,d\varphi_i\,d\varphi_j.$$

Das Oberflächenelement do auf der geodätischen Sphäre $S(Q,\rho): r(p,q)=\rho$ mit Mittelpunkt Q und Radius ρ wird $\sqrt{a}\,d\varphi_1\cdots d\varphi_{m-1}$, wo a die Determinante der a_{ij} ist. Das Volumenselement auf X wird $\sqrt{aa_0}\,dr\,d\varphi_1\cdots d\varphi_{m-1}$. Die Differentiation $\frac{\partial}{\partial n}$ wird in $S(Q,\rho)$ gleich $\frac{1}{\sqrt{a_0}}\frac{\partial}{\partial r}$. Auf S ist $\frac{1}{\rho^{m-1}}\sqrt{\frac{a}{a_0}}$ eine Funktion $E(q)=\Phi(\varphi_1,\ldots,\varphi_{m-1})$, also unabhängig von ρ. Weiter wird der Beltramioperator Δ in Polarkoordinaten für Funktionen U, die nur von r abhängen:

$$\Delta U=\frac{d^2U}{dr^2}+\frac{m-1}{r}\frac{dU}{dr}+\frac{d(\log\sqrt{g})}{dr}\frac{dU}{dr}.$$

Wir gehen nun von der Greenschen Formel

$$\int_{r(p,q)\geqq\rho}(G\Delta f-f\Delta G)\,dV(p)=\int_{r(p,q)=\rho}\left(G\frac{\partial f}{\partial n}-f\frac{\partial G}{\partial n}\right)do$$

aus, wo f eine zweimal stetig differenzierbare Funktion auf X ist. Da $\Delta G=1$ ist, erhalten wir

$$\int_{r(p,q)\geqq\rho}G\Delta f\,dV=\int_{r(p,q)\geqq\rho}f\,dV+\int_{r(p,q)=\rho}\left(G\frac{\partial f}{\partial n}-f\frac{\partial G}{\partial n}\right)do.$$

Nun ist, wenn $m>2$ ist:

$$\int_{r(p,q)=\rho}G\frac{\partial f}{\partial n}do=\frac{V}{(m-2)\,E(q)}\int G(p,q)\frac{\partial f}{\partial n}\sqrt{\frac{a}{a_0}}\,d\varphi_1\cdots d\varphi_{m-1}.$$

Nun geht $\int\frac{1}{\rho^{m-2}}\sqrt{\frac{a}{a_0}}\frac{\partial f}{\partial r}d\varphi_1\cdots d\varphi_{m-1}$, wenn f differenzierbar ist, für $\rho\to0$ gegen Null. Weiter ist

$$\int_{r(p,q)=\rho}f\frac{\partial G}{\partial n}do=\int_{r(p,q)=\rho}f\frac{\partial G}{\partial r}\sqrt{\frac{a}{a_0}}\,d\varphi_1\cdots d\varphi_{m-1}.$$

Nun ist bis auf einen Fehler $O(r^{m-2})$ bzw. für $m=2$ von der Ordnung $O(\log r)$ $\frac{\partial G}{\partial r}=\frac{V(2-m)\,r^{1-m}}{E(q)}$ und wir erhalten

$$\int f\frac{\partial G}{\partial n}do=\frac{V(2-m)}{(m-2)\,E(q)}\int_{r(p,q)=\rho}f(p)\sqrt{\frac{a}{a_0}}\frac{1}{\rho^{m-1}}\,d\varphi_1\cdots d\varphi_{m-1};$$

daraus folgt, daß für $\rho\to0$ das Integral links gegen $Vf(q)$ strebt. Wenn $m>2$ ist, haben wir daher $\int_X G\Delta f\,dV(p)=\int_X f\,dv-Vf(q)$ oder

$$f(q)=\frac{1}{V}\left(\int_X f\,dV-\int_X G(p,q)\Delta f\right)dV(p). \tag{5}$$

Ist $m=2$, dann ist $\frac{\partial G}{\partial r}=\frac{V}{E(q)}\frac{d\lg r}{dr}=\frac{V}{r}\frac{1}{E(q)}$ und wir erhalten analog

$$Vf(q)=\int fdV-\int_X G\Delta fdV.$$

Wir normieren noch das Volumen durch $d\mu=\frac{1}{V}dV$ und erhalten die Formel

$$f(q)=\int_X f(p)\,d\mu(p)-\int_X G\Delta f(p)\,d\mu(p). \tag{6}$$

Wir können nun überlegen, daß G selbst durch die Bedingungen (2), (3), (4) eindeutig bestimmt ist. Haben wir nämlich zwei solche Funktionen G, G_1, so erhalten wir $\int\limits_{r(p,q)\geqq\rho}(G_1\Delta G-G\Delta G_1)\,dV=\int\limits_{r(p,q)=\rho}\left(\frac{\partial G}{\partial n}G_1-\frac{\partial G_1}{\partial n}G\right)do$, und wir schließen wie oben, wobei jetzt (4) wesentlich ist, daß $G(p,q)=G_1(p,q)$ ist. Weiters zeigt man in üblicher Weise, daß G symmetrisch ist, also daß $G(p,q)=G(q,p)$ ist. Man schließt aus X die Punkte p, q durch Kugeln von Radius ρ aus, dann hat man mit $G(x,q)=G_1(x)$:

$$\int\limits_{r(x,p)\geqq\rho,\,r(x,q)\geqq\rho}(G_1\Delta G-G\Delta G_1)\,dV(x)$$

$$=\int\limits_{r(x,p)=\rho}do\left(\frac{\partial G}{\partial n}G_1-\frac{\partial G_1}{\partial n}G\right)+\int\limits_{r(x,q)=\rho}do\left(\quad\right).$$

Die linke Seite geht wegen (3) und (4) gegen Null. Die rechte Seite geht gegen $G_1(p,q)-G(q,p)$ für $\rho\to 0$. Es ist also $G(p,q)=G(q,p)$.

Wir betrachten nun die Eigenwerte λ und die Eigenfunktionen φ der Gleichung $\Delta\varphi+\lambda\varphi=0$. Es ist bekannt, daß es eine Folge von Eigenwerten $\lambda_0=0, \lambda_1,\dots$ und zugehörige Eigenfunktionen (φ_n) gibt, die ein orthogonales System $\int_X \varphi_i\varphi_j\,d\mu=\delta_{ij}$ bilden. Dabei gehört zu $\lambda_0=0$ die Eigenfunktion 1. Die anderen Eigenwerte sind positiv und wir denken sie uns der Größe nach geordnet. Mehrfache Eigenwerte werden nach ihrer Vielfachheit gezählt. Es sind, wie auch aus (6) hervorgeht, wenn wir

$$\Delta\varphi_n+\lambda_n\varphi_n=0 \tag{7}$$

schreiben, die λ_n Eigenwerte, bzw. die φ_n Eigenfunktionen der Gleichung

$$\varphi_n(x)+\lambda_n\int_X G(x,y)\,\varphi_n(y)\,d\mu=0.$$

Der Greenschen Funktion G ist die Fourierreihe $\sum\limits_{n=1}^{\infty}\frac{\varphi_n(x)\,\varphi_n(y)}{\lambda_n}$ zugeordnet. Wir betrachten nun die iterierten Kerne $G^{(r)}(x,y)$ definiert durch

$$G^{(r)}(x,y)=\int_X G^{(r-1)}(x,z)\,G(z,y)\,dV(z),$$

die die Fourierentwicklung $\sum\limits_{n=1}^{\infty}\frac{\varphi_n(x)\,\varphi_n(y)}{\lambda_n^r}$ haben.

34 Wenn nun r genügend groß ist, so ist die Reihe absolut und gleichmäßig konvergent, und es gilt

$$G^{(r)}(x,y)=\sum_{n=1}^{\infty}\frac{\varphi_n(x)\,\varphi_n(y)}{\lambda_n^r}.$$

Liegt X in einem euklidischen Raum, so kann man sich überlegen, daß dies für $r\geqq\left[\frac{m}{2}\right]+1$ der Fall ist, da dann auch die $G^{(r)}(x,y)$ regulär auf X^2 sind. Wir können dies aus der Tatsache folgern, daß $|G(p,q)|\leqq C\,r(p,q)^{2-m}$ ist. Es gilt nämlich folgender

Hilfssatz. *Es sei* $0<a<m$, $0<b<m$, *dann gilt für* $J(a,b)=\int\frac{dV(z)}{r_{pz}^a\,r_{zq}^b}$:

(8) $\quad J(a,b)=O(r_{pq}^{m-a-b}),\quad$ *wenn* $a+b>m$,

(8′) $\quad J(a,b)=O\,(\lg(r_{pq})),\quad$ *wenn* $a+b=m$, *und*

(8″) $\quad J(a,b)=O(1),\quad$ *wenn* $a+b<m$ *ist.*

Da wir annehmen, daß X in R^{m+1} liegt, so gilt mit positiven Konstanten c_1, c_2 $c_1|p,q|\leqq r_{pq}\leqq c_2|p,q|$, wo $|p,q|$ der euklidische Abstand ist.

Man kann auch das Volumselement dV auf X durch das m-dimensionale euklidische Volumselement ersetzen, da es nur auf die Größenordnung von $J(a,b)$ ankommt. Es ist nun

(a) $$J(a,b)=J_1+J_2,$$

wo in J_1 über alle z mit $|z,q|<2|p,q|$ integriert wird. In J_2 wird über alle z mit $|z,q|\geqq 2|p,q|$ integriert. Wir können dabei r_{pq} durch $|p,q|$ ersetzen. In J_2 ist nun, da nach der Dreiecksungleichung $|z,p|+|q,p|\geqq|z,q|$ und $|z,q|\geqq 2|p,q|$ gilt, auch $|z,p|\geqq\frac{1}{2}|z,q|$, also gilt

(b) $$J_2\leqq\int_{|z,q|>|p,q|}\frac{dV}{|z,q|^{a+b}}.$$

Führen wir Polarkoordinaten ein, also $z=q+r\xi$, so ist

(c) $$J_2=O\left(\int_{r\geqq 2|pq|}\frac{r^{m-1}\,dr}{r^{a+b}}\right)$$

und daraus folgen für J_2 die Behauptungen (8′), (8″), (8‴). Jetzt betrachten wir J_1: Wir setzen in J_1 $z=q+2dy$, wo $d=|p,q|$ sein soll. Es ist dann $|z,q|=2d|y|<2|p,q|$, also $|y|<1$. Dann ist

(d) $$J_1=d^{m-(a+b)}J_3,$$

(e) $$J_3=\int_{|y|\leqq 1}\frac{dv(y)}{|y|^a(w+y)^b},$$

wobei $w=\frac{p-q}{2d}$ gesetzt wurde. Daher ist $|w|=\frac{1}{2}$. Wir zerlegen J_3 in J_4+J_5, wo in J_4 der Integrationsbereich $|y|\leqq\frac{|w|}{2}$, in J_5: $\frac{|w|}{2}\leqq|y|\leqq 1$ ist. In J_4 führen wir wieder Polarkoordinaten $y=r\xi$ ein und erhalten, da $|w+y|\geqq|w|-|y|\geqq|w|-\frac{|w|}{2}=\frac{1}{4}$ ist,

$$J_4=O\Big(\int\limits_{r<\frac{1}{2}} r^{m-a-1}\,dr\Big)=O(1). \tag{f}$$

In J_5 setzen wir $y=-w+r\xi$. Es ist $r\leqq|w|+|y|<\frac{3}{2}$

$$J_5=O\left(\int\limits_{r\leqq\frac{3}{2}}\frac{r^{m-1}\,dr}{r^b}\right)=O(1). \tag{g}$$

Damit ist also $J_4=O(|pq|^{m-(a+b)})$, also ist $J_1=O(1)$ für $a+b\leqq m$ und daraus folgt die Behauptung für J.

Aus dem Hilfssatz folgt: Es ist für $k\geqq 1$

$$G^{(k)}(p,q)=O\left(\frac{1}{\|p,q\|^{m-2k}}\right), \quad \text{wenn } m>2 \text{ ist, und} \tag{9}$$

$$G^{(k)}(p,q)=O(1) \quad \text{für } m=2 \text{ und } k>1. \tag{9'}$$

Den Beweis von (9) führen wir durch vollständige Induktion nach k. Für $k=1$ ist er nach der Definition der Greenschen Funktion richtig. Wenn (9) für k richtig ist, so folgt

$$G^{(k+1)}(p,q)=\int G^{(k)}(p,z)\,G(z,q)\,dV(z).$$

Wenden wir den Hilfssatz auf $G^{(k)}(p,z)$ mit $a=m-2k$ und auf $G(z,q)$ mit $b=m-2$ an, so ist $G^{(k+1)}(p,q)=O(\|p,q\|^{m-(a+b)})=O(\|p,q\|^{2k+2-m})$. Damit ist (8) bewiesen.

Im Falle $m=2$ ist $G(p,q)=O(\lg|p,q|)=O(|p,q|^{-\varepsilon})$, daher

$$G^{(2)}(p,q)=O(|p,q|^{2-2\varepsilon})=O(1);$$

dies gilt für alle $G^{(k)}$ mit $k>1$. Nehmen wir in (9) $k=\left[\frac{m}{2}\right]+1$, so ist

$$m-2k=m-2\left(\left[\frac{m}{2}\right]+1\right)<0,$$

also gilt der

Satz. *Es ist*

$$G^{(r)}(p,q)=O(1) \tag{10}$$

für $r\geqq\left[\frac{m}{2}\right]+1$.

Der Satz bleibt auch für $m=2$ richtig.

Wir haben nun die Formel (wir ersetzen p, q durch x, y)

$$f(x)=\int_X f\,d\mu+(-1)^r\int_X G^{(r)}(x,y)\,\Delta^{(r)}f(y)\,d\mu(y), \tag{6'}$$

wo $\Delta^{(r)}$ der r-fach iterierte Beltramioperator ist. Wir zeigen dies leicht durch vollständige Induktion nach r. Für $r=0$ ist dies nach (6) richtig. Aus (10) folgt, daß

$$\Delta^{(r)}f(y)=-\int_X G^{(r)}(x,y)\,\Delta^{(r+1)}f(z)\,d\mu(z)$$

ist, daher $f(x)=\int f\,d\mu+(-1)^{r+1}\int G^{(r)}(x,y)\,d\mu(y)\int G(y,z)\,\Delta^{(r+1)}f(z)\,d\mu(z)$, und dies ist die Behauptung.

Es sei nun $\omega_N=(x_1,\ldots,x_N)$ eine Folge auf X, dann ist $\mu_N(f)=\frac{1}{N}\sum_{n=1}^{N}f(x_n)$. Sei $\mu(f)=\int f\,d\mu$ gesetzt, dann gilt $\mu_N(f)-\mu(f)=(-1)^r\int_X(\mu_N(G^{(r)},y)-\mu(G^{(r)},y))\,\Delta^r f(y)\,dy$. Dabei bedeutet

$$\mu_N(G^{(r)},y)=\frac{1}{N}\sum_{n=1}^{N}G^{(r)}(x_n,y),$$

analog ist $\mu(G^{(r)},y)$ definiert. Es ist dann

$$|\varepsilon_N(f)|\leqq\int|\varepsilon_N(G^{(r)},y)|\,|\Delta^{(r)}f(y)|\,d\mu(y).$$

Wir definieren nun als *r-te Diskrepanz* der Folge ω_N

$$D^{(r)}(\omega_N)=\sup_y|\varepsilon_N(G^{(r)},y)|. \tag{11}$$

Dann ist

$$|\varepsilon_N(f)|\leqq D^{(r)}(\omega_N)\int_X|\Delta^{(r)}f(y)|\,d\mu(y). \tag{12}$$

Dabei sei stets $r\geqq\left[\frac{m}{2}\right]+1$.

Wir stellen nun die Frage, ob die verschiedenen Diskrepanzen von gleichen Folgen vergleichbar sind. Zu diesem Zweck führen wir die Weylschen Summen ein:

$$W_k(\omega_N)=\frac{1}{N}\sum_{n=1}^{N}\varphi(x_n)=\mu_N(\varphi_k), \tag{13}$$

wo φ_k die k-te Eigenfunktion ist. (Es ist ja $\mu(\varphi_n)=0$, da 1 Eigenfunktion ist.) Es sind die φ_k auf X beschränkt, also

$$|\varphi_k(x)|\leqq C(k). \tag{14}$$

Nach (10) ist nun $\varphi_k(x)=(-1)^r\int G^{(r)}(xy)\,\Delta^r\varphi_k(y)\,dy$ und $\Delta^{(r)}\varphi_k=\pm\lambda_k^r\varphi_k$. Daher ist nach (12)

$$|W_k|\leqq\lambda_k^r\,C(k)\,D^{(r)}(\omega_N). \tag{15}$$

Nehmen wir statt φ_k ein lineares Kompositum $f(x)=\sum_{k=1}^{L}C_k\varphi_k(x)$, dann erhalten wir analog $|\varepsilon(f)|\leqq D^{(r)}\int|\sum_k c_k\lambda_k^{(r)}\varphi_k|\,d\mu$, daher $|\varepsilon(f)|\leqq D^{(r)}(\sum_k|c_k|^2\,\lambda_k^{2r})$.

Es ist nun $G^{(r)}(x, y) = \sum_{k=1}^{\infty} \varphi_k(x)\, \varphi_k(y)\, \lambda_k^{-r}$. Es ist $\mu_N(G^{(r)}, y) = \sum_{k=1}^{\infty} \frac{\varphi_k(y)}{\lambda_k^r} W_k$. Führen wir eine natürliche Zahl $L \geqq 1$ ein, so ist

$$(15') \qquad |\mu_N(G^{(r)}, y)| \leqq \sum_{k=1}^{L} \frac{C(k)}{\lambda_k^r} |W_k| + \sum_{k=L+1}^{\infty} \frac{C^2(k)}{\lambda_k^r}.$$

Daher ist auch $D^{(r)}(\omega_N) \leqq \sum_{k=1}^{L} \frac{C(k)}{\lambda_k^r} |W_k| + R(L)$, wo $R(L) = \sum_{k=L+1}^{\infty} \frac{C^2(k)}{\lambda_k^r}$ ist. Daher ist auch $G^{(S)}(x, y) = \sum_k \frac{\varphi_k(x)\, \varphi_k(y)}{\lambda_k^S}$. Aus der Orthogonalität der φ_k folgt, daß

$$\frac{\varphi_k(x)}{\lambda_k^S} = \int G^{(S)}(x, y)\, \varphi_k(y)\, dy = J_{S,k}$$

ist. Es ist also $\frac{W_k}{\lambda_k^S} = \int (\mu_N(G^{(S)}, y) - \mu(G^{(S)}, y)\, \varphi_k(y)\, dy$, daher $\frac{|W_k|}{\lambda_k^S} \leqq D^{(S)}\, C(k)$.

Nach (15) haben wir also $D^{(r)} \leqq D^{(S)} \sum_{k=1}^{L} C^2(k)\, \lambda_k^{S-r} + R(L)$. Wir nehmen nun die größte

35 Zahl $L(D^{(S)}) = L_S$, so daß $D^{(S)} \sum_{k=1}^{L_S} C^2(k)\, \lambda_k^{s-r} \leqq R(L)$ ist[1]); dann ist also

36 $$D^{(r)} \leqq D^{(S)} \sum_{k=1}^{L_S} C^2(k)\, \lambda_k^{s-r}.$$

Ist daher $\lim_{N\to\infty} D^{(S)}(\omega_N) = 0$, dann ist auch $\lim_{N\to\infty} D^{(r)}(\omega_N) = 0$.

Wir schreiben die Formel (6′) in der Form

$$(6'') \qquad f(x) = \int f(y)\, d\mu(y) - \int K(x, y)\, \Omega(y)\, \big(f(y)\big)\, d\mu(y).$$

37 Dabei wurde $G^{(r)}(x, y) = K(x, y)$, $\Omega(y) = \Delta^{(r)}$ gesetzt. Es bedeutet $\Omega(y)$ den Operator $\Delta^{(r)}$ auf $f(y)$ angewendet. Die Bezeichnung $\Omega(y)$ soll andeuten, daß sich die Differentiation auf die Variable y bezieht. Die Integration erstreckt sich über ganz X.

Wir wollen nun ein Gegenstück zur Formel von Koksma-Hlawka für ein Produkt Π_t von t Mannigfaltigkeiten $X_1, \ldots, X_t$ aufstellen. Es sei f eine Funktion auf Π_t, also $f(x) = f\big(x(1), \ldots, x(t)\big)$. Um nun die Formel bequem formulieren zu können, seien einige Definitionen vorausgeschickt. Mit $C(j) = \big(c(1), \ldots, c(j)\big)$ bezeichnen wir eine Kombination von j Elementen aus $\{1, \ldots, t\}$. Mit $\Pi_{C(j)}$ bezeichnen wir die Mannigfaltigkeit $X_{c(1)} \times \cdots \times X_{c(j)}$, mit $d\Pi_c$ sein normiertes Volumselement $d\mu_{c(1)} \cdots d\mu_{c(j)}$. Weiters sei $K(x, y, \Pi_c) = K\big(x(c(1)), y(c(1))\big) \cdots K\big(x(c(j)), y(c(j))\big)$ und

$$\Omega\big(\Pi_c(y)\big) = \Omega\big(y(c(1))\big) \cdots \Omega\big(y(c(j))\big).$$

Wir nehmen an, daß stets $\Omega\big(\Pi_c(y)\big)\, f(y)$ vorhanden und stetig sind, für alle Kombinationen aus $\{1, \ldots, t\}$. Dann gilt die Formel

$$(16) \qquad f(x) = \sum_{j=0}^{t} (-1)^j \sum_{c(j)} \int K(x, y, \Pi_c)\, \Omega(\Pi_c)\, (y)\, f(y)\, d\Pi_c(y).$$

---

[1]) Ist die Reihe $\sum_{k=1}^{\infty} C^2(k)\, \lambda_k{}^{S-r}$ konvergent, so ist $L_S = \infty$ zu nehmen.

Dabei erstreckt sich die Summation rechts bei festem j über alle Kombinationen $C(j)$. Der Beweis erfolgt durch vollständige Induktion von t. Für $t=1$ ist es Formel (6″). Nehmen wir (11) als richtig an, so betrachten wir $\Pi_{t+1}=\Pi\times X_{t+1}$. Wir schreiben $x=(x(1)\cdots x(t),\ x(t+1))=(x', x'')$, wo $x'=(x(1)\cdots x(t))$ und $x''=x(t+1)$ ist.

Wir wenden nun (6″) auf die Funktionen $\Omega(\Pi_C(y))\,f(y', x'')$ mit $y'=(y(1)\cdots y(t))$, $y''=y(t+1)$ an und erhalten $\Omega(\pi_c(y))\,f(y', x'')=J_1-J_2$, wo

$$J_1=\int \Omega(\Pi_C(y))\,f(y', y'')\,d\mu(y'') \quad\text{und}\quad J_2=\int K(x'', y'')\,\Omega(y'')\,\Omega(\Pi_C y'')\,d\mu(y'')$$

ist, und wenden (16) auf die Funktion $g(x')=f(x', x'')=f(x)$ an und erhalten $f(x)=\sum\limits_1+\sum\limits_2$, wo

$$\sum_1=\sum_{j=0}^{t}(-1)^j\sum_{C(j)}\int K(x', y'', \Pi_C)\,\Omega(\Pi_C y')\,f(y)\,d\Pi_C(y)$$

und

$$\sum_2=\sum_{j=0}^{t}(-1)^j\sum_{C(j)}\int K(x', y'', \Pi_C)\,K(x'', y'')\,\Omega(y'')\,\Omega(\Pi_C y')\,f(y)\,d\Pi(y).$$

Nun ist jede Kombination von $\{1,\ldots, t+1\}$ entweder eine Kombination von $\{1,\ldots, t\}$ oder eine Kombination von der Gestalt $(C(j),\ t+1)$, damit ist (16) gezeigt.

Es sei nun eine Folge ω_N auf $\Pi=\Pi_t$ gegeben. Wir definieren dann

$$D_N(C(j), \omega_N)=\sup|\varepsilon_N(K, y, \Pi_C)|;$$

somit ist

$$|\varepsilon_N(f)|\leqq\sum_{j=1}^{t}\sum_{C(j)}D_N(C(j), \omega_N)\int|\Omega(\Pi_{C(j)}, y)\,f(y)|\,d\Pi(y).$$

Führen wir die L^2-Diskrepanz $\hat{D}_N(C(j), \omega(N))=\left(\int|\varepsilon_N(Ky, \Pi_{C(j)})|^2\,d\Pi(y)\right)^{\frac{1}{2}}$ ein, so erhalten wir $|\varepsilon_N(f)|\leqq\sum\limits_{j=1}^{t}\sum\limits_{C(j)}\hat{D}_N(C(j), \omega_N)\,J(c(j))$, wo

$$J(C(j))^2=\int(\Omega(\Pi_{C(j)}y)\,f(y))^2\,d\Pi(y)$$

ist.

Besitzt $K(x(j), y(j))$ auf X_j die Bilinearentwicklung

$$K(x(j), y(j))=\sum_L\frac{\varphi_L(x(j))\,\varphi_L(y(j))}{\lambda_L(j)},\quad\text{so ist}\quad K(x, y, \Pi_C)=\sum_L\frac{\varphi_L(x, \Pi_C)\,\varphi(y, \Pi_C)}{\lambda_L(\Pi_C)},$$

wo $\varphi_L(x, \Pi_C)=\varphi_L(x(C(1)))\cdots\varphi_L(x(C(j)))$ und $\lambda_L(\pi_C)=\lambda_L(C(1))\cdots\lambda_L(C(j))$ ist.

§ 4. Wir wollen die Entwicklung in § 3 auf die Sphäre S^m bzw. auf das Produkt von solchen Sphären spezialisieren. In diesem Fall sind die Eigenwerte des Beltramioperators von S^m gleich

$$\lambda_k=k(k+m-1), \tag{1}$$

und die Eigenfunktionen sind gerade die Kugelfunktionen $H_{k,j}$ von der Vielfachheit $Z(m, k)$, wo $Z(m, k)$ von der Größenordnung k^{m-1} ist.

Es ist

$$G^{(r)}(x,y)=\sum_{k=1}^{\infty}\frac{1}{\lambda_k^r}\sum_{j=0}^{Z(m,k)}H_{k,j}(x)\,H_{k,j}(y). \tag{2}$$

Weiters ist

$$\sum_{j=0}^{Z(m,k)}H_{k,j}(x)\,H_{k,j}(y)=\frac{Z(m,k)}{V(S^m)}\,P_k(xy). \tag{3}$$

Es ist also das k-te Glied der Reihe (1) von der Größenordnung

$$O(k^{m-2r-1}). \tag{4}$$

Daher ist die Reihe für $r>\frac{m}{2}$ absolut und gleichmäßig konvergent. Man kann auch die Greensche Funktion explizit aufstellen. (Dies gilt auch, wenn X eine Mannigfaltigkeit konstanter positiver Krümmung ist.)

Es sei y der Pol (Unendlichkeitsstelle) der Greenschen Funktion $G(x,y)$. Wir führen als neue Variable $t=xy=\cos\vartheta$ ein $\left(\text{ist } \Delta \text{ der Laplaceoperator, so ist mit } x=rt,\ \Delta=\frac{\partial^2}{\partial r^2}+mr\frac{\partial}{\partial r}+\frac{1}{r^2}\Delta_m\right)$. So hat dann der Beltramioperator Δ_m auf S^m die Gestalt

$$\Delta_m=(1-t^2)\frac{\partial^2}{\partial t^2}-mt\frac{\partial}{\partial t}+\frac{1}{1-t^2}\Delta_{m-1},$$

wo Δ_{m-1} der Beltramioperator auf S^{m-1} ist.

Wir suchen nun eine Lösung U von $\Delta_m U=1$, die nur von t allein abhängt. Es ist dann die Gleichung $(1-t^2)\frac{d^2U}{dt^2}-mt\frac{dU}{dt}=1$ zu lösen. Wir setzen $\frac{dU}{dt}=V$ und lösen zuerst die homogene Gleichung (wenn $m>2$ ist) $(1-t^2)V'-mtV=0$ mit $V=C(1-t^2)^{-\frac{m}{2}}$. Für die inhomogene Gleichung $(1-t^2)V'-mtV=1$ erhalten wir $V(t)=C(t)(1-t^2)^{-\frac{m}{2}}$ mit $C(t)=\int(1-t^2)^{\frac{m-2}{2}}dt$. Daher lautet die allgemeine Lösung

$$V(t)=\frac{1}{(1-t^2)^{\frac{m}{2}}}\int_0^t(1-t_1^2)^{\frac{m}{2}-1}dt_1+\frac{C_1}{(1-t^2)^{\frac{m}{2}}} \tag{4'}$$

und es ist

$$U(t)=\int\frac{dt}{(1-t^2)^{\frac{m}{2}}}\int(1-t^2)^{\frac{m}{2}-1}dt+C_1\int\frac{dt}{(1-t^2)^{\frac{m}{2}}}+C_2. \tag{5}$$

Die Integrale können elementar ausgewertet werden. Es ist U eine Funktion von xy, genauer eine Funktion in $1-xy$. Sie ist also symmetrisch in x und y. Man muß noch C_1, C_2 so bestimmen, daß sie die vorgeschriebene Singularität $\dfrac{1}{(1-xy)^{\frac{m-2}{2}}}$ hat und $\int\limits_S U(xy)\,dy=0$ ist.

Es ist ja hier die Funktion E konstant. Im Falle $m=2$ erhält man

$$U(t)=-\lg(1-t)+C_1, \tag{6}$$

daher ist

$$G(x,y)=-\lg(1-xy)+C_1. \tag{6'}$$

Nach (2) und (3) ist

$$G^{(r)}(xy)=\sum_{k=1}^{\infty}\frac{Z(m,k)}{V(S^m)}\frac{P_k(xy)}{\lambda_k^r}. \tag{7}$$

Es sei nun $M\geqq 1$ eine natürliche Zahl, dann gilt für

$$R_M=\sum_{k=M+1}^{\infty}\frac{Z(m,k)}{V(S^m)}\frac{P_k(xy)}{\lambda_k^r} \tag{8}$$

nach (1), da $|P_k|\leqq 1$ ist, $R_M=O\left(\sum\limits_{k=M}^{\infty}k^{m-1}k^{-2r}\right)$; daher ist für $r\geqq\frac{m}{2}$,

$$R_M=O(M^{m-2r}). \tag{9}$$

Es ist also

$$G^{(r)}(xy)=\sum_{k=1}^{M}\frac{Z(m,k)}{V(S^m)}\frac{P_k(xy)}{\lambda_k^r}+O\left(\frac{1}{M^{2r-m}}\right). \tag{10}$$

Haben wir statt S^m ein Produkt $\Pi=S(1)\times\cdots\times S(t)$ von t Sphären (wir nehmen der Einfachheit halber an, daß alle Sphären die gleiche Dimension m haben) und nehmen weiters bei jeder Sphäre den r-ten Beltramioperator, so erhalten wir

38 $$G^{(r)}(x,y)=\sum_{k_1,\dots,k_r=1}^{n}\frac{Z(m,k)}{V(m)}P_k(xy)+O\left(\frac{2}{M^{2r-m}}\right). \tag{10'}$$

Dabei haben wir $(x=(x(1),\dots,x(t)))$ $G^{(r)}(x,y)=G^{(r)}(x(1),y(1))\cdots G^{(r)}(x(t),y(t))$ sowie

$$Z(m,k)=Z(m,k_1)\,Z(m,k_2)\cdots Z(m,k_t),\quad V(m)=V(S^m)^t,$$

$$P_k(z)=P_{k_1}(z(1))\cdots P_{k_t}(z(t))\quad\text{und}\quad \lambda_k=\lambda_{k_1}\lambda_{k_2}\cdots\lambda_{k_t}.$$

Wenn wir wieder eine Folge $\omega_N(x_1,\ldots,x_N)$ auf Π_t betrachten, dann ist

$$\mu_N(P_k, y)=\frac{1}{N}\sum_{n=1}^{N} P_k(x_n y)=\sum_{j=0}^{Z(m,k)} H_{kj}(y)\frac{V(m)}{Z(m,k)}W_{kj}. \tag{11}$$

Trivial ist $|\mu_N(P_k, y)|\leqq 1$. Dabei ist wieder $W_{kj}(\omega_N)=\frac{1}{N}\sum_{n=1}^{N} H_{kj}(x_n)$ die Weylsche Summe zur Folge ω_N. Weiter gelte wieder die Konvention $H_{kj}(x)=H_{k_1 j_1}(x(1))\cdots H_{k_t j_t}(x(t))$. Wenn wir auf (11) die Schwarz'sche Ungleichung anwenden, erhalten wir

$$|\mu_N(P_k, y)|\leqq\frac{V(m)}{Z(m,k)}\sqrt{\sum H_{kj}^2(y)}\cdot\sqrt{\sum_{j=0}|W_{kj}|^2}.$$

Setzen wir noch, wie schon früher

$$W_k=\sqrt{\frac{\sum|W_{kj}|^2}{Z(m,k)}}, \tag{11'}$$

so erhalten wir, wenn wir noch (3) berücksichtigen,

$$|\mu_N(P_k, y)|\leqq|W_k|. \tag{11''}$$

Damit erhalten wir nach (10) und nach Definition von $D_N(C(j),\omega_N)$ in § 3 (12), wenn wir beachten, daß $K(x,y)=G^{(r)}(x,y)$ ist (wir fügen jetzt zur Definition von D_N noch den Index r hinzu) und $k^{m-1}=k_1^{m-1}\cdots k_t^{m-1}$ bedeutet:

39 $$D_N^{(r)}((c(j)),\omega_N)\leqq\sum_{k}^{M} k^{m-1}\frac{|W_k(\omega_N)|}{\lambda_k^r}+O\left(\frac{1}{M^{2r-m}}\right). \tag{12}$$

Dabei erstreckt sich die Summation über alle $c(1),\ldots,c(j)$ mit $1\leqq c(j)\leqq M$.

Wir wollen den Vergleich der Diskrepanzen in § 3 (13), (14), (15) im Fall der Sphäre vertiefen. Wir haben nach § 3 (13) und mit (1) $W_k\leqq c(m)k^{2s}D^{(s)}$, also ist

40 $$D^{(r)}\leqq D^{(s)}\sum_{k=1}^{L} k^{2(s-r)+m-1}+O(L^{m-2s-1}).$$

Wir unterscheiden zwei Fälle:

41 1. *Fall.* $2(s-r)+m<0$, dann ist $D^{(r)}\leqq c(m)D^{(s)}$;

42 2. *Fall.* $2(s-r)+m>0$, dann wählen wir L so, daß $D^{(r)}\leqq(D^{(s)})^{\sigma}$ ist, mit $\sigma=\dfrac{2s+1-m}{4s-2r+1}$.

Mittels $D_N^{(r)}$ können wir nach § 3 (12), $|\varepsilon_N(f)|$ abschätzen, wenn $\Delta^r f$ vorhanden und stetig ist. Dabei muß $r\geqq\left[\frac{m}{2}\right]+1$ sein. Man kann $\varepsilon_N(f)$ auch abschätzen, wenn $\Delta^r f$ nur für $r\leqq\left[\frac{m}{2}\right]$, also insbesondere nur für $r=1$ vorhanden ist. Wir gehen von § 3 (6) aus:

43 $$f(x)=\mu(f)+\int G(x,y)\,\Delta f(y)\,d\mu(y), \tag{15}$$

wobei sich jetzt die Integration auf S^m erstreckt. Wir zerlegen S^m in die Kappe $C(y,\rho)\colon\|xy\|\leqq\rho$ und in C'. (C' bezeichnet das Komplement von C auf S^m.) Es sei

$\iota(C(y,\rho))$ die Indikatorfunktion. Wir schreiben kurz $\iota(y)$. Dann haben wir für das Integral in (15) $I_1(x)+I_2(x)$, wobei

(16) $$I_1(x)=\int G(x,y)\,\iota(x,y)\,\Delta f\,d\mu,$$

(16′) $$I_2(x)=\int G(x,y)\,(1-\iota(xy))\,\Delta f\,d\mu$$

ist. Es ist $\iota'=1-\iota$ die Indikatorfunktion von C'. Es ist $|I_1(x)|\leqq c(m)\,\rho M(f,\rho)$, wobei $M(f,\rho)$ das Maximum von $|\Delta f|$ auf $C(y,\rho)$ ist. Es ist $c(m)$ eine Konstante, die nur von m abhängt. Zum Beweis vergleiche man § 3 (3).

Wir erhalten also nach (15) $|\varepsilon_N(f)|\leqq\int\varepsilon_N(G\iota',y)|\Delta f|\,d\mu+c(m)\,\rho M(f,\rho)$, wo $|\varepsilon_N(G\iota',y)|=\left|\frac{1}{N}\sum_{n=1}^{N}G(x_n,y)\,\iota'(x_n,y)-\int G(x,y)\,\iota'(x,y)\,d\mu(x)\right|$ ist. Da nun $\|x_n,y\|\geqq\rho$ auf C' ist, so folgt aus dem Beispiel in § 1 (66) mit $\rho=L^{-\frac{1}{m+1}}$ (es ist ja $\alpha=m-2$):

44 (17) $$|\varepsilon_N(f)|\leqq O\Big(ML^{-\frac{1}{m+1}}+L^{\frac{m-2}{m+1}}\sum_{k=1}^{L}|W_k(\omega_N)|\int|\Delta(f)|\,d\mu\Big).$$

Man kann die Sache noch weiter verfolgen, wenn wir die Diskrepanz $D_N(\omega_N,C)$ einführen. Diese wird wie folgt definiert: Betrachten wir alle Kappen $C(y,\rho)$ auf S^m. Es ist $C(y,\rho)$ im Sinne von § 1 ein Normalbereich, und es ist $\mu(I_3(\delta))\leqq k_m\delta$, wobei k_m unabhängig von ρ ist, also nur von m abhängt. Es ist ja das Volumen

45 $$C(y,\rho+\delta)-C(y,\rho-\delta)=\bar{k}_m\big((\rho+\delta)^m-(\rho-\delta)^m\big)\leqq\bar{k}_m\rho^{m-1}\delta,$$

da $\rho\leqq\pi$ ist. Wenden wir nun § 2 (13) an, so erhalten wir

(17′) $$|\varepsilon_N(\iota(C))|\leqq\left(\sqrt{\frac{\lg L}{L}}+\sum_{k=1}^{L}k^{\frac{m-1}{2}}W_k(\omega_N)\right)c_m.$$

Es ist also die rechte Seite von (17′) von der Kappe C unabhängig. Wir definieren nun

(18) $$D_N(\omega_N,C)=\sup_C|\varepsilon_N(\iota(C))|.$$

Wir wollen diese Diskrepanz die *Kappendiskrepanz* der Folge ω_N nennen und haben also schärfer

46 (17″) $$D(\omega_N,C)\leqq\sup_C\left(\sqrt{\frac{\lg L}{L}}+\sum_{k=1}^{L}W_k(\omega_N)\,b_{kj}(\iota(C))k^{m-1}\right).$$

Wir wollen nun die Kappendiskrepanz (wir können auch eine L^2-Kappendiskrepanz einführen) mit der Diskrepanz $D_N^{(s)}(\omega_N)$ vergleichen. Wir haben nach § 3 (15) und (18)

47 $D_N(C)\leqq c_m\left(\sqrt{\frac{\lg L}{L}}+D_N^{(s)}\sum_{k=1}^{L}k^{2s+\frac{m-1}{2}}\right)$. Nun ist die Summe rechts $O(L^{2s+m})$. Nehmen

48 wir $L=(D_N^{(s)})^{-\lambda(s)}$, wo $\lambda(s)=\dfrac{1}{2s+m+\frac{1}{2}}$, so erhalten wir mit $D=D^{(s)}$:

49 (19) $$D_N(C)\leqq C(m)\,D^{1/2\,\lambda(s)}\log D.$$

50 Da D_N stets $\geqq N^{-1}$ ist, so ist $D^{(s)}\geqq N^{-4(s+m)\lambda(s)}\log N$.

Bemerkung 1. Man kann statt der Familie aller Kappen, auch allgemeiner, Familien von normalen Mengen, für die § 1 (48) mit einer festen Schranke C gilt, betrachten. Man kann dann die entsprechenden Diskrepanzen einführen. Eine solche Familie, bei der sich $D_N(C)$ nicht ändert, ist die Familie C aller Mengen, die sich als endliche Vereinigung bzw. Durchschnitt bzw. als Differenzbildung von Kappen darstellen läßt.

Wir werden zeigen (in Bemerkung 5), daß

$$W_k \leqq c(m)\, D_n(C)\, k^m \tag{20}$$

ist. Dann erhalten wir aus (12) $D_N^{(r)} = O(D_N)$ für $r > m$, $D_N^{(r)} = O(D_N \log D_N)$ für $r = m$ und $D_N^{(r)} = O\big(D_N^{\frac{2r}{m}-1}\big)$, wenn $\frac{m}{2} < r < m$. Es ist dabei $D_N = D_N(c)$.

51 **Bemerkung 2.** Verwendet man (20) in § 2 (12′), so erhält man $|\varepsilon_N(f)| \leqq M(f)\, S_L + R(L)$, wo $S_L = L^{2m} D_N(C)$ ist. Ist E eine Kappe, so kommt

52 $$|\varepsilon_N(f)| \leqq M(f, E)\, L^{2m} D_N(C) + R(L)$$

(für $m=2$ ist die entsprechende Modifikation vorzunehmen). Ist f auf C noch eine Lipschitzfunktion mit einer Lipschitzkonstanten $k(f)$, so ist

$$|\varepsilon_N(f)| \leqq k\left(\sqrt{\frac{\lg L}{L}} + M(f, C)\, L^{2m} D_N(C)\right). \tag{$*$}$$

Bemerkung 3. Die Abschätzungen in § 2 (12) beruhten auf der Approximation von f durch $T_L(f)$. Es war $T_L(f) = \sum_{k,j} C_{kj}(f)\, H_{kj}(x)$. Nehmen wir nun an, daß $\Delta^{(r)} f$ existiert und stetig ist, dann haben wir $T_L(\Delta^{(r)} f) = \sum_{k,j} C_{kj}(\Delta^r f)\, H_{kj}(x)$. Nun betrachten wir die stetige Funktion $g = f - \sum_{k,j} c_{kj} \lambda_k^{-r} H_{kj}(x) = f - T_L(f)$, wo wir $C_{kj}(\Delta^{(r)} f) = C'_{kj}$ gesetzt haben. Dann ist $\Delta^{(r)} g = \Delta^{(r)} f - \sum C'_{kj} H_{kj}(x)$, daher ist nach § 3 (10), angewendet auf g, $|g(x)| \leqq \sup_y |\Delta^{(r)} f - T_L(\Delta^{(r)} f)| \int |G^{(r)}(x, y)|\, d\mu(y)$. Wir haben dann, wenn wir auf g den Satz 1 in § 1 anwenden, nach § 1 (2) $|g(x) - T_L(g)| \leqq R(L)$, wo

$$R(L) \leqq \omega\left(\omega(\Delta^{(r)} f)\, \sqrt{\frac{\lg L}{L}}, \sqrt{\frac{\lg L}{L}}\right)$$

ist.

Erfüllt also $\Delta^{(r)} f$ eine Lipschitzbedingung, so ist $|R(L)| \leqq C \frac{\lg L}{L}$. Es wird dann $|f(x) - T_L^*(f)| \leqq c \frac{\lg L}{L}$, wo $T_L^* = T_L(f) + T_L(g)$ ist.

Man kann dann mit T_L^* genauso operieren, wie wir dies mit T_L in § 2 gemacht haben. Natürlich kann man den Prozeß wiederholen, falls $\Omega_m f = \Delta^{(rs)} f$ für $s \geqq 1$ existiert.

Erfüllt dann $\Omega_m f$ eine Lipschitzbedingung, so haben wir $|R(L)| \leqq c \left(\frac{\lg L}{L}\right)^{\frac{1}{2}}$. Man kann übrigens T_L durch $\frac{1}{N} W_L^{-1} \sum_{n=1}^{N} f(y_n)(1+y_n x)^L$ ersetzen, wo (y_n) eine gleichverteilte Folge auf S^m bzw. Π mit kleiner Diskrepanz $D_N(C)$ ist. Dies dürfte für praktische Anwendungen von Nutzen sein. (Für den Fall des Torus wurde dies vom Verfasser in [7] durchgeführt.) In einer Fortsetzung dieser Arbeit soll dies neben anderen Anwendungen ausgeführt werden.

Bemerkung 4. Nach den Entwicklungen in § 2 und § 4 kommt es bei den Abschätzungen der Diskrepanzen auf gute Abschätzungen der Weylschen Summen bzw. der W_k hinaus. Es ist oft zweckmäßig, auf § 2 (11) zurückzugreifen, wie schon dort bemerkt wurde. Dies gilt besonders für den Fall eines Produktes $\Pi = \Pi_t$. Nehmen wir an, daß $\Pi = \Pi_1 \times \Pi_2$ ist und die Folge ω dieser Zerlegung angepaßt ist, also $\omega_N = \omega_{N(1)} \times \omega_{N(2)}$ ist, wo $\omega_{N(j)}$ $(j=1, 2)$ Folgen auf Π_1 bzw. Π_2 sind und $N = N(1)\, N(2)$ ist. Es ist dann $\mu_N(f) = \frac{1}{N(1)\, N(2)} \sum_{n_1, n_2}^{N(1)\, N(2)} f(x_{n_1}, x_{n_2})$. Wir wollen dies gleich an einem konkreten Beispiel anwenden. Wir betrachten für eine natürliche Zahl s die Menge aller ganzzahligen Lösungen $k = (k_1, \ldots, k_{m+1})$ von $k^2 = k_1^2 + \cdots + k_{m+1}^2 = s$. Ihre Anzahl wird bekanntlich mit $r(s)$ bezeichnet. Dann betrachten wir mit $N = r(s)$ die Folge der Punkte $\left(\frac{k}{\sqrt{s}}\right)$ auf S^m. Es wurde nun von Malyschew [8], [9] und Pommerenke [11] gezeigt, daß

$$W_k \leqq c_1(m)\, k^m s^{-\frac{m}{4}} \tag{21}$$

ist, wenn $m \geqq 4$ ist. (Für $m = 2$ und 3 und passendes s wurden von R. Arenstorf und D. Johnson, [2], analoge Abschätzungen gezeigt.)

Es ist also für jedes L

$$\sum_{k=1}^{L} k^{m-1} W_k \leqq c_1(m)\, L^{2m} s^{-\frac{m}{4}}. \tag{22}$$

Damit folgt aus § 2 (12)

53 $$|\varepsilon_N(f)| \leqq c_2(m)\, \omega_f \left(\sqrt{\frac{\lg L}{L}}\right) (1 + s^{-\frac{m}{4}} L^{2m+\frac{1}{2}}). \tag{23}$$

Nehmen wir $L = s^\alpha$ mit

$$\alpha = \frac{m}{8m+2}, \tag{24}$$

dann erhalten wir, wenn f noch eine Lipschitzbedingung erfüllt,

$$|\varepsilon_N(f)| \leqq c(m)\, s^{-\frac{\alpha}{2}} \lg s. \tag{25}$$

Weiters erhalten wir für den allgemeinen Fall (E Normalbereich, f auf E eine Lipschitzfunktion)

$$(26)\qquad |\varepsilon_N(f)| \leqq c_3(m)\, s^{-\frac{\alpha}{2}} \lg s\, M(f),$$

wo immer $M(f)$ das Maximum von f auf E ist. Es ist insbesondere die Kappendiskrepanz

54 $$(27)\qquad D_N(C) \leqq c_4(m)\, s^{-\frac{\alpha}{2}} \lg s.$$

Benützt man (17′), so kommt schärfer $D_N(C) \leqq s^{-\frac{m}{4(3m+2)}} \lg s$. Berechnen wir noch $D_N^{(r)}$. Es ist nach § 3 (12)

$$D_N^{(r)} \leqq O\left(\sum_{k=1}^{M} \frac{k^{2m-2r-1}}{s^{m/4}}\right) + O\left(\frac{1}{M^{2r-m}}\right),$$

also

55 $$\leqq O\left(\frac{\operatorname{Min}(1, M^{2m-2r})}{s^{m/4}}\right) + O\left(\frac{1}{M^{2r-m}}\right).$$

Es ist daher

$$(28)\qquad D_N^{(r)} = O(s^{-\alpha}) \quad \text{wo} \quad \alpha = \frac{1}{4}(2r-m),$$

wenn $r > \frac{m}{2}$ ist. Es ist $D_N^{(r)}$ gleich $O(s^{-\frac{m}{4}})$, wenn $r > m$ ist. Die Konstante hängt natürlich von r ab. Wenn $\Delta^{(r)} f$ existiert, dann haben wir

$$(29)\qquad |\varepsilon_N(f)| \leqq O(s^{-\frac{m}{4}}) \int |\Delta^{(r)} f|\, d\mu.$$

Haben wir jetzt ein Produkt Π von t Sphären S^m und ist ω die Folge $\left(\frac{k_1}{\sqrt{s}}, \ldots, \frac{k_t}{\sqrt{s}}\right)$, so erhalten wir das gleiche Resultat, allerdings hängen die O-Abschätzungen noch von t ab (vgl. § 2 (9‴)). Es gilt also für eine Lipschitzfunktion f

$$(30)\qquad \left|\frac{1}{r(s)^t} \sum_{k_1 \ldots k_t} f\left(\frac{k_1}{\sqrt{s}}, \ldots, \frac{k_t}{\sqrt{s}}\right) - \int_{\Pi} f\, d\Pi\right| \leqq \frac{c(m^t)}{s^{1/20}},$$

wenn $m \neq 4$ ist, da ja dann $\alpha \geqq \frac{1}{10}$ ist.

Wir können die Formeln § 3 (6) noch anders schreiben. Es ist

$$\int \operatorname{grad} G \operatorname{grad} f\, dV(y) = -\int G \Delta f\, dV(y).$$

Wir erhalten also (sei $\int f d\mu = 0$) $f(x) = \int \operatorname{grad} G(x, y) \operatorname{grad} f(y)\, dV(y)$. Teilen wir wieder in $|x, y| \geqq \rho$ und $|x, y| \leqq \rho$, so ist $f(x) = \int_{|x,y|=\rho} \operatorname{grad} G(x, y) \operatorname{grad} f(y)\, dy + M(|\operatorname{grad} f|, \rho)$, wo $M = M(|\operatorname{grad} f|, \rho)$ das Maximum von $|\operatorname{grad} f|$ auf $|x, y| \leqq \rho$ ist. Es ist ja $|\operatorname{grad} G| \leqq |x, y|^{-m+1}$, also ist $\int |\operatorname{grad} G| \leqq C(m)$. Wir setzen nun $(\iota(|x, y| \geqq \rho)) = \iota'(x, y, \rho)$ und haben $|\varepsilon_N(f)| \leqq \int_{|x,y| \geqq \rho} \varepsilon_N \operatorname{grad} G \iota'(\rho)\, |\operatorname{grad} f|\, d\mu + M$.

Aus (*) und § 1 (67) folgt nun, da ja $|\operatorname{grad} G| \leqq \frac{C}{\rho^{m-1}}$ ist, (jetzt ist aber $\alpha = m-1$)

$$|\varepsilon_N(\operatorname{grad} G\,\iota)| \leqq O\left(\rho^{-m}\sqrt{\frac{\lg L}{L}} + \rho^{-m+1} D_N(C)\, L^{2m}\right).$$

Wir nehmen $L = D^{-\frac{1}{4m}}$, $\rho = L^{-\frac{1}{4m}}$ und erhalten

(31) $$|\varepsilon_N(f)| \leqq c\, D^{\frac{1}{17m}} \int |\operatorname{grad} f|\, d\mu(y) + M(|\operatorname{grad} f|, \rho),$$

wo $\rho = D^{-1/(16m^2)}$ ist.

Bemerkung 5. Wir wollen nun § 2 (13) bzw. die mit ihr identische Formel (17) zeigen. Es sei $\iota(C(p, \rho))$, kurz $\iota(C)$, die Kappe $\|p, y\| \leqq \rho$. Dann wollen wir zeigen, daß für

(32) $$b_{kj}(\iota) = \int \iota(C_y)\, H_{kj}(y)\, d\mu(y)$$

die Abschätzung

(33) $$|b_{kj}(\iota)| \leqq c\, \frac{|H_{kj}(p)|}{k^{\frac{m-1}{2}}}$$

gilt.

Beweis. Setzen wir $\|p, y\| = \vartheta$, so ist $\cos\vartheta = py$, daher $\vartheta = \arccos(py)$. Wir haben alle y mit $|\arccos(py)| \leqq \rho$ zu betrachten. Es sei g die Indikatorfunktion des Intervalls $|\xi| \leqq \rho$. Daher ist

(32′) $$b_{kj}(\iota) = \int_S g(\arccos py)\, H_{kj}(y)\, d\mu(y);$$

dies ist nach der Formel von Funk-Hecke

(34) $$\lambda_k H_{kj}(p).$$

Dabei ist

$$\lambda_k = \frac{V(S(m))}{V_\rho(S(m-1))} \int_{-1}^{1} g(t)(1-t^2)^{\frac{m}{2}-1} P_k(t)\, dt$$
$$= \frac{V(S(m))}{V(S(m-1))} \int_{-\rho}^{\rho} P_k(\cos\vartheta)\, |\sin\vartheta|^{m-1}\, d\vartheta.$$

Spezialisieren wir die Jacobischen Polynome (ausnahmsweise in der Bezeichnung nach Szegö, [12], p. 81) mittels $\alpha = \beta = \frac{m}{2} - 1$ zu den ultrasphärischen Polynomen mit der Benennung $C_k^{(\frac{m-1}{2})}$ (Szegö schreibt dafür $P_k^{(\frac{m-1}{2})}$), so gilt:

$$C_k^{(\frac{m-1}{2})}(\cos\delta) = \frac{\Gamma\left(\frac{m}{2}\right)\Gamma(k+m-1)}{\Gamma(m-1)\,\Gamma\left(k+\frac{m}{2}\right)}\, P_k^{(\frac{m}{2}-1, \frac{m}{2}-1)}(\cos\delta).$$

Andererseits wird der Konnex zwischen den $C_k^{(\frac{m-1}{2})}$ nach Szegö und den Gegenbauerpolynomen P_k nach C. Müller [10] durch die folgende Umnormierung hergestellt:

$$\frac{\Gamma(k+m+1)}{\Gamma(m-1)\,\Gamma(k+1)}\,P_k(\cos\delta)=C_k^{(\frac{m-1}{2})}(\cos\delta).$$

Elimination von $C_k^{(\frac{m-1}{2})}$ aus beiden Formeln ergibt:

$$P_k(\cos\delta)=\frac{\Gamma\left(\frac{m}{2}\right)\Gamma(k+1)}{\Gamma\left(k+\frac{m}{2}\right)}\,P_k^{(\frac{m}{2}-1,\frac{m}{2}-1)}(\cos\delta). \tag{35}$$

Das asymptotische Verhalten nach der Ordnung k bei ultrasphärischen Polynomen wird nach Szegö [12], p. 184 (8. 21. 10) mit $\alpha=\beta=\frac{m}{2}-1$ durch

$$\left(\sin\frac{\delta}{2}\right)^{\frac{m-1}{2}}\left(\cos\frac{\delta}{2}\right)^{\frac{m-1}{2}}P_k^{(\frac{m}{2}-1,\frac{m}{2}-1)}(\cos\delta)=O(k^{-\frac{1}{2}})$$

beschrieben. Einsetzen von (35), Berücksichtigung der Größenordnung $O(1)$ der beiden Winkelfunktionen sowie der Größenordnung der Gammafunktionen nach Stirling ergibt

$$P_k(\cos\delta)=\frac{\Gamma\left(\frac{m}{2}\right)\Gamma(k+1)}{\Gamma\left(k+\frac{m}{2}\right)}\,O(k^{-\frac{1}{2}})=O(k^{-\frac{m-1}{2}}). \tag{36}$$

Für alle ρ gilt somit

$$|\lambda_k(\rho)|\leqq c(m)\,k^{-(\frac{m-1}{2})}. \tag{37}$$

Damit ist die vorhergehende Behauptung bewiesen. Es gibt andererseits stets ein ρ_1 von der Gestalt $\frac{c_1(m)}{k}$ bei passendem $c_1(m)$, so daß

$$\left|\lambda_k\left(\frac{c_1}{k}\right)\right|\geqq c_2(m)\,k^{-m}. \tag{38}$$

Es ist nämlich nach Szegö [12], Kap. 8 gleichmäßig in ϑ

$$\lim_{k\to\infty} k^{-\frac{m}{2}+1}\,P_k^{(\frac{m}{2}-1,\frac{m}{2}-1)}\left(\cos\frac{\delta}{k}\right)=\left(\frac{\delta}{2}\right)^{-\frac{m}{2}-1}J_{\frac{m}{2}-1}(\delta), \tag{39}$$

wo $J_{\frac{m}{2}-1}$ die Besselfunktion der Ordnung $\frac{m}{2}-1$ ist. Wir wählen $|\delta|<$ die erste positive Nullstelle von $J_{\frac{m}{2}-1}$. Es ist daher

$$\int_{-\rho_1}^{\rho_1}|\sin\delta|^{m-1}\,P_k(\cos\delta)\,d\delta\geqq c_2 k^{m-1}.$$

Aus (32) und (34) folgt

(40) $$\lambda_k(\rho)\, H_{k,j}(p) = \int \iota(p, y, c)\, H_{k,j}(y)\, d\mu(y).$$

Wir setzen $p = x_n$ $(n = 1, \ldots, N)$ und $\lambda_k(\rho)\, \varepsilon_N(H_{k,j}) = \int \varepsilon_N(\iota C)\, H_{k,j}(y)\, d\mu(y)$, also $|\lambda_k(\rho)|\, |W_{k,j}| \leqq D_N(c) \int |H_{k,j}|\, d\mu(y)$, somit für $\rho = \frac{c_1}{k}$, $|W_{k,j}| \leqq k^m D_N(c) \int |H_{k,j}(y)\, d\mu(y)$, also $W_k = \sqrt{\frac{\sum W_{k,j}^2}{Z}} \leqq k^m D_N(c)$, und das ist Formel (20).

§ 5. Wir benützen nun den Satz, dessen Beweis wir gleich nachtragen werden.

Satz. *Ist $\Pi = S$ von der Dimension m, so gibt es isotrope Vektoren c, so daß $c^2 = 0$ ist (ihre Anzahl ist $Z(m, k)$), so daß sich jede Kugelfunktion H_{kj} vom Grade k in der Gestalt:*

(1) $$H_{kj}(x) = \sum_{z=0}^{Z(m,k)} \alpha_{zkj} (c_z x)^k$$

darstellen läßt.

Dies wurde schon von E. Hecke und von Pommerenke [11] bewiesen. Man findet dies auch bei Appell-Kampé de Feriét, [1]. Wir können aber noch mehr sagen. Bei passender Wahl von H_{kj} können wir die c_z in der Gestalt

(2) $$c_z = (1, i e_z)$$

annehmen, wo die e_z reelle Vektoren im R_m mit $e^2 = 1$ sind, also auf $S^{(m-1)}$ liegen. Dabei muß nun die Determinante

56 (3) $$D(e) = |\mathrm{Det}\,(H_{j,r}(V_{e_z}))| \neq 0$$

sein. Dabei ist $r = Z(j, m-1)$ und $0 \leqq j \leqq k$.

Die Anzahl dieser Kugelfunktionen ist ja gerade $Z(m, k)$. Wir werden zeigen (siehe (12)), daß es sogar solche Z Punkte auf $S^{(m-1)}$ gibt, daß

(4) $$D(e) \geqq \frac{Z(m, k)}{V^Z(S^{(m-1)})}$$

gilt. Man kann die α_z so wählen, daß

57 (5) $$|\alpha_{zkj}| \leqq \frac{k^{\frac{Z-1}{2}}}{D(e)}$$

wird.

Liegt der allgemeine Fall vor, daß Π ein Produkt von Sphären ist, dann gilt analog $H_{kj}(x) = \sum \alpha_{zkj} (c_k x)^k$, wobei gilt $1 \leqq z \leqq Z$ und

(6) $$(cx)^k = (c_1 x_1)^{k_1} \cdots (c_t x_t)^{k_t},$$

58 wenn $x = (x_1, \ldots, x_t)$ ist. Weiters ist $c_L = (1, k e_{z_L}(m_L))$, wo $e_{z_L}(m_L)$ ein Vektor auf $S^{(m_L-1)}$ ist mit

(7) $$D_j = D_j(e) = |\mathrm{Det}\, H_{j_L, r_L}(e_{z_L})|.$$

Es gilt, wenn wieder $\alpha_{Zkj}=\alpha_{Z_1k_1j_1}\cdots\alpha_{Z_tk_tj_t}$,

59 (8) $$|\alpha_{Zkj}|\leqq\frac{k^{\frac{Z-1}{2}}}{D(e)},$$

wobei $k^{\frac{Z-1}{2}}=k^{\frac{Z_1-1}{2}}\cdots k_t^{\frac{Z_t-1}{2}}$ ist, mit $D(e)=D_1\cdots D_t$. Wir erhalten dann aus § 2 (10)

$$\varepsilon_N(T_L(f))\leqq\sum_k|b_{kj}|\,|\sum\alpha_{Zkj}\varepsilon_N((c_{Zkj})^k)|.$$

Es gilt also

Satz. *Die Folge* (x_n) *ist genau dann gleichverteilt, wenn für alle isotropen Vektoren* $(c_1,\ldots,c_t)$ *und alle* $(k_1,\ldots,k_t)\neq(0,\ldots,0)$

(9) $$\lim_{N\to\infty}\frac{1}{N}\sum(c_1p(x_n))^{k_1}\cdots(c_tp(x_n))^{k_t}=0$$

gilt.

Beim *Beweis* der vorhergehenden Behauptung gehen wir von der Laplace'schen Darstellung für Kugelfunktionen

(10) $$H_k^{(m)}(x)=c(m,k)\int_{S^{m-1}}(x_1+i\sqrt{1-x_1^2}\,x''e)^kH_{j,r}^{(m-1)}(e)\,dV(S^{m-1},e)$$

aus. Dabei ist $x=(x_1,\ldots,x_{m+1})=(x_1,\sqrt{1-x_1^2}\,x'')$ mit $e=e(m-1)$, Einheitsvektor auf dem $S^{(m-1)}$. Wenn nun $0\leqq j\leqq k$, $1\leqq r\leqq Z(j,m-1)$ ist, dann ist die Menge aller $(H_{j,r})$ (ihre Anzahl ist $Z(k,m)$), eine linear unabhängige Menge von Kugelfunktionen vom Grade k auf dem S^m. Wir betrachten mit $c(e)=(1,ie)$ den Vektor

$$x=(x_1,x')=(x_1,\sqrt{1-x_1^2}\,x'').$$

Es ist $(cx)=(x_1+iex')=(x_1+i\sqrt{1-x_1^2}\,ex'')$ und

$$(cx)^k=(x_1+i\sqrt{1-x_1^2}\,ex'')^k=\sum_{j=0}^{k}\sum_{r=0}^{Z(j,m-1)}c_{jr}H_{j,r}(e).$$

Es ist nun

(11) $$c_{jr}=\int_{S^{m-1}}(x_1+i\sqrt{1-x_1^2}\,ex'')^kH_{j,r}(e)\,dV(S^{m-1},e),$$

also nach der Darstellung von Laplace $=H_k^{(m)}(x)$. Wir schreiben $\hat H_{i,r}^{(m)}$ und erhalten $(cx)^k=\sum_{j,r}\hat H_{j,r}(x)\,H_{j,r}(e)$. Wir wählen nun $e_1,\ldots,e_Z$ auf S^{m-1} so, daß $D=\operatorname{Det}(H_{j,r}(e))\neq0$ ist. Dies ist möglich, da die $(H_{j,r}(e))$ ein Orthogonalsystem bilden. Wir betrachten nun die zugehörigen Vektoren $C(e)=C_L$ und haben also $(C_Lx)^k=\sum_{j,r}\hat H_{j,r}(x)\,H_{j,r}(e_L)$.

Wir betrachten die inverse Matrix zu $(H_{j,r}(e_L))=(\alpha_{j,r})$. Nach Funk-Hecke ist das
60 Integral (11) gleich $H_{j,r}^{(m-1)}\int_{-1}^{1}(x_1+i\sqrt{1-x_1^2}\,t)^k\,t^{\frac{m-3}{2}}P_j^{(m-1)}(t)\,dt$. Also liefert die Formel (11) ein orthogonales System von Kugelfunktionen, da ja die $H_{j,r}^{(m-1)}$ orthogonal sind, das allerdings nicht normiert ist. Daher wurde ein Faktor $c(m,k)$ in Formel (10) hinzugefügt. Es ist $\alpha_{j,r}=\frac{p_{j,r}}{D}$, wo $p_{j,r}$ der (j,r)te Minor in der Matrix $(H_{j,r}(e_L))$ ist.

Wenden wir auf einen solchen Minor die Hadamard'sche Ungleichung an, so erhalten wir, daß $|p_{ir}| \leqq \Pi' \sqrt{\sum_{j,r} |H_{j,r}(e_L)|^2}$ ist, dabei erstreckt sich das Produkt auf $Z-1$ Zeilen der Determinante D. Nun ist ja

$$\sum_r |H_{j,r}(e_L)|^2 \leqq \frac{Z(j, m-1)}{V(S_{m-1})} P_j^{(m-1)}(e_L^2) = \frac{Z(j, m-1)}{V(S_{m-1})},$$

also ist $\sum_{j,r} |H_{j,r}(e_L)|^2 \leqq \sum_{j=1}^{k} \frac{Z(j, m-1)}{V(S_{m-1})}$ und damit $|p_{j,r}| \leqq \left(\sum_{j=1}^{k} \frac{Z(j, m-1)}{V(S_{m-1})}\right)^{\frac{Z-1}{2}}$. Damit ist die Behauptung (2) bewiesen.

Jetzt erhalten wir sofort

$$H_{k,j}(x) = \sum \alpha_{k,j,Z}(c_Z x)^k.$$

Wir zeigen nun, daß wir die $e_1, \ldots, e_Z$ immer so wählen können, daß auch

$$D \geqq \sqrt{\frac{Z!}{V^Z(S^{m-1})}} \tag{12}$$

gilt. Wir bilden nun das $Z(k, m)$-Produkt der $(m-1)$-dimensionalen Sphäre und betrachten

$$J = \int_{\Pi} |\mathrm{Det}\, H_{j,r}(e)|^2 \, dV_1 \cdots dV_Z, \tag{13}$$

wobei dV_j das Volumen auf S^{m-1} ist. Entwickeln wir die Determinante, so erhalten wir

$$f(e) = \sum S(i_1, \ldots, i_Z)\, S(k_1, \ldots, k_Z) \prod H_{ik}(e) \quad (S = \text{signum der Permutation } i_1, \ldots, i_Z).$$

Integrieren wir darüber, so erhalten wir $J = \sum S^2(i_1, \ldots, i_Z) = Z!$. Es muß daher $e_1, \ldots, e_Z$ geben, so daß $V^Z(S^{m-1})\, |\mathrm{Det}\, e|^2 \geqq Z!$, also $|\mathrm{Det}\, e| \geqq \sqrt{\frac{Z!}{V^Z(S^{m-1})}}$ ist. Damit sind die Behauptungen ganz bewiesen. Man kann nun versuchen, gute Punkte $e_1, \ldots, e_Z$ zu finden. Man betrachtet zu diesem Zweck eine Z-fache Folge $(e_1(s_1), \ldots, e_Z(s_Z))$, so daß $(e(s))$ auf S gut gleichverteilt ist. Dann ist, wenn wir $f(e_1, \ldots, e_Z) = |\mathrm{Det}\,(e)|^2$ nehmen,

$$\frac{1}{M(S)^Z} \sum_{s_1=1, \ldots, s_Z=1}^{M} f(e_1(s_1), \ldots, e_Z(s_Z)) - \int f d\Pi \quad \text{klein.}$$

Nimmt man z.B. die Folge $\left(\frac{k}{\sqrt{s}}\right)$, dann bietet sich $\frac{1}{M(S)^Z} \sum_{k_1, \ldots, k_Z} f\left(\frac{k_1}{\sqrt{s}}, \frac{k_2}{\sqrt{s}}, \ldots, \frac{k_Z}{\sqrt{s}}\right)$ mit $M = r(S)$ an. Die Diskrepanz ist dann in diesem Fall $D = O(S^{-\frac{\alpha}{2}} \lg s)$ (vgl. § 4 (28)).

Es muß noch nach § 3 (12) Δf bzw. $\Delta^{(r)}(f)$ mit $r = \left[\frac{m}{2}\right] + 1$ berechnet werden. Es ist $f = \sum S(i_1, \ldots, i_z)\, S(k_1, \ldots, k_z)\, H_{i_1 j_1}((e_1), \ldots, H_{iZ\,jZ}(e_Z))$. Nun ist $\Delta^{(r)} H_{i_k} = \lambda_j H_{j_k}$, daher $\Delta^{(r)} H_{ir}(e_1) = \lambda_j^r H_{jr}$. Wir erhalten also

$$|\Delta^{(r)} f| \leqq \lambda_1^{Z_1 r} \cdots \lambda_k^{Z_k r} \left(\sum_{j=1}^{k} \frac{Z(j, m-1)}{V(S_{m-1})}\right)^{Z/2}.$$

Damit erhalten wir

$$(14)\qquad \left|\frac{1}{(M(S))^Z}\sum_{k_1,\ldots,k_Z} f\left(\frac{k_1}{\sqrt{S}},\ldots,\frac{k_Z}{\sqrt{S}}\right)-Z!\right| \leqq \frac{1}{S^{\alpha/2}}\,\lambda_1^{Z_1 r}\cdots\lambda_k^{Z_k}\left(\sum_{j=1}^{k}\frac{Z(j,m-1)}{V(S_{m-1})}\right)^{Z/2},$$

wo $\lambda_i = m(m-1+j)$, $Z_j = Z(j, m-1)$ ist. Wählt man S so groß, daß die rechte Seite $< Z!$ ist, so folgt, daß $\left|\operatorname{Det}\left(H_{jr}\left(\frac{k}{\sqrt{S}}\right)\right)\right| \geqq 1$ ist.

Es gibt auch eine einfache Methode, um gute Punkte zu bestimmen, und zwar mit Hilfe der Gleichverteilung auf dem Torus. Wir stellen dazu die S^{m-1} in Polarkoordinaten mit den Winkeln $\delta_1, \ldots, \delta_{m-1}$, modulo 1 gerechnet, dar:

61 $$x_1 = \cos\pi\delta_1 \cos\pi\delta_2 \cdots \cos\pi\delta_{m-1}, \quad x_2 = \cos\pi\delta_1 \cos\pi\delta_2 \cdots \cos\pi\delta_{m-1}$$
$$\sin\pi\delta_m, \ldots, x_{m-1} = \cos\pi\delta_1, \quad x_m = \sin\pi\delta_1 \qquad 0 < \delta_j < 1\,.$$

Es ist dann das Volumselement von der Gestalt

$$dV(S^{m-1}) = c\,\sin^{m-2}\pi\,d\delta_1\cdots d\delta_{m-1} = u(\delta_1,\ldots,\delta_{m-1})\,d\delta_1\cdots d\delta_{m-1}\,.$$

Ist $x \in S^{m-1}$, so ist

$$f(x)\,dV(x) = g(\delta_1,\ldots,\delta_{m-1})\,u(\delta_1,\ldots,\delta_{m-1})\,d\delta_1\cdots d\delta_{m-1}$$
$$= h(\delta_1,\ldots,\delta_{m-1})\,d\delta_1\cdots d\delta_{m-1}\,.$$

Wir schreiben kurz mit $\delta = (\delta_1, \ldots, \delta_{m-1})$

$$f(x)\,dV(x) = h(\delta)\,d\delta, \quad \text{wo} \quad h(\delta) = g(\delta)\,u(\delta).$$

Haben wir eine Funktion f auf einem Produkt Π von Sphären gleicher Dimension, so wird

$$f(x)\,d\Pi(x) = h(\delta^1,\ldots,\delta^Z)\,d\delta^1\cdots d\delta^Z.$$

Wir betrachten nun die Z Vektoren

$$\delta_1 = \frac{k_1}{M},\ \delta_2 = \frac{k_2}{M},\ldots,\ \delta_Z = \frac{k_Z}{M} \quad (k_i = (k_{i1},\ldots,k_{im}))\,.$$

Dann gilt für h

$$\lim_{M\to\infty}\frac{1}{M^Z}\sum_{k_1,\ldots,k_Z}^{M} h\left(\frac{k_1}{M},\ldots,\frac{k_Z}{M}\right) = \int h(\delta)\,d\delta = \int f(x)\,d\Pi(x).$$

Die Diskrepanz der Folge ist bekanntlich $\leqq \frac{1}{M}$, also ist

$$\frac{1}{M^Z}\sum h\left(\frac{k_1}{M},\ldots,\frac{k_Z}{M}\right) - \int f(x)\,d\Pi(x) \leqq \frac{1}{M}\int\left|\frac{\partial^{mZ} f}{\partial\delta_1\cdots\partial\delta_Z}\right|\,d\delta_1\cdots d\delta_Z.$$

Wendet man dies auf $|\mathrm{Det}(e_1,\ldots,e_Z)|^2=|D|^2$ an und beachtet, daß

$$|D|^2=\operatorname*{Det}_{i,k}\left(\sum_j \frac{Z(m,j)}{V(S^m)}P_j(e_{z_i}e_{z_k})\right)$$

ist, dann hat man die Ableitungen der P_j zu betrachten. Nun ist bekannt, daß die s-te Ableitung von P_j ja dem Betrag nach kleiner als $\binom{j+2m+s}{j}$ ist, und wir erhalten eine Abschätzung für das Integral $\leqq cm^m$, und man kann schließen, daß

$$\frac{1}{M^Z}\left|\sum_{k \bmod M}\left(\mathrm{Det}\left(\frac{k}{M}\right)\right)\right|^2 \geqq Z!-\frac{m^m}{M}$$

ist. Daraus folgt wieder die Existenz von guten Punkten. Es gibt noch andere Wege, solche gute Punkte zu finden. Eine nützliche Aufgabe wäre es, ein einfaches System guter Punkte zu finden. Wir wollen noch in anderer Weise zeigen, wie man gleichverteilte Folgen auf den Torus dazu benützen kann, gleichverteilte Folgen auf der Sphäre (und auf einem Produkt von Sphären) zu konstruieren. Dies soll in dem folgenden Anhang geschehen.

Anhang. Es wurde von Box und Muller, [3], die Transformation

$$\xi=\sqrt{-\ln x}\cos 2\pi y,\quad \eta=\sqrt{-\ln x}\sin 2\pi y$$

eingeführt mit $0<x<1$ und $0<y<1$. Es ist $\xi^2+\eta^2=-\ln x$, also $x=\exp(-(\xi^2+\eta^2))$. Es ist

62 $$\frac{\partial(x,y)}{\partial(\xi,\eta)}=\exp(-(\xi^2+y^2))=x,\quad \text{da}\quad \frac{\partial(\xi,\eta)}{\partial(x,y)}=\frac{1}{x}.$$

Für $x=1$ ist $\xi=\eta=0$ und für $x=0$ ist ξ oder η unendlich, da

$$\xi^2+\eta^2=-\ln x.$$

Es ist, wenn E: $0<x<1$, $0<y<1$,

$$\int_E f(\sqrt{-\ln x}\cos 2\pi y, \sqrt{-\ln x}\sin 2\pi y)\,dx\,dy=\int f(\xi,\eta)\frac{\partial(x,y)}{\partial(\xi,\eta)}\,d\xi\,d\eta, \tag{1}$$

wo sich die Integration rechts in (1) über den ganzen R^2 erstreckt. Dabei wollen wir f als zweimal stetig differenzierbar voraussetzen und annehmen, daß die Integrale in (1) existieren. Man stellt folgende Formel fest, wo wir

$$F(x,y)=f(\sqrt{-\ln x}\cos 2\pi y, \sqrt{-\ln x}\sin 2\pi y)$$

setzen:

$$F_x=f_\xi\xi_x+f_\eta\eta_x=(x\sqrt{-\ln x})^{-1}(f_\xi\cos 2\pi y+f_\eta\sin 2\pi y)$$

und

$$\begin{aligned}F_{xy}&=(x\sqrt{-\ln x})^{-1}[(f_{\xi\xi}\xi_y+f_{\xi\eta}\eta_y)\cos 2\pi y+(f_{\xi\eta}\xi_y+f_{\eta\eta}\eta_y)\sin 2\pi y\\&\quad+(-f_\xi\sin 2\pi y+f_\eta\cos 2\pi y)\,2\pi]\\&=x^{-1}[f_{\xi\eta}+\sin 2\pi y\cos 2\pi y(f_{\xi\xi}+f_{\eta\eta})]\\&\quad+(x\sqrt{-\ln x})^{-1}\,2\pi(f_\xi\sin 2\pi y+f_\eta\cos 2\pi y).\end{aligned}$$

3

Es ist

64 $$\int_E |F_{xy}| dx\, dy = \int |F_{xy}| \left|\frac{\partial(x, y)}{\partial(\xi, \eta)}\right| d\xi\, d\eta = \int |F_{xy}| x\, d\xi\, d\eta$$
$$\leqq \int \{[|f_{\xi\eta}| + |f_{\xi\xi} + f_{\eta\eta}|] + 2\pi(-\ln x)^{-1/2}(|f_\xi| + |f_\eta|)\}\, d\xi\, d\eta,$$

also

$$\int_E |F_{xy}|\, dx\, dy \leqq I_1 + I_2, \tag{2}$$

wo

$$I_1 = \int (|f_{\xi\xi} + f_{\eta\eta}| + |f_{\xi\eta}|)\, d\xi\, d\eta \tag{3}$$

und

65 $$I_2 = 2\pi \int (\xi^2 + \eta^2)^{-1/2}(|f_\xi| + |f_\eta|)\, d\xi\, d\eta. \tag{4}$$

Die Integrationen erstrecken sich über den ganzen R^2. Nun ist die bisher entwickelte Rechnung nur formaler Natur, da die Transformation Singularitäten besitzt. Wir betrachten statt E zunächst nur den Bereich

$$E(\varepsilon, \varepsilon_1): \varepsilon \leqq x \leqq 1 - \varepsilon_1,\ 0 \leqq y \leqq 1.$$

Dabei ist $\varepsilon > 0$ und $\varepsilon_1 > 0$. Bei der Mullerschen Transformation geht $E(\varepsilon, \varepsilon_1)$ in den Bereich $B: \ln \frac{1}{1-\varepsilon_1} \leqq \xi^2 + \eta^2 \leqq \ln \frac{1}{\varepsilon}$ über. Dann erhalten wir, wenn wir noch den Operator $\Delta = |\partial_\xi| + |\partial_\eta|$ und damit $\Delta^2 = |\partial_{\xi\xi}| + |\partial_{\eta\eta}| + 2\,|\partial_{\xi\eta}|$ einführen, statt (2)

66 $$\int_{E(\varepsilon, \varepsilon_1)} |F_{xy}| dx\, dy \leqq \int_B (\Delta^2 f + \Delta f (\xi^2 + \eta^2)^{-1/2})\, d\xi\, d\eta.$$

Betrachten wir nun eine gleichverteilte Folge $\omega = (x_n, y_n)$ in E, dann sei, wenn $\bar{\chi}$ die Indikatorfunktion des Intervalls $\langle \varepsilon, 1 - \varepsilon_1 \rangle$ bedeutet,

$$\lambda_N(f, \bar{\chi}) = \frac{1}{N} \sum_{n=1}^{N} f(\sqrt{-\ln x_n} \cos 2\pi y_n, \sqrt{-\ln x_n} \sin 2\pi y_n)\, \bar{\chi}(x_n).$$

Es gilt dann

67 $$|\lambda_N(f) - \int_B f(\xi, \eta)\, e^{-(\xi^2+\eta^2)}\, d\xi\, d\eta| \leqq D_N \int_B (\Delta^2 f + (\xi^2 + \eta^2)^{-1/2} \Delta f)\, d\xi\, d\eta.$$

Es ist also

$$|\lambda_N(f) - \int f(\xi, \eta)\, e^{-(\xi^2+\eta^2)}\, d\xi\, d\eta| \leqq I_3 + D_N I_4,$$

wo

$$I_3 = \int_{B^*} |f(\xi, \eta)|\, e^{-(\xi^2+\eta^2)}\, d\xi\, d\eta$$

und

68 $$I_4 = \int_{B^*} (\Delta^2 f + (\xi^2 + \eta^2)^{-1/2} \Delta f)\, d\xi\, d\eta.$$

Dabei ist B^* das Komplement von B in R^2. Es sei nun $M(\xi, \eta)$ eine obere Schranke von $|f|$ in B^*. Es ist

$$B^* = \left(\xi^2 + \eta^2 \leqq \ln \frac{1}{1-\varepsilon_1}\right) \cup \left(\xi^2 + \eta^2 \geqq \ln \frac{1}{\varepsilon}\right).$$

Dann ist

$$\int\limits_{B^*} |f| \, e^{-(\xi^2+\eta^2)} \, d\xi \, d\eta \leqq M \Big(\int\limits_{\xi^2+\eta^2 \leqq \ln \frac{1}{1-\varepsilon_1}} e^{-(\xi^2+\eta^2)} \, d\xi \, d\eta + \int\limits_{\ln \frac{1}{\varepsilon} \leqq \xi^2+\eta^2} e^{-(\xi^2+\eta^2)} \, d\xi \, d\eta \Big).$$

Es ist

$$\int\limits_{\xi^2+\eta^2 \leqq \ln \frac{1}{1-\varepsilon_1}} e^{-(\xi^2+\eta^2)} \, d\xi \, d\eta = \pi \left(-1 + \frac{1}{1-\varepsilon_1}\right)$$

und

$$\int\limits_{\ln \frac{1}{\varepsilon} \leqq \xi^2+\eta^2} e^{-(\xi^2+\eta^2)} \, d\xi \, d\eta = \varepsilon \pi,$$

also ist $\iint\limits_{B^*} |f| \, e^{-(\xi^2+\eta^2)} \, d\xi \, d\eta \leqq \pi M \left(\frac{\varepsilon_1}{1-\varepsilon_1} + \varepsilon\right)$. Wir nehmen nun $\varepsilon_1 = \frac{\varepsilon}{1+\varepsilon}$. Setzen wir $M_\varepsilon = \sup |f|$ im Bereich

$$(\xi^2 + \eta^2 \leqq \ln(1+\varepsilon)) \cup \left(\xi^2 + \eta^2 \geqq \ln \frac{1}{\varepsilon}\right),$$

dann ist $\int\limits_{B^*} |f| \, e^{-(\xi^2+\eta^2)} \, d\xi \, d\eta \leqq M_\varepsilon 2 \varepsilon \pi$. Wir haben also für

$$\mathfrak{F}_N = \left| \lambda_N(f) - \frac{1}{\pi} \iint f(\xi, \eta) \, e^{-(\xi^2+\eta^2)} \, d\xi \, d\eta \right|,$$

wobei sich das Integral über die ganze Ebene erstreckt, die Abschätzung

$$\mathfrak{F}_N \leqq \varepsilon M_\varepsilon \pi + D_N \iint^* L(\xi, \eta) \, d\xi \, d\eta, \tag{6}$$

wobei

$$L = U + (\xi^2 + \eta^2)^{-1/2} \, V \tag{7}$$

mit

$$U = |f_{\xi\xi}| + 2 |f_{\xi\eta}| + |f_{\eta\eta}| \tag{8}$$

und

$$V = |f_\xi| + |f_\eta| \tag{9}$$

ist. Das Integral in (6) erstreckt sich über den Bereich

$$\ln(1+\varepsilon)^{1/2} \leqq (\xi^2 + \eta^2)^{1/2} \leqq (-\ln \varepsilon)^{1/2}. \tag{10}$$

Wir wählen nun $\varepsilon_0 = D_N$ und bezeichnen M_{ε_0} mit K_N. Es ist also

$$K_N = \sup |f| \tag{11}$$

im Bereich

$$(\xi^2 + \eta^2 \leqq \ln(1+D_N)) \cup \left(\xi^2 + \eta^2 \geqq \ln \frac{1}{D_N}\right). \tag{12}$$

Wir erhalten also

69 (13) $$\mathfrak{F}_N \leqq D_N\left(K_N + \iint^* I\, d\xi\, d\eta\right),$$

wo sich das Integral nun über den Bereich

(14) $$\log(1+D_N) \leqq (\xi^2+\eta^2) \leqq -\log D_N$$

erstreckt.

Wir wollen nun über U und V spezielle Annahmen machen und zwar, daß es eine positive Konstante C gibt, so daß

(15) $$\operatorname{Max}(U, V) \leqq C e^{-(\xi^2+\eta^2)}$$

gilt. Es ist dann

$$\iint^* L\, d\xi\, d\eta \leqq C \iint^* e^{-(\xi^2+\eta^2)}\left(1+(\xi^2+\eta^2)^{-1/2}\right) d\xi\, d\eta.$$

Das obige rechtsstehende Integral ist

$$\leqq 2\pi C \int_0^\infty e^{-r^2}(r+1)\, dr = 2\pi \frac{1}{2}(1+\sqrt{\pi}) \leqq 4\pi.$$

Weiters folgt aus dem Mittelwertsatz der Differentialrechnung

$$|f(\xi, \eta) - f(0,0)| \leqq C e^{-(\xi^2+\eta^2)}(|\xi| + |\eta|),$$

also ist

(16) $$K_N \leqq |f(0,0)| + C\left(D_N \log \frac{1}{D_N} + \left(\frac{\log(1+D_N)}{1+D_N}\right)\right).$$

Die Bedingungen (15) sind natürlich sehr einschränkend. Wir wollen nun die obigen Überlegungen auf mehr als zwei Dimensionen verallgemeinern. Es sei eine Funktion f in $2s$ Variablen $\xi_1, \eta_1, \xi_2, \eta_2, \ldots, \xi_s, \eta_s$ im R^{2s} vorgelegt, und wir machen die Transformation

(17) $$\xi_i = \sqrt{-\log x_i}\cos 2\pi y_i, \quad \eta_i = \sqrt{-\log x_i}\sin 2\pi y_i$$

für $i = 1, \ldots, s$. Wir schreiben kurz

$$\xi_i = \varphi_i(x_i, y_i), \quad \eta_i = \psi_i(x_i, y_i).$$

Wir wollen annehmen, daß f alle Ableitungen bis zur Ordnung $2s$ in $R^{2s} - 0$ besitzt, welche stetig sein sollen.

Wir benützen nun (3), welches wir abkürzend in der Form schreiben

$$\frac{\partial^2 f}{\partial x_1\, \partial y_1}(\varphi_1, \psi_1) = S_1\left(f(\xi_1, \eta_1, \varphi_2(x_2, y_2), \psi_2(x_2, y_2), \ldots)\right).$$

Daraus folgt sofort

$$\frac{\partial^4 f(\varphi_1, \varphi_2, \varphi_3, \varphi_4)}{\partial x_1\, \partial y_1\, \partial x_2\, \partial y_2} = \frac{\partial^2}{\partial x_2\, \partial y_2} S_1 f = S_1 \frac{\partial^2 f}{\partial x_2\, \partial y_2} = S_1 S_2 f.$$

Allgemein erhalten wir so

(18) $$\frac{\partial^{2s} f(\varphi_1, \psi_1, \ldots, \varphi_s, \psi_s)}{\partial x_1\, \partial y_1 \cdots \partial x_s\, \partial y_s} = S_1 \cdots S_s f(\xi_1, \eta_1, \ldots, \xi_s, \eta_s).$$

Wir erhalten also analog zu (13) für

(19) $$\mathfrak{F}_n^{(s)} = \left| \lambda_N(\mathfrak{F}) - \frac{1}{\pi^s} \int \mathfrak{F}(\xi_1, \eta_1, \ldots) e^{-\sum_{i=1}^{s} (\xi_i^2 + \eta_i^2)} d\xi_1 \cdots d\eta_s \right|,$$

wobei sich das Integral über den ganzen R^{2s} erstreckt:

(19') $$\mathfrak{F}_n^{(s)} \leqq D_N^{(2s)} (K_N + \int^* L \, d\xi_1 \cdots d\eta_s),$$

wobei

$$L = |S_1 \cdots S_s \mathfrak{F}(\xi_1, \ldots, \eta_s)|$$

und sich das Integral über dem Bereich

(20) $$\prod_{i=1} (\log(1 + D_N)) \leqq (\xi_i^2 + \eta_i^2) \leqq -\log D_N$$

erstreckt, und es ist

(21) $$K_N = \sup |f|$$

erstreckt über den Bereich

$$\prod_{i=1} (\xi_i^2 + \eta_i^2) \leqq \log(1 + D_N) \cup \left(\xi_i^2 + \eta_i^2 \geqq \log \frac{1}{D_N} \right).$$

Weiters ist, wenn (x_n) eine Folge im E^{2s} ist, also in Koordinaten

$$(p_1(x_n), q_1(x_n), \ldots, p_s(x_n), q_s(x_n))$$

$$\lambda_N(f) = \frac{1}{N} \sum_{n=1}^{N} H(n) \, G(n),$$

wo

$$H(n) = f(\sqrt{-\log p_1(x_n)} \cos 2\pi q_1(x_n), \sqrt{-\log p_2(x_n)} \sin 2\pi q_2(x_n), \ldots)$$

und

$$G(n) = \chi(p_1(x_n), \ldots, p_s(x_n)),$$

wo χ die Indikatorfunktion von

$$\prod_{i=1}^{s} \log(1 + D_N) \leqq p_i(x) \leqq -\log D_N$$

ist.

Wir wollen uns wieder auf den Fall $s = 1$ beschränken und betrachten nun eine Funktion

(22) $$f(\xi, \eta) = \Phi\left(\frac{\xi}{\sqrt{\xi^2 + \eta^2}}, \frac{\eta}{\sqrt{\xi^2 + \eta^2}} \right),$$

wo Φ zweimal stetig differenzierbar auf dem Einheitskreis sein soll. Die Funktion (22) ist stetig differenzierbar in $R^2 - 0$. Wir können also (13) anwenden und haben nur das Integral

(23) $$\iint^* L \, d\xi \, d\eta$$

im Bereich (14) abzuschätzen, wobei L nach (7) mittels (8) und (9) definiert ist. Wir setzen noch

(24) $$r^2 = \xi^2 + \eta^2, \quad O_1 = \xi / r, \quad O_2 = \eta / r.$$

Es ist $O_1^2+O_2^2=1$ und

$$\frac{\partial O_1}{\partial \xi}=\frac{O_2^2}{r},\ \frac{\partial O_1}{\partial \eta}=\frac{-\xi\eta}{r^3}=\frac{\partial O_2}{\partial \xi},\ \frac{\partial O_2}{\partial \eta}=\frac{O_1^2}{r},\ \frac{\partial^2 O_1}{\partial \xi^2}=-3\xi=\frac{-3O_1O_2^2}{r^4},\dots .$$

Es ist z.B. $\dfrac{\partial^2 \Phi}{\partial \xi^2}=\dfrac{\partial^2 \Phi}{\partial O_1^2}\left(\dfrac{\partial O_1}{\partial \xi}\right)^2+\cdots$.

Weiters ist

$$\iint^* \left|\frac{\partial^2 f}{\partial \xi^2}\right| d\xi\, d\eta=\iint r\, dr \left|\frac{\partial^2 f}{\partial \xi^2}\right| dO,$$

wo dO das Oberflächenelement der Einheitskugel E ist. Berücksichtigt man dies, so folgt, daß mit einer Konstanten C_1, die, numerisch ausgewertet, sicher unter 2000 liegt, gilt

$$\iint^* L\, d\xi\, d\eta \leqq C_1 \int^* dr\left(\frac{1}{r}+\frac{1}{r^3}\right)\iint_E \Omega\, dO,$$

wobei

$$\Omega=\left|\frac{\partial^2 \Phi}{\partial O_1^2}\right|+2\left|\frac{\partial^2 \Phi}{\partial O_1\,\partial O_2}\right|+\left|\frac{\partial^2 \Phi}{\partial O_2^2}\right|+\left|\frac{\partial \Phi}{\partial O_1}\right|+\left|\frac{\partial \Phi}{\partial O_2}\right|$$

ist. Das Integral $\int dr\left(\dfrac{1}{r}+\dfrac{1}{r^3}\right)$ ist nach (14) $\leqq 2\log\log\dfrac{1}{D_N}$.

Es ist also

$$\iint^* L\, d\xi\, d\eta \leqq C_1 \log\log\frac{1}{D_N}\iint_E \Omega\, dO. \tag{25}$$

Andererseits ist das Integral

$$\iint f e^{-(\xi^2+\eta^2)}\, d\xi\, d\eta=\int e^{-r^2} r \iint_E \phi\, dO.$$

Setzen wir

$$\mathfrak{F}_N^2=\left|\lambda_N(\phi)-\frac{1}{2\pi}\iint_E \phi\, dO\right|, \tag{26}$$

so ist also nach (13)

70 $$\mathfrak{F}_N^2 \leqq D_N\left(K_N+C_1\log\log\frac{1}{D_N}\iint_E \Omega\, dO\right), \tag{27}$$

wo $K_N=\sup\limits_E |\phi|$.

Wir haben also eine Quadraturformel für Funktionen auf dem Einheitskreis. Das hätten wir natürlich einfacher haben können. Wir können dies nun aber auch für die $2s$-dimensionale Einheitssphäre S^{2s} durchführen. Es sei ϕ eine $2s$-mal stetig differenzierbare Funktion auf S^{2s} und wir definieren $f(\xi_1,\dots,\eta_s)=\phi\left(\dfrac{\xi_1}{r},\dots,\dfrac{\eta_s}{r}\right)$, wo $r^2=\xi_1^2+\cdots+\eta_s^2$ ist. Wir setzen wieder $O_1=\dfrac{\xi_1}{r}$, $O_2=\dfrac{\eta_1}{r},\dots,O_{2s-1}=\dfrac{\xi_s}{r}$, $O_{2s}=\dfrac{\eta_s}{r}$.

Wir können alles von früher übertragen, nur müssen wir beachten, daß wir jetzt statt (14) den Bereich (20) zu betrachten haben. Es ist z.B., wenn wir $\varepsilon=\log(1+D_N)$ und $\eta=-\log D_N$ setzen, für $i=1,\ldots,s$:

$$\varepsilon\leqq\xi_i^2+\eta_1^2=r^2(O_1^2+O_2^2)\leqq\eta,$$

also $s\varepsilon\leqq r^2\leqq\eta s$, und wir erhalten analog folgendes Resultat:

Es sei mit $k=2s$

$$\mathfrak{F}_N^*=\left|\lambda_N(\phi)-\frac{1}{V(S^k)}\int\phi\,dO\right|, \tag{28}$$

dann ist

$$\mathfrak{F}_N^*\leqq C_s D_N\left(K_N+\left(\log\frac{1}{D_N}\right)^k\int_{E_k}\Omega\,dO\right), \tag{29}$$

wo C_s nur von s abhängt. Man kann $C_s=4000^s$ nehmen. Weiter ist Ω die Summe der Absolutbeträge aller Ableitungen auf E_{2s} bis zur Ordnung $2s$.

Es ist wieder $K_N=\sup_{E_{2s}}|\phi|$.

In (29) haben wir eine Quadraturformel für Integrale über die k-dimensionale Einheitssphäre, wo $k=2s$, also gerade ist. Wir können eine solche Formel auch für ungerades k von der Form $2s-1$ ableiten. Dazu brauchen wir in (19) nur Funktionen f betrachten, welche von $\eta_1,\ldots,\xi_s,\eta_s$, aber nicht von ξ_1 abhängen. Dann erhalten wir sofort aus (19):

Ist $\hat{\mathfrak{F}}_N^{(s)}=\left|\hat\lambda_N(f)-\frac{\sqrt{\pi}}{\pi^s}\int f(\xi_1,\ldots,\xi_s,\eta_s)\,e^{-\sum_{i=2}(\xi_i^2+\eta_i^2)-\xi_1^2}\,d\eta_1\cdot d\eta_s\right|$, dann gilt

$$\hat{\mathfrak{F}}_N^{(s)}\leqq D_N^{(2s)}(K_N+\int^*\hat{L}\,d\eta_1\cdots d\xi_s\,d\eta_s), \tag{30}$$

wobei $\hat{L}(f)=L(f)$ ist. Es ist jetzt

$$\hat\lambda_N(f)=\frac{1}{N}\sum_{n=1}^{N}f(\sqrt{-\log p_1(x_n)}\sin 2\pi q_1(x_n),\sqrt{-\log p_2(x_n)}\cos 2\pi q_2(x_n),\ldots).$$

Betrachtet man jetzt eine Funktion ϕ auf S^{2s-1} und definiert

$$f(\eta_1,\xi_2,\eta_2,\ldots,\xi_s,\eta_s)=\Phi\left(\frac{\eta_1}{r},\frac{\xi_2}{r},\frac{\eta_2}{r},\ldots\right),$$

wo $r^2=\eta_1^2+\xi_2^2+\eta_2^2+\cdots+\eta_s^2$ ist, so erhält man eine analoge Formel wie in (30) auch für ungerades k.

Man kann die Formel (19) auch anwenden, um Quadraturformeln für das kartesische Produkt von Einheitssphären zu gewinnen. Wir wollen der Einfachheit halber annehmen, daß die Einheitssphären $S_l^{(j)}$ alle einander gleich, sagen wir S^m sind, wo $m=2l$ gerade ist. Wenn die Anzahl der Faktoren t ist, dann nehmen wir in (19) $s=lt$.

Es sei ϕ eine Funktion auf $\prod_{j=1}^{t} S_l^j$. Dann betrachten wir die Funktion

$$f(\xi_1,\ldots,\xi_s,\eta_s)=\left(\frac{\xi_1^{(1)}}{r_1},\ldots,\frac{\eta_1^{(1)}}{r_1},\ldots,\frac{\xi_1^{(2)}}{r_2},\ldots,\frac{\eta_1^{(2)}}{r_2},\ldots,\frac{\xi_1^{(t)}}{r_t},\ldots,\frac{\eta_1^{(t)}}{r_t}\right),$$

wo

$$r_1^2=(\xi_1^{(1)})^2+\cdots+(\eta_t^{(1)})^2,\ r_2^2=(\xi_1^{(2)})^2+\cdots+(\eta_t^{(2)})^2,\ldots.$$

Wir erhalten analog wie (26), wenn Φ $2s$-mal stetig differenzierbar ist, eine Quadraturformel für

$$\int_l\cdots\int_l \phi(O_1,\ldots,O_{lm})\,dO_1\cdots dO_l.$$

Literatur

[1] *P. Appell* und *J. Kampe de Feriet*, Fonctions hypergéometriques et hyperspheriques. Polynomes d'Hermite, Paris 1926.

[2] *R. Arenstorf* and *D. Johnson*, Uniform Distribution of Integral Points on 3-Dimensional Spheres via Modular Forms, J. Number Theory **11** (1979), 218—238.

[3] *G. Box-M. Muller*, A note on the generation of random normal deviates, Annals of Math. Statistics, **29** (1958), 610—616.

[4] *R. Courant* und *D. Hilbert*, Methoden der Mathematischen Physik. I, Berlin-Heidelberg-New York 1924.

[5] *W. Feller*, Über die Lösungen der linearen partiellen Differentialgleichungen zweiter Ordnung vom elliptischen Typus, Math. Ann. **102** (1930), 633—649.

[6] *W. Freeden*, Über eine Klasse von Integralformeln der mathematischen Geodäsie, Veröffentl. d. geodätischen Instituts Aachen 1979.

[7] *E. Hlawka*, Weierstraßscher Approximationssatz und Gleichverteilung, Monatsh. f. Math. **88** (1979), 137—170.

[8] *A. V. Malysev*, The distribution of integer points in a four-dimensional sphere (Russisch), Doklady Akad. Nauk SSSR **114** (1957), 25—28.

[9] *A. V. Malysev*, Über die Darstellung ganzer Zahlen durch positive quadratische Formen mit vier und mehr Veränderlichen (Russisch), I. Izvestija Akad. Nauk SSSR, Ser. mat. **23** (1959), 337—364.

[10] *C. Müller*, Spherical Harmonics, Lecture notes in mathematics **17**, Berlin-Heidelberg-New York 1966.

[11] *C. Pommerenke*, Über die Gleichverteilung von Gitterpunkten auf *m*-dimensionalen Ellipsoiden, Acta Arith. **5** (1959), 227—257.

[12] *G. Szegö*, Orthogonal polynomials, New York 1939.

Österreichische Akademie der Wissenschaften, Institut für Informationsverarbeitung, Fleischmarkt 20, A-1010 Wien

Eingegangen 2. März 1981

Corrections

2.1 $\geq 2^L(2V^{-1}(K_m)L^{\frac{m}{2}})^{-1}$

2.2 $\geq 2^{L-1}$

2.3 $\leq e^{-\frac{L\delta^2}{16}}V(S^m)2^L,$

4.4 $c_m = 2V^{-1}(K^m)V(S^m)$

4.5 $(4\sqrt{\frac{m+1}{2}}\sqrt{\frac{\lg L}{L}})$

4.6 $\leq 2\omega_f(4\sqrt{\frac{m+1}{2}}\sqrt{\frac{\lg L}{L}})$

4.7 $k_m \leq 8\sqrt{\frac{m+1}{2}} \quad (L \geq L_0)$

5.8 $\int_{\|x,y\|\geq\delta} K(xy,L)\,dy \leq 2V^{-1}(K^m)V(S^m)e^{-\frac{L\delta^2}{16}}L^{\frac{m}{2}}$

6.9 $= 4\sqrt{\frac{m_j+1}{2}}\sqrt{\frac{\lg L_j}{L_j}}$

6.10 $\leq 2\omega_f(4\sqrt{\frac{m+1}{2}}\sqrt{\frac{\lg L}{L}}) \quad (L \geq L_0)$

6.11 $= \frac{1}{N}\sum_{n=1}^{N} f(x_n)$

7.12 $\le 2\omega_f(4\sqrt{\frac{m+1}{2}}\sqrt{\frac{\lg L}{L}}) + |\varepsilon_N(T_L)|$

8.13 $= \omega_g(\sigma) + A(f)e^{-\frac{L\sigma^2}{16}} L^{\frac{m}{2}} c_m$

8.14 $= M(g)e^{-\frac{L\delta^2}{16}} L^{\frac{m}{2}} c_m$

8.15 $\delta = \sigma = 4\sqrt{\frac{m+1}{2}}\sqrt{\frac{\lg L}{L}}$

8.16 $= \omega_g(4\sqrt{\frac{m+1}{2}}\sqrt{\frac{\lg L}{L}}) + (A(g) + 2M(g))\frac{c_m}{\sqrt{L}}$

10.17 $\le \varepsilon_N(f_2) + \mu(f_2 - f_1),$

10.18 $\ge \varepsilon_N(f_1) - \mu(f_2 - f_1).$

10.19 $\delta = 4\sqrt{\frac{m+1}{2}}\sqrt{\frac{\lg L}{L}}$

10.20 $\le |\varepsilon_N(T_L(f))| + 4M\varepsilon_N(T_L(\iota(E_3))) + R$

10.21 $= M(g)V(E_3)k_{m_1}$

10.22 $= k_{m_2}(\omega_g(4\sqrt{\frac{m+1}{2}}\sqrt{\frac{\lg L}{L}}) + M(g)\sqrt{\frac{\lg L}{L}})$

10.23 $|\varepsilon(g_\rho, \omega_N)| \le |\varepsilon_N(T_L(g_\rho))|$
$+ 4O(\rho^{-\alpha})|\varepsilon_N(T_L(\iota(E_3)))| +$

11.24 $\le \varepsilon_N(T_L(f_\rho)) + O(\rho^{-\alpha})\varepsilon_N(T_L(\iota(E_3)))$
$+ O(\rho^{m-\alpha}\lg\rho)$

11.25 Größenordnung k^{m-1}.

11.26 $= L_k(L)H_{kr}(q),$

11.27 $L_k = \frac{V(S)}{W_L V(S^{m-1})} J(L,k)$

13.28 $= |\varepsilon_N(T_L(f))| \le$

13.29 $\omega_f(4\sqrt{\frac{m+1}{2}}\sqrt{\frac{\lg L}{L}})$

13.30 $\le (M(g) + V(L))\mathrm{Min}(1, \sum_{k=1}^{L} k^{m-1}W_k)$
$+ R(L)$
$\le (M(g) + V(L))\mathrm{Min}(1, \sum |b_{kj}(f)|\,|W_{kj}|)$
$+ R(L)$

13.31 $\le c_1(\cdots) + c_2(\cdots)$

13.32 $\le c_1\sum\cdots + c_2\surd$

14.33 $|\varepsilon_N(T_L(f))| \le$

18.34 und gleichmäßig konvergent, wie man mit dem Satz von Merzer sieht, und es gilt

21.35 $\le R(L_S)$ ist

21.36 $\le D^{(S)}\sum_{k=1}^{L_S} C^2(k)\lambda_k^{S-r} + R(L_S) \le 2R(L_S).$

21.37 wurde $(-1)^{r-1}G^{(r)}(x,y) =$

24.38 $= \sum_{k_1,\dots,k_r=1}^{M} \frac{Z(m,k)}{V(m)\lambda_k^r} P_k(xy) + O(\frac{2}{M^{2r-m}})$

25.39 $\le c_1 \sum_{k}^{M}$

25.40 $\le D^{(s)}c_2 \sum_{k=1}^{L} k^{2(s-r)+m-1} + O(L^{m-2r})$

25.41 $D^{(r)} \le c_2 D^{(s)}$

25.42 daß $D^{(r)} \le c(D^{(s)})^\sigma$ ist, mit $\sigma = \frac{2r-m}{2s}$.

25.43 $= \mu(f) - \int G(x,y)\Delta f(y)\,d\mu(y)$

26.44 $\sum_{k=1}^{L} k^{m-1}|W_k(\omega_N)| \int |\Delta(f)|\,d\mu)$

26.45 $C(y, \rho+\delta) - C(y, \rho-\delta) \le \bar{k}_m\rho^{m-1}\delta$

26.46 $\sum_{k=1}^{L} W_k(\omega_N)|b_k|(\iota(C))k^{m-1})c_m.$

26.47 $O(L^{2s+(m+1)/2})$

26.48 $\lambda(s) = \frac{1}{2s+\frac{m}{2}+1}$

26.49 $\le C(m)D^{1/2\lambda(s)}\log D^{-1}$

26.50 $D^{(s)} \ge c(N\log N)^{-2/\lambda(s)}$

27.51 $\le (M(f) + c\sqrt{\frac{\lg L}{L}})S_L + R(L)$

27.52 $\le (M(f,E) + c\sqrt{\frac{\lg L}{L}})L^{2m}D_N(C) + R(L)$

28.53 $\omega_f(4\sqrt{\frac{m+1}{2}}\sqrt{\frac{\lg L}{L}})$

29.54 $\le cs^{-\frac{m}{4(3m+2)}}\lg s$

29.55 $\le O(\frac{\mathrm{Max}(1, M^{2m-2r})}{s^{m/4}}) + O(\frac{1}{M^{2r-m}})$

32.56 $= |\mathrm{Det}(H_{j,r}(e_z))| \ne 0$

32.57 $|\alpha_{zkj}| \le \frac{k^{(m-1)\frac{Z-1}{2}}}{D(e)}$

32.58 $c_L = (1, ie_{z_L}(m_L))$

33.59 wie bei 32.57 jeweils den Faktor $(m-1)$ bzw. $(m_j - 1)$ mit $1 \le j \le t$ passend einfügen.

33.60 gleich $H_{j,r}^{(m-1)}(x'')\int_{-1}^{1}$

35.61 "$x_1 = \cos\dots\dots < \delta_j < 1.$" weglassen.

36.62 $\frac{\partial(x,y)}{\partial(\xi,\eta)} = \frac{1}{\pi}\exp(-(\xi^2+\eta^2)) = x,$
da $\frac{\partial(\xi,\eta)}{\partial(x,y)} = \frac{\pi}{x}.$

36.63 $= -\frac{\pi}{x}[(f_{\eta\eta} - f_{\xi\xi})\sin 2\pi y\cos 2\pi y$
$+ f_{\xi\eta}(\cos^2 2\pi y - \sin^2 2\pi y)]$
$+\pi(x\sqrt{-\ln x})^{-1}(f_\xi \sin 2\pi y - f_\eta \cos 2\pi y).$

37.64 $= \frac{1}{\pi}\int |F_{xy}|x\,d\xi d\eta$
$\le \int\{[|f_{\xi\eta}| + |f_{\xi\xi} - f_{\eta\eta}|]$
$+ (-\ln x)^{-1/2}(|f_\xi| + |f_\eta|)\}\,d\xi d\eta$

37.65 $I_2 = \int$

37.66 $\le \int_B (\Delta^2 f + (\xi^2+\eta^2)^{-1/2}\Delta f)\,d\xi d\eta$

37.67 $|\lambda_N(f) - \frac{1}{\pi}\int_B$

37.68 $I_4 = \int_B (\Delta^2 f + (\xi^2+\eta^2)^{-1/2}\Delta f)\,d\xi d\eta$

39.69 $\le D_N(2\pi K_N + \int\int^* I\,d\xi d\eta)$

41.70 wie bei 39.69 2π einfügen.

Eine Bemerkung zur Theorie der Gleichverteilung

Dem Andenken Paul Turáns gewidmet

Studies in Pure Mathematics, Akadémiai Kiado, Budapest, 1983, pp. 337–345

Es sei ω eine Dreiecksfolge;

$$(x_{N1}, x_{N2}, \ldots, x_{NN}) \tag{1}$$

für $N = 1, 2, \ldots$ Die x_{Nk} sind reelle Zahlen in einem Intervall $I = {<}a, b{<}$. Es sei ω zur Dichte ρ gleichverteilt, wo ρ eine im Riemannschen Sinne integrierbare Funktion auf $I = {<}a, b{>}$ ist. Es sei ρ nicht negativ, wobei

$$\lambda(I) = \int_a^b \rho\, dx > 0$$

ist. Dann gelte also für jede (im Riemannschen Sinne) integrierbare Funktion g auf I

$$\lim_{N\to\infty} \frac{1}{N} \sum_{k=1}^{N} g(x_{Nk}) = \frac{1}{\lambda(I)} \int_I g(x)\, \rho(x)\, dx\,. \tag{2}$$

Wir ordnen nun für jedes feste N die Zahlen in (1) der Größe nach

$$z_{N1}, z_{N2}, \ldots, z_{NN}\,. \tag{3}$$

Wir wollen die so für $N = 1, 2, \ldots$ entstehende Folge mit $\hat{\omega}$ bezeichnen.

Es sei nun weiter eine Folge (α_k) komplexer Zahlen gegeben. Sie sei beschränkt, d. h. es existiert ein M, so daß für alle k

$$|\alpha_k| \leqq M\,. \tag{4}$$

(Wenn alle α_k reell und nicht negativ sind, so setzen wir (4) nicht voraus.) Weiters existiere

$$\lim_{N\to\infty} \frac{1}{N} \sum_{k=1}^{N} \alpha_k = \alpha\,. \tag{5}$$

Dann behaupten wir folgenden

Satz: *Für jede im Riemannschen Sinne integrierbare Funktion existiert*

$$\lim_{N\to\infty} \frac{1}{N} \sum_{k=1}^{N} \alpha_k f(z_{Nk})$$

und es ist

$$\lim_{N\to\infty} \frac{1}{N} \sum_{k=1}^{N} \alpha_k f(z_{Nk}) = \frac{\alpha}{\lambda(I)} \int_a^b f(x)\,\rho(x)\,dx\,. \tag{6}$$

Wir wollen folgenden Spezialfall betrachten. Es sei ρ auf I stets gleich 1, weiter sei

$$\bar{x}_{Nk} = a + \frac{k}{N}(b-a) \quad \text{für} \quad k = 0, 1, \ldots, N\,.$$

Es sei (α_k) eine Folge von Zahlen der Gestalt 0 oder 1, welcher man die Zahl

1 $$t = 0, \alpha_1, \alpha_2, \ldots$$

zuordnen kann. Für diese Folge ist (4) erfüllt. Weiters gelte (5). Betrachten wir nun eine Folge $\omega = (x_{Nk})$ $(k = 1, \ldots, N)$, so daß stets x_{Nk} im Intervall

$$< \bar{x}_{Nk-1}, \bar{x}_{Nk} <$$

liegt, so ist $\omega^* = \omega$; nach Definition des Riemannschen Integrals gilt (2) mit $\rho = 1$, und es ist dann noch (6), mit $\lambda(I) = b - a$

$$\lim_{N\to\infty} \frac{1}{N} \sum_{k=1}^{N} \alpha_k f(x_{Nk}) = \frac{\alpha}{\lambda(I)} \int_a^b f(x)\,dx\,. \tag{6'}$$

Diese Formel (6') wurde erstmals von CARR und HILL 1951 in *Am. Proc.*, **2** (242–245) gezeigt. Sie nennen die linke Seite von (6') das *Pattern integral* von f. Das „Muster" wird dabei durch die Zahl t gegeben.

Wir wollen, bevor wir weitere Bemerkungen zu (6) geben, zunächst (6) beweisen. Wir werden dabei eine Methode benützen, wie sie in der Theorie der Gleichverteilung üblich ist. Sie dürfte sogar im Falle (6') etwas einfacher sein, als die Methode von CARR und HILL. Wir können ohne Beschränkung der Allgemeinheit annehmen, daß I das Intervall $E = 0, 1<$ und $\lambda(I) = 1$ ist.

Wenn f auf E eine Konstante k ist, dann ist (6) richtig und folgt sofort aus (5). Wir setzen nun zunächst voraus, daß die α_k alle nicht negativ sind. Es sei nun f die Indikatorfunktion i des Teilintervalls $<0, c<$, wo $0 < c < 1$ ist. Es sei nun die natürliche Zahl $l = l(N)$, so gewählt, daß

$$z_{Nl} \leqq c < z_{N,l+1} \tag{7}$$

ist. Dann ist

$$\frac{1}{N}\sum_{k=1}^{N}\alpha_k i(z_{Nk})=\frac{1}{N}\sum_{k-1}^{l(N)}\alpha_k. \tag{8}$$

Nun ist

$$\frac{1}{N}\sum_{k=1}^{l(N)}\alpha_k=\frac{l(N)}{N}\,\frac{1}{l(N)}\sum_{k=1}^{l(N)}\alpha_k.$$

Es ist nun, nach Definition von 1 (vgl. (7))

$$l(N)=\sum_{k=1}^{N} i(z_{Nk})=\sum_{k=1}^{N} i(x_{Nk}) \tag{9}$$

denn es ist (3) nur eine Umordnung von (1) und die Summe in (9) ist eine symmetrische Funktion. Nach (2) gilt aber

$$\lim_{N\to\infty}\frac{l(N)}{N}=\lim_{N\to\infty}\frac{1}{N}\sum_{k=1}^{N} i(x_{Nk})=\int_0^1 i(x)\rho(x)dx. \tag{2'}$$

Weiters ist nach (5)

$$\lim_{N\to\infty}\frac{1}{l(N)}\sum_{k=1}^{l(N)}\alpha_k=\alpha$$

und damit ist (6) für f bewiesen, wenn $l(N)$ nicht beschränkt ist. Ist aber von einem N_0 an $l(N)=l_0$ fest, dann ist nach (2')

$$0=\lim_{N\to\infty}\frac{l_0}{N}=\int_0^1 i(x)\rho(x)dx=\int_0^c \sigma(x)dx.$$

Dann ist aber (6) auch richtig.

Es gilt (6) auch dann, wenn i_1 die Indikatorfunktion des Intervalls $<c, 1<$ ist, denn es ist doch, wenn i die gleiche Bedeutung wie vorher hat.

$$i_1=1-i.$$

Es ist ja (6) linear und homogen in f. Liegt ein Intervall $<c_1, c_2<$ aus E vor und ist i_3 seine Indikatorfunktion so ist doch $i_3=i_2-i_1$, wo i_2 die Indikatorfunktion von $<0, c_2<$ und i_1 jene von $<0, c_1<$ ist. Daraus folgt nun, daß (6) für alle Treppenfunktionen bewiesen ist.

Um nun (6) für alle im Riemannschen Sinne integrierbare Funktionen f zu beweisen, benützen wir die Tatsache, daß es zu jedem solchen f und zu jedem $\varepsilon>0$

Treppenfunktionen f_1 und f_2 gibt, so daß in E

$$(10)\qquad f_1 \leqq f \leqq f_2$$

ist und

$$(11)\qquad \int_0^1 (f_2 - f_1)\, dx < \varepsilon$$

ist. Ist M eine obere Schranke von ρ in E, so ist daher auch

$$(11')\qquad \int_0^1 (f_2 - f)\, \rho(x)\, dx \leqq \int_0^1 (f_2 - f_1)\, \rho\, dx \leqq M\varepsilon$$

und

$$(11'')\qquad \int_0^1 (f - f_1)\, \rho\, dx \leqq M\varepsilon$$

Nun benützen wir *erstmals*, daß die α_k nicht negativ sind. Es ist dann nämlich, wenn wir

$$\lambda_N(f) = \frac{1}{N} \sum_{k=1}^{N} \alpha_k f(z_{Nk}) \quad \text{setzen}$$

$$(12)\qquad \lambda_N(f_1) \leqq \lambda_N(f) \leqq \lambda_N(f_2)\,.$$

Weiter ist nach (6) angewendet auf f_1 und f_2 (für Treppenfunktionen haben wir ja schon (6) bewiesen) für großes N

$$(13)\qquad |\lambda_N(f_k) - \alpha\lambda(f_k)| < \varepsilon$$

wo

$$\lambda(g) = \int_0^1 g\rho\, dx$$

ist. Es ist also nach (13) für großes N und nach (11′)

2 $$\lambda_N(f_2) \leqq \alpha\lambda(f_2) + \varepsilon \leqq \alpha\lambda(f) + (M+1)\,\varepsilon$$

und nach (13) und (11″)

3 $$\lambda_N(f_1) \geqq \alpha\lambda(f_1) - \varepsilon \geqq \alpha\lambda(f) - (M+1)\,\varepsilon\,.$$

Es ist also und (12) für großes N

4 $$|\lambda_N(f) - \alpha\lambda(f)| \leqq (M+1)\,\varepsilon$$

und dies ist die Behauptung (6).

Es sei nun die Folge (α_k) eine beliebige reelle Folge mit (4) und (5), dann ist die Folge $(\alpha'_k) = (\alpha_k + M)$ eine nicht negative Folge mit

$$\lim_{N\to\infty} \frac{1}{N} \sum_{k=1}^{N} \alpha'_k = \alpha + M$$

und wir erhalten nach dem eben Bewiesenen

$$\lim_{N\to\infty} \frac{1}{N} \sum_{k=1}^{N} \alpha'_k f(z_{Nk}) = (\alpha + M)\,\lambda(f)\,.$$

Da

$$\lim_{N\to\infty} \frac{1}{N} \sum f(z_{Nk}) = \lambda(f)$$

wieder nach (6), mit $\alpha_k = 1$ für alle k, so folgt durch Subtraktion jetzt die Behauptung für jede reelle Folge (α_k). Ist jetzt allgemein (α_k) eine beliebige Folge mit (4) und (5), so folgt daß die Folge (Re *al* α_k) und (Im α_k) die Behauptung (6) erfüllen und damit (α_k) selbst. Damit ist der Satz vollkommen bewiesen.

Wir wollen gleich noch auf eine Verallgemeinerung hinweisen.

Satz 2: *Es seien die Voraussetzung* (1), (2), (3) *erfüllt. Es sei weiter* (a_k) *eine Folge von Vektoren in* C^s, *so daß ein* M *existiert, so daß*

(14) $$|a_k| \leqq M$$

und

(15) $$\lim_{k\to\infty} \frac{1}{N} \sum_{k=1}^{N} a_k = a$$

dann gilt

5 (16) $$\lim_{N\to\infty} \frac{1}{N} \sum_{k=1}^{N} a_k f(z_{Nk}) = a \int_I f\rho\, dx\,.$$

Der Beweis folgt aus (6) angewendet auf die Komponenten der Vektorenfolge.

Eine weitere Verallgemeinerung erhalten wir dadurch, daß wir statt dem eindimensionalen Intervall I ein l-dimensionales Intervall I^l nehmen.

Weiters seien jetzt l-Folgen $\omega^{(g)}$ $(g=1, \ldots, l)$ durch

(17) $$\omega^{(g)}=(x_{N1}^{(g)}, \ldots, x_{NN}^{(g)})$$

gegeben.

Es sei

(18) $$\lambda_{N_1 \ldots N_l}(g)=\frac{1}{N_1, N_2, \ldots, N_l} \sum_{k_1=1, \ldots, k_l=1}^{N_1, \ldots, N_l} g(x_{N_1, k_1}, \ldots, x_{N_l, k_l}).$$

Es gelte, wo

$$\lambda(I)=\int_{I^l} \rho \, dx_1, \ldots, dx_l$$

(19) $$\lim_{(N_1 \ldots N_l) \to \infty} \lambda_{N_1 \ldots N_l}(g)=\frac{1}{\lambda(I^l)} \int_{I^l} g(x_1 \ldots x_l) \, \rho(x_1 \ldots x_l) \, dx_1 \ldots dx_l$$

6 wo ρ eine Dichtefunktion auf I^l ist. Es gelte (18) und (19) für jede auf I^l (Hülle von I^l) im Riemannschen Sinn integrierbare Funktion. Wir bilden uns nach (3) zu den Folge $\omega^{(j)}$ die Folgen $\hat{\omega}^{(j)}$. Es sei weiter eine l-fache beschränkte Folge

$$(\alpha_{k_1, \ldots, k_l})$$

gegeben mit

(20) $$\lim_{(N_1 \ldots N_l) \to \infty} \frac{1}{N_1 \ldots N_l} \sum_{k_1=1, \ldots, k_l=1}^{N_1, \ldots, N_l} \alpha_{k_1 \ldots k_l}=\alpha$$

dann gilt wieder

7 (21) $$\lim_{(N_1, \ldots, N_l) \to \infty} \frac{1}{N_1 \ldots N_l} \sum_{k_1 \ldots k_l}^{N_1 \ldots N_l} \alpha_{k_1 \ldots k_l} f(z_{Nk_1}^{(1)}, \ldots, z_{Nk_l}^{(2)})=\alpha \int_{I^l} f\rho \, dx_1 \ldots dx_l$$

Der Beweis erfolgt wie vorher.

Wir wollen nun einige Folgerungen aus (6) besprechen. Aus (6) folgt zunächst einmal daß mit (x_{Nk}) auch (z_{Nk}) zur Dichte ρ gleichverteilt ist. Nehmen wir weiters eine Folge

8 (y_k) welche zur Dichte σ gleichverteilt ist, d. h. für welche jetzt gilt (g komplexwertig)

9 $$\lim_{N \to \infty} \frac{1}{N} \sum_{k=1}^{N} g(y_k)=\int_I g(x) \, \sigma(x) \, dx$$

und setzen wir in (6)

10 $$\alpha_k=\bar{g}(y_k)$$

($\bar{g}$ komplex konjugiert zu g).

So erhalten wir

$$(21)\qquad \lim_{N\to\infty}\frac{1}{N}\sum_{k=1}^{N}\bar{g}(x_k)f(z_{Nk})=\int \bar{g}(x)\,\sigma(x)\,dx\int f\rho\,dx\,.$$

Interessant erscheint der Fall, daß die Folge (x_{Nk}) gerade eine Folge (x_k) ist, dann entsteht die Folge (z_{Nk}) dadurch daß für jedes N die N Zahlen $x_1,\ldots,x_N$ nach der *Größe geordnet* werden. Wir haben dann

$$(22)\qquad \lim_{N\to\infty}\frac{1}{N}\sum_{k=1}^{N}\bar{g}(x_k)f(z_{Nk})=\int \bar{g}\,\rho\,dx\int f\rho\,dx$$

und insbesondere

$$(23)\qquad \lim_{N\to\infty}\frac{1}{N}\sum_{k=1}^{N}\bar{f}(x_n)f(z_{Nk})=\left|\int f\rho\,dx\right|^2.$$

Interessant ist auch der Fall in (20) daß $\sigma=1$ ist und wir

$$g(x)=e^{2\pi ihx}$$

nehmen ($h\neq 0$, $h\in\mathbf{Z}$). Dann ist

$$\int_0^1 g\,dx=0$$

und wir haben für jede gleichverteilte Folge (x_{Nk})

$$(24)\qquad \lim_{N\to\infty}\frac{1}{N}\sum_{k=1}^{N}e^{2\pi ihx_k}f(z_{Nk})=0\,.$$

Nehmen wir z. B. $x_k=\beta k$ mod 1, wo β irrational ist, so haben wir für die Folge (1)

$$(25)\qquad \lim_{N\to\infty}\frac{1}{N}\sum_{k=1}^{N}e^{2\pi ih\beta k}f(z_{Nk})=0\,.$$

Es ist also z. B.

$$\lim_{N\to\infty}\frac{1}{N}\sum_{k=1}^{N}e^{2\pi i\beta k}f\left(\frac{k}{N}\right)=0\,.$$

Wir bemerken noch folgendes: Es kann (5) nicht durch

$$\lim_{N\to\infty}\sum_{k=1}^{N}\alpha_{Nk}=\alpha$$

ersetzt werden, so daß (6) richtig bleibt. Wäre dies der Fall, so könnte man $\alpha_{Nk}=\frac{1}{N}f(z_{Nk})$ nehmen und wir hätten

$$\lim_{N\to\infty}\frac{1}{N}\sum_{k=1}^{N}f^2(z_{Nk})=\left(\int_0^1 f\rho\right)^2 dx .$$

Andererseits wäre nach (6) ebenfalls mit f^2 statt f

$$\lim\frac{1}{N}\sum_{k=1}^{N}f^2(z_n)=\int_0^1 f^2\,\rho\,dx .$$

Für $\rho=1$ und $f(x)=x$ für alle x in E hätten wir sofort $3=4$. Aus (20) folgt, wenn

$$\alpha_{k_1\ldots k_l}=e^{2\pi i(\beta_1 k_1+\ldots+\beta_l k_l)}$$

(*ein* β irrational), daß analog zu (24)

$$\lim_{(N_1\ldots N_l)\to\infty}\frac{1}{N_1\ldots N_l}\sum_{k_1\ldots k_l}^{N_1\ldots N_l}e^{2\pi i(\beta_1 k_1+\ldots+\beta_l k_l)}f(z^{(1)}_{Nk_1},\ldots,z^{(l)}_{Nk_l})=0 ,$$

also insbesondere

$$\lim_{(N_1\ldots N_l)\to\infty}\frac{1}{N_1\ldots N_l}\sum_{k_1\ldots k_l}^{N_1\ldots N_l}e^{2\pi i(\beta_1 k_1+\ldots+\beta_l k_l)}f\left(\frac{k_1}{N_1},\ldots,\frac{k_l}{N_l}\right)=0 .$$

Bemerkung: Wir können statt l-fachen Folgen auch den Fall betrachten, daß in (18) $N_1=N_2=\ldots=N_l$ ist, dann gilt wenn (20) beibehalten wird, daß

$$\lim_{N\to\infty}\frac{1}{N}\sum_{k_1\ldots k_l}^{N\ldots N}\alpha_{k_1\ldots k_l}f(z^{(1)}_{Nk_1},\ldots,z^{(2)}_{Nk_l})=\alpha\int_{I^l}f\rho\,dx_1\ldots dx_l .$$

Schlußbemerkung: Es sei (y_k) eine Folge eines kompakten, metrischen Raumes X, welcher zu einem Wahrscheinlichkeitsmaß auf X gleichverteilt ist. Es gilt also für jede stetige Funktion g auf X

$$\lim_{N\to\infty}\frac{1}{N}\sum_{k=1}^{N}g(y_k)=\int_X g(x)\,d\mu .$$

Aus (6) folgt mit $\alpha_k = \bar{g}(y_k)$

$$\lim_{N\to\infty} \frac{1}{N} \sum_{k=1}^{N} \bar{g}(y_k) f(z_{Nk}) = \frac{1}{\lambda(I)} \int_X g(x)\, d\mu \int_a^b f(x)\, \rho(x)\, dx$$

11 Ist $X = \{0, 1\}$, $\int_X g(x)\, d\mu = \frac{g(0)+g(1)}{2}$, dann erhält das Resultat von Carr und Hill.

Corrections

338.1 $t = 0, \alpha_1\alpha_2\cdots$

340.2 $(\alpha M + 1)$

340.3 wie 340.2

341.4 wie 340.2

341.5 $= \frac{a}{\lambda(I)} \int_I f\rho\, dx.$

342.6 auf $\bar{I}^l$ (Hülle von I^l)

342.7 $= \frac{\alpha}{\lambda(I^l)} \int_{I^l} f\rho\, dx_1 \cdots dx_l$

342.8 x_k statt y_k

342.9 wie 342.8

342.10 wie 342.8

345.11 $\int_X g(x)\, d\mu = (1-\alpha) g(0) + \alpha g(1)$, dann erhält man das

Carl Ludwig Siegel

31.12.1896 – 4.4.1981

J. Number Theory **20** (1985), 373–404

Received October 5, 1984

Carl Ludwig Siegel wurde als einziges Kind seiner Eltern am 31/12/1896 in Berlin geboren. Nach dem Besuch der dreijährigen Volksschule und von 3 Jahren Realschule machte er das Eintrittsexamen an der Oberrealschule. Darüber berichtet er: "Besonders der Unterricht in Mathematik war schlecht, ich fürchtete beim Übertritt in die Oberrealschule deswegen in Schwierigkeiten zu kommen. Mit dem Alter von 14 Jahren hatte ich begonnen, selbstständig Mathematik zu lernen und geometrische Aufgaben zu lösen. Andererseits hatte ich schon frühzeitig Freude und Geschick beim Zeichnen, weshalb ich bei einem guten Mathematikunterricht vielleicht doch Maler geworden wäre!"

Nach seiner Erzählung hatte er auch ein schlechtes Verhältnis zu seinem Mathematiklehrer. Bei Schularbeiten begann er sofort mit dem schwierigsten Beispiel und erreichte immer die Lösung des Problems. Dafür blieb für die leichten Beispiele wenig Zeit, der Lehrer aber bewertete alle Beispiele gleich. Nach seinem Abitur 1915 nahm Siegel das Studium der Mathematik, der Theoretischen Physik, und Astronomie an der Berliner Universität auf. Zur besseren Vorbereitung wollte er von der Städtischen Bücherei das Buch von Weber "Algebra" entlehnen. Da aber die beiden ersten Bände bereits entliehen waren, erhielt er nur den 3.Band. Der Inhalt war wohl wichtig, aber sehr schwer. Oft erzählte er, wie er im 1. Semester zur Vorlesung über Zahlentheorie von Frobenius kam. Im 3. Semester—Frühjahr 1917—war er durch eine Bemerkung in einem Vortrag von I. Schur zum Studium der Arbeit von Thue im Crelle'schen Journal 1909 angeregt worden. Die Bemerkung bestand darin, daß im Gegensatz zur Pellschen Gleichung, auf der Kurve mit der Gleichung $x^3 - dy^3 = 1$ nur endlich viele Gitterpunkte liegen, obwohl sich die Kurve ins Unendliche erstreckt. Diese von I. Schur zitierte Arbeit von Thue war nur von ganz wenigen Mathematikern studiert worden. Thue, der geniale Außenseiter, war auf diese Probleme gekommen durch den Versuch, die Fermatsche Vermutung zu beweisen. Wie Siegel erzählte, hat sich Thue noch in seinem letzten Lebensjahr, im Krankenhaus, damit beschäftigt. Siegel sagte selbst, daß er beim Studium zunächst durch die vielen Parameter ganz verwirrt war. Aber mit eiserner Energie gelang es ihm, durch Einführung eines weiteren Parameters, eine Verschärfung des Resultats von Thue zu erzielen. Er schrieb sein Ergebnis in konzentrierter Form nieder, dabei so daß ihm I. Schur die Arbeit ohne genaues Studium zurückgab, was ihn sehr enttäuschte. Inzwischen wurde Siegel zum Militär nach Straßburg einberufen, allerdings nach fünf—für ihn sehr schweren Wochen—wieder entlassen. Zunächst war er dann als Hauslehrer tätig, nahm aber im Sommersemester 1919 wieder sein Studium an der Universität Göttingen auf. Für E. Landau mußte er seine Entdeckung nochmals umarbeiten, bis dieser endlich zufrieden war. Er promovierte damit am 2/6/1920. Landau hob stolz hervor, daß er die Entdeckung Siegels unterstützt habe, während die Berliner Kollegen die Arbeit nicht verstanden hätten. Dieser Seitenhieb Landaus erklärt sich daraus, daß es immer Landaus Wunsch war, in einer Großstadt tätig zu sein. 1920/21 bekam Siegel einen Lehrauftrag bei E. Hecke, nahm an dessen Seminar teil, hielt sich aber nach Aussagen H. Behnkes vom Mathematischen Leben in Hamburg fern. Mit Begeisterung kehrte er nach Göttingen zurück und war 3 Semester lang Assistent bei R. Courant. Darüber sagte er in seinem Lebenslauf "diesem gütigen Menschen und tüchtigen Gelehrten bin ich während der kommenden fünfzig Jahre bis zu seinem Tod in Dankbarkeit verbunden geblieben."

Mit den Arbeiten zur additiven Zahlentheorie algebraischer Zahlkörper habilitierte er sich am 10/12/1921, wurde aber schon am 1/8/1922 als Nachfolger von Schönfliess Professor in Frankfurt. Hier schloß er enge Freundschaft mit Dehn, Hellinger, Szasz, und Epstein und er nahm mit Begeisterung am mathematischen, historischen Seminar teil, das ihn sehr beeinflußte. Auch A. Weil war Teilnehmer dieses Seminars. Wegen Überanstrengung mußte Siegel aber seine wissenschaftliche Tätigkeit unterbrechen. Er hatte 1921 und 1922 dreizehn Publikationen veröffentlicht. 1926 begann er wieder wissenschaftlich zu arbeiten. 1930 war er ein Semester in Göttingen. Im Jahre 1933 bewirkten die politischen Ereignisse, daß viele seiner Kollegen ihre Tätigkeit an der Hochschule einstellen mußten. Obwohl er davon nicht betroffen war, erschütterten ihn diese Vorgänge so, daß er 1935 in die USA ging. Am 1/1/1938 kehrte er nach Deutschland zurück und nahm eine Professur in Göttingen an. Diese Berufung war auf Betreiben des damaligen Direktors H. Hasse erfolgt. Bald aber traten zwischen diesen beiden Persönlichkeiten, die sich schon seit ihrem Studium kannten und schätzten, Spannungen auf. Dies zeigte sich bereits im gemeinsamen Seminar, das dem Studium der Siegelschen großen Arbeit aus den Denkschriften der Preußischen Akademie gewidmet war. Eine Habilitationsangelegenheit—die berühmte Affäre Ziegenbein—erhöhte die persönlichen Differenzen, die dann zu einer grundsätzlichen Aussprache und zum endgültigen Bruch der beiden führte. Siegel entschloß sich, Deutschland zu verlassen und ging im Frühjahr 1940 über Norwegen an das Institute for Advanced Study in Princeton. Bis 1945 hatte er hier ein Forschungsstipendium und wurde ab 1945 Permanent Member. Im Winter 1946/47 kam er mit dem ersten Soldatenzug nach Göttingen als Gast zurück. 1951 wurde er wieder Professor in Göttingen, ließ sich 1959 emeritieren, übte aber bis 1967 seine Vorlesungstätigkeit aus. In dieser Zeit hielt er mehrmals Gastvorlesungen am Tata Institute of Fundamental Research in Bombay. Er sagte immer, daß er sich in Indien sehr wohl gefühlt hat. Am 4/4/1981 starb er in Göttingen.

Er hat viele Ehrungen erhalten, war mehrfacher Ehrendoktor, so eben auch an der ETH-Zürich und in Basel. Die Ehrenmitgliedschaft am Tata Institute schätzte er besonders, da damals außer ihm nur Dirac diese Ehrung hatte. 1963 wurde ihm der Orden Pour le Mérite verliehen. Er ist gerne gereist, hielt an vielen Hochschulen Vorträge, beklagte sich aber in späteren Jahren, daß es ihm seine Gesundheit nicht mehr erlaube, große Reisen zu unternehmen. Er liebte die Schweizer Berge, verbrachte immer seine Ferien in diesem schönen Land, war jahrelang in Pontresina und zuletzt in Flims. In den letzten Jahren mußte er sich allerdings mit Spaziergängen in Göttingen begnügen.

Jetzt möchte ich kurz die wissenschaftlichen Arbeiten von Siegel besprechen. Vor allem möchte ich die Arbeiten des jungen Siegel behandeln, da diese für sein späteres Werk doch entscheidend geblieben sind. Er hat selbst zum Ausdruck gebracht—ungefähr 1975—daß er seine wissenschaftliche Konzeption schon vor dreißig Jahren ins Auge gefaßt hat, deren Ausführungen er später verwirklicht hat. Er war aktiv tätig, bis zum 80. Lebensjahr in dem er sich entschloß, nichts mehr zu publizieren. Die Arbeiten die noch nicht vollständig ausgeführt waren und nur in Ansätzen vorlagen, hat er vernichtet, sodaß kein wissenschaftlicher Nachlaß vorhanden ist.

1. Thue–Siegel

Wir wollen nun zum Besprechen des Werkes von Siegel übergehen und dieses in mehrere Abschnitte gliedern: Der erste Abschnitt soll sich mit dem Problemkreis "Thue–Siegel," wie er heute genannt wird, beschäftigen.

Er gliedert sich in zwei Teile: (1) *Approximation algebraischer Zahlen durch rationale Zahlen*, bzw. allgemeiner! durch algebraische Zahlen nach unten und (2) *Anwendung auf Diophantische Gleichungen.*

Wir wollen uns zunächst der ersten Frage zuwenden!

Es ist wohl bekannt, daß sich jede irrationale Zahl durch rationale

Zahlen p/q bis auf den Fehler q^{-2} approximieren läßt. Liouville hat nun gezeigt, daß sich jede reelle algebraische Zahl von einem Grad $g \geqslant 2$ nicht besser approximieren läßt, als cq^{-g} (Zahlen, die sich besser approximieren lassen, sind dann notwendigerweise transzendente Zahlen, die heute auch Liouvillesche Zahlen genannt werden). Man sieht sofort, daß dies für $g = 2$, für quadratische Irrationalitäten, auch die beste Aussage ist. A. Thue hat in seiner, schon vorher zitierten, berühmten Arbeit 1909 gezeigt, daß man diese untere Abschätzung zu $g/2 + 1 + \varepsilon$ $(\varepsilon > 0)$ verbessern kann, wenn $g > 2$ ist. Bei dem Beweis dieser Abschätzung muß er das Schubfach-prinzip benützen, sodaß es nicht möglich ist, die zugehörige Konstante c die jetzt auch von ε abhängt, konstruktiv zu bestimmen. Allerdings war es ihm in vorhergehenden Arbeiten für gewisse Irrationalitäten gelungen, auch ohne das Schubfachprinzip auszukommen. Siegel gelingt es nun, in seiner Dissertation (Arbeit Nr. 1 und 2)[1] die Zahl $g/2 + 1$ durch $2\sqrt{g}$ zu ersetzen, ja sogar etwas noch Besseres zu erreichen, allerdings wieder unter Verwendung des nicht konstruktiven Schubfachprinzipes. Er betrachtet darüber hinaus in diesen Arbeiten auch Verallgemeinerungen, wo die rationale Zahl p/q durch eine algebraische Zahl, natürlich von einen Grad $< g$ ersetzt wird. Es gelingt ihm außerdem in diesen Arbeiten unter zusätzlichen Voraussetzungen die vorher angegebene Schranke zu verkleinern. Diese zusätzlichen Voraussetzungen konnten allerdings erst von F. J. Dyson 1947 beseitigt werden, indem er die Siegelsche Schranke durch $\sqrt{2g}$ ersetzen konnte. Diese Verschärfung erscheint geringfügig, wenn man das Ziel vor Augen hat, die Schranke 2 zu erreichen, wie es schon frühzeitig vermutet wurde.

Die neue Technik von Dyson—kräftig verallgemeinert von K. F. Roth—ermöglichte es nun 1955 dieses Ziel zu erreichen. Sie ist aber wieder von nicht konstruktiver Art. Man muß aber beachten, daß zu dieser Schranke von Roth noch das ε hinzukommt. Es ist sicher nicht möglich für algebraische Zahlen vom Grad $g \geqslant 3$ das ε zu streichen. Vermutet wurde schon seit längerer Zeit aus maßtheoretischen Gründen, daß man $q^{2+\varepsilon}$ durch $q^2 \log q$ verbessern kann. Eine weitere Verschärfung wäre dann im allgemeinen unmöglich. Der russische Mathematiker Buluta hat nun in mehreren Vorträgen angekündigt, dieses schärfere Resultat erreichen zu können und sogar in konstruktiver Weise; ob das stimmt, kann im Augenblick nicht gesagt werden. Man bemüht sich, wie die letzte Tagung in Oberwolfach gezeigt hat, die Schranken von Siegel für gewisse Irrationalzahlen konstruktiv, ja sogar die Schranke von Dyson zu erreichen. Hier ist vor allem das Werk von Bombieri hervorzuheben. Wenn man nur ein etwas stärkeres Resultat, als das Liouvillesche erreichen will,

[1] Die Arbeiten von C. L. Siegel sind zitiert gemäss der in Bd. 4 der *Gesammelten Werke* (Springer, 1979), S. 331–336 gegebenen Liste.

so hat schon Thue gezeigt, wie man das in Spezialfällen erreichen kann; Gelfond hat dieses Programm weiter verfolgt und A. Baker hat dies auch erreicht. Er zeigt ($g \geqslant 3$) konstruktiv $|\alpha - p/q| > q^{-g} \exp(\log^{1/(g+2)} q)$ in berühmten Arbeiten, die ihm die Fields–Medaille eingebracht haben. Man kann nun fragen und damit kommen wir zum 2. Teil dieses Fragekreises, warum diese Verschärfungen so interessant sind. Nun ist die Approximation von Irrationalzahlen durch rationale Zahlen stets von grundsätzlicher Bedeutung, wie schon die Pythagoräer gewußt haben, die sich mit der Approximation von Quadratwurzeln beschäftigt haben und zwar mit Hilfe der Lösungen Diophantischer Gleichungen, also der Pellschen Gleichung und verwandter Gleichungen $x^2 - dy^2 = m$, wobei das m dem Betrag nach klein gewählt wurde. Sie finden dies sehr ausführlich in den Büchern von van der Waerden und auch in dem neuen Buch von A. Weil behandelt. Die Verschärfungen von Thue und Siegel usw. dienen nun dazu, aufzuzeigen, daß gewisse Diophantische Gleichungen höheren Grades nur endlich viele ganzzahlige Lösungen haben. Betrachten wir z.B. die Gleichung

$$\begin{aligned} y^g f\left(\frac{x}{y}\right) &= a_0 x^g + a_1 x^{g-1} y + \cdots + a_g y^g \\ &= a_0(x - \alpha_1 y) \cdots (x - \alpha_g y) = C, \end{aligned}$$

wobei $f(\xi)$ irreduzibel ist. Unter der Annahme, daß diese Gleichung unendlich viele Lösungen hat, folgt, daß die Ungleichung (für ein α_j)

$$\left| \frac{x}{y} - \alpha_j \right| \leqslant \frac{c}{y^g}$$

unendlich viele Lösungen hat, was z.B. der Ungleichung von Thue widerspricht. Für die Anzahl A der Lösungen erhält man $A \leqslant (4H)^{2g^2} e^3 + \exp(643 g^2)$ (H = Höhe von f). Mit der Methode von Baker erhält man für die Lösungen (x, y) der Gleichung die Abschätzung: $\max(|x|, |y|) \leqslant \exp[(gH)^{(10g)^5} + (\log |c|)^{2g+2}]$. Für die Verschärfung des Thueschen Satzes konnte Siegel für eine großere Klasse von Gleichungen nachweisen, daß derselbe Sachverhalt gilt. 1926 konnte er nun durch Kombination dieser Methode in Verbindung mit einem Mordellschen Satz zeigen, daß auf jeder elliptischen Kurve über $\mathbb{Q}$, also auf Kurven vom Geschlecht 1, nur endlich viele ganzzahlige Punkte liegen. In Verbindung mit der These von A. Weil konnte er dies auch zeigen für Kurven vom Geschlecht größer als 1. Dazu ist nun folgendes historisch zu bemerken! Mordell hat damals, als er diesen Satz bewies, die Vermutung ausgesprochen, daß für solche Kurven vom Geschlecht größer als 1 mehr gilt, daß nämlich auf solchen Kurven nur endlich viele rationale Punkte liegen. (Für Kurven vom Geschlecht 1 kann es für z.B. $y^2 = x^3 - 2x$; $x^3 + y^3 = 9 : x_0 = 2,\ y_0 = 1$; $x_{n+1} = x_n(x_n^3 + 2y_n^3)/$

$(x_n^3 - y_n^3)$, $y_{n+1} = y_n(y_n^3 + 2x_n^3)/(y_n^3 - x_n^3)$ vorkommen, daß unendlich viele rationale Punkte darauf liegen). Aus dem Beweis der Mordellschen Vermutung folgt, daß z.B. die Fermatsche Gleichung $x^n + y^n = z^n$ für $n \geqslant 5$ nur endlich viele ganzzahlige Lösungen hat (von den trivialen Lösungen abgesehen). Siegel konnte diese Mordellsche Vermutung nicht voll beweisen und dies war wohl auch der Grund, warum er diese Arbeit nur in Manuskriptform in der Bibliothek des Frankfurter Seminars auflegte, mit der Begründung (wir sind im Jahre 1926): "Es sind schon zu viele Arbeiten erschienen!" Landau wurden seine Ergebnisse mitgeteilt, z.B. (meiner Erinnerung nach) von A. Weil, der es ablehnte, dies zur Kenntnis zu nehmen, denn etwas, das nicht publiziert sei, existiere für ihn nicht, da man ja die Richtigkeit einer nicht publizierten Arbeit nicht kontrollieren könne. I. Schur lud Siegel zu einem Vortrag nach Berlin ein, mit dem Wunsch, daß er über seine Resultate berichten sollte, und es wurden zwei Vorträge von Siegel gehalten, der erste behandelte den Fall $g = 1$. Bei der Vorbereitung für den zweiten Vortrag am nächsten Tag, der den Fall $g > 1$ behandeln sollte, entdeckte Siegel eine Lücke im Beweis, und in dem Vortrag sprach er dann nur—nach seinen Erzählungen—über die Lücke in seinem Beweis. Auf der Heimfahrt nach Frankfurt sah er sofort, wie sich diese Lücke schließen liesse und er teilte dies I. Schur schriftlich mit. I. Schur erwiderte daß es doch zweckmäßig wäre, diese Arbeit zu veröffentlichen. Siegel stimmte zu, meinte aber, um seine Absicht, nichts mehr zu publizieren nicht aufzugeben, daß die Arbeit in einer möglichst wenig gelesenen Zeitschrift erscheinen sollte. Es wurden dafür die Denkschriften der Preußischen Akademie in Aussicht genommen. Nun erscheint diese ganze Affäre nicht mehr von so großem Interesse, nachdem Faltings die Mordellsche Vermutung bewiesen hat und sogar den Siegelschen Satz für $g = 1$ auch beweisen kann und wie er behauptet, sogar in konstruktiver Weise. Es sei noch bemerkt, daß Baker auch den Fall $g = 1$ in konstruktiver Weise behandeln konnte, dies bedeutet natürlich, daß man für die Lösungen obere Schranken angeben kann. Ist n der Grad des absolut irreduziblen Polynoms $f(x, y)$ vom Geschlecht g, dann gilt für jede Lösung (x, y): $\mathrm{Max}(|x|, |y|) \leqslant \exp\exp\exp[(2H)^{\exp(3n)^{10}}]$ (H = Höhe von f), und dies kann mit dem Computer berechnet werden. Der Satz von Faltings ist ein Triumph der algebraischen Geometrie und der abstrakten Methoden in der Algebra, wie G. Faltings ja in dem Jahresbericht der DMV Bd. 86 (1984) scharf hervorgehoben hat. Nun ist der Beweis, an dessen Richtigkeit kein Zweifel besteht, noch nicht vollständig publiziert. Dies wird allerdings in Kürze geschehen; er wird aber ungefähr, wie mir mitgeteilt wurde, 200 Seiten umfassen. Darüber hinaus hofft man sogar, zeigen zu können, daß die Koordinaten der Torsionspunkte, also der Punkte endlicher Ordnung auf elliptischen Kurven, unter einer Schranke liegen, die nicht von den Kurven abhängt. Es wurde in Oberwolfach in einem Vortrag erläutert, wie man die Ergebnisse von Faltings auf Approximation algebraischer Irrationalitäten durch rationale Zahlen anwenden kann, daß man also den Spieß umdrehen kann, so wie dies die Pythagoräer bei den Quadratwurzeln gemacht haben.

Meine Hoffnung besteht darin, daß man doch einfache elementare Beweise für den Satz von Faltings finden wird. Es ist doch auch gelungen, den Weilschen Beweis der Riemannschen Vermutung für die Kongruenz-Zeta-Funktion, der ja Methoden der algebraischen Geometrie benützt, durch den elementaren Beweis von Stepanow zu ersetzen. Übrigens wird dieser Weilsche Satz im Beweis des Satzes von Faltings benützt. Selbstverständlich behalten die Methoden von Thue und Siegel ihren grundsätzlichen Wert. Es sei noch nachträglich bemerkt, daß Siegel für die Thueschen Methoden eine sehr schöne Darstellung in der Arbeit Nr. 91 gegeben hat.

Ergänzend sei noch bemerkt daß man schon frühzeitig die Gleichung $ax^n - by^n = c$ studiert hat, ich führe nur die Namen Nagell, und Delone an, und Siegel selbst hat 1937 eine Arbeit darüber veröffentlicht (Nr. 28). Hier weiß man, daß solche Gleichungen, ja sogar $|ax^n - by^n| \leqslant c$, höchstens eine Lösung in teilerfremden natürlichen Zahlen besitzen können, wenn $|ab|^{n/2-1} \geqslant \lambda_n c^{2n-2}$ ist, wobei $n \geqslant 3$, a, b, c ganz $c > 0$ und $\lambda_n = 4(n \prod_{p/n} p^{1/(p-1)})^n$ ist. Es wäre noch sehr viel über diesen ganzen Problemkreis zu sagen.

Mit Diophantischen Gleichungen hat Siegel sich auch später beschäftigt. Ich hebe nur die Arbeit mit der Nr. 97 hervor, und zwar wurde diese Arbeit durch A. Weil angeregt. Gegeben ist ein algebraischer Zahlkörper H und über ihm ein Körper G und es wird nun die Gleichung

$$N_{G/H}(\xi) = m$$

untersucht, wobei m eine ganze Zahl aus H ist. Dabei kann die unbekannte Größe ξ auch eine gebrochene Zahl aus G sein. Es wird die Frage untersucht, ob man Schranken für diese unbekannte Größe ξ angeben kann, unter der Voraussetzung, daß die Gleichung überhaupt eine Lösung besitzt. Bei der Lösung dieser Aufgabe muß Siegel woraussetzen, daß G über H normal ist. Ich weiß nicht, ob die Frage weiter verfolgt worden ist und von wem, ich erinnere mich nur an Folgendes: Siegel hat mir erzählt, daß ein damaliger Student ihn aufgesucht hat und ihm eine Lösung für den Fall, daß G über H nicht normal ist, mitgeteilt hat.

2. Transzendenz-Untersuchungen

Wir wollen jetzt zum nächsten Gebiet übergehen, das den ersten Teil der vorher erwähnten großen Abhandlung Nr. 16 ausmacht. Es behandelt die

Transzendenz von Irrationalzahlen. Wir haben vorher besprochen, wie sich die algebraischen Zahlen durch rationale Zahlen und durch algebraische Zahlen niederen Grades approximieren lassen. Dies ordnet sich in die allgemeinere Frage ein, wie sich eine beliebige Irrationalzahl durch eine rationale Zahl nach unten approximieren läßt und allgemeiner eine beliebige Irrationalzahl sich durch algebraische Zahlen approximieren läßt. Diese letzte Fragestellung ist besonders wichtig, wenn es sich um transzendente Zahlen handelt und damit ergibt sich das Problem, festzustellen, wann eine Zahl transzendent ist. Hermite und Lindemann waren ja die ersten, die solche Transzendenzuntersuchungen durchgeführt haben und Borel war der erste der z.B. studiert hat, wie groß die Differenz der Zahl e zu algebraischen Zahlen ist, d.h. er hat—im modernen Sprachgebrauch—ein Transzendenzmaß für e bestimmt. Dazu benötigt man den Begriff der Höhe H einer algebraischen Zahl x, das kann man definieren als die Höhe $H(P)$ des zugehörigen irreduziblen Polynoms P über Z: $h_0 + h_1 x + \cdots + h_n x^n$ welche gegeben ist durch $|h_0| + |h_1| + \cdots + |h_n|$. Oft ist es einfacher, so hat es Borel gemacht, daß man sich ein Polynom mit ganzzahligen Koeffizienten mit der Höhe H vorgibt und dann eine Ungleichung in der Form angibt, sagen wir für die Zahl e ($n =$ Grad von P) gilt nach I. Popken (1928)

$$|P(e)| > H^{-n - c_1/\log\log H}$$

(c_1 hängt nur von H ab).

Die Beweise von Ch. Hermite und F. Lindemann wurden immer mehr und mehr vereinfacht und elementarisiert. Es hat aber Hessenberg in seinem Buch über e und π bemerkt, daß dabei der Grundgedanke der Beweise nicht mehr ersichtlich war. Siegel ist nun auf den Grundgedanken von Hermite zurückgegangen, hat ihn aber durch eigene Ideen bereichert, sodaß er nicht nur neue Resultate erzielt hat, die ich dann gleich anführen werde, sondern auch Transzendenzmaße bestimmt hat. Darüber hinaus zeigt er die algebraische Abhängigkeit von verschiedenen transzendenten Zahlen. Eine Basis für seine Überlegungen ist ein Hilfssatz über die Lösung linearer Ungleichungen, das heute mit Recht Siegelsches Lemma heißt, obwohl es schon vorher in der Literatur vorkommt. Dieses Lemma lautet folgendermaßen: Es seien m Linearformen $L_1(x),\ldots, L_m(x)$ in $x = (x_1,\ldots, x_n)$ gegeben, wobei die Koeffizienten alle ganzzahlig sind und dem Betrage nach nicht größer als die ganze Zahl A liegen. Dann gibt es, wenn $n > m$, ganze Zahlen $x_1,\ldots, x_n$ nicht alle 0, welche das homogene Gleichungssystem $L_1(x) = 0,\ldots, L_m(x) = 0$ lösen und die dem Betrag nach kleiner als

$$1 + (nA)^{m/(n-m)}$$

sind.

Dieses Lemma beweist er wieder mit dem Schubfachprinzip. Es wird nun ein System von Funktionen $(E_1, \ldots, E_m)$ betrachtet, die er E-Funktionen nennt. Es sind dies ganze Funktionen, die einem System von linearen homogenen Differentialgleichungen genügen, deren Koeffizienten rationale Funktionen in x sind. Eine E-Funktion hat die Gestalt

$$E(x) = \sum_{n=0}^{\infty} \frac{c_n}{n!} x^n,$$

wobei die Koeffizienten aus einem festen algebraischen Zahlkörper sind und für alle $\varepsilon > 0$

$$\overline{|c_n|} = O(e^{\varepsilon n \log n})$$

($\overline{|c|}$ bezeichnet das Maximum der Beträge der Konjugierten von c) gilt; weiter gibt es eine Folge natürlicher Zahlen q_n, sodaß

$$q_n = O(e^{\varepsilon n \log n}) \text{ sowie } q_n c_k \text{ ganz für } 0 \leqslant k \leqslant n \text{ ist.}$$

Der Name E-Funktion rührt daher, daß als Beispiel die Funktion e^x die Voraussetzungen erfüllt. Es wird die algebraische Abhängigkeit des Systems $(E_1(x), \ldots, E_m(x))$ an algebraischen Stellen untersucht.

Es ist unmöglich, in dieser kurzen Zeit die weiteren Gedankengänge, die eine Kombination der Methoden von Liouville und Riemann und seiner eigenen scharfsinnigen Überlegungen sind, auch nur zu skizzieren. Ich möchte mir erlauben ein konkretes Resultat anzugeben, das sich auf die nullte Bessel-Funktion bezieht: Es ist $|P_n(J_0(\xi), J_0'(\xi))| > cH^{-123n^2g^3}$ (ξ algebraisch vom Grad g). Die Untersuchungen wurden später von A. Shidlofsky verallgemeinert (1967) und die Voraussetzungen in bedeutendem Maße abgeschwächt, allerding gestatten sie nicht, wie Siegel (Nr. 96) später bemerkt hat, quantitative Resultate, wie z.B. Transzendenzmaße anzugeben, da indirekte Schlüsse benützt werden.

Im letzten Abschnitt seiner großen Arbeit leitet er auch Irrationalitätsmaße, für Zahlen von denen man nicht erwarten kann, daß sie transzendent sind, her. Dabei werden die sogenannten G-Funktionen $\sum c_n x^n$, die einen endlichen Konvergenzradius haben betrachtet.

Dieser Abschnitt ist sehr kurz gefaßt und die Resultate sind zum Teil nur angegeben und nicht bewiesen, wie auch schon Th. Schneider in seinem Nachruf in den Sitzungsberichten der Göttinger Akademie bemerkt hat. Einen Teil davon hat Schneider bewiesen, aber ich weiß nicht, ob er sie publiziert hat. H. Hasse hat mir gegenüber einmal die Meinung vertreten, daß vielleicht das Ziel dieses Abschnittes ein Beweis der Fermatschen Vermutung war, vielleicht nicht für die ganze Vermutung, aber doch für Teile davon, also zu zeigen, daß $\sqrt[n]{1 - x^n}$ irrational ist, wenn x rational ist. 1929 trat A. Gelfond mit einer neuen Methode hervor, die an eine Arbeit von G. Polya anschließt. Siegel kombinierte in der Arbeit Nr. 17 (1932) seine Methode mit der Methode von Gelfond und er erhielt neue Resultate über transzendente Zahlen, die dann von Th. Schneider (1934) verallgemeinert wurden. Es sei noch historisch bemerkt, daß Siegel über die Theorie der transzendenten Zahlen und Irrationalzahlen drei Dissertationen vergeben hat, und zwar als erste Dissertation überhaupt die an W. Maier (1925). C. L. Siegel hat stets das Verdienst von W. Maier an der Abfassung der großen Arbeit Nr. 16 hervorgehoben. Maier war ja später Professor in Jena und er hat sich dann vor allem mit der Untersuchung von Eigenschaften gewisser Zeta-Funktionen beschäftigt. Siegel sagte zu mir einmal in Flims während einer Wanderung: "... Maier zetert viel." Aber Siegel war ihm stets, bis an sein Lebensende, wohl gesinnt. Maier lebt heute in der BRD, 80 jährig.

Siegels dritter Dissertant Karl Böhle arbeitete in seiner Dissertation (1932) auf diesem Gebiet und sein 5. Dissertant Th. Schneider verfaßte seine Dissertation 1934. Es gelang ihm, unabhängig von Gelfond, die Eulersche Vermutung (es ist das das 7. Hilbertsche Problem) zu beweisen. Es ist z.B. $2^{\sqrt{2}}$ transzendent (dies wurde zuerst 1930 von O. R. Kusmin gezeigt) also auch irrational. Hilbert hielt den Beweis der letzten Tatsache für schwieriger als den Beweis der Riemannschen Vermutung, oder der Fermatschen Vermutung. Siegel bemerkt dazu: "Man kann die Schwierigkeiten eines Problems erst dann richtig erkennen, wenn man es gelöst hat. Die eine Tatsache ist bewiesen, die Vermutungen noch nicht." Diese Bemerkung machte er 1949. Siegel selbst hat dann 1949 ein Buch über transzendente Zahlen veröffentlicht, zunächst in Englisch, das dann in deutscher Sprache als Taschenbuch erschienen ist. Obwohl er später nichts mehr über dieses Thema veröffentlicht hat, hat er sich doch immer wieder für die Frage interessiert hat, ob die Eulersche Konstante eine irrationale bzw. eine transzendente Zahl ist. Es sind ja von verschiedenen Seiten Beweisversuche unternommen worden, bei denen es sich immer wieder herausstellte, daß sie falsch sind. Siegel hat großes Interesse für den Satz von Apéry über die Irrationalität von $\zeta(3)$ gezeigt. Er war an der Frage interessiert, ob sich das Verfahren auf $\zeta(2k+1)$ $(k>2)$ verallgemeinern läßt. Es läßt sich meines Wissens nach die Beweismethode auf andere Zahlen verallgemeinern, aber anscheinend nicht auf den Fall $\zeta(5)$, allgemeiner $\zeta(2k+1)$.

3. Zeta-Funktion

Als nächstes Gebiet will ich nun seine Arbeiten über die *Zeta-Funktion* besprechen. Wie ich schon bei der biographischen Skizze angeführt habe, war er ja in Hamburg bei Hecke tätig (er hatte dort einen Lehrauftrag). Wenn er auch später auf Hecke nicht sehr gut zu sprechen war, so stand er zeitlebens unter dem Eindruck der Heckeschen Arbeiten. In den 20er Jahren waren es vor allem die aufsehenerregenden Arbeiten von Hecke zur Dedekindschen Zeta-Funktion, von der Hecke bewiesen hatte, daß sie auf der ganzen komplexen Zahlenebene meromorph fortsetzbar sind. Siegel hat selbst dann mehrere andere Beweise für diesen Satz gegeben. Der eine Beweis benützt eine Formel, die ihm—wie er in der Einleitung hervorhebt, von F. Bernstein gezeigt wurde. Diese Formel ist eine Verallgemeinerung der Formel von Lipschitz, wie sich später gezeigt hat, und sie lautet

$$\sum_{k=1}^{\infty} k^{\rho-1}e^{-ky} = \Gamma(\rho)\sum_{l}\frac{1}{(y+2\pi il)^{\rho}}\ (\rho>1,\ \mathrm{Re}(y)>0).$$

Diese Formel von Lipschitz wurde von Siegel mehrmals verwendet, deshalb schreibe ich sie explizit hin. In einer weiteren Arbeit vermeidet er möglichst den Satz von Dirichlet über die Einheiten. Er hat dies später auch bei Zetafunktionen von Algebren versucht, allerdings nicht publiziert. Man kann aber ein entsprechendes Zitat bei Deuring im Buch "Algebren" finden. Es wird dort versucht, aus den Eigenschaften der Zeta-Funktion von Algebren, über algebraischen Zahlkörpern, Folgerungen auf die Theorie der Einheiten zu ziehen. Er hat sich aber auch mit der gewöhnlichen Zetafunktion und mit den L-Reihen beschäftigt. Am bekanntesten von diesen Untersuchungen ist wohl seine Arbeit Nr. 18 (1932), die sich an den Nachlaß von Riemann über die Zeta-Funktion anschließt. Riemann hat auf einer Menge von Zetteln, die nur Formeln enthielten, asymptotische Formen aufgestellt, die als Spezialfall die sog. approximative Funktionalgleichung von Hardy–Littlewood enthalten, die erst im Jahre 1914 entdeckt worden war. Diese Formeln sind—wie sich später herausstellte—alle ohne Fehlerabschätzung und ohne Beweis. Es gelang nun Siegel in diesen Rechnungen einen Zusammenhang herauszufinden, und Restabschätzungen für diese Formeln zu finden und sie zu einer asymptotischen Reihe zu erweitern und man nennt sie daher mit Recht Riemann–Siegelsche Formel. Ich erlaube mir, eine solche Formel aufzuschreiben (Eine analoge Formel gilt für $\zeta(\sigma+it)$). Es sei $t>0$, $\nu=[\sqrt{t/2\pi}]$,

$$\vartheta = -\frac{t}{2}\log\pi + \arg\Gamma\left(\frac{1}{4}+\frac{it}{2}\right) = \frac{t}{2}\log\frac{t}{2\pi e}-\frac{\pi}{8}+O(t^{-1})$$

$$e^{i\vartheta}\zeta\left(\frac{1}{2}+it\right) = 2\sum_{n=1}^{\nu}\frac{1}{\sqrt{n}}\cos(\vartheta - t\log n)+R.$$

Es gilt (k ganz $\geqslant 0$)

$$R=(-1)^{\nu-1}\left(\frac{2\pi}{t}\right)^{1/4}(C_0+C_1t^{-1/2}+\cdots+C_kt^{-k/2}+R_k),$$
$$R_k=O(t^{-(k+1)/2}),$$

wobei die C_j beschränkte Funktionen sind, z.B.

$$C_0=\cos\left(2\pi u^2+\frac{3\pi}{8}\right)\Big/\cos 2\pi u \qquad \left(u=\sqrt{\frac{t}{2\pi}}-\nu-\frac{1}{2}\right).$$

Siegel ist (Nr. 12) später (1943) auf diese Sache zurückgekommen und er hat sie auf die L-Reihen, angeregt durch eine Arbeit von R. O. Kusin, 1934, ausgedehnt und vereinfacht. Darüber hinaus dehnte er diese Untersuchungen auf die Epsteinsche Zetafunktion aus. Er hat in einer Arbeit Nr. 51 (1945) die Untersuchungen von R.E.A.C. Paley (1931) über die L-Reihen fortgesetzt und z.B. einen Satz von G. Hoheisel (1926) verallgemeinert. Diese Untersuchungen hat er in einer seiner letzten Arbeiten Nr. 97 auf die Heckeschen Zetafunktionen mit Größencharakteren ausgedehnt. In Bezug auf die Riemannsche Vermutung war er der Meinung, daß sie richtig für die gewöhnliche Riemannsche Zetafunktion (Ein Beweis wurde von Matsumoto angekündigt), aber nicht für die allgemeinen Heckeschen Zetafunktionen sei. Für die Epsteinsche Zetafunktion ist sie bekanntlich nicht richtig. In den letzten Jahren hat er sich vor allem mit der Verallgemeinerung des Eulerschen Resultates, das die gewöhnliche Zetafunktion an geraden Stellen abgesehen von einer Potenz von π immer eine rationale Zahl ist, beschäftigt und er hat sie für L-Reihen und auf die Dedekindsche Zetafunktion an diesen Stellen ausgedehnt. Dabei muß man natürlich noch ($d=$ Diskrimante) von einer Potenz von d absehen.

Die letzte Arbeit Nr. 99 beschäftigt sich mit der Summierung von Reihen der Gestalt

$$L(s)=\sum_{m,n} e^{2\pi i(mu+nv)}(Q(m,n))^{-s} \qquad (s=1,2,\ldots),$$

$Q(x,y)=Ax^2+Bxy+Cy^2$ indefinit, $A, B, C,$ ganz, $B^2=4AC=D$ keine Quadratzahl (u, v rational). Es sei noch $A>0$ $(A,B,C)=1$. C. L. Siegel zeigt $\pi^{-2s}\sqrt{D}\,L(s)$ ist rational. Er hatte sich schon vorher (Nr. 84, 1968) damit beschäftigt. Damals verwendete er im Anschluß an Hecke die komplexe Funktionentheorie und in dieser allerletzten Arbeit Nr. 99 bemüht er sich, nur mit Hilfsmitteln der reellen Analysis auszukommen.

Es ist jetzt noch ein Satz über L-Reihen zu besprechen, der Siegelsche Satz, der allerdings den Titel: "Über die Classenzahl quadratischer Zahlkörper" hat (Nr. 21′), und 1935 in *Acta Arithmetica* (S. 83–86) erschienen ist. Die Angelegenheit hat folgende Vorgeschichte. Bereits Gauß hat vermutet, daß die Klassenzahl $h(d)$ imaginär quadratischer Zahlkörper gegen Unendlich geht. Dies wurde nach Vorarbeitten von Mordell, Hecke, und Deuring von Heilbronn bewiesen. Es gelang nun Siegel folgende Verschärfung zu zeigen, nämlich:

$$\lim_{d \to \infty} \frac{\log h(d)}{\log \sqrt{|d|}} = 1.$$

Dies zeigt er unter Verwendung der Klassenkörpertheorie, und einen analogen Satz zeigte er bei reellen quadratischen Zahlenkörpern, wo $\varepsilon(d)$ Pell'sche Einheit ist: $\log h(d) \log \varepsilon(d)/\log \sqrt{d} \to 1$ für $d \to \infty$. Aus dem engen Zusammenhang, (es ist ja $\pi\,|d|^{-1/2}\,h(d) = L_d(1)$, wenn $d < 0$, $2d^{-1/2}h(d) \log \varepsilon(d) = L_d(1)$, wenn $d > 0$), der Klassenzahl mit den L-Reihen ist es ein Satz über $L_d(1) = \sum_{n=1}^{\infty}(d/n)(1/n)$. Wie Walfisz entdeckte, wird in dieser Arbeit von Siegel folgendes bewiesen, heute als *Siegelscher Satz* bekannt: Zu jedem $\varepsilon > 0$ existiert eine positive Zahl $C(\varepsilon)$, so daß, wenn χ ein reeller Charakter mod q, aber nicht der Haupcharakter, $L(s, \chi) \neq 0$ ist für alle $s \geqslant 1 - c(\varepsilon)\, q^{-6}$.

Allerdings ist dieser Satz erneut nicht konstruktiv, sodaß man das $c(\varepsilon)$ nicht explizit als Funktion von ε angeben kann, dagegen ist die schwächere Abschätzung von A. Page $1/\sqrt{q} \log q$ konstruktiv. A. Walfisz zeigte (1936) mit Hilfe des Siegelschen Satzes folgenden Satz für die Anzahl $\pi(x, q, a)$ der Primzahlen in einer arithmetischen Reihe: Es sei $N \geqslant 1$. Dann existiert ein $C(N)$, sodaß gleichmäßig für alle q mit $q \leqslant (\log x)^N$: $\pi(x, q, a) = li(x)/\varphi(q) + O(xe^{-C(N)\sqrt{\log x}})$ für alle a mit $(a, q) = 1$ ist. Es sei noch bemerkt, daß Walfisz, der dann später Professor in Tiflis war, gleichzeitig mit Siegel und Hasse studiert hat. Er fiel nach Auskunft von Hasse durch seinen schönen Spazierstock auf, den er nie aus der Hand ließ. Es sei mir auch noch gestattet, eine kleine Anekdote, die für Siegel und für Landau charakteristisch ist, zu erzählen. Landau pflegte vor allem auch seine Dissertanden, am Tag ihrer Prüfung nach der Prüfung zum Mittagessen, aber auch andere Gelehrte, die gerade in Göttingen waren, einzuladen. Beim Essen hielt er dann meist eine Lobesrede auf seine Kandidaten. Siegel, der diesen Brauch kannte, erklärte, daß er nur dann zum Mittagessen kommen würde, wenn von seinem Rigorosum, das bei Runge nicht besonders gut verlaufen war, nichts erwähnt werden würde. Landau, der Siegel gut kannte und ihn als seine Entdeckung ansah, stimmte ihm zu. Am gleichen Tag legte auch Walfisz seine Prüfungen ab. Beim Mittagessen erhob sich nun Landau und sagte: "Unter dem Siegel der Verschwiegenheit teile ich mit, daß Walfisz heute seine Prüfungen abgelegt hat." Es war dann immer üblich, daß nach dem Mittagessen verschiedene Spiele gespielt wurden, z.B. auch Schach, bei denen der Hausherr gerne den Sieg davon trug. So waren einmal bei einem solchen Essen auch Siegel und Gelfond eingeladen und nach dem Mittagessen schlug Landau Gelfond eine Schachpartie vor, die Gelfond akzeptierte. Aber zur großen, unangenehmen Überraschung des Hausherrn war er nach drei Zügen matt. Landau dachte, das sei nur Zufall und er schlug eine zweite Partie vor, die er ebenfalls in kurzer Zeit verlor. Es stellte sich nach Befragen des jungen, aber sehr schweigsamen Gelfond heraus, daß er der erste russische Schachmeister auf akademischen Boden war. Der Hausherr fand dann endlich ein Spiel, das Gelfond unbekannt war und dabei siegte Landau triumphierend.

Um nun auf den Satz von Siegel zurückzukommen, so sei noch bemerkt, daß man heute in der analytischen Zahlentheorie den elementaren Beweis von Estermann zu Grunde legt, der ohne algebraische Zahlentheorie auskommt. Ein konstruktiver Beweis liegt bisher nur in Spezialfällen vor. Es waren andere Methoden erforderlich, um alle Körper mit $h(d)=1$ zu bestimmen. Dies gelang zuerst K. Heegner, dann H. M. Stark. Mit einer anderen Methode, die von A. Gelfond herrührt, gelang es A. Baker nicht nur den Fall $h(d)=1$, sondern auch $h(d)=2$ zu erledigen. Nach einer Mitteilung von O. Taussky ist es B. Gross und Don Zagier nach Vorarbeiten von Goldfeld gelungen, den allgemeinen Fall der Klassenzahl imaginär quadratischer Zahlkörper zu erledigen "unter gigantischer Verwendung der Theorie der elliptischen Kurven," wie Olga sagt. C. L. Siegel hat in Nr. 85 einen anderen Beweis des Satzes von Heegner–Stark gegeben.

Es sei noch bemerkt, daß dieser Satz von Siegel an wichtiger Stelle im Beweis des Goldbachschen Satzes von Vinogradow eingeht.

Diese letzten Bemerkungen haben uns somit zur algebraischen Zahlentheorie geführt.

4. Algebraische Zahlen und Geometrie der Zahlen

Das Werk von Hecke, Algebraische Zahlen, hat Siegel auf das Stärkste beeinflußt. Wenn wir die Begriffe: Diskriminante, Klassenzahl, Regulator, Einheiten, Primzahlzerlegung, Reziprozitätsgesetz, u.s.w. ins Auge fassen und die Chronologie beiseite lassen, so hat er in der Arbeit Nr. 47 eine Vermutung von Gauß bewiesen: Es sei $h(d)$ die Klassenzahl primitiver binärer quadratischer Formen mit positiver Diskriminante d, welche keine Quadratzahl ist. Es sei $\varepsilon(d)$ die zugehörige kleinste Pellsche Einheit, dann ist

$$\sum_{k\leqslant N} h(4k)\,\varepsilon(4k)\sim\frac{4\pi^2}{21\zeta(3)}N^{3/2}$$

während eine analoge Vermutung von Gauß für Formen negativer Diskriminante bereits von Mertens bewiesen wurde. Er stützt sich dabei auf den Satz von G. Polya über Charaktersummen. In dieser Arbeit werden dann diese Sätze auf quadratische Formen in mehreren Variablen verallgemeinert. In die 20er Jahre zurück führt uns eine Reihenentwicklung für die Diskriminante total reeller Zahlkörper, die in den Göttinger Sitzungsberichten (1922) erschienen ist (Nr. 10). Diese Entwicklungen wurden dann später von Ramanathan auf hyperkomplexe Systeme und von Deuring weiter verallgemeinert. Siegel hat den Gedankengang in Nr. 10 dazu benützt, sie auf die Geometrie der Zahlen anzuwenden. Es ist, wie Zassenhaus zurecht bemerkt hat, Siegel wohl in den 20er Jahren in Deutschland der einzige, der sich mit der Geometrie der Zahlen, dieser Theorie, die von Minkowski geschaffen wurde, beschäftigt hat. So gibt er einen neuen analytischen Beweis des Minkowskischen Linearformensatzes, der sich auf folgende Formel stützt (Re $s_k > 0$, $D = \det(a_{ik})$, (A_{ik}) reziprok zu (a_{ik}), a_{ik} reelle Zahlen):

$$D \sum_{l_1,\ldots,l_n=-\infty}^{\infty} e^{-\sum_{k=1}^{n} |a_{k1}l_1 + \cdots + a_{kn}l_n| s_k}$$

$$= \sum_{l_1,\ldots,l_n} \prod_{k=1}^{n} \left\{ \frac{1}{s_k - 2\pi i(A_{k1}l_1 + \cdots + A_{kn}l_n)} + \frac{1}{s_k + 2\pi i(A_{k1}l_1 + \cdots + A_{kn}l_n)} \right\}.$$

Daraus folgt die folgende Gleichung (*)

$$D \sum_{\text{Klammern} > 0}' \prod_{k=1}^{n} (t_k - |a_{k1}l_1 + \cdots + a_{kn}l_n|)$$

$$= \sum_{l_1,\ldots,l_n} \prod_{k=1}^{n} \left(\frac{\sin t_k \pi(A_{k1}l_1 + \cdots + A_{kn}l_n)}{\pi(A_{k1}l_1 + \cdots + A_{kn}l_n)} \right)^2$$

und daraus folgt der Linearformensatz: Ist $|\;| < t_k$ $(k = 1,\ldots, n)$ nur für $l_1 = \cdots = l_n = 0$ lösbar, so ist die linke Seite $D(t_1,\ldots, t_n)$. Die rechte Seite ist $\geqslant (t_1,\ldots, t_n)^2$, also ist $D \geqslant \prod_{k=1}^{n} t_k$.

Wir haben also statt Ungleichungen jetzt Gleichungen, etwas was z.B. Hasse und A. Weil immer gewünscht haben, bzw. fordern.

Zassenhaus zitiert gerne Weil: "Es muß hinter der Ungleichung eine Gleichung stecken." Siegel hoffte damit auch eine Vermutung von Minkowski beweisen zu können. Dies hat er später mit den Worten im 4. Band zurückgezogen, mit dem Wort "... Hochstapelei." Der Beweis dieser Minkowskischen Vermutung wurde dann von Hajós geführt, aber dieser

Ausdruck von Siegel ist eine Übertreibung, da ja der Beweis von Hajós eine Folgerung aus dieser Identität benützt.

Siegel selbst hat dann den 1. Fundamentalsatz von Minkowski durch eine Gleichung ersetzt, nämlich durch folgende: Enthält der konvexe Körper K mit Mittelpunkt 0 keinen Gitterpunkt $g \neq 0$, so ist

$$2^n = V + V^{-1} \sum_{g \neq 0} \left| \int_K e^{\pi i \langle g, x \rangle} \, dx \right|^2.$$

H. Weyl und andere wie z.B. meine Wenigkeit haben an diese Gleichung angeknüpft. Ich hatte weiters 1942 folgende Ungleichung gezeigt: Es gibt eine Matrix A mit Determinante $=1$, sodaß zu jeder Funktion f und zu jedem $\varepsilon > 0$ gilt:

$$\sum_{g \neq 0} f(Ag) < \int_{R^n} f(x) \, dx + \varepsilon$$

($f \equiv 0$ außerhalb eines Würfels c, integrierbar im Riemannschen Sinn). Daraus folgt eine andere Vermutung von Minkowski. Siegel ersetzt nun auch hier die Ungleichung durch eine Gleichung, indem er folgende Gleichung zeigt (Arbeit Nr. 50):

$$\int_F \left(\sum_{g \neq 0} f(Ag) \right) d\omega = \int_{R^n} f(x) \, dx.$$

Dabei ist Ω_1 die multiplikative Gruppe aller A von Determinante 1 auf welchem es ein invariantes Maß $d\omega$ gibt, eindeutig bestimmt bis auf einen konstanten Faktor. Es sei $\Gamma 1$ die unimodulare ganzzahlige Gruppe. Es ist dann F der Fundamentalbereich von $\Gamma 1$ in Ω_1. Sein Volumen ist endlich und wird nun zu 1 normiert. Er hat über die Geometrie der Zahlen auch 1944/45 eine Vorlesung am Courant-Institut in New York gehalten. In der späteren Arbeit Nr. 88 hat er sich bemüht, für die Einheiten in algebraischen Zahlkörpern konstruktive Methoden zu entwickeln, beeinflußt durch die Arbeiten von Remak. Eine erste Arbeit von Remak in dieser Hinsicht findet sich im selben Band des Crelleschen Journals, in der sich auch die große Arbeit von Thue befindet und in der Remak für die Lösung der Pellschen Gleichung obere Schranken angibt. Diese Untersuchungen von Siegel sind wohl durch die Untersuchungen von Baker veranlaßt worden. Aber im Alter pflegte Siegel überhaupt mehr und mehr den konstruktiven Standpunkt; Er findet sogar den Begriff der unendlichen Menge suspekt. Dieser konstruktive Standpunkt der sich aber auch bei Hasse in den späteren Jahren findet, stellt eine Abkehrung von den Methoden von Hilbert dar, der in der algebraischen Zahlentheorie, die abstrakten, axiomatischen, nicht konstruktiven Methoden gegenüber den expliziten Formeln z.B. von Kummer vorzieht. Zassenhaus, der einer der lebhaftesten Verfechter der konstruktiven Methoden der algebraischen Zahlentheorie ist, hat hervorgehoben, daß es sich bei den Einheiten um ein grundsätzlich einfaches Problem handelt, nämlich um die Auflösung von linearen Ungleichungssystemen in mehreren Unbekannten. Aber diese grundsätzlichen Aufgaben sind eben sehr schwer zu lösen.

Siegel selbst hat auch eine Arbeit Nr. 74 zum quadratischen Reziprozitätsgesetz in algebraischen Zahlkörpern geschrieben, worin er im engen Anschluß an Dirichlet und Hecke mit Gaußschen Summen operiert und zwar beweist er die explizite Formel, die Hasse 1924 entwickelt hatte

$$\left(\frac{\alpha}{\beta}\right)\left(\frac{\beta}{\beta}\right)=(-1)^{\varepsilon(\alpha,\beta)}$$

wobei $\varepsilon(\alpha,\beta)=(-1)^{g+S((\alpha-1)/2(\beta-1)/2)}$, wobei $\alpha\equiv\beta\equiv1 \pmod 2$,

$$g=\sum_{p=1}^{m}\frac{\operatorname{sgn}\alpha^{(p)}-1}{2}\cdot\frac{\operatorname{sgn}\beta^{(p)}-1}{2}$$

($m=$ Grad von K, sgn nur für die reellen konjugierten Zahlen, $S=$ Spur). Dabei weist er darauf hin, daß der Beweis des Reziprozitätsgesetzes von Gaußschen Summen von Krazer durch Fouriersche Reihen im Fall von mehreren Variablen nicht korrekt ist, da die Reihen in diesem Fall divergent sind. Siegel selbst führt den Beweis durch sehr geschickte vollständige Induktion aus. Es hat sich später, und zwar unter Benützung der vorher zitierten Siegelschen Formeln und des Minkowskischen Linearformensatz gezeigt, daß man den Krazerschen Beweis allerdings durch sehr mühevolle Überlegungen und passende Interpretation der Fourierkoeffizienten retten kann.

Das Hauptanliegen von Siegel in den 20er Jahren war aber die Entwicklung einer additiven Zahlentheorie in algebraischen Zahlkörpern, insbesondere die Darstellbarkeit einer algebraischen Zahl als Summe von Quadraten. A. Weil hat in seinem Buch über die Geschichte der Zahlentheorie bemerkt, daß das Hauptanliegen von Diophant und Fermat die Behandlung von Diophantischen Gleichungen höheren Grades in zwei Variablen und die Darstellung einer natürlichen Zahl als Summe von einer vorgegebenen festen Anzahl von Quadraten ist. Eine Verallgemeinerung stellt das Waringsche Problem dar, wobei die Quadrate durch höhere Potenzen ersetzt werden. Dieses Problem wurde zum ersten Mal von Hilbert gelöst. Seine Methode liefert aber für die zugehörige Anzahl m der Potenzen, die notwendig sind, daß sich jede natürliche Zahl b als Summe von n-ten Potenzen darstellen läßt, schlechte Schranken. (Es wird oft die Ansicht geäußert, daß die Hilbert-Methode es nicht gestattet, ein solches m zu bestimmen. Das ist nicht richtig, wie z.B. Rieger gezeigt hat.)

5. Methoden von Hardy–Littlewood

Hardy, Littlewood, und Ramanujan entwickelten in den 20er Jahren in sensationellen Arbeiten eine analytische Methode, um gute m zu bestimmen. Diese Methode nannten sie die Kreismethode. Sie beruht auf folgender Idee (Landau hat sie scherzhaft eine schülerhafte Idee genannt). Es sei eine Folge von natürlichen Zahlen (N_r)... gegeben und eine Zahl m und man fragt nach der Anzahl $A_m(t)$ der Lösungen der Gleichung

$$N_1 + N_2 + \cdots + N_m = t$$

die von der Folge, von m und t abhängt. Wir betrachten dazu die Potenzreihe $f(z) = \sum_{k=1}^{\infty} z^{N_k}$ die im Inneren des Einheitskreises konvergent ist. Dann ist die m-te Potenz $f^m(z)$ gleich $\sum_t A_m(t)\, z^t$ und nach der Cauchyschen Integralformel ist

$$A_m(t) = \frac{1}{2\pi i} \int_C \frac{f^m(z)}{z^{t+1}}\, dz$$

wobei C ein Kreis ist, der im Inneren des Einheitskreises liegt.

Nun ist die Funktion $f(z)$ im allgemeinen über den Einheitskreis hinaus nicht analytisch fortsetzbar. Die Idee dieser drei Mathematiker war nun, sozusagen, die schwersten Singularitäten zu umgehen. Die Vermutung war nun, daß diese Stellen die Punkte $e^{2\pi i\alpha}$ sind, wobei α in der Nähe einer rationalen Stelle d/q liegt. Diese Stellen werden nun mit Bögen umgeben, mit größeren, bzw. kleineren Bögen major arcs und minor arcs genannt. Die major arcs liefern in ihrer Gesamtheit den Hauptanteil der asymptotischen Entwicklung für $A_m(t)$, die aus einer unendlichen Reihe S, singular series genannt, noch mit einer Potenz von t versehen, besteht. Die Gesamtheit der minor arcs liefert das Restglied. So sind dann zwei Dinge zu untersuchen: (1) ob S von 0 verschieden ist und (2) ob das Restglied gegenüber dem Hauptglied klein ist. Ich möchte nun das Verfahren ein bißchen expliziter schildern, in der Fassung, wie es von Vinogradow modifiziert wurde, der das Rechnen mit unendlichen Reihen und das Operieren mit einem kleineren Kreis C als dem Einheitskreis erspart. Soll die Gleichung

$$n_1^k + \cdots + n_m^k = t$$

in natürlichen Zahlen lösbar sein, so müssen die $n \leqslant [t^{1/k}] = P$ sein. Es wird also daher der Ansatz gemacht

$$T(\alpha) = \sum_{n=1}^{P} e^{2\pi i \alpha n^k},$$

dann ist

$$A_m(t) = \int_0^1 (T(\alpha))^m e^{-2\pi i t\alpha} d\alpha$$

und dann werden die major arcs, so definiert $I_{d,q}$: $|\alpha - d/q| \leqslant 1/q\tau$, wobei $(d, q) = 1$ und $\tau \leqslant P^\beta$ ist. (Die Wahl von β ist nur von technischem Interesse) und nun wird folgendes gemacht:

Es wird in $I_{d,q}$ in der Funktion $T(\alpha)$ die Zahl α durch die Zahl d/q ersetzt und da die Funktion $e^{2\pi i(d/q)n^k}$ die Periode q hat, so ist der Wert dieser Summe ungefähr gleich

$$\frac{p}{q}\sum_{d=0}^{q-1} e^{2\pi i(d/q)n^k},$$

also erhalten wir, wenn wir gleich über alle Bögen summieren $St^{m/k-1}$. (Wir müssen nämlich das ganze noch korrigieren, indem wir durch t dividieren) und dazu kommt dann noch das Restglied. Dabei ist

$$S = \sum_{q=1}^{\infty} \sum_{d(d,q)=1}^{q} \left(\frac{1}{q}\sum_{n=1}^{q} e^{2\pi i(d/q)n^k}\right)^m e^{-2\pi i(d/q)t}$$

die Singular Series. Wenden wir das auf $k=2$ an, so stehen hier in der Klammer die Gaußschen Summen, und die m-te Potenz davon ist wieder eine Gaußsche Summe zur quadratischen Einheitsform E_m in m Variablen.

Hardy und Mordell zeigen darüber hinaus für $5 \leqslant m \leqslant 8$ mit Hilfe der elliptischen Modulfunktionen, daß diese asymptotischen Formeln sogar exakte Formeln sind, das Restglied also Null ist.

Diese Erkenntnisse, die hier aufgezeigt wurden, waren für die weitere Arbeitsrichtung von Siegel entscheidend. Die Methoden von Hardy–Littlewood, die ich hier skizziert habe, wurden im Seminar von Hecke nach den Erinnerungen von Behnke ausführlich behandelt, allerdings sah Hecke nicht, wie diese auf algebraische Zahlkörper anzuwenden sind, und wandte sich dann der Beschäftigung mit der Klassenkörpertheorie, wie sie von Takagi veröffentlicht worden war, zu. Siegel erzählte mir einmal, daß er Artin und Hecke auf diese Arbeit von Takagi aufmerksam gemacht hätte, die ihm durchsichtiger erschien, als die Theorie von Hilbert und Furtwängler. Siegel ließ aber nicht locker und es gelang ihm in seiner Habilitationsschrift (Nr. 3, 1921) zunächst den Fall $m \geqslant 5$ zu behandeln und in der Arbeit Nr. 8 sogar den Fall $m=4$. Dabei ist t eine total positive Zahl aus dem betrachteten reell quadratischen Zahlkörper bzw. in Nr. 14 ist es sogar ein algebraischer Zahlkörper höheren Grades.

Dabei gelang es ihm auch einen Satz von Hilbert zu beweisen und, wie ich schon vorher erzählt habe, war Siegel zunächst enttäuscht, daß Hilbert darauf nicht reagierte, aber Hilbert schenkte ihm dann die Gesammelten Werke von Minkowski. Die Arbeit Nr. 12 ist ein Bericht über seinen Vortrag, den er bei der Jahresversammlung 1922 der deutschen Mathematikervereinigung gehalten hat. Diese Versammlung ist wohl die einzige, die er besucht hat. Von dieser Versammlung hatte er hauptsächlich einen Vortrag, bzw. eine Diskussion von Brouwer, über den Intuitionismus, aber vor allem, daß Brouwer und die Tafel plötzlich zu Boden gefallen war, in Erinnerung. Es waren in der Arbeit von Siegel mehrere Schwierigkeiten zu überwinden: Die eine bestand darin, die Methode der großen und kleinen Bögen auf algebraische Zahlkörper zu übertragen. Dies gelang ihm mit Hilfe des Minkowskischen Linearformensatzes. Die zweite Schwierigkeit liegt natürlich in dem Studium der singulären Reihe und hier benützt er wieder die Formel, die ihm Felix Bernstein mitgeteilt hatte. Damals erschien es ganz hoffnungslos. Das Waringsche Problem auf algebraische Zahlenkörper zu verallgemeinern, dies gelang ihm erst spät, 1944 in der Arbeit Nr. 44.

Die Hauptschwierigkeit liegt darin, daß im Bereich der ganzen Zahlen, jede natürliche Zahl n eine Summe von k-ten Potenzen der Zahl 1 ist, deren Anzahl allerdings von n abhängt. Hierzu gibt es kein Gegenstück in algebraischen Zahlkörpern. Siegel führt den Ring J_k aller Summen

$$a_1\lambda_1^k + \cdots + a_n\lambda_n^k \qquad (\lambda_i \text{ ganz aus } K)$$

ein, erzeugt von den k-ten Potenzen der ganzen Zahlen aus dem algebraischen Körper K, wobei die Koeffizienten a_k ganz rationale Zahlen sind, und dann kann er den folgenden Satz aussprechen:

Ist der Körper K nicht total reell, so läßt sich jede total positive Zahl v aus diesem Ring (wobei man bei der Darstellung $v = \lambda_1^k + \cdots + \lambda_m^k$ verlangen muß $|\lambda_r^{(\tau)}|^k \leqslant |v^{(\tau)}|$ für $r = 1, 2, \ldots, m$, wenn der konjugierte Körper $K^{(\tau)}$ komplex ist. Der Ausdruck totalpositiv bezieht sich nur auf die reellen konjugierten Körper $K^{(\tau)}$) als Summe von $m > (2^{k-1} + g)\, gk$ k-ten Potenzen aus J_r darstellen, wobei der Grad von K über $\mathbb{Q}$ gleich g ist.

Ist K total reell, dann gilt der Satz nur, wenn die total positive Zahl eine *genügend große Norm besitzt.* Auf Anregung von Siegel hat der amerikanische Mathematiker Peck diesen Satz verallgemeinert. Siegel selbst hat sich später mit der Kreismethode nur mehr indirekt beschäftigt, durch Anregung von Dissertationen, so durch die Dissertation von O. Körner 1960, der das Goldbachsche Problem auf algebraische Zahlkörper erweitert hat.

Es sei noch bemerkt, daß Siegel nicht nur Dissertationen aus seinen Arbeitsgebieten gegeben hat; so beschäftigt sich eine Dissertation mit projektiver Geometrie, bzw. mit den Grundlagen der Geometrie, eine andere mit periodischen Minimalflächen.

6. Quadratische Formen

Wir haben schon von den asymptotischen Gesetzen für die Anzahl der Darstellungen durch eine quadratische Form gesprochen. Nun gibt es aber seit Lagrange, Gauß, Jacobi exakte Formeln für $m=2$, 3, 4 und Hardy hat ja sogar exakte Formeln für die übrigen Fälle bis $m=8$. Man kann jetzt fragen, wie steht es für größere m und allgemeiner mit der Anzahl $A(S, t)$ der Darstellungen einer natürlichen Zahl t durch eine ganzzählige quadratische Form S. Mit dieser Fragestellung für den Fall, daß die quadratische Form S eine positiv definite ist, haben sich Gauß und Eisenstein für ternäre Formen beschäftigt, allgemeiner Minkowski. Nun hat schon Gauß bemerkt, daß man diese Fragestellung der Darstellung verallgemeinern muß, durch Untersuchungen der Darstellungen $A(S, T)$ einer ganzzähligen symmetrischen Matrix T durch die quadratische Form S. Es handelt sich also um die Auflösung der Gleichung

$$X'SX=S[X]=T \text{(Dabei sei } S\ m\text{-reihig, } T\ n\text{-reihig, } X'=\text{transponierte Matrix zu } X)$$

in ganzzahligen Matrizen X, wo $m \geqslant n$ ist. Im Fall $T=S$ schreibt man man seit Eisenstein $A(S, S)=E(S)$. Es ändert sich $A(S, T)$ klarerweise nicht, wenn man S durch eine zu S äquivalente Matrix S' ersetzt, d.g. die Anzahl hängt nur von der Klasse von S ab.

Einfache Beispiele zeigen, daß man für diese A keine einfachen Formel erwarten kann. Dies ändert sich aber, wenn man zu dem Geschlecht von S übergeht. Man nennt zwei quadratische Formen S, S_1 verwandt, wenn für jeden Modul q die Kongruenzen $S[X] \equiv S_1 \pmod q$ und $S_1[X_1] \equiv S \pmod q$ lösbar sind und vom gleichen Trägheitsindex sind, wenn S und S_1 nicht positiv definit sind). Die Menge aller Formen die zu S verwandt sind, bildet das Geschlecht von S. Man weiß seit Hermite, daß sich jedes Geschlecht aus endlich vielen Klassen zusammensetzt. Es seien nun die Repräsentanten dieses Geschlechts $S_1=S, S_2, \ldots, S_h$, dann bildet Siegel den Mittelwert

$$\bar{A}(S, T)=\sum_{i=1}^{h} \frac{A(S_i, T)}{E(S_i)} \Bigg/ \sum_{i=1}^{h} \frac{1}{E(S_i)}.$$

Es sei nun weiter

$$A_q(S, T)=\sum_{S[X] \equiv T \pmod q} 1$$

und dann zeigt er die folgende grundlegende Formel

$$\frac{\bar{A}(S, T)}{A_\infty(S, T)}=\varepsilon \lim_{(q) \to \infty} \frac{A_q(S, T)}{q^{mn-n(n+1)/2}},$$

wo der lim der rechten Seite ein eingeschränkter lim ist. So werden nur solche Folgen in lim zugelassen, wo $q_i \mid q_{i+1}$. Eine solche zugelassene Folge ist die Folge $n!$ Der Ausdruck $A_\infty(S, T)$, zunächst von Siegel mit A_0 bezeichnet, hängt nur von den Determinanten von S und T ab. Der Faktor ε ist im Fall $m > n+1$ gleich 1; im Fall $m = n+1$ gleich $\frac{1}{2}$ und im Fall $m = n$ ebenfalls $\frac{1}{2}$. Weiters ist, wenn $m = n$ auf der rechten Seite A_q noch durch $2^{-\omega(q)}$ zu dividieren, wobei $\omega(q)$ die Anzahl der Primteiler von q ist. Diese grundlegende Formel wird die *Siegelsche Maßformel* genannt, deshalb, da seit Eisenstein und Minkowski $M(s) = \sum_{i=1}^{h} 1/E(S_i)$ das Maß von S genannt wird. Diese Hauptformel schreibt Siegel noch in einer anderen Art, in der berücksichtigt wird, daß A_q eine multiplikative Funktion ist. Wir setzen $\alpha_q = A_q(S, T) \mid q^{mn-n(n+1)/2}$ bzw. für $m = n, \alpha_q = A_q(S, T) \mid 2^{\omega(q)} q^{mn-n(n+1)/2}$. Es existiert dann $\lim_{n \to \infty} \alpha_p n = d_p$.

Dann lautet die Formel folgendermaßen (p durchläuft alle Primzahlen)

$$\frac{\bar{A}(S, T)}{A_\infty(S, T)} = \prod_p d_p(S, T).$$

Für den Fall daß $T = S$ erhält man die Maßformel von Minkowski, wobei Siegel ausdrücklich darauf hinweist, daß die Potenz von 2 bei Minkowski falsch bestimmt ist. Er betrachtet nun Beispiele, um zu zeigen, daß darin die klassischen Resultate enthalten sind, z.B. das Ergebnis von Jacobi für $S = E_4$, $t \equiv 1 \pmod 2$,

$$d_2 = 1, \qquad d_p = \left(1 - \frac{1}{p^2}\right)\left(1 + \frac{1}{p} + \cdots + \frac{1}{p^\tau}\right) \cdots \text{wenn } p \equiv 1 \pmod 2,$$

$p^\tau \| t$. Es ist dann

$$\prod_p d_p = \prod_{p>2} \left(1 - \frac{1}{p^2}\right) \sum_{g \mid t} \frac{1}{g},$$

$$A(E_4, t) = 8 \sum_{g \mid t} g = 8\sigma(t).$$

Andere Fälle sind in dem schönen Buch von van der Waerden und H. Groß "Studien zur Theorie der quadratischen Formen, 1968" enthalten.

Zum Beweis, der außerordentlich umfangreich ist (Tamagawa und A. Weil haben andere Beweise mit Hilfe der Adele gegeben und in die Theorie der algebraischen Gruppen eingeordnet), und sich auf die Untersuchungen von Gauß und Minkowski stützt, sei nur folgendes bemerkt: Der Ausgangspunkt besteht darin, daß man sich zunächst eine rationale Lösung x_0 der Gleichung $S[X] = T$ verschafft und zwar mit Hilfe des Satzes von Minkowski und Hasse, der ja aussagt, daß diese Gleichung eine rationale Lösung besitzt, wenn die zugehörigen Kongruenzen mod q lösbar sind (Hassesches Prinzip). Nun will man ja die Lösungen in ganzzahligen X bestimmen, dazu bracht man eine Parameterdarstellung für die Lösungen der Gleichung. Sie enthält als Spezialfall die Parameter darstellung für die Erzeugenden einer Fläche 2. Ordnung, bzw. die Parameterdarstellung von Cayley. Sie lautet folgendermaßn

$$X = X_0 + B(A - \tfrac{1}{2}B'SB)^{-1} B'SX_0 \qquad (B \text{ beliebig, } A \text{ schief}),$$
$$\operatorname{Det}(A - \tfrac{1}{2}B'SB) \neq 0, \qquad \operatorname{Det}(X'SX_0 - T) \neq 0.$$

Jetzt wird in sehr schwierigen Überlegungen, die auch eine Verschärfung des Hasseschen Satzes enthalten, der Beweis der Hauptformel vorbereitet, der durch vollständige Induktion nach m erfolgt. Der Beweis besteht dann aus zwei Teilen, aus einem rein zahlentheoretischen Teil, wo gezeigt wird, daß sich die beiden Seiten der Hauptformel nur um eine Faktor $\gamma(S)$ unterscheiden. Mit analytischen Methoden, wie sie von Dirichlet und Minkowski entwickelt wurden, wird nun gezeigt, daß $\gamma(S) = 1$ ist.

Hier bieten die niederdimensionalen Fälle die Hauptschwierigkeit.

Nun setzt eine grundlegende Umformung dieser Siegelschen Maßformel ein, die ich jetzt etwas näher schildern will. Wir betrachten den Fall daß $T = t$ eine natürliche Zahl ist, dann läßt sich in üblicher Weise A_q durch Gaußsche Summen (Siegel verwendet diesen Namen nur wenn $(d, q) = 1$ ist) darstellen:

$$\left(G\left(S, \frac{d}{q}\right) = \sum_{x \bmod q} e^{-2\pi i (d/q) S(x)} \right),$$

$$A_q(S, t) = \sum_{S(x) \equiv t \,(\mathrm{mod}\, q)1} = \frac{1}{q} \sum_{d \bmod q} e^{-2\pi i (dt/q)} G(S, d/q),$$

wenn man benützt, daß

$$\frac{1}{q} \sum_{d=0}^{q-1} e^{2\pi i (hd/q)} = 1 \quad \text{wenn } h \equiv 0 \ (\mathrm{mod}\ q),$$
$$= 0 \quad \text{sonst}$$

ist.

Um nun zu dem Limes $q \to \infty$ über zugehen, wird d/q in reduzierter Form a/b mit $(a, b) = 1$, $b \mid q$ geschrieben und man erhält

$$x_q = \sum_{\substack{0 < a/b \leq 1 \\ (a,b) = 1, b \mid q}} \frac{1}{q^m} G\left(S, \frac{a}{b}\right) e^{-2\pi i (a/b) t} \left(\frac{q}{b}\right)^m;$$

für $q \to \infty$ erhält man $\lim \alpha_q$, wenn er existiert, und insgesamt

$$\bar{A}(S, t) = \frac{\pi^{m/2}}{\sqrt{\operatorname{Det} S}} t^{m/2-1} \sum_{\substack{0 < a/b < 1 \\ (a,b)=1}} \frac{G(S, a/b)}{b^m} e^{-2\pi i (a/b) t},$$

was für $m > 4$ absolut konvergent ist, und für $E_5, \ldots, E_8$ mit der Formel von Hardy übereinstimmt und mit $A(S, t)$ identisch ist, da in diesem Fall das Geschlecht nur aus einer Klasse besteht. Und jetzt geht er entgegengesetzt wie Hardy vor. Er bildet die erzeugende θ-Funktion ($\tau = x + iy$, $y > 0$),

$$f(s, \tau) = \sum_x e^{\pi i S(x)} = 1 + \sum_{t=1} A(S, t) e^{2\pi i t},$$

bildet die Funktion $F(S, \tau)$:

$$F(S, \tau) = \sum_{k=1}^{h} \frac{f(S_k, t)}{E(S_k)} \Bigg| \sum_{k=1}^{h} \frac{1}{E(S_k)} = 1 + \sum_{t=1} \bar{A}(S, t) e^{2\pi i t \tau}.$$

Es wird $F(s, \tau) = 1 + \pi^{m/2}$

$$(\operatorname{Det} S)^{-1/2} \sum_{0 < a/b < 1} \frac{G(S, a/b)}{b^m} \sum_{t=1}^{\infty} t^{m/2-1} e^{\pi i t(\tau - 2a/b)}.$$

Jetzt wendet er die Formel von Lipschitz an und er erhält folgende Partialbruchzerlegung

$$F(s, \tau) = 1 + \frac{i^{m/2}}{\sqrt{\operatorname{Det} S}} \sum_{\tau} \sum_{0 < a/b < 1, b > 0} \frac{G(S, a/b)}{b^m} \left(\tau - 2\frac{a}{b} - 2l\right)^{-m/2};$$

als Beispiel führt er jetzt den Fall $S = E_8$ vor und er erhält die Formel

$$F(E_8, \tau) = \left(\sum_k e^{\pi i k^2 \tau}\right)^8 = \sum_{\substack{(a,b)=1 \\ ab \equiv 0 \ (\mathrm{mod}\ 2) \\ a > 0 \text{ oder } a = 0, b > 0}} (a\tau + b)^{-4},$$

die schon von Hardy aufgestellt und wie Siegel in einer späteren Arbeit Nr. 66 sagt, war diese Identität Ausgangspunkt seiner Überlegungen. Es sei noch zu dieser Mittelbildung bemerkt, daß Siegel sie auch in einer anderen Arbeit (Nr. 51) auf die Epsteinsche Zetafunktion angewendet hat. Nach diesen Überlegungen ist es für ihn ganz naheliegend t durch $Z = X + iY$ mit $Y = \operatorname{Im} Z > 0$ zu ersetzen, die θ-Funktion

$$f(S, Z) = \sum_C e^{\pi i \operatorname{Sp}(S(C)Z)} = \sum_{T \geqslant 0} A(S, T) e^{\pi i \operatorname{Sp}(TZ)}$$

($T > 0$ bedeutet T positiv definit) einzuführen, F zu bilden und jetzt braucht man in den vorher gegebenen Entwicklungen nur einige Modifikationen vorzunehmen. Allerdings sind die Beweise sehr schwierig, vor allem muß ein Analog zur Lipschitzschen Formel gefunden werden und er erhält dann folgende allgemeine Formel (Die Reihe ist sicher konvergent für $m > 2(n+1)$)

$$F(S, Z) = \sum_{[A,B]} h(S, A, B)(\mathrm{Det}(AZ+B))^{-m/2},$$

wobei A, B symmetrisch, teilerfremd, nicht assoziierte Matrizen durchlaufen, d.h. $AB' =$ symmetrisch und die n-zeiligen Unter-Determinanten von (A, B) sind teilerfremd und verschiedene Paare sind nicht linksassoziiert, d.h. es gilt nicht $(A, B) = U(A, B)$ (U unitär).

Als Beispiel für diese Formel: $m = 8$, $n = 2$, $S = E_8$, $Z = \begin{pmatrix} x & y \\ y & z \end{pmatrix}$,

$$F(E_8, Z) = \left(\sum_{a,b} e^{\pi i(xa^2+2yab+zb^2)}\right)^8 = \sum_{A,B} (\mathrm{Det}(AZ+B))^{-4}$$

wobei er bemerkt, daß diese θ-Funktion eine Riemann'sche θ-Funktion ist. Damit war für ihn das Programm gegeben, nämlich eine Theorie der Modulfunktionen zu entwickeln, aber davon später.

Siegel hat nun diese vorliegende Theorie auch auf indefinite Formen und auf quadratische Formen über algebraische Zahlkörper ausgedehnt. Hier müssen nun verschiedene Modifikationen vorgenommen werden, da z.B. bei indefiniten Formen $E(S)$, die Anzahl der Automorphismen oder wie Siegel sagt, die Anzahl der Einheiten, unendlich wird. Bei der Theorie über algebraischen Zahlkörpern müssen auch ausgeartete Formen mit Determinante 0 zugrunde gelegt werden. Ich möchte hier auf diese Modifikationen nicht näher eingehen, aber nur einige Fragen besprechen, welche mit der Theorie der quadratischen Formen eng zusammenhängen. Zunächst ist schon im Fall der positiv definiten Form die Reduktionstheorie der quadratischen Formen heranzuziehen. Bei den positiv definiten Formen gibt es ja verschiedene Reduktionsmethoden, am meisten bewährt hat sich die Reduktionstheorie von Minkowski, die Siegel zugrunde legt. Sie besteht aus folgendem: Will man zu einer Form S, sie sei jetzt positiv definit, eine reduzierte Form $S[U]$ finden, wobei U ganzzahlig unimodular ist, so legt man die erste Spalte U_1 von U durch die Minimalbedingung fest, daß $S(U_1)$ ein Minimum sein soll, die zweite Spalte U_2 durch die Minimalbedingung, daß bei festem U_1, $S(U_2)$ ein Minimum sein soll (wobei eventuell noch U_1 durch $-U_1$ ersetzt wird) usw. Die reduzierten Formen bilden im Raum P aller positiv definiten Formen einen konvexen Kegel K mit der Spitze im Nullpunkt der durch endlich viele Ebenen begrenzt ist, und dessen Bilder bei Anwendung der unimodularen

Transformationen U den ganzen Raum P überdeckt. Damit erhält man, wie schon Minkowski gezeigt hat, auch einen Fundamentalbereich im Raum aller nicht singulären Matrizen X in Bezug auf die unimodulare Gruppe. Dabei wird zwischen U und $-U$ nicht unterschieden. In der großen Arbeit Nr. 72 (kurz als japanische Arbeit bezeichnet) hat er den Raum P kompaktifiziert: jede Form $S[x]$ läßt sich in der Form schreiben

$$S[x]=\sum_{k=1}^{m} t_k y_k^2, \qquad y_k=x_k+\sum_{\tau=k+1}^{m} d_{k\tau}x_{k\tau}.$$

Wir betrachten $u_k=t_k/t_{k+1}$ $(k=1,\ldots,m-1)$ und die $d_{k\tau}$, für die $k<\tau$, und nennen sie die Normalkoordinaten von S in P.

Man betrachtet nun den Raum $\bar{P}$ aller $(u_k, d_{k\tau})$ wobei nur $u_k \geqslant 0$ verlangt wird, und die $d_{k\tau}$ beliebig reell sind und der P enthält, den Siegel ausführlich untersucht. Zur Reduktionstheorie zurückkommend sei noch folgendes Historische bemerkt: Die Theorie wurde durch Anwendung der Geometrie der Zahlen von K. Mahler und unabhängig davon von H. Weyl einerseits und andererseits in der gemeinsamen Arbeit von Bieberbach und I. Schur bedeutend vereinfacht. H. Weyl hat über diese gemeinsame Arbeit Jahre später das Wort tempi passati hinzugefügt. Eine schöne Darstellung der Minkowskischen Reduktionstheorie hat van der Waerden in seinem oben genannten Buch gegeben. Siegel hat verschiedene Modifikationen an der Minkowskischen Theorie angebracht, die die Handhabung der Theorie erleichtern, sie sind aber von technischer Natur und ich möchte daher nicht weiter darauf eingehen. Bei der Reduktionstheorie und Einheitstheorie von indefiniten Formen liegt die Sache bedeutend komplizierter. Bei den binären quadratischen Formen liegt die Theorie von Lagrange vor, wo man Kettenbrüche benützen kann und die Pellsche Gleichung nur von der Diskriminante abhängt.

Hermite hat im Fall der ternären quadratischen Formen folgende Reduktionstheorie indefinite Formen vorgeschlagen.

Ist die quadratische Form von der Gestalt $y_1^2+y_2^2-y_3^2$, so soll sie reduziert heißen, wenn die zugehörige positiv definite Form $y_1^2+y_2^2+y_2^3$ reduziert ist. Der Beweis, daß es in der Menge aller ternären quadratischen Formen mit gegebener Diskriminante nur endlich viele reduzierte Formen gibt, ist Hermite nicht gelungen und wurde erst in einer umfangreichen Arbeit von Stouff erbracht. In der Theorie der Einheiten im ternären Fall gibt es auch eine Pellsche Gleichung, die aber von der zugrunde gelegten Form abhängt. Siegel legt nun in der berühmten Abhandlung Nr. 33 die in den Hamburger Abhandlungen 1940 erschienen ist, die allgemeine Theorie dar. Ist S eine indefinite Form, so betrachtet er alle positiv definiten Formen H mit der Gleichung

$$H'S^{-1}H=S^{-1}[H]=S$$

und es wird nun definiert: S ist reduziert, wenn H im Minkowskischen Sinn reduziert ist. Siegel kann dann in einfacher Weise zeigen, daß der Satz von Stouff in dieser Definition richtig ist. Er betrachtet nun die Menge aller H, die diese Gleichung erfüllen. Sie bilden eine endlich dimensionale algebraische Mannigfaltigkeit M. Man zeigt nun, daß die Einheiten von S auf M eine diskontinuierliche Gruppe bilden, die endlich viele Erzeugende besitzt. Er ordnet weiterhin der Einheitengruppe ein Maß zu, das er als Volumen eines Fundamentalbereiches in einem zugeordneten Raum Y darstellt.

Ich möchte nicht auf weitere Arbeiten von Siegel über die Theorie der quadratischen Formen eingehen, aber eine seiner letzten Arbeiten besprechen, so die Arbeit Nr. 96, die eine Verallgemeinerung der Diophantischen Gleichung $Ax^2+2Bxy+Cy^2+2Dx+2Ey+F=0$ ist, nämlich

$$S[X]+B'X+X'B+C=0.$$

Die Gleichung wurde von C. Jordan 1881 behandelt. Die Beweise von Jordan enthalten aber, wie J. Dieudonné, Herausgeber der gesammelten Werke von Jordan bemerkt, eine Lücke. So war es für Siegel eine große Befriedigung, daran kann ich mich noch lebhaft erinnern, daß es ihm gelungen ist, mit Hilfe seiner Methoden diese Gleichung zu behandeln und die Lücke zu schließen.

Wenden wir uns nun der Theorie der Modulfunktionen zu, wie sie von Siegel in Nr. 32 (1939) entwickelt wurde. Führen wir gleich die Bezeichnungen ein, um uns kurz ausdrücken zu können. Zunächst definieren wir einmal die sogenannte Siegelsche Halbebene H_n. Sie besteht aus allen komplexen n-reihigen $(n \geqslant 1)$ symmetrischen Matrizen. $Z=X+iY$ wobei Y die Matrix einer positiv definiten quadratischen Form ist. X, Y reelle Matrizen. Es wird nun die symplektische Gruppe

$$\mathrm{Sp}(n, R)$$

eingeführt, bestehend aus allen $2n$-zeiligen Matrizen (A, B, C, D reell)

$$M=\begin{pmatrix} A & B \\ C & D \end{pmatrix},$$

welche die Gleichung erfüllen

$$M'JM=J$$

wobei $J=\left(\begin{smallmatrix} 0 & E \\ -E & 0 \end{smallmatrix}\right)$ ist (E Einheitsmatrix) und es werden nun die Transformationen der Halbebene betrachtet,

$$Z \to M\langle Z\rangle=(AZ+B)(CZ+D)^{-1}.$$

Das zahlentheoretische Element kommt nun dadurch hinein, daß man

$Sp(n, Z)$ betrachtet, wobei also A, B, C, D ganzzahlige Matrizen sind. Diese Untergruppe der Siegelschen Modulgruppe Γ_n ist diskontinuierlich in H_n und besitzt einen Fundamentalbereich. Meist benützt Siegel den folgenden Fundamentalbereich $F(\Gamma_n)$ $|\mathrm{Det}(C|z|+D)| \geqslant 1$, Y reduziert nach Minkowski, $|x_{kL}| \leqslant \frac{1}{2}$. Wie Siegel bemerkt, hat schon Picard diese Gruppe Γ_n ins Auge gefaßt, aber nicht weiter verfolgt, wohl deshalb, weil die Transformationsformel nicht linear ist. Bevor ich weiter auf die Siegelsche Modulgruppe eingehe, sei nur auf die große Arbeit Nr. 41, in der die geometrischen Eigenschaften der symplektischen Gruppe studiert wird, hingewiesen. Er führt in H_n das Bogenelement $ds^2 = \mathrm{Sp}(Y^{-1}\,dZ\,Y^{-1}\,d\bar{Z})$ ein. Das Volumselement ist

$$dv = \frac{dX\,dY}{(\mathrm{Det}\ Y)^{n+1}}.$$

Siegel bestimmt die Kroneckersche Krümmung als rationale Zahl mit Hilfe des verallgemeinerten Satzes von Gauß und Bonnet, wie sie von Allendörffer und unabhängig davon von Fenchel aufgestellt wurde (damit kann die Eulersche Charakteristik von $F(\Gamma_n)$ bestimmt werden). Es sei noch bemerkt, daß Siegel in seinem Buch über die Himmelsmechanik 1956 auf die Bedeutung der symplektischen Gruppe in der Mechanik, insbesondere in der Theorie der kanonischen Transformationen als erster hingewiesen hat. In der Arbeit Nr. 41 behandelt er verschiedene diskrete Untergruppen und ihre Charakterisierungen, aber darauf soll jetzt nicht näher eingegangen werden, sondern wir beschränken uns auf die Siegelsche Modulgruppe. Er definiert nun unter einer Modulform n-ten Grades vom ganzzahligen Gewicht $k \geqslant 0$ jede holomorphe Funktion (beschränkt im Fundamentalbereich) mit der Eigenschaft

$$f(M\langle Z\rangle) = \det(CZ+D)^{+k} f(Z) \qquad \text{für alle } M \text{ aus } \mathrm{Sp}(n, Z).$$

Es wurde erst später von M. Köcher, 1954, erkannt, daß man diese Bedingung in der Klammer für $n > 1$ weglassen kann, Siegel zeigt nun folgende Sätze:

1. Die Modulformen mit gegebenem Gewicht k bilden einen endlichen Vektorraum von der Dimension $O(k^{n(n+1)/2})$.

2. Je $n(n+1)/2+2$ Modulformen (sie können auch von verschiedenem Gewicht sein), sind algebraisch abhängig.

3. Die Eisensteinreihe (absolut konvergent für $k > n+1$ nach einem Satz von Hel Braun)

$$E_n^k(Z) = \sum_{[C,D]} \mathrm{Det}(CZ+D)^{-k}$$

ist Modulform $\not\equiv 0$ vom Gewicht k.

Er definiert nun Modulfunktionen als Quotienten von Modulformen gleichen Gewichts und zeigt, daß sie einen algebraischen Funktionenkörper vom Transzendenzgrad $n(n+1)/2$ bilden. Dies wird mit Hilfe der Eisensteinreihen gezeigt. Das Problem, das sich in der weiteren Forschung ergeben hat, war, diese Modulfunktion funktionentheoretisch zu charakterisieren.

Dies ist nun durch eine Reihe von wichtigen Arbeiten von verschiedenen Mathematikern geschehen.

Ich erlaube mir nur einige Namen anzuführen: z.B. W. L. Baily, A. Borel, H. Grauert, U. Christian, I. Satake, J. P. Serre und Siegel selbst, wobei verschiedene Kompaktifizierungen benützt wurden, bis sich dann herausstellte, daß sich diese an sich enorm wichtigen Dinge für die Charakterisierung der Modulfunktionen als nicht notwendig erwiesen haben; Sie sind für $n>1$ meromorph, Γ_n-invariante Funktionen (Arbeit Nr. 75, 1960).

Man kann vielleicht ganz grob sagen, daß es ein wichtiges Ziel in der Funktionentheorie mehrerer Variabler ist, aus der analytischen Abhängigkeit von meromorphen Funktionen auf die algebraische Abhängigkeit zu schließen. Siegel selbst setzte in den fünfziger und sechziger Jahren seinen Ehrgeiz darin, in möglichst elementarer Weise, ohne Benützung von tiefliegenden Sätzen aus der Funktionentheorie von mehreren komplexen Variablen, diese Resultate, wie sie z.B. von Serre gefunden wurden, zu beweisen. So sagt er in einer Arbeit (Nr. 64, 1955) über den Satz von Serre (Es sei M eine kompakte analytische Mannigfaltigkeit von n komplexen Dimensionen, K der Körper der auf M eindeutigen meromorphen Funktionen. Dann sind je $n+1$ Funktionen algebraisch abhängig. Sind $f,\ldots,f_n$ algebraisch unabhängig, dann ist K eine endliche algebraische Erweiterung von $C(f_1,\ldots,f_n)$: Er läßt sich in einer Kollegstunde bequem vortragen. Es sei noch bemerkt, daß die Modulfunktionen, die bei Siegel in der Theorie der quadratischen Formen auftreten, nicht zur vollen Modulgruppe gehören, sondern zu einer Kongruenzuntergruppe. Ohne auf die umfangreichen Theorien, die durch diese Arbeiten von Siegel entstanden sind, einzugehen, sei auf das ausführliche Referat von H. Klingen, das in den Jahresbericht der DMV 1983 erschienen ist, verwiesen.

Ich möchte besonders auf die Fourierentwicklung

$$E_n^k(z) = \sum_{T \text{ positiv semidefinit}} a(T)\, e^{2\pi i \,\mathrm{Sp}(TZ)}$$

der Eisensteinreihen hinweisen. Die Koeffizienten $a(T)$ sind rationale Zahlen, wie Siegel entdeckt hat. Ihre Nenner wurden von Witt und Siegel selbst untersucht. In diesem Zusammenhang möchte ich auch auf die

ζ-Funktionen indefiniter, quadratischer Formen hinweisen, die Siegel in mehreren Arbeiten studiert hat.

Siegel hat sich auch mit anderen diskontinuierlichen Gruppen beschäftigt und auch zur allgemeinen Theorie dieser Gruppen wichtiges beigetragen. Auch auf diesen Gegenstand will ich nicht näher eingehen, aber eine kleine Arbeit (Nr. 56) sei hervorgehoben, die an einen Satz von J. Nielsen anschließt, sie ist 1940 erschienen und sie besagt, daß eine nichtabelsche Gruppe F von Transformationen $(az+b)/(cz+d)$, welche das Innere des Einheitskreises (E) in sich überführen und abgesehen von der Einheit aus hyperbolischen Transformationen bestehen, notwendigerweise im Einheitskreis diskontinuierlich ist.

7. Himmelsmechanik

Wie schon in der Einleitung gesagt, hat sich Siegel immer für Astronomie, besonders für Himmelsmechanik und für die Hamilton–Jacobische Theorie, interessiert, allerdings war sie zunächst Anlaß für ein Mißgeschick. Er hatte sich Runge, der damals nurmehr Mathematikvorlesungen hielt, soviel ich weiß, der aber auch durch seine physikalischen Untersuchungen bekannt ist, als Prüfer gewünscht und hatte erwartet, daß er in theoretischer Physik geprüft werde, besonders in Hamilton–Jacobischen Theorie, in der er so gut vorbereitet war. Tatsächlich prüfte ihn Runge nur nach physikalischen Maßkonstanten und numerischen Daten, auf die Siegel in keiner Weise gefaßt war; nach den Akten soll Runge indigniert gewesen sein, daß er überhaupt als physikalischer Prüfer fungieren sollte. Wie es auch sein mag, diese Prüfung bewirkte, daß das gesamte Rigorosum mit "rite" beurteilt wurde, statt wie Siegel gehofft hatte mit "cum laude." Aber glücklicherweise hat dies Siegel nicht davon abgehalten, sich weiterhin mit Himmelsmechanik zu beschäftigen. 1936 bestimmte er in der Arbeit Nr. 23 alle algebraischen Integrale des restringierten Drei–Körper-Problems. Als Vorbild diente ihm der Satz von Bruns, der ja beim allgemeinen Drei–Körper-Problem gezeigt hatte, daß ausser den zehn bekannten algebraischen Integralen (Schwerpunktsätze, Flächensätze, Energiesatz usw.) keine weiteren algebraischen Integrale vorhanden sind. Daraus folgt aber nicht das Resultat von Siegel in der zitierten Arbeit, weil es beim eingeschränkten Drei–Körper-Problem noch ein elftes, das sogenannte Jacobische Integral gibt. (Bei dem eingeschränkten Drei–Körper-Problem findet bekanntlich die Bewegung in einer Ebene statt, wobei zwei Körper sich auf einem Kreis bewegen und der dritte Körper die Masse 0 hat). 1941 behandelte er im Anschluß an die Arbeit von Sondmann 1907 den Dreierstoß und zeigte, daß sich die Lösungen nach Potenzen von $t-t_0$ (t die Zeit), wo die Exponenten ganzzahlige Vielfache von $\frac{2}{3}$ sind entwickeln läßt. Über seine Arbeiten in der Himmelsmechanik schrieb er sogar einen populären Artikel in der *Amer. Math. Monthly* 1948. 1951 veröffentliche er in dem Band, der zum 75. Geburtstag von Erhard Schmidt erschienen ist, eine Arbeit mit dem Titel "Über periodische Lösungen im ebenen Drei–Körper-Problem." Er gab eine Modifikation der Hillschen Theorie des Mondes. Er erzielte eine große Vereinfachung dadurch, daß er die Entwicklung nach trigonometrischen Funktionen durch eine Potenzreihenentwicklung ersetzte und in ganz genialer Weise die Cauchysche Majorantenmethode anwendete. In der Arbeit Nr. 39, 1942 beschäftigte er sich mit der sogenannten Schröderschen Funktionalgleichung: Ist $f(z)$ eine Potenzreihe von der Gestalt $a_1 z + a_2 z^2 + \cdots +$, welche einen positiven Konvergenzradius besitzt, dann sucht man eine Reihe $\varphi(\zeta) = \zeta + \sum_{h=2}^{\infty} c_k \zeta^k$ (Schrödersche Reihe), sodaß

$$\varphi(a_1 \zeta) = f(\varphi(\zeta)).$$

Wenn a_1 nicht vom Betrage 1 ist, dann war es schon lange bekannt, daß diese Schrödersche Reihe konvergent ist. So war nur der Fall offen, daß a_1 von der Gestalt $e^{2\pi i\alpha}$ ist. Dabei ist α irrational, G. Julia hatte die Vermutung ausgesprochen, daß in diesem Fall die Schrödersche Reihe divergent ist und ein Schüler von Birkhoff, Pfeifer, hatte dies an einem Beispiel auch nachgewiesen. Der deutsche Mathematiker H. Cremer, Professor in Aachen, durch seine mathematischen Gedichte wohl bekannt, hatte sich jahrelang damit beschäftigt, diese Vermutung zu beweisen und er hatte in den *Mathematischen Annalen* 115, 1938 ein Kriterium für a_1 aufgestellt, mit der Eigenschaft, daß die zugehörige Schrödersche Reihe divergent war. Siegel erkannte sofort, daß diese Bedingung mit der Bedingung gleichbedeutend war, daß α eine Liouvillesche Zahl ist. Solche Zahlen waren ihm ja aus seinen Anfangsarbeiten wohl vertraut. Er nahm nun an, α sei keine Liouvillesche Zahl und zeigte, daß die zugehörige Schrödersche Reihe konvergent ist, damit war gezeigt, daß die Vermutung von Julia falsch ist. Er stellte sich nun die größere Aufgabe, solche Voraussetzungen auch in der Himmelsmechanik, wo das Problem der kleinen Nenner auftritt, anzuwenden. Seine Untersuchungen hatte er in seinem Buch über die Himmelsmechanik zusammengestellt. Der volle Erfolg stellte sich erst durch die Arbeiten von Kolmogorow, Arnold und Moser ein.

Von den Arbeiten, die die allgemeine Theorie der Mechanik betreffen, möchte ich nur aufzählen die Arbeit Nr. 37 "Über die Integrale kanonischer Systeme" und die Arbeit Nr. 52 "Über Differentialgleichungen auf dem Torus," im Anschluß an Denjoy. Siegel hat sich drei Ziele gesetzt: (1) die Stabilität von Lösungen von Differentialgleichungen in der Nähe einer Gleichgewichtslage zu untersuchen, vor allem gelang es ihm, manche

Behauptungen, die Cherry 1926 ohne stichhaltige Beweise aufgestellt hat, zu beweisen, dann war er bemüht, periodische Lösungen von Differentialgleichungen zu studieren und das gilt vor allem für die vorletzte Arbeit Nr. 98. Dann studierte er die ganze Mannigfaltigkeit der Lösungen in der Nähe einer Gleichgewichtslage. Vorbilder waren für ihn: Bohl, Perron, Poincaré, und Ljapunow. Ich will hier auf Details nicht weiter eingehen, Weiter hat einer seiner letzten Schüler, H. Rüßmann, eine genaue Beschreibung der Siegelschen Arbeiten in den *Jahresberichten Bd.* 85 der DMV 1983 gegeben.

8. Schlussbemerkungen

Ich bin nun am Schluß meines Berichtes über die Arbeiten Siegels angelangt. Es konnten nur Andeutungen über sein Werk gegeben werden, vieles mußte unberücksichtigt bleiben. Siegel strebte in seinen Arbeiten eine möglichst abgeschlossene Darstellung an, er sagte immer, ein Werk müsse reifen. Obwohl er ein Schüler von Landau war, pflegte er eigentlich nicht den sogenannten 2. Landau-Stil: Voraussetzung, Behauptung, Beweis.

Ich glaube, daß für ihn der Stil von I. Schur Vorbild war.

Ich möchte jetzt noch ein wenig über seine Vorlesungen sprechen, und zwar möchte ich die Vorlesung erwähnen, die er 1945 gehalten hat, wo er den θ-Satz beweist, den Weierstraß und Riemann ohne Beweis aufgestellt hatten. Poincaré und Wirtinger hatten einen Beweis skizziert. Appel hatte für den Fall $n=2$ schon einen vollständigen Beweis gegeben, Siegel gab nun einen allgemeinen Beweis für diesen Satz, Conforto einen anderen Beweis, es handelt sich hier um die Auflösung eines Systems von Differenzengleichungen.

Ich möchte aber auch noch einige Vorlesungen, die Siegel am Tata-Institut gehalten hat, hervorheben:

(1) On Quadratic Forms 1984.

(2) On Advanced Analytic Number Theory 1980.

(3) On Riemann Matrices 1963.

Man hört jetzt schon manchmal, daß das Werk von Siegel outdated ist, dasselbe hat man nach dem Tod von Poincaré, aber auch bei Hecke gesagt: Wie lebendig ist es jetzt? Ich bin davon überzeugt, daß das Werk von Siegel noch für viele Generationen eine Fundgrube von Ideen sein wird, das gilt nicht nur für seine Arbeiten, sondern auch für seine Vorlesungen und Bücher.

List of publications

Books

Topics in the Theory of Numbers. California Institute of Technology, Pasadena, California, 1967, 124 pp.

Theorie der Gleichverteilung. Bibliographisches Institut Wissenschaftsverlag, Mannheim Wien Zürich, 1979, 142 pp.

Zahlentheorie – eine Einführung (with J. Schoissengeier). Manz, Wien, 1979, 159 pp.

Grundbegriffe der Mathematik (with C. Binder and P. Schmitt). Prugg, Wien, 1979, 196 pp.

Zahlentheoretische Methoden in der Numerischen Mathematik. Schriftenreihe der Österr. Computer Gesellschaft, Band 12. Oldenbourg, Wien München, 1981, 149 pp.

The Theory of Uniform Distribution (translated by Henry Orde). AB Academic Publishers, Berkhamsted, 1984, 141 pp.

Zahlentheoretische Analysis. Seminar Wien, Lecture Notes in Mathematics, vol. 1114. Springer, Berlin Heidelberg New York Tokyo, 1985, 157 pp. (Editor)

Geometrische und analytische Zahlentheorie (with J. Schoissengeier and R. J. Taschner). Manz, Wien, 1986, 194 pp.

Zahlentheoretische Analysis II. Seminar Wien, Lecture Notes in Mathematics, vol. 1262. Springer, Berlin Heidelberg New York Tokyo, 1987, 158 pp. (Editor)

Johann Radon – Collected Papers, vols. 1 and 2. Birkhäuser, Basel Boston, 392 and 496 pp. (Edited jointly with P. M. Gruber, W. Nöbauer, and L. Schmetterer)

Articles

[1] Eine asymptotische Formel der Laguerreschen Polynome. Monatsh. Math. Phys. **42** (1935) 275–278

[2] Über asymptotische Entwicklungen von Lösungen linearer Differentialgleichungen 2. Ordnung. Monatsh. Math. Phys. **46** (1937) 34–37

[3]* Über die Approximation von zwei komplexen inhomogenen Linearformen. Monatsh. Math. Phys. **46** (1938) 324–334 Page 9

---

* Indicates a paper contained in these Selecta for which the page number is given at the end of the citation.

[4]* Über die Approximation von inhomogenen Linearformen. Monatsh. Math. Phys. **47** (1938) 181–185 Page 20

[5]* Über komplexe homogene Linearformen. Monatsh. Math. Phys. **49** (1941) 321–326 Page 25

[6]* Eine Ungleichung für Vektorlängen. Enthalten in: H. Hornich: Eine Ungleichung für Vektorlängen. Math. Z. **48** (1942) 268–274 Page 31

[7]* Zur Geometrie der Zahlen. Math. Z. **49** (1943) 285–312 Page 38

[8] Über Potenzsummen von Linearformen. Sitzungsber., Abt. II, Österr. Akad. Wiss., Math.-Naturwiss. Kl. **154** (1945) 50–58

[9] Inhomogene Linearformen in algebraischen Zahlkörpern. Sitzungsber., Abt. II, Österr. Akad. Wiss., Math.-Naturwiss. Kl. **155** (1947) 63–73

[10] Über einen Satz aus der Geometrie der Zahlen. Sitzungsber., Abt. II, Österr. Akad. Wiss., Math.-Naturwiss. Kl. **155** (1947) 75–82

[11] Über Gitterpunkte in Zylindern. Sitzungsber., Abt. II, Österr. Akad. Wiss., Math.-Naturwiss. Kl. **156** (1947) 203–217

[12] Über Potenzsummen von Linearformen II. Sitzungsber., Abt. II, Österr. Akad. Wiss., Math.-Naturwiss. Kl. **156** (1947) 247–254

[13] Über Folgen von Quadratwurzeln komplexer Zahlen. Sitzungsber., Abt. II, Österr. Akad. Wiss., Math.-Naturwiss. Kl. **156** (1947) 255–262

[14] Eine asymptotische Formel für Potenzsummen komplexer Linearformen. Monatsh. Math. **52** (1948) 248–254

[15] Ausfüllung und Überdeckung durch Zylinder. Anz. Österr. Akad. Wiss., Math.-Naturwiss. Kl. **85** (1948) 116–119

[16]* Ausfüllung und Überdeckung konvexer Körper durch konvexe Körper. Monatsh. Math. **53** (1949) 81–131 Page 66

[17] Über eine Verallgemeinerung des Satzes von Mordell. Ann. Math. **50** (1949) 314–317

[18]* Über Gitterpunkte in Parallelepipeden. J. Reine Angew. Math. **187** (1950) 246–252 Page 118

[19]* Über Integrale auf konvexen Körpern I. Monatsh. Math. **54** (1950) 1–36 Page 125

[20] Integrale auf konvexen Körpern II. Monatsh. Math. **54** (1950) 81–99

[21] Über die Zeta-Funktion konvexer Körper. Monatsh. Math. **54** (1950) 100–107. (Herrn Geheimrat O. Perron zum 70. Geburtstag)

[22] Bemerkungen zu einem Satz von R. Rado. Anz. Österr. Akad. Wiss., Math.-Naturwiss. Kl. **87** (1950) 219–226

[23] Ein Satz über additive Mengenfunktionen. Math. Nachr. **4** (1950–51) 150–155. (Herrn Prof. Dr. Erhard Schmidt zum 75. Geburtstag gewidmet)

[24] Integrale auf konvexen Körpern III. Monatsh. Math. **55** (1951) 105–137. (Herrn Prof. G. Herglotz zum 70. Geburtstag)

[25] Zur Theorie des Figurengitters. Math. Ann. **125** (1952) 183–207

[26]* Über eine Klasse von mehrfachen Integralen. Abh. Math. Semin. Univ. Hamb. **18** (1952) 53–69 Page 161

[27] Grundbegriffe der Geometrie der Zahlen. Jahresber. Dtsch. Math.-Ver. **57** (1954) 37–55

[28] Zur Theorie der Überdeckung durch konvexe Körper. Monatsh. Math. **58** (1954) 287–291

[29] Inhomogene Minima von Sternkörpern. Monatsh. Math. **58** (1954) 292–305

[30]* Das inhomogene Problem in der Geometrie der Zahlen. Proc. Intern. Congr. Math., Amsterdam, vol. III, 1954, pp. 20–27 Page 178

[31]* Über einen Satz von van der Corput. Archiv Math. **6** (1955) 115–120 Page 186

[32]* Zur formalen Theorie der Gleichverteilung in kompakten Gruppen. Rend. Circ. Mat. Palermo, Ser. II, **4** (1955) 1–15 Page 192

[33]* Folgen auf kompakten Räumen. Abh. Math. Semin. Univ. Hamb. **20** (1956) 223–241. (Wilhelm Blaschke zum 70. Geburtstag gewidmet) Page 207

[34]* Zur Überdeckung der Ebene durch konvexe Scheiben. Anz. Österr. Akad. Wiss., Math.-Naturwiss. Kl. **94** (1957) 79–82 Page 226

[35] Normal gleichverteilte Folgen auf kompakten Räumen. Anz. Österr. Akad. Wiss., Math.-Naturwiss. Kl. **94** (1957) 94–96

[36]* Zum Hauptsatz der Theorie der Gleichverteilung. Anz. Österr. Akad. Wiss., Math.-Naturwiss. Kl. **94** (1957) 313–317 Page 230

[37] Folgen auf kompakten Räumen II. Math. Nachr. **18** (1958) 188–202. (H. L. Schmid gewidmet)

[38] Zur Theorie der diophantischen Approximationen. Anz. Österr. Akad. Wiss., Math.-Naturwiss. Kl. **95** (1958) 41–48

[39] Statistik auf kompakten Gruppen. Anz. Österr. Akad. Wiss., Math.-Naturwiss. Kl. **96** (1959) 64–76

[40]* Erbliche Eigenschaften in der Theorie der Gleichverteilung. Publ. Math. **7** (1960) 181–186. (Herrn Prof. Dr. O. Varga zum 50. Geburtstag gewidmet) Page 235

[41] Über C-Gleichverteilung. Ann. Mat. Pura Appl., IV. Ser. **49** (1960) 311–326. (Giovanni Sansone zu seinem 70. Geburtstag gewidmet)

[42] Cremonatransformation von Folgen modulo 1. Monatsh. Math. **65** (1961), 227–232. (Herrn Professor E. Kruppa zum 75. Geburtstag gewidmet)

[43]* Funktionen von beschränkter Variation in der Theorie der Gleichverteilung. Ann. Mat. Pura Appl., IV. Ser. **54** (1961) 325–334. (Enrico Bompiani zu seinem wissenschaftlichen Jubiläum) Page 241

[44] Über die Diskrepanz mehrdimensionaler Folgen mod. 1. Math. Z. **77** (1961) 273–284. (Herrn Friedrich Karl Schmidt zum 60. Geburtstag am 22. 9. 1961 gewidmet)

[45]* Zur angenäherten Berechnung mehrfacher Integrale. Monatsh. Math. **66** (1962) 140–151. (Professor Paul Funk zum 75. Geburtstag gewidmet) Page 250

[46]* Rhythmische Folgen auf kompakten Gruppen I. Sitzungsber., Abt. II, Österr. Akad. Wiss., Math.-Naturwiss. Kl. **171** (1963) 67–74 Page 262

[47] Lösung von Integralgleichungen mittels zahlentheoretischer Methoden I. Sitzungsber., Abt. II, Österr. Akad. Wiss., Math.-Naturwiss. Kl. **171** (1963) 103–123

[48] Geordnete Schätzfunktionen und Diskrepanz. Math. Ann. **150** (1963) 259–267. (Herrn B. L. van der Waerden zum 60. Geburtstag am 2. 2. 1963 gewidmet)

[49]* Zur Geometrie der Zahlen. Jahrbuch der Akademie der Wissenschaften zu Göttingen, 1963, pp. 23–29 Page 270

[50] Lösungen von Integralgleichungen mittels zahlentheoretischer Methoden II (jointly with K. Kreiter). Sitzungsber., Abt. II, Österr. Akad. Wiss., Math.-Naturwiss. Kl. **172** (1963) 229–250

[51] Überdeckung durch konvexe Scheiben. Sitzungsber. Berliner Math. Ges., Jahrgang 1961–64, pp. 28–36

[52]* Discrepancy and uniform distribution of sequences. Compos. Math. **16** (1964) 83–91 Page 277

[53]* Uniform distribution modulo 1 and numerical analysis. Compos. Math. **16** (1964) 92–105 Page 286

[54] Trigonometrische Interpolation bei Funktionen von mehreren Variablen. Acta Arith. **9** (1964) 305–320. (Herrn Professor L. J. Mordell zum 75. Geburtstag)

[55] Rhythmische Folgen auf kompakten Gruppen II (jointly with W. Henhapl). Sitzungsber., Abt. II, Österr. Akad. Wiss., Math.-Naturwiss. Kl. **174** (1965) 139–173

[56] Geordnete Schätzfunktionen und Diskrepanz II (jointly with W. Kuich). Sitzungsber., Abt. II, Österr. Akad. Wiss., Math.-Naturwiss. Kl. **174** (1965) 235–286

[57] Mathematische Modelle zur kinetischen Gastheorie. Sitzungsber., Abt. II, Österr. Akad. Wiss., Math.-Naturwiss. Kl. **174** (1965) 287–307. (Prof. Paul Urban zum 60. Geburtstag gewidmet)

[58] Geometrie der Zahlen und trigonometrische Interpolation bei Funktionen von mehreren Variablen. Les Tendances Géométriques en Algèbre et Théorie des Nombres 1966, Colloques Internationaux du Centre National de la Recherche Scientifique No. 143, Clermont-Ferrand 1964, (1966), pp. 83–86

[59] Leibniz als Mathematiker. Philosophia Naturalis, Archiv für Naturphilosophie und der philosophischen Grenzgebiete der exakten Wissenschaften und Wissenschaftsgeschichte **10** (1968) 146–158

[60]* Interpolation analytischer Funktionen auf dem Einheitskreis. Abhandlungen aus Zahlentheorie und Analysis (E. Landau Gedenkband), VEB Deutscher Verlag der Wissenschaften, Berlin, 1968, pp. 99–118 Page 300

[61] Diskrepanz in kompakten abelschen Gruppen I (jointly with H. Niederreiter). Manuscr. Math. **1** (1969) 259–288

[62] Karl Mayrhofer – ein Siebziger. IMN **92** (1969) 57–58

[63] Die mengentheoretische Topologie. Beiträge zur Lehrerfortbildung, Schriftenreihe zur Fortbildung der Lehrer an allgemeinbildenden Höheren Schulen, Band 3, Mathematik, 1969, pp. 48–71

[64] Mathematische Modelle der kinetischen Gastheorie II. Symp. Math. IV (1970) 81–97

[65] Mathematische Modelle der kinetischen Gastheorie III. Sitzungsber., Abt. II, Österr. Akad. Wiss., Math.-Naturwiss. Kl. **178** (1970) 1–12. (Herr. Prof. H. Hadwiger zum 60. Geburtstag gewidmet)

[66] Bemerkungen zum großen Sieb von Linnik. Sitzungsber., Abt. II, Österr. Akad. Wiss., Math.-Naturwiss. Kl. **178** (1970) 13–18

[67]* Ein metrischer Satz in der Theorie der C-Gleichverteilung. Monatsh. Math. **74** (1970) 108–118. (Herrn Professor N. Hofreiter zum 65. Geburtstag gewidmet) Page 320

[68] Differentiable manifolds. Acta Physica Austr., Suppl. VII, (1970) 265–307

[69]* Discrepancy and Riemann Integration. Studies in Pure Mathematics (Papers presented to Richard Rado). Academic Press, London New York, 1971, pp. 121–129 Page 331

[70] Zur Definition der Diskrepanz. Acta Arith. **18** (1971) 233–241. (Gewidmet dem Andenken H. Davenports)

[71] A Transformation of Equidistributed Sequences (jointly with R. Mück). Applications of Number Theory to Numerical Analysis. Academic Press, London, New York, 1972, pp. 371–388

[72] Ein metrisches Gegenstück zu einem Satz von W. A. Veech. Monatsh. Math. **76** (1972) 436–447

[73] Über eine Transformation von gleichverteilten Folgen II (jointly with R. Mück). Computing **9** (1972) 127–138

[74]* Über eine Methode von E. Hecke in der Theorie der Gleichverteilung. Acta Arith. **24** (1973) 11–31. (Herrn Prof. C. L. Siegel zum 75. Geburtstag gewidmet in Dankbarkeit und Verehrung) Page 340

[75]* Mathematische Modelle der kinetischen Gastheorie. Rheinisch-Westfälische Akademie der Wissenschaften, Natur-, Ingenieur- und Wirtschaftswiss., Vorträge 240 (1974), 15 pp. Page 361

[76] On some concepts, theorems and problems in the theory of uniform distribution. Colloq. Math. Soc. János Bolyai **13** (1974) 97–109

[77] Zum großen Sieb von Linnik. Acta Arith. **27** (1975) 89–100. (Dem Andenken von Yu. V. Linnik gewidmet)

[78] Zur Theorie der Gleichverteilung I. Anz. Österr. Akad. Wiss., Math.-Naturwiss. Kl. **112** (1975) 13–14

[79] Zur Theorie der Gleichverteilung II. Anz. Österr. Akad. Wiss., Math.-Naturwiss. Kl. **112** (1975) 23–24

[80]* Anwendung zahlentheoretischer Methoden auf Probleme der numerischen Mathematik I. Sitzungsber., Abt. II, Österr. Akad. Wiss., Math.-Naturwiss. Kl. **184** (1975) 217–225 Page 376

[81] Numerische analytische Fortsetzung in Polyzylindern. Sitzungsber., Abt. II, Österr. Akad. Wiss., Math.-Naturwiss. Kl. **184** (1975) 307–331. (Herrn Prof. Dr. P. Urban in alter Freundschaft zum 70. Geburtstag gewidmet)

[82] Zur quantitativen Theorie der Gleichverteilung. Sitzungsber., Abt. II, Österr. Akad. Wiss., Math.-Naturwiss. Kl. **184** (1975) 355–365. (Herrn Prof. Theodor Schneider zum 65. Geburtstag gewidmet)

[83]* Über die Gleichverteilung gewisser Folgen, welche mit den Nullstellen der Zetafunktion zusammenhängen. Sitzungsber., Abt. II, Österr. Akad. Wiss., Math.-Naturwiss. Kl. **184** (1975) 459–471. (Prof. C. L. Siegel zum 80. Geburtstag in Verehrung gewidmet) Page 385

[84] Abschätzung von trigonometrischen Summen mittels diophantischer Approximation. Sitzungsber., Abt. II, Österr. Akad. Wiss., Math.-Naturwiss. Kl. **185** (1976) 43–50. (Hans Hornich zum 70. Geburtstag gewidmet)

[85] Das Werk Perrons auf dem Gebiete der diophantischen Approximation. Jahresber. Dtsch. Math.-Ver. **80** (1978) 1–12

[86] Weierstraßscher Approximationssatz und Gleichverteilung. Monatsh. Math. **88** (1979) 137–170. (Meinem Freund Prof. L. Schmetterer zum 60. Geburtstag gewidmet)

[87] Mathematica – quo vadis? Wissenschaftliche Nachrichten, April 1979, pp. 25–29

[88] Bemerkungen zur Ausbildung von Lehrern. Zeitschrift für Hochschuldidaktik, Sonderheft **53** (1979) 107–125

[89] Über einige Reihen, welche mit den Vielfachen von Irrationalzahlen zusammenhängen. Acta Arith. **37** (1980) 285–306. (Dem Andenken von Paul Turán gewidmet)

[90] Gleichverteilung und mathematische Linguistik. Sitzungsber., Abt. II, Österr. Akad. Wiss., Math.-Naturwiss. Kl. **189** (1980) 437–490

[91] Über einige Sätze, Begriffe und Probleme in der Theorie der Gleichverteilung. Sitzungsber., Abt. II, Österr. Akad. Wiss., Math.-Naturwiss. Kl. **189** (1980) 209–248

[92]* 90 Jahre Geometrie der Zahlen. Jahrbuch Überblicke Mathematik, Bibliographisches Institut, Mannheim, 1980, pp. 9–41 Page 398

[93] Zur Geschichte des Inhaltsbegriffs. ÖMG Didaktikreihe, Heft 2 (1980) 1–56

[94]* Approximation von Irrationalzahlen und pythagoräische Tripel. Bonner Mathematische Schriften **121** (1980) 1–32. (Vortrag am 9. Dez. 1977 in Bonn anläßlich des 70. Geburtstages von Prof. E. Peschl) Page 431

[95] Nachruf auf Prof. Hans Hornich. Monatsh. Math. **89** (1980) 1–8

[96] Gleichverteilung und Quadratwurzelschnecke. Monatsh. Math. **89** (1980), 19–44. (Herrn Prof. R. M. Redheffer in Freundschaft zum 60. Geburtstag gewidmet)

[97] Regiomontan-Studien, Einleitung (Editor G. Hamann). Sitzungsber., Österr. Akad. Wiss., Phil.-Hist. Kl. **364** (1980) 7–10

[98] Hans Hornich, Nachruf. Almanach, Österr. Akad. Wiss. **130** (1980) 309–314

[99] Anwendung einer zahlentheoretischen Methode von C. L. Siegel auf Probleme der Analysis. Comment. Math. Helv. **56** (1981) 66–82. (Prof. Chandrasekharan zum 60. Geburtstag gewidmet)

[100] Carl Ludwig Siegel, Nachruf. Almanach, Österr. Akad. Wiss. **131** (1981) 289–293

[101] Über einige Reihen, die mit den Vielfachen von Irrationalzahlen zusammenhängen II. Sitzungsber., Abt. II, Österr. Akad. Wiss., Math.-Naturwiss. Kl. **190** (1981) 33–61

[102]* Gleichverteilung auf Produkten von Sphären. J. Reine Angew. Math. **330** (1982) 1–43 Page 456

[103] Nachruf auf Carl Ludwig Siegel. Jahrbuch Überblicke Mathematik, Bibliographisches Institut, Mannheim, 1982, pp. 159–168. Leserbrief 1983, pp. 215–216

[104] Zum Zahlbegriff. Philosophia naturalis **19** (1982) 413–470

[105] Mathematica – quo vadis? ÖMG Didaktikreihe, Heft 8 (1982) 33–54

[106] Trigonometrische Interpolation bei Funktionen von mehreren Variablen II. Sitzungsber., Abt. II, Österr. Akad. Wiss., Math.-Naturwiss. Kl. **191** (1982) 187–201

[107] Lineare Differenzengleichungen in mehreren Variablen. J. Reine Angew. Math. **339** (1983) 166–178. (Hans Rohrbach zum 80. Geburtstag gewidmet)

[108] Gleichverteilung und das Konvergenzverhalten von Potenzreihen am Rande des Konvergenzkreises. Manuscr. Math. **44** (1983) 231–263

[109]* Eine Bemerkung zur Theorie der Gleichverteilung. Studies in Pure Mathematics, Akadémiai Kiado, Budapest, 1983, pp. 337–345. (Dem Andenken Paul Turáns gewidmet) Page 500

[110] Pavel Aleksandrov, Nachruf. Almanach, Österr. Akad. Wiss. **133** (1983) 329–334

[111] Über einen Satz von C. Radoux. Mathematical Structures – Computational Mathematics – Mathematical Modelling. Papers dedicated to Prof L. Iliev's 70th Anniversary, Sofia, 1984, pp. 208–215

[112] Über einige Gitterreihen und Gitterfunktionen. Sitzungsber., Abt. II, Österr. Akad. Wiss., Math.-Naturwiss. Kl. **193** (1984) 247–287

[113] Näherungslösungen der Wellengleichung und verwandter Gebiete durch zahlentheoretische Methoden. Sitzungsber., Abt. II, Österr. Akad. Wiss., Math.-Naturwiss. Kl. **193** (1984) 359–442

[114] Über ein Produkt, das in der Interpolation analytischer Funktionen im Einheitskreis auftritt. In: Zahlentheoretische Analysis. Seminar Wien, Editor E. Hlawka, Lecture Notes in Mathematics, vol. 1114. Springer, Berlin Heidelberg New York Tokyo, 1985, pp. 19–25

[115] Bemerkung zum Lemma von Du Bois-Reymond. In: Zahlentheoretische Analysis. Seminar Wien, Editor E. Hlawka, Lecture Notes in Mathematics, vol. 1114. Springer, Berlin Heidelberg New York Tokyo, 1985, pp. 26–29

[116] Bemerkung zum Lemma von Du Bois-Reymond II. In: Zahlentheoretische Analysis. Seminar Wien, Editor E. Hlawka, Lecture Notes in Mathematics, vol. 1114. Springer, Berlin Heidelberg New York Tokyo, 1985, pp. 30–39

[117]* Carl Ludwig Siegel (31.12.1896 - 4.4.1981). J. Number Theory **20** (1985) 373–404 Page 509

[118] Gleichverteilung und ein Satz von Müntz. J. Number Theory **24** (1986) 35–46

[119] Erinnerungen an Kurt Gödel. IMN **142/43** (1986) 2–6

[120] Über die Entwicklung der Theorie der Gleichverteilung in den Jahren 1909 bis 1916 (jointly with C. Binder). Arch. Hist. Exact Sci. **36** (1986) 197–249

[121] Gábor Szegö, Nachruf. Almanach, Österr. Akad. Wiss. **136** (1986) 335–342

[122] Karl Menger, Nachruf. Almanach, Österr. Akad. Wiss. **136** (1986) 343–361

[123] Josef Lense, Nachruf. Almanach, Österr. Akad. Wiss. **136** (1986) 367–374

[124] Ernst Ferdinand Peschl, Nachruf. Almanach, Österr. Akad. Wiss. **136** (1986) 387–397

[125] Über die direkten Methoden der Variationsrechnung und Gleichverteilung. In: Zahlentheoretische Analysis II. Seminar Wien, Editor E. Hlawka, Lecture Notes in Mathematics, vol. 1262. Springer, Berlin Heidelberg New York Tokyo, 1987, pp. 50–85

[126] Über eine geometrische Reihe von E. Hecke. Acta Arith. **49** (1987) 113–125. (Gewidmet Paul Erdös zum 75. Geburtstag)

[127] Eine Anwendung diophantischer Approximationen auf die Theorie von Differentialgleichungen. Aequationes Math. **35** (1988) 232–253. (Herrn Prof. Dr. Dr. h.c. mult. Otto Haupt mit den besten Wünschen zum 100. Geburtstag gewidmet)

[128] Mathematik auf dem Weg durch die Zeit. ÖMG Didaktikheft 15, Nov. 1987, pp. 88–145

[129] On a class of multiple Lambert series. Journ. Math. Phys. Sci. **22** (1988) 45–66. (To the memory of Ramanujan)

[130] Erinnerungen an Johann Radon. In: Johann Radon – Collected Works, vol. 1. Birkhäuser, Basel Boston, 1987, pp. 3–26

[131] Näherungsformeln zur Berechnung von mehrfachen Integralen mit Anwendungen auf die Berechnung von Potentialen, Induktionskoeffizienten und Lösungen von Gleichungssystemen (in print)

[132] Beiträge zur Theorie der Gleichverteilung und ihren Anwendungen I. Einleitung. Sitzungsber., Abt. II, Österr. Akad. Wiss., Math.-Naturwiss. Kl. **197** (in print). (Dem Andenken an W. Nöbauer gewidmet)

[133] Beiträge zur Theorie der Gleichverteilung und ihren Anwendungen II. Der euklidische Fall. Sitzungsber., Abt. II, Österr. Akad. Wiss., Math.-Naturwiss. Kl. (in print). (A. Florian zum 60. Geburtstag gewidmet)

[134] Beiträge zur Theorie der Gleichverteilung und ihren Anwendungen III. Der nichteuklidische Fall. Sitzungsber., Abt. II, Österr. Akad. Wiss., Math.-Naturwiss. Kl. (in print)

[135] Beiträge zur Theorie der Gleichverteilung und ihren Anwendungen IV. Der sphärische Fall. Sitzungsber., Abt. II, Österr. Akad. Wiss., Math.-Naturwiss. Kl. (in print).

[136] Beiträge zur Theorie der Gleichverteilung und ihren Anwendungen V. Der Fall der unitären Gruppe. Sitzungsber., Abt. II, Österr. Akad. Wiss., Math.-Naturwiss. Kl. (in print). (G. Helmberg zum 60. Geburtstag freundschaftlich gewidmet)

[137] Wilfried Nöbauer, Nachruf. Almanach, Österr. Akad. Wiss. **138** (1988) 307–315

[138] Gleichverteilung und Entropie (Das Entropiespiel von T. und P. Ehrenfest) (to appear)

Unpublished Manuscripts

Ungleichungen. 50 pp.
Euklidische Geometrie der Ebene. 1965, 52 pp.
Diophantische Gleichungen. 1952/53, 150 pp.
Diophantische Ungleichungen. 1952/53, 200 pp.
Erinnerungen eines österreichischen Mathematikers. 140 pp.
The work of C. A. Rogers in the geometry of numbers. 1986, 20 pp.
Mathematik von 1945 bis 1985 in Österreich. 1987, 12 pp.

Acknowledgements

We thank the publishers and copyright holders of Edmund Hlawka's papers for the permission to reprint them here.

The numbers following each source refer to the numbered articles in the List of publications.

Academic Press: 69, 117
Akadémiai Kiado: 109
Akademie der Wissenschaften zu Göttingen: 49
Akademische Verlagsgesellschaft: 3, 4, 5
Bibliographisches Institut: 92
Birkhäuser: 31
Rendiconti del Circolo Matematica di Palermo: 32
Fondazione Annali di Matematica Pura ed Applicata: 43
Institutum Mathematicum Universitatis Debreceniensis: 40
Kluwer Academic Publishers: 52, 53
Mathematisches Institut der Universität Bonn: 94
Österreichische Akademie der Wissenschaften: 34, 36, 46, 80, 83
PWN Polish Scientific Publishers: 74
VEB Deutscher Verlag der Wissenschaften: 60
Walter de Gruyter & Co: 18, 102
Westdeutscher Verlag: 75
Wolters-Noordhoff: 30